SPACECRAFT DYNAMICS AND CONTROL

SPACECRAFT DYNAMICS AND CONTROL
AN INTRODUCTION

Anton H.J. de Ruiter
Ryerson University, Canada
Christopher J. Damaren
University of Toronto, Canada
James R. Forbes
McGill University, Canada

WILEY
A John Wiley & Sons, Ltd., Publication

This edition first published 2013

Registered office
John Wiley & Sons Ltd, The Atrium, Southern Gate, Chichester, West Sussex, PO19 8SQ, United Kingdom

For details of our global editorial offices, for customer services and for information about how to apply for permission to reuse the copyright material in this book please see our website at www.wiley.com.

Library of Congress Cataloging-in-Publication Data

De Ruiter, Anton H. J.
Spacecraft dynamics and control : an introduction / Anton H.J. de Ruiter, Christopher Damaren, James R. Forbes.
pages cm
Includes bibliographical references and index.
ISBN 978-1-118-34236-7 (hardback)
1. Space vehicles–Attitude control systems. 2. Space vehicles–Dynamics. I. Damaren, Christopher. II. Forbes, James R. III. Title.
TL3260.D33 2013
629.4′1–dc23

2012033616

A catalogue record for this book is available from the British Library.

ISBN: 9781118342367

Typeset in 10/12pt Times by Aptara Inc., New Delhi, India

To Janice, Thomas, Benjamin, Therese and Marie
A.dR

To Yvonne, Gwen, and Georgia
C.J.D

For Allison
J.R.F

Contents

Preface

This book presents a fundamental introduction to spacecraft orbital and attitude dynamics as well as its control. There are several excellent books related to these subjects. It is not our intention to compete with these well-established texts. However, many of them assume relatively significant backgrounds on behalf of the reader, which can make them difficult to follow for the beginner. It is our hope that this book will fill that void, and that by studying this book, more advanced texts on the subject will become more accessible to the reader. This book is suitable for first courses in spacecraft dynamics and control at the upper undergraduate level or at the beginning graduate level. The book is naturally split between orbital mechanics, and spacecraft attitude dynamics and control. It could therefore be used for two one semester courses, one on each subject. It could also be used for self-study.

The primary objective of this book is to educate, and the structure of the book reflects this. This book could also be used by the professional looking to refresh some of the fundamentals. We have made this book as self-contained as possible. In each chapter we develop a subject at a fundamental level (perhaps drawing on results from previous chapters). As a result, the reader should not only understand what the key mathematical results are, but also how they were obtained and what their limitations are (if applicable). At the end of each chapter we provide a few recommended references should the reader have interest in exploring the subject further.

The assumed reader background is minimal. Junior undergraduate level mathematics and mechanics taught in standard engineering programs should be sufficient. While a background in classical control theory would help, it is not necessary to have it in order to be able to follow the treatment of spacecraft attitude control. The presentation in this book on spacecraft attitude control is completely self-contained, and it could in fact be used as a substitute for a complete first (undergraduate level) course in classical control. The reader without a control background will learn classical control theory motivated by a real system to be controlled, namely, a spacecraft (as opposed to some abstract transfer functions). The reader with a prior control background may gain new appreciation of the theory by seeing it presented in the context of an application.

In Chapters 1 and 2, we present the vector kinematics and rigid body dynamics required to be able to describe spacecraft motion. Chapters 3 to 10 contain the orbital mechanics component of this book. Topics include the two-body problem, preliminary orbit determination, orbital and interplanetary maneuvers, orbital perturbations, low-thrust trajectory design, spacecraft formation flying, and the restricted three-body problem. Chapter 11 presents a high level overview of both passive and active means of spacecraft attitude stabilization, and provides an introduction to control systems. Chapters 12 to 16 present aspects of spacecraft attitude dynamics (disturbance torques and a solution for torque-free motion), and more detailed treatments of passive means of spacecraft attitude stabilization. Chapters 17 to 23 present active means of spacecraft attitude control using classical control techniques. Chapters 24 and 25 present introductions to some more advanced topics, namely nonlinear spacecraft attitude control and spacecraft attitude determination. These chapters also provide a brief introduction to nonlinear control theory and state estimation. Chapter 26 presents an overview of practical issues that must be dealt with in designing a spacecraft attitude control system, namely different spacecraft attitude sensor and actuator types, digital control implementation issues and effects of unmodeled dynamics on spacecraft attitude control systems. Finally, Appendices A and B contain some background reference material.

After finishing this book, the reader should have a strong understanding of the fundamentals of spacecraft orbital and attitude dynamics and control, and should be aware of important practical issues that need to be accounted for in spacecraft attitude control design. The reader will be well-prepared for further study in the subject.

The first author would like to express his deep gratitude to the Department of Mechanical and Aerospace Engineering at Carleton University in Ottawa, Canada, for the opportunity to develop and teach courses in orbital mechanics and spacecraft dynamics and control. The notes developed for these courses were the starting point for much of this book.

The reader will notice that this book contains no exercises. This was a decision made in order to keep the page count down. However, the reader will find a full set of exercises accompanying the book, as well as other supplementary material on the book's companion website: *http://arrow.utias.utoronto.ca/damaren/book/*.

Anton H.J. de Ruiter
Christopher J. Damaren
James R. Forbes

1

Kinematics

Spacecraft are free bodies, possessing both translational and rotational motion. The translational component is the subject of *orbital dynamics*, the rotational component is the subject of *attitude dynamics*. It will be seen that the two classes of motion are essentially uncoupled, and can be treated separately.

To be able to study the motion of a spacecraft mathematically, we need a framework for describing it. For this purpose, we need to have a solid understanding of vectors and reference frames, and the associated calculus.

1.1 Physical Vectors

A *physical vector* is a three-dimensional quantity that possesses a *magnitude* and a *direction*. A physical vector will be denoted as $\vec{\mathbf{r}}$, for example. It can be represented graphically by an arrow. Vector addition is defined head-to-tail as shown in Figure 1.1. Multiplication of a vector $\vec{\mathbf{r}}$ by a scalar a scales the magnitude by $|a|$. If a is positive, the direction is unchanged, and if a is negative, the direction is reversed. It is also useful to define a zero-vector denoted by $\vec{\mathbf{0}}$, which has magnitude 0, but no specified direction.

Under these definitions, physical vectors satisfy the following rules for addition:

$$\begin{aligned}
(\vec{\mathbf{a}} + \vec{\mathbf{b}}) + \vec{\mathbf{c}} &= \vec{\mathbf{a}} + (\vec{\mathbf{b}} + \vec{\mathbf{c}}),\\
\vec{\mathbf{a}} + \vec{\mathbf{b}} &= \vec{\mathbf{b}} + \vec{\mathbf{a}},\\
\vec{\mathbf{a}} + \vec{\mathbf{0}} &= \vec{\mathbf{a}},\\
\vec{\mathbf{a}} + (-\vec{\mathbf{a}}) &= \vec{\mathbf{0}},
\end{aligned}$$

Spacecraft Dynamics and Control: An Introduction, First Edition.
Anton H.J. de Ruiter, Christopher J. Damaren and James R. Forbes.

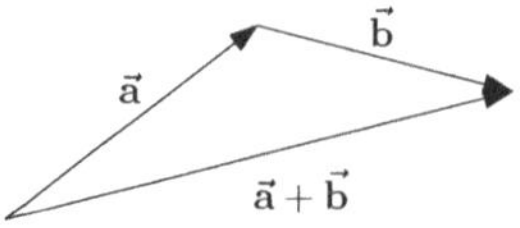

Figure 1.1 Physical vector addition

and the following rules for scalar multiplication:

$$a(b\vec{\mathbf{c}}) = (ab)\vec{\mathbf{c}},$$
$$(a+b)\vec{\mathbf{c}} = a\vec{\mathbf{c}} + b\vec{\mathbf{c}},$$
$$a(\vec{\mathbf{b}}+\vec{\mathbf{c}}) = a\vec{\mathbf{b}} + a\vec{\mathbf{c}},$$
$$1\vec{\mathbf{a}} = \vec{\mathbf{a}},$$
$$0\vec{\mathbf{a}} = \vec{\mathbf{0}}.$$

It is very important to note that the concept of a physical vector is *independent* of a coordinate system.

1.1.1 Scalar Product

Given vectors $\vec{\mathbf{a}}$ and $\vec{\mathbf{b}}$, the scalar (or dot) product between the two vectors is defined as

$$\vec{\mathbf{a}} \cdot \vec{\mathbf{b}} \triangleq |\vec{\mathbf{a}}|\,|\vec{\mathbf{b}}| \cos\theta,$$

where $0 \leq \theta \leq 180°$ is the small angle between the two vectors, as shown in Figure 1.2. By this definition, the scalar product is commutative, that is

$$\vec{\mathbf{a}} \cdot \vec{\mathbf{b}} = \vec{\mathbf{b}} \cdot \vec{\mathbf{a}}.$$

As demonstrated in Figure 1.2, the scalar product $\vec{\mathbf{a}} \cdot \vec{\mathbf{b}}$ is just the projection of $\vec{\mathbf{a}}$ onto $\vec{\mathbf{b}}$ multiplied by $|\vec{\mathbf{b}}|$. Projections are additive, as shown in Figure 1.3, therefore, the scalar product is also distributive, that is

$$(\vec{\mathbf{a}} + \vec{\mathbf{b}}) \cdot \vec{\mathbf{c}} = \vec{\mathbf{a}} \cdot \vec{\mathbf{c}} + \vec{\mathbf{b}} \cdot \vec{\mathbf{c}}. \tag{1.1}$$

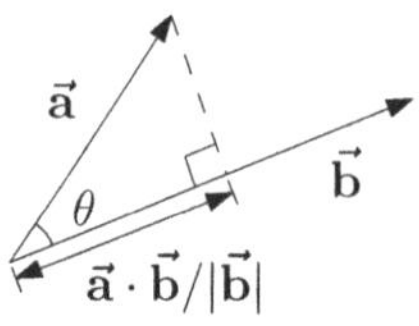

Figure 1.2 Scalar product geometry

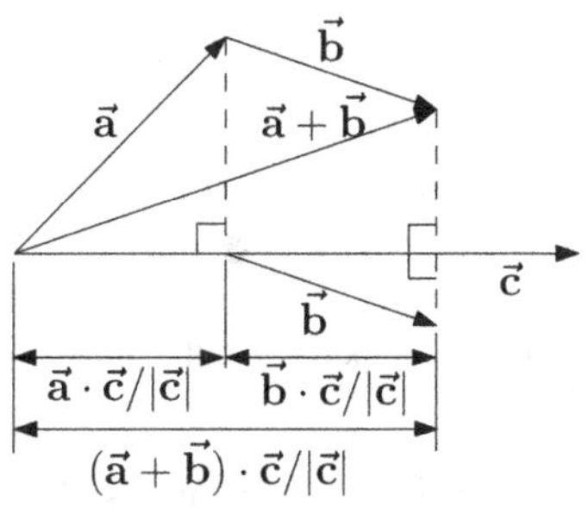

Figure 1.3 Distributivity of scalar product

The following properties are also readily verified from the definition

$$\vec{\mathbf{a}} \cdot \vec{\mathbf{a}} = |\vec{\mathbf{a}}|^2 \geq 0, \tag{1.2}$$

$$\vec{\mathbf{a}} \cdot \vec{\mathbf{a}} = 0 \Leftrightarrow \vec{\mathbf{a}} = \vec{\mathbf{0}}, \tag{1.3}$$

$$\vec{\mathbf{a}} \cdot (c\vec{\mathbf{b}}) = c\vec{\mathbf{a}} \cdot \vec{\mathbf{b}}, \tag{1.4}$$

$$\vec{\mathbf{a}} \cdot \vec{\mathbf{b}} = 0 \Leftrightarrow \vec{\mathbf{a}} \perp \vec{\mathbf{b}} \text{ or } \vec{\mathbf{a}} = \vec{\mathbf{0}} \text{ or } \vec{\mathbf{b}} = \vec{\mathbf{0}}. \tag{1.5}$$

1.1.2 Vector Cross Product

Given vectors $\vec{\mathbf{a}}$ and $\vec{\mathbf{b}}$, the cross-product is defined as a vector $\vec{\mathbf{c}}$, denoted by $\vec{\mathbf{c}} = \vec{\mathbf{a}} \times \vec{\mathbf{b}}$ with magnitude

$$|\vec{\mathbf{c}}| = |\vec{\mathbf{a}}|\, |\vec{\mathbf{b}}| \sin\theta,$$

with a direction perpendicular to both $\vec{\mathbf{a}}$ and $\vec{\mathbf{b}}$, chosen according to the right-hand rule, as shown in Figure 1.4. Note that $0 \leq \theta \leq 180^\circ$ is again the small angle between the two vectors.

From the definition of the cross-product, it is clear that changing the order simply reverses the direction of the cross-product, that is

$$\vec{\mathbf{a}} \times \vec{\mathbf{b}} = -\vec{\mathbf{b}} \times \vec{\mathbf{a}}.$$

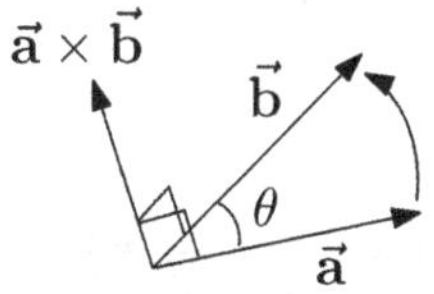

Figure 1.4 Vector cross product

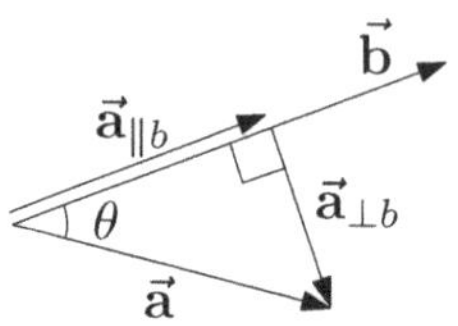

Figure 1.5 Parallel and perpendicular vector components

Now, as shown in Figure 1.5, the vector $\vec{\mathbf{a}}$ can be decomposed into two mutually perpendicular vectors $\vec{\mathbf{a}} = \vec{\mathbf{a}}_{\perp b} + \vec{\mathbf{a}}_{\|b}$, where $\vec{\mathbf{a}}_{\perp b}$ is perpendicular to $\vec{\mathbf{b}}$, and $\vec{\mathbf{a}}_{\|b}$ is parallel to $\vec{\mathbf{b}}$. These components are given by

$$\vec{\mathbf{a}}_{\|b} = \frac{(\vec{\mathbf{a}} \cdot \vec{\mathbf{b}})}{|\vec{\mathbf{b}}|^2}\vec{\mathbf{b}},$$

which is the projection of $\vec{\mathbf{a}}$ onto the direction of $\vec{\mathbf{b}}$, and

$$\vec{\mathbf{a}}_{\perp b} = \vec{\mathbf{a}} - \vec{\mathbf{a}}_{\|b} = \vec{\mathbf{a}} - \frac{(\vec{\mathbf{a}} \cdot \vec{\mathbf{b}})}{|\vec{\mathbf{b}}|^2}\vec{\mathbf{b}}.$$

Since $|\vec{\mathbf{a}}_{\perp b}| = |\vec{\mathbf{a}}| \sin\theta$ (see Figure 1.5), and $\vec{\mathbf{a}}_{\perp b}$ is perpendicular to $\vec{\mathbf{b}}$, $|\vec{\mathbf{a}}_{\perp b} \times \vec{\mathbf{b}}| = |\vec{\mathbf{a}}|\,|\vec{\mathbf{b}}| \sin\theta$. Since $\vec{\mathbf{a}}_{\perp b}$ lies in the plane defined by $\vec{\mathbf{a}}$ and $\vec{\mathbf{b}}$, and points to the same side of $\vec{\mathbf{b}}$ as $\vec{\mathbf{a}}$, $\vec{\mathbf{a}}_{\perp b} \times \vec{\mathbf{b}}$ has the same direction as $\vec{\mathbf{a}} \times \vec{\mathbf{b}}$. Therefore,

$$\vec{\mathbf{a}}_{\perp b} \times \vec{\mathbf{b}} = \vec{\mathbf{a}} \times \vec{\mathbf{b}}. \tag{1.6}$$

Now, we are in a position to show a distributive property of the cross-product. Consider three vectors $\vec{\mathbf{a}}$, $\vec{\mathbf{b}}$ and $\vec{\mathbf{c}}$. First of all, note that

$$\begin{aligned}
(\vec{\mathbf{a}} + \vec{\mathbf{b}})_{\perp c} &= (\vec{\mathbf{a}} + \vec{\mathbf{b}}) - \frac{\left((\vec{\mathbf{a}} + \vec{\mathbf{b}}) \cdot \vec{\mathbf{c}}\right)}{|\vec{\mathbf{c}}|^2}\vec{\mathbf{c}} \\
&= (\vec{\mathbf{a}} + \vec{\mathbf{b}}) - \frac{\left(\vec{\mathbf{a}} \cdot \vec{\mathbf{c}} + \vec{\mathbf{b}} \cdot \vec{\mathbf{c}}\right)}{|\vec{\mathbf{c}}|^2}\vec{\mathbf{c}} \\
&= \left(\vec{\mathbf{a}} - \frac{(\vec{\mathbf{a}} \cdot \vec{\mathbf{c}})}{|\vec{\mathbf{c}}|^2}\vec{\mathbf{c}}\right) + \left(\vec{\mathbf{b}} - \frac{(\vec{\mathbf{b}} \cdot \vec{\mathbf{c}})}{|\vec{\mathbf{c}}|^2}\vec{\mathbf{c}}\right) \\
&= \vec{\mathbf{a}}_{\perp c} + \vec{\mathbf{b}}_{\perp c}
\end{aligned}$$

Therefore, we have

$$\begin{aligned}
\left(\vec{\mathbf{a}} + \vec{\mathbf{b}}\right) \times \vec{\mathbf{c}} &= \left(\vec{\mathbf{a}} + \vec{\mathbf{b}}\right)_{\perp c} \times \vec{\mathbf{c}} \\
&= \left(\vec{\mathbf{a}}_{\perp c} + \vec{\mathbf{b}}_{\perp c}\right) \times \vec{\mathbf{c}}.
\end{aligned}$$

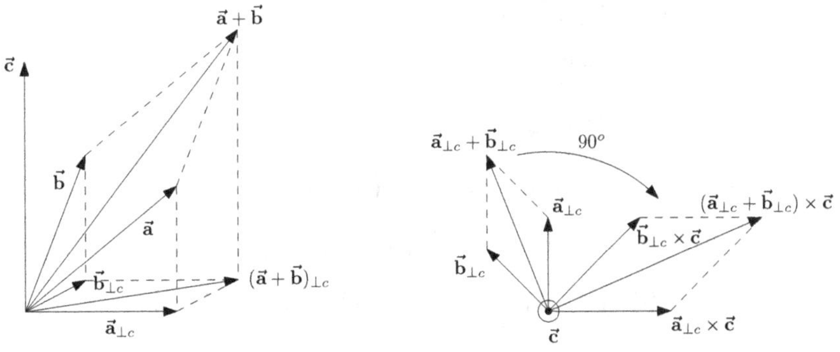

Figure 1.6 Distributivity of vector cross product

Now, the vectors $\vec{\mathbf{a}}_{\perp c}$, $\vec{\mathbf{b}}_{\perp c}$ and $\vec{\mathbf{a}}_{\perp c}+\vec{\mathbf{b}}_{\perp c}$ all are perpendicular to $\vec{\mathbf{c}}$. Therefore,

$$|\vec{\mathbf{a}}_{\perp c}\times\vec{\mathbf{c}}| = |\vec{\mathbf{a}}_{\perp c}|\,|\vec{\mathbf{c}}|,$$

$$|\vec{\mathbf{b}}_{\perp c}\times\vec{\mathbf{c}}| = |\vec{\mathbf{b}}_{\perp c}|\,|\vec{\mathbf{c}}|,$$

$$\left|\left(\vec{\mathbf{a}}_{\perp c}+\vec{\mathbf{b}}_{\perp c}\right)\times\vec{\mathbf{c}}\right| = \left|\left(\vec{\mathbf{a}}_{\perp c}+\vec{\mathbf{b}}_{\perp c}\right)\right|\left|\vec{\mathbf{c}}\right|.$$

Since the vectors $\vec{\mathbf{a}}_{\perp c}$, $\vec{\mathbf{b}}_{\perp c}$ and $\vec{\mathbf{a}}_{\perp c}+\vec{\mathbf{b}}_{\perp c}$ are all perpendicular to $\vec{\mathbf{c}}$, the cross-products $\vec{\mathbf{a}}_{\perp c}\times\vec{\mathbf{c}}$, $\vec{\mathbf{b}}_{\perp c}\times\vec{\mathbf{c}}$ and $\left(\vec{\mathbf{a}}_{\perp c}+\vec{\mathbf{b}}_{\perp c}\right)\times\vec{\mathbf{c}}$ are all simply the vectors $\vec{\mathbf{a}}_{\perp c}$, $\vec{\mathbf{b}}_{\perp c}$ and $\vec{\mathbf{a}}_{\perp c}+\vec{\mathbf{b}}_{\perp c}$ rotated by 90° about the vector $\vec{\mathbf{c}}$, and then scaled by the factor $|\vec{\mathbf{c}}|$, as shown in Figure 1.6. What this shows is that

$$\left(\vec{\mathbf{a}}_{\perp c}+\vec{\mathbf{b}}_{\perp c}\right)\times\vec{\mathbf{c}} = \vec{\mathbf{a}}_{\perp c}\times\vec{\mathbf{c}}+\vec{\mathbf{b}}_{\perp c}\times\vec{\mathbf{c}},$$

and therefore by (1.6),

$$\left(\vec{\mathbf{a}}+\vec{\mathbf{b}}\right)\times\vec{\mathbf{c}} = \vec{\mathbf{a}}\times\vec{\mathbf{c}}+\vec{\mathbf{b}}\times\vec{\mathbf{c}}, \tag{1.7}$$

which is the distributive property we wanted to show. Finally, the following results are also readily derived from the definition:

$$\vec{\mathbf{a}}\times\vec{\mathbf{a}} = \vec{\mathbf{0}}, \tag{1.8}$$

$$(a\vec{\mathbf{b}})\times\vec{\mathbf{c}} = a(\vec{\mathbf{b}}\times\vec{\mathbf{c}}). \tag{1.9}$$

1.1.3 Other Useful Vector Identities

Some other useful vector identities are:

$$\begin{aligned}
&\vec{\mathbf{a}} \times \left(\vec{\mathbf{b}} \times \vec{\mathbf{c}}\right) = (\vec{\mathbf{a}} \cdot \vec{\mathbf{c}})\vec{\mathbf{b}} - (\vec{\mathbf{a}} \cdot \vec{\mathbf{b}})\vec{\mathbf{c}},\\
&\vec{\mathbf{a}} \cdot \left(\vec{\mathbf{b}} \times \vec{\mathbf{c}}\right) = \vec{\mathbf{b}} \cdot (\vec{\mathbf{c}} \times \vec{\mathbf{a}}) = \vec{\mathbf{c}} \cdot \left(\vec{\mathbf{a}} \times \vec{\mathbf{b}}\right),\\
&\vec{\mathbf{a}} \times \left(\vec{\mathbf{b}} \times \vec{\mathbf{c}}\right) + \vec{\mathbf{b}} \times (\vec{\mathbf{c}} \times \vec{\mathbf{a}}) + \vec{\mathbf{c}} \times \left(\vec{\mathbf{a}} \times \vec{\mathbf{b}}\right) = \vec{\mathbf{0}},\\
&\left(\vec{\mathbf{a}} \times \vec{\mathbf{b}}\right) \cdot \left(\vec{\mathbf{c}} \times \vec{\mathbf{d}}\right) = (\vec{\mathbf{a}} \cdot \vec{\mathbf{c}}) \left(\vec{\mathbf{b}} \cdot \vec{\mathbf{d}}\right) - \left(\vec{\mathbf{a}} \cdot \vec{\mathbf{d}}\right) \left(\vec{\mathbf{b}} \cdot \vec{\mathbf{c}}\right).
\end{aligned}$$

Note that the definitions of scalar- and cross-product and all of the associated properties and identities above are *independent* of a coordinate system.

1.2 Reference Frames and Physical Vector Coordinates

Up to this point, we have only considered physical vectors, without any mention of a frame of reference. For computational purposes we need to introduce the concept of a reference frame. Reference frames are also needed to describe the orientation of an object, and are needed for the formulation of kinematics and dynamics.

To define a reference frame, say reference frame "1" (which we will label $\mathcal{F}_1$), it is customary to identify three mutually perpendicular unit length (length of one) physical vectors, labeled as $\vec{\mathbf{x}}_1$, $\vec{\mathbf{y}}_1$ and $\vec{\mathbf{z}}_1$ respectively. The notation used here corresponds to the usual *x*-*y*-*z* axes defined for a Cartesian three-dimensional coordinate system. These three vectors then define the reference frame. The unit vectors are chosen according to the right-handed rule, as shown in Figure 1.7. Under the right-handed rule, the unit vectors satisfy

$$\begin{aligned}
\vec{\mathbf{x}}_1 \times \vec{\mathbf{y}}_1 &= \vec{\mathbf{z}}_1,\\
\vec{\mathbf{y}}_1 \times \vec{\mathbf{z}}_1 &= \vec{\mathbf{x}}_1,\\
\vec{\mathbf{z}}_1 \times \vec{\mathbf{x}}_1 &= \vec{\mathbf{y}}_1.
\end{aligned}$$

Since they are perpendicular, they also satisfy

$$\begin{aligned}
\vec{\mathbf{x}}_1 \cdot \vec{\mathbf{x}}_1 = \vec{\mathbf{y}}_1 \cdot \vec{\mathbf{y}}_1 = \vec{\mathbf{z}}_1 \cdot \vec{\mathbf{z}}_1 = 1,\\
\vec{\mathbf{x}}_1 \cdot \vec{\mathbf{y}}_1 = \vec{\mathbf{x}}_1 \cdot \vec{\mathbf{z}}_1 = \vec{\mathbf{y}}_1 \cdot \vec{\mathbf{z}}_1 = 0.
\end{aligned} \tag{1.10}$$

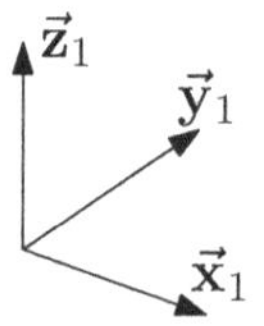

Figure 1.7 Reference frame basis vectors

Now, since the three unit vectors form a basis for physical three-dimensional space, any physical vector $\vec{\mathbf{r}}$ can be written as a linear combination of the unit vectors, that is

$$\begin{aligned}\vec{\mathbf{r}} &= r_{x,1}\vec{\mathbf{x}}_1 + r_{y,1}\vec{\mathbf{y}}_1 + r_{z,1}\vec{\mathbf{z}}_1 \\ &= \begin{bmatrix}\vec{\mathbf{x}}_1 & \vec{\mathbf{y}}_1 & \vec{\mathbf{z}}_1\end{bmatrix}\begin{bmatrix}r_{x,1}\\ r_{y,1}\\ r_{z,1}\end{bmatrix} \\ &= \vec{\mathcal{F}}_1^T\mathbf{r}_1.\end{aligned} \tag{1.11}$$

where

$$\mathbf{r}_1 = \begin{bmatrix}r_{x,1}\\ r_{y,1}\\ r_{z,1}\end{bmatrix} \tag{1.12}$$

is a column matrix containing the coordinates of the physical vector $\vec{\mathbf{r}}$ in reference frame $\mathcal{F}_1$, and

$$\vec{\mathcal{F}}_1 = \begin{bmatrix}\vec{\mathbf{x}}_1\\ \vec{\mathbf{y}}_1\\ \vec{\mathbf{z}}_1\end{bmatrix} \tag{1.13}$$

is a column matrix containing the unit physical vectors defining reference frame $\mathcal{F}_1$. We shall refer to $\vec{\mathcal{F}}_1$ as a *vectrix* (that is, a matrix of physical vectors).

To determine the coordinates of the vector $\vec{\mathbf{r}}$ in frame $\mathcal{F}_1$, we can simply take the dot product of the physical vector (1.11) with each of the unit vectors. For example,

$$\begin{aligned}\vec{\mathbf{r}}\cdot\vec{\mathbf{x}}_1 &= \left(r_{x,1}\vec{\mathbf{x}}_1 + r_{y,1}\vec{\mathbf{y}}_1 + r_{z,1}\vec{\mathbf{z}}_1\right)\cdot\vec{\mathbf{x}}_1, \\ &= r_{x,1}\vec{\mathbf{x}}_1\cdot\vec{\mathbf{x}}_1 + r_{y,1}\vec{\mathbf{y}}_1\cdot\vec{\mathbf{x}}_1 + r_{z,1}\vec{\mathbf{z}}_1\cdot\vec{\mathbf{x}}_1, \\ &= r_{x,1}.\end{aligned}$$

Here, we have made use of properties (1.1) and (1.4) of the dot product of physical vectors. In fact, these properties allow us to treat the dot product of physical vectors in the same manner as scalar multiplication. Using the vectrix notation, we can take advantage of this fact to concisely determine $\mathbf{r}_1$ by taking the dot product of the vector (1.11) with the vectrix (1.13) as follows

$$\vec{\mathcal{F}}_1 \cdot \vec{\mathbf{r}} = \vec{\mathcal{F}}_1 \cdot \left(\vec{\mathcal{F}}_1^T \mathbf{r}_1\right) = \left(\begin{bmatrix} \vec{\mathbf{x}}_1 \\ \vec{\mathbf{y}}_1 \\ \vec{\mathbf{z}}_1 \end{bmatrix} \cdot \begin{bmatrix} \vec{\mathbf{x}}_1 & \vec{\mathbf{y}}_1 & \vec{\mathbf{z}}_1 \end{bmatrix}\right) \mathbf{r}_1$$
$$= \begin{bmatrix} \vec{\mathbf{x}}_1 \cdot \vec{\mathbf{x}}_1 & \vec{\mathbf{x}}_1 \cdot \vec{\mathbf{y}}_1 & \vec{\mathbf{x}}_1 \cdot \vec{\mathbf{z}}_1 \\ \vec{\mathbf{x}}_1 \cdot \vec{\mathbf{y}}_1 & \vec{\mathbf{y}}_1 \cdot \vec{\mathbf{y}}_1 & \vec{\mathbf{y}}_1 \cdot \vec{\mathbf{z}}_1 \\ \vec{\mathbf{x}}_1 \cdot \vec{\mathbf{z}}_1 & \vec{\mathbf{y}}_1 \cdot \vec{\mathbf{z}}_1 & \vec{\mathbf{z}}_1 \cdot \vec{\mathbf{z}}_1 \end{bmatrix} \mathbf{r}_1$$
$$= \begin{bmatrix} 1 & 0 & 0 \\ 0 & 1 & 0 \\ 0 & 0 & 1 \end{bmatrix} \mathbf{r}_1$$

Note that properties (1.1) and (1.4) allowed us to treat the dot product in the same manner as scalar multiplication, and apply the associativity rule for matrix multiplication as we did above. Finally, we have

$$r_{x,1} = \vec{\mathbf{r}} \cdot \vec{\mathbf{x}}_1,$$
$$r_{y,1} = \vec{\mathbf{r}} \cdot \vec{\mathbf{y}}_1,$$
$$r_{z,1} = \vec{\mathbf{r}} \cdot \vec{\mathbf{z}}_1.$$

1.2.1 *Vector Addition and Scalar Multiplication*

We can now determine how to perform vector addition and scalar multiplication operations in terms of the coordinates of a vector in a given reference frame. To this end, let us consider two physical vectors $\vec{\mathbf{a}}$ and $\vec{\mathbf{b}}$ expressed in the same reference frame $\mathcal{F}_1$, and a scalar, c:

$$\vec{\mathbf{a}} = \begin{bmatrix} \vec{\mathbf{x}}_1 & \vec{\mathbf{y}}_1 & \vec{\mathbf{z}}_1 \end{bmatrix} \begin{bmatrix} a_{x,1} \\ a_{y,1} \\ a_{z,1} \end{bmatrix} = \vec{\mathcal{F}}_1^T \mathbf{a}, \quad \vec{\mathbf{b}} = \begin{bmatrix} \vec{\mathbf{x}}_1 & \vec{\mathbf{y}}_1 & \vec{\mathbf{z}}_1 \end{bmatrix} \begin{bmatrix} b_{x,1} \\ b_{y,1} \\ b_{z,1} \end{bmatrix} \vec{\mathcal{F}}_1^T \mathbf{b},$$

It is obvious from the rules for physical vector addition and scalar multiplication that

$$\vec{\mathbf{a}} + \vec{\mathbf{b}} = \begin{bmatrix} \vec{\mathbf{x}}_1 & \vec{\mathbf{y}}_1 & \vec{\mathbf{z}}_1 \end{bmatrix} \begin{bmatrix} a_{x,1} + b_{x,1} \\ a_{y,1} + b_{y,1} \\ a_{z,1} + b_{z,1} \end{bmatrix} = \vec{\mathcal{F}}_1^T (\mathbf{a} + \mathbf{b}),$$

and

$$c\vec{\mathbf{a}} = \begin{bmatrix} \vec{\mathbf{x}}_1 & \vec{\mathbf{y}}_1 & \vec{\mathbf{z}}_1 \end{bmatrix} \begin{bmatrix} c\,a_{x,1} \\ c\,a_{y,1} \\ c\,a_{z,1} \end{bmatrix} = \vec{\mathcal{F}}_1^T\,(c\mathbf{a}).$$

That is, vector addition and scalar multiplication operations can be directly applied to the coordinates of the vectors.

1.2.2 Scalar Product

Let us now examine how to compute the scalar (or dot) product in terms of the coordinates of the vectors in a given reference frame. To this end, let us consider two physical vectors $\vec{\mathbf{a}}$ and $\vec{\mathbf{b}}$ expressed in the same reference frame $\mathcal{F}_1$:

$$\vec{\mathbf{a}} = \begin{bmatrix} \vec{\mathbf{x}}_1 & \vec{\mathbf{y}}_1 & \vec{\mathbf{z}}_1 \end{bmatrix} \begin{bmatrix} a_{x,1} \\ a_{y,1} \\ a_{z,1} \end{bmatrix}, \quad \vec{\mathbf{b}} = \begin{bmatrix} \vec{\mathbf{x}}_1 & \vec{\mathbf{y}}_1 & \vec{\mathbf{z}}_1 \end{bmatrix} \begin{bmatrix} b_{x,1} \\ b_{y,1} \\ b_{z,1} \end{bmatrix}$$

The dot product is now given by

$$\begin{aligned}
\vec{\mathbf{a}} \cdot \vec{\mathbf{b}} &= \begin{bmatrix} a_{x,1} & a_{y,1} & a_{z,1} \end{bmatrix} \begin{bmatrix} \vec{\mathbf{x}}_1 \\ \vec{\mathbf{y}}_1 \\ \vec{\mathbf{z}}_1 \end{bmatrix} \cdot \begin{bmatrix} \vec{\mathbf{x}}_1 & \vec{\mathbf{y}}_1 & \vec{\mathbf{z}}_1 \end{bmatrix} \begin{bmatrix} b_{x,1} \\ b_{y,1} \\ b_{z,1} \end{bmatrix} \\
&= \begin{bmatrix} a_{x,1} & a_{y,1} & a_{z,1} \end{bmatrix} \begin{bmatrix} \vec{\mathbf{x}}_1 \cdot \vec{\mathbf{x}}_1 & \vec{\mathbf{x}}_1 \cdot \vec{\mathbf{y}}_1 & \vec{\mathbf{x}}_1 \cdot \vec{\mathbf{z}}_1 \\ \vec{\mathbf{x}}_1 \cdot \vec{\mathbf{y}}_1 & \vec{\mathbf{y}}_1 \cdot \vec{\mathbf{y}}_1 & \vec{\mathbf{y}}_1 \cdot \vec{\mathbf{z}}_1 \\ \vec{\mathbf{x}}_1 \cdot \vec{\mathbf{z}}_1 & \vec{\mathbf{y}}_1 \cdot \vec{\mathbf{z}}_1 & \vec{\mathbf{z}}_1 \cdot \vec{\mathbf{z}}_1 \end{bmatrix} \begin{bmatrix} b_{x,1} \\ b_{y,1} \\ b_{z,1} \end{bmatrix} \\
&= \begin{bmatrix} a_{x,1} & a_{y,1} & a_{z,1} \end{bmatrix} \begin{bmatrix} 1 & 0 & 0 \\ 0 & 1 & 0 \\ 0 & 0 & 1 \end{bmatrix} \begin{bmatrix} b_{x,1} \\ b_{y,1} \\ b_{z,1} \end{bmatrix} \\
&= \mathbf{a}_1^T \mathbf{b}_1
\end{aligned}$$

Again, properties (1.1) and (1.4) allowed us to treat the dot product in the same manner as scalar multiplication, and apply the associativity rule for matrix multiplication as we did above. Making use of identity (1.2), we can relate the length of the physical vector to the length of its coordinate representation, that is:

$$||\mathbf{a}_1|| = \sqrt{\mathbf{a}_1^T \mathbf{a}_1} = |\vec{\mathbf{a}}|,$$

where $||\mathbf{a}_1||$ is the standard Euclidean length of a column matrix.

1.2.3 Vector Cross Product

We can also determine the cross-product of two vectors in terms of the coordinates with respect to a given reference frame. Consider again the same two vectors as in Section 1.2.2. Since the vector cross product satisfies the same distributive and scalar multiplication properties (1.7) and (1.9) as the vector dot product (compare to (1.1) and (1.4)), we can concisely determine the vector cross-product in terms of the coordinates in the same manner as we determined the dot product in Section 1.2.2 (provided we respect the order in which each individual vector cross-product is taken). We have

$$
\begin{aligned}
\vec{\mathbf{a}} \times \vec{\mathbf{b}} &= \begin{bmatrix} a_{x,1} & a_{y,1} & a_{z,1} \end{bmatrix} \begin{bmatrix} \vec{\mathbf{x}}_1 \\ \vec{\mathbf{y}}_1 \\ \vec{\mathbf{z}}_1 \end{bmatrix} \times \begin{bmatrix} \vec{\mathbf{x}}_1 & \vec{\mathbf{y}}_1 & \vec{\mathbf{z}}_1 \end{bmatrix} \begin{bmatrix} b_{x,1} \\ b_{y,1} \\ b_{z,1} \end{bmatrix} \\
&= \begin{bmatrix} a_{x,1} & a_{y,1} & a_{z,1} \end{bmatrix} \begin{bmatrix} \vec{\mathbf{x}}_1 \times \vec{\mathbf{x}}_1 & \vec{\mathbf{x}}_1 \times \vec{\mathbf{y}}_1 & \vec{\mathbf{x}}_1 \times \vec{\mathbf{z}}_1 \\ \vec{\mathbf{y}}_1 \times \vec{\mathbf{x}}_1 & \vec{\mathbf{y}}_1 \times \vec{\mathbf{y}}_1 & \vec{\mathbf{y}}_1 \times \vec{\mathbf{z}}_1 \\ \vec{\mathbf{z}}_1 \times \vec{\mathbf{x}}_1 & \vec{\mathbf{z}}_1 \times \vec{\mathbf{y}}_1 & \vec{\mathbf{z}}_1 \times \vec{\mathbf{z}}_1 \end{bmatrix} \begin{bmatrix} b_{x,1} \\ b_{y,1} \\ b_{z,1} \end{bmatrix} \\
&= \begin{bmatrix} a_{x,1} & a_{y,1} & a_{z,1} \end{bmatrix} \begin{bmatrix} \vec{\mathbf{0}} & \vec{\mathbf{z}}_1 & -\vec{\mathbf{y}}_1 \\ -\vec{\mathbf{z}}_1 & \vec{\mathbf{0}} & \vec{\mathbf{x}}_1 \\ \vec{\mathbf{y}}_1 & -\vec{\mathbf{x}}_1 & \vec{\mathbf{0}} \end{bmatrix} \begin{bmatrix} b_{x,1} \\ b_{y,1} \\ b_{z,1} \end{bmatrix} \\
&= \begin{bmatrix} \vec{\mathbf{x}}_1 & \vec{\mathbf{y}}_1 & \vec{\mathbf{z}}_1 \end{bmatrix} \begin{bmatrix} 0 & -a_{z,1} & a_{y,1} \\ a_{z,1} & 0 & -a_{x,1} \\ -a_{y,1} & a_{x,1} & 0 \end{bmatrix} \begin{bmatrix} b_{x,1} \\ b_{y,1} \\ b_{z,1} \end{bmatrix} \\
&= \vec{\mathcal{F}}_1^T \mathbf{a}_1^{\times} \mathbf{b}_1,
\end{aligned}
$$

where the 3×3 matrix

$$
\mathbf{a}_1^{\times} \triangleq \begin{bmatrix} 0 & -a_{z,1} & a_{y,1} \\ a_{z,1} & 0 & -a_{x,1} \\ -a_{y,1} & a_{x,1} & 0 \end{bmatrix}
$$

is the cross-product operator matrix corresponding to the vector $\vec{\mathbf{a}}$ in reference frame $\mathcal{F}_1$ coordinates.

1.2.4 Column Matrix Identities

The vector identities presented in Sections 1.1.2 and 1.1.3 can all be rewritten in terms of column matrices. To this end, let us consider four physical vectors $\vec{\mathbf{a}}$, $\vec{\mathbf{b}}$, $\vec{\mathbf{c}}$ and $\vec{\mathbf{d}}$ expressed in the same reference frame $\mathcal{F}$, with corresponding coordinates $\mathbf{a}$, $\mathbf{b}$, $\mathbf{c}$ and $\mathbf{d}$ respectively.

Making use of the above results for the computations for the scalar- and cross-products, the following column matrix identities are automatically obtained

$$
\begin{aligned}
&\mathbf{a}^{\times}\mathbf{a} = \mathbf{0} \\
&\mathbf{a}^{\times}\mathbf{b} = -\mathbf{b}^{\times}\mathbf{a} \\
&\mathbf{a}^{\times}\mathbf{b}^{\times}\mathbf{c} = (\mathbf{a}^{T}\mathbf{c})\mathbf{b} - (\mathbf{a}^{T}\mathbf{b})\mathbf{c}, \\
&\mathbf{a}^{T}\mathbf{b}^{\times}\mathbf{c} = \mathbf{b}^{T}\mathbf{c}^{\times}\mathbf{a} = \mathbf{c}^{T}\mathbf{a}^{\times}\mathbf{b}, \\
&\mathbf{a}^{\times}\mathbf{b}^{\times}\mathbf{c} + \mathbf{b}^{\times}\mathbf{c}^{\times}\mathbf{a} + \mathbf{c}^{\times}\mathbf{a}^{\times}\mathbf{b} = \mathbf{0}, \\
&\left(\mathbf{a}^{\times}\mathbf{b}\right)^{T}\left(\mathbf{c}^{\times}\mathbf{d}\right) = \left(\mathbf{a}^{T}\mathbf{c}\right)\left(\mathbf{b}^{T}\mathbf{d}\right) - \left(\mathbf{a}^{T}\mathbf{d}\right)\left(\mathbf{b}^{T}\mathbf{c}\right).
\end{aligned}
$$

The third identity leads to a very useful result. That is,

$$
\begin{aligned}
\mathbf{a}^{\times}\mathbf{b}^{\times}\mathbf{c} &= (\mathbf{a}^{T}\mathbf{c})\mathbf{b} - (\mathbf{a}^{T}\mathbf{b})\mathbf{c} \\
&= \left(\mathbf{b}\mathbf{a}^{T} - (\mathbf{a}^{T}\mathbf{b})\mathbf{1}\right)\mathbf{c}
\end{aligned}
$$

where we denote the identity matrix by $\mathbf{1}$. Since the above result must hold for any column matrix $\mathbf{c} \in R^3$, we must have

$$
\mathbf{a}^{\times}\mathbf{b}^{\times} = \mathbf{b}\mathbf{a}^{T} - (\mathbf{a}^{T}\mathbf{b})\mathbf{1}. \tag{1.14}
$$

1.3 Rotation Matrices

Spacecraft dynamics problems generally involve the use of several reference frames. For example, to describe the orientation of the spacecraft with respect to the Earth, it makes sense to fix one reference frame (say $\mathcal{F}_1$) to the Earth, and the other (say $\mathcal{F}_2$) to the spacecraft body. The orientation of the spacecraft with respect to the Earth can then be described by the orientation of reference frame $\mathcal{F}_2$ with respect to $\mathcal{F}_1$. In spacecraft terminology, we call the orientation the *attitude*.

Since multiple reference frames will be used, it is necessary to know how to:

1. Describe the orientation of one reference frame with respect to another.
2. Transform the coordinates of a vector from one reference frame to another.

As we shall see, these two issues are directly related. Let us now consider two reference frames, $\mathcal{F}_1$ defined by the unit vectors $\vec{\mathbf{x}}_1$, $\vec{\mathbf{y}}_1$ and $\vec{\mathbf{z}}_1$, and $\mathcal{F}_2$ defined by the unit vectors $\vec{\mathbf{x}}_2$, $\vec{\mathbf{y}}_2$ and $\vec{\mathbf{z}}_2$. Let us also consider an arbitrary vector $\vec{\mathbf{r}}$, as shown in Figure 1.8. Let us now express the vector $\vec{\mathbf{r}}$ in each frame as

$$
\vec{\mathbf{r}} = \vec{\mathcal{F}}_1^T\mathbf{r}_1 = \vec{\mathcal{F}}_2^T\mathbf{r}_2.
$$

As we have shown before, the column matrices $\mathbf{r}_1$ and $\mathbf{r}_2$ contain the coordinates of the vector $\vec{\mathbf{r}}$, in the reference frames $\mathcal{F}_1$ and $\mathcal{F}_2$ respectively. To address point 1, we can find the

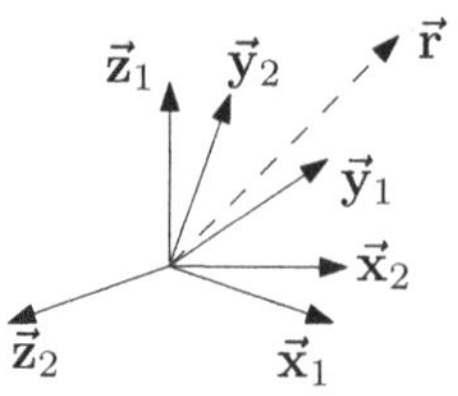

Figure 1.8 Multiple reference frames

coordinates in reference frame $\mathcal{F}_2$ of the unit vectors defining reference frame $\mathcal{F}_1$. As we have seen before, we can write

$$\vec{\mathbf{x}}_1 = \vec{\mathcal{F}}_2^T \mathbf{x}_{1,2}, \quad \vec{\mathbf{y}}_1 = \vec{\mathcal{F}}_2^T \mathbf{y}_{1,2}, \quad \text{and } \vec{\mathbf{z}}_1 = \vec{\mathcal{F}}_2^T \mathbf{z}_{1,2},$$

where

$$\mathbf{x}_{1,2} = \begin{bmatrix} \vec{\mathbf{x}}_1 \cdot \vec{\mathbf{x}}_2 \\ \vec{\mathbf{x}}_1 \cdot \vec{\mathbf{y}}_2 \\ \vec{\mathbf{x}}_1 \cdot \vec{\mathbf{z}}_2 \end{bmatrix}, \quad \mathbf{y}_{1,2} = \begin{bmatrix} \vec{\mathbf{y}}_1 \cdot \vec{\mathbf{x}}_2 \\ \vec{\mathbf{y}}_1 \cdot \vec{\mathbf{y}}_2 \\ \vec{\mathbf{y}}_1 \cdot \vec{\mathbf{z}}_2 \end{bmatrix}, \quad \text{and } \mathbf{z}_{1,2} = \begin{bmatrix} \vec{\mathbf{z}}_1 \cdot \vec{\mathbf{x}}_2 \\ \vec{\mathbf{z}}_1 \cdot \vec{\mathbf{y}}_2 \\ \vec{\mathbf{z}}_1 \cdot \vec{\mathbf{z}}_2 \end{bmatrix}. \tag{1.15}$$

To address point 2, we now obtain a relationship between coordinates of $\vec{\mathbf{r}}$ in the different reference frames as follows:

$$\begin{aligned} \vec{\mathcal{F}}_2^T \mathbf{r}_2 &= \vec{\mathcal{F}}_1^T \mathbf{r}_1 \quad \Rightarrow \\ \vec{\mathcal{F}}_2 \cdot \vec{\mathcal{F}}_2^T \mathbf{r}_2 &= \vec{\mathcal{F}}_2 \cdot \vec{\mathcal{F}}_1^T \mathbf{r}_1 \Rightarrow \\ \mathbf{r}_2 &= \mathbf{C}_{21} \mathbf{r}_1 \end{aligned} \tag{1.16}$$

where

$$\begin{aligned} \mathbf{C}_{21} &= \vec{\mathcal{F}}_2 \cdot \vec{\mathcal{F}}_1^T \\ &= \begin{bmatrix} \vec{\mathbf{x}}_2 \\ \vec{\mathbf{y}}_2 \\ \vec{\mathbf{z}}_2 \end{bmatrix} \cdot \begin{bmatrix} \vec{\mathbf{x}}_1 & \vec{\mathbf{y}}_1 & \vec{\mathbf{z}}_1 \end{bmatrix} \\ &= \begin{bmatrix} \vec{\mathbf{x}}_2 \cdot \vec{\mathbf{x}}_1 & \vec{\mathbf{x}}_2 \cdot \vec{\mathbf{y}}_1 & \vec{\mathbf{x}}_2 \cdot \vec{\mathbf{z}}_1 \\ \vec{\mathbf{y}}_2 \cdot \vec{\mathbf{x}}_1 & \vec{\mathbf{y}}_2 \cdot \vec{\mathbf{y}}_1 & \vec{\mathbf{y}}_2 \cdot \vec{\mathbf{z}}_1 \\ \vec{\mathbf{z}}_2 \cdot \vec{\mathbf{x}}_1 & \vec{\mathbf{z}}_2 \cdot \vec{\mathbf{y}}_1 & \vec{\mathbf{z}}_2 \cdot \vec{\mathbf{z}}_1 \end{bmatrix} \end{aligned} \tag{1.17}$$

The matrix $\mathbf{C}_{21}$ is called a *rotation matrix*. It is sometimes referred to as the "direction cosine matrix", since each entry is a scalar product between two unit vectors, giving the cosine of the angle between them (by definition of the scalar product). Comparing the matrix $\mathbf{C}_{21}$ in (1.17) with (1.15), it can be seen that the columns of the matrix $\mathbf{C}_{21}$ are just the unit vectors defining $\mathcal{F}_1$ expressed in $\mathcal{F}_2$. That is,

$$\mathbf{C}_{21} = \begin{bmatrix} \mathbf{x}_{1,2} & \mathbf{y}_{1,2} & \mathbf{z}_{1,2} \end{bmatrix}. \tag{1.18}$$

The rotation matrix $\mathbf{C}_{21}$ therefore contains all of the information needed to address both points 1 and 2. In particular, the rotation matrix fully describes the attitude of $\mathcal{F}_2$ with respect to $\mathcal{F}_1$.

Rotation matrices have a number of special properties. First of all, from equation (1.18), the rotation matrix is invertible, since its columns are the coordinates of mutually perpendicular unit vectors. Furthermore, making use of the relationship between the unit vectors in (1.10), we have

$$\begin{aligned}
\mathbf{C}_{21}^T\mathbf{C}_{21} &= \begin{bmatrix} \mathbf{x}_{1,2}^T \\ \mathbf{y}_{1,2}^T \\ \mathbf{z}_{1,2}^T \end{bmatrix} \begin{bmatrix} \mathbf{x}_{1,2} & \mathbf{y}_{1,2} & \mathbf{z}_{1,2} \end{bmatrix} \\
&= \begin{bmatrix} \mathbf{x}_{1,2}^T\mathbf{x}_{1,2} & \mathbf{x}_{1,2}^T\mathbf{y}_{1,2} & \mathbf{x}_{1,2}^T\mathbf{z}_{1,2} \\ \mathbf{y}_{1,2}^T\mathbf{x}_{1,2} & \mathbf{y}_{1,2}^T\mathbf{y}_{1,2} & \mathbf{y}_{1,2}^T\mathbf{z}_{1,2} \\ \mathbf{z}_{1,2}^T\mathbf{x}_{1,2} & \mathbf{z}_{1,2}^T\mathbf{y}_{1,2} & \mathbf{z}_{1,2}^T\mathbf{z}_{1,2} \end{bmatrix} \\
&= \begin{bmatrix} 1 & 0 & 0 \\ 0 & 1 & 0 \\ 0 & 0 & 1 \end{bmatrix} = \mathbf{1}.
\end{aligned}$$

This shows that $\mathbf{C}_{21}^{-1} = \mathbf{C}_{21}^T$, and that the rotation matrix is an *orthonormal matrix*, since its inverse is equal to its transpose. Making use of this result, inverting the transformation between coordinates in equation (1.16) gives

$$\mathbf{r}_1 = \mathbf{C}_{21}^T\mathbf{r}_2 = \mathbf{C}_{12}\mathbf{r}_2.$$

That is, $\mathbf{C}_{12} = \mathbf{C}_{21}^T$.

Let us now consider three reference frames, $\mathcal{F}_1$, $\mathcal{F}_2$ and $\mathcal{F}_3$. The vector $\vec{\mathbf{r}}$ is then given by

$$\vec{\mathbf{r}} = \vec{\mathcal{F}}_1^T\mathbf{r}_1 = \vec{\mathcal{F}}_2^T\mathbf{r}_2 = \vec{\mathcal{F}}_3^T\mathbf{r}_3.$$

From our results obtained so far, we have the following transformations:

$$\mathbf{r}_3 = \mathbf{C}_{31}\mathbf{r}_1,$$

and

$$\mathbf{r}_3 = \mathbf{C}_{32}\mathbf{r}_2 = \mathbf{C}_{32}\mathbf{C}_{21}\mathbf{r}_1.$$

Combining these two results leads to (since they must hold for any column matrix $\mathbf{r}_1$)

$$\mathbf{C}_{31} = \mathbf{C}_{32}\mathbf{C}_{21}. \tag{1.19}$$

That is, products of rotation matrices are also rotation matrices, and successive rotations can be combined by multiplying the rotation matrices in the reverse order of the rotations (from right to left).

An important relationship arises when considering the cross-product of two vectors, expressed in two different reference frames. Consider the vectors $\vec{\mathbf{a}} = \vec{\mathcal{F}}_1^T\mathbf{a}_1 = \vec{\mathcal{F}}_2^T\mathbf{a}_2$ and $\vec{\mathbf{b}} = \vec{\mathcal{F}}_1^T\mathbf{b}_1 = \vec{\mathcal{F}}_2^T\mathbf{b}_2$. We have the relationships $\mathbf{a}_2 = \mathbf{C}_{21}\mathbf{a}_1$ and $\mathbf{b}_2 = \mathbf{C}_{21}\mathbf{b}_1$. Let us now examine the cross product of the two vectors. We have

$$\begin{aligned}\vec{\mathbf{a}} \times \vec{\mathbf{b}} = \vec{\mathcal{F}}_2^T\mathbf{a}_2^\times\mathbf{b}_2 &= \vec{\mathcal{F}}_1^T\mathbf{a}_1^\times\mathbf{b}_1\\ &= \vec{\mathcal{F}}_2^T\mathbf{C}_{21}\mathbf{a}_1^\times\mathbf{b}_1\\ &= \vec{\mathcal{F}}_2^T\mathbf{C}_{21}\mathbf{a}_1^\times\mathbf{C}_{21}^T\mathbf{b}_2.\end{aligned}$$

From this we obtain

$$\mathbf{a}_2^\times\mathbf{b}_2 = \mathbf{C}_{21}\mathbf{a}_1^\times\mathbf{C}_{21}^T\mathbf{b}_2.$$

Since this must hold for any $\mathbf{b}_2 \in R^3$, we have

$$\mathbf{a}_2^\times = \mathbf{C}_{21}\mathbf{a}_1^\times\mathbf{C}_{21}^T. \tag{1.20}$$

Finally, since $\mathbf{a}_2 = \mathbf{C}_{21}\mathbf{a}_1$, we have the result

$$(\mathbf{C}_{21}\mathbf{a}_1)^\times = \mathbf{C}_{21}\mathbf{a}_1^\times\mathbf{C}_{21}^T. \tag{1.21}$$

1.3.1 Principal Rotations

An important class of rotations are those about one of the coordinate axes, as shown in Figure 1.9. When reference frame $\mathcal{F}_2$ is obtained from frame $\mathcal{F}_1$ by a rotation about the z-axis, the associated rotation matrix is

$$\mathbf{C}_z(\theta_z) = \begin{bmatrix} \cos\theta_z & \sin\theta_z & 0 \\ -\sin\theta_z & \cos\theta_z & 0 \\ 0 & 0 & 1 \end{bmatrix}.$$

The subscript "z" is used here instead of the subscript "21" to emphasize the fact that the rotation has occurred about the z-axis.

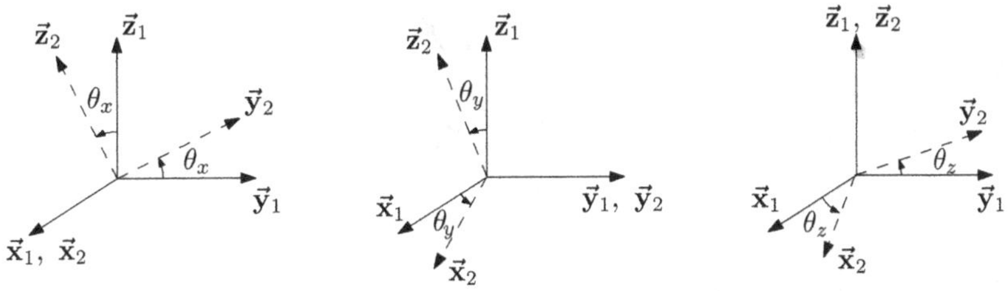

Figure 1.9 Principal rotations

For a rotation about the y-axis,

$$\mathbf{C}_y(\theta_y) = \begin{bmatrix} \cos\theta_y & 0 & -\sin\theta_y \\ 0 & 1 & 0 \\ \sin\theta_y & 0 & \cos\theta_y \end{bmatrix}.$$

For a rotation about the x-axis,

$$\mathbf{C}_x(\theta_x) = \begin{bmatrix} 1 & 0 & 0 \\ 0 & \cos\theta_x & \sin\theta_x \\ 0 & -\sin\theta_x & \cos\theta_x \end{bmatrix}.$$

1.3.2 General Rotations

Up to this point, we have given the definition of the transformation between reference frames (1.17), but we have not justified why it is called a rotation matrix. The answer is found in Euler's Theorem, which was obtained by Leonhard Euler in 1775.

Euler's Theorem: *The most general motion of a rigid body with one point fixed is a rotation about an axis through that point.*

As illustrated in Figure 1.10, what Euler's theorem means is that given any two reference frames $\mathcal{F}_1$ and $\mathcal{F}_2$, frame $\mathcal{F}_2$ can be obtained from frame $\mathcal{F}_1$ by a single rotation about some unit vector which we will denote $\vec{\mathbf{a}} = \vec{\mathcal{F}}_1^T\mathbf{a}$. We shall now demonstrate this. Consider the rotation matrix $\mathbf{C}_{12}$, which transforms coordinates of vectors in frame $\mathcal{F}_2$ to coordinates in $\mathcal{F}_1$. As we have seen, the rotation matrix is given by

$$\mathbf{C}_{12} = \begin{bmatrix} \mathbf{x}_{2,1} & \mathbf{y}_{2,1} & \mathbf{z}_{2,1} \end{bmatrix}, \tag{1.22}$$

where $\mathbf{x}_{2,1}$, $\mathbf{y}_{2,1}$ and $\mathbf{z}_{2,1}$ are the coordinates in $\mathcal{F}_1$ of the basis vectors defining $\mathcal{F}_2$, that is,

$$\vec{\mathbf{x}}_2 = \vec{\mathcal{F}}_1^T\mathbf{x}_{2,1}, \quad \vec{\mathbf{y}}_2 = \vec{\mathcal{F}}_1^T\mathbf{y}_{2,1}, \quad \vec{\mathbf{z}}_2 = \vec{\mathcal{F}}_1^T\mathbf{z}_{2,1}.$$

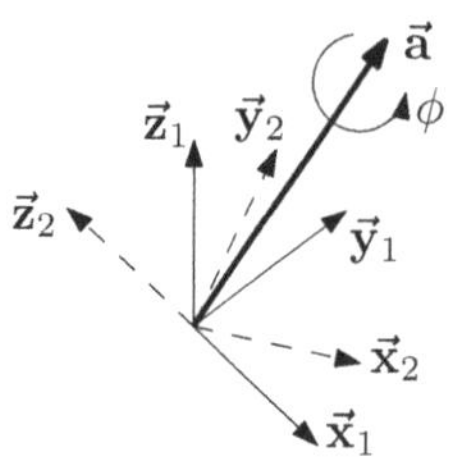

Figure 1.10 General rotation

First of all, we note that the determinant of $\mathbf{C}_{12}$ is given by

$$\det[\mathbf{C}_{12}] = \mathbf{x}_{2,1}^T \mathbf{y}_{2,1}^{\times} \mathbf{z}_{2,1}.$$

This can readily be shown by expansion of the determinant. Since $\mathcal{F}_1$ is a right-handed frame, we have seen that this means that

$$\det[\mathbf{C}_{12}] = \vec{\mathbf{x}}_2 \cdot \left(\vec{\mathbf{y}}_2 \times \vec{\mathbf{z}}_2\right).$$

Since $\mathcal{F}_2$ is a right-handed frame, we have $\vec{\mathbf{x}}_2 = \vec{\mathbf{y}}_2 \times \vec{\mathbf{z}}_2$, and therefore

$$\det[\mathbf{C}_{12}] = 1. \tag{1.23}$$

Therefore, a rotation matrix is a 3 by 3 orthonormal matrix with determinant equal to $+1$. Next, we shall show that $+1$ is an eigenvalue of the rotation matrix. We shall make use of the following facts for matrix determinants.

1. The determinant of a square matrix is equal to the determinant of its transpose,

$$\det[\mathbf{A}] = \det\left[\mathbf{A}^T\right].$$

2. The determinant of a product of square matrices is equal to the product of the individual determinants,

$$\det[\mathbf{AB}] = \det[\mathbf{A}]\det[\mathbf{B}].$$

3. Given an n by n matrix, $\mathbf{A}$ and a scalar a,

$$\det[a\mathbf{A}] = a^n \det[\mathbf{A}].$$

We shall now show that +1 is an eigenvalue of the rotation matrix $\mathbf{C}_{12}$. Consider $\det[\mathbf{C}_{12} - \mathbf{1}]$, where $\mathbf{1}$ is the 3 by 3 identity matrix. Then, since $\det[\mathbf{C}_{12}] = 1$, we have

$$\det[\mathbf{C}_{12} - \mathbf{1}] = \det\left[\mathbf{C}_{12}^T\right]\det[\mathbf{C}_{12} - \mathbf{1}],$$

by fact 1. By fact 2 and the orthonormality of $\mathbf{C}_{12}$, this becomes

$$\begin{aligned}\det\left[\mathbf{C}_{12} - \mathbf{1}\right] &= \det\left[\mathbf{C}_{12}^T\left(\mathbf{C}_{12} - \mathbf{1}\right)\right] \\ &= \det\left[\mathbf{1} - \mathbf{C}_{12}^T\right].\end{aligned}$$

By fact 1 and fact 3, this becomes

$$\begin{aligned}\det\left[\mathbf{C}_{12} - \mathbf{1}\right] &= \det\left[\mathbf{1} - \mathbf{C}_{12}\right] \\ &= (-1)^3 \det\left[\mathbf{C}_{12} - \mathbf{1}\right].\end{aligned}$$

The only way that this final equality holds is if

$$\det\left[\mathbf{C}_{12} - \mathbf{1}\right] = 0,$$

which shows that +1 is indeed an eigenvalue of $\mathbf{C}_{12}$.

Now, let $\mathbf{a}$ be a unit eigenvector of $\mathbf{C}_{12}$, corresponding to the eigenvalue +1. Then, we have

$$\mathbf{C}_{12}\mathbf{a} = \mathbf{a},$$

and the rotation matrix $\mathbf{C}_{12}$ leaves the unit eigenvector $\mathbf{a}$ unchanged. We now define the unit physical vector $\vec{\mathbf{a}} = \vec{\mathcal{F}}_1^T\mathbf{a}$. Let us now choose another unit vector $\vec{\mathbf{b}} = \vec{\mathcal{F}}_1^T\mathbf{b}$ perpendicular to $\vec{\mathbf{a}}$, that is $\mathbf{b}^T\mathbf{b} = 1$ and $\mathbf{a}^T\mathbf{b} = 0$. Transforming the coordinates of the vector $\vec{\mathbf{b}}$ through the rotation matrix yields a new vector given by

$$\vec{\mathbf{b}}' = \vec{\mathcal{F}}_1^T\mathbf{C}_{12}\mathbf{b}.$$

We see immediately that

$$|\vec{\mathbf{b}}'| = \left((\mathbf{C}_{12}\mathbf{b})^T\,\mathbf{C}_{12}\mathbf{b}\right)^{\frac{1}{2}} = 1,$$

and

$$\begin{aligned}\vec{\mathbf{a}} \cdot \vec{\mathbf{b}}' &= \mathbf{a}^T\mathbf{C}_{12}\mathbf{b} \\ &= (\mathbf{C}_{12}\mathbf{a})^T\,\mathbf{C}_{12}\mathbf{b} \\ &= 0.\end{aligned}$$

Therefore, as shown in Figure 1.11, the transformed vector $\vec{\mathbf{b}}'$ is also a unit vector perpendicular to $\vec{\mathbf{a}}$. This means that the transformation of the coordinates $\mathbf{b}$ by the matrix $\mathbf{C}_{12}$ is equivalent to a rotation of the vector $\vec{\mathbf{b}}$ about the vector $\vec{\mathbf{a}}$ through some angle ϕ. Note that we define the angle of rotation ϕ to be positive in the right hand sense about $\vec{\mathbf{a}}$. Next, let us define a third unit vector $\vec{\mathbf{c}} = \vec{\mathcal{F}}_1^T\mathbf{c} = \vec{\mathbf{a}} \times \vec{\mathbf{b}}$. In frame $\mathcal{F}_1$, the coordinates are $\mathbf{c} = \mathbf{a}^{\times}\mathbf{b}$. The vectors $\vec{\mathbf{a}}$, $\vec{\mathbf{b}}$ and $\vec{\mathbf{c}}$ therefore constitute an orthogonal triad, and any other vector can be written in terms of these. Transforming the coordinates of the vector $\vec{\mathbf{c}}$ through the rotation matrix yields a new vector given by

$$\vec{\mathbf{c}}' = \vec{\mathcal{F}}_1^T\mathbf{C}_{12}\mathbf{c}.$$

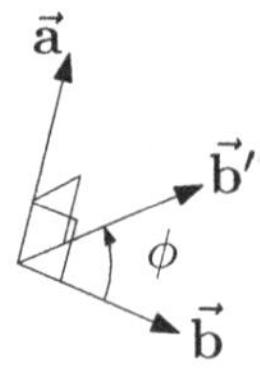

Figure 1.11 Rotation about $\vec{\mathbf{a}}$

We now see that

$$\begin{aligned}\vec{\mathbf{c}}' &= \vec{\mathcal{F}}_1^T \mathbf{C}_{12}\mathbf{a}^\times \mathbf{b},\\ &= \vec{\mathcal{F}}_1^T \mathbf{C}_{12}\mathbf{a}^\times \mathbf{C}_{12}^T\mathbf{C}_{12}\mathbf{b},\\ &= \vec{\mathcal{F}}_1^T \left(\mathbf{C}_{12}\mathbf{a}\right)^\times \mathbf{C}_{12}\mathbf{b},\\ &= \vec{\mathcal{F}}_1^T \mathbf{a}^\times \mathbf{C}_{12}\mathbf{b},\\ &= \vec{\mathbf{a}} \times \vec{\mathbf{b}}'.\end{aligned}$$

This means that the vectors $\vec{\mathbf{a}}$, $\vec{\mathbf{b}}'$ and $\vec{\mathbf{c}}'$ are also an orthogonal triad with the same sense as $\vec{\mathbf{a}}$, $\vec{\mathbf{b}}$ and $\vec{\mathbf{c}}$. The only way that this is possible is if the transformation of the coordinates $\mathbf{c}$ by the matrix $\mathbf{C}_{12}$ is equivalent to a rotation of the vector $\vec{\mathbf{c}}$ about the vector $\vec{\mathbf{a}}$ through the same angle ϕ as the vector $\vec{\mathbf{b}}$, as shown in Figure 1.12. Now, since any vector $\vec{\mathbf{g}} = \vec{\mathcal{F}}_1^T \mathbf{g}$ can be written as a linear combination of the vectors $\vec{\mathbf{a}}$, $\vec{\mathbf{b}}$ and $\vec{\mathbf{c}}$, the transformation of the coordinates $\mathbf{g}$ by the matrix $\mathbf{C}_{12}$ is equivalent to a rotation of the vector $\vec{\mathbf{g}}$ about the vector $\vec{\mathbf{a}}$ through the same angle ϕ.

Let us now see what this means in terms of the relationship between frames $\mathcal{F}_1$ and $\mathcal{F}_2$. It is easy to from the expression for the rotation matrix in (1.22) that

$$\mathbf{x}_{2,1} = \mathbf{C}_{12}\begin{bmatrix}1\\0\\0\end{bmatrix}, \quad \mathbf{y}_{2,1} = \mathbf{C}_{12}\begin{bmatrix}0\\1\\0\end{bmatrix}, \quad \mathbf{z}_{2,1} = \mathbf{C}_{12}\begin{bmatrix}0\\0\\1\end{bmatrix}.$$

Since the basis vectors defining frame $\mathcal{F}_1$ satisfy

$$\vec{\mathbf{x}}_1 = \vec{\mathcal{F}}_1^T \mathbf{x}_{1,1}, \quad \vec{\mathbf{y}}_1 = \vec{\mathcal{F}}_1^T \mathbf{y}_{1,1}, \quad \vec{\mathbf{z}}_1 = \vec{\mathcal{F}}_1^T \mathbf{y}_{1,1},$$

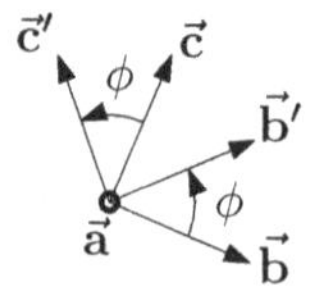

Figure 1.12 Rotation about $\vec{\mathbf{a}}$

with

$$\mathbf{x}_{1,1} = \begin{bmatrix} 1 \\ 0 \\ 0 \end{bmatrix}, \quad \mathbf{y}_{1,1} = \begin{bmatrix} 0 \\ 1 \\ 0 \end{bmatrix}, \quad \mathbf{z}_{1,1} = \begin{bmatrix} 0 \\ 0 \\ 1 \end{bmatrix}$$

it is clear now that the basis vectors defining frame $\mathcal{F}_2$ can all be obtained by rotating the basis vectors defining frame $\mathcal{F}_1$ about the unit vector $\vec{\mathbf{a}}$ through the angle ϕ. This is precisely Euler's Theorem.

1.3.2.1 Rotation Matrix in Terms of the Principal Axis and Angle of Rotation

We shall now find an expression for the rotation matrix in terms of the principal axis and angle of rotation. Let $\vec{\mathbf{a}} = \vec{\mathcal{F}}_1^T \mathbf{a}$ and ϕ be the principal axis and angle of rotation corresponding to the rotation matrix $\mathbf{C}_{12}$, respectively.

To this end, let us examine a rotation of an arbitrary vector $\vec{\mathbf{v}} = \vec{\mathcal{F}}_1^T \mathbf{v}$ about $\vec{\mathbf{a}}$ through angle ϕ in the right-hand sense, as shown in Figure 1.13. We can decompose the vector $\vec{\mathbf{v}}$ into a part parallel to $\vec{\mathbf{a}}$ (denoted $\vec{\mathbf{v}}_{\parallel}$), and a part perpendicular to $\vec{\mathbf{a}}$ (denoted $\vec{\mathbf{v}}_{\perp}$). That is,

$$\vec{\mathbf{v}} = \vec{\mathbf{v}}_{\parallel} + \vec{\mathbf{v}}_{\perp}.$$

These components are given by

$$\begin{aligned} \vec{\mathbf{v}}_{\parallel} &= (\vec{\mathbf{a}} \cdot \vec{\mathbf{v}})\vec{\mathbf{a}}, \\ \vec{\mathbf{v}}_{\perp} &= \vec{\mathbf{v}} - (\vec{\mathbf{a}} \cdot \vec{\mathbf{v}})\vec{\mathbf{a}}. \end{aligned} \tag{1.24}$$

Now, as seen in Figure 1.13, to rotate the vector $\vec{\mathbf{v}}$ about $\vec{\mathbf{a}}$ (which we will denote $\vec{\mathbf{v}}_{rot}$), we only need to rotate the perpendicular part (which we will denote $\vec{\mathbf{v}}_{\perp rot}$), i.e.

$$\vec{\mathbf{v}}_{rot} = \vec{\mathbf{v}}_{\parallel} + \vec{\mathbf{v}}_{\perp rot}.$$

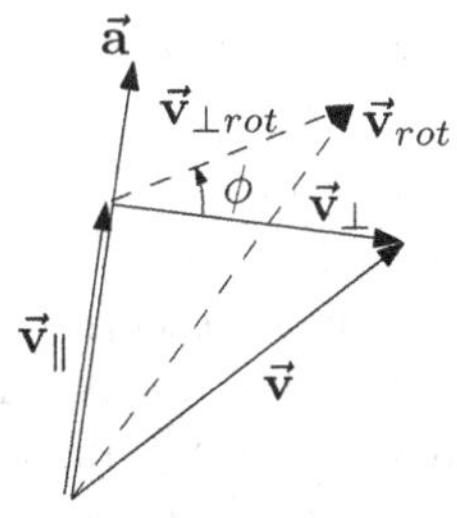

Figure 1.13 Rotation of a vector

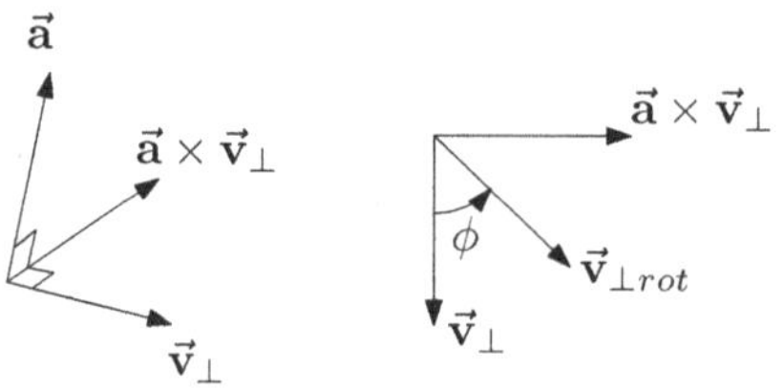

Figure 1.14 Rotation of the perpendicular component of a vector

To this end, we can set up a vector perpendicular to $\vec{\mathbf{a}}$ and $\vec{\mathbf{v}}_\perp$, given by $\vec{\mathbf{a}} \times \vec{\mathbf{v}}_\perp$ as shown in Figure 1.14. Note that $|\vec{\mathbf{a}} \times \vec{\mathbf{v}}_\perp| = |\vec{\mathbf{v}}_\perp|$. As seen in Figure 1.14, the rotated vector $\vec{\mathbf{v}}_{\perp rot}$ can now be expressed as

$$\vec{\mathbf{v}}_{\perp rot} = \vec{\mathbf{v}}_\perp \cos\phi + \vec{\mathbf{a}} \times \vec{\mathbf{v}}_\perp \sin\phi.$$

Substituting (1.24) and making use of the fact that $\vec{\mathbf{a}} \times \vec{\mathbf{a}} = \vec{\mathbf{0}}$ gives

$$\vec{\mathbf{v}}_{rot} = \vec{\mathbf{v}} \cos\phi + (\vec{\mathbf{a}} \cdot \vec{\mathbf{v}})\vec{\mathbf{a}}(1 - \cos\phi) + \vec{\mathbf{a}} \times \vec{\mathbf{v}} \sin\phi.$$

Expressing this in $\mathcal{F}_1$ coordinates, we have

$$\vec{\mathbf{v}}_{rot} = \vec{\mathcal{F}}_1^T \left[\cos\phi\mathbf{1} + (1 - \cos\phi)\mathbf{a}\mathbf{a}^T + \sin\phi\mathbf{a}^\times\right]\mathbf{v} = \vec{\mathcal{F}}_1^T \mathbf{C}_{12}\mathbf{v}. \tag{1.25}$$

Note that the second equality is due to the fact that the axis $\vec{\mathbf{a}}$ and angle ϕ correspond to the rotation matrix $\mathbf{C}_{12}$. Since the above relationship must hold for any $\mathbf{v} \in \mathbf{R}^3$, it must be that

$$\mathbf{C}_{12} = \cos\phi\mathbf{1} + (1 - \cos\phi)\mathbf{a}\mathbf{a}^T + \sin\phi\mathbf{a}^\times.$$

Taking the transpose, we finally obtain

$$\mathbf{C}_{21} = \cos\phi\mathbf{1} + (1 - \cos\phi)\mathbf{a}\mathbf{a}^T - \sin\phi\mathbf{a}^\times. \tag{1.26}$$

Note that while we have used the coordinates of $\vec{\mathbf{a}}$ in reference frame $\mathcal{F}_1$, we could have used the coordinates in $\mathcal{F}_2$, since they are the same. A simple calculation shows that

$$\mathbf{C}_{21}\mathbf{a} = \mathbf{a},$$

as it should be. Another consequence of Euler's Theorem is that while the rotation matrix $\mathbf{C}_{21}$ contains nine entries, it can be fully described by four parameters, that is the three components of the axis of rotation $\mathbf{a}$, and that angle of rotation ϕ. In fact, the three components of the axis of rotation are not independent, since $\mathbf{a}^T\mathbf{a} = 1$ ($\vec{\mathbf{a}}$ is a unit vector), so in fact, three parameters could be used to fully describe the rotation matrix.

1.3.2.2 Finding the Principal Axis and Angle from the Rotation Matrix

Let

$$\mathbf{a} = \begin{bmatrix} a_1 \\ a_2 \\ a_3 \end{bmatrix}, \text{ and } \mathbf{C}_{21} = \begin{bmatrix} c_{11} & c_{12} & c_{13} \\ c_{21} & c_{22} & c_{23} \\ c_{31} & c_{32} & c_{33} \end{bmatrix}.$$

Then, taking the trace of $\mathbf{C}_{21}$ in (1.26), we obtain

$$\text{trace}\,[\mathbf{C}_{21}] = 3\cos\phi + (1-\cos\phi)(a_1^2 + a_2^2 + a_3^2) = 1 + 2\cos\phi.$$

Note that the last equality follows since $\mathbf{a}$ is a unit column matrix. Therefore, we can find the principal angle of rotation as

$$\phi = \cos^{-1}\left(\frac{\text{trace}\,[\mathbf{C}_{21}] - 1}{2}\right). \tag{1.27}$$

An important point to note is that a rotation about the axis $\vec{\mathbf{a}}$ through the angle ϕ is equivalent to a rotation about the axis $-\vec{\mathbf{a}}$ through the angle $-\phi$. Therefore, when solving 1.27, we shall restrict ourselves to the range $0 \le \phi \le \pi$, since a rotation angle in the range $-\pi < \phi < 0$ about $\mathbf{a}$ is equivalent to a rotation angle in the range $0 \le \phi \le \pi$ about the negative axis of rotation $-\mathbf{a}$.

Having found the angle ϕ, we can now find the corresponding axis of rotation $\mathbf{a}$. From (1.26), we find that

$$\mathbf{C}_{21}^T - \mathbf{C}_{21} = 2\sin\phi\,\mathbf{a}^\times,$$

from which we can readily find that if $0 < \phi < \pi$, the axis of rotation can be obtained from

$$\begin{aligned} a_1 &= \frac{c_{23} - c_{32}}{2\sin\phi}, \\ a_2 &= \frac{c_{31} - c_{13}}{2\sin\phi}, \\ a_3 &= \frac{c_{12} - c_{21}}{2\sin\phi}. \end{aligned} \tag{1.28}$$

When $\phi = \pm\pi$, the rotation matrix becomes

$$\mathbf{C}_{21} = -\mathbf{1} + 2\mathbf{a}\mathbf{a}^T.$$

From this, we find that if $\phi = \pm\pi$,

$$|a_1| = \left(\frac{c_{11}+1}{2}\right)^{\frac{1}{2}}, \quad |a_2| = \left(\frac{c_{22}+1}{2}\right)^{\frac{1}{2}}, \quad |a_3| = \left(\frac{c_{33}+1}{2}\right)^{\frac{1}{2}}. \tag{1.29}$$

Since a rotation of either π or $-\pi$ about the axes **a** and $-$**a** are equivalent, we can arbitrarily choose the sign of one of a_1, a_2 or a_3. Some possible solutions are

- If $\phi = \pm\pi$ and $|a_1| > 0$,

$$\begin{aligned} a_1 &= |a_1|, \\ a_2 &= \text{sign}\,(c_{12})\,|a_2|, \\ a_3 &= \text{sign}\,(c_{13})\,|a_3|. \end{aligned}$$

- If $\phi = \pm\pi$ and $|a_2| > 0$,

$$\begin{aligned} a_1 &= \text{sign}\,(c_{12})\,|a_1|, \\ a_2 &= |a_2|, \\ a_3 &= \text{sign}\,(c_{23})\,|a_3|. \end{aligned}$$

- If $\phi = \pm\pi$ and $|a_3| > 0$,

$$\begin{aligned} a_1 &= \text{sign}\,(c_{13})\,|a_1|, \\ a_2 &= \text{sign}\,(c_{23})\,|a_2|, \\ a_3 &= |a_3|. \end{aligned}$$

When $\phi = 0$, the axis **a** cannot be determined uniquely. Physically this makes sense, since a zero rotation about any axis results in the same orientation.

1.3.3 Euler Angles

In the preceeding section we have found that the rotation matrix can be described by three parameters. There are in fact many parameterizations that could be used. A common set of parameters is known as the Euler Angles. Euler Angles describe three successive principal rotations. For example, a possible sequence is:

1. A rotation ψ about the original z-axis (called a "yaw" rotation).
2. A rotation θ about the intermediate y-axis (called a "pitch" rotation).
3. A rotation ϕ about the transformed x-axis (called a "roll" rotation).

This is a very common choice in aerospace applications, and is called a 3-2-1 attitude sequence, and is depicted in Figure 1.15. The terminology relates to the order of rotations. A principal z-axis (labeled 3) rotation is first, followed by a principal y-axis (labeled 2) rotation, followed

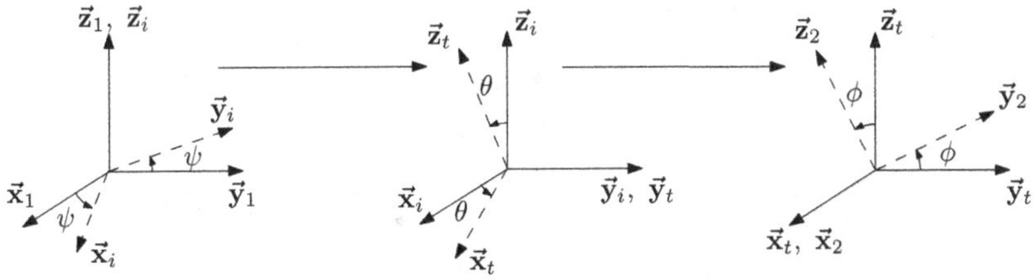

Figure 1.15 3-2-1 Euler rotation sequence

by a principal x-axis (labeled 1) rotation. In this case, the rotation matrix from frame $\mathcal{F}_1$ to frame $\mathcal{F}_2$ is given by

$$\begin{aligned} \mathbf{C}_{21}(\phi, \theta, \psi) &= \mathbf{C}_x(\phi)\mathbf{C}_y(\theta)\mathbf{C}_z(\psi) \\ &= \begin{bmatrix} c_\theta c_\psi & c_\theta s_\psi & -s_\theta \\ s_\phi s_\theta c_\psi - c_\phi s_\psi & s_\phi s_\theta s_\psi + c_\phi c_\psi & s_\phi c_\theta \\ c_\phi s_\theta c_\psi + s_\phi s_\psi & c_\phi s_\theta s_\psi - s_\phi c_\psi & c_\phi c_\theta \end{bmatrix} \end{aligned} \tag{1.30}$$

where $s_b = \sin b$ and $c_b = \cos b$.

An unfortunate consequence of using three parameters to describe the rotation matrix is that a singularity occurs. It can be shown that this will occur for any three-dimensional parameterization of the rotation matrix. For the 3-2-1 sequence above, the singularity occurs when the pitch angle is $\theta = \pm 90°$. For example, when $\theta = 90°$, the rotation matrix becomes

$$\mathbf{C}_{21}(\phi, 90°, \psi) = \begin{bmatrix} 0 & 0 & -1 \\ \sin(\phi - \psi) & \cos(\phi - \psi) & 0 \\ \cos(\phi - \psi) & -\sin(\phi - \psi) & 0 \end{bmatrix}.$$

Physically, at the singularity the first and third rotations in the sequence occur about the same axis. In this case, the roll and yaw angles (ϕ and ψ) are associated with the same rotation, and cannot be determined uniquely. Outside of the singularity however, we can uniquely determine the angles from the rotation matrix. Denoting the rotation matrix by

$$\mathbf{C} = \begin{bmatrix} c_{11} & c_{12} & c_{13} \\ c_{21} & c_{22} & c_{23} \\ c_{31} & c_{32} & c_{33} \end{bmatrix},$$

from equation (1.30), we see that

$$\begin{aligned} \phi &= \tan^{-1}(c_{23}/c_{33}), \\ \theta &= -\sin^{-1}(c_{13}), \\ \psi &= \tan^{-1}(c_{12}/c_{11}). \end{aligned}$$

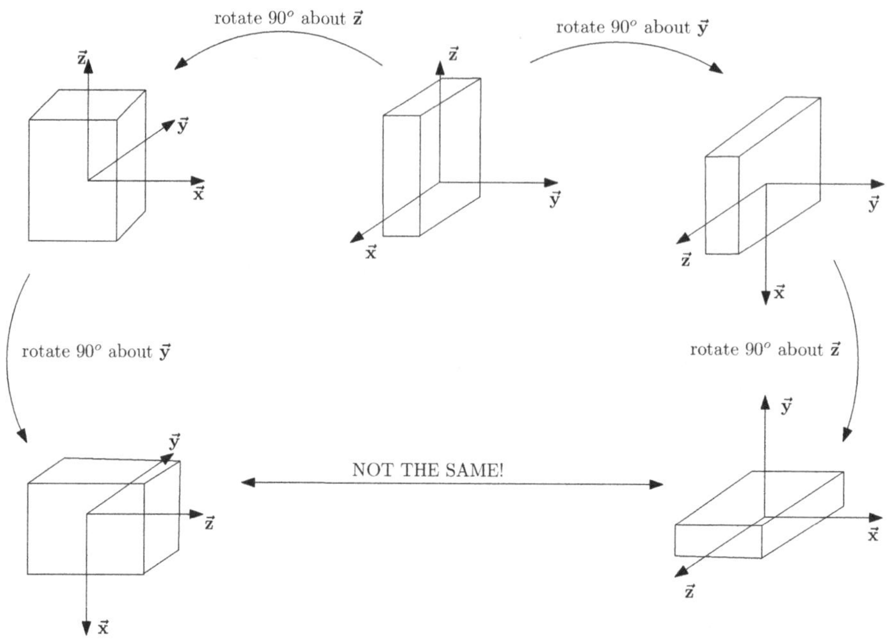

Figure 1.16 Non commuting rotations

The signs of c_{23}, c_{33} and c_{11}, c_{12} determine the quadrants of ϕ and ψ respectively.

Note that it is very important to specify the order of rotations (e.g. 3-2-1), since as seen in Figure 1.16, **rotations do NOT commute!**

1.3.4 Quaternions

As seen in the previous section, an unfortunate consequence of using a three-dimensional parameterization of the rotation matrix is the existence of a singularity, that is, a rotation for which the three parameters cannot be uniquely determined. To overcome this problem, it is necessary to add a redundant parameter to the parameterization. One possibility is the principal axis and angle of rotation. A more common and very useful choice of parameterization is the quaternion (also known as Euler parameters). The advantage of using quaternions, unlike axis-angle parameters is that the expression for the rotation matrix, and kinematics are purely algebraic (they contain no trigonometric functions). This makes them very useful and efficient for computational purposes.

To define the quaternion, we first need to reexamine the rotation matrix in terms of the principal axis $\mathbf{a}$ and angle ϕ of rotation. From (1.26), we have

$$\mathbf{C} = \cos\phi\mathbf{1} + (1 - \cos\phi)\mathbf{a}\mathbf{a}^T - \sin\phi\mathbf{a}^\times.$$

Let us now make use of the trigonometric identities

$$\sin\phi = 2\sin\frac{\phi}{2}\cos\frac{\phi}{2}, \quad \cos\phi = 2\cos^2\frac{\phi}{2} - 1 = 1 - 2\sin^2\frac{\phi}{2}.$$

We can now rewrite the rotation matrix as

$$\mathbf{C} = \left(2\cos^2\frac{\phi}{2} - 1\right) + 2\sin^2\frac{\phi}{2}\mathbf{a}\mathbf{a}^T - 2\sin\frac{\phi}{2}\cos\frac{\phi}{2}\mathbf{a}^\times.$$

Based on this, we now define a vector and scalar part of the quaternion as

$$\boldsymbol{\epsilon} = \mathbf{a}\sin\frac{\phi}{2}, \quad \eta = \cos\frac{\phi}{2}, \tag{1.31}$$

respectively. We immediately see that the quaternion satisfies a unit magnitude constraint

$$\boldsymbol{\epsilon}^T\boldsymbol{\epsilon} + \eta^2 = 1. \tag{1.32}$$

The rotation matrix in terms of the quaternion is given by

$$\mathbf{C} = \left(2\eta^2 - 1\right)\mathbf{1} + 2\boldsymbol{\epsilon}\boldsymbol{\epsilon}^T - 2\eta\boldsymbol{\epsilon}^\times. \tag{1.33}$$

1.3.4.1 Quaternion from Rotation Matrix

Let

$$\boldsymbol{\epsilon} = \begin{bmatrix} \epsilon_1 \\ \epsilon_2 \\ \epsilon_3 \end{bmatrix}, \text{ and } \mathbf{C} = \begin{bmatrix} c_{11} & c_{12} & c_{13} \\ c_{21} & c_{22} & c_{23} \\ c_{31} & c_{32} & c_{33} \end{bmatrix}.$$

Then, taking the trace of $\mathbf{C}$ in (1.33), we obtain

$$\text{trace}\,[\mathbf{C}] = 3\left(2\eta^2 - 1\right) + 2(\epsilon_1^2 + \epsilon_2^2 + \epsilon_3^2) = 4\eta^2 - 1, \tag{1.34}$$

Note that we have made use of (1.32) to achieve the last equality. Therefore, we obtain

$$\eta = \pm\frac{(\text{trace}\,[\mathbf{C}] + 1)^{\frac{1}{2}}}{2}. \tag{1.35}$$

An important point to note is that we may choose either sign for η. It can easily be seen from (1.33) that the quaternions $(\boldsymbol{\epsilon}, \eta)$ and $(-\boldsymbol{\epsilon}, -\eta)$ correspond to the same rotation matrix. Physically, a positive value for η corresponds to a principal angle in the range $-\pi < \phi < \pi$. A negative value for η corresponds to a principal angle in the range $-2\pi < \phi < -\pi$ or $\pi < \phi < 2\pi$. As shown in Figure 1.17, by appropriately choosing the direction of the principal

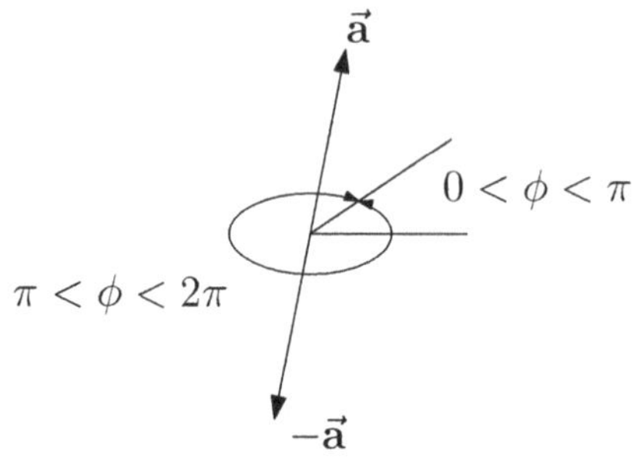

Figure 1.17 Equivalent rotations

axis of rotation, these rotations are equivalent. Having found η, we can now obtain $\boldsymbol{\epsilon}$. From (1.33),

$$\mathbf{C}^T - \mathbf{C} = 4\eta\boldsymbol{\epsilon}^\times, \tag{1.36}$$

and we see that if $\eta \neq 0$, then the corresponding $\boldsymbol{\epsilon}$ is given by

$$\begin{aligned} \epsilon_1 &= \frac{c_{23} - c_{32}}{4\eta}, \\ \epsilon_2 &= \frac{c_{31} - c_{13}}{4\eta}, \\ \epsilon_3 &= \frac{c_{12} - c_{21}}{4\eta}. \end{aligned} \tag{1.37}$$

When $\eta = 0$, the rotation matrix becomes

$$\mathbf{C} = -\mathbf{1} + 2\boldsymbol{\epsilon}\boldsymbol{\epsilon}^T.$$

Physically, $\eta = 0$ corresponds to a principal rotation angle $\phi = \pm\pi$. We now find that if $\eta = 0$,

$$\begin{aligned} |\epsilon_1| = \left(\frac{c_{11}+1}{2}\right)^{\frac{1}{2}} = |a_1|, \quad |\epsilon_2| = \left(\frac{c_{22}+1}{2}\right)^{\frac{1}{2}} = |a_2|, \quad |\epsilon_3| \\ = \left(\frac{c_{33}+1}{2}\right)^{\frac{1}{2}} = |a_3|. \end{aligned} \tag{1.38}$$

Since a rotation of either π or $-\pi$ about the axes $\mathbf{a}$ and $-\mathbf{a}$ are equivalent, we can arbitrarily choose the sign of one of ϵ_1, ϵ_2 or ϵ_3. Some possible solutions are

- If $\eta = 0$ and $|\epsilon_1| > 0$,

$$\begin{aligned} \epsilon_1 &= |\epsilon_1|, \\ \epsilon_2 &= \operatorname{sign}(c_{12})\,|\epsilon_2|, \\ \epsilon_3 &= \operatorname{sign}(c_{13})\,|\epsilon_3|. \end{aligned}$$

- If $\eta = 0$ and $|\epsilon_2| > 0$,

$$\begin{aligned} \epsilon_1 &= \operatorname{sign}(c_{12})\,|\epsilon_1|, \\ \epsilon_2 &= |\epsilon_2|, \\ \epsilon_3 &= \operatorname{sign}(c_{23})\,|\epsilon_3|. \end{aligned}$$

- If $\eta = 0$ and $|\epsilon_3| > 0$,

$$\begin{aligned} \epsilon_1 &= \operatorname{sign}(c_{13})\,|\epsilon_1|, \\ \epsilon_2 &= \operatorname{sign}(c_{23})\,|\epsilon_2|, \\ \epsilon_3 &= |\epsilon_3|. \end{aligned}$$

We now note some useful points. First of all, the quaternion parameterization of the zero rotation ($\phi = 0$ or $\mathbf{C} = \mathbf{1}$) is $(\mathbf{0}, \pm 1)$. Next, from (1.33) is immediately obvious that if $(\boldsymbol{\epsilon}, \eta)$ is the quaternion parameterization of $\mathbf{C}$, then $(-\boldsymbol{\epsilon}, \eta)$ is a quaternion parameterization of the inverse rotation, $\mathbf{C}^T$.

1.3.4.2 Successive Rotations

Let us now consider three reference frames $\mathcal{F}_1$, $\mathcal{F}_2$ and $\mathcal{F}_3$, with associated rotation matrices $\mathbf{C}_{21}$, $\mathbf{C}_{32}$ and $\mathbf{C}_{31}$, such that

$$\mathbf{C}_{31} = \mathbf{C}_{32}\mathbf{C}_{21}.$$

We parameterize each rotation matrix with a quaternion, $\mathbf{C}_{21} = \mathbf{C}(\boldsymbol{\epsilon}_1, \eta_1)$, $\mathbf{C}_{32} = \mathbf{C}(\boldsymbol{\epsilon}_2, \eta_2)$ and $\mathbf{C}_{31} = \mathbf{C}(\boldsymbol{\epsilon}_3, \eta_3)$, such that

$$\begin{aligned} \mathbf{C}_{21} &= \left(2\eta_1^2 - 1\right)\mathbf{1} + 2\boldsymbol{\epsilon}_1\boldsymbol{\epsilon}_1^T - 2\eta_1\boldsymbol{\epsilon}_1^\times, \\ \mathbf{C}_{32} &= \left(2\eta_2^2 - 1\right)\mathbf{1} + 2\boldsymbol{\epsilon}_2\boldsymbol{\epsilon}_2^T - 2\eta_2\boldsymbol{\epsilon}_2^\times, \\ \mathbf{C}_{31} &= \left(2\eta_3^2 - 1\right)\mathbf{1} + 2\boldsymbol{\epsilon}_3\boldsymbol{\epsilon}_3^T - 2\eta_3\boldsymbol{\epsilon}_3^\times. \end{aligned}$$

We shall now find $(\boldsymbol{\epsilon}_3, \eta_3)$ as a function of $(\boldsymbol{\epsilon}_1, \eta_1)$ and $(\boldsymbol{\epsilon}_2, \eta_2)$. The resulting expressions are very useful, since they allow the computation of the new quaternion without having to first form the rotation matrix and then extracting it.

To address this problem, we note that the scalar and vector parts of the quaternion satisfy (1.34) and (1.36) respectively. Let us now compute $\mathbf{C}_{31}$. We have,

$$\begin{aligned}
\mathbf{C}_{31} &= \left[\left(2\eta_2^2 - 1\right)\mathbf{1} + 2\boldsymbol{\epsilon}_2\boldsymbol{\epsilon}_2^T - 2\eta_2\boldsymbol{\epsilon}_2^\times\right]\left[\left(2\eta_1^2 - 1\right)\mathbf{1} + 2\boldsymbol{\epsilon}_1\boldsymbol{\epsilon}_1^T - 2\eta_1\boldsymbol{\epsilon}_1^\times\right] \\
&= \left(2\eta_2^2 - 1\right)\left(2\eta_1^2 - 1\right)\mathbf{1} + 2\left(2\eta_2^2 - 1\right)\boldsymbol{\epsilon}_1\boldsymbol{\epsilon}_1^T - 2\eta_1\left(2\eta_2^2 - 1\right)\boldsymbol{\epsilon}_1^\times \\
&\quad + 2\left(2\eta_1^2 - 1\right)\boldsymbol{\epsilon}_2\boldsymbol{\epsilon}_2^T + 4\boldsymbol{\epsilon}_2\boldsymbol{\epsilon}_2^T\boldsymbol{\epsilon}_1\boldsymbol{\epsilon}_1^T - 4\eta_1\boldsymbol{\epsilon}_2\boldsymbol{\epsilon}_2^T\boldsymbol{\epsilon}_1^\times \\
&\quad - 2\eta_2\left(2\eta_1^2 - 1\right)\boldsymbol{\epsilon}_2^\times - 4\eta_2\boldsymbol{\epsilon}_2^\times\boldsymbol{\epsilon}_1\boldsymbol{\epsilon}_1^T + 4\eta_1\eta_2\boldsymbol{\epsilon}_2^\times\boldsymbol{\epsilon}_1^\times.
\end{aligned}$$

Noting that

$$\boldsymbol{\epsilon}_2\boldsymbol{\epsilon}_2^T\boldsymbol{\epsilon}_1\boldsymbol{\epsilon}_1^T = \left(\boldsymbol{\epsilon}_2^T\boldsymbol{\epsilon}_1\right)\boldsymbol{\epsilon}_2\boldsymbol{\epsilon}_1^T$$

and

$$\boldsymbol{\epsilon}_2^\times\boldsymbol{\epsilon}_1^\times = \boldsymbol{\epsilon}_1\boldsymbol{\epsilon}_2^T - \left(\boldsymbol{\epsilon}_2^T\boldsymbol{\epsilon}_1\right)\mathbf{1},$$

we can rewrite the rotation matrix as

$$\begin{aligned}
\mathbf{C}_{31} &= \left(2\eta_2^2 - 1\right)\left(2\eta_1^2 - 1\right)\mathbf{1} + 2\left(2\eta_2^2 - 1\right)\boldsymbol{\epsilon}_1\boldsymbol{\epsilon}_1^T - 2\eta_1\left(2\eta_2^2 - 1\right)\boldsymbol{\epsilon}_1^\times \\
&\quad + 2\left(2\eta_1^2 - 1\right)\boldsymbol{\epsilon}_2\boldsymbol{\epsilon}_2^T + 4\left(\boldsymbol{\epsilon}_2^T\boldsymbol{\epsilon}_1\right)\boldsymbol{\epsilon}_2\boldsymbol{\epsilon}_1^T - 4\eta_1\boldsymbol{\epsilon}_2\boldsymbol{\epsilon}_2^T\boldsymbol{\epsilon}_1^\times \qquad (1.39)\\
&\quad - 2\eta_2\left(2\eta_1^2 - 1\right)\boldsymbol{\epsilon}_2^\times - 4\eta_2\boldsymbol{\epsilon}_2^\times\boldsymbol{\epsilon}_1\boldsymbol{\epsilon}_1^T + 4\eta_1\eta_2\boldsymbol{\epsilon}_1\boldsymbol{\epsilon}_2^T - 4\eta_1\eta_2\left(\boldsymbol{\epsilon}_2^T\boldsymbol{\epsilon}_1\right)\mathbf{1}.
\end{aligned}$$

First, we shall find η_3. Taking the trace of $\mathbf{C}_{31}$ in (1.39), we obtain (note that we make use of the fact that for any two matrices $\mathbf{A}$ and $\mathbf{B}$, the trace satisfies trace $[\mathbf{AB}]$ = trace $[\mathbf{BA}]$, provided the multiplication on the right makes sense),

$$\begin{aligned}
\text{trace}\left[\mathbf{C}_{31}\right] &= 3\left(2\eta_2^2 - 1\right)\left(2\eta_1^2 - 1\right) + 2\left(2\eta_2^2 - 1\right)\boldsymbol{\epsilon}_1^T\boldsymbol{\epsilon}_1 + 2\left(2\eta_1^2 - 1\right)\boldsymbol{\epsilon}_2^T\boldsymbol{\epsilon}_2 \\
&\quad + 4\left(\boldsymbol{\epsilon}_2^T\boldsymbol{\epsilon}_1\right)^2 - 4\eta_1\boldsymbol{\epsilon}_2^T\boldsymbol{\epsilon}_1^\times\boldsymbol{\epsilon}_2 - 4\eta_2\boldsymbol{\epsilon}_1^T\boldsymbol{\epsilon}_2^\times\boldsymbol{\epsilon}_1 \\
&\quad + 4\eta_1\eta_2\left(\boldsymbol{\epsilon}_2^T\boldsymbol{\epsilon}_1\right) - 12\eta_1\eta_2\left(\boldsymbol{\epsilon}_2^T\boldsymbol{\epsilon}_1\right).
\end{aligned}$$

Making use of the facts $\boldsymbol{\epsilon}_1^T\boldsymbol{\epsilon}_1 = 1 - \eta_1^2$, $\boldsymbol{\epsilon}_2^T\boldsymbol{\epsilon}_2 = 1 - \eta_2^2$, and noting that $\boldsymbol{\epsilon}_1^T\boldsymbol{\epsilon}_2^\times\boldsymbol{\epsilon}_1 = \boldsymbol{\epsilon}_2^T\boldsymbol{\epsilon}_1^\times\boldsymbol{\epsilon}_2 = 0$, we can reduce this to

$$\begin{aligned}
\text{trace}\left[\mathbf{C}_{31}\right] &= 4\left(\eta_1\eta_2\right)^2 - 8\left(\eta_1\eta_2\right)\left(\boldsymbol{\epsilon}_2^T\boldsymbol{\epsilon}_1\right) + 4\left(\boldsymbol{\epsilon}_2^T\boldsymbol{\epsilon}_1\right)^2 - 1, \\
&= 4\left[\eta_1\eta_2 - \boldsymbol{\epsilon}_2^T\boldsymbol{\epsilon}_1\right]^2 - 1.
\end{aligned}$$

Comparing this to (1.34), and recalling that we are free to choose the sign of η_3, we see that we can choose

$$\eta_3 = \eta_1\eta_2 - \boldsymbol{\epsilon}_2^T\boldsymbol{\epsilon}_1. \qquad (1.40)$$

Now, let us find the vector part of the quaternion, $\boldsymbol{\epsilon}_3$. From (1.36), we see that we must compute $\mathbf{C}_{31}^T - \mathbf{C}_{31}$. From (1.39) we have,

$$\begin{aligned}\mathbf{C}_{31}^T - \mathbf{C}_{31} &= 4\eta_1\left(2\eta_2^2 - 1\right)\boldsymbol{\epsilon}_1^\times - 4\left(\boldsymbol{\epsilon}_2^T\boldsymbol{\epsilon}_1\right)\left[\boldsymbol{\epsilon}_2\boldsymbol{\epsilon}_1^T - \boldsymbol{\epsilon}_1\boldsymbol{\epsilon}_2^T\right]\\ &+4\eta_1\left[\boldsymbol{\epsilon}_2\boldsymbol{\epsilon}_2^T\boldsymbol{\epsilon}_1^\times + \boldsymbol{\epsilon}_1^\times\boldsymbol{\epsilon}_2\boldsymbol{\epsilon}_2^T\right] + 4\eta_2\left(2\eta_1^2 - 1\right)\boldsymbol{\epsilon}_2^\times\\ &+4\eta_2\left[\boldsymbol{\epsilon}_1\boldsymbol{\epsilon}_1^T\boldsymbol{\epsilon}_2^\times + \boldsymbol{\epsilon}_2^\times\boldsymbol{\epsilon}_1\boldsymbol{\epsilon}_1^T\right] + 4\eta_1\eta_2\left[\boldsymbol{\epsilon}_2\boldsymbol{\epsilon}_1^T - \boldsymbol{\epsilon}_1\boldsymbol{\epsilon}_2^T\right].\end{aligned} \tag{1.41}$$

We now derive a new column matrix identity, which shall be very useful. We have

$$\left(\mathbf{a}^\times\mathbf{b}\right)^\times\mathbf{c} = -\mathbf{c}^\times\mathbf{a}^\times\mathbf{b} = -\left[\left(\mathbf{b}^T\mathbf{c}\right)\mathbf{a} - \left(\mathbf{a}^T\mathbf{c}\right)\mathbf{b}\right] = \left[\mathbf{b}\mathbf{a}^T - \mathbf{a}\mathbf{b}^T\right]\mathbf{c}.$$

Since this must hold for all $\mathbf{c} \in \mathbf{R}^3$, we have

$$\left(\mathbf{a}^\times\mathbf{b}\right)^\times = \mathbf{b}\mathbf{a}^T - \mathbf{a}\mathbf{b}^T. \tag{1.42}$$

Applying this identity, we have

$$\boldsymbol{\epsilon}_2\boldsymbol{\epsilon}_1^T - \boldsymbol{\epsilon}_1\boldsymbol{\epsilon}_2^T = \left(\boldsymbol{\epsilon}_1^\times\boldsymbol{\epsilon}_2\right)^\times.$$

Making use of this and noting that $2\eta^2 - 1 = \eta^2 - \boldsymbol{\epsilon}^T\boldsymbol{\epsilon}$, we can rewrite (1.41) as

$$\begin{aligned}\mathbf{C}_{31}^T - \mathbf{C}_{31} &= 4\eta_1\eta_2\left(\eta_2\boldsymbol{\epsilon}_1^\times + \eta_1\boldsymbol{\epsilon}_2^\times\right) + 4\left(\eta_1\eta_2 - \boldsymbol{\epsilon}_2^T\boldsymbol{\epsilon}_1\right)\left(\boldsymbol{\epsilon}_1^\times\boldsymbol{\epsilon}_2\right)^\times\\ &-4\eta_1\left(\boldsymbol{\epsilon}_2^T\boldsymbol{\epsilon}_2\right)\boldsymbol{\epsilon}_1^\times - 4\eta_2\left(\boldsymbol{\epsilon}_1^T\boldsymbol{\epsilon}_1\right)\boldsymbol{\epsilon}_2^\times\\ &+4\eta_1\left[\boldsymbol{\epsilon}_2\boldsymbol{\epsilon}_2^T\boldsymbol{\epsilon}_1^\times + \boldsymbol{\epsilon}_1^\times\boldsymbol{\epsilon}_2\boldsymbol{\epsilon}_2^T\right] + 4\eta_2\left[\boldsymbol{\epsilon}_1\boldsymbol{\epsilon}_1^T\boldsymbol{\epsilon}_2^\times + \boldsymbol{\epsilon}_2^\times\boldsymbol{\epsilon}_1\boldsymbol{\epsilon}_1^T\right].\end{aligned} \tag{1.43}$$

Now, making use of the identity (1.42), we shall reduce the square bracket terms. We have

$$\begin{aligned}\boldsymbol{\epsilon}_2\boldsymbol{\epsilon}_2^T\boldsymbol{\epsilon}_1^\times + \boldsymbol{\epsilon}_1^\times\boldsymbol{\epsilon}_2\boldsymbol{\epsilon}_2^T &= \left(\boldsymbol{\epsilon}_1^\times\boldsymbol{\epsilon}_2\right)\boldsymbol{\epsilon}_2^T - \boldsymbol{\epsilon}_2\left(\boldsymbol{\epsilon}_1^\times\boldsymbol{\epsilon}_2\right)^T\\ &= \left(\boldsymbol{\epsilon}_2^\times\boldsymbol{\epsilon}_1^\times\boldsymbol{\epsilon}_2\right)^\times,\end{aligned}$$

and

$$\begin{aligned}\boldsymbol{\epsilon}_1\boldsymbol{\epsilon}_1^T\boldsymbol{\epsilon}_2^\times + \boldsymbol{\epsilon}_2^\times\boldsymbol{\epsilon}_1\boldsymbol{\epsilon}_1^T &= \left(\boldsymbol{\epsilon}_2^\times\boldsymbol{\epsilon}_1\right)\boldsymbol{\epsilon}_1^T - \boldsymbol{\epsilon}_1\left(\boldsymbol{\epsilon}_2^\times\boldsymbol{\epsilon}_1\right)^T\\ &= \left(\boldsymbol{\epsilon}_1^\times\boldsymbol{\epsilon}_2^\times\boldsymbol{\epsilon}_1\right)^\times.\end{aligned}$$

Making use of the identity

$$\mathbf{a}^\times\mathbf{b}^\times\mathbf{c} = (\mathbf{a}^T\mathbf{c})\mathbf{b} - (\mathbf{a}^T\mathbf{b})\mathbf{c},$$

we can reduce these to

$$\begin{aligned}\boldsymbol{\epsilon}_2\boldsymbol{\epsilon}_2^T\boldsymbol{\epsilon}_1^\times + \boldsymbol{\epsilon}_1^\times\boldsymbol{\epsilon}_2\boldsymbol{\epsilon}_2^T &= \left(\boldsymbol{\epsilon}_2^\times\boldsymbol{\epsilon}_1^\times\boldsymbol{\epsilon}_2\right)^\times \\ &= \left[\left(\boldsymbol{\epsilon}_2^T\boldsymbol{\epsilon}_2\right)\boldsymbol{\epsilon}_1 - \left(\boldsymbol{\epsilon}_1^T\boldsymbol{\epsilon}_2\right)\boldsymbol{\epsilon}_2\right]^\times \\ &= \left(\boldsymbol{\epsilon}_2^T\boldsymbol{\epsilon}_2\right)\boldsymbol{\epsilon}_1^\times - \left(\boldsymbol{\epsilon}_1^T\boldsymbol{\epsilon}_2\right)\boldsymbol{\epsilon}_2^\times,\end{aligned}$$

and

$$\begin{aligned}\boldsymbol{\epsilon}_1\boldsymbol{\epsilon}_1^T\boldsymbol{\epsilon}_2^\times + \boldsymbol{\epsilon}_2^\times\boldsymbol{\epsilon}_1\boldsymbol{\epsilon}_1^T &= \left(\boldsymbol{\epsilon}_1^\times\boldsymbol{\epsilon}_2^\times\boldsymbol{\epsilon}_1\right)^\times \\ &= \left[\left(\boldsymbol{\epsilon}_1^T\boldsymbol{\epsilon}_1\right)\boldsymbol{\epsilon}_2 - \left(\boldsymbol{\epsilon}_2^T\boldsymbol{\epsilon}_1\right)\boldsymbol{\epsilon}_1\right]^\times \\ &= \left(\boldsymbol{\epsilon}_1^T\boldsymbol{\epsilon}_1\right)\boldsymbol{\epsilon}_2^\times - \left(\boldsymbol{\epsilon}_2^T\boldsymbol{\epsilon}_1\right)\boldsymbol{\epsilon}_1^\times.\end{aligned}$$

Making use of these in (1.43), we finally obtain

$$\mathbf{C}_{31}^T - \mathbf{C}_{31} = 4\left(\eta_1\eta_2 - \boldsymbol{\epsilon}_2^T\boldsymbol{\epsilon}_1\right)\left[\eta_2\boldsymbol{\epsilon}_1 + \eta_1\boldsymbol{\epsilon}_2 + \boldsymbol{\epsilon}_1^\times\boldsymbol{\epsilon}_2\right]^\times.$$

Recognizing η_3 from before, we have

$$\mathbf{C}_{31}^T - \mathbf{C}_{31} = 4\eta_3\left[\eta_2\boldsymbol{\epsilon}_1 + \eta_1\boldsymbol{\epsilon}_2 + \boldsymbol{\epsilon}_1^\times\boldsymbol{\epsilon}_2\right]^\times.$$

Comparing this to (1.36), we see that if $\eta_3 \neq 0$, then we can find the corresponding vector part of the quaternion as

$$\boldsymbol{\epsilon}_3 = \eta_2\boldsymbol{\epsilon}_1 + \eta_1\boldsymbol{\epsilon}_2 + \boldsymbol{\epsilon}_1^\times\boldsymbol{\epsilon}_2. \tag{1.44}$$

Now all that remains is to check the vector part of the quaternion when $\eta_3 = 0$. We note that when $\eta_3 = 0$, $\mathbf{C}_{31}$ becomes

$$\mathbf{C}_{31} = -\mathbf{1} + 2\boldsymbol{\epsilon}_3\boldsymbol{\epsilon}_3^T,$$

which is symmetric. In particular, we have

$$\mathbf{C}_{31} + \mathbf{C}_{31}^T = -2\mathbf{1} + 4\boldsymbol{\epsilon}_3\boldsymbol{\epsilon}_3^T. \tag{1.45}$$

From (1.39), we have

$$\begin{aligned}\mathbf{C}_{31} + \mathbf{C}_{31}^T = {} & 2\left(2\eta_2^2 - 1\right)\left(2\eta_1^2 - 1\right)\mathbf{1} + 4\left(2\eta_2^2 - 1\right)\boldsymbol{\epsilon}_1\boldsymbol{\epsilon}_1^T \\ & + 4\left(2\eta_1^2 - 1\right)\boldsymbol{\epsilon}_2\boldsymbol{\epsilon}_2^T + 4\left(\boldsymbol{\epsilon}_2^T\boldsymbol{\epsilon}_1\right)\left[\boldsymbol{\epsilon}_2\boldsymbol{\epsilon}_1^T + \boldsymbol{\epsilon}_1\boldsymbol{\epsilon}_2^T\right] \\ & - 4\eta_1\left[\boldsymbol{\epsilon}_2\boldsymbol{\epsilon}_2^T\boldsymbol{\epsilon}_1^\times - \boldsymbol{\epsilon}_1^\times\boldsymbol{\epsilon}_2\boldsymbol{\epsilon}_2^T\right] - 4\eta_2\left[\boldsymbol{\epsilon}_2^\times\boldsymbol{\epsilon}_1\boldsymbol{\epsilon}_1^T - \boldsymbol{\epsilon}_1\boldsymbol{\epsilon}_1^T\boldsymbol{\epsilon}_2^\times\right] \\ & 4\eta_1\eta_2\left[\boldsymbol{\epsilon}_1\boldsymbol{\epsilon}_2^T + \boldsymbol{\epsilon}_2\boldsymbol{\epsilon}_1^T\right] - 8\eta_1\eta_2\left(\boldsymbol{\epsilon}_2^T\boldsymbol{\epsilon}_1\right)\mathbf{1}\end{aligned} \tag{1.46}$$

Let us now examine this term by term. Noting from (1.40) that $\eta_1\eta_2 = \boldsymbol{\epsilon}_2^T\boldsymbol{\epsilon}_1$ when $\eta_3 = 0$, we have

$$2\left(2\eta_2^2 - 1\right)\left(2\eta_1^2 - 1\right) - 8\eta_1\eta_2\left(\boldsymbol{\epsilon}_2^T\boldsymbol{\epsilon}_1\right) = -2 + \left[4 - 4\left(\eta_1^2 + \eta_2^2\right)\right].$$

Next, since $\eta^2 + \boldsymbol{\epsilon}^T\boldsymbol{\epsilon} = 1$, we have

$$\begin{aligned}
4\left(2\eta_2^2 - 1\right)\boldsymbol{\epsilon}_1\boldsymbol{\epsilon}_1^T &= 4\left(\eta_2\boldsymbol{\epsilon}_1\right)\left(\eta_2\boldsymbol{\epsilon}_1\right)^T - 4\left(\boldsymbol{\epsilon}_2^T\boldsymbol{\epsilon}_2\right)\boldsymbol{\epsilon}_1\boldsymbol{\epsilon}_1^T,\\
4\left(2\eta_1^2 - 1\right)\boldsymbol{\epsilon}_2\boldsymbol{\epsilon}_2^T &= 4\left(\eta_1\boldsymbol{\epsilon}_2\right)\left(\eta_1\boldsymbol{\epsilon}_2\right)^T - 4\left(\boldsymbol{\epsilon}_1^T\boldsymbol{\epsilon}_1\right)\boldsymbol{\epsilon}_2\boldsymbol{\epsilon}_2^T.
\end{aligned}$$

We can also rewrite the following factors

$$\begin{aligned}
-4\eta_1\left[\boldsymbol{\epsilon}_2\boldsymbol{\epsilon}_2^T\boldsymbol{\epsilon}_1^\times - \boldsymbol{\epsilon}_1^\times\boldsymbol{\epsilon}_2\boldsymbol{\epsilon}_2^T\right] &= 4\left(\eta_1\boldsymbol{\epsilon}_2\right)\left(\boldsymbol{\epsilon}_1^\times\boldsymbol{\epsilon}_2\right)^T + 4\left(\boldsymbol{\epsilon}_1^\times\boldsymbol{\epsilon}_2\right)\left(\eta_1\boldsymbol{\epsilon}_2\right)^T,\\
-4\eta_2\left[\boldsymbol{\epsilon}_2^\times\boldsymbol{\epsilon}_1\boldsymbol{\epsilon}_1^T - \boldsymbol{\epsilon}_1\boldsymbol{\epsilon}_1^T\boldsymbol{\epsilon}_2^\times\right] &= 4\left(\boldsymbol{\epsilon}_1^\times\boldsymbol{\epsilon}_2\right)\left(\eta_2\boldsymbol{\epsilon}_1\right)^T + 4\left(\eta_2\boldsymbol{\epsilon}_1\right)\left(\boldsymbol{\epsilon}_1^\times\boldsymbol{\epsilon}_2\right)^T,\\
4\eta_1\eta_2\left[\boldsymbol{\epsilon}_1\boldsymbol{\epsilon}_2^T + \boldsymbol{\epsilon}_2\boldsymbol{\epsilon}_1^T\right] &= 4\left(\eta_2\boldsymbol{\epsilon}_1\right)\left(\eta_1\boldsymbol{\epsilon}_2\right)^T + 4\left(\eta_1\boldsymbol{\epsilon}_2\right)\left(\eta_2\boldsymbol{\epsilon}_1\right)^T.
\end{aligned}$$

Making use of these expansions, we can rewrite (1.46) as

$$\begin{aligned}
\mathbf{C}_{31} + \mathbf{C}_{31}^T = &-2\mathbf{1} + 4\left(\eta_2\boldsymbol{\epsilon}_1\right)\left(\eta_2\boldsymbol{\epsilon}_1\right)^T + 4\left(\eta_1\boldsymbol{\epsilon}_2\right)\left(\eta_1\boldsymbol{\epsilon}_2\right)^T\\
&+ 4\left(\eta_1\boldsymbol{\epsilon}_2\right)\left(\boldsymbol{\epsilon}_1^\times\boldsymbol{\epsilon}_2\right)^T + 4\left(\boldsymbol{\epsilon}_1^\times\boldsymbol{\epsilon}_2\right)\left(\eta_1\boldsymbol{\epsilon}_2\right)^T\\
&+ 4\left(\boldsymbol{\epsilon}_1^\times\boldsymbol{\epsilon}_2\right)\left(\eta_2\boldsymbol{\epsilon}_1\right)^T + 4\left(\eta_2\boldsymbol{\epsilon}_1\right)\left(\boldsymbol{\epsilon}_1^\times\boldsymbol{\epsilon}_2\right)^T\\
&+ 4\left(\eta_2\boldsymbol{\epsilon}_1\right)\left(\eta_1\boldsymbol{\epsilon}_2\right)^T + 4\left(\eta_1\boldsymbol{\epsilon}_2\right)\left(\eta_2\boldsymbol{\epsilon}_1\right)^T\\
&+ 4\Big[\left(\boldsymbol{\epsilon}_2^T\boldsymbol{\epsilon}_1\right)\left[\boldsymbol{\epsilon}_2\boldsymbol{\epsilon}_1^T + \boldsymbol{\epsilon}_1\boldsymbol{\epsilon}_2^T\right] - \left(\boldsymbol{\epsilon}_2^T\boldsymbol{\epsilon}_2\right)\boldsymbol{\epsilon}_1\boldsymbol{\epsilon}_1^T - \left(\boldsymbol{\epsilon}_1^T\boldsymbol{\epsilon}_1\right)\boldsymbol{\epsilon}_2\boldsymbol{\epsilon}_2^T\\
&+ \left[1 - \left(\eta_1^2 + \eta_2^2\right)\right]\mathbf{1}\Big]
\end{aligned} \tag{1.47}$$

Now we just need to take care of the final bracketed term. Making use of the identity $\mathbf{a}\mathbf{a}^T = \mathbf{a}^\times\mathbf{a}^\times + \mathbf{a}^T\mathbf{a}\mathbf{1}$, we have

$$\begin{aligned}
\left(\boldsymbol{\epsilon}_1^\times\boldsymbol{\epsilon}_2\right)\left(\boldsymbol{\epsilon}_1^\times\boldsymbol{\epsilon}_2\right)^T &= \left(\boldsymbol{\epsilon}_1^\times\boldsymbol{\epsilon}_2\right)^\times\left(\boldsymbol{\epsilon}_1^\times\boldsymbol{\epsilon}_2\right)^\times + \left(\boldsymbol{\epsilon}_1^\times\boldsymbol{\epsilon}_2\right)^T\left(\boldsymbol{\epsilon}_1^\times\boldsymbol{\epsilon}_2\right)\mathbf{1}\\
&= \left(\boldsymbol{\epsilon}_2\boldsymbol{\epsilon}_1^T - \boldsymbol{\epsilon}_1\boldsymbol{\epsilon}_2^T\right)\left(\boldsymbol{\epsilon}_2\boldsymbol{\epsilon}_1^T - \boldsymbol{\epsilon}_1\boldsymbol{\epsilon}_2^T\right) - \boldsymbol{\epsilon}_2^T\boldsymbol{\epsilon}_1^\times\boldsymbol{\epsilon}_1^\times\boldsymbol{\epsilon}_2\mathbf{1}\\
&= \boldsymbol{\epsilon}_2\boldsymbol{\epsilon}_1^T\boldsymbol{\epsilon}_2\boldsymbol{\epsilon}_1^T - \boldsymbol{\epsilon}_2\boldsymbol{\epsilon}_1^T\boldsymbol{\epsilon}_1\boldsymbol{\epsilon}_2^T - \boldsymbol{\epsilon}_1\boldsymbol{\epsilon}_2^T\boldsymbol{\epsilon}_2\boldsymbol{\epsilon}_1^T + \boldsymbol{\epsilon}_1\boldsymbol{\epsilon}_2^T\boldsymbol{\epsilon}_1\boldsymbol{\epsilon}_2^T\\
&\quad - \boldsymbol{\epsilon}_2^T\left(\boldsymbol{\epsilon}_1\boldsymbol{\epsilon}_1^T - \boldsymbol{\epsilon}_1^T\boldsymbol{\epsilon}_1\mathbf{1}\right)\boldsymbol{\epsilon}_2\mathbf{1}\\
&= \left(\boldsymbol{\epsilon}_2^T\boldsymbol{\epsilon}_1\right)\left[\boldsymbol{\epsilon}_2\boldsymbol{\epsilon}_1^T + \boldsymbol{\epsilon}_1\boldsymbol{\epsilon}_2^T\right] - \left(\boldsymbol{\epsilon}_1^T\boldsymbol{\epsilon}_1\right)\boldsymbol{\epsilon}_2\boldsymbol{\epsilon}_2^T - \left(\boldsymbol{\epsilon}_2^T\boldsymbol{\epsilon}_2\right)\boldsymbol{\epsilon}_1\boldsymbol{\epsilon}_1^T\\
&\quad + \left[\left(\boldsymbol{\epsilon}_1^T\boldsymbol{\epsilon}_1\right)\left(\boldsymbol{\epsilon}_2^T\boldsymbol{\epsilon}_2\right) - \left(\boldsymbol{\epsilon}_2^T\boldsymbol{\epsilon}_1\right)^2\right]\mathbf{1}.
\end{aligned}$$

Since $\eta_3 = 0$ means that $\eta_1\eta_2 = \boldsymbol{\epsilon}_2^T\boldsymbol{\epsilon}_1$, and making use of the unit magnitude constraint $\eta^2 + \boldsymbol{\epsilon}^T\boldsymbol{\epsilon} = 1$, the last term in brackets becomes

$$\left(\boldsymbol{\epsilon}_1^T\boldsymbol{\epsilon}_1\right)\left(\boldsymbol{\epsilon}_2^T\boldsymbol{\epsilon}_2\right) - \left(\boldsymbol{\epsilon}_2^T\boldsymbol{\epsilon}_1\right)^2 = 1 - \left(\eta_1^2 + \eta_2^2\right).$$

Substituting this back in, we have

$$\begin{aligned}\left(\boldsymbol{\epsilon}_1^\times\boldsymbol{\epsilon}_2\right)\left(\boldsymbol{\epsilon}_1^\times\boldsymbol{\epsilon}_2\right)^T = &\left(\boldsymbol{\epsilon}_2^T\boldsymbol{\epsilon}_1\right)\left[\boldsymbol{\epsilon}_2\boldsymbol{\epsilon}_1^T + \boldsymbol{\epsilon}_1\boldsymbol{\epsilon}_2^T\right] - \left(\boldsymbol{\epsilon}_1^T\boldsymbol{\epsilon}_1\right)\boldsymbol{\epsilon}_2\boldsymbol{\epsilon}_2^T - \left(\boldsymbol{\epsilon}_2^T\boldsymbol{\epsilon}_2\right)\boldsymbol{\epsilon}_1\boldsymbol{\epsilon}_1^T \\ &+ \left[1 - \left(\eta_1^2 + \eta_2^2\right)\right]\mathbf{1}.\end{aligned}$$

We can now see that this is precisely the last bracketed term in (1.47). Therefore, we can finally write

$$\mathbf{C}_{31} + \mathbf{C}_{31}^T = -2\mathbf{1} + 4\left[\eta_2\boldsymbol{\epsilon}_1 + \eta_1\boldsymbol{\epsilon}_2 + \boldsymbol{\epsilon}_1^\times\boldsymbol{\epsilon}_2\right]\left[\eta_2\boldsymbol{\epsilon}_1 + \eta_1\boldsymbol{\epsilon}_2 + \boldsymbol{\epsilon}_1^\times\boldsymbol{\epsilon}_2\right]^T. \tag{1.48}$$

We shall now make use of the fact that for two column matrices $\mathbf{a}$ and $\mathbf{b}$, if $\mathbf{a}\mathbf{a}^T = \mathbf{b}\mathbf{b}^T$, then $\mathbf{b} = \pm\mathbf{a}$. When $\eta_3 = 0$, the sign of the vector part of the quaternion does not matter. Comparing (1.48) with (1.45), we see therefore that we may take

$$\boldsymbol{\epsilon}_3 = \eta_2\boldsymbol{\epsilon}_1 + \eta_1\boldsymbol{\epsilon}_2 + \boldsymbol{\epsilon}_1^\times\boldsymbol{\epsilon}_2,$$

which is the same as for the case $\eta_3 \neq 0$.

In summary, given the rotation matrices $\mathbf{C}_{21} = \mathbf{C}(\boldsymbol{\epsilon}_1, \eta_1)$, $\mathbf{C}_{32} = \mathbf{C}(\boldsymbol{\epsilon}_2, \eta_2)$ and $\mathbf{C}_{31} = \mathbf{C}(\boldsymbol{\epsilon}_3, \eta_3)$ and their respective quaternion parameterizations, given that $\mathbf{C}_{31} = \mathbf{C}_{32}\mathbf{C}_{21}$, the associated quaternions are related by

$$\begin{aligned}\boldsymbol{\epsilon}_3 &= \eta_2\boldsymbol{\epsilon}_1 + \eta_1\boldsymbol{\epsilon}_2 + \boldsymbol{\epsilon}_1^\times\boldsymbol{\epsilon}_2, \\ \eta_3 &= \eta_1\eta_2 - \boldsymbol{\epsilon}_2^T\boldsymbol{\epsilon}_1.\end{aligned} \tag{1.49}$$

1.4 Derivatives of Vectors

Now that we have examined physical vectors and their representations with respect to reference frames, we are in a position to look at the evolution of physical vectors with respect to time. Reference frames are crucial for this, since the evolution of a physical vector depends entirely on the point of view of an observer. For example, an observer attached to one reference frame will see a different evolution of the physical vector compared with an observer attached to a different reference frame. For this reason, when we talk about temporal derivatives of a physical vector, we will talk about the derivative as seen in a particular reference frame.

As such, let us consider a reference frame $\mathcal{F}_1$ defined by the unit vectors $\vec{\mathbf{x}}_1$, $\vec{\mathbf{y}}_1$ and $\vec{\mathbf{z}}_1$. Given a physical vector

$$\vec{\mathbf{r}} = \vec{\mathcal{F}}_1^T\mathbf{r}_1,$$

the time derivative of $\vec{\mathbf{r}}$ as seen in reference frame $\mathcal{F}_1$ is defined to be

$$\overset{\bullet}{\vec{\mathbf{r}}} \stackrel{\Delta}{=} \lim_{\delta t \to 0} \frac{\delta \vec{\mathbf{r}}}{\delta t},$$

where

$$\delta \vec{\mathbf{r}} \stackrel{\Delta}{=} \vec{\mathcal{F}}_1^T \left(\mathbf{r}_1(t + \delta t) - \mathbf{r}_1(t) \right).$$

Therefore, the time derivative of $\vec{\mathbf{r}}$ as seen in reference frame $\mathcal{F}_1$ is given by

$$\overset{\bullet}{\vec{\mathbf{r}}} = \vec{\mathcal{F}}_1^T \dot{\mathbf{r}}_1.$$

As a simple consequence of this, if we multiply the vector $\vec{\mathbf{r}}$ by a time-varying scalar a, we have

$$\begin{aligned} \overbrace{(a\vec{\mathbf{r}})}^{\bullet} &= \vec{\mathcal{F}}_1^T (a\dot{\mathbf{r}}_1) \\ &= \vec{\mathcal{F}}_1^T \left(\dot{a}\mathbf{r}_1 + a\dot{\mathbf{r}}_1 \right) \\ &= \dot{a}\vec{\mathbf{r}} + a\overset{\bullet}{\vec{\mathbf{r}}}. \end{aligned}$$

Similarly, for sums of vectors,

$$\begin{aligned} \overbrace{\left(\vec{\mathbf{a}} + \vec{\mathbf{b}} \right)}^{\bullet} &= \vec{\mathcal{F}}_1^T \left[\dot{\mathbf{a}}_1 + \dot{\mathbf{b}}_1 \right] \\ &= \overset{\bullet}{\vec{\mathbf{a}}} + \overset{\bullet}{\vec{\mathbf{b}}}. \end{aligned}$$

We also see immediately that the time derivatives of the unit vectors defining frame $\mathcal{F}_1$ are all $\vec{\mathbf{0}}$, and that therefore

$$\overset{\bullet}{\vec{\mathcal{F}}}_1 = \begin{bmatrix} \vec{\mathbf{0}} \\ \vec{\mathbf{0}} \\ \vec{\mathbf{0}} \end{bmatrix}.$$

Finally, we can obtain product rules for scalar- and cross-products of vectors:

$$\begin{aligned} \frac{d}{dt} \left(\vec{\mathbf{a}} \cdot \vec{\mathbf{b}} \right) &= \frac{d}{dt} \left[\mathbf{a}^T \mathbf{b} \right] \\ &= \dot{\mathbf{a}}^T \mathbf{b} + \mathbf{a}^T \dot{\mathbf{b}} \\ &= \overset{\bullet}{\vec{\mathbf{a}}} \cdot \vec{\mathbf{b}} + \vec{\mathbf{a}} \cdot \overset{\bullet}{\vec{\mathbf{b}}}. \end{aligned}$$

and

$$
\begin{aligned}
\overbrace{\left(\vec{\mathbf{a}} \times \vec{\mathbf{b}}\right)}^{\cdot} &= \vec{\mathcal{F}}_1^T \frac{d}{dt}\left[\mathbf{a}^{\times}\mathbf{b}\right] \\
&= \vec{\mathcal{F}}_1^T \left[\dot{\mathbf{a}}^{\times}\mathbf{b} + \mathbf{a}^{\times}\dot{\mathbf{b}}\right] \\
&= \dot{\vec{\mathbf{a}}} \times \vec{\mathbf{b}} + \vec{\mathbf{a}} \times \dot{\vec{\mathbf{b}}}.
\end{aligned}
$$

1.4.1 Angular Velocity

Let us consider a second reference frame $\mathcal{F}_2$, which is rotating with respect to $\mathcal{F}_1$ with angular velocity $\vec{\boldsymbol{\omega}}_{21}$, as shown in Figure 1.18. The magnitude of $\vec{\boldsymbol{\omega}}_{21}$, $|\vec{\boldsymbol{\omega}}_{21}|$ is the rate of rotation, and the direction $\vec{\boldsymbol{\omega}}_{21}/|\vec{\boldsymbol{\omega}}_{21}|$ is the *instantaneous* axis of rotation. To understand what this means, consider the reference frame $\mathcal{F}_2$ at two times t and $t+\delta t$ as seen in reference frame $\mathcal{F}_1$. We know that the transformation from $\mathcal{F}_2(t)$ to $\mathcal{F}_2(t+\delta t)$ is a rotation about some axis $\vec{\mathbf{a}}(\delta t)$ through an angle $\phi(\delta t)$. We formally define the angular velocity to be

$$
\vec{\boldsymbol{\omega}}_{21} = \lim_{\delta t \to 0} \vec{\mathbf{a}}(\delta t)\frac{\phi(\delta t)}{\delta t}.
$$

Let us now consider the time-derivative of an arbitrary physical vector $\vec{\mathbf{v}}$ rotating with angular velocity $\vec{\boldsymbol{\omega}}$ as seen in reference frame $\mathcal{F}_1$. First, let us examine a finite rotation of the vector $\vec{\mathbf{v}}$ about a unit vector $\vec{\mathbf{a}}$ through an angle ϕ as seen in $\mathcal{F}_1$.

We have seen in (1.25) that the rotated vector is given in $\mathcal{F}_1$ coordinates by

$$
\vec{\mathbf{v}}_{rot} = \vec{\mathcal{F}}_1^T \left[\cos\phi\mathbf{1} + (1-\cos\phi)\mathbf{a}\mathbf{a}^T + \sin\phi\mathbf{a}^{\times}\right]\mathbf{v}. \tag{1.50}
$$

Now, let δt be very small such that the angle $\phi(\delta t)$ is very small, and $\sin\phi \approx \phi$ and $\cos\phi \approx 1$. Under these approximations, we have

$$
\vec{\mathbf{v}}(t+\delta t) = \vec{\mathcal{F}}_1^T \left[\mathbf{1} + \mathbf{a}^{\times}\phi\right]\mathbf{v}.
$$

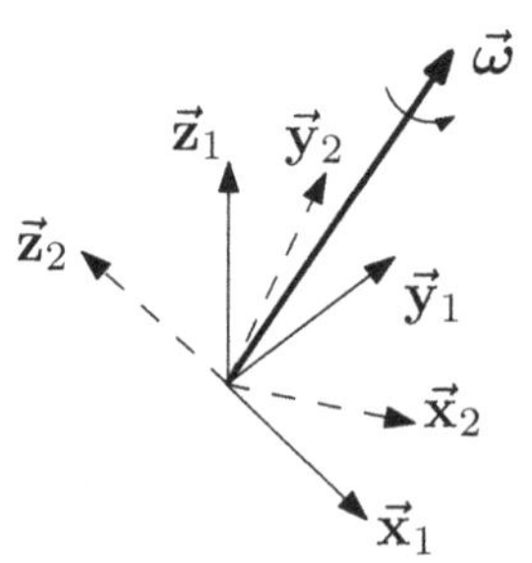

Figure 1.18 Angular velocity

From this we find that

$$\vec{\mathbf{v}}(t+\delta t) - \vec{\mathbf{v}}(t) = \vec{\mathcal{F}}_1^T \left[\mathbf{a}^\times \phi\right] \mathbf{v},$$

which leads to

$$\begin{aligned} \dot{\vec{\mathbf{v}}}(t) = \lim\nolimits_{\delta t \to 0} \frac{\vec{\mathbf{v}}(t+\delta t) - \vec{\mathbf{v}}(t)}{\delta t} &= \vec{\mathcal{F}}_1^T \boldsymbol{\omega}^\times \mathbf{v}, \\ &= \vec{\boldsymbol{\omega}} \times \vec{\mathbf{v}}. \end{aligned} \tag{1.51}$$

Let us now return to the consideration of the two reference frames $\mathcal{F}_1$ and $\mathcal{F}_2$, with $\mathcal{F}_2$ rotating with angular velocity $\vec{\boldsymbol{\omega}}_{21}$ with respect to $\mathcal{F}_1$. Using the above result, we have that the time-derivatives as seen in $\mathcal{F}_1$ of the unit vectors defining $\mathcal{F}_2$ are

$$\dot{\vec{\mathbf{x}}}_2 = \vec{\boldsymbol{\omega}}_{21} \times \vec{\mathbf{x}}_2, \quad \dot{\vec{\mathbf{y}}}_2 = \vec{\boldsymbol{\omega}}_{21} \times \vec{\mathbf{y}}_2, \quad \dot{\vec{\mathbf{z}}}_2 = \vec{\boldsymbol{\omega}}_{21} \times \vec{\mathbf{z}}_2.$$

This can be written compactly as

$$\dot{\vec{\mathcal{F}}}_2^T = \vec{\boldsymbol{\omega}}_{21} \times \vec{\mathcal{F}}_2^T.$$

As we have already discussed, observers in the two reference frames do not see the same motion, because of their own relative motions. We have already denoted the time-derivatives of vectors as seen in frame $\mathcal{F}_1$ by $(\dot{\ })$. Let us denote time derivatives of vectors as seen in frame $\mathcal{F}_2$ by $(\mathring{\ })$. We now consider the vector

$$\vec{\mathbf{r}} = \vec{\mathcal{F}}_1^T \mathbf{r}_1 = \vec{\mathcal{F}}_2^T \mathbf{r}_2.$$

By definition, the time derivatives of the vector as seen in each of the frames are

$$\dot{\vec{\mathbf{r}}} = \vec{\mathcal{F}}_1^T \dot{\mathbf{r}}_1, \quad \mathring{\vec{\mathbf{r}}} = \vec{\mathcal{F}}_2^T \dot{\mathbf{r}}_2,$$

respectively. Let us now obtain a relationship between the time-derivatives of vectors as seen in the two frames. Making use of the rules for vector differentiation we obtained earlier, we have

$$\begin{aligned} \dot{\vec{\mathbf{r}}} &= \vec{\mathcal{F}}_2^T \dot{\mathbf{r}}_2 + \dot{\vec{\mathcal{F}}}_2^T \mathbf{r}_2 \\ &= \mathring{\vec{\mathbf{r}}} + \vec{\boldsymbol{\omega}}_{21} \times \vec{\mathcal{F}}_2^T \mathbf{r}_2 \\ &= \mathring{\vec{\mathbf{r}}} + \vec{\boldsymbol{\omega}}_{21} \times \vec{\mathbf{r}}. \end{aligned} \tag{1.52}$$

This very important relationship (1.52) is true for any vector $\vec{\mathbf{r}}$, and is referred to as the *Transport Theorem*. A very important application occurs when $\vec{\mathbf{r}}$ denotes the position (of say a spacecraft), $\mathcal{F}_1$ is a nonrotating inertial reference frame, and $\mathcal{F}_2$ is a frame that rotates with the body in question. Equation (1.52) also shows that provided two reference frames are not rotating with respect to each other (if $\vec{\boldsymbol{\omega}}_{21} = \vec{\mathbf{0}}$), then the time-derivative of a vector as seen in both frames is the same, that is $\dot{\vec{\mathbf{r}}} = \mathring{\vec{\mathbf{r}}}$.

Let us express the angular velocity in frame $\mathcal{F}_2$ as

$$\vec{\omega}_{21} = \vec{\mathcal{F}}_2^T \omega_{21}.$$

Making use of this, equation (1.52) becomes

$$\begin{aligned}\dot{\vec{r}} = \vec{\mathcal{F}}_1^T \dot{\mathbf{r}}_1 &= \vec{\mathcal{F}}_2^T \left[\dot{\mathbf{r}}_2 + \omega_{21}^{\times}\mathbf{r}_2\right] \\ &= \vec{\mathcal{F}}_1^T \mathbf{C}_{12}\left[\dot{\mathbf{r}}_2 + \omega_{21}^{\times}\mathbf{r}_2\right],\end{aligned}$$

that is, the relationship between the coordinates of the time-derivatives with respect to the two frames is

$$\dot{\mathbf{r}}_1 = \mathbf{C}_{12}\left[\dot{\mathbf{r}}_2 + \omega_{21}^{\times}\mathbf{r}_2\right]. \tag{1.53}$$

We can make use of the above equation (1.53) to obtain the differential equation for the rotation matrix $\mathbf{C}_{21}$. To this end, let us consider an arbitrary constant vector $\vec{r}$ as seen in frame $\mathcal{F}_1$, that is $\dot{\mathbf{r}}_1 = \mathbf{0}$. In this case, the equation (1.53) becomes

$$\dot{\mathbf{r}}_2 + \omega_{21}^{\times}\mathbf{r}_2 = \mathbf{0}$$

Making use of the transformation of coordinates relationship $\mathbf{r}_2 = \mathbf{C}_{21}\mathbf{r}_1$, this becomes

$$\left[\dot{\mathbf{C}}_{21} + \omega_{21}^{\times}\mathbf{C}_{21}\right]\mathbf{r}_1 = \mathbf{0}.$$

Since the vector $\vec{r}$ was arbitrary, this must hold for any $\mathbf{r}_1 \in R^3$. Therefore, we have the differential equation for the rotation matrix

$$\dot{\mathbf{C}}_{21} = -\omega_{21}^{\times}\mathbf{C}_{21}. \tag{1.54}$$

This equation can be rewritten to give the angular velocity, when the rotation matrix is known as a function of time:

$$\omega_{21}^{\times} = -\dot{\mathbf{C}}_{21}\mathbf{C}_{21}^T. \tag{1.55}$$

Finally, we can demonstrate that angular velocities are additive. Let us consider three reference frames $\mathcal{F}_1$, $\mathcal{F}_2$ and $\mathcal{F}_3$. Let frame $\mathcal{F}_2$ rotate with angular velocity $\vec{\omega}_{21} = \vec{\mathcal{F}}_2^T \omega_{21}$ with respect to $\mathcal{F}_1$, let frame $\mathcal{F}_3$ rotate with angular velocity $\vec{\omega}_{32} = \vec{\mathcal{F}}_3^T \omega_{32}$ with respect to $\mathcal{F}_2$, and let frame $\mathcal{F}_3$ rotate with angular velocity $\vec{\omega}_{31} = \vec{\mathcal{F}}_3^T \omega_{31}$ with respect to $\mathcal{F}_1$. Let $\mathbf{C}_{21}$, $\mathbf{C}_{32}$ and $\mathbf{C}_{31}$ be the associated rotation matrices respectively. From equation (1.54) we have

$$\dot{\mathbf{C}}_{21} = -\omega_{21}^{\times}\mathbf{C}_{21}, \quad \dot{\mathbf{C}}_{32} = -\omega_{32}^{\times}\mathbf{C}_{32}, \quad \dot{\mathbf{C}}_{31} = -\omega_{31}^{\times}\mathbf{C}_{31}.$$

But, we also know from equation (1.19) that

$$\mathbf{C}_{31} = \mathbf{C}_{32}\mathbf{C}_{21}.$$

Differentiating this and using the above differential equations gives

$$\begin{aligned}-\boldsymbol{\omega}_{31}^{\times}\mathbf{C}_{31} = \dot{\mathbf{C}}_{31} &= \dot{\mathbf{C}}_{32}\mathbf{C}_{21} + \mathbf{C}_{32}\dot{\mathbf{C}}_{21} \\ &= -\boldsymbol{\omega}_{32}^{\times}\mathbf{C}_{32}\mathbf{C}_{21} - \mathbf{C}_{32}\boldsymbol{\omega}_{21}^{\times}\mathbf{C}_{21} \\ &= -\boldsymbol{\omega}_{32}^{\times}\mathbf{C}_{31} - \mathbf{C}_{32}\boldsymbol{\omega}_{21}^{\times}\mathbf{C}_{21}.\end{aligned}$$

Post-multiplying both sides by $\mathbf{C}_{31}^T$, and making use of the fact that $\mathbf{C}_{32} = \mathbf{C}_{31}\mathbf{C}_{21}^T$, this becomes

$$-\boldsymbol{\omega}_{31}^{\times} = -\boldsymbol{\omega}_{32}^{\times} - \mathbf{C}_{32}\boldsymbol{\omega}_{21}^{\times}\mathbf{C}_{32}^T.$$

Now, from equation (1.21) we have $(\mathbf{C}_{32}\boldsymbol{\omega}_{21})^{\times} = \mathbf{C}_{32}\boldsymbol{\omega}_{21}^{\times}\mathbf{C}_{32}^T$. Therefore, we obtain

$$-\boldsymbol{\omega}_{31}^{\times} = -\boldsymbol{\omega}_{32}^{\times} - (\mathbf{C}_{32}\boldsymbol{\omega}_{21})^{\times},$$

and

$$\boldsymbol{\omega}_{31} = \boldsymbol{\omega}_{32} + (\mathbf{C}_{32}\boldsymbol{\omega}_{21}).$$

Finally, recognizing that these are just the coordinates of the angular velocities expressed in $\mathcal{F}_3$, this leads to

$$\vec{\boldsymbol{\omega}}_{31} = \vec{\boldsymbol{\omega}}_{32} + \vec{\boldsymbol{\omega}}_{21}, \tag{1.56}$$

that is, **angular velocities are additive**. As a consequence, since a reference frame has zero angular velocity relative to itself, the angular velocity of frame $\mathcal{F}_1$ with respect to $\mathcal{F}_2$ is $\vec{\boldsymbol{\omega}}_{12} = -\vec{\boldsymbol{\omega}}_{21}$.

1.4.2 Angular Velocity in Terms of Euler Angle Rates

We have seen that any frame $\mathcal{F}_2$ can be obtained from another frame $\mathcal{F}_1$ through three successive principal rotations. The angles of these rotations are called Euler Angles. Making use of the fact that angular velocities are additive, we can now easily obtain an expression for the angular velocity in terms of the Euler Angle rates. We will demonstrate this using the 3-2-1 Euler rotation sequence presented previously. First, it will be useful to define some intermediate reference frames. Referring to Figure 1.15, let us denote frame $\mathcal{F}_i$ as the frame obtained from $\mathcal{F}_1$ by a principal rotation about the $\mathcal{F}_1$ z-axis. Next, let us denote frame $\mathcal{F}_t$ as the frame obtained from $\mathcal{F}_i$ by a principal rotation about the $\mathcal{F}_i$ y-axis. Finally, frame $\mathcal{F}_2$ is obtained from frame $\mathcal{F}_t$ by a principal rotation about the $\mathcal{F}_t$ x-axis.

It is now clear that frame $\mathcal{F}_i$ is rotating with respect to $\mathcal{F}_1$ with an angular velocity given by

$$\vec{\boldsymbol{\omega}}_{i1} = \vec{\mathcal{F}}_i^T \begin{bmatrix} 0 \\ 0 \\ \dot{\psi} \end{bmatrix} = \vec{\mathcal{F}}_2^T \mathbf{C}_x(\phi)\mathbf{C}_y(\theta) \begin{bmatrix} 0 \\ 0 \\ \dot{\psi} \end{bmatrix},$$

frame $\mathcal{F}_t$ is rotating with respect to $\mathcal{F}_i$ with an angular velocity given by

$$\vec{\boldsymbol{\omega}}_{ti} = \vec{\mathcal{F}}_t^T \begin{bmatrix} 0 \\ \dot{\theta} \\ 0 \end{bmatrix} = \vec{\mathcal{F}}_2^T \mathbf{C}_x(\phi) \begin{bmatrix} 0 \\ \dot{\theta} \\ 0 \end{bmatrix},$$

and frame $\mathcal{F}_2$ is rotating with respect to $\mathcal{F}_t$ with an angular velocity given by

$$\vec{\boldsymbol{\omega}}_{2t} = \vec{\mathcal{F}}_2^T \begin{bmatrix} \dot{\phi} \\ 0 \\ 0 \end{bmatrix}.$$

Making use of the fact that the angular velocities are additive, we find that frame $\mathcal{F}_2$ rotates with respect to frame $\mathcal{F}_1$ with angular velocity

$$\vec{\boldsymbol{\omega}}_{21} = \vec{\boldsymbol{\omega}}_{2t} + \vec{\boldsymbol{\omega}}_{ti} + \vec{\boldsymbol{\omega}}_{i1}.$$

With $\vec{\boldsymbol{\omega}}_{21} = \vec{\mathcal{F}}_2^T \boldsymbol{\omega}_{21}$, we can obtain the relationship between the Euler Angle rates and the angular velocity in the coordinates of $\mathcal{F}_2$ as

$$\boldsymbol{\omega}_{21} = \begin{bmatrix} \dot{\phi} \\ 0 \\ 0 \end{bmatrix} + \mathbf{C}_x(\phi) \begin{bmatrix} 0 \\ \dot{\theta} \\ 0 \end{bmatrix} + \mathbf{C}_x(\phi)\mathbf{C}_y(\theta) \begin{bmatrix} 0 \\ 0 \\ \dot{\psi} \end{bmatrix}. \tag{1.57}$$

Written out in full, this is

$$\boldsymbol{\omega}_{21} = \begin{bmatrix} 1 & 0 & -\sin\theta \\ 0 & \cos\phi & \sin\phi\cos\theta \\ 0 & -\sin\phi & \cos\phi\cos\theta \end{bmatrix} \begin{bmatrix} \dot{\phi} \\ \dot{\theta} \\ \dot{\psi} \end{bmatrix}. \tag{1.58}$$

This can be inverted to give the Euler Angle rates in terms of the angular velocity:

$$\begin{bmatrix} \dot{\phi} \\ \dot{\theta} \\ \dot{\psi} \end{bmatrix} = \begin{bmatrix} 1 & \sin\phi\tan\theta & \cos\phi\tan\theta \\ 0 & \cos\phi & -\sin\phi \\ 0 & \sin\phi\sec\theta & \cos\phi\sec\theta \end{bmatrix} \boldsymbol{\omega}_{21}. \tag{1.59}$$

From this, it is clear that when the 3-2-1 rotation sequence is at the singularity (recall this is when $\theta = \pm 90°$, the Euler Angle rates cannot be obtained from (1.59) due to the terms $\tan\theta$ and $\sec\theta$ being undefined.

1.4.3 Angular Velocity in Terms of Quaternion Rates

We have seen from (1.55) that if frame $\mathcal{F}_2$ is rotating with angular velocity $\vec{\boldsymbol{\omega}}_{21} = \vec{\mathcal{F}}_2^T \boldsymbol{\omega}_{21}$ relative to frame $\mathcal{F}_1$, then

$$\boldsymbol{\omega}_{21}^{\times} = -\dot{\mathbf{C}}_{21}\mathbf{C}_{21}^T. \tag{1.60}$$

It is important to note that $\boldsymbol{\omega}_{21}$ are the coordinates of $\vec{\omega}_{21}$ in frame $\mathcal{F}_2$, and $\mathbf{C}_{21}$ is the rotation matrix transforming coordinates from frame $\mathcal{F}_1$ to frame $\mathcal{F}_2$.

Making use of the identity $\boldsymbol{\epsilon}^{\times}\boldsymbol{\epsilon}^{\times} = \boldsymbol{\epsilon}\boldsymbol{\epsilon}^T - \boldsymbol{\epsilon}^T\boldsymbol{\epsilon}\mathbf{1}$ (see (1.14)), and the unit constraint (1.32), the rotation matrix $\mathbf{C}_{21}$ with quaternion parameterization $(\boldsymbol{\epsilon}, \eta)$, may be written as

$$\mathbf{C}_{21} = \mathbf{1} + 2\boldsymbol{\epsilon}^{\times}\boldsymbol{\epsilon}^{\times} - 2\eta\boldsymbol{\epsilon}^{\times}. \tag{1.61}$$

Differentiating, we have

$$\dot{\mathbf{C}}_{21} = 2\left[\left(\dot{\boldsymbol{\epsilon}}^{\times}\boldsymbol{\epsilon}^{\times} + \boldsymbol{\epsilon}^{\times}\dot{\boldsymbol{\epsilon}}^{\times}\right) - \dot{\eta}\boldsymbol{\epsilon}^{\times} - \eta\dot{\boldsymbol{\epsilon}}^{\times}\right]. \tag{1.62}$$

By post-multiplying (1.62) with (1.61) and expanding, we find that

$$\begin{aligned}\tfrac{1}{2}\dot{\mathbf{C}}_{21}\mathbf{C}_{21}^T &= \left(\dot{\boldsymbol{\epsilon}}^{\times}\boldsymbol{\epsilon}^{\times} + \boldsymbol{\epsilon}^{\times}\dot{\boldsymbol{\epsilon}}^{\times}\right) - \dot{\eta}\boldsymbol{\epsilon}^{\times} - \eta\dot{\boldsymbol{\epsilon}}^{\times} \\ &\quad + 2\left(\dot{\boldsymbol{\epsilon}}^{\times}\boldsymbol{\epsilon}^{\times} + \boldsymbol{\epsilon}^{\times}\dot{\boldsymbol{\epsilon}}^{\times}\right)\boldsymbol{\epsilon}^{\times}\boldsymbol{\epsilon}^{\times} - 2\dot{\eta}\boldsymbol{\epsilon}^{\times}\boldsymbol{\epsilon}^{\times}\boldsymbol{\epsilon}^{\times} \\ &\quad + 2\eta\boldsymbol{\epsilon}^{\times}\dot{\boldsymbol{\epsilon}}^{\times}\boldsymbol{\epsilon}^{\times} - 2\eta\dot{\eta}\boldsymbol{\epsilon}^{\times}\boldsymbol{\epsilon}^{\times} - 2\eta^2\dot{\boldsymbol{\epsilon}}^{\times}\boldsymbol{\epsilon}^{\times}.\end{aligned} \tag{1.63}$$

To reduce this expression, we need to make use of some column matrix identities. We have previously derived

$$\mathbf{a}^{\times}\mathbf{b}^{\times} = \mathbf{b}\mathbf{a}^T - \left(\mathbf{a}^T\mathbf{b}\right)\mathbf{1}.$$

Making use of this identity, we now expand the individual terms in (1.63). We have

$$\begin{aligned}\boldsymbol{\epsilon}^{\times}\boldsymbol{\epsilon}^{\times} &= \boldsymbol{\epsilon}\boldsymbol{\epsilon}^T - \left(\boldsymbol{\epsilon}^T\boldsymbol{\epsilon}\right)\mathbf{1}, \\ \boldsymbol{\epsilon}^{\times}\boldsymbol{\epsilon}^{\times}\boldsymbol{\epsilon}^{\times} &= -\left(\boldsymbol{\epsilon}^T\boldsymbol{\epsilon}\right)\boldsymbol{\epsilon}^{\times}, \\ \dot{\boldsymbol{\epsilon}}^{\times}\boldsymbol{\epsilon}^{\times} &= \boldsymbol{\epsilon}\dot{\boldsymbol{\epsilon}}^T - \left(\boldsymbol{\epsilon}^T\dot{\boldsymbol{\epsilon}}\right)\mathbf{1}, \\ \boldsymbol{\epsilon}^{\times}\dot{\boldsymbol{\epsilon}}^{\times}\boldsymbol{\epsilon}^{\times} &= -\left(\boldsymbol{\epsilon}^T\dot{\boldsymbol{\epsilon}}\right)\boldsymbol{\epsilon}^{\times}, \\ \boldsymbol{\epsilon}^{\times}\dot{\boldsymbol{\epsilon}}^{\times} &= \dot{\boldsymbol{\epsilon}}\boldsymbol{\epsilon}^T - \left(\boldsymbol{\epsilon}^T\dot{\boldsymbol{\epsilon}}\right)\mathbf{1},\end{aligned}$$

Substituting these into (1.63) and collecting like terms, we obtain

$$\begin{aligned}\tfrac{1}{2}\dot{\mathbf{C}}_{21}\mathbf{C}_{21}^T &= \dot{\boldsymbol{\epsilon}}\boldsymbol{\epsilon}^T + \left[1 - 2\left(\boldsymbol{\epsilon}^T\boldsymbol{\epsilon}\right) - 2\eta^2\right]\boldsymbol{\epsilon}\dot{\boldsymbol{\epsilon}}^T \\ &\quad + \left[-2\left(\boldsymbol{\epsilon}^T\dot{\boldsymbol{\epsilon}}\right) + 4\left(\boldsymbol{\epsilon}^T\dot{\boldsymbol{\epsilon}}\right)\left(\boldsymbol{\epsilon}^T\boldsymbol{\epsilon}\right) + 2\eta\dot{\eta}\left(\boldsymbol{\epsilon}^T\boldsymbol{\epsilon}\right) + 2\eta^2\left(\boldsymbol{\epsilon}^T\dot{\boldsymbol{\epsilon}}\right)\right]\mathbf{1} \\ &\quad + \left[-\dot{\eta} + 2\dot{\eta}\left(\boldsymbol{\epsilon}^T\boldsymbol{\epsilon}\right) - 2\eta\left(\boldsymbol{\epsilon}^T\dot{\boldsymbol{\epsilon}}\right)\right]\boldsymbol{\epsilon}^{\times} - \eta\dot{\boldsymbol{\epsilon}}^{\times} \\ &\quad + \left[-2\left(\boldsymbol{\epsilon}^T\dot{\boldsymbol{\epsilon}}\right) - 2\eta\dot{\eta}\right]\boldsymbol{\epsilon}\boldsymbol{\epsilon}^T.\end{aligned} \tag{1.64}$$

We can now further reduce each of the coefficients in (1.64), by making use of the unit magnitude constraint on the quaternion

$$\boldsymbol{\epsilon}^T\boldsymbol{\epsilon} + \eta^2 = 1, \tag{1.65}$$

and its derivative

$$\boldsymbol{\epsilon}^T\dot{\boldsymbol{\epsilon}} + \eta\dot{\eta} = 0. \tag{1.66}$$

Making use of (1.65), we find that

$$1 - 2\left(\boldsymbol{\epsilon}^T\boldsymbol{\epsilon}\right) - 2\eta^2 = -1.$$

Making use of (1.65) and (1.66), we find that

$$\begin{aligned}
&-2\left(\boldsymbol{\epsilon}^T\dot{\boldsymbol{\epsilon}}\right) + 4\left(\boldsymbol{\epsilon}^T\dot{\boldsymbol{\epsilon}}\right)\left(\boldsymbol{\epsilon}^T\boldsymbol{\epsilon}\right) + 2\eta\dot{\eta}\left(\boldsymbol{\epsilon}^T\boldsymbol{\epsilon}\right) + 2\eta^2\left(\boldsymbol{\epsilon}^T\dot{\boldsymbol{\epsilon}}\right) \\
&= -2\left(\boldsymbol{\epsilon}^T\dot{\boldsymbol{\epsilon}}\right) + 2\left(\boldsymbol{\epsilon}^T\boldsymbol{\epsilon} + \eta^2\right)\left(\boldsymbol{\epsilon}^T\dot{\boldsymbol{\epsilon}}\right) + 2\left(\boldsymbol{\epsilon}^T\dot{\boldsymbol{\epsilon}} + \eta\dot{\eta}\right)\left(\boldsymbol{\epsilon}^T\boldsymbol{\epsilon}\right) = 0.
\end{aligned}$$

$$\begin{aligned}
-\dot{\eta} + 2\dot{\eta}\left(\boldsymbol{\epsilon}^T\boldsymbol{\epsilon}\right) - 2\eta\left(\boldsymbol{\epsilon}^T\dot{\boldsymbol{\epsilon}}\right) &= \dot{\eta}\left(2\left(\boldsymbol{\epsilon}^T\boldsymbol{\epsilon}\right) - 1\right) - 2\eta\left(\boldsymbol{\epsilon}^T\dot{\boldsymbol{\epsilon}}\right) \\
&= \dot{\eta}\left(1 - 2\eta^2\right) - 2\eta\left(\boldsymbol{\epsilon}^T\dot{\boldsymbol{\epsilon}}\right) \\
&= \dot{\eta} - 2\left(\eta\dot{\eta} + \boldsymbol{\epsilon}^T\dot{\boldsymbol{\epsilon}}\right)\eta \\
&= \dot{\eta}
\end{aligned}$$

$$-2\left(\boldsymbol{\epsilon}^T\dot{\boldsymbol{\epsilon}}\right) - 2\eta\dot{\eta} = 0.$$

Therefore, (1.64) becomes

$$\frac{1}{2}\dot{\mathbf{C}}_{21}\mathbf{C}_{21}^T = \left(\dot{\boldsymbol{\epsilon}}\boldsymbol{\epsilon}^T - \boldsymbol{\epsilon}\dot{\boldsymbol{\epsilon}}^T\right) - \eta\dot{\boldsymbol{\epsilon}}^\times + \dot{\eta}\dot{\boldsymbol{\epsilon}}^\times.$$

We need one more identity to reduce the bracketed term. Making use of (1.42), we find that

$$\dot{\boldsymbol{\epsilon}}\boldsymbol{\epsilon}^T - \boldsymbol{\epsilon}\dot{\boldsymbol{\epsilon}}^T = \left(\boldsymbol{\epsilon}^\times\dot{\boldsymbol{\epsilon}}\right)^\times,$$

and we finally have

$$\frac{1}{2}\dot{\mathbf{C}}_{21}\mathbf{C}_{21}^T = \left(\boldsymbol{\epsilon}^\times\dot{\boldsymbol{\epsilon}}\right)^\times - \eta\dot{\boldsymbol{\epsilon}}^\times + \dot{\eta}\boldsymbol{\epsilon}^\times.$$

Therefore, from (1.60), we have

$$\boldsymbol{\omega}_{21}^\times = -2\left[\left(\boldsymbol{\epsilon}^\times\dot{\boldsymbol{\epsilon}}\right)^\times - \eta\dot{\boldsymbol{\epsilon}}^\times + \dot{\eta}\boldsymbol{\epsilon}^\times\right]$$

Finally, since both sides of the above equation are cross-product operators, we can extract the equation for the angular velocity as

$$\boldsymbol{\omega}_{21} = 2\left(\eta\mathbf{1} - \boldsymbol{\epsilon}^\times\right)\dot{\boldsymbol{\epsilon}} - 2\boldsymbol{\epsilon}\dot{\eta}. \tag{1.67}$$

We can now invert (1.67) to find the quaternion rates in terms of the angular velocity. To do this, we write (1.67) in matrix form, and append the derivative constraint (1.66).

$$\begin{bmatrix} \boldsymbol{\omega}_{21} \\ 0 \end{bmatrix} = 2 \begin{bmatrix} \left(\eta \mathbf{1} - \boldsymbol{\epsilon}^{\times}\right) & -\boldsymbol{\epsilon} \\ \boldsymbol{\epsilon}^{T} & \eta \end{bmatrix} \begin{bmatrix} \dot{\boldsymbol{\epsilon}} \\ \dot{\eta} \end{bmatrix}.$$

It is straightforward to show that the above matrix is orthonormal (just multiply it by its transpose and use the same identities as before). Therefore, we may invert the above relationship to obtain

$$\begin{bmatrix} \dot{\boldsymbol{\epsilon}} \\ \dot{\eta} \end{bmatrix} = \frac{1}{2} \begin{bmatrix} \left(\eta \mathbf{1} + \boldsymbol{\epsilon}^{\times}\right) & \boldsymbol{\epsilon} \\ -\boldsymbol{\epsilon}^{T} & \eta \end{bmatrix} \begin{bmatrix} \boldsymbol{\omega}_{21} \\ 0 \end{bmatrix}.$$

From this, we can extract the equations for the quaternion rates in terms of the angular velocity

$$\begin{aligned} \dot{\boldsymbol{\epsilon}} &= \frac{1}{2}\left(\eta \mathbf{1} + \boldsymbol{\epsilon}^{\times}\right) \boldsymbol{\omega}_{21}, \\ \dot{\eta} &= -\frac{1}{2} \boldsymbol{\epsilon}^{T} \boldsymbol{\omega}_{21}. \end{aligned} \tag{1.68}$$

1.5 Velocity and Acceleration

We shall now obtain expressions for the velocity and acceleration as seen in different reference frames. As before, we consider the frames $\mathcal{F}_1$ and $\mathcal{F}_2$.

Let us denote the *velocity* as seen in frame $\mathcal{F}_1$ by

$$\vec{\mathbf{v}} = \dot{\vec{\mathbf{r}}} = \mathring{\vec{\mathbf{r}}} + \vec{\boldsymbol{\omega}}_{21} \times \vec{\mathbf{r}} \tag{1.69}$$

The *acceleration* as seen in frame $\mathcal{F}_1$ can be obtained by applying the rule in (1.52) to $\vec{\mathbf{v}}$:

$$\begin{aligned} \ddot{\vec{\mathbf{r}}} = \dot{\vec{\mathbf{v}}} &= \mathring{\vec{\mathbf{v}}} + \vec{\boldsymbol{\omega}}_{21} \times \vec{\mathbf{v}} \\ &= \left(\overset{\circ\circ}{\vec{\mathbf{r}}} + \vec{\boldsymbol{\omega}}_{21} \times \mathring{\vec{\mathbf{r}}} + \mathring{\vec{\boldsymbol{\omega}}}_{21} \times \vec{\mathbf{r}}\right) + \left(\vec{\boldsymbol{\omega}}_{21} \times \mathring{\vec{\mathbf{r}}} + \vec{\boldsymbol{\omega}}_{21} \times \vec{\boldsymbol{\omega}}_{21} \times \vec{\mathbf{r}}\right) \\ &= \overset{\circ\circ}{\vec{\mathbf{r}}} + 2\vec{\boldsymbol{\omega}}_{21} \times \mathring{\vec{\mathbf{r}}} + \mathring{\vec{\boldsymbol{\omega}}}_{21} \times \vec{\mathbf{r}} + \vec{\boldsymbol{\omega}}_{21} \times \vec{\boldsymbol{\omega}}_{21} \times \vec{\mathbf{r}}. \end{aligned} \tag{1.70}$$

The different terms in the above expression for the acceleration have special names:

$\overset{\circ\circ}{\vec{\mathbf{r}}}$	acceleration as seen in $\mathcal{F}_2$,
$2\vec{\boldsymbol{\omega}}_{21} \times \mathring{\vec{\mathbf{r}}}$	coriolis acceleration,
$\mathring{\vec{\boldsymbol{\omega}}}_{21} \times \vec{\mathbf{r}}$	angular acceleration,
$\vec{\boldsymbol{\omega}}_{21} \times \vec{\boldsymbol{\omega}}_{21} \times \vec{\mathbf{r}}$	centripetal acceleration.

The coordinates for acceleration in (1.70) given in the two reference frames can be obtained by using the expressions

$$\ddot{\vec{\mathbf{r}}} = \vec{\mathcal{F}}_1^T \ddot{\mathbf{r}}_1, \quad \overset{\circ\circ}{\vec{\mathbf{r}}} = \vec{\mathcal{F}}_2^T \ddot{\mathbf{r}}_2, \quad \vec{\omega}_{21} = \vec{\mathcal{F}}_2^T \omega_{21}.$$

The result is

$$\ddot{\mathbf{r}}_1 = \mathbf{C}_{12}\left[\ddot{\mathbf{r}}_2 + 2\omega_{21}^{\times}\dot{\mathbf{r}}_2 + \dot{\omega}_{21}^{\times}\mathbf{r}_2 + \omega_{21}^{\times}\omega_{21}^{\times}\mathbf{r}_2\right].$$

1.6 More Rigorous Definition of Angular Velocity

Consider two reference frames $\mathcal{F}_2$ and $\mathcal{F}_1$. Let the unit basis vectors of frame $\mathcal{F}_2$ be $\vec{\mathbf{x}}_2$, $\vec{\mathbf{y}}_2$ and $\vec{\mathbf{z}}_2$. Recall that the basis vectors must satisfy the normality and orthogonality constraints

$$\vec{\mathbf{x}}_2 \cdot \vec{\mathbf{x}}_2 = \vec{\mathbf{y}}_2 \cdot \vec{\mathbf{y}}_2 = \vec{\mathbf{z}}_2 \cdot \vec{\mathbf{z}}_2 = 1, \tag{1.71}$$

and

$$\vec{\mathbf{x}}_2 \cdot \vec{\mathbf{y}}_2 = \vec{\mathbf{x}}_2 \cdot \vec{\mathbf{z}}_2 = \vec{\mathbf{y}}_2 \cdot \vec{\mathbf{z}}_2 = 0. \tag{1.72}$$

respectively.

We are now going to examine the time-derivative of frame $\mathcal{F}_2$ as seen in frame $\mathcal{F}_1$. Let $(\dot{\vec{\mathbf{a}}})$ denote the time-derivative of the vector $\vec{\mathbf{a}}$ as seen in $\mathcal{F}_1$.

Let us now take the derivatives of the normality constraints (1.71). We see that

$$\vec{\mathbf{x}}_2 \cdot \dot{\vec{\mathbf{x}}}_2 = 0, \quad \vec{\mathbf{y}}_2 \cdot \dot{\vec{\mathbf{y}}}_2 = 0, \quad \vec{\mathbf{z}}_2 \cdot \dot{\vec{\mathbf{z}}}_2 = 0.$$

That is,

$$\vec{\mathbf{x}}_2 \perp \dot{\vec{\mathbf{x}}}_2, \quad \vec{\mathbf{y}}_2 \perp \dot{\vec{\mathbf{y}}}_2, \quad \vec{\mathbf{z}}_2 \perp \dot{\vec{\mathbf{z}}}_2.$$

Therefore, we can find vectors $\vec{\mathbf{a}}$, $\vec{\mathbf{b}}$ and $\vec{\mathbf{c}}$ such that

$$\dot{\vec{\mathbf{x}}}_2 = \vec{\mathbf{a}} \times \vec{\mathbf{x}}_2, \quad \dot{\vec{\mathbf{y}}}_2 = \vec{\mathbf{b}} \times \vec{\mathbf{y}}_2, \quad \dot{\vec{\mathbf{z}}}_2 = \vec{\mathbf{c}} \times \vec{\mathbf{z}}_2. \tag{1.73}$$

The reason we can do this will be clear shortly. Let $\vec{\mathbf{a}}_{\perp x}$, $\vec{\mathbf{b}}_{\perp y}$ and $\vec{\mathbf{c}}_{\perp z}$ be the components of $\vec{\mathbf{a}}$, $\vec{\mathbf{b}}$ and $\vec{\mathbf{c}}$ that are perpendicular to $\vec{\mathbf{x}}_2$, $\vec{\mathbf{y}}_2$ and $\vec{\mathbf{z}}_2$ respectively. As we have seen in (1.6), we can rewrite the above derivatives as

$$\dot{\vec{\mathbf{x}}}_2 = \vec{\mathbf{a}}_{\perp x} \times \vec{\mathbf{x}}_2, \quad \dot{\vec{\mathbf{y}}}_2 = \vec{\mathbf{b}}_{\perp y} \times \vec{\mathbf{y}}_2, \quad \dot{\vec{\mathbf{z}}}_2 = \vec{\mathbf{c}}_{\perp z} \times \vec{\mathbf{z}}_2.$$

Now, let us express each of the vectors in $\mathcal{F}_2$ coordinates. We have

$$\vec{\mathbf{x}}_2 = \vec{\mathcal{F}}_2^T \begin{bmatrix} 1 \\ 0 \\ 0 \end{bmatrix}, \quad \vec{\mathbf{y}}_2 = \vec{\mathcal{F}}_2^T \begin{bmatrix} 0 \\ 1 \\ 0 \end{bmatrix}, \quad \vec{\mathbf{z}}_2 = \vec{\mathcal{F}}_2^T \begin{bmatrix} 0 \\ 0 \\ 1 \end{bmatrix},$$

and

$$\vec{\mathbf{a}}_{\perp x} = \vec{\mathcal{F}}_2^T \begin{bmatrix} 0 \\ a_y \\ a_z \end{bmatrix}, \quad \vec{\mathbf{b}}_{\perp y} = \vec{\mathcal{F}}_2^T \begin{bmatrix} b_x \\ 0 \\ b_z \end{bmatrix}, \quad \vec{\mathbf{c}}_{\perp x} = \vec{\mathcal{F}}_2^T \begin{bmatrix} c_x \\ c_y \\ 0 \end{bmatrix}.$$

Let us now see why we can always find vectors $\vec{\mathbf{a}}$, $\vec{\mathbf{b}}$ and $\vec{\mathbf{c}}$ such that our derivatives are given by (1.73). Consider $\dot{\vec{\mathbf{x}}}_2$. Since $\vec{\mathbf{x}}_2 \perp \dot{\vec{\mathbf{x}}}_2$, when $\dot{\vec{\mathbf{x}}}_2$ is expressed in $\mathcal{F}_2$ coordinates, it must have the form

$$\dot{\vec{\mathbf{x}}}_2 = \vec{\mathcal{F}}_2^T \begin{bmatrix} 0 \\ p \\ q \end{bmatrix},$$

for some p and q. Let us now look at the derivative expression in $\mathcal{F}_2$. We have

$$\dot{\vec{\mathbf{x}}}_2 = \vec{\mathbf{a}}_{\perp x} \times \vec{\mathbf{x}}_2 = -\vec{\mathbf{x}}_2 \times \vec{\mathbf{a}}_{\perp x},$$

which leads to

$$\begin{bmatrix} 0 \\ p \\ q \end{bmatrix} = \begin{bmatrix} 0 & 0 & 0 \\ 0 & 0 & 1 \\ 0 & -1 & 0 \end{bmatrix} \begin{bmatrix} 0 \\ a_y \\ a_z \end{bmatrix} = \begin{bmatrix} 0 \\ a_z \\ +a_y \end{bmatrix}.$$

which is clearly always solvable for a_y and a_z, and hence $\vec{\mathbf{a}}$. We can similarly always solve for $\vec{\mathbf{b}}$ and $\vec{\mathbf{c}}$.

Now, let us examine the orthogonality constraints (1.72). Differentiating these, we obtain

$$\begin{aligned} \dot{\vec{\mathbf{x}}}_2 \cdot \vec{\mathbf{y}}_2 + \vec{\mathbf{x}}_2 \cdot \dot{\vec{\mathbf{y}}}_2 &= 0, \\ \dot{\vec{\mathbf{x}}}_2 \cdot \vec{\mathbf{z}}_2 + \vec{\mathbf{x}}_2 \cdot \dot{\vec{\mathbf{z}}}_2 &= 0, \\ \dot{\vec{\mathbf{y}}}_2 \cdot \vec{\mathbf{z}}_2 + \vec{\mathbf{y}}_2 \cdot \dot{\vec{\mathbf{z}}}_2 &= 0. \end{aligned}$$

Substituting the derivative expressions, these become

$$\begin{aligned} \left(\vec{\mathbf{a}}_{\perp x} \times \vec{\mathbf{x}}_2\right) \cdot \vec{\mathbf{y}}_2 + \vec{\mathbf{x}}_2 \cdot \left(\vec{\mathbf{b}}_{\perp y} \times \vec{\mathbf{y}}_2\right) &= 0, \\ \left(\vec{\mathbf{a}}_{\perp x} \times \vec{\mathbf{x}}_2\right) \cdot \vec{\mathbf{z}}_2 + \vec{\mathbf{x}}_2 \cdot \left(\vec{\mathbf{c}}_{\perp z} \times \vec{\mathbf{z}}_2\right) &= 0, \\ \left(\vec{\mathbf{b}}_{\perp y} \times \vec{\mathbf{y}}_2\right) \cdot \vec{\mathbf{z}}_2 + \vec{\mathbf{y}}_2 \cdot \left(\vec{\mathbf{c}}_{\perp z} \times \vec{\mathbf{z}}_2\right) &= 0. \end{aligned}$$

Making use of the scalar triple product identity (see the end of Section 2.1), and the fact that for a right-handed coordinate system we have $\vec{\mathbf{x}}_1 \times \vec{\mathbf{y}}_1 = \vec{\mathbf{z}}_1$, $\vec{\mathbf{y}}_1 \times \vec{\mathbf{z}}_1 = \vec{\mathbf{x}}_1$ and $\vec{\mathbf{z}}_1 \times \vec{\mathbf{x}}_1 = \vec{\mathbf{y}}_1$, we obtain the constraints

$$\left(\vec{\mathbf{b}}_{\perp y} - \vec{\mathbf{a}}_{\perp x}\right) \cdot \vec{\mathbf{z}}_2 = 0,$$
$$\left(\vec{\mathbf{c}}_{\perp z} - \vec{\mathbf{a}}_{\perp x}\right) \cdot \vec{\mathbf{y}}_2 = 0,$$
$$\left(\vec{\mathbf{c}}_{\perp z} - \vec{\mathbf{b}}_{\perp y}\right) \cdot \vec{\mathbf{x}}_2 = 0.$$

These constraints lead to

$$b_z = a_z, \quad c_x = b_x, \quad c_y = a_y,$$

so that we must have

$$\vec{\mathbf{a}}_{\perp x} = \vec{\mathcal{F}}_2^T \begin{bmatrix} 0 \\ a_y \\ a_z \end{bmatrix}, \quad \vec{\mathbf{b}}_{\perp y} = \vec{\mathcal{F}}_2^T \begin{bmatrix} b_x \\ 0 \\ a_z \end{bmatrix}, \quad \vec{\mathbf{c}}_{\perp z} = \vec{\mathcal{F}}_2^T \begin{bmatrix} b_x \\ a_y \\ 0 \end{bmatrix}.$$

Now, note that the vector $\vec{\mathcal{F}}_2^T \begin{bmatrix} b_x \\ 0 \\ 0 \end{bmatrix}$ is parallel to $\vec{\mathbf{x}}_2$, the vector $\vec{\mathcal{F}}_2^T \begin{bmatrix} 0 \\ a_y \\ 0 \end{bmatrix}$ is parallel to $\vec{\mathbf{y}}_2$ and the vector $\vec{\mathcal{F}}_2^T \begin{bmatrix} 0 \\ 0 \\ a_z \end{bmatrix}$ is parallel to $\vec{\mathbf{z}}_2$. Therefore, see that we can write

$$\vec{\boldsymbol{\omega}}_{21} \times \vec{\mathbf{x}}_2 = \vec{\mathbf{a}}_{\perp x} \times \vec{\mathbf{x}}_2, \quad \vec{\boldsymbol{\omega}}_{21} \times \vec{\mathbf{y}}_2 = \vec{\mathbf{b}}_{\perp y} \times \vec{\mathbf{y}}_2, \quad \vec{\boldsymbol{\omega}}_{21} \times \vec{\mathbf{z}}_2 = \vec{\mathbf{c}}_{\perp z} \times \vec{\mathbf{z}}_2,$$

where

$$\vec{\boldsymbol{\omega}}_{21} = \vec{\mathbf{a}}_{\perp x} + \vec{\mathcal{F}}_2^T \begin{bmatrix} b_x \\ 0 \\ 0 \end{bmatrix} = \vec{\mathbf{b}}_{\perp y} + \vec{\mathcal{F}}_2^T \begin{bmatrix} 0 \\ a_y \\ 0 \end{bmatrix} = \vec{\mathbf{c}}_{\perp z} + \vec{\mathcal{F}}_2^T \begin{bmatrix} 0 \\ 0 \\ a_z \end{bmatrix} = \vec{\mathcal{F}}_2^T \begin{bmatrix} b_x \\ a_y \\ a_z \end{bmatrix}.$$

We call the vector $\vec{\boldsymbol{\omega}}_{21}$ the angular velocity of frame $\mathcal{F}_2$ relative to frame $\mathcal{F}_1$, and we have the relationships

$$\dot{\vec{\mathbf{x}}}_2 = \vec{\boldsymbol{\omega}}_{21} \times \vec{\mathbf{x}}_2, \quad \dot{\vec{\mathbf{y}}}_2 = \vec{\boldsymbol{\omega}}_{21} \times \vec{\mathbf{y}}_2, \quad \dot{\vec{\mathbf{z}}}_2 = \vec{\boldsymbol{\omega}}_{21} \times \vec{\mathbf{z}}_2.$$

These are the same as we have formally derived in Section 1.4.1.

Notes

In this chapter, we have developed the mathematics necessary to describe the kinematics of spacecraft motion. We have purposely chosen to develop the kinematics using the vectrix formalism. The use of vectrices provides a clear distinction between a physical vector and its coordinate representation in a particular reference frame. This becomes important in spacecraft dynamics problems where many different reference frames may be considered simultaneously. This then results in several different coordinate representations for the same physical vector. As will become evident in later chapters in this book, the use of vectrices allows the spacecraft equations of motion to be derived purely in terms of physical vectors, without consideration of its coordinates in any particular reference frame. Finally, once the equations of motion have been obtained in physical vector form, their coordinate representations in any particular reference frame may be obtained directly. The term vectrix was first coined by Peter C. Hughes, and the reader can find further treatment of vectrices in Hughes (2004). Similar formalisms may also be found in Likins (1973) and Wittenburg (1977). In Section 1.3, we have explained how the spacecraft attitude is described by a rotation matrix, which may be equivalently represented by a principal axis and angle of rotation, a quaternion or a set of Euler Angles. There are several other parameterizations of a rotation matrix. A detailed treatment is contained in Shuster (1993).

References

Hughes PC 2004 *Spacecraft Attitude Dynamics*. Dover Publications, Mineola, NY.

Likins PW 1973 *Elements of Engineering Mechanics*. McGraw-Hill, New York, NY.

Shuster MD 1993 A Survey of Attitude Representations. *Journal of the Astronautical Sciences* **41**(4), 439–517.

Wittenburg J 1977 *Dynamics of Systems of Rigid Bodies*. Teubner, Stuttgart.

2

Rigid Body Dynamics

Now that we have a framework for the description of the position and attitude of a spacecraft in three dimensions, we can formulate the equations of motion. Generally, spacecraft structures have some flexibility (that is, they can elastically deform). However, often times they can be approximated as *rigid bodies* (that is, they cannot deform). This will be the approach taken in this book. We will start with the dynamics of a single particle, then extend it to a system of particles, and finally extend this to a rigid body.

2.1 Dynamics of a Single Particle

The dynamics of a single particle are governed by Newton's second law of motion. First, we need to specify an *inertial* frame of reference, which we will denote $\mathcal{F}_I$. This is a frame of reference where Newton's laws hold. The specification of such a reference frame depends on the problem at hand. For our purposes, such a frame has the origin at the sun, with orientation fixed with respect to the stars. It is a property of inertial reference frames that any frame that is stationary or moving with constant velocity (but not rotating) with respect to an inertial reference frame is also inertial. This is since the acceleration with respect to those frames is the same, which is a simple consequence of the fact that we discovered in Section 1.4.1 that time-derivatives of vectors as seen in frames that are not rotating with respect to each other, are the same.

This is easy to show. Let us consider a particle of constant mass m with position $\vec{\mathbf{r}}$ with respect to the origin of $\mathcal{F}_I$, as shown in Figure 2.1. Consider a second reference frame $\mathcal{F}_2$ whose origin is located at a position $\vec{\mathbf{r}}_{F2}$ from the origin of $\mathcal{F}_I$, and is translating with constant velocity relative to $\mathcal{F}_I$, but is not rotating with respect to $\mathcal{F}_I$. The position of the particle with respect to the origin of $\mathcal{F}_2$ is given by $\vec{\mathbf{r}}_2 = \vec{\mathbf{r}} - \vec{\mathbf{r}}_{F2}$. Note that the velocity of $\mathcal{F}_2$ is given by $\vec{\mathbf{v}}_{F2} = \dot{\vec{\mathbf{r}}}_{F2}$, which is constant. Now, let us find the acceleration of the particle as seen in both frames. In frame $\mathcal{F}_I$, the velocity of the particle is

$$\dot{\vec{\mathbf{r}}} = \dot{\vec{\mathbf{r}}}_{F2} + \dot{\vec{\mathbf{r}}}_2 = \vec{\mathbf{v}}_{F2} + \dot{\vec{\mathbf{r}}}_2,$$

Spacecraft Dynamics and Control: An Introduction, First Edition.
Anton H.J. de Ruiter, Christopher J. Damaren and James R. Forbes.

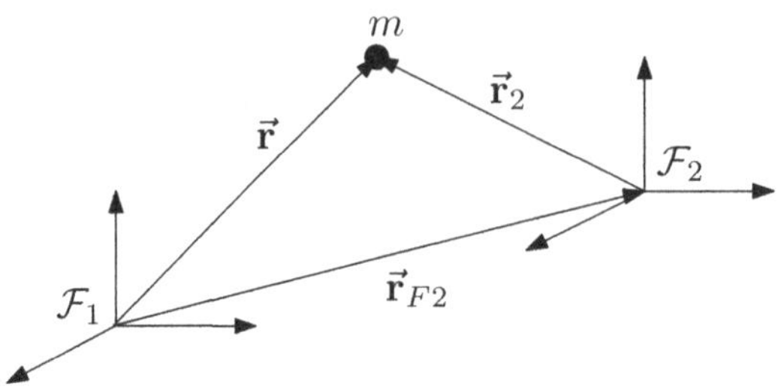

Figure 2.1 Translating reference frame

and the acceleration is

$$\ddot{\vec{\mathbf{r}}} = \dot{\vec{\mathbf{v}}}_{F2} + \ddot{\vec{\mathbf{r}}}_2 = \ddot{\vec{\mathbf{r}}}_2.$$

But, since $\mathcal{F}_2$ is not rotating with respect to $\mathcal{F}_I$, time-derivatives of vectors are the same and $\ddot{\vec{\mathbf{r}}}_2$ is just the acceleration of the particle as seen in $\mathcal{F}_2$. From this point on in this book, the time-derivative of a physical vector as seen in an inertial frame of reference shall be denoted by an overdot $(\dot{\ })$.

As a consequence of this, once we have identified one inertial frame of reference, it is possible to obtain any other inertial frame from it. Of particular interest to this book, since the earth is orbiting the sun slowly (one revolution per year), a frame of reference with origin at the center of the Earth whose orientation is fixed with respect to the stars is translating with approximately constant velocity with respect to the sun (throughout the period of a geocentric orbit, which is on the order of hours). This frame can be considered (almost) inertial (for the purposes of modeling geocentric orbits), and is called the Earth-Centered-Inertial (ECI) frame. This frame is important when describing the motion of Earth-orbiting satellites.

Having discussed all this, let us now return to the problem at hand (the dynamics of a single particle). We consider a particle of constant mass m with position $\vec{\mathbf{r}}$, as shown in Figure 2.2. The momentum of the particle is defined as $\vec{\mathbf{p}} = m\dot{\vec{\mathbf{r}}}$, and the particle is subject to the force $\vec{\mathbf{f}}$. Newton's second law of motion gives the translational dynamics of the particle:

$$m\ddot{\vec{\mathbf{r}}} = \vec{\mathbf{f}}, \tag{2.1}$$

The *angular momentum* of the particle about O is defined by

$$\vec{\mathbf{h}}_O \stackrel{\Delta}{=} \vec{\mathbf{r}} \times \vec{\mathbf{p}} = m\vec{\mathbf{r}} \times \dot{\vec{\mathbf{r}}}.$$

The inertial time-derivative of the angular momentum is

$$\dot{\vec{\mathbf{h}}}_O = m\dot{\vec{\mathbf{r}}} \times \dot{\vec{\mathbf{r}}} + m\vec{\mathbf{r}} \times \ddot{\vec{\mathbf{r}}}$$

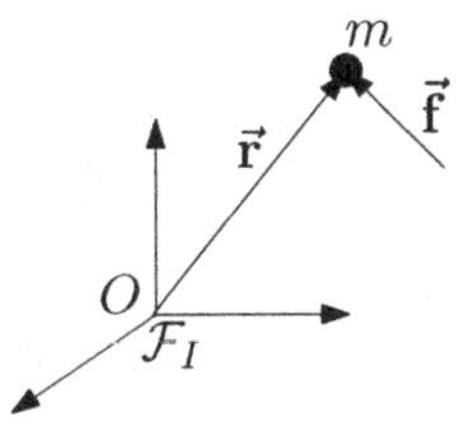

Figure 2.2 Single particle dynamics

Making use of (2.1), this gives

$$\dot{\vec{\mathbf{h}}}_O = \vec{\mathbf{t}}_O, \quad \vec{\mathbf{t}}_O = \vec{\mathbf{r}} \times \vec{\mathbf{f}}, \tag{2.2}$$

where the vector $\vec{\mathbf{t}}_O$ is the torque about O due to the force $\vec{\mathbf{f}}$. That is, the inertial time-derivative of the angular momentum is equal to the external torque.

2.2 Dynamics of a System of Particles

We can now extend the formulation to a system of N particles. As shown in Figure 2.3, we let each of the particles have mass m_i and position $\vec{\mathbf{r}}_i$ for $i = 1, \ldots, N$. Each particle has forces acting upon them. They can be divided into two classes:

1. Forces from external sources, denoted by $\vec{\mathbf{F}}_i$, $i = 1, \ldots, N$.
2. Forces from internal sources, that is forces from the other $N - 1$ masses. These will be denoted by $\vec{\mathbf{f}}_{ij}$, meaning the force exerted on particle i by particle j. Since a particle does not exert a force on itself, we set $\vec{\mathbf{f}}_{ii} = \vec{\mathbf{0}}$.

From Newton's third law, if particle j exerts a force on particle i, then particle i exerts an opposing force on particle j, equal in magnitude, but opposite in direction. That is,

$$\vec{\mathbf{f}}_{ij} = -\vec{\mathbf{f}}_{ji} \tag{2.3}$$

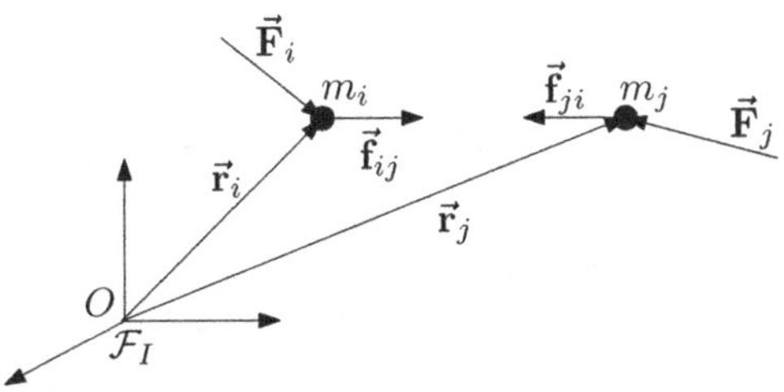

Figure 2.3 Dynamics of a system of particles

Now, applying Newton's second law to the ith particle gives

$$m_i \ddot{\vec{\mathbf{r}}}_i = \vec{\mathbf{F}}_i + \sum_{j=1}^{N} \vec{\mathbf{f}}_{ij}. \tag{2.4}$$

Summing all of the equations of motion in (2.4) for the N masses gives

$$\sum_{i=1}^{N} m_i \ddot{\vec{\mathbf{r}}}_i = \sum_{i=1}^{N} \vec{\mathbf{F}}_i + \sum_{i=1}^{N} \sum_{j=1}^{N} \vec{\mathbf{f}}_{ij}.$$

From Newton's third law (2.3), the last term is $\sum_{i=1}^{N} \sum_{j=1}^{N} \vec{\mathbf{f}}_{ij} = \vec{\mathbf{0}}$. Therefore, we have

$$\sum_{i=1}^{N} m_i \ddot{\vec{\mathbf{r}}}_i = \sum_{i=1}^{N} \vec{\mathbf{F}}_i. \tag{2.5}$$

Now, it will be useful to define the center of mass of the system of particles. This is defined as

$$\vec{\mathbf{r}}_c = \frac{\sum_{i=1}^{N} m_i \vec{\mathbf{r}}_i}{m}, \tag{2.6}$$

where $m = \sum_{i=1}^{N} m_i$ is the total mass of the system. Making use of this definition, we can obtain the translational equation for the center of mass from (2.5) as

$$m\ddot{\vec{\mathbf{r}}}_c = \vec{\mathbf{F}}, \quad \vec{\mathbf{F}} = \sum_{i=1}^{N} \vec{\mathbf{F}}_i, \tag{2.7}$$

where $\vec{\mathbf{F}}$ is the total external force acting on the system of particles. Comparing this to (2.1), we can see that the equation of motion for the center of mass of a system of particles is the same as for a single particle. It will be useful to write the angular equation of motion about an arbitrary point O^*, which may or may not be moving, as shown in Figure 2.4. It is clear that the position of the ith particle is given by

$$\vec{\mathbf{r}}_i = \vec{\mathbf{r}}_{O^*} + \vec{\boldsymbol{\rho}}_i, \quad i = 1, \ldots, N.$$

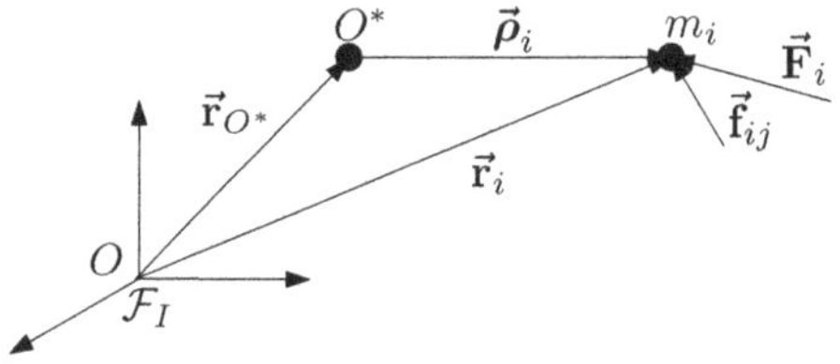

Figure 2.4 Dynamics of a system of particles about point O^*

The equation of motion of the particle (2.4) becomes

$$m_i\ddot{\vec{\boldsymbol{\rho}}}_i = \vec{\mathbf{F}}_i + \sum_{j=1}^{N}\vec{\mathbf{f}}_{ij} - m_i\ddot{\vec{\mathbf{r}}}_{O^*}. \tag{2.8}$$

The angular momentum of the ith particle about O^* is defined as

$$\vec{\mathbf{h}}_{i,O^*} \triangleq m_i\vec{\boldsymbol{\rho}}_i \times \dot{\vec{\boldsymbol{\rho}}}_i, \quad i = 1, \ldots, N.$$

The angular momentum of the system of particles about O^* is then

$$\vec{\mathbf{h}}_{O^*} = \sum_{i=1}^{N}\vec{\mathbf{h}}_{i,O^*} = \sum_{i=1}^{N} m_i\vec{\boldsymbol{\rho}}_i \times \dot{\vec{\boldsymbol{\rho}}}_i. \tag{2.9}$$

Taking the inertial time-derivative of (2.9) and making use of (2.8) leads to

$$\begin{aligned}\dot{\vec{\mathbf{h}}}_{O^*} &= \sum_{i=1}^{N}\vec{\boldsymbol{\rho}}_i \times \vec{\mathbf{F}}_i + \sum_{i=1}^{N}\vec{\boldsymbol{\rho}}_i \times \sum_{j=1}^{N}\vec{\mathbf{f}}_{ij} - \sum_{i=1}^{N} m_i\vec{\boldsymbol{\rho}}_i \times \ddot{\vec{\mathbf{r}}}_{O^*},\\ &= \sum_{i=1}^{N}\vec{\boldsymbol{\rho}}_i \times \vec{\mathbf{F}}_i + \sum_{i=1}^{N}\sum_{j=1}^{N}\vec{\boldsymbol{\rho}}_i \times \vec{\mathbf{f}}_{ij} - \vec{\mathbf{c}}_{O^*} \times \ddot{\vec{\mathbf{r}}}_{O^*},\end{aligned} \tag{2.10}$$

where we have defined the *first moment of mass* of the system about O^* as

$$\vec{\mathbf{c}}_{O^*} \triangleq \sum_{i=1}^{N} m_i\vec{\boldsymbol{\rho}}_i.$$

Now, making use of Newton's third law (2.3), the double sum in (2.10) becomes

$$\sum_{i=1}^{N}\sum_{j=1}^{N}\vec{\boldsymbol{\rho}}_i \times \vec{\mathbf{f}}_{ij} = \frac{1}{2}\sum_{i=1}^{N}\sum_{j=1}^{N}(\vec{\boldsymbol{\rho}}_i - \vec{\boldsymbol{\rho}}_j) \times \vec{\mathbf{f}}_{ij}.$$

But, the vector $(\vec{\boldsymbol{\rho}}_i - \vec{\boldsymbol{\rho}}_j)$ is just the vector pointing from particle j to particle i. Assuming that the force $\vec{\mathbf{f}}_{ij}$ acts along the line joining particle i and j as shown in Figure 2.5, that is assuming $(\vec{\boldsymbol{\rho}}_i - \vec{\boldsymbol{\rho}}_j)$ and $\vec{\mathbf{f}}_{ij}$ are parallel, we have

$$\sum_{i=1}^{N}\sum_{j=1}^{N}\vec{\boldsymbol{\rho}}_i \times \vec{\mathbf{f}}_{ij} = \vec{\mathbf{0}}.$$

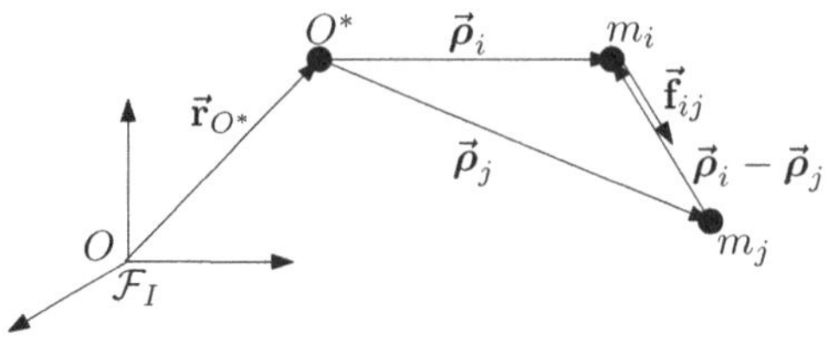

Figure 2.5 Internal forces acting along the line between particles

The angular equation of motion (2.10) then becomes

$$\dot{\vec{\mathbf{h}}}_{O^*} = \vec{\mathbf{T}}_{O^*} - \vec{\mathbf{c}}_{O^*} \times \ddot{\vec{\mathbf{r}}}_{O^*}, \quad \vec{\mathbf{T}}_{O^*} = \sum_{i=1}^{N} \vec{\rho}_i \times \vec{\mathbf{F}}_i. \tag{2.11}$$

where $\vec{\mathbf{T}}_{O^*}$ is the total external torque applied to the system about the point O^*.

A special situation arises when we choose the point O^* to be the system center of mass, c. In this case, the first moment of mass about c is $\vec{\mathbf{c}}_c = \vec{\mathbf{0}}$ by definition, and the angular equation of motion becomes

$$\dot{\vec{\mathbf{h}}}_c = \vec{\mathbf{T}}_c, \quad \vec{\mathbf{T}}_c = \sum_{i=1}^{N} \vec{\rho}_i \times \vec{\mathbf{F}}_i. \tag{2.12}$$

2.3 Rigid Body Dynamics

We will now extend the formulation for a system of particles to a rigid body.

Definition: A *rigid body* is a continuum in which the distance between any two points on the body remains fixed.

This means that the body does not deform. We shall treat a rigid body as the limiting case of a system of particles where the number of particles becomes infinite, and the mass of each particle becomes infinitesimal. To do this, we make the following changes to the formulation for a system of particles (refer to Figure 2.6):

$$\begin{aligned} \vec{\rho}_i &\rightarrow \vec{\rho} \\ m_i &\rightarrow dm = \sigma(\vec{\rho})dV \\ \vec{\mathbf{F}}_i &\rightarrow \vec{\mathbf{f}}(\vec{\rho})dV \\ \sum_{i=1}^{N} &\rightarrow \int_V \end{aligned} \tag{2.13}$$

Here, dm is an infinitesimal mass element, dV is an infinitesimal volume element, and $\sigma(\vec{\rho})$ is the mass density per unit volume at a point $\vec{\rho}$. The vector $\vec{\mathbf{f}}(\vec{\rho})$ is the force due to external sources per unit volume at a point $\vec{\rho}$. Note that the integral is taken over the entire volume V of the body.

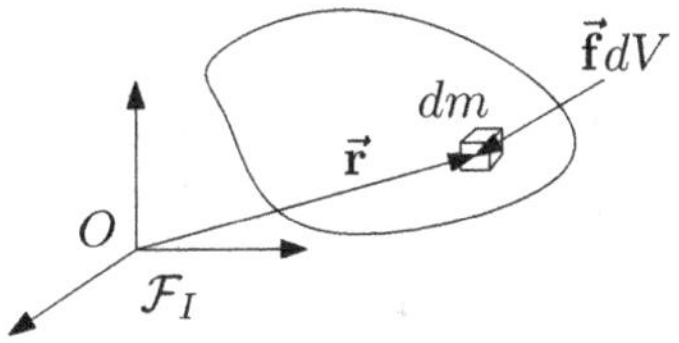

Figure 2.6 Modeling continuum dynamics

2.3.1 Translational Dynamics

For a system of point masses, the total mass of the system, and the center of mass are given by

$$m = \sum_{i=1}^{N} m_i \quad \text{and} \quad \vec{\mathbf{r}}_c = \frac{\sum_{i=1}^{N} m_i \vec{\mathbf{r}}_i}{m},$$

respectively. Analogously, making use of the changes in (2.13), for a rigid body they become

$$m = \int_V dm \quad \text{and} \quad \vec{\mathbf{r}}_c = \frac{\int_V \vec{\mathbf{r}} dm}{m}. \tag{2.14}$$

For a system of point masses, the translational equation (2.5) is:

$$\sum_{i=1}^{N} m_i \ddot{\vec{\mathbf{r}}}_i = \sum_{i=1}^{N} \vec{\mathbf{F}}_i.$$

Applying the changes in (2.13) to this, we have

$$\int_V \ddot{\vec{\mathbf{r}}} dm = \int_V \vec{\mathbf{f}} dV. \tag{2.15}$$

Now, taking the inertial derivative of (2.14) gives

$$m\ddot{\vec{\mathbf{r}}}_c = \int_V \ddot{\vec{\mathbf{r}}} dm. \tag{2.16}$$

In summary, combining (2.15) and (2.16), the translational dynamics of the center of mass of a rigid body is

$$m\ddot{\vec{\mathbf{r}}}_c = \vec{\mathbf{F}}, \quad \vec{\mathbf{F}} \triangleq \int_V \vec{\mathbf{f}} dV, \tag{2.17}$$

where $\vec{\mathbf{F}}$ is the total force on the body due to external forces. What equation (2.17) shows is that the center of mass of a rigid body behaves like a point mass. That is, provided we know

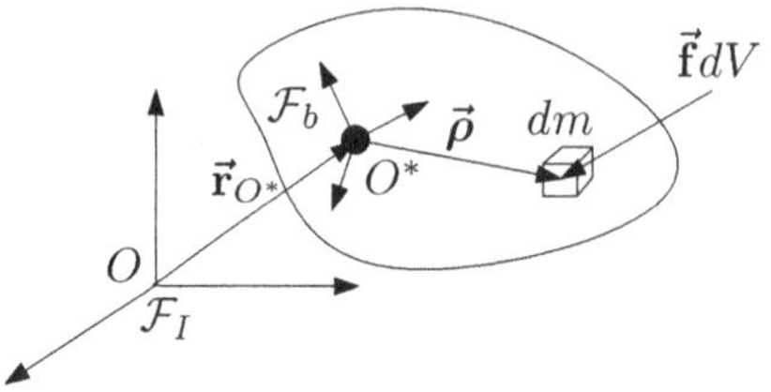

Figure 2.7 Modeling rotational dynamics of a rigid body

the total force acting on the body, and we know the total mass, we can determine the motion of the body center of mass. Also, note that nowhere did we apply the rigid body condition (points remain fixed relative to each other), so in fact, the translational results derived here apply to *any continuum*, rigid or not.

2.3.2 Rotational Dynamics

Since we are considering a rigid body, it will be useful to attach a reference frame to the body, as shown in Figure 2.7. We shall call this frame a *body-fixed frame*, and it will be denoted $\mathcal{F}_b$. The origin of this frame will be located at a point O^* within the body. Since the reference frame is attached to the body, all points within the body are fixed as seen in this frame due to the rigid body assumption. We shall denote time derivatives of vectors as seen in $\mathcal{F}_b$ by an overcircle ($\mathring{}$). Now, the angular momentum of a system of particles about O^* is

$$\vec{\mathbf{h}}_{O^*} = \sum_{i=1}^{N} m_i \vec{\rho}_i \times \dot{\vec{\rho}}_i.$$

As before, making the changes in (2.13) the angular momentum of a rigid body about O^* is

$$\vec{\mathbf{h}}_{O^*} = \int_V \vec{\rho} \times \dot{\vec{\rho}} dm. \tag{2.18}$$

Now, we can make use of the rigid body hypothesis. The inertial time-derivative of $\vec{\rho}$ is related to the time-derivative of $\vec{\rho}$ as seen in $\mathcal{F}_b$ by

$$\dot{\vec{\rho}} = \mathring{\vec{\rho}} + \vec{\omega} \times \vec{\rho}, \tag{2.19}$$

where $\vec{\omega}$ is the angular velocity of $\mathcal{F}_b$ (and hence the rigid body) with respect to the inertial frame $\mathcal{F}_I$. Now, due to the rigid body assumption, $\mathring{\vec{\rho}} = \vec{\mathbf{0}}$. Therefore, we can write the angular momentum of a rigid body as

$$\begin{aligned} \vec{\mathbf{h}}_{O^*} &= \int_V \vec{\rho} \times (\vec{\omega} \times \vec{\rho}) dm \\ &= -\int_V \vec{\rho} \times (\vec{\rho} \times \vec{\omega}) dm. \end{aligned} \tag{2.20}$$

Let us now obtain the coordinates of the angular momentum in $\mathcal{F}_b$. The position, angular velocity and angular momentum vectors become

$$\vec{\rho} = \vec{\mathcal{F}}_b^T \boldsymbol{\rho}, \quad \vec{\omega} = \vec{\mathcal{F}}_b^T \boldsymbol{\omega}, \quad \vec{\mathbf{h}}_{O^*} = \vec{\mathcal{F}}_b^T \mathbf{h}_{O^*}. \tag{2.21}$$

Making use of this in (2.20), we obtain

$$\mathbf{h}_{O^*} = \left[-\int_V \boldsymbol{\rho}^\times \boldsymbol{\rho}^\times dm \right] \boldsymbol{\omega}. \tag{2.22}$$

The quantity

$$\begin{aligned} \mathbf{J} &\triangleq -\int_V \boldsymbol{\rho}^\times \boldsymbol{\rho}^\times dm \\ &= \int_V \begin{bmatrix} (\rho_y^2 + \rho_z^2) & -\rho_x \rho_y & -\rho_x \rho_z \\ -\rho_x \rho_y & (\rho_x^2 + \rho_z^2) & -\rho_y \rho_z \\ -\rho_x \rho_z & -\rho_y \rho_z & (\rho_x^2 + \rho_y^2) \end{bmatrix} \sigma(\rho_x, \rho_y, \rho_z) dV, \end{aligned} \tag{2.23}$$

is called the *moment of inertia matrix about* O^*. In the special case that the point O^* is chosen to be the center of mass, c, we use the label $\mathbf{I}$ instead of $\mathbf{J}$. We can now write the angular momentum of the body about O^* as

$$\mathbf{h}_{O^*} = \mathbf{J}\boldsymbol{\omega}. \tag{2.24}$$

We have seen for a system of particles that the rotational equation of motion is given by

$$\dot{\vec{\mathbf{h}}}_{O^*} = \vec{\mathbf{T}}_{O^*} - \vec{\mathbf{c}}_{O^*} \times \ddot{\vec{\mathbf{r}}}_{O^*}, \quad \vec{\mathbf{T}}_{O^*} = \sum_{i=1}^{N} \vec{\rho}_i \times \vec{\mathbf{F}}_i,$$

where $\vec{\mathbf{c}}_{O^*} = \sum_{i=1}^N m_i \vec{\rho}_i$ is the first moment of mass about the point O^*. Making the changes in (2.13), we obtain for a rigid body

$$\dot{\vec{\mathbf{h}}}_{O^*} = \vec{\mathbf{T}}_{O^*} - \vec{\mathbf{c}}_{O^*} \times \ddot{\vec{\mathbf{r}}}_{O^*}, \quad \vec{\mathbf{T}}_{O^*} = \int_V \vec{\rho} \times \vec{\mathbf{f}} dV, \tag{2.25}$$

where $\vec{\mathbf{T}}_{O^*}$ is the total external torque applied to the body about the point O^*, and $\vec{\mathbf{c}}_{O^*} = \int_V \vec{\rho} dm$ is the body's first moment of mass about O^*. In particular, when O^* is chosen as the center of mass, c, we have

$$\dot{\vec{\mathbf{h}}}_c = \vec{\mathbf{T}}_c, \quad \vec{\mathbf{T}}_c = \int_V \vec{\rho} \times \vec{\mathbf{f}} dV, \tag{2.26}$$

and $\vec{\mathbf{T}}_c$ is the total external torque about the center of mass. For a single body, such as a spacecraft, the center of mass is the most useful point, since this decouples the rotational dynamics from the translational dynamics (we do not have to consider the translational motion of the center of mass). For a multi-bodied system, such as a robot arm with multiple links, the center of mass is not always a convenient origin. That is why we have presented the more general case.

Now, let us return to the dynamical equation. The relationship between the inertial time-derivative of the angular momentum $\vec{\mathbf{h}}_c$ and the time-derivative as seen in $\mathcal{F}_b$ is

$$\dot{\vec{\mathbf{h}}}_c = \overset{\circ}{\vec{\mathbf{h}}}_c + \vec{\omega} \times \vec{\mathbf{h}}_c.$$

Therefore, from (2.26), we obtain

$$\overset{\circ}{\vec{\mathbf{h}}}_c + \vec{\omega} \times \vec{\mathbf{h}}_c = \vec{\mathbf{T}}_c. \tag{2.27}$$

In terms of coordinates in the body-fixed frame $\mathcal{F}_b$, this is

$$\dot{\mathbf{h}}_c + \boldsymbol{\omega}^{\times}\mathbf{h}_c = \mathbf{T}_c. \tag{2.28}$$

Substituting the expression for the angular momentum $\mathbf{h}_c = \mathbf{I}\boldsymbol{\omega}$, we obtain

$$\mathbf{I}\dot{\boldsymbol{\omega}} + \boldsymbol{\omega}^{\times}\mathbf{I}\boldsymbol{\omega} = \mathbf{T}_c. \tag{2.29}$$

This is known as *Euler's equation*, and fully describes the dynamics of the angular velocity of the rigid body (i.e. $\mathcal{F}_b$). The attitude of the rigid body (orientation of $\mathcal{F}_b$ with respect to $\mathcal{F}_I$) is fully described by the rotation matrix $\mathbf{C}_{bI}$. Therefore, the full set of equations for the rotational motion of a rigid body are

$$\begin{aligned} \dot{\mathbf{C}}_{bI} &= -\boldsymbol{\omega}^{\times}\mathbf{C}_{bI}, \\ \mathbf{I}\dot{\boldsymbol{\omega}} + \boldsymbol{\omega}^{\times}\mathbf{I}\boldsymbol{\omega} &= \mathbf{T}_c. \end{aligned} \tag{2.30}$$

2.4 The Inertia Matrix

We have introduced the concept of an inertia matrix. These have several properties that are worth examining.

First of all, the inertia matrix is *real and symmetric* (see (2.23)), that is

$$\mathbf{J} = \mathbf{J}^T.$$

Next, the inertia matrix is *positive definite*, that is given any $\mathbf{x} \neq \mathbf{0} \in R^3$,

$$\mathbf{x}^T\mathbf{J}\mathbf{x} > 0.$$

As a consequence of these two properties, the eigenvalues of $\mathbf{J}$ are real and positive. The eigenvalues actually have a special meaning, which we will discuss later.

In any body-fixed frame, the moment of inertia matrix is constant for a rigid body.

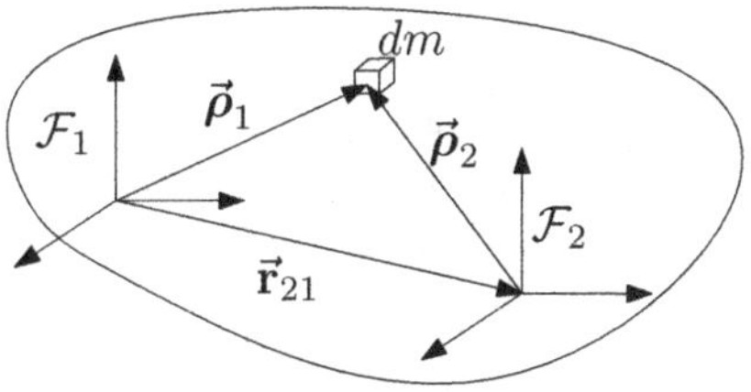

Figure 2.8 Parallel axis theorem

2.4.1 A Parallel Axis Theorem

The only requirement for a reference frame to be called a body-fixed frame is for it to be stationary with respect to the body. The origin of the reference frame may be defined at any point within the body. As such, the moment of inertia matrix may be defined with respect to any body-fixed reference frame. Consider now two body-fixed frames $\mathcal{F}_1$ and $\mathcal{F}_2$, which have the same orientation, but different origins, as shown in Figure 2.8. Specifically, the origin of $\mathcal{F}_2$ is located at $\vec{\mathbf{r}}_{21}$ from the origin of $\mathcal{F}_1$. Now, the location of a mass element, dm, within the body is located at $\vec{\boldsymbol{\rho}}_1$ from the origin of $\mathcal{F}_1$ and at $\vec{\boldsymbol{\rho}}_2$ from the origin of $\mathcal{F}_2$. We clearly have the relationship

$$\vec{\boldsymbol{\rho}}_2 = \vec{\boldsymbol{\rho}}_1 - \vec{\mathbf{r}}_{21},$$

or in terms of the coordinates in $\mathcal{F}_1$ and $\mathcal{F}_2$,

$$\boldsymbol{\rho}_2 = \boldsymbol{\rho}_1 - \mathbf{r}_{21}.$$

The moment of inertia matrix with respect to frame $\mathcal{F}_1$ is given by

$$\mathbf{J}_1 = -\int_V \boldsymbol{\rho}_1^\times \boldsymbol{\rho}_1^\times dm. \tag{2.31}$$

The moment of inertia matrix with respect to frame $\mathcal{F}_2$ is

$$\begin{aligned}
\mathbf{J}_2 &= -\int_V \boldsymbol{\rho}_2^\times \boldsymbol{\rho}_2^\times dm \\
&= -\int_V (\boldsymbol{\rho}_1 - \mathbf{r}_{21})^\times (\boldsymbol{\rho}_1 - \mathbf{r}_{21})^\times dm \\
&= -\int_V \boldsymbol{\rho}_1^\times \boldsymbol{\rho}_1^\times dm + \int_V \mathbf{r}_{21}^\times \boldsymbol{\rho}_1^\times dm + \int_V \boldsymbol{\rho}_1^\times \mathbf{r}_{21}^\times dm - \int_V \mathbf{r}_{21}^\times \mathbf{r}_{21}^\times dm.
\end{aligned} \tag{2.32}$$

Recognizing that $\int_V dm = m$ is the mass and $\int_V \boldsymbol{\rho}_1 dm = \mathbf{c}_1$ is the first moment of mass about the origin of $\mathcal{F}_1$, we have the parallel axis theorem for inertia matrices:

$$\mathbf{J}_2 = \mathbf{J}_1 + \mathbf{r}_{21}^\times \mathbf{c}_1^\times + \mathbf{c}_1^\times \mathbf{r}_{21}^\times - m\mathbf{r}_{21}^\times \mathbf{r}_{21}^\times. \tag{2.33}$$

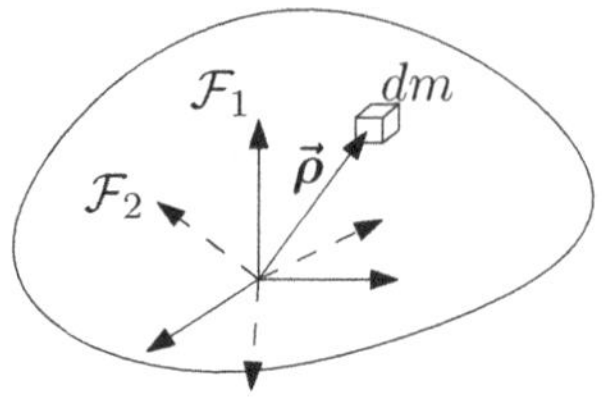

Figure 2.9 Rotational transformation theorem

In particular, if the origin of $\mathcal{F}_1$ is the center of mass, then $\mathbf{c}_1 = \mathbf{0}$ and

$$\mathbf{J}_2 = \mathbf{I} - m\mathbf{r}_{21}^{\times}\mathbf{r}_{21}^{\times}. \tag{2.34}$$

2.4.2 A Rotational Transformation Theorem

Just as the origin of a body-fixed frame can be chosen arbitrarily, so can the orientation. Consider now two body-fixed frames $\mathcal{F}_1$ and $\mathcal{F}_2$ whose origins coincide, but orientations differ, as shown in Figure 2.9. The transformation from $\mathcal{F}_1$ to $\mathcal{F}_2$ can be described by the rotation matrix $\mathbf{C}_{21}$. That is, for any vector,

$$\vec{\rho} = \vec{\mathcal{F}}_1^T \boldsymbol{\rho}_1 = \vec{\mathcal{F}}_2^T \boldsymbol{\rho}_2,$$

we have the relationship $\boldsymbol{\rho}_2 = \mathbf{C}_{21}\boldsymbol{\rho}_1$. The moment of inertia matrix with respect to $\mathcal{F}_1$ is

$$\mathbf{J}_1 = -\int_V \boldsymbol{\rho}_1^{\times}\boldsymbol{\rho}_1^{\times} dm. \tag{2.35}$$

Now, making use of the identities $\mathbf{C}_{21}^{-1} = \mathbf{C}_{21}^T$ and $(\mathbf{C}_{21}\boldsymbol{\rho}_1)^{\times} = \mathbf{C}_{21}\boldsymbol{\rho}_1^{\times}\mathbf{C}_{21}^T$, the moment of inertia matrix with respect to $\mathcal{F}_2$ is

$$\begin{aligned}\mathbf{J}_2 &= -\int_V \boldsymbol{\rho}_2^{\times}\boldsymbol{\rho}_2^{\times} dm \\ &= -\int_V (\mathbf{C}_{21}\boldsymbol{\rho}_1)^{\times}(\mathbf{C}_{21}\boldsymbol{\rho}_1)^{\times} dm \\ &= -\int_V \mathbf{C}_{21}\boldsymbol{\rho}_1^{\times}\mathbf{C}_{21}^T\mathbf{C}_{21}\boldsymbol{\rho}_1^{\times}\mathbf{C}_{21}^T dm \\ &= -\int_V \mathbf{C}_{21}\boldsymbol{\rho}_1^{\times}\boldsymbol{\rho}_1^{\times}\mathbf{C}_{21}^T dm.\end{aligned} \tag{2.36}$$

Therefore, we have the rotational transformation theorem for inertia matrices:

$$\mathbf{J}_2 = \mathbf{C}_{21}\mathbf{J}_1\mathbf{C}_{21}^T. \tag{2.37}$$

2.4.3 Principal Axes

Definition: *A* principal axes frame *is a body-fixed frame in which the moment of inertia matrix is diagonal, i.e.*

$$\mathbf{J} = \begin{bmatrix} J_1 & 0 & 0 \\ 0 & J_2 & 0 \\ 0 & 0 & J_3 \end{bmatrix}.$$

The diagonal elements of **J** are called the *principal moments of inertia.*

Now, as already discussed, the moment of inertia matrix is real, symmetric and positive definite. Let us consider the eigenvalue problem for the inertia matrix:

$$\mathbf{J}\mathbf{e}_i = \lambda_i \mathbf{e}_i, \quad i = 1, 2, 3. \tag{2.38}$$

where λ_i are the eigenvalues, and $\mathbf{e}_i \neq \mathbf{0}$ are the eigenvectors. We now make use of the following result from matrix theory:

Theorem: *The eigenvectors for any real symmetric matrix can be chosen to be real, satisfying*

$$\mathbf{e}_i^T \mathbf{e}_i = \begin{cases} 1, \ i = j, \\ 0, \ i \neq j \end{cases} \tag{2.39}$$

What this means is that the eigenvectors $\mathbf{e}_i$ of the inertia matrix **J**, represent the coordinates of three mutually perpendicular vectors. Since $-\mathbf{e}_i$ is also an eigenvector, the eigenvectors $\mathbf{e}_i$ may be chosen to be the coordinates of the unit vectors defining a right-handed coordinate system. This means that the matrix

$$\mathbf{E} = \begin{bmatrix} \mathbf{e}_1 & \mathbf{e}_2 & \mathbf{e}_3 \end{bmatrix}$$

is actually a rotation matrix, and the relationship (2.39) can be written compactly as

$$\mathbf{E}^T \mathbf{E} = \mathbf{1}. \tag{2.40}$$

Returning to the eigenvalue problem (2.38), this can also be written compactly as

$$\mathbf{J}\mathbf{E} = \mathbf{E}\mathbf{\Lambda}, \quad \mathbf{\Lambda} = \begin{bmatrix} \lambda_1 & 0 & 0 \\ 0 & \lambda_2 & 0 \\ 0 & 0 & \lambda_3 \end{bmatrix}.$$

Making use of (2.40), this can be rewritten as

$$\mathbf{\Lambda} = \mathbf{E}^T \mathbf{J}\mathbf{E} \quad \text{or equivalently} \quad \mathbf{J} = \mathbf{E}\mathbf{\Lambda}\mathbf{E}^T. \tag{2.41}$$

Since we have identified that $\mathbf{E}$ is a rotation matrix, equation (2.41) and (2.37) show that $\mathbf{\Lambda}$ is itself a moment of inertia matrix. Since $\mathbf{\Lambda}$ is diagonal, $\mathbf{E}^T$ describes the rotation from the original frame used to compute $\mathbf{J}$ to a principal axes frame. The columns of $\mathbf{E}$ (i.e. the eigenvectors of $\mathbf{J}$) are just the coordinates (in the original frame) of the unit vectors defining the principal axes frame. The diagonal entries of $\mathbf{\Lambda}$ (i.e. the eigenvalues of $\mathbf{J}$) are the principal moments of inertia.

What we have shown is that for *any* rigid body, we can always find a principal axes frame. The fact that we can do this allows us to simplify the rotational dynamics considerably. Assume that the chosen body-fixed frame $\mathcal{F}_b$ is a principal axes frame, and let

$$\mathbf{I} = \begin{bmatrix} I_x & 0 & 0 \\ 0 & I_y & 0 \\ 0 & 0 & I_z \end{bmatrix}, \quad \boldsymbol{\omega} = \begin{bmatrix} \omega_x \\ \omega_y \\ \omega_z \end{bmatrix}, \quad \mathbf{T}_c = \begin{bmatrix} T_x \\ T_y \\ T_z \end{bmatrix}.$$

Then, Euler's equation (17.2) becomes

$$\begin{aligned} I_x\dot{\omega}_x + (I_z - I_y)\omega_y\omega_z &= T_x \\ I_y\dot{\omega}_y + (I_x - I_z)\omega_x\omega_z &= T_y \\ I_z\dot{\omega}_z + (I_y - I_x)\omega_x\omega_y &= T_z \end{aligned} \tag{2.42}$$

2.5 Kinetic Energy of a Rigid Body

Consider the rigid body as shown in Figure 2.10. We embed a body-fixed frame $\mathcal{F}_b$ in the rigid body, with origin at the center of mass. The vector $\vec{\mathbf{r}}$ denotes the position of the rigid body center of mass relative to the origin of an inertial frame $\mathcal{F}_I$. The mass element dm is located by vector $\vec{\rho}$ from the rigid body center of mass. Let us denote the inertial angular velocity of the rigid body by the vector $\vec{\omega}$.

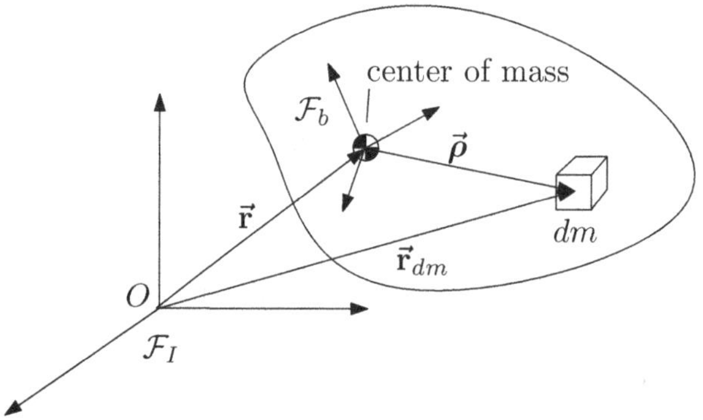

Figure 2.10 Rigid body

As shown in Figure 2.10, the mass element dm is located relative to the origin of $\mathcal{F}_I$ by the vector

$$\vec{\mathbf{r}}_{dm} = \vec{\mathbf{r}} + \vec{\rho}.$$

Therefore, mass element dm has inertial velocity

$$\dot{\vec{\mathbf{r}}}_{dm} = \dot{\vec{\mathbf{r}}} + \dot{\vec{\rho}},$$

where $(\dot{\ })$ denotes the time-derivative as seen in $\mathcal{F}_I$. Since the body is rigid, we have

$$\dot{\vec{\rho}} = \vec{\omega} \times \vec{\rho},$$

as seen in Section 2.3.2. Therefore, the inertial velocity of dm is given by

$$\dot{\vec{\mathbf{r}}}_{dm} = \vec{\mathbf{v}} + \vec{\omega} \times \vec{\rho}, \tag{2.43}$$

where $\vec{\mathbf{v}} = \dot{\vec{\mathbf{r}}}$ is the inertial velocity of the rigid body center of mass.

The kinetic energy of mass element dm is defined as

$$dT = \frac{1}{2}\dot{\vec{\mathbf{r}}}_{dm} \cdot \dot{\vec{\mathbf{r}}}_{dm} dm.$$

Making use of (2.43), this becomes

$$dT = \frac{1}{2}\left(\vec{\mathbf{v}} + \vec{\omega} \times \vec{\rho}\right) \cdot \left(\vec{\mathbf{v}} + \vec{\omega} \times \vec{\rho}\right) dm.$$

Expanding, we have

$$dT = \frac{1}{2}\vec{\mathbf{v}} \cdot \vec{\mathbf{v}} dm + \vec{\mathbf{v}} \cdot \left(\vec{\omega} \times \vec{\rho}\right) dm + \frac{1}{2}\left(\vec{\omega} \times \vec{\rho}\right) \cdot \left(\vec{\omega} \times \vec{\rho}\right) dm. \tag{2.44}$$

The total kinetic energy of the rigid body is now obtained by integrating (2.44) over the entire rigid body

$$T = \frac{1}{2}\int_B \vec{\mathbf{v}} \cdot \vec{\mathbf{v}} dm + \int_B \vec{\mathbf{v}} \cdot \left(\vec{\omega} \times \vec{\rho}\right) dm + \frac{1}{2}\int_B \left(\vec{\omega} \times \vec{\rho}\right) \cdot \left(\vec{\omega} \times \vec{\rho}\right) dm. \tag{2.45}$$

Let us now examine this term-by-term. First of all, we recognize that the total mass of the rigid body is given by

$$m = \int_B dm.$$

Therefore, we have

$$\frac{1}{2}\int_B \vec{\mathbf{v}} \cdot \vec{\mathbf{v}} dm = \frac{1}{2} m \vec{\mathbf{v}} \cdot \vec{\mathbf{v}} = \frac{1}{2} m v^2, \quad v = |\vec{\mathbf{v}}|, \tag{2.46}$$

since $\vec{\mathbf{v}}$ is constant with respect to the integration. Next, we recognize that since the origin of $\mathcal{F}_B$ is at the rigid body center of mass, the first moment of mass is zero, that is

$$\int_B \vec{\rho} dm = \vec{\mathbf{0}}.$$

Therefore, we have

$$\int_B \vec{\mathbf{v}} \cdot (\vec{\omega} \times \vec{\rho}) \, dm = 0, \tag{2.47}$$

since $\vec{\mathbf{v}}$ and $\vec{\omega}$ are constant with respect to the integration. Finally, let us evaluate the third term in (2.45). First, let us represent $\vec{\omega}$ and $\vec{\rho}$ in body coordinates as

$$\vec{\omega} = \vec{\mathcal{F}}_B^T \boldsymbol{\omega} \quad \text{and} \quad \vec{\rho} = \vec{\mathcal{F}}_B^T \boldsymbol{\rho},$$

respectively. We then have

$$\begin{aligned} \frac{1}{2}\int_B (\vec{\omega} \times \vec{\rho}) \cdot (\vec{\omega} \times \rho) \, dm &= \frac{1}{2}\int_B (\boldsymbol{\omega}^\times \boldsymbol{\rho})^T (\boldsymbol{\omega}^\times \boldsymbol{\rho}) \, dm \\ &= \frac{1}{2}\int_B (\boldsymbol{\rho}^\times \boldsymbol{\omega})^T (\boldsymbol{\rho}^\times \boldsymbol{\omega}) \, dm \\ &= \frac{1}{2}\boldsymbol{\omega}^T \left[-\int_B \boldsymbol{\rho}^\times \boldsymbol{\rho}^\times dm \right] \boldsymbol{\omega} \end{aligned}$$

We recognize the final integral from Section 2.3.2 as the moment of inertia matrix about the rigid body center of mass, $\mathbf{I}$. Therefore, we have

$$\frac{1}{2}\int_B (\vec{\omega} \times \vec{\rho}) \cdot (\vec{\omega} \times \vec{\rho}) \, dm = \frac{1}{2}\boldsymbol{\omega}^T \mathbf{I} \boldsymbol{\omega}. \tag{2.48}$$

From equation (2.45) and equations (2.46) to (2.48), we see that the total kinetic energy of the rigid body is given by

$$T = T_t + T_r, \tag{2.49}$$

where

$$T_t = \frac{1}{2} m v^2, \tag{2.50}$$

is the rigid body *translational kinetic energy*, and

$$T_r = \frac{1}{2}\boldsymbol{\omega}^T \mathbf{I} \boldsymbol{\omega}, \tag{2.51}$$

is the rigid body *rotational kinetic energy*.

We note that when $\mathcal{F}_B$ is a principal axes frame (see Section 2.4.3), that is,

$$\mathbf{I} = \begin{bmatrix} I_x & 0 & 0 \\ 0 & I_y & 0 \\ 0 & 0 & I_z \end{bmatrix}, \quad \boldsymbol{\omega} = \begin{bmatrix} \omega_x \\ \omega_y \\ \omega_z \end{bmatrix},$$

then the rotational kinetic energy (2.51) becomes

$$T_r = \frac{1}{2}\left(I_x\omega_x^2 + I_y\omega_y^2 + I_z\omega_z^2\right).$$

Notes

This chapter contains a restricted presentation of the mechanics required to describe the types of spacecraft motion considered in this book. For example, we have not considered spacecraft made up of multiple bodies. We have adopted the Newton-Euler approach (as opposed to the Lagrangian or Hamiltonian approach), since it best suits our purpose. For a more complete presentation of classical mechanics geared toward the description of spacecraft motion, the reader is referred to Hughes (2004). There are several books containing comprehensive treatments of classical mechanics in general. One such book is Goldstein et al. (2002).

References

Goldstein H, Poole C and Safko J 2002 *Classical Mechanics* 3rd edn. Addison-Wesley, San Francisco, CA.
Hughes PC 2004 *Spacecraft Attitude Dynamics*. Dover Publications, Mineola, NY.

3

The Keplerian Two-Body Problem

We are now going to look at the Keplerian two-body problem. That is, the orbital problem for two point masses. It turns out that this is a reasonable model for spacecraft orbiting planets, planets orbiting the sun, and others. We will see in Chapter 7 that the finite size of the planets does have an effect on the orbit of a spacecraft. However, treating the planet and spacecraft as point masses, as in the Keplerian two-body problem provides a great deal of insight into the behavior of a spacecraft in orbit about a planet.

3.1 Equations of Motion

Consider two point masses of mass m_1 and m_2, as shown in Figure 3.1. As we have seen, in order to apply Newton's laws of motion, we require an inertial reference frame $\mathcal{F}_I$. Let $\vec{\mathbf{r}}_1$ and $\vec{\mathbf{r}}_2$ be the positions of the two point masses from the origin of $\mathcal{F}_I$ respectively. The position of mass m_2 with respect to mass m_1 is given by

$$\vec{\mathbf{r}}_{21} = \vec{\mathbf{r}}_2 - \vec{\mathbf{r}}_1.$$

By Newton's law of gravitation, mass m_1 exerts a force on m_2 given by

$$\vec{\mathbf{F}}_{21} = -\frac{Gm_1m_2}{|\vec{\mathbf{r}}_{21}|^3}\vec{\mathbf{r}}_{21},$$

and likewise, mass m_2 exerts a force on m_1 given by

$$\vec{\mathbf{F}}_{12} = \frac{Gm_1m_2}{|\vec{\mathbf{r}}_{21}|^3}\vec{\mathbf{r}}_{21}.$$

Note that G is Newton's universal gravitational constant.

Spacecraft Dynamics and Control: An Introduction, First Edition.
Anton H.J. de Ruiter, Christopher J. Damaren and James R. Forbes.

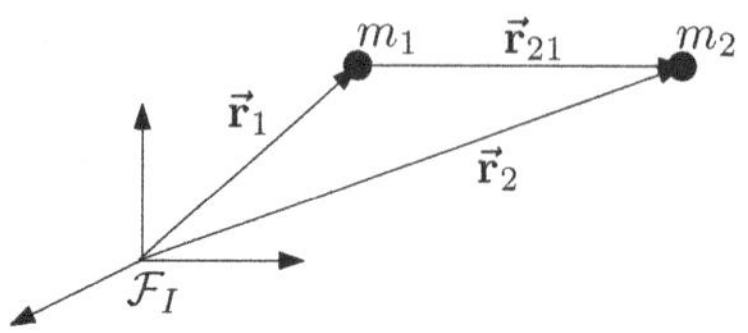

Figure 3.1 Two-body problem

The equations of motion for m_1 and m_2 are given by Newton's second law, and are

$$m_1\ddot{\vec{\mathbf{r}}}_1 = \vec{\mathbf{F}}_{12} = \frac{Gm_1m_2}{|\vec{\mathbf{r}}_{21}|^3}\vec{\mathbf{r}}_{21},$$

$$m_2\ddot{\vec{\mathbf{r}}}_2 = \vec{\mathbf{F}}_{21} = -\frac{Gm_1m_2}{|\vec{\mathbf{r}}_{21}|^3}\vec{\mathbf{r}}_{21}.$$

These lead to

$$\ddot{\vec{\mathbf{r}}}_1 = \frac{Gm_2}{|\vec{\mathbf{r}}_{21}|^3}\vec{\mathbf{r}}_{21}, \tag{3.1}$$

$$\ddot{\vec{\mathbf{r}}}_2 = -\frac{Gm_1}{|\vec{\mathbf{r}}_{21}|^3}\vec{\mathbf{r}}_{21}. \tag{3.2}$$

Taking the difference between (3.2) and (3.1) gives the equation of motion for the position of m_2 with respect to m_1.

$$\ddot{\vec{\mathbf{r}}}_{21} = \ddot{\vec{\mathbf{r}}}_2 - \ddot{\vec{\mathbf{r}}}_1 = -\frac{G(m_1 + m_2)}{|\vec{\mathbf{r}}_{21}|^3}\vec{\mathbf{r}}_{21} \tag{3.3}$$

Now, we will assume that $m_1 \gg m_2$ as is the case for example when m_1 is a planet and m_2 is a spacecraft, or when m_1 is the sun and m_2 is a planet. Under this assumption, $m_1 + m_2 \approx m_1$, and (3.3) becomes

$$\ddot{\vec{\mathbf{r}}}_{21} = -\frac{Gm_1}{|\vec{\mathbf{r}}_{21}|^3}\vec{\mathbf{r}}_{21}.$$

From this point on, we will drop the subscript "21", and $\vec{\mathbf{r}} = \vec{\mathbf{r}}_{21}$ will refer to the position of m_2 with respect to m_1. The mass m_1 will be referred to as the *primary body*, due to the fact that $m_1 \gg m_2$. We also define the constant $\mu = Gm_1$. This will prove much more useful, since the universal gravitational constant G is not known accurately, and neither is the mass m_1 of each planet. The gravitational constant μ on the other hand can be determined very accurately. It is specific to the primary body.

Finally, we have the Keplerian two-body orbital equation of motion

$$\ddot{\vec{\mathbf{r}}} = -\frac{\mu}{|\vec{\mathbf{r}}|^3}\vec{\mathbf{r}}. \tag{3.4}$$

Equation (3.4) gives the motion of the mass m_2 about the primary body m_1. It can be seen from (3.4) that this motion is independent of the mass m_2. This is due to the assumption $m_1 \gg m_2$.

Let us now examine some of the properties of a Keplerian orbit.

3.2 Constants of the Motion

3.2.1 *Orbital Angular Momentum*

First of all, let us examine the *orbital angular momentum*. This is defined as

$$\vec{\mathbf{h}} \stackrel{\Delta}{=} \vec{\mathbf{r}} \times \vec{\mathbf{v}}, \tag{3.5}$$

where $\vec{\mathbf{v}} = \dot{\vec{\mathbf{r}}}$ is the orbital velocity. Let us now take the inertial time-derivative of (3.5). We have

$$\begin{aligned}\dot{\vec{\mathbf{h}}} &= \dot{\vec{\mathbf{r}}} \times \vec{\mathbf{v}} + \vec{\mathbf{r}} \times \dot{\vec{\mathbf{v}}} \\ &= \vec{\mathbf{v}} \times \vec{\mathbf{v}} + \vec{\mathbf{r}} \times \ddot{\vec{\mathbf{r}}} \\ &= \vec{\mathbf{r}} \times \ddot{\vec{\mathbf{r}}}.\end{aligned}$$

Making use of the equation of motion (3.4), we have

$$\dot{\vec{\mathbf{h}}} = -\frac{\mu}{|\vec{\mathbf{r}}|^3}\vec{\mathbf{r}} \times \vec{\mathbf{r}} = \vec{\mathbf{0}}. \tag{3.6}$$

This shows that the orbital angular momentum is constant with respect to $\mathcal{F}_I$. By definition of $\vec{\mathbf{h}}$ (3.5), and the definition of the vector cross-product, the position vector $\vec{\mathbf{r}}$ is always perpendicular to $\vec{\mathbf{h}}$. Since $\vec{\mathbf{h}}$ is constant, the position vector $\vec{\mathbf{r}}$ evolves on a plane in inertial space that is perpendicular to $\vec{\mathbf{h}}$.

Thus, we have discovered that Keplerian orbital motion is planar, and the orbital angular momentum $\vec{\mathbf{h}}$ is constant and specifies the orbital plane.

3.2.2 *Orbital Energy*

Next, let us examine the *orbital energy*. This is defined as

$$\mathcal{E} \stackrel{\Delta}{=} \frac{v^2}{2} - \frac{\mu}{r} = \frac{\vec{\mathbf{v}} \cdot \vec{\mathbf{v}}}{2} - \frac{\mu}{r}, \tag{3.7}$$

where $v = |\vec{\mathbf{v}}|$ is the orbital speed, and $r = |\vec{\mathbf{r}}|$ is the distance of m_2 from m_1. As such, $\frac{v^2}{2}$ is the *kinetic energy per unit mass*, and $-\frac{\mu}{r}$ is the *two-body gravitational potential energy per unit mass*. The reason for the terminology for $-\frac{\mu}{r}$ is that it is a potential function from which the gravitational two-body acceleration can be obtained, that is:

$$-\frac{\mu}{|\vec{\mathbf{r}}|^3}\vec{\mathbf{r}} = \vec{\nabla}\left(\frac{\mu}{r}\right).$$

Let us now compute the time-derivative of (3.7):

$$\dot{\mathcal{E}} = \vec{\mathbf{v}} \cdot \dot{\vec{\mathbf{v}}} + \frac{\dot{r}\mu}{r^2}. \tag{3.8}$$

Now,

$$\begin{aligned} r^2 &= \vec{\mathbf{r}} \cdot \vec{\mathbf{r}}, \quad \Rightarrow \\ r\dot{r} &= \vec{\mathbf{r}} \cdot \dot{\vec{\mathbf{r}}}, \quad \Rightarrow \\ \dot{r} &= \frac{\vec{\mathbf{r}} \cdot \vec{\mathbf{v}}}{r} \end{aligned} \tag{3.9}$$

Substituting this into (3.8) and making use of (3.4) as well, we have

$$\dot{\mathcal{E}} = \vec{\mathbf{v}} \cdot \left(-\frac{\mu}{r^3}\vec{\mathbf{r}}\right) + \frac{\vec{\mathbf{r}} \cdot \vec{\mathbf{v}}\mu}{r^3} = 0. \tag{3.10}$$

Therefore, we have discovered that the Keplerian two-body orbital energy is constant.

3.2.3 The Eccentricity Vector

Let us examine the vectors $\vec{\mathbf{v}} \times \vec{\mathbf{h}}$ and $\frac{\vec{\mathbf{r}}}{r}$. Both of these vectors lie in the orbital plane, since $\vec{\mathbf{v}} \times \vec{\mathbf{h}}$ is perpendicular to $\vec{\mathbf{h}}$. Let us now take the inertial time-derivative of these two vectors. Making use of the rules for vector differentiation (see Section 1.4), we have:

$$\overbrace{\left(\vec{\mathbf{v}} \times \vec{\mathbf{h}}\right)}^{\cdot} = \dot{\vec{\mathbf{v}}} \times \vec{\mathbf{h}},$$

since $\dot{\vec{\mathbf{h}}} = \vec{\mathbf{0}}$. Making use of the identity for the vector triple product (see the end of Section 1.1), this becomes

$$\dot{\vec{\mathbf{v}}} \times \vec{\mathbf{h}} = \dot{\vec{\mathbf{v}}} \times (\vec{\mathbf{r}} \times \vec{\mathbf{v}}) = \vec{\mathbf{r}}(\dot{\vec{\mathbf{v}}} \cdot \vec{\mathbf{v}}) - \vec{\mathbf{v}}(\dot{\vec{\mathbf{v}}} \cdot \vec{\mathbf{r}})$$

Substituting the equation of motion (3.4) into the above expression leads to

$$\begin{aligned} \overbrace{\left(\vec{\mathbf{v}} \times \vec{\mathbf{h}}\right)}^{\cdot} &= \vec{\mathbf{r}}\left(\vec{\mathbf{v}} \cdot \left(-\frac{\mu\vec{\mathbf{r}}}{r^3}\right)\right) - \vec{\mathbf{v}}(\vec{\mathbf{r}} \cdot \left(-\frac{\mu\vec{\mathbf{r}}}{r^3}\right)) \\ &= -\frac{\mu}{r^3}\vec{\mathbf{r}}(\vec{\mathbf{r}} \cdot \vec{\mathbf{v}}) + \frac{\mu}{r}\vec{\mathbf{v}}. \end{aligned} \tag{3.11}$$

Similarly, recognizing that $\vec{\mathbf{v}} = \dot{\vec{\mathbf{r}}}$ we have

$$\overset{\bullet}{\overbrace{\left(\frac{\vec{\mathbf{r}}}{r}\right)}} = \frac{\dot{\vec{\mathbf{r}}}}{r} - \frac{\dot{r}\vec{\mathbf{r}}}{r^2} \\ = \frac{\vec{\mathbf{v}}}{r} - \frac{\vec{\mathbf{r}}(\vec{\mathbf{r}} \cdot \vec{\mathbf{v}})}{r^3}, \tag{3.12}$$

where expression (3.9) was used for $\dot{r}$.

Comparing expressions (3.11) and (3.12), we see that

$$\overset{\bullet}{\overbrace{\left(\frac{\vec{\mathbf{r}}}{r}\right)}} = \frac{1}{\mu}\overset{\bullet}{\overbrace{\left(\vec{\mathbf{v}} \times \vec{\mathbf{h}}\right)}}.$$

Making use of this fact, we define the *eccentricity vector* as

$$\vec{\mathbf{e}} \triangleq \frac{\vec{\mathbf{v}} \times \vec{\mathbf{h}}}{\mu} - \frac{\vec{\mathbf{r}}}{r}. \tag{3.13}$$

Thus, by our above results, the inertial time-derivative is

$$\dot{\vec{\mathbf{e}}} = \vec{\mathbf{0}}. \tag{3.14}$$

That is, the eccentricity vector is inertially constant. Since $\vec{\mathbf{v}} \times \vec{\mathbf{h}}$ and $\vec{\mathbf{r}}/r$ are perpendicular to $\vec{\mathbf{h}}$, it must be that the eccentricity vector $\vec{\mathbf{e}}$ also lies in the orbital plane.

Therefore, we have discovered that the eccentricity vector $\vec{\mathbf{e}}$ lies in the orbital plane, and is constant.

3.3 Shape of a Keplerian Orbit

We have established that Keplerian orbital motion is planar, and we can specify the orbital plane by the angular momentum vector $\vec{\mathbf{h}}$. We have also found a vector that is fixed in the orbital plane, that is the eccentricity vector $\vec{\mathbf{e}}$. Now, all that is needed to describe the position of m_2 with respect to m_1 within the orbital plane, is the distance $r = |\vec{\mathbf{r}}|$ and the angle θ the position vector $\vec{\mathbf{r}}$ makes with $\vec{\mathbf{e}}$ (see Figure 3.2). Additionally, if we can find the relationship between r and θ, then we will obtain the shape of the orbit. Before continuing, it will be useful to define the *eccentricity* of the orbit as the magnitude of the eccentricity vector, that is $e = |\vec{\mathbf{e}}|$. Clearly, the eccentricity is non-negative ($e \geq 0$). This will be important in describing the shape of the orbit.

Let us now take the scalar product of the eccentricity vector $\vec{\mathbf{e}}$ as defined in (7.77) with the position vector $\vec{\mathbf{r}}$. We have

$$\vec{\mathbf{r}} \cdot \vec{\mathbf{e}} = \frac{\vec{\mathbf{r}} \cdot (\vec{\mathbf{v}} \times \vec{\mathbf{h}})}{\mu} - \frac{\vec{\mathbf{r}} \cdot \vec{\mathbf{r}}}{r}.$$

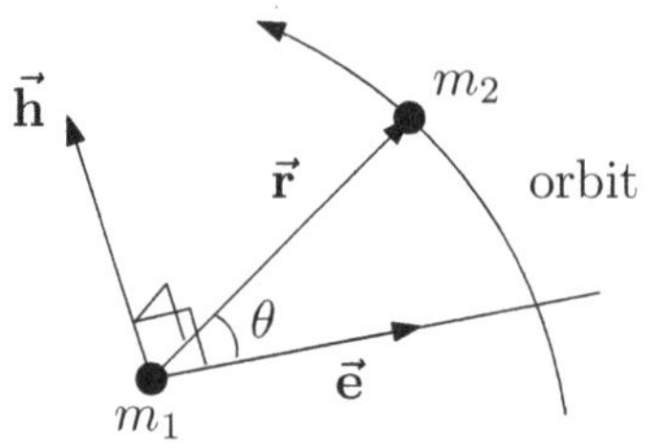

Figure 3.2 Orbit description

Now, using the vector identity for the scalar triple product (see Section 1.1.3),

$$\vec{\mathbf{r}} \cdot (\vec{\mathbf{v}} \times \vec{\mathbf{h}}) = \vec{\mathbf{h}} \cdot (\vec{\mathbf{r}} \times \vec{\mathbf{v}}) = \vec{\mathbf{h}} \cdot \vec{\mathbf{h}} = h^2, \tag{3.15}$$

where $h = |\vec{\mathbf{h}}|$ is the magnitude of the orbital angular momentum. Making use of this, and the definition of the scalar product, we have

$$re\cos\theta = \frac{h^2}{\mu} - r,$$

which can be rearranged to give

$$r = \frac{(h^2/\mu)}{1 + e\cos\theta}. \tag{3.16}$$

It is common to define the parameter

$$p = \frac{h^2}{\mu}, \tag{3.17}$$

which is known as the *semilatus rectum*, and is constant for a given orbit (since h is constant). Hence, (3.16) becomes

$$r = \frac{p}{1 + e\cos\theta}. \tag{3.18}$$

This gives the distance from m_1 to m_2 as a function of the angle θ. The angle θ has a special name, it is called the *true anomaly*. Therefore, (3.18) completely describes the shape of the orbit within the orbital plane. We can immediately see from (3.18) that r is a minimum when $\theta = 0$. That is, the closest point of m_2 to m_1 lies along the eccentricity vector $\vec{\mathbf{e}}$. The closest point is called *periapsis*, and has distance

$$r_{min} = \frac{p}{1 + e}. \tag{3.19}$$

We also see immediately from (3.18) that if the eccentricity $e = 0$, then $r = p = constant$. That is, when $e = 0$, the orbit is circular.

It is clear from (3.18) that the eccentricity e is critical in determining the shape of the orbit. Let us now examine e further. From equation (7.77), we have

$$\begin{aligned} e^2 = \vec{\mathbf{e}} \cdot \vec{\mathbf{e}} &= \left(\frac{\vec{\mathbf{v}} \times \vec{\mathbf{h}}}{\mu} - \frac{\vec{\mathbf{r}}}{r} \right) \cdot \left(\frac{\vec{\mathbf{v}} \times \vec{\mathbf{h}}}{\mu} - \frac{\vec{\mathbf{r}}}{r} \right) \\ &= \frac{(\vec{\mathbf{v}} \times \vec{\mathbf{h}}) \cdot (\vec{\mathbf{v}} \times \vec{\mathbf{h}})}{\mu^2} - 2\frac{\vec{\mathbf{r}} \cdot (\vec{\mathbf{v}} \times \vec{\mathbf{h}})}{\mu r} + \frac{\vec{\mathbf{r}} \cdot \vec{\mathbf{r}}}{r^2}. \end{aligned} \tag{3.20}$$

From the vector identities in Section 1.1.3,

$$\begin{aligned} (\vec{\mathbf{v}} \times \vec{\mathbf{h}}) \cdot (\vec{\mathbf{v}} \times \vec{\mathbf{h}}) &= (\vec{\mathbf{v}} \cdot \vec{\mathbf{v}})(\vec{\mathbf{h}} \cdot \vec{\mathbf{h}}) - (\vec{\mathbf{v}} \cdot \vec{\mathbf{h}})(\vec{\mathbf{h}} \cdot \vec{\mathbf{v}}) \\ &= v^2 h^2, \end{aligned}$$

where the second term on the top right was eliminated since $\vec{\mathbf{v}}$ is perpendicular to $\vec{\mathbf{h}}$. Substituting this and (3.15) into (3.20), we obtain

$$e^2 = \frac{v^2 h^2}{\mu^2} - \frac{2h^2}{\mu r} + 1.$$

This can be rearranged into the form

$$\begin{aligned} 1 - e^2 &= \frac{h^2}{\mu} \left(\frac{2}{r} - \frac{v^2}{\mu} \right) \\ &= p \left(\frac{2}{r} - \frac{v^2}{\mu} \right). \end{aligned} \tag{3.21}$$

It will be useful to define the parameter

$$a = \left(\frac{2}{r} - \frac{v^2}{\mu} \right)^{-1}. \tag{3.22}$$

This parameter will also have a special significance in describing the shape of the orbit, as we shall see shortly. With this definition, we obtain from (3.21) the relationship between a, p and e, namely

$$a = \frac{p}{1 - e^2}. \tag{3.23}$$

Since $p > 0$ (because $p = h^2/\mu$), it is clear from (3.23) that $a < 0$ when $e > 1$, and $a > 0$ when $e < 1$. When $e = 1$, a is undefined. Also, since p and e are constant for a given orbit, a is also constant. To obtain further physical significance of a, from (3.22) we find that

$$a^{-1} = \frac{2}{r} - \frac{v^2}{\mu} = \frac{2}{\mu}\left(\frac{\mu}{r} - \frac{v^2}{2}\right) = \frac{2}{\mu}(-\mathcal{E}). \tag{3.24}$$

where the expression for the orbital energy (3.7) has been used. Therefore, we have an alternative expression for the orbital energy in terms of a:

$$\mathcal{E} = -\frac{\mu}{2a}. \tag{3.25}$$

Substituting (3.24) into (3.21) and rearranging, we can also obtain an expression for the orbital energy in terms of the eccentricity:

$$\mathcal{E} = \frac{\mu^2(e^2 - 1)}{2h^2}. \tag{3.26}$$

From this, we see that

$$\begin{aligned} e > 1 &\Rightarrow \mathcal{E} > 0, \\ e = 1 &\Rightarrow \mathcal{E} = 0, \\ e < 1 &\Rightarrow \mathcal{E} < 0. \end{aligned} \tag{3.27}$$

A useful result is found when we equate (3.25) with (3.7). After some work, we have

$$v = \sqrt{\mu\left(\frac{2}{r} - \frac{1}{a}\right)}. \tag{3.28}$$

This is known as the *vis-viva equation*, and it relates the orbital speed in terms of the energy. This equation will be useful when we consider transfers from one orbit to another.

3.3.1 Perifocal Coordinate System

As we have seen in section 2.1, once we have one inertial frame of reference, it is possible to define another inertial frame of reference, provided it is 1) not rotating with respect to the original inertial frame, and 2) it's origin is either not translating, or translating with constant velocity with respect to the original frame. Now, we have seen that the Keplerian orbital motion evolves on a plane (normal to $\vec{\mathbf{h}}$), and that this plane is inertially fixed. To simplify the description of the motion, it will be useful to have an inertial frame, with two of the unit vectors lying in that plane. The position of $\vec{\mathbf{r}}$ can then be represented in terms of those two

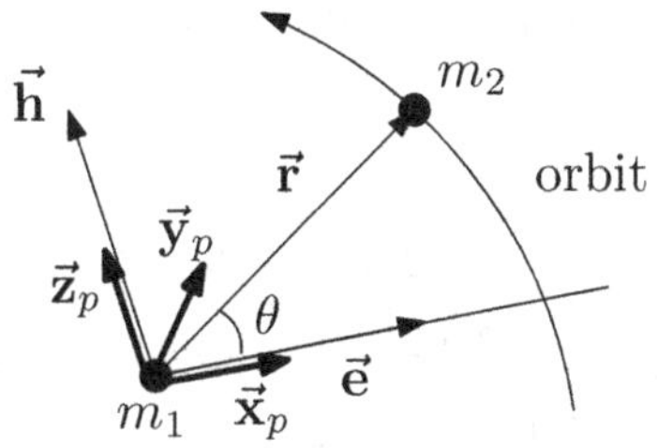

Figure 3.3 Perifocal coordinate system

vectors. We have found a vector within this plane that is inertially fixed, that is $\vec{\mathbf{e}}$. We can therefore define a new inertial reference frame $\mathcal{F}_p$ by the unit vectors

$$\vec{\mathbf{x}}_p = \frac{\vec{\mathbf{e}}}{e}, \quad \vec{\mathbf{z}}_p = \frac{\vec{\mathbf{h}}}{h}, \quad \vec{\mathbf{y}}_p = \vec{\mathbf{z}}_p \times \vec{\mathbf{x}}_p. \tag{3.29}$$

This reference frame is called the *perifocal reference frame*, and is shown in Figure 3.3. The reason for the name will become clear shortly. Note that when $\vec{\mathbf{e}} = \vec{\mathbf{0}}$, we can arbitrarily choose $\vec{\mathbf{x}}_p$ in (3.29) by any inertially fixed vector in the orbital plane. In this coordinate system, the position of m_2 with respect to m_1 is given by

$$\vec{\mathbf{r}} = \vec{\mathcal{F}}_p^T \begin{bmatrix} x_p \\ y_p \\ z_p \end{bmatrix} = \vec{\mathcal{F}}_p^T \begin{bmatrix} r\cos\theta \\ r\sin\theta \\ 0 \end{bmatrix}. \tag{3.30}$$

Let us now return to the question of the shape of the orbit. The orbit is given in polar form by equation (3.18). Now, let us examine the cartesian form with $x_p = r\cos\theta$ and $y_p = r\sin\theta$. Rearranging (3.18) gives

$$p = r(1 + e\cos\theta) = r + ex_p,$$

which leads to

$$r = p - ex_p. \tag{3.31}$$

By trigonometry, $r^2 = x_p^2 + y_p^2$. Therefore, we have

$$\begin{aligned} y_p^2 = r^2 - x_p^2 &= (p - ex_p)^2 - x_p^2, \\ &= p^2 - 2pex_p + (e^2 - 1)x_p^2. \end{aligned} \tag{3.32}$$

Now, when $e \neq 1$, we can substitute (3.23) in (3.32) to obtain

$$\begin{aligned} y_p^2 &= (p - ex_p)^2 - x_p^2, \\ &= a^2(1-e^2)^2 - 2a(1-e^2)ex_p - (1-e^2)x_p^2, \\ &= (1-e^2)\left[a^2(1-e^2) - 2aex_p - x_p^2\right], \\ &= (1-e^2)\left[a^2 - (x_p + ae)^2\right]. \end{aligned}$$

This can be rearranged to give

$$\frac{(x_p + ae)^2}{a^2} + \frac{y_p^2}{(1-e^2)a^2} = 1. \tag{3.33}$$

When $e = 1$, (3.32) becomes

$$y_p^2 = p^2 - 2px_p,$$

which rearranges to become

$$x_p = \frac{p}{2} - \frac{y_p^2}{2p}. \tag{3.34}$$

Let us now look at the different cases for e.

Case 1, $e < 1$
When the eccentricity $e < 1$, the energy of the orbit is negative, $\mathcal{E} < 0$, see (3.27). In this case, $a^2(1-e^2) > 0$, and we define $b = a\sqrt{(1-e^2)}$. Equation (3.33) becomes

$$\frac{(x_p + ae)^2}{a^2} + \frac{y_p^2}{b^2} = 1. \tag{3.35}$$

This is the equation for an ellipse, centered at $x_p = -ae$, $y_p = 0$ with semi-major axis a, and semi-minor axis b, as seen in Figure 3.4. Since an ellipse is finite in size, it is clear that when the orbital energy is negative, the mass m_2 can never escape from m_1.

For any orbit, we can compute the shortest distance from m_1 to m_2. As we have already explained, the corresponding location is called periapsis . For an elliptical orbit, it is given by (substituting (3.23) into (3.18)),

$$r_{min} = \frac{a(1-e^2)}{1+e} = a(1-e).$$

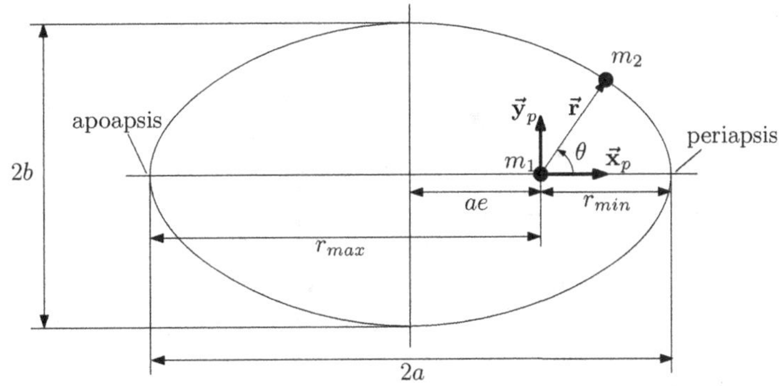

Figure 3.4 Elliptical orbit geometry

Due to the finite size of an elliptical orbit, we can also compute the maximum distance from m_1 to m_2. From equation (3.18), the maximum value of r is obtained when $\theta = 180°$. This point is called *apoapsis*. Setting $\theta = 180°$ in (3.18) and making use of (3.23), we have

$$r_{max} = \frac{a(1-e^2)}{1-e} = a(1+e).$$

Therefore, we see that both the minimum and the maximum distances lie on the same line (which is parallel to the eccentricity vector $\vec{\mathbf{e}}$). This line is called the *line of apsides.*

Ellipses have several properties. We shall not dwell on these too much, and only mention one property. That property is:

An ellipse has two focii located along the semimajor axis at equal distances from the ellipse center, such that the sum of the distances to any point on the ellipse from each of the focii is constant, and equal to twice the semi-major axis. The distance of each focus from the ellipse center is ae, where a is the semi-major axis, and $0 < e < 1$ *is the eccentricity. Referring to Figure* 3.5, $r + \bar{r} = 2a$.

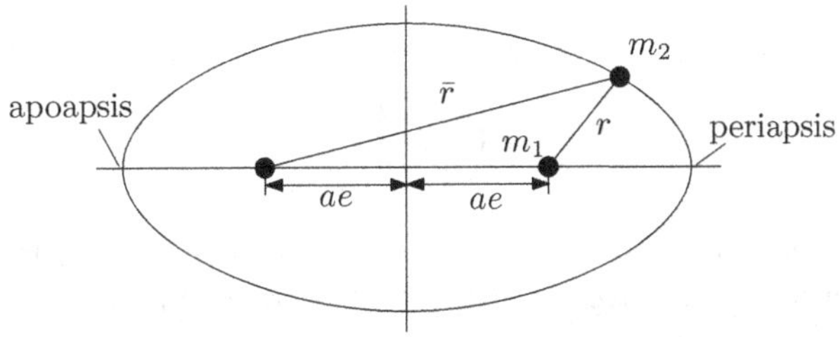

Figure 3.5 Ellipse focal property

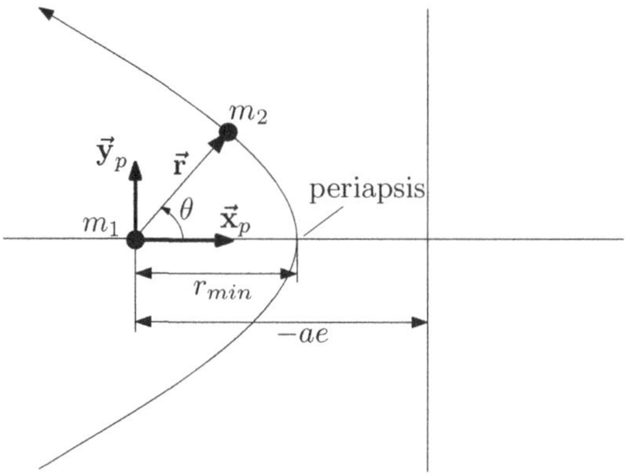

Figure 3.6 Hyperbolic orbit geometry

In summary, we have shown that when $e < 1$, equivalently, $\mathcal{E} < 0$ or $a > 0$, the mass m_2 is in an elliptical orbit about m_1, with m_1 located at the focus of the ellipse closest to periapsis. Hence, the term "perifocal reference frame".

Case 2, $e > 1$
When the eccentricity $e > 1$, the energy of the orbit is positive, $\mathcal{E} > 0$, see (3.27). In this case, $a^2(1 - e^2) < 0$, and we define $b = a\sqrt{(e^2 - 1)}$. Equation (3.33) becomes

$$\frac{(x_p + ae)^2}{a^2} - \frac{y_p^2}{b^2} = 1. \tag{3.36}$$

This is the equation for a hyperbola, with origin at $x_p = -ae$, $y = 0$, as seen in Figure 3.6. Note that in this case, $a = \frac{p}{1-e^2} < 0$ is negative. A hyperbola is infinite in size. Therefore, when the energy of the orbit is positive, the mass m_2 eventually escapes from m_1. Let us now look at the limiting speed of m_2. Rearranging equation for the orbital energy (3.7) gives

$$v = \sqrt{2\left(\mathcal{E} + \frac{\mu}{r}\right)}. \tag{3.37}$$

Letting $r \to \infty$, we have

$$v_\infty \triangleq \lim_{r\to\infty} \sqrt{2\left(\mathcal{E} + \frac{\mu}{r}\right)} = \sqrt{2\mathcal{E}} > 0. \tag{3.38}$$

This is called the *hyperbolic excess speed*, the speed m_2 has upon escaping from m_1. Therefore, when the orbital energy is positive, m_2 escapes from m_1 with a non-zero velocity. This is important for interplanetary travel, when it is necessary to escape from one planet in order to travel to another. We will consider this in more detail later.

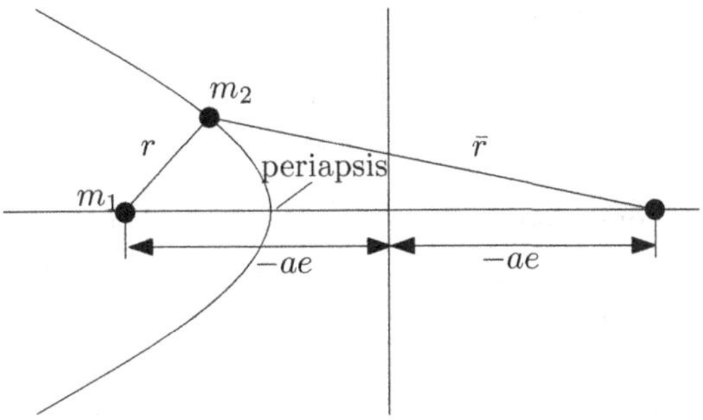

Figure 3.7 Hyperbola focal property

Hyperbolae have many properties. A hyperbola as a similar focal property as the ellipse, that is:

A hyperbola has two focii located at equal distances from hyperbola origin, such that the difference of the distances to any point on the hyperbola from each of the focii is constant. The distance of each focus from the hyperbola origin is $-ae$, and the difference in distances is -2a. Referring to Figure 3.7, $\bar{r} - r = -2a$

In summary, we have shown that when $e > 1$, equivalently, $\mathcal{E} > 0$ or $a < 0$, the mass m_2 is in an hyperbolic orbit about m_1, with m_1 located at the focus of the hyperbola closest to periapsis.

Another important property of hyperbolae is:

A hyperbola has a pair of asymptotes, as shown in Figure 3.8.

The general equation for a hyperbola with origin at $x = y = 0$ is given by

$$\frac{x^2}{a^2} - \frac{y^2}{b^2} = 1.$$

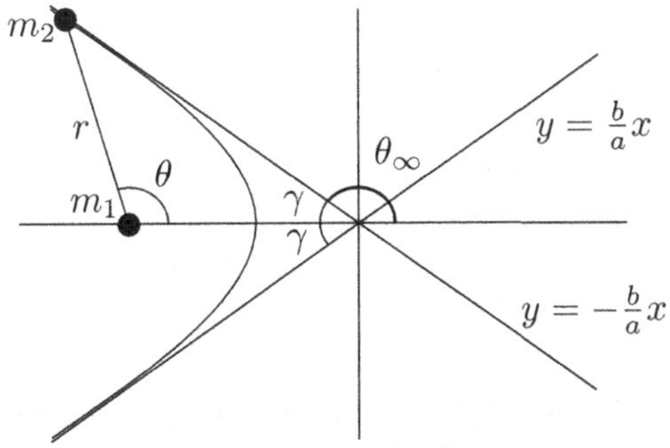

Figure 3.8 Hyperbola asymptotes

This can be rearranged into the form

$$y = \pm\frac{b}{a}\sqrt{x^2 - a^2}.$$

We see then that as $|x|$ becomes large, $y \rightarrow \pm\frac{b}{a}x$. That is, the hyperbola has asymptotes given by

$$y = \pm\frac{b}{a}x. \tag{3.39}$$

Now, a point of significant interest is the angle the asymptotes make with the x-axis, denoted by γ as seen in Figure 3.8. We clearly have $0 < \gamma < 90°$, and

$$\tan\gamma = \left|\frac{b}{a}\right|.$$

In terms of the hyperbolic orbit, where $b = a\sqrt{(e^2 - 1)}$, we have

$$\tan\gamma = \sqrt{(e^2 - 1)}.$$

Note that this angle could also have been obtained directly from the polar form of the hyperbola,

$$r = \frac{a(1 - e^2)}{1 + e\cos\theta}.$$

The asymptotic behavior is obtained by allowing $r \rightarrow \infty$. Clearly, this occurs when $1 + e\cos\theta \rightarrow 0$, or

$$\cos\theta \rightarrow -\frac{1}{e}.$$

Let $\theta_\infty = \cos^{-1}\left(-\frac{1}{e}\right)$. Because γ is the angle the asymptotes make with the x-axis, $\gamma = 180° - \theta_\infty$ (see Figure 3.8). Therefore,

$$\cos\gamma = \cos(180° - \theta_\infty) = -\cos\theta_\infty = \frac{1}{e}.$$

Making use of the right-angled triangle shown in Figure 3.9, we can see that

$$\tan\gamma = \sqrt{(e^2 - 1)},$$

as before.

The angle γ is important, since it shows that when a mass m_2 approaches m_1 from far away with $e > 1$, it will be turned through an angle $\delta = 180° - 2\gamma$. This is relevant when planning interplanetary fly-by or sling-shot maneuvers.

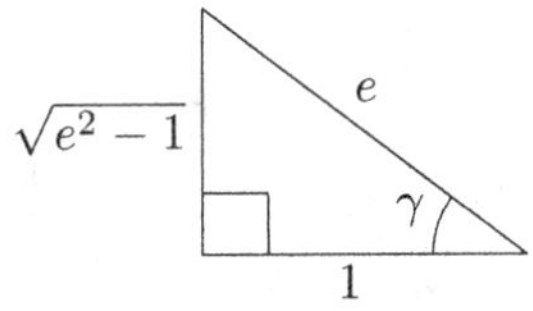

Figure 3.9 Trigonometric relationship between gamma and e

Case 3 $e = 1$

When the eccentricity $e = 1$, the energy of the orbit is zero, $\mathcal{E} = 0$, see (3.27). In this case, a is undefined, and the orbit is described by equation (3.34). This can be recognized as a parabola, with origin at $x_o = \frac{p}{2}$, $y = 0$, as shown in Figure 3.10.

A parabola is infinite in size. Therefore, when the energy of the orbit is zero, the mass m_2 eventually escapes from m_1. As for the hyperbolic case, let us examine the limiting speed. Making use of (3.37), we have

$$v_\infty \triangleq \lim_{r\to\infty} \sqrt{\frac{2\mu}{r}} = 0. \tag{3.40}$$

That is, when the orbital energy is zero, the mass m_2 escapes from m_1, but the velocity eventually goes to zero. The parabolic orbit represents the transition from a bounded elliptical orbit to an unbounded hyperbolic orbit. The orbital energy $\mathcal{E} = 0$ is the minimum energy required for mass m_2 to escape from m_1. Using (3.37), given an orbital radius r, the minimum orbital speed required for m_2 to escape from m_1 is

$$v = \sqrt{\frac{2\mu}{r}}.$$

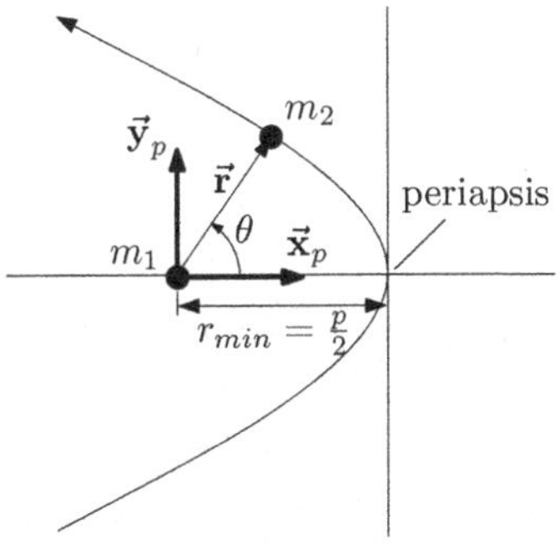

Figure 3.10 Parabolic orbit geometry

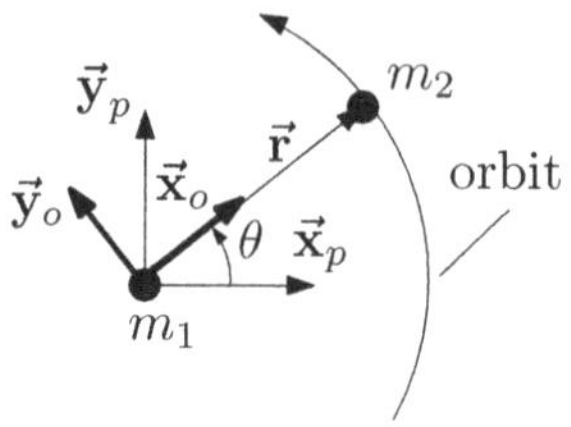

Figure 3.11 Cylindrical coordinate system

When m_2 is at periapsis ($\theta = 0$), this is called the *escape velocity*, and is given by

$$v_{esc} = \sqrt{\frac{2\mu}{r_{min}}}, \tag{3.41}$$

where $r_{min} = \frac{p}{2}$.

In conclusion, the shape of a Keplerian orbit is an ellipse when the orbital energy is negative ($\mathcal{E} < 0$), a parabola when the orbital energy is zero ($\mathcal{E} = 0$), and a hyperbola when the orbital energy is positive ($\mathcal{E} > 0$). These three shapes are all *conic sections*. Therefore, the most general description of the shape Keplerian orbit is a conic section.

3.4 Kepler's Laws

Let us now further develop the expressions for r and θ. It will be useful to define a new reference frame. This frame will be a cylindrical coordinate system, denoted $\mathcal{F}_o$ that can be obtained from the perifocal coordinate system, $\mathcal{F}_p$, by a rotation about $\vec{\mathbf{z}}_p$ through an angle θ, as shown in Figure 3.11. Therefore we have $\mathbf{C}_{op} = \mathbf{C}_z(\theta)$. The subscript "o" is chosen to reflect the fact that $\mathcal{F}_o$ rotates with the orbit. The angular velocity of $\mathcal{F}_o$ with respect to $\mathcal{F}_p$ is given by

$$\vec{\boldsymbol{\omega}}_{op} = \vec{\mathcal{F}}_o^T \begin{bmatrix} 0 \\ 0 \\ \dot{\theta} \end{bmatrix}.$$

Represented in $\mathcal{F}_o$ coordinates, the position vector is

$$\vec{\mathbf{r}} = \vec{\mathcal{F}}_o^T \begin{bmatrix} r \\ 0 \\ 0 \end{bmatrix}. \tag{3.42}$$

Taking the inertial time-derivative of this gives the velocity

$$\begin{aligned}\vec{\mathbf{v}} &= \vec{\mathcal{F}}_o^T \begin{bmatrix} \dot{r} \\ 0 \\ 0 \end{bmatrix} + \vec{\mathcal{F}}_o^T \begin{bmatrix} 0 \\ 0 \\ \dot{\theta} \end{bmatrix}^{\times} \begin{bmatrix} r \\ 0 \\ 0 \end{bmatrix}, \\ &= \vec{\mathcal{F}}_o^T \begin{bmatrix} \dot{r} \\ 0 \\ 0 \end{bmatrix} + \vec{\mathcal{F}}_o^T \begin{bmatrix} 0 & -\dot{\theta} & 0 \\ \dot{\theta} & 0 & 0 \\ 0 & 0 & 0 \end{bmatrix} \begin{bmatrix} r \\ 0 \\ 0 \end{bmatrix}, \\ &= \vec{\mathcal{F}}_o^T \begin{bmatrix} \dot{r} \\ r\dot{\theta} \\ 0 \end{bmatrix}\end{aligned} \tag{3.43}$$

Let us now compute the angular momentum vector:

$$\begin{aligned}\vec{\mathbf{h}} &= \vec{\mathbf{r}} \times \vec{\mathbf{v}}, \\ &= \vec{\mathcal{F}}_o^T \begin{bmatrix} r \\ 0 \\ 0 \end{bmatrix}^{\times} \begin{bmatrix} \dot{r} \\ r\dot{\theta} \\ 0 \end{bmatrix}, \\ &= \vec{\mathcal{F}}_o^T \begin{bmatrix} 0 & 0 & 0 \\ 0 & 0 & -r \\ 0 & r & 0 \end{bmatrix} \begin{bmatrix} \dot{r} \\ r\dot{\theta} \\ 0 \end{bmatrix}, \\ &= \vec{\mathcal{F}}_o^T \begin{bmatrix} 0 \\ 0 \\ r^2\dot{\theta} \end{bmatrix}.\end{aligned} \tag{3.44}$$

Since the angular momentum vector $\vec{\mathbf{h}}$ points in the same direction as the z-axis ($\vec{\mathbf{z}}_o = \vec{\mathbf{z}}_p$) of the frame $\mathcal{F}_o$ by the definition (3.29), the angular rate must be positive, that is $\dot{\theta} > 0$. We can therefore conclude from (3.44) that the magnitude of the orbital angular momentum is

$$h = r^2\dot{\theta}. \tag{3.45}$$

We are now in a position to state and prove Kepler's Laws.

Kepler's Laws

Johannes Kepler postulated the following laws based on empirical data obtained by Tycho Brahe:

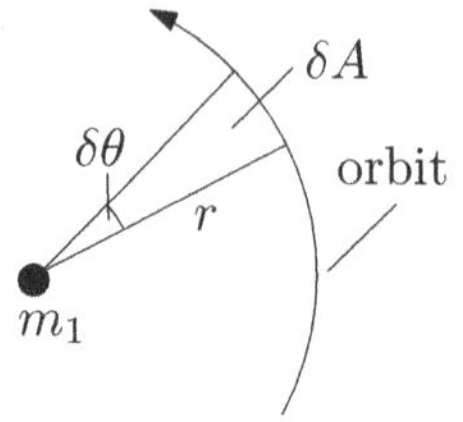

Figure 3.12 Area swept out by orbital position vector

1. *The orbit of each planet is an ellipse with the sun at one focus.*
2. *The radius vector drawn from the sun to the planet sweeps out equal areas in equal times.*
3. *The squares of the periods of the planetary orbits are proportional to the cubes of the semi-major axes of the orbits.*

Newton later derived these laws using his law of gravitation and laws of motion. We shall now prove these laws. We have already derived Kepler's first law, which is valid for $e < 1$ (which is the case for all of the planets). To prove the second law, consider the diagram in Figure 3.12. We see that the infinitessimal area segment is given by

$$\delta A = \frac{1}{2} r^2 \delta\theta.$$

Now, $\delta\theta = \dot{\theta}\delta t$. Therefore, $\delta A = \frac{1}{2} r^2 \dot{\theta} \delta t$, ie.

$$\frac{\delta A}{\delta t} = \frac{1}{2} r^2 \dot{\theta}.$$

But, we now know from (3.45) that $r^2\dot{\theta} = h$, which is constant. Letting $\delta t \to 0$, we obtain the the result

$$\frac{dA}{dt} = \frac{h}{2} = constant. \tag{3.46}$$

Note that Kepler's second law is valid for all three orbit shapes, elliptical, parabolic and hyperbolic.

Kepler's third law is only makes sense for elliptical orbits (since parabolic and hyperbolic orbits do not have a period). To derive it, let T be the orbital period, that is the time it takes to complete one revolution of the ellipse. Then, the area of the ellipse is given by

$$A = \int_0^T \frac{dA}{dt} dt.$$

By Kepler's second law (3.46), this becomes

$$A = \frac{h}{2} T. \tag{3.47}$$

Now, in terms of the semi-major axis a and semi-minor axis b of an ellipse, the area is given by

$$A = \pi ab.$$

Using this in (3.47) gives

$$T = \frac{2\pi ab}{h}. \tag{3.48}$$

Noting that for an elliptical orbit, $h = \sqrt{\mu p} = \sqrt{\mu a(1 - e^2)}$ (see (3.17) and (3.23)), and that $b = a\sqrt{1 - e^2}$, we have

$$T = \frac{2\pi a^2\sqrt{1 - e^2}}{\sqrt{\mu a(1 - e^2)}} = \frac{2\pi a^{\frac{3}{2}}}{\sqrt{\mu}}, \tag{3.49}$$

or equivalently

$$T^2 = \frac{4\pi^2}{\mu}a^3, \tag{3.50}$$

which is Kepler's third law.

Related to the orbital period, the *mean orbital motion* is defined as

$$n \triangleq \frac{2\pi}{T} = \sqrt{\frac{\mu}{a^3}}. \tag{3.51}$$

This is the average angular rate of the orbit.

3.5 Time of Flight

So far we have completely defined the shape of the orbit. However, to fully solve the orbital problem, we need to determine how m_2 traverses the orbit with respect to time. What we are missing is the true anomaly θ as a function of time, t.

3.5.1 Circular Orbits

For a circular orbit ($e = 0$), this is trivial. From (3.45), we have

$$\dot{\theta} = \frac{h}{r^2}. \tag{3.52}$$

But, for a circular orbit, we have that $r = a$ and $h = \sqrt{\mu a}$ (see (3.17), (3.18) and (3.23)). Therefore,

$$\dot{\theta} = \sqrt{\frac{\mu}{a^3}}. \tag{3.53}$$

Comparing this to (3.51), we see that **for a circular orbit, the orbital angular rate is simply the mean orbital motion**. That is,

$$\dot{\theta} = n. \tag{3.54}$$

Integrating this leads to the relationship between θ and t:

$$\theta(t) - \theta(t_0) = n(t - t_0). \tag{3.55}$$

3.5.2 Elliptical Orbits

Unfortunately, in the more general case, it is very difficult to obtain a direct relationship between the true anomaly, θ and time, t. We shall only treat the elliptical ($e < 1$) case fully. The parabolic and hyperbolic cases can be treated in similar fashion.

3.5.2.1 Eccentric Anomaly

Because it is difficult to obtain a direct relationship between θ and t, we shall instead make use of an auxiliary variable called the *eccentric anomaly*. To define this, consider the elliptical orbit described by (3.35),

$$\frac{(x_p + ae)^2}{a^2} + \frac{y_p^2}{b^2} = 1.$$

where $b = a\sqrt{1 - e^2}$. We note that this ellipse may also be represented in polar form by

$$x_p + ae = a\cos E, \quad y_p = b\sin E, \tag{3.56}$$

for some angle E. It is readily verified that

$$\frac{(x_p + ae)^2}{a^2} + \frac{y_p^2}{b^2} = \cos^2 E + \sin^2 E = 1,$$

as required. We call the angle E, the *eccentric anomaly*. The geometric interpretation of the eccentric anomaly is shown in Figure 3.13. Now, we can find relationships between E and θ. Equating the expressions for x_p and y_p in (3.56) and (3.30), we obtain

$$x_p = r\cos\theta = a\,(\cos E - e)\,, \quad y_p = r\sin\theta = b\sin E,$$

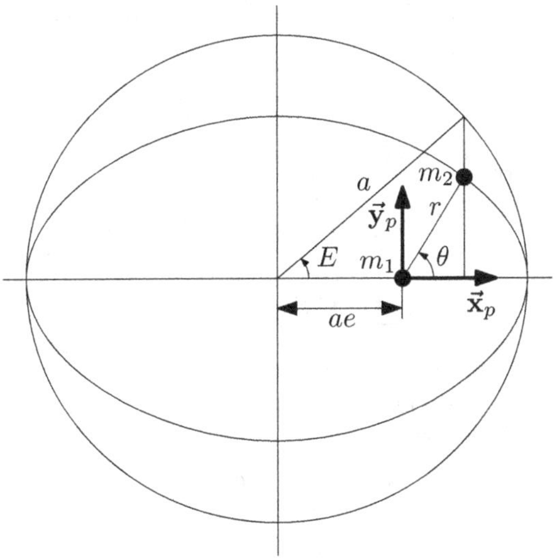

Figure 3.13 Geometric interpretation of eccentric anomaly

which leads to

$$\cos\theta = \frac{a\,(\cos E - e)}{r}, \tag{3.57}$$

and

$$\sin\theta = \frac{a\sqrt{1-e^2}\sin E}{r}, \tag{3.58}$$

where we have substituted for b. Next, we note from (3.31) and (3.23) that for any point on the ellipse,

$$r = a(1-e^2) - e x_p. \tag{3.59}$$

Substituting from (3.56) for x_p, this leads to

$$r = a(1-e^2) - ae(\cos E - e) = a(1 - e\cos E). \tag{3.60}$$

Now let us make use of some double angle formulae, specifically:

$$\cos u = 2\cos^2\frac{u}{2} - 1, \tag{3.61}$$

and

$$\sin u = 2\sin\frac{u}{2}\cos\frac{u}{2}. \tag{3.62}$$

First, let us apply (3.61) to the left-hand side of (3.57). Doing so leads to

$$2\cos^2\frac{\theta}{2} - 1 = \frac{a\,(\cos E - e)}{r} \Rightarrow$$
$$2\cos^2\frac{\theta}{2} = \frac{a\,(\cos E - e) + r}{r} \Rightarrow$$
$$2\cos^2\frac{\theta}{2} = \frac{a\,(\cos E - e) + a(1 - e\cos E)}{r} \Rightarrow$$
$$2\cos^2\frac{\theta}{2} = \frac{a(1-e)\,(1+\cos E)}{r}.$$

Note that we have made use of (3.60). Now, applying (3.61) to the right-hand side, we have

$$\cos^2\frac{\theta}{2} = \frac{a(1-e)\cos^2\frac{E}{2}}{r}. \tag{3.63}$$

Next, let us apply formula (3.62) to (3.58). Doing so leads to

$$\sin\frac{\theta}{2}\cos\frac{\theta}{2} = \frac{a\sqrt{1-e^2}\sin\frac{E}{2}\cos\frac{E}{2}}{r}. \tag{3.64}$$

Finally, dividing (3.64) by (3.63) gives

$$\tan\frac{\theta}{2} = \frac{\sqrt{1-e^2}}{1-e}\tan\frac{E}{2} = \sqrt{\frac{(1-e)(1+e)}{(1-e)^2}}\tan\frac{E}{2},$$

which leads to

$$\tan\frac{\theta}{2} = \sqrt{\frac{1+e}{1-e}}\tan\frac{E}{2}. \tag{3.65}$$

This last relationship (3.65) is the most useful, since $\frac{E}{2}$ and $\frac{\theta}{2}$ share the same quadrant, as is clear from Figure 3.13. Restricting $0 \leq E \leq 360^\circ$, we have $0 \leq \frac{E}{2} \leq 180^\circ$. From Figure 3.13, if $0 \leq E \leq 180^\circ$, then $0 \leq \theta \leq 180^\circ$, i.e. if $0 \leq \frac{E}{2} \leq 90^\circ$, then $0 \leq \frac{\theta}{2} < 90^\circ$. Similarly, if $180^\circ \leq E \leq 360^\circ$, then $180 \leq \theta \leq 360^\circ$, i.e. if $90 \leq \frac{E}{2} \leq 180^\circ$, then $90 \leq \frac{\theta}{2} < 180^\circ$. Therefore, given the eccentric anomaly E, we can uniquely determine the true anomaly θ using (3.65).

Now, having established a one-to-one relationship between the eccentric anomaly E and the true anomaly θ, in order to find the relationship between θ and t, we only need to find the relationship between E and t.

To do this, we now return to the expression for the magnitude of the angular momentum (3.45), which is

$$h = r^2\frac{d\theta}{dt}. \tag{3.66}$$

Let t_0 be the time of periapsis passage, that is $\theta(t_0) = 0$. Integrating (3.45) gives

$$\int_0^\theta r^2 d\theta = \int_{t_0}^t h\,dt = h(t - t_0). \tag{3.67}$$

Let us now make the change of variables, $E = E(\theta)$. Differentiating (3.65), we have

$$\frac{1}{2\cos^2\frac{\theta}{2}}\frac{d\theta}{dE} = \frac{1}{2\cos^2\frac{E}{2}}\sqrt{\frac{1+e}{1-e}}, \Rightarrow$$

$$\frac{d\theta}{dE} = \sqrt{\frac{1+e}{1-e}}\frac{\cos^2\frac{\theta}{2}}{\cos^2\frac{E}{2}}.$$

Making use of (3.63), this becomes

$$\begin{aligned}\frac{d\theta}{dE} &= \sqrt{\frac{1+e}{1-e}}\frac{a(1-e)}{r},\\ &= \frac{a\sqrt{1-e^2}}{r}.\end{aligned}$$

Therefore, (3.67) becomes

$$h(t - t_0) = \int_0^E r^2\frac{d\theta}{dE}dE = a\sqrt{1-e^2}\int_0^E r\,dE.$$

Since $r = a(1 - e\cos E)$ (from (3.60)), the above integral becomes

$$\begin{aligned}h(t - t_0) &= a^2\sqrt{1-e^2}\textstyle\int_0^E(1 - e\cos E)dE,\\ &= a^2\sqrt{1-e^2}(E - e\sin E).\end{aligned}$$

Substituting $h = \sqrt{\mu a(1-e^2)}$, the above equation becomes

$$E - e\sin E = \sqrt{\frac{\mu}{a^3}}(t - t_0).$$

Now, from (3.51), we recognize that $\sqrt{\frac{\mu}{a^3}}$ is just the mean motion. Defining the *mean anomaly* as

$$M = n(t - t_0), \tag{3.68}$$

the equation for the eccentric anomaly above becomes

$$E - e\sin E = M. \tag{3.69}$$

This is the famous Kepler's equation.

To summarize, for an elliptical orbit, in order to find the true anomaly, θ given the time since periapsis passage, $t - t_0$, we follow the procedure:

1. **Compute the mean anomaly, M from** (3.68).
2. **Solve Kepler's equation** (3.69) **for the eccentric anomaly, E.**
3. **Compute the true anomaly, θ from** (3.65).

We can also reverse the procedure. Given the true anomaly, θ, we can determine the time since periapsis passage $t - t_0$ by the following:

1. **Compute the eccentric anomaly, E from** (3.65).
2. **Compute the mean anomaly, M from** (3.69).
3. **Compute the time since periapsis passage, $t - t_0$ from** (3.68).

3.5.2.2 Solution for Kepler's Equation

There is no analytical solution for E given M in Kepler's equation. It must be solved numerically. There are several methods that can be used. One such method is known as Newton's method. This is an iterative scheme that converges quite quickly:

$$E_{k+1} = E_k - \frac{g(E_k)}{g'(E_k)}, \quad k = 0, 1, 2, \ldots \tag{3.70}$$

where

$$g(E) \triangleq E - e \sin E - M, \quad g'(E) = 1 - e \cos E.$$

An initial guess for E that usually works well is $E_0 = M$. The iteration (3.70) is repeated until either $|E_{k+1} - E_k| < \epsilon$ or $|g(E_k)| < \delta$ where δ and ϵ are appropriately chosen small, positive numbers.

3.5.3 Parabolic Orbits

It can be shown by direct integration of (3.67) that the time-of-flight equation for parabolic orbits is

$$6\sqrt{\frac{\mu}{p^3}}(t - t_0) = \left[3 \tan\frac{\theta}{2} + \tan^3\frac{\theta}{2}\right]. \tag{3.71}$$

3.5.4 Hyperbolic Orbits

Similar to the elliptical case, a *hyperbolic eccentric anomaly*, H can be found, satisfying

$$\tanh\frac{H}{2}=\sqrt{\frac{e-1}{e+1}}\tan\frac{\theta}{2}. \tag{3.72}$$

Defining the *hyperbolic mean anomaly* as

$$N\triangleq\sqrt{\frac{\mu}{-a^3}}(t-t_0), \tag{3.73}$$

it is possible to obtain a hyperbolic version of Kepler's equation, given by

$$e\sinh H-H=N. \tag{3.74}$$

The procedure for obtaining the true anomaly θ given the time since periapsis passage $t-t_0$ is the same as for the elliptical case.

3.6 Orbital Elements

Thus far, we have established that two-body orbital motion is planar, and that the shape of the orbit is either circular, elliptical, parabolic or hyperbolic. The size and shape are completely determined by the semi-latus rectum, p and the eccentricity, e (for elliptical orbits it is more common to use the semi-major axis, a, instead of p). These in turn, are completely determined by the orbital angular momentum, h and energy, $\mathcal{E}$. We have also established how to determine our location on the orbit (the true anomaly θ) given the time since periapsis passage $t-t_0$, using the time-of-flight equations.

All that is remaining to completely specify the two-body orbital motion is to locate the orbit in three-dimensional space. Specifically, we need to specify the plane of the orbit, and the direction of periapsis within that plane.

In order to be able to do this, we need an inertial reference frame within which we can specify these parameters. For orbits around the sun, the heliocentric-ecliptic coordinate system is useful. For orbits around the Earth, the geocentric-equatorial system is useful.

3.6.1 Heliocentric-Ecliptic Coordinate System

As shown in Figure 3.14, the *heliocentric-ecliptic* system, denoted $\mathcal{F}_{\mathcal{E}}$ is defined using the Earth's orbit about the sun. The origin is at the center of the sun. The fundamental plane of the system (x-y plane) is the ecliptic plane, which is the plane of the Earth's orbit around the sun. The $\vec{\mathbf{z}}_{\mathcal{E}}$ direction is normal to the ecliptic plane, such that it has the same direction as the Earth's orbital angular momentum vector about the sun. The $\vec{\mathbf{x}}_{\mathcal{E}}$ direction is determined by the line of intersection between the ecliptic plane, and the Earth's equatorial plane. The direction of $\vec{\mathbf{x}}_{\mathcal{E}}$ along the line of intersection is chosen such that on the first day of autumn, a vector pointing from the sun to the Earth lies along the positive $\vec{\mathbf{x}}_{\mathcal{E}}$ direction. This is called the

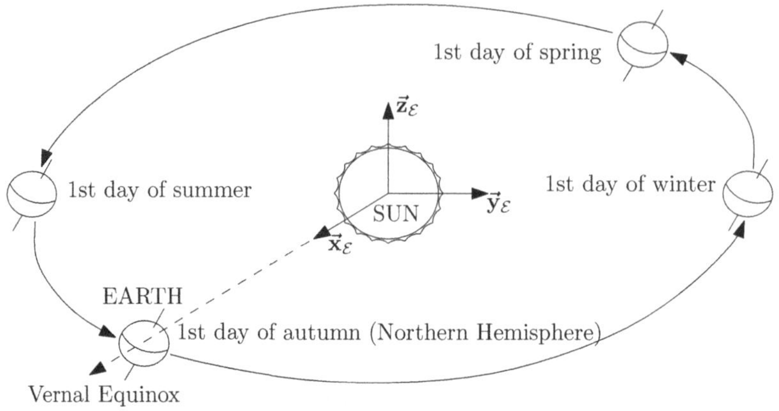

Figure 3.14 Heliocentric-ecliptic coordinate system

vernal equinox direction. The $\vec{\mathbf{y}}_\mathcal{E}$ direction is chosen to complete the right-handed coordinate system, that is $\vec{\mathbf{y}}_\mathcal{E} = \vec{\mathbf{z}}_\mathcal{E} \times \vec{\mathbf{x}}_\mathcal{E}$.

Note that the Earth's equatorial plane is the plane perpendicular to the Earth's axis of rotation. Now, the Earth's axis of rotation is not inertially fixed, but is slowly precessing. Hence, the line of intersection between the equatorial plane and the ecliptic moves. Therefore, the reference frame just defined is not truly inertial. To obtain an inertial frame, we must specify the vernal equinox direction ($\vec{\mathbf{x}}_\mathcal{E}$) at a particular *epoch* (date).

3.6.2 *Geocentric-Equatorial Coordinate System*

As shown in Figure 3.15, the *geocentric-equatorial* system, denoted $\mathcal{F}_G$ is defined by the Earth's axis of rotation. The origin of the system is at the center of the Earth. The fundamental plane (x-y-plane) is the equatorial plane. The $\vec{\mathbf{x}}_G$ direction is chosen to point towards the vernal equinox ($\vec{\mathbf{x}}_G = \vec{\mathbf{x}}_\mathcal{E}$). The $\vec{\mathbf{z}}_G$ direction points towards the north pole. The $\vec{\mathbf{y}}_G$ is chosen to complete the right-handed coordinate system, that is $\vec{\mathbf{y}}_G = \vec{\mathbf{z}}_G \times \vec{\mathbf{x}}_G$. The geocentric-equatorial system is often referred to as the *Earth-Centered Inertial (ECI)* coordinate system. Note that while the origin of the system is at the center of the Earth, the coordinate system does not rotate with the Earth. Since the origin moves with the Earth's orbit around the sun, this system is not truly inertial, however for practical purposes, it is sufficient.

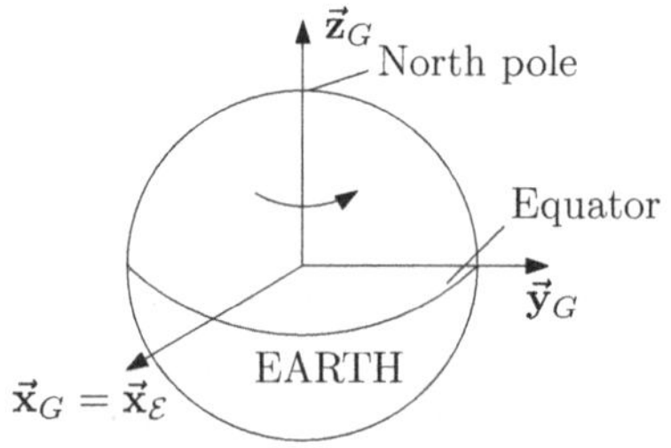

Figure 3.15 Geocentric-equatorial coordinate system

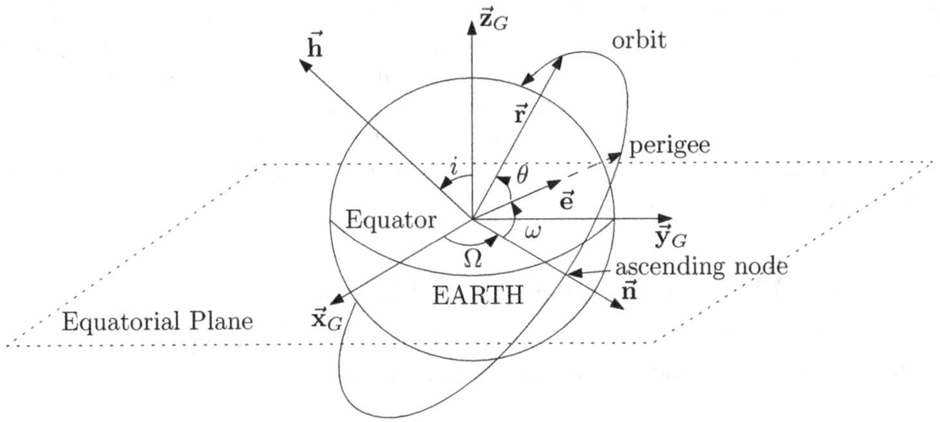

Figure 3.16 Orbital elements

Now that we have an inertial coordinate system for orbits around the Earth, Figure 3.16 shows how to locate the orbit relative to it.

To specify the orbital plane, we can use the angle that the plane makes with the equatorial plane, and the line of intersection between the two planes.

The angle between the two planes is given by the angle between the two positive normals, that is the angle between $\vec{\mathbf{h}}$ and $\vec{\mathbf{z}}_G$. This angle is called the *inclination*, and is denoted i.

The line of intersection between the equatorial and orbital planes is called the *line of nodes*. In particular, the point where the orbit crosses from the negative to positive hemisphere (the orbit's coordinate along $\vec{\mathbf{z}}_G$ goes from negative to positive), is called the *ascending node*. The line of intersection between the two planes can now be specified by the angle between the $\vec{\mathbf{x}}_G$ direction, and the ascending node direction. This angle is called the *right ascension of the ascending node (RAAN)*, and is denoted Ω.

The two angles i and Ω fully specify the orbital plane relative to $\mathcal{F}_G$. As we have discussed, the final quantity that is remaining to fully specify the orbit is the direction of periapsis. This can be described in the orbital plane by the angle between the ascending node, and periapsis. This angle is called the *argument of perigee*, and is denoted by ω. Note that periapsis is the generic term for closest distance of an orbit. When referring to orbits around the Earth, it becomes *perigee*.

Summary

To fully describe the position of an object in a two-body orbit about the Earth, we can use the six quantities:

$$\begin{aligned}
a &- \text{semi-major axis,}\\
e &- \text{eccentricity,}\\
i &- \text{inclination,}\\
\Omega &- \text{right ascension of the ascending node,}\\
\omega &- \text{argument of perigee,}\\
t_0 &- \text{time of perigee passage.}
\end{aligned} \tag{3.75}$$

These six quantities are called the *classical orbital elements*. Even though we have defined them for objects orbiting the Earth, we could similarly define them for objects orbiting other planets or stars (or moons, etc). The classical elements are by no means the only set of quantities that can be used to fully describe the orbit. In fact, there are many other sets of elements. Since the orbit is the solution of a second order differential equation, we could even have specified the orbit in terms of the initial conditions $\vec{\mathbf{r}}_0$ and $\vec{\mathbf{v}}_0$. However, the elements are much more useful from a conceptual point of view.

3.7 Orbital Elements given Position and Velocity

Now that we have defined the orbital elements, we need a means to compute them. Assume that the inertial position $\vec{\mathbf{r}}$ and velocity $\vec{\mathbf{v}}$ are given at time t.

1. The semi-major axis of the orbit, a can be obtained from the orbital energy. Therefore, we compute

$$\mathcal{E} = \frac{\vec{\mathbf{v}} \cdot \vec{\mathbf{v}}}{2} - \frac{\mu}{|\vec{\mathbf{r}}|}.$$

   We can now obtain the semi-major axis from (3.25) as

$$a = -\frac{\mu}{2\mathcal{E}}.$$

2. Next, the eccentricity can be obtained directly from the eccentricity vector (7.77). Therefore, we compute

$$\vec{\mathbf{h}} = \vec{\mathbf{r}} \times \vec{\mathbf{v}}$$
$$\vec{\mathbf{e}} = \frac{\vec{\mathbf{v}} \times \vec{\mathbf{h}}}{\mu} - \frac{\vec{\mathbf{r}}}{|\vec{\mathbf{r}}|},$$
$$e = |\vec{\mathbf{e}}|.$$

3. The inclination can now be obtained from the orbital angular momentum vector $\vec{\mathbf{h}}$ and the $\vec{\mathbf{z}}_G$ axis as follows

$$\vec{\mathbf{h}} \cdot \mathbf{z}_G = |\vec{\mathbf{h}}| \cos i,$$

   which leads to

$$i = \cos^{-1}\left(\frac{\vec{\mathbf{h}} \cdot \vec{\mathbf{z}}_G}{|\vec{\mathbf{h}}|}\right).$$

4. The right ascension of the ascending node can be obtained from the $\vec{\mathbf{x}}_G$ axis and a vector $\vec{\mathbf{n}}$ which points toward the ascending node. From Figure 3.16, such a vector is

$$\vec{\mathbf{n}} = \vec{\mathbf{z}}_G \times \vec{\mathbf{h}}.$$

Making use of this, we have

$$\vec{\mathbf{n}} \cdot \vec{\mathbf{x}}_G = |\vec{\mathbf{n}}| \cos \Omega,$$

which leads to

$$\Omega = \cos^{-1}\left(\frac{\vec{\mathbf{n}} \cdot \vec{\mathbf{x}}_G}{|\vec{\mathbf{n}}|}\right).$$

To determine the quadrant, we see from Figure 3.16, that if

$$\vec{\mathbf{n}} \cdot \vec{\mathbf{y}}_G \geq 0, \quad 0 \leq \Omega \leq 180^\circ,$$

and if

$$\vec{\mathbf{n}} \cdot \vec{\mathbf{y}}_G < 0, \quad 180 < \Omega < 360^\circ.$$

5. The argument of perigee can now be obtained from the vectors $\vec{\mathbf{n}}$ and $\vec{\mathbf{e}}$ as

$$\omega = \cos^{-1}\left(\frac{\vec{\mathbf{n}} \cdot \vec{\mathbf{e}}}{|\vec{\mathbf{n}}| e}\right).$$

To determine the quadrant, we see from Figure 3.16, that if

$$\vec{\mathbf{e}} \cdot \vec{\mathbf{z}}_G \geq 0, \quad 0 \leq \omega \leq 180^\circ,$$

and if

$$\vec{\mathbf{e}} \cdot \vec{\mathbf{z}}_G < 0, \quad 180 < \omega < 360^\circ.$$

6. To find the time of perigee passage, we first need to determine our location on the orbit (true anomaly, θ). We will make use of the argument of latitude, defined as

$$u \triangleq \theta + \omega.$$

That is, the argument of latitude is the angle the position vector $\vec{\mathbf{r}}$ makes with the ascending node vector $\vec{\mathbf{n}}$. Therefore, we have

$$u = \cos^{-1}\left(\frac{\vec{\mathbf{n}} \cdot \vec{\mathbf{r}}}{|\vec{\mathbf{n}}||\vec{\mathbf{r}}|}\right).$$

To determine the quadrant, we see from the diagram that if

$$\vec{\mathbf{r}} \cdot \vec{\mathbf{z}}_G \geq 0, \quad 0 \leq u \leq 180^\circ,$$

and if

$$\vec{\mathbf{r}} \cdot \vec{\mathbf{z}}_G < 0, \quad 180 < u < 360^\circ.$$

Now, we can compute the true anomaly from

$$\theta = u - \omega.$$

Using the time-of-flight equation, we can obtain the time of perigee passage. First, we need to determine the eccentric anomaly. From (3.65), we have

$$E = 2\tan^{-1}\left(\sqrt{\frac{1-e}{1+e}}\tan\frac{\theta}{2}\right).$$

The time of perigee passage can now be obtained from Kepler's equation (3.69) as

$$t_0 = t - (E - e\sin E)\sqrt{\frac{a^3}{\mu}}.$$

Note that in the above computations of the orbital elements, the only thing that matters is that $\vec{\mathbf{r}}$ represents the position relative to the center of the Earth, and that $\vec{\mathbf{v}}$ represents the inertial velocity. The vectors themselves (including $\vec{\mathbf{x}}_G$, $\vec{\mathbf{y}}_G$ and $\vec{\mathbf{z}}_G$) can be expressed in any reference frame, inertial or not, provided all vectors are expressed in the same reference frame.

3.8 Position and Velocity given Orbital Elements

Suppose now that the orbital elements are given, and that we want to obtain the inertial position and velocity at a given time t. To do this, we are going to make use of the perifocal coordinate system, and the transformation between the perifocal reference frame and the Earth-Centered-Inertial (ECI) frame. First, we are going to determine the position and velocity in perifocal coordinates. From (3.30), the position vector in perifocal coordinates is given by

$$\vec{\mathbf{r}} = \vec{\mathcal{F}}_p^T \begin{bmatrix} r\cos\theta \\ r\sin\theta \\ 0 \end{bmatrix},$$

where from (3.18),

$$r = \frac{a(1-e^2)}{1+e\cos\theta}. \tag{3.76}$$

This leads to

$$\vec{\mathbf{r}} = \vec{\mathcal{F}}_p^T \begin{bmatrix} \dfrac{a(1-e^2)\cos\theta}{1+e\cos\theta} \\ \dfrac{a(1-e^2)\sin\theta}{1+e\cos\theta} \\ 0 \end{bmatrix},$$

Taking the inertial time-derivative of the position vector, the inertial velocity is given by

$$\vec{\mathbf{v}} = \vec{\mathcal{F}}_p^T \begin{bmatrix} \dot{r}\cos\theta - r\dot{\theta}\sin\theta \\ \dot{r}\sin\theta + r\dot{\theta}\cos\theta \\ 0 \end{bmatrix}. \tag{3.77}$$

Now, let us differentiate (3.76) to get

$$\dot{r} = \frac{ae(1-e^2)\dot{\theta}\sin\theta}{(1+e\cos\theta)^2}. \tag{3.78}$$

From (3.45), we have

$$\dot{\theta} = \frac{h}{r^2}.$$

By definition of $p = a(1-e^2)$ in (3.17), we have

$$h = \sqrt{\mu a(1-e^2)}.$$

Therefore, we have

$$\dot{\theta} = \frac{\sqrt{\mu a(1-e^2)}}{r^2} = \frac{\sqrt{\mu a(1-e^2)}(1+e\cos\theta)^2}{a^2(1-e^2)^2},$$

which leads to

$$\dot{\theta} = \sqrt{\frac{\mu}{a^3}}\frac{(1+e\cos\theta)^2}{(1-e^2)^{\frac{3}{2}}}. \tag{3.79}$$

Substituting this into (3.78) gives

$$\dot{r} = \sqrt{\frac{\mu}{a(1-e^2)}}e\sin\theta. \tag{3.80}$$

Multiplying (3.76) with (3.79), we have

$$r\dot{\theta} = \sqrt{\frac{\mu}{a(1-e^2)}}(1 + e\cos\theta). \tag{3.81}$$

Substituting (7.18) and (7.19) into (3.77) gives

$$\vec{\mathbf{v}} = \vec{\mathcal{F}}_p^T \begin{bmatrix} -\sqrt{\frac{\mu}{a(1-e^2)}}\sin\theta \\ \sqrt{\frac{\mu}{a(1-e^2)}}(e + \cos\theta) \\ 0 \end{bmatrix}. \tag{3.82}$$

We have now found the position and inertial velocity in perifocal coordinates. However, we ultimately want to obtain them in ECI coordinates. That is, we seek $\mathbf{r}_G$ and $\mathbf{v}_G$ such that $\vec{\mathbf{r}} = \vec{\mathcal{F}}_G^T \mathbf{r}_G$ and $\vec{\mathbf{v}} = \vec{\mathcal{F}}_G^T \mathbf{v}_G$. As we have seen in Section 1.3, this transformation is given by

$$\mathbf{r}_G = \mathbf{C}_{Gp} \begin{bmatrix} \frac{a(1-e^2)\cos\theta}{1+e\cos\theta} \\ \frac{a(1-e^2)\sin\theta}{1+e\cos\theta} \\ 0 \end{bmatrix},$$

and

$$\mathbf{v}_G = \mathbf{C}_{Gp} \begin{bmatrix} -\sqrt{\frac{\mu}{a(1-e^2)}}\sin\theta \\ \sqrt{\frac{\mu}{a(1-e^2)}}(e + \cos\theta) \\ 0 \end{bmatrix}.$$

Therefore, all we need to do is find the transformation from perifocal to inertial coordinates.

As shown in Figure 3.17, it can be seen that the perifocal coordinate system is obtained by three successive principal rotations from the ECI frame (refer also to Figure 3.16 to see why).

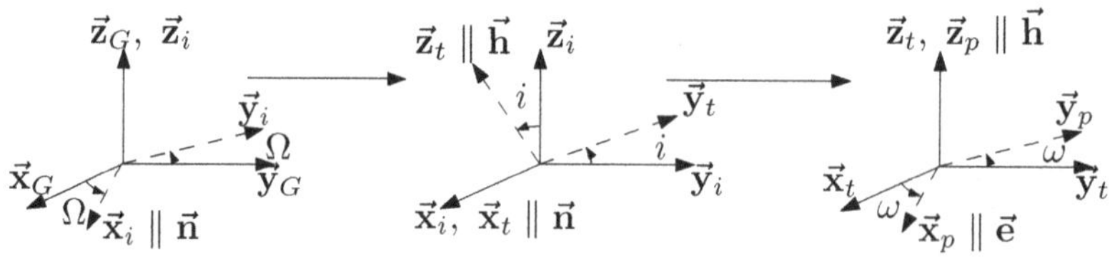

Figure 3.17 Earth-centered-inertial to perifocal transformation

The first is a rotation about the $\vec{\mathbf{z}}_G$ axis through an angle of Ω. The second is a rotation about the intermediate $\vec{\mathbf{x}}$ axis through an angle i. The third is a rotation about the transformed $\vec{\mathbf{z}}$ axis through and angle ω. Mathematically, this transformation is expressed by the rotation matrix

$$\mathbf{C}_{pG} = \mathbf{C}_z(\omega)\mathbf{C}_x(i)\mathbf{C}_z(\Omega).$$

However, we want the inverse transformation. This is just the transpose, given by

$$\mathbf{C}_{Gp} = \mathbf{C}_{pG}^T = \mathbf{C}_z^T(\Omega)\mathbf{C}_x^T(i)\mathbf{C}_z^T(\omega).$$

Making use of the principal rotation matrices in Section 1.3.1, this becomes

$$\mathbf{C}_{Gp} = \begin{bmatrix} c_\Omega c_\omega - s_\Omega c_i s_\omega & -c_\Omega s_\omega - s_\Omega c_i c_\omega & s_\Omega s_i \\ s_\Omega c_\omega + c_\Omega c_i s_\omega & -s_\Omega s_\omega + c_\Omega c_i c_\omega & -c_\Omega s_i \\ s_i s_\omega & s_i c_\omega & c_i \end{bmatrix}, \tag{3.83}$$

where $c_x = \cos x$ and $s_x = \sin x$.

Summary

To summarize, to obtain the position and velocity at a time t, given the orbital elements, first the true anomaly is computed using Kepler's equation (3.69) and (3.65). That is,

$$\begin{aligned} M &= n(t - t_0), \\ E - e\sin E &= M, \\ \theta &= 2\tan^{-1}\left(\sqrt{\frac{1+e}{1-e}}\tan\frac{E}{2}\right). \end{aligned}$$

The position and inertial velocity coordinates in the ECI frame are computed as

$$\mathbf{r}_G = \begin{bmatrix} c_\Omega c_\omega - s_\Omega c_i s_\omega & -c_\Omega s_\omega - s_\Omega c_i c_\omega & s_\Omega s_i \\ s_\Omega c_\omega + c_\Omega c_i s_\omega & -s_\Omega s_\omega + c_\Omega c_i c_\omega & -c_\Omega s_i \\ s_i s_\omega & s_i c_\omega & c_i \end{bmatrix} \begin{bmatrix} \dfrac{a(1-e^2)\cos\theta}{1+e\cos\theta} \\ \dfrac{a(1-e^2)\sin\theta}{1+e\cos\theta} \\ 0 \end{bmatrix},$$

$$\mathbf{v}_G = \begin{bmatrix} c_\Omega c_\omega - s_\Omega c_i s_\omega & -c_\Omega s_\omega - s_\Omega c_i c_\omega & s_\Omega s_i \\ s_\Omega c_\omega + c_\Omega c_i s_\omega & -s_\Omega s_\omega + c_\Omega c_i c_\omega & -c_\Omega s_i \\ s_i s_\omega & s_i c_\omega & c_i \end{bmatrix} \begin{bmatrix} -\sqrt{\dfrac{\mu}{a(1-e^2)}}\sin\theta \\ \sqrt{\dfrac{\mu}{a(1-e^2)}}(e+\cos\theta) \\ 0 \end{bmatrix}.$$

Important Remarks

In some cases, not all orbital elements are defined. For example, for a circular orbit, there is no periapsis (since r is constant). Therefore, ω is undefined. In this case, the argument of latitude u (the angle from the ascending node) can be used instead of the true anomaly. For elliptical equatorial orbits, there is no line of nodes (the orbital and equatorial planes coincide). In this case, Ω is undefined, and the longitude of periapsis, Π (the angle between periapsis and the vernal equinox) is used instead of ω. For an equatorial circular orbit, there is no line of nodes, and no periapsis. In this case, Ω and ω are undefined, as is Π. The true longitude, l (the angle from the vernal equinox) may be used instead of the true anomaly.

Notes

In this chapter, we have presented the two-body problem, and its solution. A more comprehensive treatment of the two-body problem may be found in Battin (1999).

We have introduced the concept of classical orbital elements in order to represent the orbit. However, there are other sets of elements that can also be used. The reader is referred to Vallado (2007) for more details. In presenting the orbital elements, we have loosely defined two coordinate systems, the Heliocentric-Ecliptic system and the Geocentric-Equatorial system. In addition, there are other important coordinate systems used in astrodynamics. In this chapter, we have not discussed different time systems used in astrodynamics. However, since we are interested in describing orbital motion as a function of time, it is important that the correct time system be used (a time system in which the orbital equations of motion are valid). For a more detailed presentation of different coordinate and time systems used in astrodynamics, the reader may wish to consult Vallado (2007), Montenbruck and Gill (2000) or Seidelmann (2006).

References

Battin RH 1999 *An Introduction to the Mathematics and Methods of Astrodynamics, Revised Edition.* American Institute of Aeronautics and Astronautics, Reston, VA.

Montenbruck O and Gill E 2000 *Satellite Orbits: Models, Methods, Applications.* Springer, Berlin Heidelberg.

Seidelmann PK 2006 *Explanatory Supplement to the Astronomical Almanac.* University Science Books.

Vallado DA 2007 *Fundamentals of Astrodynamics and Applications* 3rd edn. Microcosm Press, Hawthorne, CA.

4

Preliminary Orbit Determination

Now that we know how to describe the motion of an object in orbit, a significant question is how do we actually determine it. There are very sophisticated methods of orbit determination, based upon statistical approaches, in fact, entire books have been written about the subject. However, all of these sophisticated methods are based upon correcting preliminary orbit estimates. Therefore, methods of initially determining the orbit of an object are relevant. There are several methods (and variations upon these methods) to accomplish this, and they depend upon the measurements that are available. We shall examine three methods.

4.1 Orbit Determination from Three Position Vectors

Suppose we are given three successive position vectors, $\vec{\mathbf{r}}_1$, $\vec{\mathbf{r}}_2$ and $\vec{\mathbf{r}}_3$, as shown in Figure 4.1, and that the time for at least one of them is given, that is we have t_i for some $i = 1, 2, 3$. Since the three vectors lie on the same orbit, they must be coplanar. The objective is to obtain the orbital elements from this information. There are different solutions to this problem. The one we are going to examine is a vector-based method, most suitable for automated computer implementation, called Gibb's method. We shall assume that the true anomalies θ_1 and θ_3 corresponding to $\vec{\mathbf{r}}_1$ and $\vec{\mathbf{r}}_3$ respectively, satisfy $\theta_3 - \theta_1 < 180°$. From Section 3.7, we know that if we have the inertial velocity corresponding to one of these position vectors, then we have completely determined the orbit, and the orbital elements can be computed. This is what we are going to do. We are going to find the velocity vector $\vec{\mathbf{v}}_i$ corresponding to the position $\vec{\mathbf{r}}_i$ that we have a time fix t_i for.

The first point we are going to note is that since the position vectors are coplanar, we can obtain any one of them as a linear combination of the other two (provided the other two are not parallel), that is

$$\vec{\mathbf{r}}_2 = \alpha\vec{\mathbf{r}}_1 + \beta\vec{\mathbf{r}}_3, \tag{4.1}$$

provided $\vec{\mathbf{r}}_1$ and $\vec{\mathbf{r}}_3$ are not parallel.

Spacecraft Dynamics and Control: An Introduction, First Edition.
Anton H.J. de Ruiter, Christopher J. Damaren and James R. Forbes.

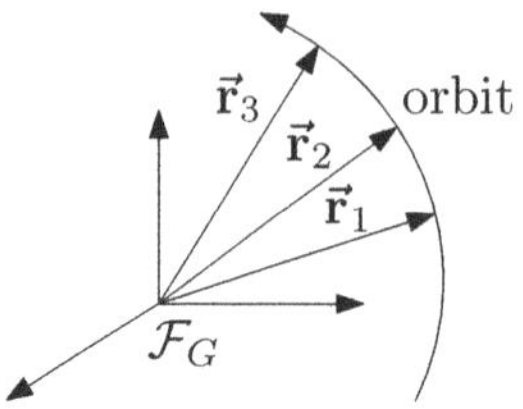

Figure 4.1 Orbit determination from three position vectors

We now determine the scalars α and β. Taking the cross-product of (4.1) with $\vec{\mathbf{r}}_1$ and $\vec{\mathbf{r}}_3$ separately, we find

$$\vec{\mathbf{r}}_1 \times \vec{\mathbf{r}}_2 = \alpha \vec{\mathbf{r}}_1 \times \vec{\mathbf{r}}_1 + \beta \vec{\mathbf{r}}_1 \times \vec{\mathbf{r}}_3 = \beta \vec{\mathbf{r}}_1 \times \vec{\mathbf{r}}_3, \tag{4.2}$$

$$\vec{\mathbf{r}}_3 \times \vec{\mathbf{r}}_2 = \alpha \vec{\mathbf{r}}_3 \times \vec{\mathbf{r}}_1 + \beta \vec{\mathbf{r}}_3 \times \vec{\mathbf{r}}_3 = \alpha \vec{\mathbf{r}}_3 \times \vec{\mathbf{r}}_1. \tag{4.3}$$

Since $\vec{\mathbf{r}}_1$, $\vec{\mathbf{r}}_2$ and $\vec{\mathbf{r}}_3$ are coplanar, the vectors $\vec{\mathbf{r}}_1 \times \vec{\mathbf{r}}_2$, $\vec{\mathbf{r}}_1 \times \vec{\mathbf{r}}_3$ and $\vec{\mathbf{r}}_2 \times \vec{\mathbf{r}}_3$ are parallel. Let us define

$$\vec{\mathbf{n}} = \vec{\mathbf{r}}_1 \times \vec{\mathbf{r}}_3, \quad n = \sqrt{\vec{\mathbf{n}} \cdot \vec{\mathbf{n}}}, \tag{4.4}$$

as shown in Figure 4.2. Taking the dot product of $\vec{\mathbf{n}}$ with each of (4.2) and (4.3), we have

$$\beta = \frac{\vec{\mathbf{n}} \cdot (\vec{\mathbf{r}}_1 \times \vec{\mathbf{r}}_2)}{n^2}, \tag{4.5}$$

$$\alpha = \frac{\vec{\mathbf{n}} \cdot (\vec{\mathbf{r}}_2 \times \vec{\mathbf{r}}_3)}{n^2}. \tag{4.6}$$

Now, we know from Section 3.3, that for any point on the orbit,

$$\vec{\mathbf{r}} \cdot \vec{\mathbf{e}} = \frac{h^2}{\mu} - r, \tag{4.7}$$

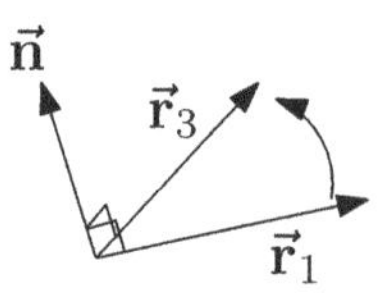

Figure 4.2 Orbit normal vector

where $r = \sqrt{\vec{\mathbf{r}} \cdot \vec{\mathbf{r}}}$, and h is the magnitude of the orbital angular momentum. Therefore, taking the dot product of (4.1) with the eccentricity vector $\vec{\mathbf{e}}$, gives

$$\frac{h^2}{\mu} - r_2 = \alpha \left(\frac{h^2}{\mu} - r_1 \right) + \beta \left(\frac{h^2}{\mu} - r_3 \right),$$

which can be rearranged to give

$$\frac{h^2}{\mu}(1 - \alpha - \beta) = r_2 - \alpha r_1 - \beta r_3,$$

leading to

$$h = \sqrt{\frac{\mu(r_2 - \alpha r_1 - \beta r_3)}{1 - \alpha - \beta}}. \tag{4.8}$$

Now, since the vectors $\vec{\mathbf{r}}_1$ and $\vec{\mathbf{r}}_3$ are successive ($\vec{\mathbf{r}}_3$ follows $\vec{\mathbf{r}}_1$), and lie in the orbital plane, the orbital angular momentum vector $\vec{\mathbf{h}}$ must be parallel to the vector $\vec{\mathbf{n}}$, and point in the same direction (since $\theta_3 - \theta_1 < 180°$). Therefore, the orbital angular momentum vector is given by

$$\vec{\mathbf{h}} = h \frac{\vec{\mathbf{n}}}{n}, \tag{4.9}$$

and using (4.4) and (4.8) we can compute the orbit normal vector.

Let us now examine the vector cross product of $\vec{\mathbf{n}}$ with $\vec{\mathbf{e}}$,

$$\vec{\mathbf{n}} \times \vec{\mathbf{e}} = (\vec{\mathbf{r}}_1 \times \vec{\mathbf{r}}_3) \times \vec{\mathbf{e}}.$$

Making use of the vector triple product identity in Section 1.1.3, this becomes

$$\vec{\mathbf{n}} \times \vec{\mathbf{e}} = -\left(\vec{\mathbf{e}} \cdot \vec{\mathbf{r}}_3\right) \vec{\mathbf{r}}_1 + \left(\vec{\mathbf{e}} \cdot \vec{\mathbf{r}}_1\right) \vec{\mathbf{r}}_3.$$

Making use of (4.7), we have

$$\vec{\mathbf{n}} \times \vec{\mathbf{e}} = \left(\frac{h^2}{\mu} - r_1 \right) \vec{\mathbf{r}}_3 - \left(\frac{h^2}{\mu} - r_3 \right) \vec{\mathbf{r}}_1. \tag{4.10}$$

Next, taking the vector cross product of $\vec{\mathbf{n}} \times \vec{\mathbf{e}}$ with $\vec{\mathbf{n}}$, and making use of the vector triple product identity, we have

$$(\vec{\mathbf{n}} \times \vec{\mathbf{e}}) \times \vec{\mathbf{n}} = -(\vec{\mathbf{n}} \cdot \vec{\mathbf{e}})\vec{\mathbf{n}} + (\vec{\mathbf{n}} \cdot \vec{\mathbf{n}})\vec{\mathbf{e}} = n^2 \vec{\mathbf{e}}, \tag{4.11}$$

where the fact that $\vec{\mathbf{n}} \cdot \vec{\mathbf{e}} = 0$ has been used, since the eccentricity vector $\vec{\mathbf{e}}$ lies in the orbital plane. Combining (4.10) and (4.11), we obtain the eccentricity vector $\vec{\mathbf{e}}$,

$$\vec{\mathbf{e}} = \frac{1}{n^2}\left[\left(\frac{h^2}{\mu} - r_1\right)\vec{\mathbf{r}}_3 \times \vec{\mathbf{n}} - \left(\frac{h^2}{\mu} - r_3\right)\vec{\mathbf{r}}_1 \times \vec{\mathbf{n}}\right]. \tag{4.12}$$

Finally, let us obtain an expression for the velocity vector, $\vec{\mathbf{v}}$. Recall from Section 3.2.3 that the eccentricity vector is defined as

$$\vec{\mathbf{e}} \overset{\Delta}{=} \frac{\vec{\mathbf{v}} \times \vec{\mathbf{h}}}{\mu} - \frac{\vec{\mathbf{r}}}{r}.$$

Taking the vector cross-product with the angular momentum vector $\vec{\mathbf{h}}$ leads to

$$\vec{\mathbf{h}} \times \vec{\mathbf{e}} = \frac{\vec{\mathbf{h}} \times (\vec{\mathbf{v}} \times \vec{\mathbf{h}})}{\mu} - \frac{\vec{\mathbf{h}} \times \vec{\mathbf{r}}}{r}. \tag{4.13}$$

Making use of the vector triple product identity in Section 1.1.3, we have

$$\vec{\mathbf{h}} \times (\vec{\mathbf{v}} \times \vec{\mathbf{h}}) = (\vec{\mathbf{h}} \cdot \vec{\mathbf{h}})\vec{\mathbf{v}} - (\vec{\mathbf{h}} \cdot \vec{\mathbf{v}})\vec{\mathbf{h}} = h^2\vec{\mathbf{v}},$$

since $\vec{\mathbf{h}} \cdot \vec{\mathbf{v}} = 0$. Substituting this into (4.13) gives

$$\vec{\mathbf{h}} \times \vec{\mathbf{e}} = \frac{h^2}{\mu}\vec{\mathbf{v}} - \frac{\vec{\mathbf{h}} \times \vec{\mathbf{r}}}{r},$$

which leads to

$$\vec{\mathbf{v}} = \frac{\mu}{h^2}\vec{\mathbf{h}} \times \left(\vec{\mathbf{e}} + \frac{\vec{\mathbf{r}}}{r}\right). \tag{4.14}$$

Since we have already obtained $\vec{\mathbf{h}}$ and $\vec{\mathbf{e}}$, we can compute $\vec{\mathbf{v}}_i$ corresponding to any of the $\vec{\mathbf{r}}_i$.

Given $\vec{\mathbf{r}}_i$ and $\vec{\mathbf{v}}_i$ with corresponding time t_i, the orbital elements can be obtained from Section 3.7.

Summary

Given are position vectors $\vec{\mathbf{r}}_1, \vec{\mathbf{r}}_2, \vec{\mathbf{r}}_3$, and a time t_i for some $i = 1, 2, 3$. To obtain the orbital elements:

1. Compute $r_1 = \sqrt{\vec{\mathbf{r}}_1 \cdot \vec{\mathbf{r}}_1}$, $r_2 = \sqrt{\vec{\mathbf{r}}_2 \cdot \vec{\mathbf{r}}_2}$ and $r_3 = \sqrt{\vec{\mathbf{r}}_3 \cdot \vec{\mathbf{r}}_3}$.
2. Compute $\vec{\mathbf{n}} = \vec{\mathbf{r}}_1 \times \vec{\mathbf{r}}_3$ and $n = \sqrt{\vec{\mathbf{n}} \cdot \vec{\mathbf{n}}}$.
3. Compute α and β

$$\alpha = \frac{\vec{\mathbf{n}} \cdot (\vec{\mathbf{r}}_2 \times \vec{\mathbf{r}}_3)}{n^2}, \quad \beta = \frac{\vec{\mathbf{n}} \cdot (\vec{\mathbf{r}}_1 \times \vec{\mathbf{r}}_2)}{n^2}.$$

4. Compute the angular momentum vector

$$h = \sqrt{\frac{\mu(r_2 - \alpha r_1 - \beta r_3)}{1 - \alpha - \beta}}, \quad \vec{\mathbf{h}} = h\frac{\vec{\mathbf{n}}}{n}.$$

5. Compute the eccentricity vector

$$\vec{\mathbf{e}} = \frac{1}{n^2}\left[\left(\frac{h^2}{\mu} - r_1\right)\vec{\mathbf{r}}_3 \times \vec{\mathbf{n}} - \left(\frac{h^2}{\mu} - r_3\right)\vec{\mathbf{r}}_1 \times \vec{\mathbf{n}}\right].$$

6. Compute the velocity $\vec{\mathbf{v}}_i$ corresponding to time t_i

$$\vec{\mathbf{v}}_i = \frac{\mu}{h^2}\vec{\mathbf{h}} \times \left(\vec{\mathbf{e}} + \frac{\vec{\mathbf{r}}_i}{r_i}\right)$$

7. Using $\vec{\mathbf{r}}_i$, $\vec{\mathbf{v}}_i$ and t_i, compute the orbital elements.

4.2 Orbit Determination from Three Line-of-Sight Vectors

Suppose we only have optical measurements available. In this case, we can only obtain the direction of the object (whose orbit we are trying to determine) from our observation location, but not the range (distance). Therefore, we cannot compute the position vector of the object. We must determine the orbit on the basis of line-of-sight vectors only. This is a very old problem, with famous solutions by Laplace and Gauss, among others, who developed their methods to determine the orbits of celestial objects. Gauss famously used his method in 1801 to obtain the orbit of the minor planet Ceres, and successfully predicted the date and location of its next sighting.

There are several methods for orbit determination from line-of-sight measurements, each of which has their own merits and limitations. Because these methods are each more suitable for different situations, someone working in the field requires knowledge of them all. For our purposes in this book, we are going to restrict ourselves to a simplified Gauss' method. As we shall see, this method is more suitable when the observations are relatively closely spaced.

Let $\vec{\mathbf{R}}_i$ be the (known) position of the observer, $\vec{\ell}_i$ the (measured) unit line-of-sight vector, and ρ_i the (unknown) range (distance to the object) at times t_i for $i = 1, 2, 3$, where $t_1 < t_2 < t_3$ (see Figure 4.3). The object orbital position at time t_i is $\vec{\mathbf{r}}_i$. For Earth orbiting objects, $\vec{\mathbf{R}}_i$ is typically the location of the observer relative to the center of the Earth (it is a point on the surface of the Earth). For objects orbiting the sun, $\vec{\mathbf{R}}_i$ is typically the orbital position of the Earth relative to the sun.

We shall now assume that the true anomalies θ_1 and θ_3 corresponding to $\vec{\mathbf{r}}_1$ and $\vec{\mathbf{r}}_3$ respectively, satisfy $\theta_3 - \theta_1 < 90°$, and that the orbit is elliptical. The orbital position of the satellite is clearly given by

$$\vec{\mathbf{r}}_i = \vec{\mathbf{R}}_i + \rho_i \vec{\ell}_i. \tag{4.15}$$

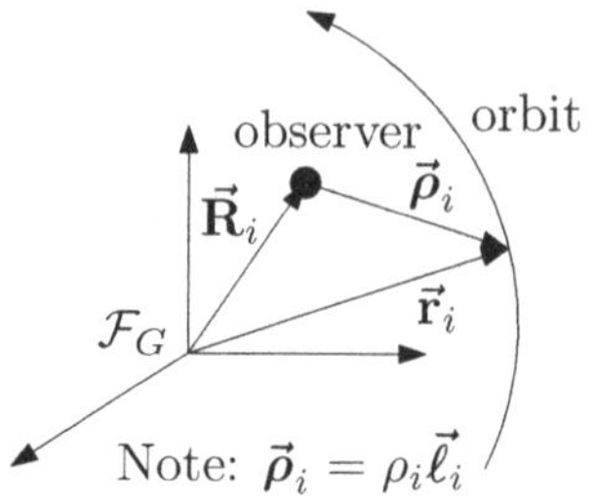

Figure 4.3 Orbit determination from three line-of-sight vectors

As in Section 4.1, because the object's orbit lies in a plane, we can write

$$\vec{\mathbf{r}}_2 = \alpha \vec{\mathbf{r}}_1 + \beta \vec{\mathbf{r}}_3. \tag{4.16}$$

Substituting this into (4.15) leads to

$$\vec{\mathbf{R}}_2 + \rho_2 \vec{\ell}_2 = \alpha(\vec{\mathbf{R}}_1 + \rho_1 \vec{\ell}_1) + \beta(\vec{\mathbf{R}}_3 + \rho_3 \vec{\ell}_3),$$

which rearranges as

$$\rho_2 \vec{\ell}_2 - \alpha \rho_1 \vec{\ell}_1 - \beta \rho_3 \vec{\ell}_3 = \alpha \vec{\mathbf{R}}_1 + \beta \vec{\mathbf{R}}_3 - \vec{\mathbf{R}}_2. \tag{4.17}$$

We now define the vectors

$$\vec{\mathbf{d}}_1 = \vec{\ell}_2 \times \vec{\ell}_3, \quad \vec{\mathbf{d}}_2 = \vec{\ell}_3 \times \vec{\ell}_1, \quad \vec{\mathbf{d}}_3 = \vec{\ell}_1 \times \vec{\ell}_2, \tag{4.18}$$

We note that

$$\begin{aligned} \vec{\mathbf{d}}_1 \cdot \vec{\ell}_2 &= 0, & \vec{\mathbf{d}}_1 \cdot \vec{\ell}_3 &= 0, \\ \vec{\mathbf{d}}_2 \cdot \vec{\ell}_1 &= 0, & \vec{\mathbf{d}}_2 \cdot \vec{\ell}_3 &= 0, \\ \vec{\mathbf{d}}_3 \cdot \vec{\ell}_1 &= 0, & \vec{\mathbf{d}}_3 \cdot \vec{\ell}_2 &= 0. \end{aligned}$$

Taking the dot product of each of $\vec{\mathbf{d}}_1$, $\vec{\mathbf{d}}_2$, $\vec{\mathbf{d}}_3$ with (4.17), leads to

$$\begin{aligned} -\alpha \rho_1 \vec{\mathbf{d}}_1 \cdot \vec{\ell}_1 &= \left[\alpha \vec{\mathbf{R}}_1 + \beta \vec{\mathbf{R}}_3 - \vec{\mathbf{R}}_2\right] \cdot \vec{\mathbf{d}}_1, \\ \rho_2 \vec{\mathbf{d}}_2 \cdot \vec{\ell}_2 &= \left[\alpha \vec{\mathbf{R}}_1 + \beta \vec{\mathbf{R}}_3 - \vec{\mathbf{R}}_2\right] \cdot \vec{\mathbf{d}}_2, \\ -\beta \rho_3 \vec{\mathbf{d}}_3 \cdot \vec{\ell}_3 &= \left[\alpha \vec{\mathbf{R}}_1 + \beta \vec{\mathbf{R}}_3 - \vec{\mathbf{R}}_2\right] \cdot \vec{\mathbf{d}}_3. \end{aligned} \tag{4.19}$$

Now, making use of the identity for the scalar triple product in Section 1.1.3, we find that $\vec{\mathbf{d}}_1 \cdot \vec{\mathbf{l}}_1 = \vec{\mathbf{d}}_2 \cdot \vec{\mathbf{l}}_2 = \vec{\mathbf{d}}_3 \cdot \vec{\mathbf{l}}_3$ Let us call this quantity D, that is

$$D = \vec{\mathbf{d}}_1 \cdot \vec{\mathbf{l}}_1 = (\vec{\mathbf{l}}_2 \times \vec{\mathbf{l}}_3) \cdot \vec{\mathbf{l}}_1.$$

Therefore, we can rewrite (4.19) as

$$\begin{aligned}
\rho_1 &= -\frac{1}{\alpha D}\left[\alpha \vec{\mathbf{R}}_1 \cdot \vec{\mathbf{d}}_1 + \beta \vec{\mathbf{R}}_3 \cdot \vec{\mathbf{d}}_1 - \vec{\mathbf{R}}_2 \cdot \vec{\mathbf{d}}_1\right], \\
\rho_2 &= \frac{1}{D}\left[\alpha \vec{\mathbf{R}}_1 \cdot \vec{\mathbf{d}}_2 + \beta \vec{\mathbf{R}}_3 \cdot \vec{\mathbf{d}}_2 - \vec{\mathbf{R}}_2 \cdot \vec{\mathbf{d}}_2\right], \\
\rho_3 &= -\frac{1}{\beta D}\left[\alpha \vec{\mathbf{R}}_1 \cdot \vec{\mathbf{d}}_3 + \beta \vec{\mathbf{R}}_3 \cdot \vec{\mathbf{d}}_3 - \vec{\mathbf{R}}_2 \cdot \vec{\mathbf{d}}_3\right].
\end{aligned} \tag{4.20}$$

That is, the ranges ρ_1, ρ_2 and ρ_3 are functions of α and β. By imposing the coplanar condition (4.16), we have reduced the number of unknowns from three down to two.

Now, taking the cross-product of (4.16) with $\vec{\mathbf{r}}_1$ and $\vec{\mathbf{r}}_3$, we have

$$\begin{aligned}
\vec{\mathbf{r}}_1 \times \vec{\mathbf{r}}_2 &= \beta \vec{\mathbf{r}}_1 \times \vec{\mathbf{r}}_3, \\
\vec{\mathbf{r}}_2 \times \vec{\mathbf{r}}_3 &= \alpha \vec{\mathbf{r}}_1 \times \vec{\mathbf{r}}_3.
\end{aligned} \tag{4.21}$$

Since the vectors $\vec{\mathbf{r}}_1$, $\vec{\mathbf{r}}_2$ and $\vec{\mathbf{r}}_3$ are coplanar and successive, the vectors $\vec{\mathbf{r}}_1 \times \vec{\mathbf{r}}_2$, $\vec{\mathbf{r}}_1 \times \vec{\mathbf{r}}_3$ and $\vec{\mathbf{r}}_2 \times \vec{\mathbf{r}}_3$ are parallel and have the same direction (since $\theta_3 - \theta_1 < 90°$). Therefore, we can obtain the following expressions from (4.21)

$$\alpha = \frac{|\vec{\mathbf{r}}_2 \times \vec{\mathbf{r}}_3|}{|\vec{\mathbf{r}}_1 \times \vec{\mathbf{r}}_3|}, \quad \beta = \frac{|\vec{\mathbf{r}}_1 \times \vec{\mathbf{r}}_2|}{|\vec{\mathbf{r}}_1 \times \vec{\mathbf{r}}_3|}. \tag{4.22}$$

Let us take a look at what $|\vec{\mathbf{a}} \times \vec{\mathbf{b}}|$ physically means. From the definition of the cross-product in Section 1.1.2, we have that $|\vec{\mathbf{a}} \times \vec{\mathbf{b}}| = |\vec{\mathbf{a}}||\vec{\mathbf{b}}| \sin\theta$, where θ is the angle between the two vectors. Figure 4.4 shows a graphical interpretation. The area of the triangle is given by $\frac{1}{2}|\vec{\mathbf{a}}||\vec{\mathbf{b}}| \sin\theta = \frac{1}{2}|\vec{\mathbf{a}} \times \vec{\mathbf{b}}|$. That is, $|\vec{\mathbf{a}} \times \vec{\mathbf{b}}|$ is equal to twice the area of the triangle enclosed by the two vectors $\vec{\mathbf{a}}$ and $\vec{\mathbf{b}}$.

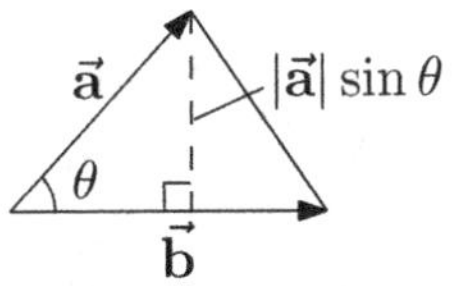

Figure 4.4 Triangle geometry

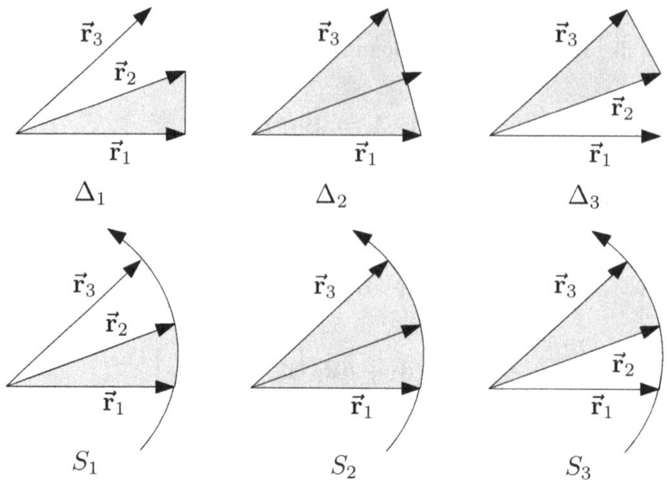

Figure 4.5 Triangle and sector areas

Referring to Figure 4.5, let us now define the areas of the three triangles generated by $\vec{\mathbf{r}}_1$, $\vec{\mathbf{r}}_2$ and $\vec{\mathbf{r}}_3$ as

$$\Delta_1 = \frac{1}{2}|\vec{\mathbf{r}}_1 \times \vec{\mathbf{r}}_2|, \quad \Delta_2 = \frac{1}{2}|\vec{\mathbf{r}}_1 \times \vec{\mathbf{r}}_3|, \quad \Delta_3 = \frac{1}{2}|\vec{\mathbf{r}}_2 \times \vec{\mathbf{r}}_3|. \tag{4.23}$$

Then, from (4.22), we see that α and β are just the ratios of these triangular areas, that is

$$\alpha = \frac{\Delta_3}{\Delta_2}, \quad \beta = \frac{\Delta_1}{\Delta_2}. \tag{4.24}$$

Related to the areas of the triangles are the areas of the sectors of the orbit generated by the vectors $\vec{\mathbf{r}}_1$, $\vec{\mathbf{r}}_2$ and $\vec{\mathbf{r}}_3$, denoted S_1, S_2 and S_3, as shown in Figure 4.5. The areas of the sectors can be obtained from Kepler's 2nd law as

$$S_1 = \frac{h}{2}(t_2 - t_1), \quad S_2 = \frac{h}{2}(t_3 - t_1), \quad S_3 = \frac{h}{2}(t_3 - t_2), \tag{4.25}$$

where h is the orbital angular momentum. The ratio of the sector to triangle areas, denoted by η_i is given by

$$\eta_1 = \frac{S_1}{\Delta_1} = \frac{h(t_2 - t_1)}{|\vec{\mathbf{r}}_1 \times \vec{\mathbf{r}}_2|},$$
$$\eta_2 = \frac{S_2}{\Delta_2} = \frac{h(t_3 - t_1)}{|\vec{\mathbf{r}}_1 \times \vec{\mathbf{r}}_3|},$$
$$\eta_3 = \frac{S_3}{\Delta_3} = \frac{h(t_3 - t_2)}{|\vec{\mathbf{r}}_2 \times \vec{\mathbf{r}}_3|}.$$

Therefore, the constants α and β may be expressed as

$$\alpha = \frac{\eta_2 S_3}{\eta_3 S_2} = \frac{\eta_2(t_3 - t_2)}{\eta_3(t_3 - t_1)}, \quad \beta = \frac{\eta_2 S_1}{\eta_1 S_2} = \frac{\eta_2(t_2 - t_1)}{\eta_1(t_3 - t_1)}. \tag{4.26}$$

Since the measurement times are known, if we can determine the sector to triangle area ratios, we can compute α and β, from which we can obtain ρ_1, ρ_2 and ρ_3 and determine the orbit.

It can be shown that the sector to triangle ratio for a two-body elliptical orbit between the position vectors $\vec{\mathbf{r}}_a$ and $\vec{\mathbf{r}}_b$ satisfies

$$\eta = 1 + \frac{m}{\eta^2} W\left(\frac{m}{\eta^2} - l\right), \tag{4.27}$$

where

$$m = \frac{\mu(t_b - t_a)^2}{\left[2\sqrt{r_a r_b}\cos((\theta_b - \theta_a)/2)\right]^3}, \quad l = \frac{r_a + r_b}{4\sqrt{r_a r_b}\cos((\theta_b - \theta_a)/2)} - \frac{1}{2},$$

and

$$W(w) = \frac{2g - \sin(2g)}{\sin^3 g}, \quad g = 2\sin^{-1}\sqrt{w}.$$

This is a transcendental equation for η, and must be solved iteratively. One such method is the secant procedure, given by

$$\eta_{i+1} = \eta_i - f(\eta_i)\frac{\eta_i - \eta_{i-1}}{f(\eta_i) - f(\eta_{i-1})},$$

where

$$f(x) = 1 - x + \frac{m}{x^2} W\left(\frac{m}{x^2} - l\right).$$

Suitable starting values for this iteration are

$$\eta_1 = \eta_H + 0.1, \quad \eta_2 = \eta_H,$$

where η_H is known as the Hansen Approximation

$$\eta_H = \frac{12}{22} + \frac{10}{22}\sqrt{1 + \frac{44}{9}\frac{m}{l + 5/6}}.$$

Now, for closely spaced measurements, the sector to triangle ratio is approximately one ($\eta_i \approx 1$, see Figure 4.5). Therefore, as a first approximation, we may take

$$\alpha = \frac{\eta_2 S_3}{\eta_3 S_2} \approx \frac{t_3 - t_2}{t_3 - t_1}, \quad \beta = \frac{\eta_2 S_1}{\eta_1 S_2} \approx \frac{t_2 - t_1}{t_3 - t_1}. \tag{4.28}$$

We can now substitute these approximations into (4.20) to obtain approximate ranges. Approximate position vectors are then obtained using (4.15). Now, we can iterate to improve our solution. With the computed position vectors, we can determine new sector triangle area ratios using the above method. Using these, we can compute new values for α and β from (4.28), and so on. Once our iterations have converged, and we have computed converged position vectors $\vec{\mathbf{r}}_1$, $\vec{\mathbf{r}}_2$ and $\vec{\mathbf{r}}_3$, we can determine the orbit from these using the method in Section 4.1.

Summary

We are given line-of-sight measurements $\vec{\ell}_1$, $\vec{\ell}_2$, $\vec{\ell}_3$ at times t_1, t_2 and t_3. The position of observation is known at each of these times, and is given by $\vec{\mathbf{R}}_1$, $\vec{\mathbf{R}}_2$ and $\vec{\mathbf{R}}_3$.

1. Set $\alpha = (t_3 - t_2)/(t_3 - t_1)$ and $\beta = (t_2 - t_1)/(t_3 - t_1)$, and select a tolerance $\epsilon > 0$
2. Compute the vectors $\vec{\mathbf{d}}_1 = \vec{\ell}_2 \times \vec{\ell}_3$, $\vec{\mathbf{d}}_2 = \vec{\ell}_3 \times \vec{\ell}_1$ and $\vec{\mathbf{d}}_3 = \vec{\ell}_1 \times \vec{\ell}_2$, and the scalar $D = \vec{\mathbf{d}}_1 \cdot \vec{\ell}_1$.
3. Compute the ranges

$$\rho_1 = -\frac{1}{\alpha D}\left[\alpha\vec{\mathbf{R}}_1 \cdot \vec{\mathbf{d}}_1 + \beta\vec{\mathbf{R}}_3 \cdot \vec{\mathbf{d}}_1 - \vec{\mathbf{R}}_2 \cdot \vec{\mathbf{d}}_1\right],$$

$$\rho_2 = \frac{1}{D}\left[\alpha\vec{\mathbf{R}}_1 \cdot \vec{\mathbf{d}}_2 + \beta\vec{\mathbf{R}}_3 \cdot \vec{\mathbf{d}}_2 - \vec{\mathbf{R}}_2 \cdot \vec{\mathbf{d}}_2\right],$$

$$\rho_3 = -\frac{1}{\beta D}\left[\alpha\vec{\mathbf{R}}_1 \cdot \vec{\mathbf{d}}_3 + \beta\vec{\mathbf{R}}_3 \cdot \vec{\mathbf{d}}_3 - \vec{\mathbf{R}}_2 \cdot \vec{\mathbf{d}}_3\right].$$

4. Compute the position vectors

$$\vec{\mathbf{r}}_1 = \vec{\mathbf{R}}_1 + \rho_1\vec{\ell}_1,$$
$$\vec{\mathbf{r}}_2 = \vec{\mathbf{R}}_2 + \rho_2\vec{\ell}_2,$$
$$\vec{\mathbf{r}}_3 = \vec{\mathbf{R}}_3 + \rho_3\vec{\ell}_3.$$

5. Compute the sector triangle area ratios η_1, η_2 and η_3 by solving

$$\eta = 1 + \frac{m}{\eta^2} W\left(\frac{m}{\eta^2} - l\right),$$

where

$$m = \frac{\mu(t_b - t_a)^2}{\left[2\sqrt{r_a r_b}\cos((\theta_b - \theta_a)/2)\right]^3}, \quad l = \frac{r_a + r_b}{4\sqrt{r_a r_b}\cos((\theta_b - \theta_a)/2)} - \frac{1}{2},$$

and

$$W(w) = \frac{2g - \sin(2g)}{\sin^3 g}, \quad g = 2\sin^{-1}\sqrt{w},$$

using the iterative procedure

$$\eta_{i+1} = \eta_i - f(\eta_i)\frac{\eta_i - \eta_{i-1}}{f(\eta_i) - f(\eta_{i-1})},$$

where

$$f(x) = 1 - x + \frac{m}{x^2}W\left(\frac{m}{x^2} - l\right),$$

with starting values

$$\eta_1 = \eta_H + 0.1, \quad \eta_2 = \eta_H,$$

where

$$\eta_H = \frac{12}{22} + \frac{10}{22}\sqrt{1 + \frac{44}{9}\frac{m}{l + 5/6}}.$$

6. Compute new approximations α_{new} and β_{new} by solving

$$\alpha_{new} = \frac{\eta_2 S_3}{\eta_3 S_2} = \frac{\eta_2(t_3 - t_2)}{\eta_3(t_3 - t_1)}, \quad \beta_{new} = \frac{\eta_2 S_1}{\eta_1 S_2} = \frac{\eta_2(t_2 - t_1)}{\eta_1(t_3 - t_1)}.$$

7. If $|\alpha - \alpha_{new}| < \epsilon$ and $|\beta - \beta_{new}| < \epsilon$, go to step 8. Otherwise, set $\alpha = \alpha_{new}$ and $\beta = \beta_{new}$ and go to step 3.
8. Using $\vec{\mathbf{r}}_1$, $\vec{\mathbf{r}}_2$ and $\vec{\mathbf{r}}_3$, compute the orbit as in Section 4.1.

Warning: The method outlined here only performs well if the orbit is elliptical and the condition $\theta_3 - \theta_1 < 90^\circ$ is satisfied.

4.3 Orbit Determination from Two Position Vectors and Time (Lambert's Problem)

It is possible to determine the orbit given two position vectors, and the time taken to travel between them. This is a very important problem, and has been the subject of much study by some of the greatest mathematicians in history. It has been given the name *Lambert's Problem*, because Lambert was the first to form a solution. This problem basically consists of finding the orbit required to achieve a given transit time between two position vectors. As such, it can be thought of as a preliminary orbit determination problem. On the other hand, it has important applications in orbit transfer, rendezvous and targeting problems. Specifically, given an initial position $\vec{\mathbf{r}}_1$, determine the orbit required to get to position $\vec{\mathbf{r}}_2$ in the time $t_2 - t_1$. For example, $\vec{\mathbf{r}}_1$ might be the position of the Earth at time t_1, and $\vec{\mathbf{r}}_2$ the position of another planet, say Mars at time t_2. The solution to Lambert's problem then provides the transfer orbit. As we have seen, the orbit is fully specified once the position and velocity vectors are simultaneously specified. The orbital elements can then be obtained from them using the procedure in Section 3.7. We can find either $\vec{\mathbf{v}}_1$ or $\vec{\mathbf{v}}_2$. Of more interest to the orbital transfer problem is the velocity $\vec{\mathbf{v}}_1$, since this lets us know what velocity we need to give the spacecraft at time t_1 to be able to get to $\vec{\mathbf{r}}_2$ at time t_2.

There are several solutions to this problem, each of which has applicability in different situations. The solution we are going to present is due to Gauss. We will restrict ourselves to elliptical orbits, with the transfer angle being less than 90°.

4.3.1 *The Lagrangian Coefficients*

Since the motion of an object in orbit is planar, and the position and velocity vectors $\vec{\mathbf{r}}$ and $\vec{\mathbf{v}}$ are not parallel, we are able to relate the position and velocity at one point in an orbit to another by the relations

$$\begin{aligned} \vec{\mathbf{r}}_2 &= F\vec{\mathbf{r}}_1 + G\vec{\mathbf{v}}_1, \\ \vec{\mathbf{v}}_2 &= F_t\vec{\mathbf{r}}_1 + G_t\vec{\mathbf{v}}_1. \end{aligned} \tag{4.29}$$

The scalars F, G, F_t and G_t are known as the *Lagrangian Coefficients*. They are very useful for the orbit determination problem in general. For our purpose, we only need the first of (4.29), and we see that if we know F and G for a given transfer orbit between $\vec{\mathbf{r}}_1$ and $\vec{\mathbf{r}}_2$, then we can obtain the velocity $\vec{\mathbf{v}}_1$ by

$$\vec{\mathbf{v}}_1 = \frac{\vec{\mathbf{r}}_2 - F\vec{\mathbf{r}}_1}{G}. \tag{4.30}$$

We will now obtain expressions for F, G. To do this, we are going to make use of the expressions for the position and velocity in the perifocal coordinate system (see Section 3.8). They are:

$$\begin{bmatrix} \vec{\mathbf{r}} \\ \vec{\mathbf{v}} \end{bmatrix} = \begin{bmatrix} r\cos\theta & r\sin\theta \\ -\sqrt{\dfrac{\mu}{p}}\sin\theta & \sqrt{\dfrac{\mu}{p}}(e+\cos\theta) \end{bmatrix} \begin{bmatrix} \vec{\mathbf{x}}_p \\ \vec{\mathbf{y}}_p \end{bmatrix}. \tag{4.31}$$

The determinant of the matrix in (4.31) is given by

$$\begin{aligned} \det\left(\begin{bmatrix} r\cos\theta & r\sin\theta \\ -\sqrt{\dfrac{\mu}{p}}\sin\theta & \sqrt{\dfrac{\mu}{p}}(e+\cos\theta) \end{bmatrix}\right) &= r\sqrt{\frac{\mu}{p}}\cos\theta(e+\cos\theta) + r\sqrt{\frac{\mu}{p}}\sin^2\theta, \\ &= r\sqrt{\frac{\mu}{p}}(1+e\cos\theta). \end{aligned}$$

Making use of the polar equation for the orbit

$$r = \frac{p}{1+e\cos\theta}, \tag{4.32}$$

this becomes

$$\det\left(\begin{bmatrix} r\cos\theta & r\sin\theta \\ -\sqrt{\frac{\mu}{p}}\sin\theta & \sqrt{\frac{\mu}{p}}(e+\cos\theta) \end{bmatrix}\right) = \sqrt{\mu p} = h.$$

Since $h \neq 0$ for an orbit, the matrix is invertible, with inverse given by

$$\begin{bmatrix} r\cos\theta & r\sin\theta \\ -\sqrt{\frac{\mu}{p}}\sin\theta & \sqrt{\frac{\mu}{p}}(e+\cos\theta) \end{bmatrix}^{-1} = \begin{bmatrix} \frac{1}{p}(e+\cos\theta) & -\frac{r}{\sqrt{\mu p}}\sin\theta \\ \frac{1}{p}\sin\theta & \frac{r}{\sqrt{\mu p}}\cos\theta \end{bmatrix}.$$

Therefore, the unit vectors of the perifocal coordinate system can be obtained from the position and velocity at any point in the orbit, in particular $\vec{\mathbf{r}}_1$ and $\vec{\mathbf{v}}_1$ by

$$\begin{bmatrix} \vec{\mathbf{x}}_p \\ \vec{\mathbf{y}}_p \end{bmatrix} = \begin{bmatrix} \frac{1}{p}(e+\cos\theta_1) & -\frac{r_1}{\sqrt{\mu p}}\sin\theta_1 \\ \frac{1}{p}\sin\theta_1 & \frac{r_1}{\sqrt{\mu p}}\cos\theta_1 \end{bmatrix} \begin{bmatrix} \vec{\mathbf{r}}_1 \\ \vec{\mathbf{v}}_1 \end{bmatrix}. \tag{4.33}$$

Substituting this into equation (4.31) for $\vec{\mathbf{r}}_2$ and $\vec{\mathbf{v}}_2$, we have

$$\begin{aligned} \begin{bmatrix} \vec{\mathbf{r}}_2 \\ \vec{\mathbf{v}}_2 \end{bmatrix} &= \begin{bmatrix} r_2\cos\theta_2 & r_2\sin\theta_2 \\ -\sqrt{\frac{\mu}{p}}\sin\theta_2 & \sqrt{\frac{\mu}{p}}(e+\cos\theta_2) \end{bmatrix} \\ &\times \begin{bmatrix} \frac{1}{p}(e+\cos\theta_1) & -\frac{r_1}{\sqrt{\mu p}}\sin\theta_1 \\ \frac{1}{p}\sin\theta_1 & \frac{r_1}{\sqrt{\mu p}}\cos\theta_1 \end{bmatrix} \begin{bmatrix} \vec{\mathbf{r}}_1 \\ \vec{\mathbf{v}}_1 \end{bmatrix}. \end{aligned} \tag{4.34}$$

From equation (4.34), we find that

$$\begin{aligned} F &= \frac{r_2}{p}\cos\theta_2(e+\cos\theta_1) + \frac{r_2}{p}\sin\theta_2\sin\theta_1, \\ &= \frac{r_2}{p}\left[e\cos\theta_2 + \cos\theta_2\cos\theta_1 + \sin\theta_2\sin\theta_1\right]. \end{aligned}$$

and

$$\begin{aligned} G &= -\frac{r_1 r_2}{\sqrt{\mu p}}\cos\theta_2\sin\theta_1 + \frac{r_1 r_2}{\sqrt{\mu p}}\sin\theta_2\cos\theta_1, \\ &= \frac{r_1 r_2}{\sqrt{\mu p}}\left[\sin\theta_2\cos\theta_1 - \cos\theta_2\sin\theta_1\right]. \end{aligned}$$

Now, from the polar equation for the orbit (4.32), we have

$$e\cos\theta_2 = \frac{p}{r_2} - 1.$$

Making use of this as well as the trigonometric identities

$$\cos(\theta_2 - \theta_1) = \cos\theta_2\cos\theta_1 + \sin\theta_2\sin\theta_1,$$
$$\sin(\theta_2 - \theta_1) = \sin\theta_2\cos\theta_1 - \cos\theta_2\sin\theta_1,$$

we obtain

$$F = 1 - \frac{r_2}{p}\left[1 - \cos(\theta_2 - \theta_1)\right], \tag{4.35}$$

and

$$G = \frac{r_1 r_2}{\sqrt{\mu p}}\sin(\theta_2 - \theta_1). \tag{4.36}$$

From these expressions, we see that the only item that is not known is the semi-parameter p. We know $r_1 = |\vec{\mathbf{r}}_1|$ and $r_2 = |\vec{\mathbf{r}}_2|$, and $\theta_2 - \theta_1$ is just the angle between the vectors $\vec{\mathbf{r}}_1$ and $\vec{\mathbf{r}}_2$. Therefore, in order to fully determine the orbit, we only need to obtain p.

From Section 4.2, using the fact that $h = \sqrt{\mu p}$ the semi-parameter p is directly related to the sector-triangle area ratio η by

$$p = \frac{\eta^2|\vec{\mathbf{r}}_1 \times \vec{\mathbf{r}}_2|^2}{\mu(t_2 - t_1)^2}. \tag{4.37}$$

Therefore, all we need to do is find η given the transfer time. We have already seen how to do this in Section 4.2.

Summary

We are given positions $\vec{\mathbf{r}}_1$ and $\vec{\mathbf{r}}_2$, and the transfer time $t_2 - t_1$. We want to find the transfer orbit between $\vec{\mathbf{r}}_1$ and $\vec{\mathbf{r}}_2$. It is assumed that $\theta_2 - \theta_1 < 90^\circ$, and that the transfer orbit is elliptical.

1. Compute the sector to triangle area ratio by solving

$$\eta = 1 + \frac{m}{\eta^2}W\left(\frac{m}{\eta^2} - l\right),$$

where

$$m = \frac{\mu(t_2 - t_1)^2}{\left[2\sqrt{r_1 r_2}\cos((\theta_2 - \theta_1)/2)\right]^3}, \quad l = \frac{r_1 + r_2}{4\sqrt{r_1 r_2}\cos((\theta_2 - \theta_1)/2)} - \frac{1}{2},$$

and

$$W(w) = \frac{2g - \sin(2g)}{\sin^3 g}, \qquad g = 2\sin^{-1}\sqrt{w},$$

using the iterative procedure

$$\eta_{i+1} = \eta_i - f(\eta_i)\frac{\eta_i - \eta_{i-1}}{f(\eta_i) - f(\eta_{i-1})},$$

where

$$f(x) = 1 - x + \frac{m}{x^2}W\left(\frac{m}{x^2} - l\right),$$

with starting values

$$\eta_1 = \eta_H + 0.1, \qquad \eta_2 = \eta_H,$$

where

$$\eta_H = \frac{12}{22} + \frac{10}{22}\sqrt{1 + \frac{44}{9}\frac{m}{l + 5/6}}.$$

2. Compute the semi-parameter from

$$p = \frac{\eta^2 |\vec{\mathbf{r}}_1 \times \vec{\mathbf{r}}_2|^2}{\mu(t_2 - t_1)^2}.$$

3. Compute the Lagrangian coefficients

$$F = 1 - \frac{r_2}{p}\left[1 - \cos(\theta_2 - \theta_1)\right],$$

and

$$G = \frac{r_1 r_2}{\sqrt{\mu p}}\sin(\theta_2 - \theta_1).$$

4. Compute the velocity $\vec{\mathbf{v}}_1$ using

$$\vec{\mathbf{v}}_1 = \frac{\vec{\mathbf{r}}_2 - F\vec{\mathbf{r}}_1}{G}.$$

5. If required, compute the orbital elements using Section 3.7.

Note This procedure has been shown to work very efficiently in practise for transfer angles less than 90°. For larger transfer angles, the iterative scheme for the sector-triangle area ratio

η has instabilities. There are other solutions to Lambert's problem that work better for large transfer angles. See for example Vallado (2007), Escobal (1976) or Battin (1999).

Notes

This chapter has presented an introduction to preliminary orbit determination. Our presentation is by no means complete. We have focused on three relatively straightforward methods of preliminary orbit determination. However, there are several other methods in the literature. Our presentation of Gibb's method in Section 4.1 is based on that given in Battin (1999) and Bate et al. (1971). Our presentations on orbit determination using three line of sight vectors in Section 4.2 and Lambert's problem in Section 4.3 are based Montenbruck and Gill (2000). A derivation of equation (4.27) for the sector to triangle area ratio may be found in Escobal (1976). For more comprehensive treatments of preliminary orbit determination, readers are recommended to consult Escobal (1976), Vallado (2007) or Battin (1999). As explained in the introduction to this chapter, the techniques presented here may be used to initially determine an orbit given some measurements. Other techniques may then be applied to improve the estimated orbit once additional measurements are available. See for example Vallado (2007), Montenbruck and Gill (2000) or Tapley et al. (2004) for details.

References

Bate RR, Mueller DD and White JE 1971 *Fundamentals of Astrodynamics*. Dover Publications, Mineola, NY.

Battin RH 1999 *An Introduction to the Mathematics and Methods of Astrodynamics, Revised Edition*. American Institute of Aeronautics and Astronautics, Reston, VA.

Escobal PR 1976 *Methods of Orbit Determination*. Krieger Publishing, Malabar, FL.

Montenbruck O and Gill E 2000 *Satellite Orbits: Models, Methods, Applications*. Springer, Berlin Heidelberg.

Tapley BD, Schutz BE and Born GH 2004 *Statistical Orbit Determination*. Elsevier, Burlington, MA.

Vallado DA 2007 *Fundamentals of Astrodynamics and Applications* 3rd edn. Microcosm Press, Hawthorne, CA.

5

Orbital Maneuvers

When a satellite is launched, we typically want a specific orbit depending on the mission objectives. This orbit may or may not be achievable directly from launch, given the launch site or type of launch vehicle. Therefore, *orbital maneuvers* are often required. Maneuvers are also required for interplanetary flight, but we will deal with these separately in Chapter 6.

Maneuvers are performed using onboard thrusters, typically (although not always) in a sequence of short duration bursts. For our purposes, we will assume that these bursts are impulsive, causing an instantaneous change to the spacecraft velocity vector $\vec{\mathbf{v}}$ (without affecting the position $\vec{\mathbf{r}}$).

The most general type of orbital maneuvers require the solution of Lambert's problem (see Section 4.3). However, there are a number of simpler maneuvers that can allow us to achieve our goal.

5.1 Simple Impulsive Maneuvers

The simplest type of maneuver requires only a single impulse. That is, at a particular point in the orbit, we instantaneously change the velocity from say $\vec{\mathbf{v}}_1$ to $\vec{\mathbf{v}}_2$.

Recall that an orbit is fully specified by the position $\vec{\mathbf{r}}$ and the velocity $\vec{\mathbf{v}}$ at any point in the orbit. Therefore, as shown in Figure 5.1, a single impulsive maneuver from one orbit to another is only possible if the two orbits share a common point. The change in velocity is given by

$$\Delta\vec{\mathbf{v}} = \vec{\mathbf{v}}_2 - \vec{\mathbf{v}}_1.$$

An important parameter of the maneuver is the magnitude of the velocity change,

$$\Delta v = |\Delta\vec{\mathbf{v}}|,$$

which is a measure of the fuel consumption. Minimum fuel maneuvers require minimum Δv.

Spacecraft Dynamics and Control: An Introduction, First Edition.
Anton H.J. de Ruiter, Christopher J. Damaren and James R. Forbes.

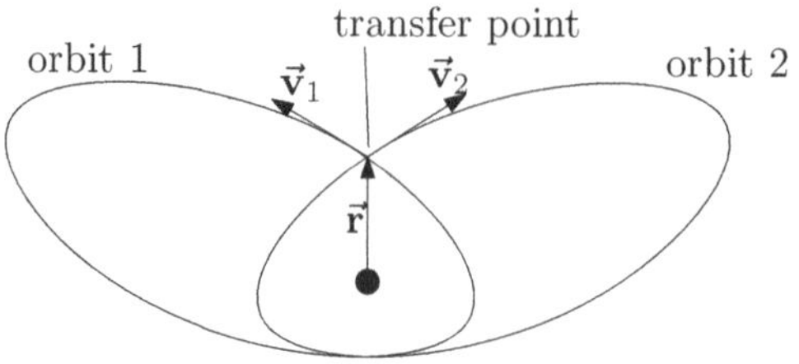

Figure 5.1 Single impulse maneuver

While the fuel consumption is very important, the *time of flight* (*TOF*), that is, the time required to complete a maneuver is also important. Obviously, for a single impulse transfer, $TOF = 0$. However, for maneuvers requiring more than one impulse, the time of flight is not zero. It may be that a minimum Δv transfer takes too long for other mission requirements. Hence, a compromise may be required.

5.2 Coplanar Maneuvers

Let us assume that the initial and final orbits lie in the same plane. In this case, we only need to change the orbital elements a, e and ω, that is, the size, shape and orientation of the orbit within the plane.

We will restrict ourselves to tangential velocity changes, that is, only changing the velocity magnitude, but not its direction.

Before continuing, let us recall the expressions for orbital position and velocity in cylindrical coordinates from Section 3.4 (see Figure 5.2). We have

$$\vec{\mathbf{r}} = r\vec{\mathbf{x}}_o, \quad \vec{\mathbf{v}} = \dot{r}\vec{\mathbf{x}}_o + r\dot{\theta}\vec{\mathbf{y}}_o. \tag{5.1}$$

Recall from Section 3.8 that

$$\dot{r} = \sqrt{\frac{\mu}{a(1-e^2)}} e \sin\theta. \tag{5.2}$$

This shows that for circular orbits (i.e. $e = 0$), $\dot{r} = 0$, which means from (5.1) that $\vec{\mathbf{r}}$ and $\vec{\mathbf{v}}$ are perpendicular. Equation (5.2) also shows that for elliptical orbits (i.e. $0 < e < 1$), $\dot{r} = 0$ only when $\theta = 0, 180^\circ$, i.e., at periapsis and apoapsis. Therefore, (5.1) shows that $\vec{\mathbf{r}}$ and $\vec{\mathbf{v}}$ are perpendicular only at periapsis and apoapsis.

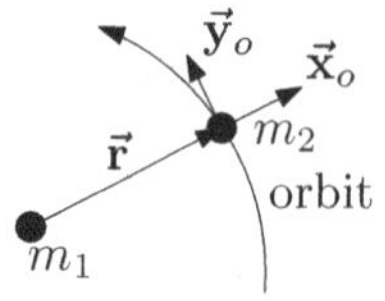

Figure 5.2 Cylindrical coordinate system

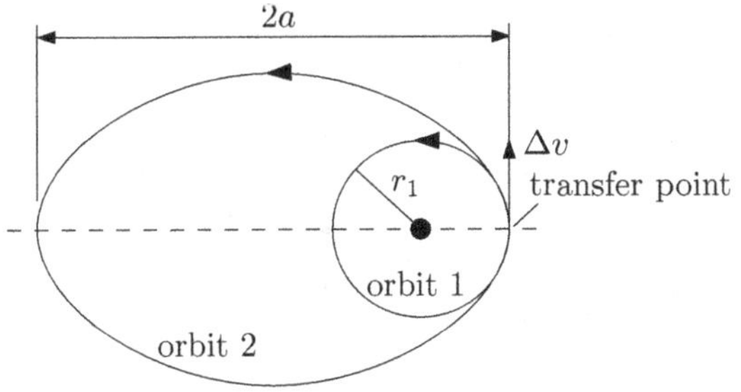

Figure 5.3 Circular to elliptical transfer

This means that any tangential velocity change at any point on a circular orbit results in that point becoming either periapsis or apoapsis of the new orbit. For an elliptical orbit, any tangential velocity change at periapsis results in a change of apoapsis height, and any tangential velocity change at apoapsis results in a change in height of periapsis. If we wish to circularize an elliptical orbit using a tangential transfer, we can only do so at periapsis or apoapsis.

A fundamental equation for tangential transfers is the *vis-viva equation* from Section 3.3, relating the orbital speed v to the orbital radius r.

$$v = \sqrt{\mu\left(\frac{2}{r} - \frac{1}{a}\right)}. \tag{5.3}$$

Circular to Elliptical Transfer

Consider a transfer from a circular orbit (orbit 1) to an elliptical orbit (orbit 2), as shown in Figure 5.3. The orbital speeds of the two orbits at the transfer point are

$$v_1 = \sqrt{\mu\left(\frac{2}{r_1} - \frac{1}{r_1}\right)} = \sqrt{\frac{\mu}{r_1}},$$

$$v_2 = \sqrt{\mu\left(\frac{2}{r_1} - \frac{1}{a}\right)},$$

where the fact that $a = r$ for a circular orbit has been used. If $a > r_1$, we have $v_2 > v_1$, and the total velocity change is

$$\Delta v = v_2 - v_1 = \sqrt{\mu\left(\frac{2}{r_1} - \frac{1}{a}\right)} - \sqrt{\frac{\mu}{r_1}}.$$

If $a < r_1$, we have $v_1 > v_2$, and the total velocity change is

$$\Delta v = v_1 - v_2 = \sqrt{\frac{\mu}{r_1}} - \sqrt{\mu\left(\frac{2}{r_1} - \frac{1}{a}\right)}.$$

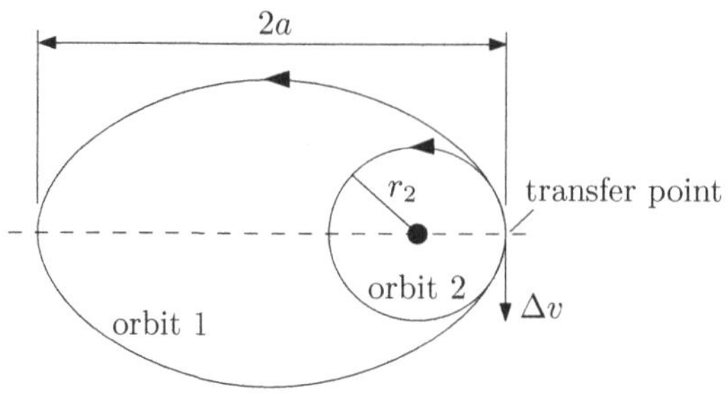

Figure 5.4 Elliptical to circular transfer

Elliptical to Circular Transfer

Now, consider a transfer from an elliptical orbit (orbit 1) to a circular orbit (orbit 2), as shown in Figure 5.4. As explained before, the transfer point must be either periapsis or apoapsis of orbit 1. The orbital speeds of the two orbits at the transfer point are

$$v_1 = \sqrt{\mu\left(\frac{2}{r_2} - \frac{1}{a}\right)},$$

$$v_2 = \sqrt{\frac{\mu}{r_2}}.$$

If $a > r_2$ (transfer at periapsis of orbit 1), we have $v_1 > v_2$, and the total velocity change is

$$\Delta v = v_1 - v_2 = \sqrt{\mu\left(\frac{2}{r_2} - \frac{1}{a}\right)} - \sqrt{\frac{\mu}{r_2}}.$$

If $a < r_2$ (transfer at apoapsis of orbit 1), we have $v_2 > v_1$, and the total velocity change is

$$\Delta v = v_2 - v_1 = \sqrt{\frac{\mu}{r_2}} - \sqrt{\mu\left(\frac{2}{r_2} - \frac{1}{a}\right)}.$$

5.2.1 Hohmann Transfers

Consider now a transfer between two circular orbits of radius r_1 and r_2. These orbits do not have a point in common, and therefore a single impulse transfer is not possible. Walter Hohmann proposed a double impulse transfer in 1925 for this purpose. It consists of two tangential maneuvers: a circular to elliptical transfer followed by an elliptical to circular transfer. That is, a transfer from orbit 1 to orbit 2 is obtained through an elliptical transfer orbit, as shown in Figure 5.5. It turns out that a Hohmann transfer is the minimum Δv double-impulse maneuver

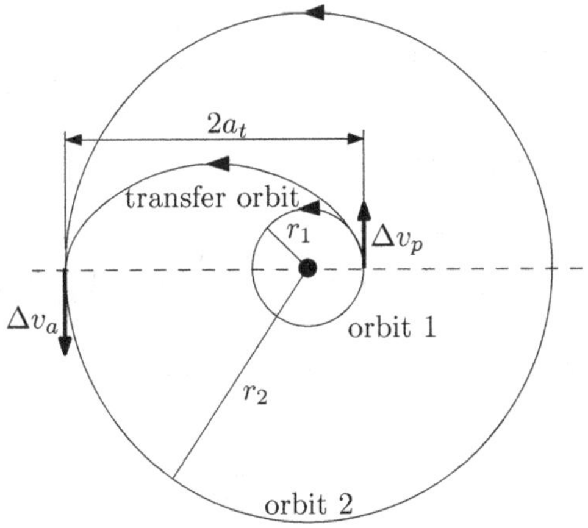

Figure 5.5 Hohmann transfer

between coplanar circular orbits. We see that the semimajor axis of the transfer orbit is given by

$$a_t = \frac{r_1 + r_2}{2}.$$

In the case considered in Figure 5.5, $r_2 > r_1$. Therefore, the first transfer occurs at periapsis of the transfer orbit, with velocity change denoted Δv_p. The second transfer occurs at apoapsis of the transfer orbit, with velocity change Δv_a.

The orbital speeds at the transfer points are

$$v_1 = \sqrt{\frac{\mu}{r_1}},$$

$$v_{tp} = \sqrt{\mu\left(\frac{2}{r_1} - \frac{1}{a_t}\right)},$$

$$v_{ta} = \sqrt{\mu\left(\frac{2}{r_2} - \frac{1}{a_t}\right)},$$

$$v_2 = \sqrt{\frac{\mu}{r_2}}.$$

Since $r_1 < a_t < r_2$, we have $v_{tp} > v_1$ and $v_2 > v_{ta}$, and the velocity changes at each transfer point are

$$\Delta v_p = v_{tp} - v_1 = \sqrt{\mu\left(\frac{2}{r_1} - \frac{1}{a_t}\right)} - \sqrt{\frac{\mu}{r_1}},$$

$$\Delta v_a = v_2 - v_{ta} = \sqrt{\frac{\mu}{r_2}} - \sqrt{\mu\left(\frac{2}{r_2} - \frac{1}{a_t}\right)}.$$

The total velocity change for the maneuver is given by

$$\Delta v = \Delta v_p + \Delta v_a = \sqrt{\frac{\mu}{r_2}} - \sqrt{\mu\left(\frac{2}{r_2} - \frac{1}{a_t}\right)} + \sqrt{\mu\left(\frac{2}{r_1} - \frac{1}{a_t}\right)} - \sqrt{\frac{\mu}{r_1}}. \tag{5.4}$$

The time of flight for the maneuver is half the period of the transfer ellipse, given by (see Section 3.4)

$$TOF = \pi\sqrt{\frac{a_t^3}{\mu}}. \tag{5.5}$$

We now obtain an expression for the total velocity change ratio $\Delta v/v_1$ in terms of the ratio of the orbital radii r_2/r_1. Dividing (5.4) by $v_1 = \sqrt{\mu/r_1}$, and rearranging leads to

$$\frac{\Delta v}{v_1} = \sqrt{\frac{r_1}{r_2}} - 1 - \sqrt{\left(\frac{r_1}{r_2}\right)\frac{2}{[1 + (r_2/r_1)]}} + \sqrt{\left(\frac{r_2}{r_1}\right)\frac{2}{[1 + (r_2/r_1)]}}. \tag{5.6}$$

This equation will be useful when we compare Hohmann transfers with Bi-Elliptic transfers in Section 5.2.2.

Note that Hohmann transfers can also be performed between co-planar elliptical orbits, provided they have the same line of apsides.

5.2.2 *Bi-Elliptic Transfers*

A variant of the Hohmann transfer is the *Bi-Elliptical* transfer, which involves two Hohmann transfers, as shown in Figure 5.6. The first transfer is to a circular orbit of radius r_3, which is larger than the target orbit. The second is a transfer inward to the target orbit. Thus, it is a triple impulse maneuver. We will see that this sometimes results in fuel savings compared with a single Hohmann transfer. We will again consider $r_2 > r_1$. For the two transfer orbits, we have the semi-major axes

$$a_{t1} = \frac{r_1 + r_3}{2},$$
$$a_{t2} = \frac{r_2 + r_3}{2}.$$

The orbital speeds of the initial and target circular orbits are

$$v_1 = \sqrt{\frac{\mu}{r_1}}, \quad v_2 = \sqrt{\frac{\mu}{r_2}}.$$

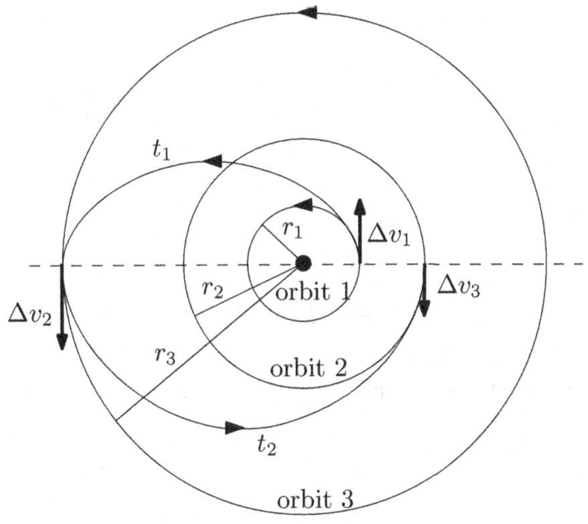

Figure 5.6 Bi-Elliptic transfer

For transfer orbit 1, the speeds at the transfer points are

$$v_{t1p} = \sqrt{\mu\left(\frac{2}{r_1} - \frac{1}{a_{t1}}\right)} : \quad \text{periapsis,}$$

$$v_{t1a} = \sqrt{\mu\left(\frac{2}{r_3} - \frac{1}{a_{t1}}\right)} : \quad \text{apoapsis.}$$

For transfer orbit 2, we have

$$v_{t2a} = \sqrt{\mu\left(\frac{2}{r_3} - \frac{1}{a_{t2}}\right)} : \quad \text{apoapsis,}$$

$$v_{t2p} = \sqrt{\mu\left(\frac{2}{r_2} - \frac{1}{a_{t2}}\right)} : \quad \text{periapsis.}$$

Since $a_{t1} > r_1$, we have $v_{t1p} > v_1$, which leads to

$$\Delta v_1 = v_{t1p} - v_1 = \sqrt{\mu\left(\frac{2}{r_1} - \frac{1}{a_{t1}}\right)} - \sqrt{\frac{\mu}{r_1}}.$$

Since $a_{t2} > a_{t1}$, we have $v_{t2a} > v_{t1a}$, which leads to

$$\Delta v_2 = v_{t2a} - v_{t1a} = \sqrt{\mu\left(\frac{2}{r_3} - \frac{1}{a_{t2}}\right)} - \sqrt{\mu\left(\frac{2}{r_3} - \frac{1}{a_{t1}}\right)}.$$

Since $a_{t2} > r_2$, we have $v_{t2p} > v_2$, which leads to

$$\Delta v_3 = v_{t2p} - v_2 = \sqrt{\mu\left(\frac{2}{r_2} - \frac{1}{a_{t2}}\right)} - \sqrt{\frac{\mu}{r_2}}.$$

The total velocity change for the maneuver is

$$\Delta v = \Delta v_1 + \Delta v_2 + \Delta v_3.$$

The time of flight for the maneuver is half the period of transfer orbit 1 plus half the period of transfer orbit 2, that is,

$$TOF = \pi\left(\sqrt{\frac{a_{t1}^3}{\mu}} + \sqrt{\frac{a_{t2}^3}{\mu}}\right). \tag{5.7}$$

Comparing (5.5) with (5.7), it is clear that a bi-elliptic transfer takes much longer than a Hohmann transfer.

The total velocity change for a bi-elliptic maneuver can now be obtained as

$$\begin{aligned}\Delta v = &\sqrt{\mu\left(\frac{2}{r_1} - \frac{1}{a_{t1}}\right)} - \sqrt{\frac{\mu}{r_1}} + \sqrt{\mu\left(\frac{2}{r_3} - \frac{1}{a_{t2}}\right)} - \sqrt{\mu\left(\frac{2}{r_3} - \frac{1}{a_{t1}}\right)} \\ &+ \sqrt{\mu\left(\frac{2}{r_2} - \frac{1}{a_{t2}}\right)} - \sqrt{\frac{\mu}{r_2}}.\end{aligned} \tag{5.8}$$

As for the Hohmann transfer, we can obtain the ratio $\Delta v/v_1$ as a function of the ratio r_2/r_1 for a given ratio r_3/r_1.

$$\begin{aligned}\frac{\Delta v}{v_1} = &\sqrt{\frac{2(r_3/r_1)}{1+(r_3/r_1)}} - 1 + \sqrt{2\left(\frac{r_1}{r_3}\right)}\left[\sqrt{\frac{(r_2/r_1)}{(r_2/r_1)+(r_3/r_1)}} - \sqrt{\frac{1}{1+(r_3/r_1)}}\right] \\ &+ \sqrt{\frac{r_1}{r_2}}\left[\sqrt{\frac{2(r_3/r_1)}{(r_2/r_1)+(r_3/r_1)}} - 1\right]\end{aligned} \tag{5.9}$$

Equations (5.6) and (5.9) can now be used to determine which maneuver (Hohmann or Bi-Elliptic) is more fuel efficient. Plotting the two for a given ratio r_3/r_1 leads to plots of the form shown in Figure 5.7. It can be shown that when $r_2/r_1 < 11.94$, a Hohmann maneuver is more efficient, and that when $r_2/r_1 > 15.58$, a Bi-Elliptic transfer is more efficient (see Vallado (2007) for details).

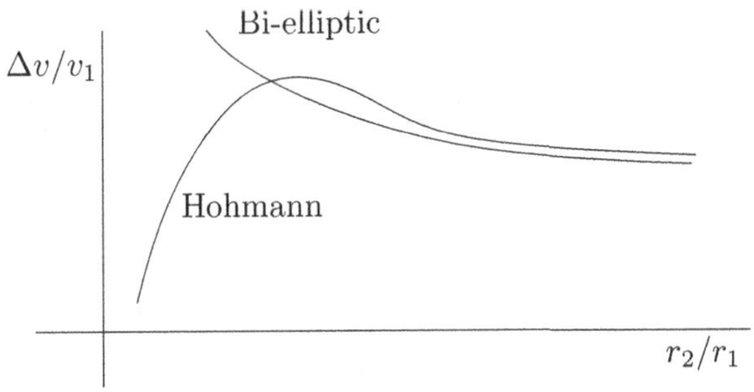

Figure 5.7 Hohmann and bi-elliptic transfer efficiencies

5.3 Plane Change Maneuvers

The two orbital elements associated with the plane of the orbit are the inclination, i, and the right ascension of the ascending node, Ω. To change these without affecting the size or shape of the orbit, we simply rotate the velocity vector $\vec{\mathbf{v}}$ about the position vector $\vec{\mathbf{r}}$, as shown in Figure 5.8. By doing this, the orbital energy, $\mathcal{E}$ and magnitude of the orbital angular momentum h are unaffected, and hence the size and shape of the orbit are unaffected also.

If the plane change maneuver occurs when $\vec{\mathbf{r}}$ and $\vec{\mathbf{v}}$ are perpendicular, that is, anywhere on a circular orbit, or for an elliptical orbit at periapsis or apoapsis, then as shown in Figure 5.9, the velocity change is given by

$$\Delta v = 2v \sin \frac{\theta}{2}. \tag{5.10}$$

This shows that the velocity change can be very large. For example, if $\theta = 60°$, then $\Delta v = v$. For a typical low earth orbiting spacecraft, this is approximately 7.5 km/s! Therefore, if possible, the plane change should be performed at the point of minimum orbital velocity.

Now, in general, a plane change will alter both i and Ω, as shown in Figure 5.10. If a pure inclination change is needed, the plane change must occur at the point where the orbit crosses the equatorial plane, as shown in Figure 5.11.

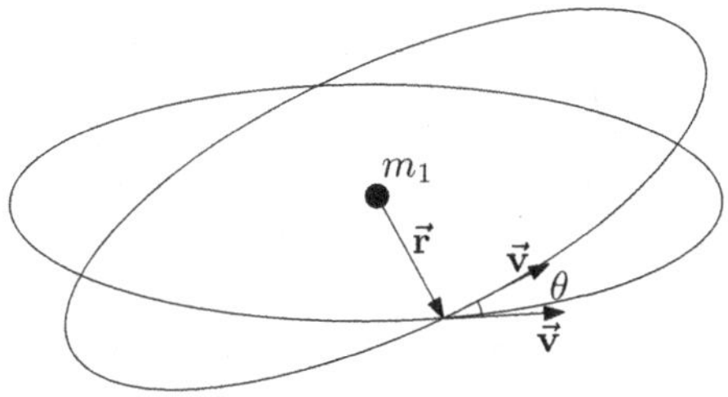

Figure 5.8 Plane change maneuver

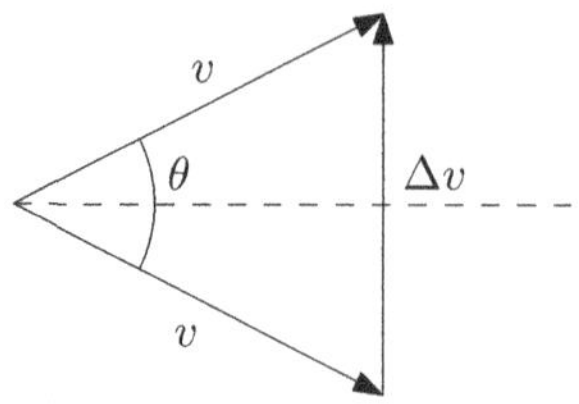

Figure 5.9 Velocity change for plane change maneuver

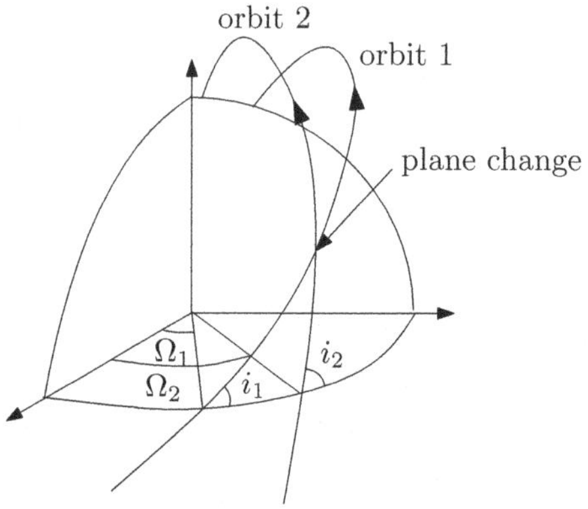

Figure 5.10 General plane change maneuver

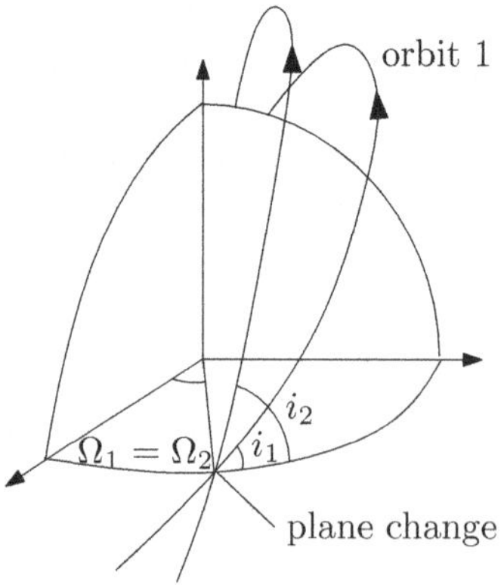

Figure 5.11 Plane change maneuver at equatorial crossing

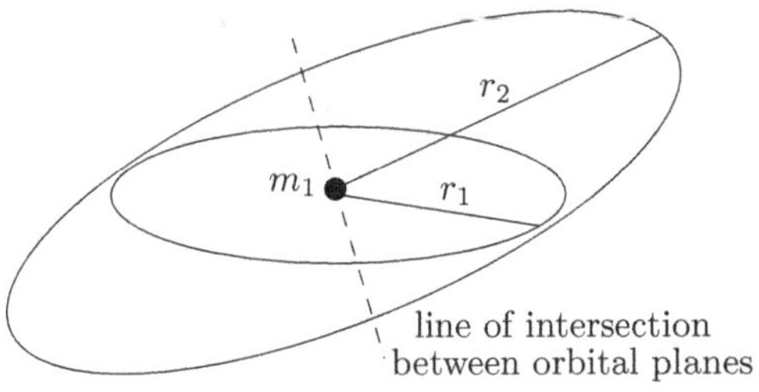

Figure 5.12 Combined maneuver

5.4 Combined Maneuvers

In general, if we wish to change the size, shape and plane of the orbit, combinations of the previous maneuvers are required. As an example, let us consider a transfer between two non-coplanar circular orbits, as shown in Figure 5.12. To achieve this, we can use a plane change and a Hohmann transfer. We can combine the plane change with one of the Hohmann maneuvers, that is, either at periapsis or apoapsis of the elliptical transfer orbit. Given equation (5.10), it makes sense to perform the plane change at apoapsis where the orbital velocity is lowest.

We will consider the case where $r_2 > r_1$. Therefore, the first Hohmann maneuver is at periapsis of the transfer ellipse. Since the plane change occurs at apoapsis of the transfer ellipse, the first maneuver is in the plane of orbit 1.

Following Section 5.2, the semi-major axis of the transfer orbit is

$$a_t = \frac{r_1 + r_2}{2},$$

with change in orbital speed at periapsis

$$\Delta v_p = \sqrt{\mu\left(\frac{2}{r_1} - \frac{1}{a_t}\right)} - \sqrt{\frac{\mu}{r_1}}.$$

The second maneuver in the Hohmann transfer sequence would be to circularize the transfer orbit at its apoapsis. However, we also want to change the plane of the orbit. Now, at apoapsis of the transfer orbit, the orbital speed is

$$v_{ta} = \sqrt{\mu\left(\frac{2}{r_2} - \frac{1}{a_t}\right)},$$

and the orbital speed on orbit 2 is

$$v_2 = \sqrt{\frac{\mu}{r_2}}.$$

Given the desired orbital plane change, θ, the orbital speed change at apoapsis can be obtained from the cosine rule as (see Figure 5.13)

$$\Delta v_a = \sqrt{v_{ta}^2 + v_2^2 - 2v_{ta}v_2\cos\theta}.$$

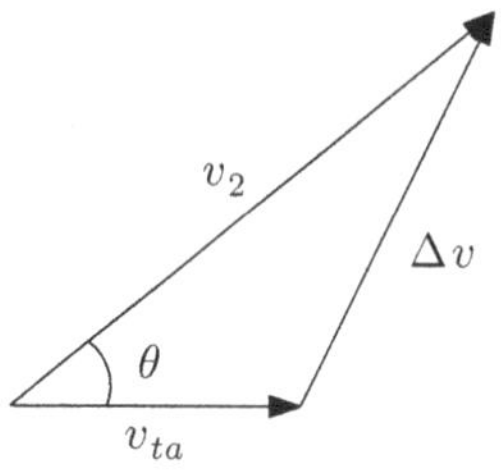

Figure 5.13 Velocity change

Therefore, the total Δv for the combined maneuver is

$$\Delta v = \Delta v_p + \Delta v_a = \sqrt{\mu\left(\frac{2}{r_1} - \frac{1}{a_t}\right)} - \sqrt{\frac{\mu}{r_1}} + \sqrt{v_{ta}^2 + v_2^2 - 2v_{ta}v_2\cos\theta}.$$

Note that the line of apsides of the transfer orbit must coincide with the line of intersection between the two orbital planes.

Application

When a satellite is injected into orbit using some launch vehicle, the point of injection is usually roughly above the launch site. Since the orbital plane must contain the center of the earth and the insertion point (which is roughly above the launch site), the minimum inclination obtainable is roughly equal to the angle between the launch site and the equatorial plane (also known as the launch site latitude, refer to Figure 5.14). Therefore, it is very difficult to launch a satellite directly into an equatorial orbit (unless the launch site is very close to the equator). A common application for the above combined maneuver is transfer into an equatorial orbit.

A very useful type of equatorial orbit is the geostationary orbit, whose period equals the rotational period of the Earth. From the point of view of an observer on Earth, the satellite remains fixed above a point on the equator, making it very useful for communication purposes.

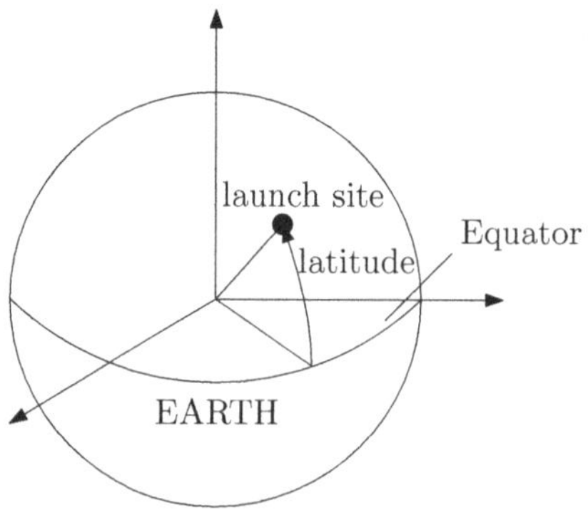

Figure 5.14 Launch site location

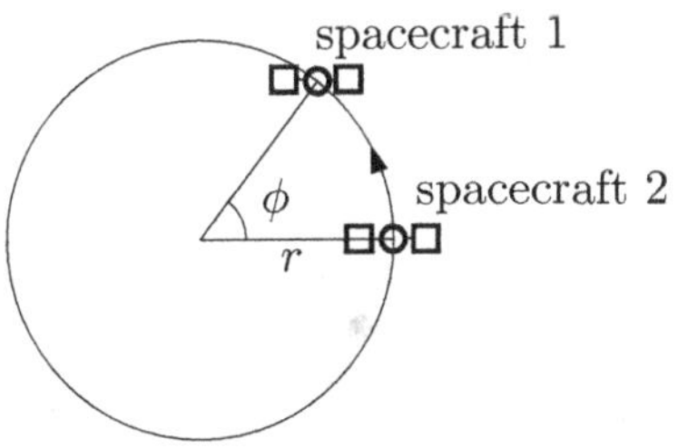

Figure 5.15 Rendezvous problem

5.5 Rendezvous

There are situations when it is desirable to match the orbit of another object. In this case, a *rendezvous* maneuver is required. We shall consider a simplified case, however more general cases are straightforward to determine.

Consider two spacecraft in the same circular orbit of radius r, separated by a phase angle ϕ, as shown in Figure 5.15. We will assume that spacecraft 1 is ahead of spacecraft 2, that is, $\phi > 0$. Suppose spacecraft 2 wishes to rendezvous with spacecraft 1. Since the orbit is circular, the angular rate is constant. Therefore, spacecraft 1 is

$$\frac{\phi T}{2\pi}$$

seconds ahead of spacecraft 2, where $T = 2\pi\sqrt{r^3/\mu}$ is the orbital period. This means that in

$$T - \frac{\phi T}{2\pi}$$

seconds, spacecraft 1 will be at the current location of spacecraft 2. Therefore, if spacecraft 2 transfers into an elliptical orbit of period

$$T_{trans} = T - \frac{\phi T}{2\pi},$$

it will rendezvous with spacecraft 1 after one revolution of the transfer orbit, as shown in Figure 5.16. Spacecraft 2 must then perform a second maneuver to match the orbital velocity of spacecraft 1. This maneuver is also called a *phasing* maneuver. The period of the transfer orbit is given by $T_{trans} = 2\pi\sqrt{a_t^3/\mu}$. Therefore, semi-major axis of the transfer orbit can be computed from

$$a_t = \left[\mu\left(\frac{T_{trans}}{2\pi}\right)^2\right]^{\frac{1}{3}}.$$

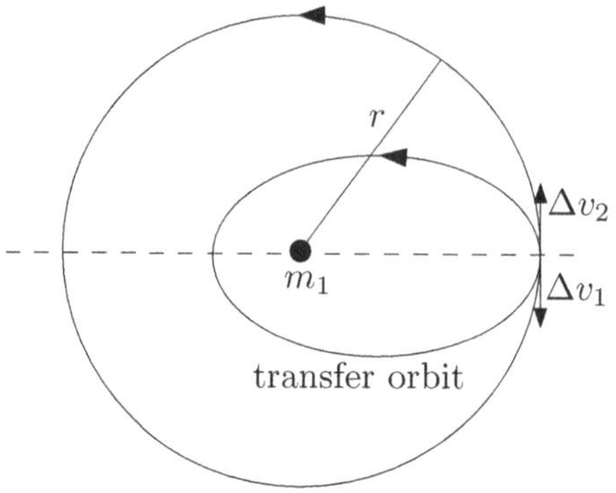

Figure 5.16 Phasing maneuver

The first velocity change is a tangential reduction in velocity

$$\Delta v_1 = \sqrt{\frac{\mu}{r}} - \sqrt{\mu\left(\frac{2}{r} - \frac{1}{a_t}\right)}.$$

The second velocity change is a tangential increase in velocity

$$\Delta v_2 = \Delta v_1.$$

If spacecraft 1 was behind spacecraft 2 ($\phi < 0$), then the opposite would be true, i.e. the first velocity change would be a tangential increase, and the second velocity change would be a tangential decrease. Regardless of the sign of ϕ, the total Δv for the rendezvous maneuver is

$$\Delta v = 2\left|\sqrt{\frac{\mu}{r}} - \sqrt{\mu\left(\frac{2}{r} - \frac{1}{a_t}\right)}\right|.$$

This procedure can easily be generalized to elliptical orbits by using the mean anomaly difference between the two spacecraft (rather than the true anomaly difference) to obtain the period of the transfer orbit. The maneuver is then either executed at periapsis or apoapsis.

Notes

This chapter has presented a discussion of various impulsive maneuvers for changing the orbit. Further treatment of this subject may be found in Vallado (2007).

Reference

Vallado DA 2007 *Fundamentals of Astrodynamics and Applications* 3rd edn. Microcosm Press, Hawthorne, CA.

6

Interplanetary Trajectories

We will now consider interplanetary travel. Consider a trajectory from Earth to Mars. Near the Earth, the main influence on the spacecraft is the Earth's gravitational field. Near Mars, the main influence is Mars' gravitational field. For most of the trip, the main influence on the spacecraft is the sun. In general, the spacecraft is simultaneously influenced by the sun, Earth and Mars (as well as all other celestial bodies). However, for the purposes of initial mission planning, it is useful to decouple those effects, considering the spacecraft to be influenced by only one body at a time. This is called the method of "patched conics". For our Earth to Mars example, we have

1. Near Earth: the spacecraft is in a hyperbolic geocentric (Earth-centered) orbit (while the spacecraft escapes from Earth).
2. Between the Earth and Mars: the spacecraft is in a heliocentric (sun-centered) elliptical orbit (traveling toward Mars).
3. Near Mars: the spacecraft is in a two-body orbit about Mars.

To be able to effectively perform this approximation, we need a criterion for when to turn each gravitational field on and off. For this purpose, it is useful to define a *sphere of influence* for each planet, within which the spacecraft is considered to be in a two-body orbit about that planet, and outside of which the spacecraft is considered to be in a two-body orbit about the sun.

6.1 Sphere of Influence

We will derive an expression for the sphere of influence based upon the equations of motion of the spacecraft about the sun, and the spacecraft about the planet. We will then define the sphere of influence based upon whether the sun has a larger influence over the motion about the planet, or the planet has a larger influence over the motion about the sun.

Let $\vec{\mathbf{R}}_p$ and $\vec{\mathbf{R}}$ denote the position of the planet and spacecraft relative to the sun respectively, and let $\vec{\mathbf{r}}$ denote the position of the spacecraft relative to the planet (see Figure 6.1). Also,

Spacecraft Dynamics and Control: An Introduction, First Edition.
Anton H.J. de Ruiter, Christopher J. Damaren and James R. Forbes.

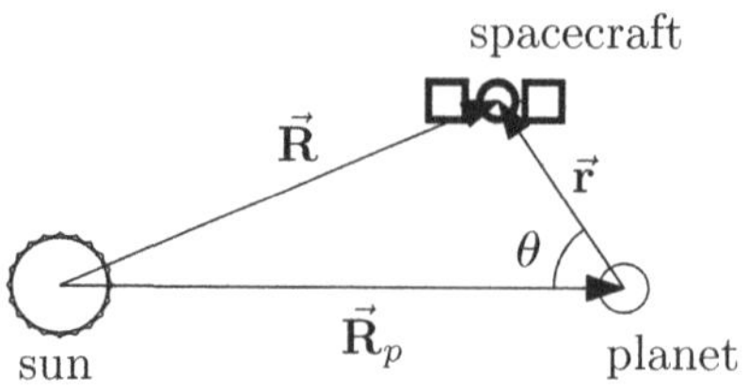

Figure 6.1 Sun, planet, spacecraft system

let m_s, m_p and m denote the mass of the sun, planet and spacecraft respectively. It is clear that $\vec{\mathbf{R}} = \vec{\mathbf{R}}_p + \vec{\mathbf{r}}$. Referring to Figure 6.1, we find that the magnitudes of the vectors are related by

$$R = \left(R_p^2 + r^2 - 2R_p r\cos\theta\right)^{1/2} = R_p\left(1 + \frac{r^2}{R_p^2} - 2\frac{r}{R_p}\cos\theta\right)^{1/2}. \tag{6.1}$$

We expect that within the sphere of influence of the planet, $r \ll R_p$, so that based on (6.1), we can use the approximation

$$R \approx R_p.$$

The gravitational force of the sun and planet on the spacecraft are

$$\vec{\mathbf{f}}_{g,s} = -\frac{Gm_s m}{R^3}\vec{\mathbf{R}}, \quad \vec{\mathbf{f}}_{g,p} = -\frac{Gm_p m}{r^3}\vec{\mathbf{r}},$$

respectively. The gravitational force of the sun on the planet is

$$\vec{\mathbf{f}}_{g,s}^p = -\frac{Gm_s m_p}{R_p^3}\vec{\mathbf{R}}_p.$$

Assuming the sun to be inertially fixed (which is almost true since the mass of the planets are so much smaller than that of the sun), the equation of motion of the spacecraft relative to the sun is

$$m\ddot{\vec{\mathbf{R}}} = \vec{\mathbf{f}}_{g,s} + \vec{\mathbf{f}}_{g,p} = -\frac{Gm_s m}{R^3}\vec{\mathbf{R}} - \frac{Gm_p m}{r^3}\vec{\mathbf{r}},$$

which can be rewritten as

$$\ddot{\vec{\mathbf{R}}} = \vec{\mathbf{A}}_s + \vec{\mathbf{P}}_p, \tag{6.2}$$

where

$$\vec{\mathbf{A}}_s = -\frac{Gm_s}{R^3}\vec{\mathbf{R}}$$

is the acceleration of the spacecraft relative to the sun due to the sun's gravitation and

$$\vec{\mathbf{P}}_p = -\frac{Gm_p}{r^3}\vec{\mathbf{r}}$$

is the acceleration of the spacecraft relative to the sun due to the planet's gravitation. The magnitudes of $\vec{\mathbf{A}}_s$ and $\vec{\mathbf{P}}_p$ are

$$A_s = \frac{Gm_s}{R^2} \approx \frac{Gm_s}{R_p^2}, \quad P_p = \frac{Gm_p}{r^2}.$$

The ratio of P_p to A_s is given by

$$\frac{P_p}{A_s} = \frac{m_p}{m_s}\left(\frac{R_p}{r}\right)^2, \tag{6.3}$$

and is a measure of the influence of the planet on the spacecraft's orbit about the sun.

We would now like to obtain a similar ratio for the disturbing effect of the sun on the spacecraft's orbit about the planet. Now, the equation of motion of the planet about the sun is

$$m_p\ddot{\vec{\mathbf{R}}}_p = -\frac{Gm_s m_p}{R_p^3}\vec{\mathbf{R}}_p,$$

which leads to

$$\ddot{\vec{\mathbf{R}}}_p = -\frac{Gm_s}{R_p^3}\vec{\mathbf{R}}_p. \tag{6.4}$$

Subtracting (6.4) from (6.2) leads to the equation of motion of the spacecraft about the planet, that is

$$\begin{aligned}
\ddot{\vec{\mathbf{r}}} = \ddot{\vec{\mathbf{R}}} - \ddot{\vec{\mathbf{R}}}_p &= -\frac{Gm_s}{R^3}\vec{\mathbf{R}} - \frac{Gm_p}{r^3}\vec{\mathbf{r}} + \frac{Gm_s}{R_p^3}\vec{\mathbf{R}}_p, \\
&\approx -\frac{Gm_s}{R_p^3}(\vec{\mathbf{R}} - \vec{\mathbf{R}}_p) - \frac{Gm_p}{r^3}\vec{\mathbf{r}}, \\
&= -\frac{Gm_s}{R_p^3}\vec{\mathbf{r}} - \frac{Gm_p}{r^3}\vec{\mathbf{r}}, \\
&= \vec{\mathbf{a}}_p + \vec{\mathbf{p}}_s,
\end{aligned}$$

where we have identified

$$\vec{\mathbf{a}}_p = -\frac{Gm_p}{r^3}\vec{\mathbf{r}},$$

as the acceleration of the spacecraft relative to the planet due to the planet's gravitation, and

$$\vec{\mathbf{p}}_s = -\frac{Gm_s}{R_p^3}\vec{\mathbf{r}},$$

as the acceleration of the spacecraft relative to the planet due to the sun's gravitation. The magnitudes of $\vec{\mathbf{a}}_p$ and $\vec{\mathbf{p}}_s$ are

$$a_p = \frac{Gm_p}{r^2}, \qquad p_s = \frac{Gm_s r}{R_p^3}.$$

Similar to equation (6.3), the ratio of p_s to a_p is obtained as

$$\frac{p_s}{a_p} = \frac{m_s}{m_p}\left(\frac{r}{R_p}\right)^3, \tag{6.5}$$

and is a measure of the influence of the sun on the spacecraft's orbit about the planet.

Now, for us to consider the spacecraft to be primarily under the influence of the planet, it seems reasonable that the influence of the sun on the orbit about the planet be less than the influence of the planet on the spacecraft's orbit about the sun. In terms of the ratios (6.3) and (6.5), this is

$$\frac{p_s}{a_p} < \frac{P_p}{A_s}, \text{ or equivalently } \frac{m_s}{m_p}\left(\frac{r}{R_p}\right)^3 < \frac{m_p}{m_s}\left(\frac{R_p}{r}\right)^2.$$

Based on this, the sphere of influence of a planet is defined as the boundary of this region, satisfying

$$\frac{p_s}{a_p} = \frac{P_p}{A_s},$$

which leads to

$$r_{SOI} = R_p\left(\frac{m_p}{m_s}\right)^{2/5}. \tag{6.6}$$

To get a sense of the size of the sphere of influence for a planet, let us compute it for the Earth. The masses of the Earth and the sun are:

$$m_{earth} = 5.974 \times 10^{24}\ kg, \qquad m_{sun} = 1.989 \times 10^{30}\ kg.$$

The mean orbital radius of the Earth about the sun is

$$R_{earth} = 149.6 \times 10^6\ km.$$

Using equation (6.6), the size of the sphere of influence of the Earth is

$$r_{SOI,Earth} = 9.25 \times 10^5\ km.$$

It is clear that compared to the distance to the sun R_{earth}, the sphere of influence is very small, that is

$$\frac{r_{SOI,Earth}}{R_{earth}} = 0.0062.$$

However, relative to the size of the Earth itself, the sphere of influence is very large. Namely, given that the radius of the Earth is approximately 6378 km, the sphere of influence is 145 times the radius of the Earth.

Note that the sphere of influence is not an exact quantity. It is simply a reasonable estimate of when the planet's gravitational field dominates that of the sun.

6.2 Interplanetary Hohmann Transfers

For the majority of the planets in our solar system, the orbits about the sun have very low eccentricities, and lie in approximately the same plane. In practise, we need to account for the eccentricities, and differences in orbital planes, but as a first approach, it will be useful to consider the planets to be in coplanar circular orbits about the sun.

Given that the spheres of influence of the planets are so small compared to their orbital radii about the sun, we will treat the starting and finishing points of an interplanetary transfer as coinciding with the location of each planet respectively, with the transfers occurring impulsively (as in Chapter 5). In Chapter 5, we found that we can transfer from one circular orbit to another using a Hohmann transfer. We have seen that the Hohmann transfer is quite slow (half the period of the transfer orbit). We could use other double impulse maneuvers to reduce travel time, but they would cost a lot more fuel.

Consider two planets, planet 1 and planet 2 respectively. Let the orbits of planet 1 and planet 2 have radii r_1 and r_2 respectively. We are going to consider Hohmann transfers from planet 1 to 2, as shown in Figure 6.2. As in Section 5.2, the semimajor axis of the transfer orbit is

$$a_t = \frac{r_1 + r_2}{2},$$

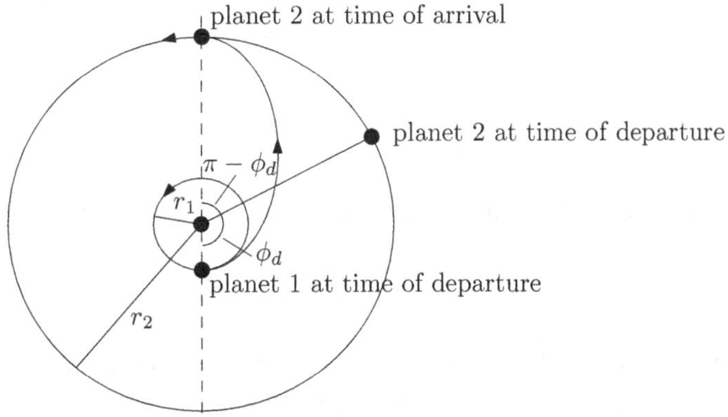

Figure 6.2 Interplanetary Hohmann transfer

and the time of flight is

$$T_{12} = \pi\sqrt{\frac{a_t^3}{\mu_s}},$$

where μ_s is the gravitational parameter of the sun.

Now, an additional constraint on the transfer is that planet 2 must be there when the spacecraft arrives at its orbit. Therefore, there is a constraint on the location of planet 2 at the time of spacecraft departure from planet 1 (see Figure 6.2). Specifically, planet 2 must have a certain phase, ϕ_d, with respect to planet 1 at the time of departure. This phase angle can be determined from the time of flight. Referring to Figure 6.2, it is clear that in order for the spacecraft to intercept planet 2 upon it's arrival, planet 2 must advance its orbital location by the angle

$$\pi - \phi_d.$$

Given that the orbit of planet 2 is circular, the orbital angular velocity is constant and is given by (see Section 3.5.1)

$$n_2 = \sqrt{\frac{\mu_s}{r_2^3}}.$$

Therefore, we must have $\pi - \phi_d = n_2 T_{12}$, which can be rearranged for the required phase angle

$$\phi_d = \pi - n_2 T_{12}. \tag{6.7}$$

When planet 2 has a phase angle ϕ_d with respect to planet 1, we say that a launch window exists.

Now, what if we miss a launch window. How long do we need to wait until the next one? Since both planets are circular with respective orbital angular velocities

$$n_1 = \sqrt{\frac{\mu_s}{r_1^3}}, \quad n_2 = \sqrt{\frac{\mu_s}{r_2^3}},$$

the true anomalies of the planets with respect to a common datum are given by (see Figure 6.3)

$$\theta_1 = \theta_{1,0} + n_1 t, \quad \theta_2 = \theta_{2,0} + n_2 t,$$

where time t is measured from the first launch window, and $\theta_{1,0}$ and $\theta_{2,0}$ are the true anomalies at $t = 0$, such that $\phi_d = \theta_{2,0} - \theta_{1,0}$. The phase angle of planet 2 relative to planet 1 is then

$$\phi = \theta_2 - \theta_1 = \phi_d + (n_2 - n_1)t. \tag{6.8}$$

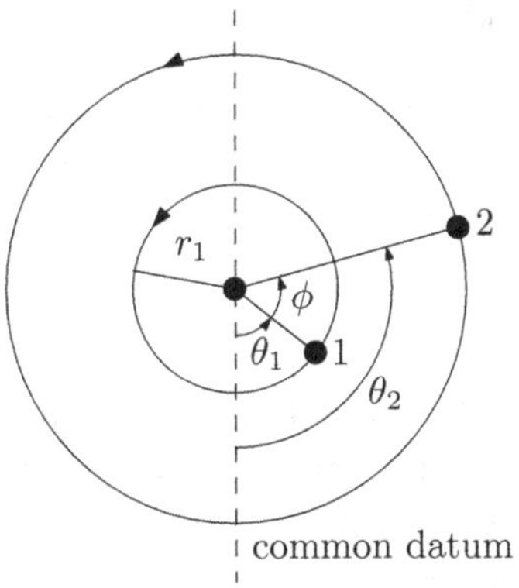

Figure 6.3 Phase angle between planets

If $n_2 > n_1$, that is $r_1 > r_2$, the next launch opportunity occurs when $\phi = \phi_d + 2\pi$. Solving for t gives

$$t = \frac{2\pi}{n_2 - n_1}.$$

If $n_1 > n_2$, that is $r_2 > r_1$, the next launch opportunity occurs when $\phi = \phi_d - 2\pi$. Solving for t gives

$$t = \frac{2\pi}{n_1 - n_2}.$$

Based upon this, the time between launch windows is always given by

$$T_{syn} = \frac{2\pi}{|n_1 - n_2|}, \tag{6.9}$$

which is known as the *synodic period.*

Suppose we have successfully transferred to planet 2, but now want to return to planet 1. We now have to wait until we have the required phase of planet 1 with respect to planet 2. Because the transfer orbit to get back is the same, the time of flight is the same as it was to get there, namely T_{12}. Based upon the same arguments as before, the required phase of planet 1 with respect to planet 2 is

$$-\phi_r = \pi - n_1 T_{12}. \tag{6.10}$$

Note that the negative sign is due to the fact that we have defined ϕ as the phase of planet 2 relative to planet 1, which is the negative of the phase of planet 1 relative to planet 2.

Using (6.8), the phase of planet 2 relative to planet 1 upon arrival at planet 2 is

$$\phi_{arr} = \phi_d + (n_2 - n_1)T_{12}.$$

Substituting for ϕ_d from (6.7) leads to

$$\phi_{arr} = \pi - n_2 T_{12} + (n_2 - n_1)T_{12} = \pi - n_1 T_{12}.$$

Comparing this with (6.10), we see that

$$\phi_r = -\phi_{arr}. \tag{6.11}$$

We can now use equation (6.8) to determine the time we need to wait after arrival for a launch window to return. We reset $t = 0$ at the time of arrival at planet 2, and replace ϕ_d with ϕ_{arr}. Denoting the time required to wait by T_{wait}, we have

$$\phi_r = \phi_{arr} + (n_2 - n_1)T_{wait}.$$

Making use of (6.11), this becomes

$$-\phi_{arr} = \phi_{arr} + (n_2 - n_1)T_{wait},$$

which can be rearranged to give

$$T_{wait} = \frac{-2\phi_{arr}}{n_2 - n_1}.$$

However, a problem with the above equation exists, namely the solution obtained for a launch window may be in the past (T_{wait} is negative). We may therefore need to add or subtract multiples of 2π to make T_{wait} positive. Specifically, we have

$$T_{wait} = \frac{-2\phi_{arr} \pm N2\pi}{n_2 - n_1}, \tag{6.12}$$

where N is the integer with smallest magnitude that makes T_{wait} positive.

The total time for a round trip from planet 1 to planet 2 is the time taken to get there plus the wait time plus the time taken to get back. Mathematically, this is

$$T_{trip} = T_{wait} + 2T_{12}. \tag{6.13}$$

The times for trips from Earth to some of the planets are listed in the table below.

Planet	T_{syn} (days)	T_{12} (days)	T_{wait} (days)	T_{trip} (days)
Mercury	115.8	105.4	66.9	277.9
Venus	583.9	146.1	467.0	759.2
Mars	779.9	258.8	454.3	972.1
Jupiter	398.8	997.5	214.6	2209.6
Saturn	378.1	2209.1	363.2	4454.5

6.3 Patched Conics

As we have described, within the sphere of influence of a planet, the spacecraft is considered to be in a two-body orbit about that planet. Outside, it is considered to be in a two-body orbit about the sun. We have already covered this latter part with the interplanetary Hohmann transfer in the previous section. In the remaining sections, we will deal with the parts within the spheres of influence and the transitions in and out of them. The relationships between the spacecraft position and velocity at the boundary of a sphere of influence are called the "patch conditions." They are:

$$\vec{\mathbf{R}} = \vec{\mathbf{R}}_p + \vec{\mathbf{r}}, \tag{6.14}$$

and

$$\vec{\mathbf{V}} = \vec{\mathbf{V}}_p + \vec{\mathbf{v}}, \tag{6.15}$$

where

$\vec{\mathbf{R}}$ = spacecraft position relative to the sun,
$\vec{\mathbf{R}}_p$ = planet position relative to the sun,
$\vec{\mathbf{r}}$ = spacecraft position relative to the planet,
$\vec{\mathbf{V}}$ = spacecraft velocity relative to the sun,
$\vec{\mathbf{V}}_p$ = planet velocity relative to the sun,
$\vec{\mathbf{v}}$ = spacecraft velocity relative to the planet.

A typical orbit for departure from a planet, or arrival at a planet, is a hyperbolic orbit. That is, the eccentricity satisfies $e > 1$. We now review the relevant features of hyperbolic orbits from Chapter 3. Refer to Figure 6.4. The orbital energy is given by

$$\mathcal{E} = \frac{v^2}{2} - \frac{\mu_p}{r} = -\frac{\mu_p}{2a}, \tag{6.16}$$

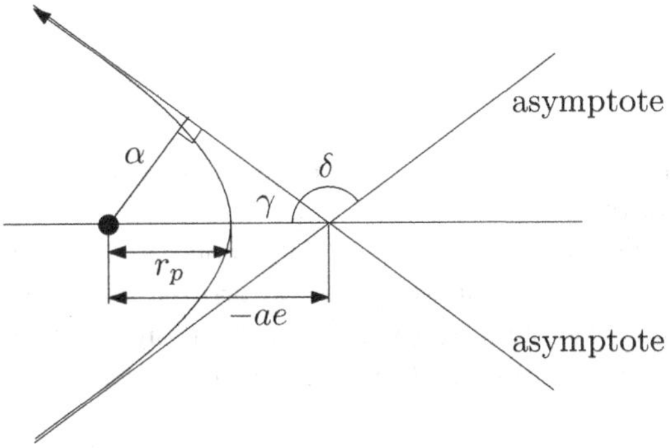

Figure 6.4 Hyperbola geometry

where $\mu_p = Gm_p$ is the gravitational parameter of the planet, and a is the semi-parameter of the orbit, which is negative for a hyperbola (see Section 3.3.1). The angle of the asymptote satisfies

$$\cos\gamma = \frac{1}{e}, \quad \tan\gamma = \sqrt{e^2 - 1}, \quad \sin\gamma = \frac{\sqrt{e^2 - 1}}{e}. \tag{6.17}$$

From the vis-viva equation ($v = \sqrt{\mu_p (2/r - 1/a)}$), the hyperbolic excess speed (as $r \to \infty$) is

$$v_\infty = \sqrt{-\frac{\mu_p}{a}}. \tag{6.18}$$

From the polar equation of the orbit (3.76), the radius at periapsis is ($\theta = 0$)

$$r_p = a(1 - e). \tag{6.19}$$

From the energy equation (6.16) and equation (6.18), we can determine the velocity at periapsis as

$$v_p = \sqrt{v_\infty^2 + \frac{2\mu_p}{r_p}}. \tag{6.20}$$

Recall that for a hyperbolic orbit, the parameter b is defined as

$$b = a\sqrt{e^2 - 1}. \tag{6.21}$$

We can now obtain a physical interpretation of b. Consider the perpendicular distance from the center of the planet to one of the asymptotes, denoted α in Figure 6.4. Referring to Figure 6.4 and making use of (6.17), we have

$$\alpha = -ae\sin\gamma = -a\sqrt{e^2 - 1} = -b.$$

Therefore, the parameter b is the negative of the perpendicular distance from the center of the planet to one of the asymptotes.

Since the sphere of influence is typically very large compared with the radius of periapsis, we will assume for our purposes that the spacecraft arrives and departs from the sphere of influence along an orbit asymptote with velocity v_∞. Therefore, the spacecraft velocity vector relative to the planet in a flyby is turned through an angle of

$$\delta = \pi - 2\gamma = \pi - 2\cos^{-1}\left(\frac{1}{e}\right). \tag{6.22}$$

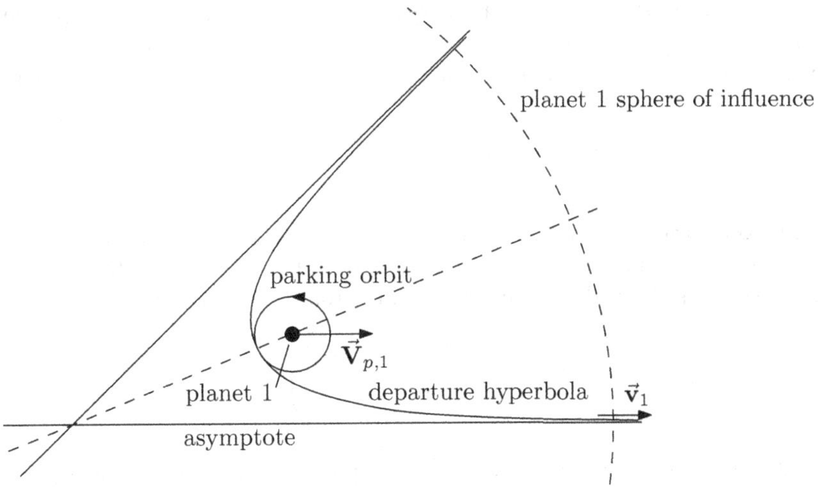

Figure 6.5 Planetary departure

6.3.1 Departure Hyperbola

Recall that for an interplanetary Hohmann transfer, the spacecraft velocity on the transfer orbit leaving planet 1 must be tangent to the orbit of planet 1 about the sun. To accomplish this, we must make the velocity vector $\vec{\mathbf{v}}$ departing from the sphere of influence parallel to the velocity vector of the planet $\vec{\mathbf{V}}_p$. This also gives the spacecraft the maximum velocity boost from planet 1. To accomplish this, we align the asymptote of our departure hyperbola with the velocity vector of planet 1, as shown in Figure 6.5. The spacecraft velocity relative to the sun upon exit from the sphere of influence is then given by the patch condition

$$\vec{\mathbf{V}}_1 = \vec{\mathbf{V}}_{p,1} + \vec{\mathbf{v}}_1. \tag{6.23}$$

The required magnitude of the spacecraft heliocentric velocity at this location is the velocity at the start of the Hohmann transfer ellipse, and can be obtained from the vis-viva equation as

$$V_1 = \sqrt{\mu_s\left(\frac{2}{R_{p1}} - \frac{1}{a_t}\right)}. \tag{6.24}$$

From equations (6.23) and (6.24), we can determine the required hyperbolic excess speed

$$v_{\infty,1} = V_1 - V_{p,1}. \tag{6.25}$$

Now, the semi-parameter a_1 for the departure hyperbola can be computed from (6.18)

$$a_1 = -\frac{\mu_{p1}}{v_{\infty,1}^2}. \tag{6.26}$$

Before initiating departure from planet 1, the spacecraft will initially be in some parking orbit about planet 1, as shown in Figure 6.5. Let us assume that the parking orbit is circular of radius r_{park}. The corresponding orbital velocity can be obtained from the vis-viva equation as

$$v_{park} = \sqrt{\frac{\mu_{p1}}{r_{park}}}. \tag{6.27}$$

If we use a tangential velocity change to enter the departure hyperbola, the radius of the parking orbit will be the radius of periapsis of the departure hyperbola. The associated velocity of the spacecraft at periapsis of the departure hyperbola is (from the vis-viva equation)

$$v_d = \sqrt{\mu_{p1}\left(\frac{2}{r_{park}} - \frac{1}{a_1}\right)}. \tag{6.28}$$

Therefore, the required Δv to inject the spacecraft into the departure hyperbola is

$$\Delta v = v_d - v_{park} = \sqrt{\mu_{p1}\left(\frac{2}{r_{park}} - \frac{1}{a_1}\right)} - \sqrt{\frac{\mu_{p1}}{r_{park}}}. \tag{6.29}$$

We now need to determine the location on the parking orbit for transfer to the departure hyperbola, such that the asymptote is parallel to the velocity vector of the planet. Since r_{park} is the radius of periapsis of the departure hyperbola, we can use equation (6.19) to obtain the departure hyperbola eccentricity

$$e_1 = 1 - \frac{r_{park}}{a_1}. \tag{6.30}$$

We can now determine the location of transfer from the parking orbit to the departure hyperbola as the phase ϕ_1 relative planet 1's velocity vector. From Figure 6.6, we see that the required

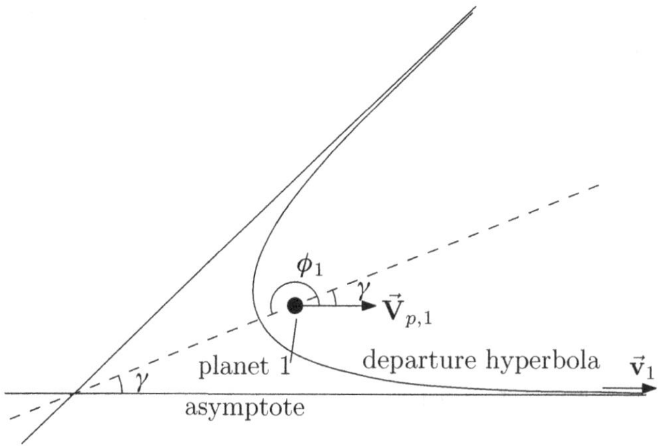

Figure 6.6 Phase for planetary departure

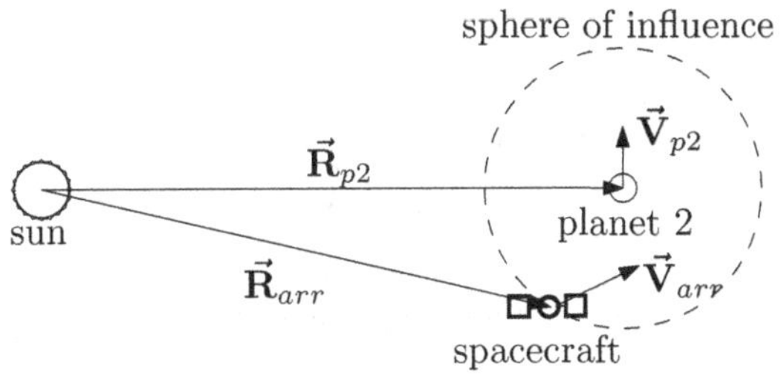

Figure 6.7 Planetary arrival

phase relative to $\vec{\mathbf{V}}_{p,1}$ is

$$\phi_1 = \pi + \gamma = \pi + \cos^{-1}\left(\frac{1}{e_1}\right). \tag{6.31}$$

Important: Equation (6.31) provides the required phase when $v_{\infty,1}$ computed in (6.25) is positive, which means that the spacecraft velocity relative to the sun must be larger than the planet's velocity, which is the case when traveling to an orbit with larger orbital radius about the sun. If $v_{\infty,1}$ computed in (6.25) is negative, which means the transfer is to a planet with smaller orbital radius, then the departure asymptote must point in the opposite direction of $\vec{\mathbf{V}}_{p,1}$. The required phase is then $\phi_1 = \gamma$.

6.3.2 *Arrival Hyperbola*

Let the position and velocity of the spacecraft with respect to the sun be denoted $\vec{\mathbf{R}}_{arr}$ and $\vec{\mathbf{V}}_{arr}$ respectively, at the time when the spacecraft enters the sphere of influence of planet 2. Let the corresponding position and velocity of planet 2 be $\vec{\mathbf{R}}_{p2}$ and $\vec{\mathbf{V}}_{p2}$ respectively (see Figure 6.7). From the patch conditions, the position and velocity of the spacecraft relative to planet 2 upon entry to the sphere of influence are

$$\vec{\mathbf{r}}_{arr} = \vec{\mathbf{R}}_{arr} - \vec{\mathbf{R}}_{p2}. \tag{6.32}$$

and

$$\vec{\mathbf{v}}_{arr} = \vec{\mathbf{V}}_{arr} - \vec{\mathbf{V}}_{p2}. \tag{6.33}$$

This information can now be used to determine the arrival hyperbola. The speed at which the spacecraft enters the sphere of influence is taken to be the hyperbolic excess speed, that is

$$v_{\infty,arr} = |\vec{\mathbf{v}}_{arr}|. \tag{6.34}$$

Using equation (6.18), this immediately gives the arrival hyperbola semi-parameter

$$a_2 = \frac{-\mu_{p2}}{v^2_{\infty,arr}}. \tag{6.35}$$

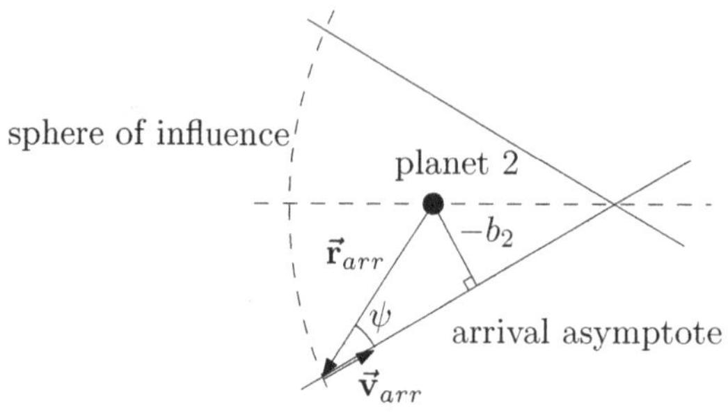

Figure 6.8 Arrival geometry

Next, as we have seen, the parameter b is the negative of the perpendicular distance from the center of the planet to each asymptote. Now, the asymptote is aligned with $\vec{\mathbf{v}}_{arr}$, so that b_2 can be determined from geometry using $\vec{\mathbf{r}}_{arr}$ and $\vec{\mathbf{v}}_{arr}$. Consider the diagram shown in Figure 6.8. The angle ψ is given by

$$\psi = \cos^{-1}\left(\frac{-\vec{\mathbf{r}}_{arr}\cdot\vec{\mathbf{v}}_{arr}}{r_{arr}v_{\infty,arr}}\right). \tag{6.36}$$

By trigonometry then,

$$b_2 = -r_{arr}\sin\psi. \tag{6.37}$$

Note that $r_{arr} = r_{SOI,2}$, the radius of the sphere of influence of planet 2. Now, from equation (6.21) we can determine the eccentricity of the arrival hyperbola

$$e_2 = \sqrt{1+\left(\frac{b_2}{a_2}\right)^2}. \tag{6.38}$$

From equation (6.19), we can now determine the closest pass of the spacecraft with respect to the center of planet 2 (the radius of periapsis of the arrival hyperbola), namely

$$r_{p2} = a_2\left[1-\sqrt{1+\left(\frac{b_2}{a_2}\right)^2}\right]. \tag{6.39}$$

Of course, planet 2 has a finite size. Let us assume that planet 2 is spherical with radius $\bar{R}_{p2}$. In order to avoid a collision with the planet, we must have

$$r_{p2} > \bar{R}_{p2}, \tag{6.40}$$

as shown in Figure 6.9. Making use of equations (6.39) and (6.40), we have

$$a_2\left[1-\sqrt{1+\left(\frac{b_2}{a_2}\right)^2}\right] > \bar{R}_{p2},$$

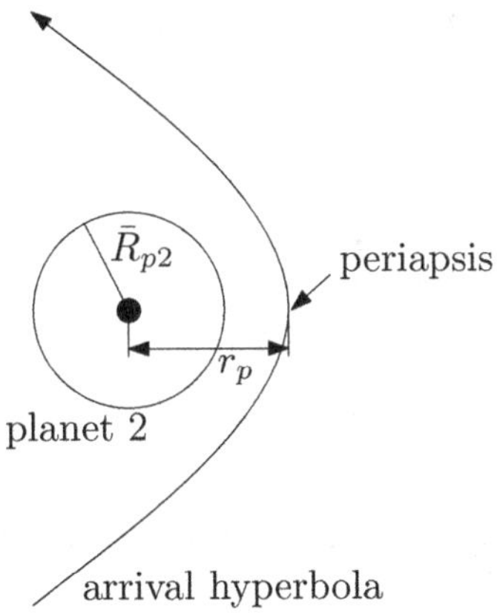

Figure 6.9 Collision avoidance

which leads to

$$1 - \frac{\bar{R}_{p2}}{a_2} < \sqrt{1 + \left(\frac{b_2}{a_2}\right)^2}.$$

Note that the reverse in the inequality is due to the fact that we have divided by a_2, which is negative for a hyperbola. Finally, after squaring the above expression and rearranging, we obtain

$$|b_2| > |a_2| \left[\left(1 - \frac{\bar{R}_{p2}}{a_2}\right)^2 - 1 \right]^{1/2}. \tag{6.41}$$

Therefore, in order to avoid a collision, inequality (6.41) must be satisfied. Note that equation (6.41) is useful, since the parameter b_2 can be readily adjusted by making small changes to the spacecraft velocity vector while the spacecraft is still far away from planet 2.

6.4 Planetary Flyby

We have already determined the parameters of the arrival hyperbola, namely a_2 and b_2. We also have $\vec{\mathbf{v}}_{arr}$ and $v_{\infty,arr}$. Without any further maneuvers, the spacecraft will exit the sphere of influence of planet 2 along the other hyperbola asymptote, with velocity vector $\vec{\mathbf{v}}_{dep}$ relative to the planet with magnitude $v_{\infty,arr}$, as shown in Figure 6.10. That is, $|\vec{\mathbf{v}}_{arr}| = |\vec{\mathbf{v}}_{dep}| = v_{\infty,arr}$. From equation (6.22), the angle through which the velocity vector is turned is

$$\delta_2 = \pi - 2\gamma_2 = \pi - 2\cos^{-1}\left(\frac{1}{e_2}\right). \tag{6.42}$$

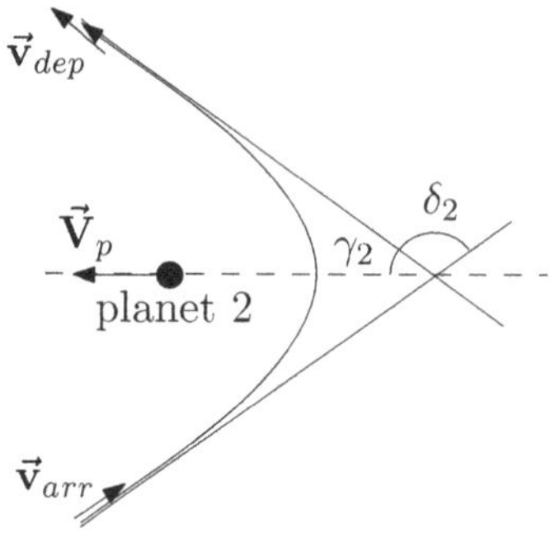

Figure 6.10 Planetary flyby (trailing edge)

Therefore, $\vec{\mathbf{v}}_{arr} \neq \vec{\mathbf{v}}_{dep}$, and the planet exerts a $\Delta\vec{\mathbf{v}}$ on the spacecraft. The spacecraft velocity relative to the sun upon exit from the sphere of influence is

$$\vec{\mathbf{V}}_{dep} = \vec{\mathbf{V}}_p + \vec{\mathbf{v}}_{dep}. \tag{6.43}$$

Treating the size of the sphere of influence of planet 2, $r_{SOI,2}$, as negligible compared to the orbital radius of planet 2, that is $|\vec{\mathbf{R}}_{arr}| \approx |\vec{\mathbf{R}}_{dep}| \approx |\vec{\mathbf{R}}_{p2}|$, and also treating the time of transit of the spacecraft through the sphere of influence as negligible (such that the planet's velocity vector is unchanged during this time), the change in spacecraft heliocentric orbital energy is

$$\Delta\mathcal{E} = \frac{1}{2}\left(V_{dep}^2 - V_{arr}^2\right), \tag{6.44}$$

which may be positive or negative. That is, planet 2 either increases or decreases the velocity of the spacecraft relative to the sun. A positive $\Delta\mathcal{E}$ occurs as a result of a trailing edge flyby, that is, the spacecraft passes behind the planet, as shown in Figure 6.10.

This increase is a "gravitational slingshot", and can be used as an alternative to the Hohmann transfer for travel to the outer planets. For example, to get to Saturn via a Jupiter fly-by.

A negative $\Delta\mathcal{E}$ occurs as a result of a leading edge flyby, that is, the spacecraft passes in front of the planet, as shown in Figure 6.11. This can be used to reduce the spacecraft velocity relative to the sun.

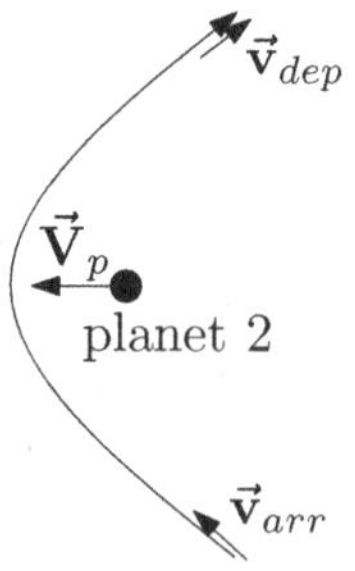

Figure 6.11 Leading edge planetary flyby

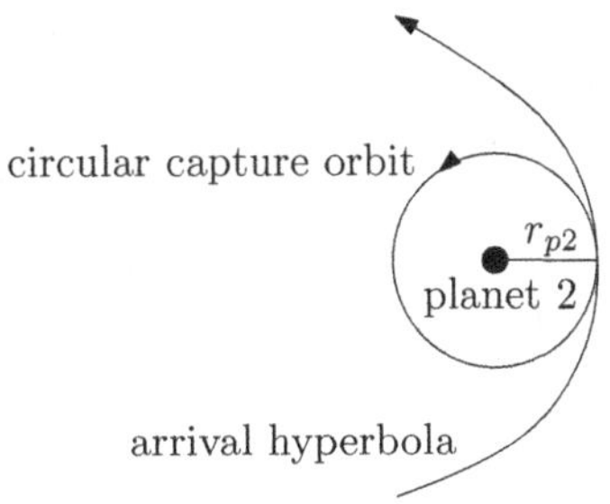

Figure 6.12 Planetary capture

As some examples, Voyager II used this method to perform flybys of Jupiter, Saturn, Uranus and Neptune. Galileo went to Jupiter using the Venus-Earth-Earth-Gravity-Assist (VEEGA) trajectory.

6.5 Planetary Capture

If we wish to stay at the target planet, we need to perform a maneuver to convert the arrival hyperbola into a bound (elliptical or circular) orbit about the planet. We have seen that if no maneuvers are performed, the spacecraft will exit from the sphere of influence. In general, the arrival velocity $v_{\infty,arr}$ cannot be controlled easily, but the entry point into the sphere of influence (and hence b_2) can be, by small corrections to the interplanetary transfer orbit.

We have seen that the radius of periapsis of the arrival hyperbola is given by

$$r_{p2} = a_2\left[1 - \sqrt{1 + \left(\frac{b_2}{a_2}\right)^2}\right]. \tag{6.45}$$

Now, one way of capturing the spacecraft about planet 2 is to perform a circularization of the arrival hyperbola at its periapsis, as shown in Figure 6.12. From the vis-viva equation, the orbital velocity on the circular capture orbit is

$$v_{cap} = \sqrt{\frac{\mu_{p2}}{r_{p2}}}. \tag{6.46}$$

From equation (6.20), the orbital velocity at periapsis of the arrival hyperbola is

$$v_{p2} = \sqrt{v_{\infty,arr}^2 + \frac{2\mu_{p2}}{r_{p2}}}. \tag{6.47}$$

Therefore, the Δv required for capture is

$$\Delta v_{cap} = v_{p2} - v_{cap} = \sqrt{v_{\infty,arr}^2 + \frac{2\mu_{p2}}{r_{p2}}} - \sqrt{\frac{\mu_{p2}}{r_{p2}}}. \tag{6.48}$$

Since we have control over b_2, we have control over r_{p2} through equation (6.39). Let us now determine the radius for optimal capture. For minimum Δv_{cap}, we set the derivative of (6.48) with respect to r_{p2} equal to zero. Namely,

$$0 = \frac{d\Delta v_{cap}}{dr_{p2}} = -\frac{2\mu_{p2}}{r_{p2}^2}\left(\frac{1}{2}\right)\left(v_{\infty,arr}^2 + \frac{2\mu_{p2}}{r_{p2}}\right)^{-\frac{1}{2}} - \sqrt{\mu_{p2}}\left(-\frac{1}{2}\right)\frac{1}{r_{p2}^{\frac{3}{2}}},$$

$$= \frac{-2\mu_{p2} + \sqrt{\mu_{p2}r_{p2}\left(v_{\infty,arr}^2 + \frac{2\mu_{p2}}{r_{p2}}\right)}}{2r_{p2}^2\sqrt{v_{\infty,arr}^2 + \frac{2\mu_{p2}}{r_{p2}}}}.$$

This leads to

$$\sqrt{\mu_{p2}r_{p2}\left(v_{\infty,arr}^2 + \frac{2\mu_{p2}}{r_{p2}}\right)} = 2\mu_{p2},$$

and upon squaring and rearranging, we have the radius for optimal capture

$$r_{p2,opt} = \frac{2\mu_{p2}}{v_{\infty,arr}^2}. \tag{6.49}$$

The associated Δv is obtained by substituting (6.49) into (6.48)

$$\Delta v_{cap,opt} = \sqrt{v_{\infty,arr}^2 + \frac{2\mu_{p2}v_{\infty,arr}^2}{2\mu_{p2}}} - \sqrt{\frac{\mu_{p2}v_{\infty,arr}^2}{2\mu_{p2}}} = \sqrt{2}v_{\infty,arr} - \frac{v_{\infty,arr}}{\sqrt{2}},$$

which becomes

$$\Delta v_{cap,opt} = \frac{v_{\infty,arr}}{\sqrt{2}}. \tag{6.50}$$

If the planet has an atmosphere, we can reduce the Δv_{cap} requirement by aerobraking using the atmosphere.

Notes

This chapter has presented simple techniques that can be used for initial interplanetary mission design. Since the patched conic approach is an approximation (gravity fields are continuous), initially designed interplanetary trajectories must be corrected to account for the continuous gravity fields of all celestial bodies. Methods are presented in Battin (1999) for how to determine these differential corrections. Readers interested in an in depth treatment of interplanetary trajectory design are encouraged to consult Kemble (2006).

References

Battin RH 1999 *An Introduction to the Mathematics and Methods of Astrodynamics, Revised Edition*. American Institute of Aeronautics and Astronautics, Reston, VA.

Kemble S 2006 *Interplanetary Mission Analysis and Design*. Springer, Chichester, UK.

7

Orbital Perturbations

Up to now, we have considered only Keplerian two-body motion of point masses m_1 and m_2, where m_1 is taken to be the primary with $m_1 \gg m_2$. As we have seen, the differential equation governing the motion of m_2 with respect to m_1 is

$$\ddot{\vec{\mathbf{r}}} = -\frac{\mu}{r^3}\vec{\mathbf{r}}, \tag{7.1}$$

with initial conditions

$$\vec{\mathbf{r}}(0) = \vec{\mathbf{r}}_0, \quad \dot{\vec{\mathbf{r}}}(0) = \vec{\mathbf{v}}_0.$$

For example, for an Earth orbiting satellite, m_1 represents the Earth, and m_2 the satellite. For a planet orbiting the sun, m_1 represents the sun, and m_2 the planet.

The Keplerian two-body orbit is very much an idealized motion. It works well for short periods of time, but there are several factors that cause the actual motion to deviate from the Keplerian orbit, and these factors must be accounted for in practise. For a geocentric orbit, these factors include:

1. Non-sphericity of the primary (the Earth is not perfectly spherical).
   In the Keplerian orbit, masses m_1 and m_2 are assumed to be point masses. Due to the small size of m_2 (compared with the distance from the center of m_1), this is a reasonable approximation for m_2. As we shall see later, if the primary (m_1) is spherically symmetric, then it can also be treated as a point mass. However, this is not the case, and the non-sphericity of the primary must be taken into account.
2. Presence of other bodies and their gravitational fields.
   For a spacecraft orbiting the Earth, the spacecraft is also influenced by the sun and moon (and to a much lesser extent, the other planets).
3. Atmospheric drag.
   For near-Earth orbits, there is still some residual atmosphere, creating drag on the spacecraft. This results in a gradual orbit decay.

Spacecraft Dynamics and Control: An Introduction, First Edition.
Anton H.J. de Ruiter, Christopher J. Damaren and James R. Forbes.

4. Solar radiation pressure.
   Light from the sun (photons) creates pressure on the lit surface of the spacecraft, which is caused by momentum transfer from the photons to the spacecraft surface.

These effects are *perturbations* to the Keplerian orbit, and this Chapter shows how we may deal with them.

Let $\vec{\mathbf{f}}_p$ be the *perturbative acceleration* due to the perturbing effects. Then, the true equation of motion is (compared to (7.1))

$$\ddot{\vec{\mathbf{r}}} = -\frac{\mu}{r^3}\vec{\mathbf{r}} + \vec{\mathbf{f}}_p, \tag{7.2}$$

with initial conditions

$$\vec{\mathbf{r}}(0) = \vec{\mathbf{r}}_0, \quad \dot{\vec{\mathbf{r}}}(0) = \vec{\mathbf{v}}_0.$$

There are two approaches for dealing with the perturbations. These are classed as *special perturbations* and *general perturbations*.

Special perturbations determine the effects numerically, by performing some kind of numerical integration of equation (7.2). They are called "special", since the solution is only valid for one set of initial conditions. The procedure must be repeated once a different set of initial conditions is given.

General perturbations determine the effects *analytically*. They deal with variations in the orbital elements (which are constant for a Keplerian orbit) due to the perturbative acceleration $\vec{\mathbf{f}}_p$. Specifically, expressions for

$$\frac{da}{dt}, \quad \frac{de}{dt}, \quad \frac{di}{dt}, \quad \frac{d\Omega}{dt}, \quad \frac{d\omega}{dt},$$

are obtained. These solutions are called general, since they are valid for any set of initial conditions. They provide a great deal of physical insight, and are useful for mission planning. However, approximations are often made in their derivations, so they are not as accurate as special perturbation techniques can potentially be.

7.1 Special Perturbations

As we have mentioned, special perturbation techniques involve *numerical integration* of the equations of motion. As discussed in Appendix B, this requires writing the equations of motion in the first order form

$$\dot{\mathbf{x}} = \mathbf{F}(\mathbf{x}, t), \quad \mathbf{x}(t_0) = \mathbf{x}_0.$$

Having written the equations of motion in this form, numerical integration techniques (such as that presented in Appendix B) may be applied to determine $\mathbf{x}(t)$ for $t \geq t_0$ numerically.

7.1.1 Cowell's Method

This method was developed by P.H. Cowell in the early 20th century. It is essentially a brute-force approach, where the equations of motion (7.2) are numerically integrated directly.

Specifically, if we represent all quantities in (7.2) in an inertial frame of reference $\mathcal{F}_I$, such that

$$\vec{\mathbf{r}} = \vec{\mathcal{F}}_I^T \mathbf{r}, \quad \vec{\mathbf{v}} = \dot{\vec{\mathbf{r}}} = \vec{\mathcal{F}}_I^T \mathbf{v}, \quad \vec{\mathbf{f}}_p = \vec{\mathcal{F}}_I^T \mathbf{f}_p,$$
$$\vec{\mathbf{r}}_0 = \vec{\mathcal{F}}_I^T \mathbf{r}_0, \quad \vec{\mathbf{v}}_0 = \vec{\mathcal{F}}_I^T \mathbf{v}_0,$$

we can write the equations of motion (7.2) in the first order form appropriate for numerical integration, namely

$$\begin{bmatrix} \dot{\mathbf{r}} \\ \dot{\mathbf{v}} \end{bmatrix} = \begin{bmatrix} \mathbf{v} \\ -\dfrac{\mu}{(\mathbf{r}^T\mathbf{r})^{\frac{3}{2}}}\mathbf{r} + \mathbf{f}_p \end{bmatrix}, \tag{7.3}$$

with initial conditions

$$\begin{bmatrix} \mathbf{r}(0) \\ \mathbf{v}(0) \end{bmatrix} = \begin{bmatrix} \mathbf{r}_0 \\ \mathbf{v}_0 \end{bmatrix}.$$

Cowell's method then numerically integrates equations (7.3) directly. The advantage of this method is that any perturbations $\mathbf{f}_p$ can be included without modification to this method. However, to achieve the required accuracy, small time-steps are typically required, making it computationally expensive. Also, the round-off errors due to numerical integration can accumulate quite rapidly, making long term solutions inaccurate. The required small time-step compounds this effect. This is less of an issue today than it was in the past, due to the computational power and sophisticated techniques available.

7.1.2 Encke's Method

Encke's method is quite a bit more sophisticated than Cowell's method, even though it was developed much earlier, in 1857. The reason is the lack of computational power available at the time.

Encke's method works by numerically integrating the deviation of the true (perturbed) orbit satisfying

$$\ddot{\vec{\mathbf{r}}} = -\frac{\mu}{r^3}\vec{\mathbf{r}} + \vec{\mathbf{f}}_p, \quad \vec{\mathbf{r}}(0) = \vec{\mathbf{r}}_0, \quad \dot{\vec{\mathbf{r}}}(0) = \vec{\mathbf{v}}_0, \tag{7.4}$$

from a reference Keplerian orbit satisfying

$$\ddot{\vec{\boldsymbol{\rho}}} = -\frac{\mu}{\rho^3}\vec{\boldsymbol{\rho}}, \quad \vec{\boldsymbol{\rho}}(0) = \vec{\mathbf{r}}_0, \quad \dot{\vec{\boldsymbol{\rho}}}(0) = \vec{\mathbf{v}}_0. \tag{7.5}$$

The reference orbit $\vec{\rho}(t)$ is called the *osculating orbit*, and can be determined analytically using the two-body solution we found in Chapter 3. Let us define the deviation from the osculating orbit as

$$\delta\vec{\mathbf{r}} \stackrel{\Delta}{=} \vec{\mathbf{r}} - \vec{\rho}.$$

Taking the difference between (7.4) and (7.5), we have

$$\begin{aligned}
\delta\ddot{\vec{\mathbf{r}}} = \ddot{\vec{\mathbf{r}}} - \ddot{\vec{\rho}} &= -\frac{\mu}{r^3}\vec{\mathbf{r}} + \frac{\mu}{\rho^3}\vec{\rho} + \vec{\mathbf{f}}_p, \\
&= -\mu\left[\frac{\vec{\mathbf{r}}}{r^3} - \frac{(\vec{\mathbf{r}} - \delta\vec{\mathbf{r}})}{\rho^3}\right] + \vec{\mathbf{f}}_p, \\
&= -\frac{\mu}{\rho^3}\left[\delta\vec{\mathbf{r}} - \left(1 - \frac{\rho^3}{r^3}\right)\vec{\mathbf{r}}\right] + \vec{\mathbf{f}}_p.
\end{aligned}$$

Taking the difference in initial conditions of (7.4) and (7.5), we have

$$\delta\vec{\mathbf{r}}(0) = \vec{\mathbf{0}}, \quad \delta\dot{\vec{\mathbf{r}}}(0) = \vec{\mathbf{0}}.$$

Encke's method then numerically integrates

$$\delta\ddot{\vec{\mathbf{r}}} = -\frac{\mu}{\rho^3}\left[\delta\vec{\mathbf{r}} - \left(1 - \frac{\rho^3}{r^3}\right)\vec{\mathbf{r}}\right] + \vec{\mathbf{f}}_p, \tag{7.6}$$

with initial conditions

$$\delta\vec{\mathbf{r}}(0) = \vec{\mathbf{0}}, \quad \delta\dot{\vec{\mathbf{r}}}(0) = \vec{\mathbf{0}}.$$

Since $\vec{\rho}$ is a known function of time, obtained analytically from the Keplerian orbit, the true position vector $\vec{\mathbf{r}}$ can be obtained as

$$\vec{\mathbf{r}} = \vec{\rho} + \delta\vec{\mathbf{r}}. \tag{7.7}$$

However, the formulation is not yet complete. Equation (7.6) has the term $1 - \frac{\rho^3}{r^3}$, which is the difference between two almost equal quantities for small $\delta\vec{\mathbf{r}}$, which can cause numerical inaccuracies. In order to deal with this problem, a small variable q is defined by

$$2q = 1 - \frac{r^2}{\rho^2}, \tag{7.8}$$

leading to

$$1 - \frac{\rho^3}{r^3} = 1 - (1 - 2q)^{-\frac{3}{2}}.$$

Expanding this in a Taylor scrics, we obtain

$$1 - \frac{\rho^3}{r^3} = 3q - \frac{3 \times 5}{2!} q^2 + \cdots \tag{7.9}$$

which converges very rapidly for small q, and avoids the loss of precision associated with direct computation of $1 - \frac{\rho^3}{r^3}$. Now, all we need is an expression for q. From its definition (7.8), we can find that

$$\begin{aligned} q = \frac{1}{2}\left(\frac{\rho^2 - r^2}{\rho^2}\right) &= \frac{1}{2}\left(\frac{\vec{\rho}\cdot\vec{\rho} - \vec{\mathbf{r}}\cdot\vec{\mathbf{r}}}{\rho^2}\right), \\ &= \frac{1}{2}\left(\frac{\vec{\rho}\cdot\vec{\rho} - (\vec{\rho}+\delta\vec{\mathbf{r}})\cdot(\vec{\rho}+\delta\vec{\mathbf{r}})}{\rho^2}\right), \\ &= -\frac{\delta\vec{\mathbf{r}}\cdot\left(\vec{\rho}+\frac{\delta\vec{\mathbf{r}}}{2}\right)}{\rho^2}. \end{aligned}$$

For very small $\delta\vec{\mathbf{r}}$ compared to $\vec{\rho}$, we can use

$$q \approx -\frac{\vec{\rho}\cdot\delta\vec{\mathbf{r}}}{\rho^2}.$$

Now, since the quantity $\delta\vec{\mathbf{r}}$ changes much more slowly than $\vec{\mathbf{r}}$, the numerical integration scheme can use much larger time-steps for the same accuracy, leading to reduced computational demand. It has been found to be up to ten times faster than Cowell's method.

Finally, since $\delta\vec{\mathbf{r}}$ deviates from $\vec{\rho}$, there comes a time when $\delta\vec{\mathbf{r}}$ is no longer small. At this time, Encke's method requires a *rectification*. That is, a new osculating orbit is defined using the initial conditions of the true orbit at the time of rectification. This process is shown in Figure 7.1.

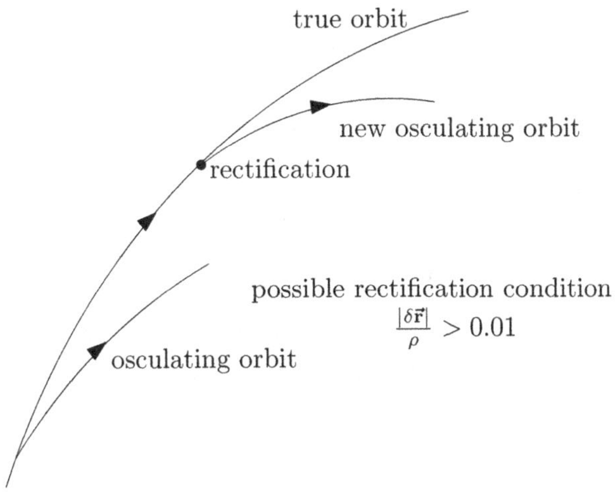

Figure 7.1 Encke's method and recification

7.2 General Perturbations

As we have briefly described earlier, general perturbation methods are concerned with obtaining analytical expressions for the changes in the orbital elements with time, specifically

$$\frac{da}{dt},\quad \frac{de}{dt},\quad \frac{di}{dt},\quad \frac{d\Omega}{dt},\quad \frac{d\omega}{dt}.$$

The expressions we will present are due to Gauss.

The starting point is the perturbed two-body equations of motion

$$\ddot{\vec{\mathbf{r}}} = -\frac{\mu}{r^3}\vec{\mathbf{r}} + \vec{\mathbf{f}}_p. \tag{7.10}$$

We will make use of the cylindrical coordinate frame $\mathcal{F}_o$ (see the start of Section 3.4), which has basis vectors $\vec{\mathbf{x}}_o$, $\vec{\mathbf{y}}_o$ and $\vec{\mathbf{z}}_o$.

Recall that in this coordinate frame, the position and velocity are given by

$$\vec{\mathbf{r}} = r\vec{\mathbf{x}}_o, \quad \vec{\mathbf{v}} = \dot{r}\vec{\mathbf{x}}_o + r\dot{\theta}\vec{\mathbf{y}}_o, \tag{7.11}$$

and that the angular momentum is given by

$$\vec{\mathbf{h}} = h\vec{\mathbf{z}}_o = r^2\dot{\theta}\vec{\mathbf{z}}_o. \tag{7.12}$$

Let us express the perturbing force in $\mathcal{F}_o$ coordinates as

$$\vec{\mathbf{f}}_p = f_r\vec{\mathbf{x}}_o + f_\theta\vec{\mathbf{y}}_o + f_z\vec{\mathbf{z}}_o. \tag{7.13}$$

We will now find an expression for $\frac{da}{dt}$. Recall that the orbital energy is defined by

$$\mathcal{E} = \frac{\vec{\mathbf{v}}\cdot\vec{\mathbf{v}}}{2} - \frac{\mu}{r}, \tag{7.14}$$

and is related to the orbit semi-major axis by (see Section 3.3)

$$\mathcal{E} = -\frac{\mu}{2a}. \tag{7.15}$$

Differentiating this expression with respect to time gives

$$\frac{d\mathcal{E}}{dt} = \frac{da}{dt}\frac{\mu}{2a^2},$$

which leads to

$$\frac{da}{dt} = \frac{2a^2}{\mu}\frac{d\mathcal{E}}{dt}. \tag{7.16}$$

Therefore, we need to find $\frac{d\mathcal{E}}{dt}$. Differentiating (7.14), we find

$$\frac{d\mathcal{E}}{dt} = \vec{\mathbf{v}} \cdot \dot{\vec{\mathbf{v}}} + \dot{r}\frac{\mu}{r^2}.$$

Recalling that $\dot{r} = \frac{\vec{\mathbf{r}} \cdot \vec{\mathbf{v}}}{r}$, and substituting from the equation of motion (7.10), gives

$$\frac{d\mathcal{E}}{dt} = \vec{\mathbf{v}} \cdot \left(-\frac{\mu}{r^3}\vec{\mathbf{r}}\right) + \vec{\mathbf{v}} \cdot \vec{\mathbf{f}}_p + \mu\frac{\vec{\mathbf{r}} \cdot \vec{\mathbf{v}}}{r^3} = \vec{\mathbf{v}} \cdot \vec{\mathbf{f}}_p.$$

Using the expressions in cylindrical coordinates (7.11) and (7.13), we obtain

$$\begin{aligned} \frac{d\mathcal{E}}{dt} &= \left[\dot{r}\vec{\mathbf{x}}_o + r\dot{\theta}\vec{\mathbf{y}}_o\right] \cdot \left[f_r\vec{\mathbf{x}}_o + f_\theta\vec{\mathbf{y}}_o + f_z\vec{\mathbf{z}}_o\right], \\ &= \dot{r}f_r + r\dot{\theta}f_\theta. \end{aligned} \tag{7.17}$$

Now, we need expressions for $\dot{r}$ and $r\dot{\theta}$. These are given by their instantaneous Keplerian two-body quantities, which are (see Section 3.8)

$$\dot{r} = \sqrt{\frac{\mu}{a(1-e^2)}}e\sin\theta. \tag{7.18}$$

and

$$r\dot{\theta} = \sqrt{\frac{\mu}{a(1-e^2)}}(1 + e\cos\theta). \tag{7.19}$$

Therefore, substituting these into (7.17), and then making use of (7.16),we have

$$\frac{d\mathcal{E}}{dt} = \sqrt{\frac{\mu}{a(1-e^2)}}\left[e\sin\theta f_r + (1 + e\cos\theta)f_\theta\right],$$

which leads to the desired expression for $\frac{da}{dt}$

$$\frac{da}{dt} = \frac{2a^2}{\sqrt{\mu a(1-e^2)}}\left[e\sin\theta f_r + (1 + e\cos\theta)f_\theta\right]. \tag{7.20}$$

Following a similar procedure, we can obtain variational equations for all of the orbital elements, which are:

$$\begin{aligned} \frac{de}{dt} &= \sqrt{\frac{a(1-e^2)}{\mu}}\left[\sin\theta f_r + \frac{2\cos\theta + e(1+\cos^2\theta)}{(1+e\cos\theta)}f_\theta\right], \\ \frac{di}{dt} &= \sqrt{\frac{a(1-e^2)}{\mu}}\frac{\cos(\omega+\theta)}{1+e\cos\theta}f_z, \end{aligned}$$

$$\frac{d\Omega}{dt} = \sqrt{\frac{a(1-e^2)}{\mu}} \frac{\sin(\omega+\theta)}{\sin i(1+e\cos\theta)} f_z,$$

$$\frac{d\omega}{dt} = \sqrt{\frac{a(1-e^2)}{\mu}} \left[-\frac{\cos\theta}{e} f_r + \frac{(2+e\cos\theta)\sin\theta}{e(1+e\cos\theta)} f_\theta - \frac{\sin(\omega+\theta)}{\tan i(1+e\cos\theta)} f_z \right].$$

For reference, the derivations of these are contained in Section 7.7.

7.3 Gravitational Perturbations due to a Non-Spherical Primary Body

Now that the special and general perturbation methods have been formulated, we need expressions for the perturbative accelerations $\vec{\mathbf{f}}_p$. As has been described, there are several sources of perturbative accelerations. We will focus only on the effects due to a non-spherical primary. These effects are very important for Earth-orbiting satellites, and as we will see they lead to some favorable effects which can be exploited during orbit design.

As we have seen in Section 3.1, the gravitational force due to a point mass m_1 acting on a point mass m_2 is

$$\vec{\mathbf{F}} = -\frac{Gm_1m_2}{r^3}\vec{\mathbf{r}},$$

where $\vec{\mathbf{r}}$ is the position of m_2 relative to m_1. The resulting force per unit mass is

$$\vec{\mathbf{f}} = -\frac{Gm_1}{r^3}\vec{\mathbf{r}} = -\frac{\mu}{r^3}\vec{\mathbf{r}}.$$

We briefly mentioned that the force per unit mass could be obtained from a potential function

$$\phi = \frac{Gm_1}{r},$$

such that

$$\vec{\mathbf{f}} = \vec{\nabla}\phi,$$

where $\vec{\nabla}$ is the gradient operator, which is given in Cartesian coordinates by

$$\vec{\nabla}(\cdot) = \frac{\partial}{\partial x}(\cdot)\vec{\mathbf{x}} + \frac{\partial}{\partial y}(\cdot)\vec{\mathbf{y}} + \frac{\partial}{\partial z}(\cdot)\vec{\mathbf{z}},$$

where (x, y, z) are the coordinates of a vector in some reference frame $\mathcal{F}$ with unit basis vectors $\vec{\mathbf{x}}$, $\vec{\mathbf{y}}$ and $\vec{\mathbf{z}}$.

The force per unit mass acting on m_2 due to a series of point masses is given by

$$\vec{\mathbf{f}} = -\sum_i \frac{Gm_i}{r_i^3}\vec{\mathbf{r}}_i,$$

which can likewise be obtained from the potential function

$$\phi = \sum_i \frac{Gm_i}{r_i}. \tag{7.21}$$

It is natural to extend this to the gravitational potential of a continuous body by setting

$$m_i \to dm = \rho dV, \quad \sum_i \to \int_V$$

where dm is an infinitessimal mass element (where ρ is the mass density and dV is an infinitessimal volume element), and the integral is taken over the entire body.

Let us now derive the gravitational potential for an arbitrary body of total mass m_1 (refer to Figure 7.2). Let $\mathcal{F}$ be some reference frame, with unit basis vectors $\vec{\mathbf{x}}$, $\vec{\mathbf{y}}$ and $\vec{\mathbf{z}}$. Let $dm(\vec{\mathbf{r}}')$ be a mass element in the body located at position

$$\vec{\mathbf{r}}' = x'\vec{\mathbf{x}} + y'\vec{\mathbf{y}} + z'\vec{\mathbf{z}}. \tag{7.22}$$

We will use spherical coordinates (r', λ', δ') such that

$$x' = r' \cos\delta' \cos\lambda', \quad y' = r' \cos\delta' \sin\lambda', \quad z' = r' \sin\delta'.$$

We will assume that our frame $\mathcal{F}$ has origin at the body center of mass.

Now, let us find the gravitational potential at a point $\vec{\mathbf{r}}$ *outside* the body, with coordinates

$$\vec{\mathbf{r}} = x\vec{\mathbf{x}} + y\vec{\mathbf{y}} + z\vec{\mathbf{z}}, \tag{7.23}$$

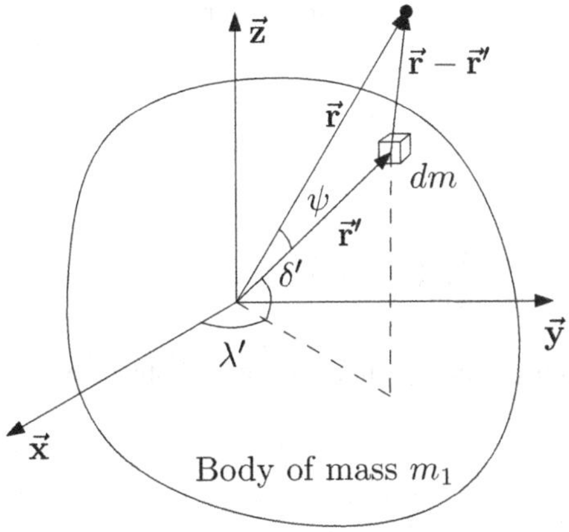

Figure 7.2 Location of mass element and test point $\vec{\mathbf{r}}$

with associated spherical coordinates

$$x = r\cos\delta\cos\lambda, \quad y = r\cos\delta\sin\lambda, \quad z = r\sin\delta.$$

The potential at $\vec{\mathbf{r}}$ due to the mass element $dm(\vec{\mathbf{r}}')$ is

$$d\phi(\vec{\mathbf{r}}) = \frac{G\rho(\vec{\mathbf{r}}')dV(\vec{\mathbf{r}}')}{|\vec{\mathbf{r}} - \vec{\mathbf{r}}'|},$$

which leads to the total potential at $\vec{\mathbf{r}}$

$$\phi(\vec{\mathbf{r}}) = \int_V d\phi(\vec{\mathbf{r}}) = \int_V \frac{G\rho(\vec{\mathbf{r}}')dV(\vec{\mathbf{r}}')}{|\vec{\mathbf{r}} - \vec{\mathbf{r}}'|}, \tag{7.24}$$

where the integral is taken over the entire volume of the body V. Now,

$$\begin{aligned}
|\vec{\mathbf{r}} - \vec{\mathbf{r}}'| &= \sqrt{(\vec{\mathbf{r}} - \vec{\mathbf{r}}') \cdot (\vec{\mathbf{r}} - \vec{\mathbf{r}}')}, \\
&= \sqrt{\vec{\mathbf{r}} \cdot \vec{\mathbf{r}} + \vec{\mathbf{r}}' \cdot \vec{\mathbf{r}}' - 2\vec{\mathbf{r}} \cdot \vec{\mathbf{r}}'}, \\
&= \sqrt{r^2 + r'^2 - 2rr'\cos\psi}, \\
&= r\sqrt{1 - 2\frac{r'}{r}\cos\psi + \left(\frac{r'}{r}\right)^2}.
\end{aligned}$$

where ψ is the angle between $\vec{\mathbf{r}}$ and $\vec{\mathbf{r}}'$. Therefore, the gravitational potential in (7.24) becomes

$$\phi(\vec{\mathbf{r}}) = \int_V d\phi(\vec{\mathbf{r}}) = \frac{G}{r}\int_V \frac{\rho(\vec{\mathbf{r}}')dV(\vec{\mathbf{r}}')}{\sqrt{1 - 2\frac{r'}{r}\cos\psi + \left(\frac{r'}{r}\right)^2}}, \tag{7.25}$$

Assuming that $r' < r$, the reciprocal of the square root expression can be expanded in a power series in $\frac{r'}{r}$ using the Legendre polynomials in the variable $\cos\psi$:

$$\begin{aligned}
\frac{1}{\sqrt{1 - 2\frac{r'}{r}\cos\psi + \left(\frac{r'}{r}\right)^2}} &= P_0(\cos\psi) + \left(\frac{r'}{r}\right)P_1(\cos\psi) + \left(\frac{r'}{r}\right)^2 P_2(\cos\psi) + \cdots \\
&= \sum_{n=0}^{\infty}\left(\frac{r'}{r}\right)^n P_n(\cos\psi),
\end{aligned}$$

where $P_n(x)$ are the Legendre polynomials, with the first three given by

$$\begin{aligned}
P_0(x) &= 1, \\
P_1(x) &= x, \\
P_2(x) &= \frac{3}{2}x^2 - \frac{1}{2},
\end{aligned} \tag{7.26}$$

and in general,

$$(n+1)P_{n+1}(x) = (2n+1)xP_n(x) - nP_{n-1}(x).$$

Therefore, the potential at $\vec{\mathbf{r}}$ in (7.25) becomes

$$\phi(\vec{\mathbf{r}}) = \frac{G}{r}\sum_{n=0}^{\infty}\int_V \rho(\vec{\mathbf{r}}')\left(\frac{r'}{r}\right)^n P_n(\cos\psi)dV. \tag{7.27}$$

Let us now expand the potential (7.27) in terms of the first two terms (using (7.26))

$$\phi(\vec{\mathbf{r}}) = \frac{G}{r}\int_V \rho(\vec{\mathbf{r}}')dV + \frac{G}{r^2}\int_V \rho(\vec{\mathbf{r}}')r'\cos\psi dV + \frac{G}{r}\sum_{n=2}^{\infty}\int_V \rho(\vec{\mathbf{r}}')\left(\frac{r'}{r}\right)^n P_n(\cos\psi)dV. \tag{7.28}$$

We immediately recognize the first integral as the total mass of the body, m_1,

$$\int_V \rho(\vec{\mathbf{r}}')dV = m_1.$$

For the second integral, we can use the fact that

$$\cos\psi = \frac{\vec{\mathbf{r}}\cdot\vec{\mathbf{r}}'}{rr'},$$

to rewrite the integral as

$$\int_V \rho(\vec{\mathbf{r}}')r'\cos\psi dV = \frac{1}{r}\int_V \rho(\vec{\mathbf{r}}')\vec{\mathbf{r}}\cdot\vec{\mathbf{r}}'dV = \frac{1}{r}\int_V \rho(\vec{\mathbf{r}}')\vec{\mathbf{r}}'dV\cdot\vec{\mathbf{r}}.$$

But, as we saw in Section 2.3, the integral

$$\frac{1}{m_1}\int_V \rho(\vec{\mathbf{r}}')\vec{\mathbf{r}}'dV,$$

defines the center of mass of the body. Since we have chosen the origin of $\mathcal{F}$ to coincide with the center of mass, this integral is zero

$$\int_V \rho(\vec{\mathbf{r}}')\vec{\mathbf{r}}'dV = \vec{\mathbf{0}},$$

which leads to

$$\int_V \rho(\vec{\mathbf{r}}')r'\cos\psi dV = \frac{1}{r}\vec{\mathbf{0}}\cdot\vec{\mathbf{r}} = 0.$$

Therefore, the potential due to the body (7.28) can be expressed as

$$\phi(\vec{\mathbf{r}}) = \frac{Gm_1}{r} + \frac{G}{r}\sum_{n=2}^{\infty}\int_V \rho(\vec{\mathbf{r}}')\left(\frac{r'}{r}\right)^n P_n(\cos\psi)dV, \tag{7.29}$$

and we see that since the first term $\frac{Gm_1}{r}$ is just the two-body potential for a point mass. The potential resulting in the perturbative force per unit mass is

$$\phi_p(\vec{\mathbf{r}}) = \frac{G}{r}\sum_{n=2}^{\infty}\int_V \rho(\vec{\mathbf{r}}')\left(\frac{r'}{r}\right)^n P_n(\cos\psi)dV. \tag{7.30}$$

Now, in order to evaluate the integrals in (7.30), it would be useful to write it in terms of the spherical coordinates rather than $\cos\psi$. As we have mentioned,

$$\vec{\mathbf{r}}\cdot\vec{\mathbf{r}}' = rr'\cos\psi.$$

Making use of the spherical coordinates for $\vec{\mathbf{r}}$ and $\vec{\mathbf{r}}'$, this becomes

$$\begin{aligned} rr'\cos\psi &= rr'\left[\cos\delta\cos\delta'\cos\lambda\cos\lambda' + \cos\delta\cos\delta'\sin\lambda\sin\lambda' + \sin\delta\sin\delta'\right], \\ &= rr'\left[\cos\delta\cos\delta'\cos(\lambda-\lambda') + \sin\delta\sin\delta'\right], \end{aligned}$$

which leads to

$$\cos\psi = \cos\delta\cos\delta'\cos(\lambda-\lambda') + \sin\delta\sin\delta'.$$

Now, there is an *addition theorem* for spherical harmonics that states

$$P_n(\cos\psi) = P_n(\sin\delta)P_n(\sin\delta') + 2\sum_{m=1}^{n}\frac{(n-m)!}{(n+m)!}\left[A_{n,m}A'_{n,m} + B_{n,m}B'_{n,m}\right], \tag{7.31}$$

where

$$\begin{aligned} A_{n,m} &= P_{n,m}(\sin\delta)\cos(m\lambda), & A'_{n,m} &= P_{n,m}(\sin\delta')\cos(m\lambda'), \\ B_{n,m} &= P_{n,m}(\sin\delta)\sin(m\lambda), & B'_{n,m} &= P_{n,m}(\sin\delta')\sin(m\lambda'), \end{aligned}$$

and the *associated Legendre functions* $P_{n,m}(x)$ are

$$P_{n,m}(x) = (1-x^2)^{\frac{m}{2}}\frac{d^m}{dx^m}P_n(x).$$

Substituting (7.31) into (7.30) leads to

$$\begin{aligned}
\phi_p(\vec{\mathbf{r}}) &= \frac{G}{r}\sum_{n=2}^{\infty}\int_V \rho(\vec{\mathbf{r}}')\left(\frac{r'}{r}\right)^n P_n(\sin\delta)P_n(\sin\delta')dV, \\
&+ \frac{G}{r}\sum_{n=2}^{\infty}\sum_{m=1}^{n} 2\frac{(n-m)!}{(n+m)!}\int_V \rho(\vec{\mathbf{r}}')\left(\frac{r'}{r}\right)^n \left[P_{n,m}(\sin\delta')\cos(m\lambda')\right. \\
&\times P_{n,m}(\sin\delta)\cos(m\lambda) + P_{n,m}(\sin\delta')\sin(m\lambda')P_{n,m}(\sin\delta)\sin(m\lambda)\left.\right] dV.
\end{aligned} \tag{7.32}$$

Based on this, let us define the factors

$$J_n = -\frac{1}{R_e^n m_1}\int_V \rho(\vec{\mathbf{r}}')(r')^n P_n(\sin\delta')dV, \tag{7.33}$$

$$C_{n,m} = \frac{1}{R_e^n m_1} 2\frac{(n-m)!}{(n+m)!}\int_V \rho(\vec{\mathbf{r}}')(r')^n P_{n,m}(\sin\delta')\cos(m\lambda')dV, \tag{7.34}$$

and

$$S_{n,m} = \frac{1}{R_e^n m_1} 2\frac{(n-m)!}{(n+m)!}\int_V \rho(\vec{\mathbf{r}}')(r')^n P_{n,m}(\sin\delta')\sin(m\lambda')dV, \tag{7.35}$$

where R_e is some normalizing radius of the body m_1. For the Earth, R_e is taken to be the equatorial radius. Then, the perturbing gravitational potential of the body is given by

$$\begin{aligned}
\phi_p(\vec{\mathbf{r}}) = \frac{Gm_1}{r}&\left[-\sum_{n=2}^{\infty} J_n\left(\frac{R_e}{r}\right)^n P_n(\sin\delta)\right. \\
&\left.+\sum_{n=2}^{\infty}\sum_{m=1}^{n}\left(\frac{R_e}{r}\right)^n P_{n,m}(\sin\delta)\left[C_{n,m}\cos(m\lambda) + S_{n,m}\sin(m\lambda)\right]\right].
\end{aligned} \tag{7.36}$$

In practice, the coefficients J_n, $C_{n,m}$ and $S_{n,m}$ are determined experimentally from satellite observations, and can be obtained from tables.

Let us now note a few interesting points. First, if the body is rotationally symmetric about $\vec{\mathbf{z}}$, then the density function is independent of the longitude λ', that is $\rho(r', \lambda', \delta') = \rho(r', \delta')$. Using spherical coordinates, the volume element is $dV = (r')^2\cos\delta' d\lambda' dr' d\delta'$, and the integrals in $C_{n,m}$ and $S_{n,m}$ (7.34) and (7.35) become

$$\begin{aligned}
&\int_V \rho(\vec{\mathbf{r}}')(r')^n P_{n,m}(\sin\delta')\cos(m\lambda')dV \\
&= \int_{-\frac{\pi}{2}}^{\frac{\pi}{2}}\int_0^{R(\delta')}\int_0^{2\pi} \rho(r', \delta')(r')^{n+2} P_{n,m}(\sin\delta')\cos\delta'\cos(m\lambda')d\lambda' dr' d\delta' = 0,
\end{aligned}$$

and

$$\int_V \rho(\vec{\mathbf{r}}')(r')^n P_{n,m}(\sin\delta')\sin(m\lambda')dV$$
$$= \int_{-\frac{\pi}{2}}^{\frac{\pi}{2}} \int_0^{R(\delta')} \int_0^{2\pi} \rho(r',\delta')(r')^{n+2} P_{n,m}(\sin\delta')\cos\delta' \sin(m\lambda')d\lambda' dr' d\delta' = 0,$$

since

$$\int_0^{2\pi} \cos(m\lambda')d\lambda' = \int_0^{2\pi} \sin(m\lambda')d\lambda' = 0 \text{ for } m \geq 1.$$

Therefore, **for rotationally symmetric bodies about $\vec{\mathbf{z}}$, we have** $C_{n,m} = S_{n,m} = 0$, and

$$\phi_p(\vec{\mathbf{r}}) = -\frac{Gm_1}{r}\sum_{n=2}^{\infty} J_n \left(\frac{R_e}{r}\right)^n P_n(\sin\delta).$$

A property of the Legendre polynomials is that they satisfy an orthogonality property, that is

$$\int_{-1}^{1} P_n(x)P_m(x)dx = 0, \quad n \neq m.$$

Since $P_0(x) = 1$, this gives

$$\int_{-1}^{1} P_n(x)dx = 0, \quad n \geq 1.$$

If we now make the change of variables $x = \sin\delta'$, the integral above becomes

$$\int_{-\frac{\pi}{2}}^{\frac{\pi}{2}} P_n(\sin\delta')\cos\delta' d\delta' = \int_{-1}^{1} P_n(x)dx = 0.$$

This now shows that if the body is spherically symmetric, and therefore $\rho(r',\lambda',\delta') = \rho(r')$ is independent of both λ' and δ', then the integrals in the J_n terms (7.33) become

$$\int_V \rho(\vec{\mathbf{r}}')(r')^n P_n(\sin\delta')dV = \int_{-\frac{\pi}{2}}^{\frac{\pi}{2}} \int_0^{R} \int_0^{2\pi} \rho(r')(r')^{n+2} P_n(\sin\delta')\cos\delta' d\lambda' dr' d\delta' = 0.$$

Therefore, for a spherically symmetric body, the perturbing potential is $\phi_p = 0$, and the resulting force per unit mass on m_2 is the same as for a point mass m_1 located at the center of mass of the body. Since planets are nearly spherically symmetric, this justifies our initial treatment of orbits using point masses.

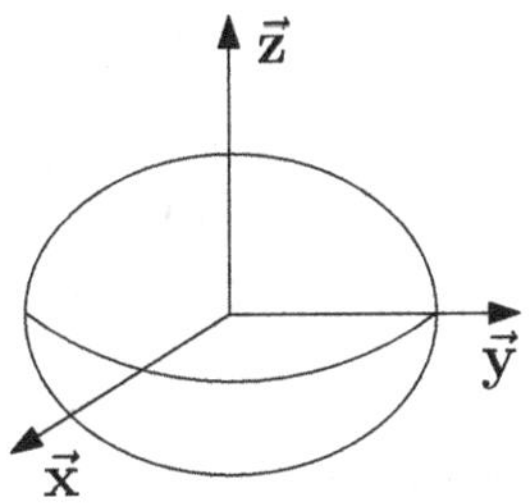

Figure 7.3 Earth flattening at the poles

For the Earth, the most dominant perturbing effect is the J_2 term, which is a result of the Earth's oblate shape (it is flattened at the poles, as shown in Figure 7.3). For the Earth,

$$J_2 = 1.083 \times 10^{-3}.$$

The perturbing potential including J_2 effects only is

$$\phi_p = -\frac{\mu}{r^3} J_2 R_e^2 \left(\frac{3}{2} \sin^2 \delta - \frac{1}{2} \right). \tag{7.37}$$

7.3.1 *The Perturbative Force Per Unit Mass Due to J_2*

Now that we have the potential ϕ_p, we can obtain the perturbative force per unit mass by taking its gradient. Let us do this in ECI coordinates. In ECI coordinates, the position vector of a spacecraft is given by

$$\vec{\mathbf{r}} = x\vec{\mathbf{x}}_G + y\vec{\mathbf{y}}_G + z\vec{\mathbf{z}}_G, \tag{7.38}$$

with magnitude $r = \sqrt{x^2 + y^2 + z^2}$.

Since $z = r \sin \delta$, we have $\sin \delta = \frac{z}{r}$. Therefore, the perturbing potential due to J_2 in equation (7.37) becomes

$$\phi_p = -\frac{\mu}{r^3} J_2 R_e^2 \left(\frac{3}{2} \left(\frac{z}{r} \right)^2 - \frac{1}{2} \right). \tag{7.39}$$

Let us now take the gradient of this term by term. We start with

$$\frac{\partial}{\partial x}\left(\frac{1}{r^3} \right) = \frac{\partial}{\partial x}\left(\frac{1}{(x^2 + y^2 + z^2)^{\frac{3}{2}}} \right) = -\frac{3}{2}\frac{2x}{(x^2 + y^2 + z^2)^{\frac{5}{2}}} = -\frac{3x}{r^5}.$$

Likewise,

$$\frac{\partial}{\partial y}\left(\frac{1}{r^3} \right) = -\frac{3y}{r^5}, \quad \frac{\partial}{\partial z}\left(\frac{1}{r^3} \right) = -\frac{3z}{r^5}.$$

Next,

$$\frac{\partial}{\partial x}\left(\frac{z^2}{r^5}\right) = -\frac{5xz^2}{r^7}, \quad \frac{\partial}{\partial y}\left(\frac{z^2}{r^5}\right) = -\frac{5yz^2}{r^7}.$$

and

$$\frac{\partial}{\partial z}\left(\frac{z^2}{r^5}\right) = \frac{2z}{r^5} - \frac{5z^3}{r^7}.$$

Therefore,

$$\begin{aligned}\vec{\mathbf{f}}_p = \vec{\nabla}\phi_p &= -\mu J_2 R_e^2\left[\left(-\frac{15xz^2}{2r^7} + \frac{3x}{2r^5}\right)\vec{\mathbf{x}}_G + \left(-\frac{15yz^2}{2r^7} + \frac{3y}{2r^5}\right)\vec{\mathbf{y}}_G\right.\\ &\quad \left. + \left(-\frac{15z^3}{2r^7} + \frac{3z}{r^5} + \frac{3z}{2r^5}\right)\vec{\mathbf{z}}_G\right],\\ &= \frac{3\mu J_2 R_e^2}{2r^5}\left[\left(5\frac{z^2}{r^2} - 1\right)(x\vec{\mathbf{x}}_G + y\vec{\mathbf{y}}_G + z\vec{\mathbf{z}}_G) - 2z\vec{\mathbf{z}}_G\right].\end{aligned}$$

Recognizing that $\vec{\mathbf{r}} = x\vec{\mathbf{x}}_G + y\vec{\mathbf{y}}_G + z\vec{\mathbf{z}}_G$, and that $z = \vec{\mathbf{r}} \cdot \vec{\mathbf{z}}_G$, we can write the perturbative force per unit mass purely in terms of physical vectors

$$\vec{\mathbf{f}}_p = \frac{3\mu J_2 R_e^2}{2r^5}\left[\left(5\frac{(\vec{\mathbf{r}} \cdot \vec{\mathbf{z}}_G)^2}{r^2} - 1\right)\vec{\mathbf{r}} - 2(\vec{\mathbf{r}} \cdot \vec{\mathbf{z}}_G)\vec{\mathbf{z}}_G\right]. \tag{7.40}$$

7.4 Effect of J_2 on the Orbital Elements

Now that we have an expression for the perturbative force per unit mass due to J_2, we can use the Gauss variational equations to determine its effect. To do this, we need to express the perturbative force per unit mass in cylindrical coordinates. Now, in cylindrical coordinates, the position vector is given by

$$\vec{\mathbf{r}} = r\vec{\mathbf{x}}_o.$$

As we can see from equation (7.40), we need an expression for $\vec{\mathbf{z}}_G$. This has been obtained in Section 7.7 as (see equation (7.59))

$$\vec{\mathbf{z}}_G = \sin i \sin(\omega + \theta)\vec{\mathbf{x}}_o + \sin i \cos(\omega + \theta)\vec{\mathbf{y}}_o + \cos i \vec{\mathbf{z}}_o.$$

We therefore have

$$\vec{\mathbf{z}}_G \cdot \vec{\mathbf{r}} = r \sin i \sin(\omega + \theta),$$

and we can obtain the perturbative force per unit mass in cylindrical coordinates directly from (7.40) as

$$\begin{aligned}\vec{\mathbf{f}}_p &= \frac{3\mu J_2 R_e^2}{2r^5}\left[\left(5\frac{(r\sin i\sin(\omega+\theta))^2}{r^2}-1\right)r\vec{\mathbf{x}}_o\right.\\ &\quad \left.-2r\sin i\sin(\omega+\theta)\left(\sin i\sin(\omega+\theta)\vec{\mathbf{x}}_o+\sin i\cos(\omega+\theta)\vec{\mathbf{y}}_o+\cos i\vec{\mathbf{z}}_o\right)\right],\\ &= \frac{3\mu J_2 R_e^2}{2r^4}\left[\left(3\sin^2 i\sin^2(\omega+\theta)-1\right)\vec{\mathbf{x}}_o-\sin^2 i\sin(2(\omega+\theta))\vec{\mathbf{y}}_o\right.\\ &\quad \left.-\sin 2i\sin(\omega+\theta)\vec{\mathbf{z}}_o\right],\end{aligned} \tag{7.41}$$

from which we can identify

$$f_r = \frac{3\mu J_2 R_e^2}{2r^4}\left(3\sin^2 i\sin^2(\omega+\theta)-1\right), \tag{7.42}$$

$$f_\theta = -\frac{3\mu J_2 R_e^2}{2r^4}\sin^2 i\sin(2(\omega+\theta)), \tag{7.43}$$

and

$$f_z = -\frac{3\mu J_2 R_e^2}{2r^4}\sin 2i\sin(\omega+\theta). \tag{7.44}$$

Typically, a perturbed orbital element will have secular and periodic variations, as shown in Figure 7.4. The secular variation consists of long term changes in the element, while the periodic variation leads to no net change in the element. Let us now use the Gauss variational equations to determine the secular variation due to J_2. We shall only examine the secular variation in Ω. We have seen that the Gauss variational equation for Ω is

$$\frac{d\Omega}{dt} = \sqrt{\frac{a(1-e^2)}{\mu}}\frac{\sin(\omega+\theta)}{\sin i(1+e\cos\theta)}f_z.$$

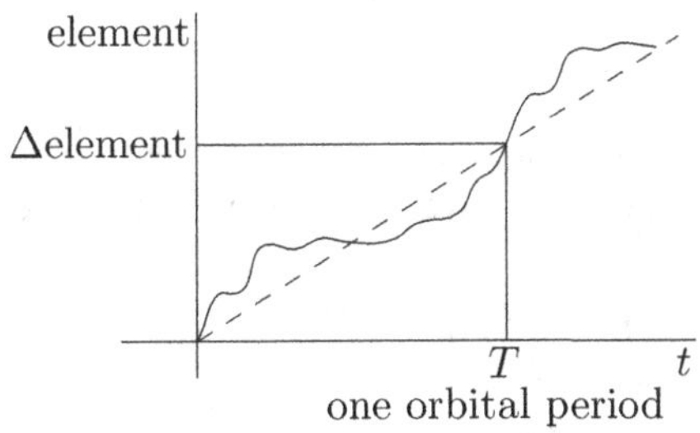

Figure 7.4 Time history of an element

It will be more useful to find the variation in terms of the true anomaly θ. To this end, the chain rule gives us

$$\frac{d\Omega}{dt} = \frac{d\Omega}{d\theta}\dot{\theta},$$

which leads to

$$\frac{d\Omega}{d\theta} = \frac{1}{\dot{\theta}}\frac{d\Omega}{dt}. \tag{7.45}$$

From Section 3.8, we have for a two-body orbit

$$\dot{\theta} = \sqrt{\frac{\mu}{a^3}}\frac{(1 + e\cos\theta)^2}{(1 - e^2)^{\frac{3}{2}}},$$

which is sufficiently accurate assuming the effects of $\vec{\mathbf{f}}_p$ are small. This leads to

$$\frac{d\Omega}{d\theta} = \frac{a^2(1 - e^2)}{\mu}\frac{\sin(\omega + \theta)}{\sin i(1 + e\cos\theta)^3}f_z. \tag{7.46}$$

Making use of the polar equation for the orbit (equation (3.76) in Section 3.8), and the identity $2\sin i\cos i = \sin 2i$, in equation (7.44), f_z becomes

$$f_z = -\frac{3\mu J_2 R_e^2}{a^4(1 - e^2)^4}\sin i\cos i\sin(\omega + \theta)(1 + e\cos\theta)^4. \tag{7.47}$$

Substituting this into (7.46) leads to

$$\frac{d\Omega}{d\theta} = -\frac{3J_2 R_e^2}{a^2(1 - e^2)^2}\cos i\sin^2(\omega + \theta)(1 + e\cos\theta). \tag{7.48}$$

To determine the secular change in Ω, we can find the average over one orbit,

$$\begin{aligned}\Delta\Omega = \int_0^{\Delta\Omega} d\Omega &= \int_0^{2\pi}\frac{d\Omega}{d\theta}d\theta,\\ &= -\frac{3J_2 R_e^2}{a^2(1 - e^2)^2}\int_0^{2\pi}\cos i\sin^2(\omega + \theta)(1 + e\cos\theta)d\theta.\end{aligned}$$

Now, we expect changes over an orbit in each of the elements to be small, such that they can be considered constant in the integrand (we vary θ only). Therefore, we have

$$\begin{aligned}\Delta\Omega = \int_0^{\Delta\Omega} d\Omega &= \int_0^{2\pi}\frac{d\Omega}{d\theta}d\theta,\\ &= -\frac{3J_2 R_e^2}{a^2(1 - e^2)^2}\cos i\int_0^{2\pi}\sin^2(\omega + \theta)(1 + e\cos\theta)d\theta.\end{aligned}$$

Now,

$$\int_0^{2\pi} \sin^2(\omega+\theta)(1+e\cos\theta)d\theta = \int_0^{2\pi} \sin^2(\omega+\theta)d\theta + e\int_0^{2\pi} \sin^2(\omega+\theta)\cos\theta d\theta.$$

The first integral is simply

$$\int_0^{2\pi} \sin^2(\omega+\theta)d\theta = \int_0^{2\pi} \frac{1}{2} - \frac{1}{2}\cos(2(\omega+\theta))d\theta = \pi.$$

To evaluate the second integral, we must make use of some trigonometric identities. Specifically,

$$\begin{aligned}
\sin^2(\omega+\theta)\cos\theta &= \cos\theta\left(\frac{1-\cos(2(\omega+\theta))}{2}\right),\\
&= \frac{\cos\theta}{2} - \frac{\cos\theta}{2}\left[\cos 2\omega\cos 2\theta - \sin 2\omega \sin 2\theta\right],\\
&= \frac{\cos\theta}{2} - \frac{\cos 2\omega\cos\theta(1-2\sin^2\theta)}{2} + \frac{\cos\theta\sin 2\omega\, 2\sin\theta\cos\theta}{2},\\
&= \frac{\cos\theta}{2}(1-\cos 2\omega) + \cos 2\omega\cos\theta\sin^2\theta + \sin 2\omega\sin\theta\cos^2\theta.
\end{aligned}$$

Now,

$$\int_0^{2\pi} \cos\theta d\theta = 0, \quad \int_0^{2\pi} \cos\theta\sin^2\theta d\theta = \left.\frac{\sin^3\theta}{3}\right|_0^{2\pi} = 0,$$

$$\int_0^{2\pi} \sin\theta\cos^2\theta d\theta = -\left.\frac{\cos^3\theta}{3}\right|_0^{2\pi} = 0.$$

Therefore,

$$\int_0^{2\pi} \sin^2(\omega+\theta)\cos\theta d\theta = 0,$$

and

$$\int_0^{2\pi} \sin^2(\omega+\theta)(1+e\cos\theta)d\theta = \pi,$$

from which we obtain the change in Ω as

$$\Delta\Omega = \int_0^{\Delta\Omega} d\Omega = \int_0^{2\pi} \frac{d\Omega}{d\theta} d\theta,$$
$$= -\frac{3\pi J_2 R_e^2}{a^2(1-e^2)^2} \cos i.$$

To obtain the secular (average) rate of change of Ω (which we will denote $\langle\dot{\Omega}\rangle$), we divide by the orbital period ΔT, which is given by

$$\Delta T = 2\pi\sqrt{\frac{a^3}{\mu}},$$

to obtain

$$\langle\dot{\Omega}\rangle = \frac{\Delta\Omega}{\Delta T} = -\frac{3J_2 R_e^2}{2(1-e^2)^2}\sqrt{\frac{\mu}{a^7}} \cos i. \tag{7.49}$$

This secular change in Ω is called *nodal regression.*

If we repeated this procedure for the other orbital elements, we would find that

$$\begin{aligned} \langle\dot{a}\rangle &= 0, \\ \langle\dot{e}\rangle &= 0, \\ \langle\dot{i}\rangle &= 0, \\ \langle\dot{\omega}\rangle &= \frac{3J_2 R_e^2}{4(1-e^2)^2}\sqrt{\frac{\mu}{a^7}}(5\cos^2 i - 1). \end{aligned} \tag{7.50}$$

Therefore, the oblateness effects of the earth (through J_2) only affect Ω and ω in the long-term, with the orbital plane rotating about the Earth's spin axis at an average rate of $\langle\dot{\Omega}\rangle$, and the argument of perigee rotating about the orbit normal at an average rate of $\langle\dot{\omega}\rangle$.

7.5 Special Types of Orbits

The secular variations of Ω and ω provide useful types of orbits. We shall present two such types.

7.5.1 Sun-Synchronous Orbits

We see from equation (7.49), that for a given semi-major axis a, and eccentricity e, we can appropriately select the inclination to achieve a given average nodal regression $\langle\dot{\Omega}\rangle$.

In particular, if we choose i such that

$$\langle\dot{\Omega}\rangle = 360^\circ/year,$$

then the orbital plane rotates at the same rate as the Earth orbits around the sun. Therefore, the orbit's ascending node is always the same relative to the sun, as shown in Figure 7.5. This

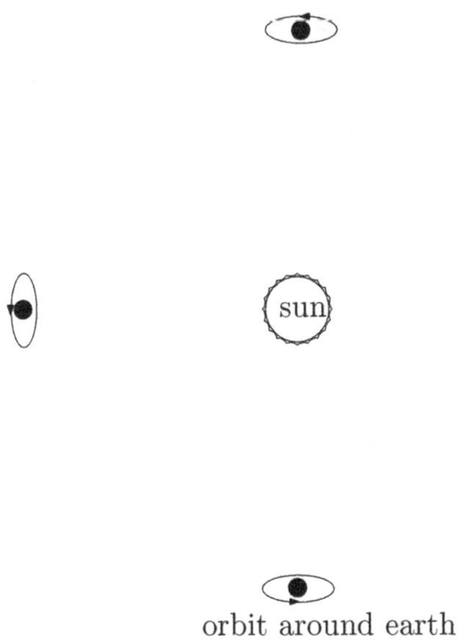

Figure 7.5 Sun-synchronous orbit

is called a *sun-synchronous* orbit, and it finds many applications. It is particularly useful for Earth-sensing missions, since the lighting conditions of the Earth are always the same every orbit.

7.5.2 *Molniya Orbits*

From the last equation in (7.50), we see that by selecting $5\cos^2 i - 1 = 0$, or $i = 63.4°,\ 116.6°$, the perigee advance is zero ($\langle\dot{\omega}\rangle = 0$). This is called a frozen orbit.

A Molniya orbit is a highly eccentric frozen orbit with a twelve-hour period. This is very useful for communications with high latitude regions, since the apogee remains over the same latitude (see Figure 7.6). For highly-eccentric orbits, the spacecraft spends the majority of its orbit close to apogee. By having a number of satellites in Molniya-type orbits, continuous coverage of high latitude regions is possible.

7.6 Small Impulse Form of the Gauss Variational Equations

We have seen in Chapter 5 that we can create orbit changes by impulsively changing the velocity Δv, at an appropriate point in the orbit. If the velocity change is small, we can use Gauss' variational equations to observe the change in the orbital elements.

Let $\vec{\mathbf{f}}_p$ be a small constant thrust applied over a short interval δt, such that a velocity change (from the two-body velocity) $\delta\vec{\mathbf{v}}$ results. From equation (7.2),

$$\dot{\vec{\mathbf{v}}} = -\frac{\mu}{r^3}\vec{\mathbf{r}} + \vec{\mathbf{f}}_p.$$

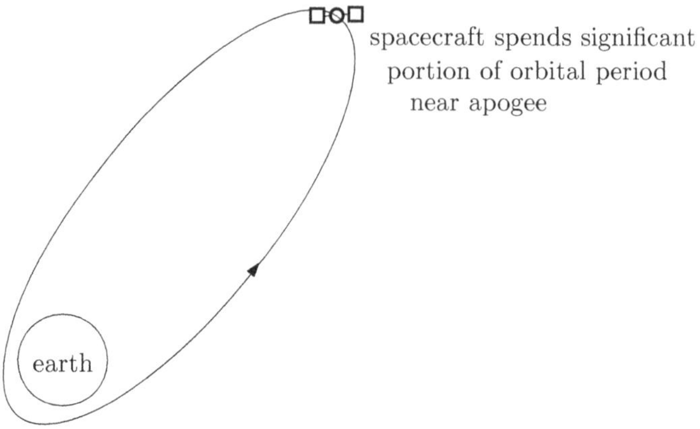
spacecraft spends significant
portion of orbital period
near apogee
earth

Figure 7.6 Molniya orbit

For two-body motion, the velocity change over a short interval is (setting $\vec{\mathbf{f}}_p = \vec{\mathbf{0}}$)

$$\delta \vec{\mathbf{v}}_{tb} = \dot{\vec{\mathbf{v}}} \delta t = -\frac{\mu}{r^3} \vec{\mathbf{r}} \delta t.$$

For the true motion, we have

$$\delta \vec{\mathbf{v}}_{tb} + \delta \vec{\mathbf{v}} = -\frac{\mu}{r^3} \vec{\mathbf{r}} \delta t + \vec{\mathbf{f}}_p \delta t,$$

from which we can identify the required thrust

$$\vec{\mathbf{f}}_p = \frac{\delta \vec{\mathbf{v}}}{\delta t}.$$

Writing $\delta \vec{\mathbf{v}}$ in cylindrical coordinates

$$\delta \vec{\mathbf{v}} = \vec{\mathcal{F}}_o^T \begin{bmatrix} \delta v_r \\ \delta v_\theta \\ \delta v_z \end{bmatrix},$$

and substituting this into Gauss variational equations at the end of Section 7.2, the small impulse form of the Gauss variational equations can be obtained:

$$\delta a = \frac{2a^2}{\sqrt{\mu a(1-e^2)}} \left[e \sin\theta \delta v_r + (1 + e\cos\theta) \delta v_\theta \right],$$

$$\delta e = \sqrt{\frac{a(1-e^2)}{\mu}} \left[\sin\theta \delta v_r + \frac{2\cos\theta + e(1+\cos^2\theta)}{(1+e\cos\theta)} \delta v_\theta \right],$$

$$\delta i = \sqrt{\frac{a(1-e^2)}{\mu}} \frac{\cos(\omega+\theta)}{1+e\cos\theta}\delta v_z,$$

$$\delta\Omega = \sqrt{\frac{a(1-e^2)}{\mu}} \frac{\sin(\omega+\theta)}{\sin i(1+e\cos\theta)}\delta v_z,$$

$$\delta\omega = \sqrt{\frac{a(1-e^2)}{\mu}} \left[-\frac{\cos\theta}{e}\delta v_r + \frac{(2+e\cos\theta)\sin\theta}{e(1+e\cos\theta)}\delta v_\theta - \frac{\sin(\omega+\theta)}{\tan i(1+e\cos\theta)}\delta v_z\right].$$

The above small impulse form of the Gauss variational equations can be used to determine the orbital maneuvers required to make small changes to the spacecraft orbit.

7.7 Derivation of the Remaining Gauss Variational Equations

Equation for $\frac{de}{dt}$

For the two-body solution, the elements a and e are related to the orbital angular momentum by (see Section 3.3)

$$\frac{h^2}{\mu} = a(1-e^2). \tag{7.51}$$

Differentiating this leads to

$$\frac{2h\dot{h}}{\mu} = \dot{a}(1-e^2) - 2ae\dot{e},$$

which can be rearranged for $\dot{e} = \frac{de}{dt}$ as

$$\frac{de}{dt} = \frac{\dot{a}(1-e^2)}{2ae} - \frac{h\dot{h}}{\mu ae}. \tag{7.52}$$

Now, we already have $\dot{a} = \frac{da}{dt}$ (equation (7.20)). The remaining quantity we need is $\frac{dh}{dt}$, which is given by (differentiate $h^2 = \vec{\mathbf{h}} \cdot \vec{\mathbf{h}}$ and rearrange)

$$\frac{dh}{dt} = \frac{\vec{\mathbf{h}} \cdot \dot{\vec{\mathbf{h}}}}{h}. \tag{7.53}$$

Now, recall that the angular momentum vector is defined as $\vec{\mathbf{r}} \times \vec{\mathbf{v}}$. Differentiation gives

$$\dot{\vec{\mathbf{h}}} = \dot{\vec{\mathbf{r}}} \times \vec{\mathbf{v}} + \vec{\mathbf{r}} \times \dot{\vec{\mathbf{v}}} = \vec{\mathbf{r}} \times \dot{\vec{\mathbf{v}}}.$$

Substituting the equation of motion (7.10), gives

$$\dot{\vec{\mathbf{h}}} = \vec{\mathbf{r}} \times \left(-\frac{\mu}{r^3}\vec{\mathbf{r}} + \vec{\mathbf{f}}_p\right) = \vec{\mathbf{r}} \times \vec{\mathbf{f}}_p.$$

Making use of the expressions in cylindrical coordinates (7.11) and (7.13), this becomes

$$\begin{aligned}\dot{\vec{\mathbf{h}}} &= r\vec{\mathbf{x}}_o \times \left[f_r\vec{\mathbf{x}}_o + f_\theta\vec{\mathbf{y}}_o + f_z\vec{\mathbf{z}}_o\right], \\ &= rf_\theta\vec{\mathbf{z}}_o - rf_z\vec{\mathbf{y}}_o.\end{aligned} \tag{7.54}$$

Now,

$$\vec{\mathbf{h}} \cdot \dot{\vec{\mathbf{h}}} = h\vec{\mathbf{z}}_o \cdot \left[rf_\theta\vec{\mathbf{z}}_o - rf_z\vec{\mathbf{y}}_o\right] = hrf_\theta.$$

Substituting this into (7.53) gives

$$\frac{dh}{dt} = rf_\theta = \frac{a(1-e^2)}{1+e\cos\theta}f_\theta. \tag{7.55}$$

where the polar equation of the instantaneous two-body orbit,

$$r = \frac{a(1-e^2)}{1+e\cos\theta},$$

has been used. Substituting (7.55) and the expression for $\frac{da}{dt}$ (equation (7.20)) into (7.52) gives

$$\frac{de}{dt} = \frac{(1-e^2)}{2ae}\frac{2a^2}{\sqrt{\mu a(1-e^2)}}\left[e\sin\theta f_r + (1+e\cos\theta)f_\theta\right] - \frac{h}{\mu ae}\frac{a(1-e^2)}{1+e\cos\theta}f_\theta.$$

which can be rearranged (using equation (7.51) for h) to give

$$\frac{de}{dt} = \sqrt{\frac{a(1-e^2)}{\mu}}\left[\sin\theta f_r + \frac{2\cos\theta + e(1+\cos^2\theta)}{(1+e\cos\theta)}f_\theta\right]. \tag{7.56}$$

Equation for $\frac{di}{dt}$

Recall that the inclination satisfies (see Section 3.7)

$$\cos i = \frac{\vec{\mathbf{h}} \cdot \vec{\mathbf{z}}_G}{h}.$$

Differentiating this leads to

$$-\sin i\frac{di}{dt} = \frac{\dot{\vec{\mathbf{h}}}\cdot\vec{\mathbf{z}}_G}{h} - \frac{dh}{dt}\frac{\vec{\mathbf{h}}\cdot\vec{\mathbf{z}}_G}{h^2}, \\ = \frac{\dot{\vec{\mathbf{h}}}\cdot\vec{\mathbf{z}}_G}{h} - \frac{dh}{dt}\frac{\cos i}{h}. \tag{7.57}$$

Now, we already have expressions for $\dot{\vec{\mathbf{h}}}$ and $\dot{h}$ in terms of the cylindrical coordinates (equations (7.54) and (7.55)). We now need an expression for $\vec{\mathbf{z}}_G$ in cylindrical coordinates. To do this, we can make use of the rotation matrix from the ECI frame $\mathcal{F}_G$ to the cylindrical frame $\mathcal{F}_o$, which is given by (see Section 3.8) a rotation Ω about $\vec{\mathbf{z}}_G$, followed by a rotation i about the intermediate $\vec{\mathbf{x}}$ axis, followed by a rotation $\omega+\theta$ about $\vec{\mathbf{z}}_o$. The rotation matrix is obtained from equation (3.83) in Section 3.8 by replacing ω by $\omega+\theta$.

$$\mathbf{C}_{oG} = \mathbf{C}_{Go}^T,$$

where

$$\mathbf{C}_{Go} = \begin{bmatrix} c_\Omega c_{\omega+\theta} - s_\Omega c_i s_{\omega+\theta} & -c_\Omega s_{\omega+\theta} - s_\Omega c_i c_{\omega+\theta} & s_\Omega s_i \\ s_\Omega c_{\omega+\theta} + c_\Omega c_i s_{\omega+\theta} & -s_\Omega s_{\omega+\theta} + c_\Omega c_i c_{\omega+\theta} & -c_\Omega s_i \\ s_i s_{\omega+\theta} & s_i c_{\omega+\theta} & c_i \end{bmatrix}, \tag{7.58}$$

where $c_x = \cos x$ and $s_x = \sin x$. We can now express the vector $\vec{\mathbf{z}}_G$ in both reference frames as

$$\vec{\mathbf{z}}_G = \vec{\mathcal{F}}_G^T \begin{bmatrix} 0 \\ 0 \\ 1 \end{bmatrix} = \vec{\mathcal{F}}_o^T \mathbf{C}_{Go}^T \begin{bmatrix} 0 \\ 0 \\ 1 \end{bmatrix} = \vec{\mathcal{F}}_o^T \begin{bmatrix} \sin i \sin(\omega+\theta) \\ \sin i \cos(\omega+\theta) \\ \cos i \end{bmatrix}.$$

We therefore have

$$\vec{\mathbf{z}}_G = \sin i \sin(\omega+\theta)\vec{\mathbf{x}}_o + \sin i \cos(\omega+\theta)\vec{\mathbf{y}}_o + \cos i\,\vec{\mathbf{z}}_o. \tag{7.59}$$

Now, making use of (7.54) and (7.59), we can compute

$$\dot{\vec{\mathbf{h}}}\cdot\vec{\mathbf{z}}_G = \left[rf_\theta\vec{\mathbf{z}}_o - rf_z\vec{\mathbf{y}}_o\right]\cdot\left[\sin i \sin(\omega+\theta)\vec{\mathbf{x}}_o + \sin i\cos(\omega+\theta)\vec{\mathbf{y}}_o + \cos i\,\vec{\mathbf{z}}_o\right], \\ = rf_\theta\cos i - rf_z \sin i \cos(\omega+\theta).$$

Substituting this and (7.55) into (7.57), we obtain

$$-\sin i\frac{di}{dt} = \frac{rf_\theta\cos i - rf_z\sin i\cos(\omega+\theta) - rf_\theta\cos i}{h} = -\frac{rf_z\sin i\cos(\omega+\theta)}{h}.$$

Making use of the polar equation for the instantaneous two-body orbit, this becomes

$$\frac{di}{dt} = \sqrt{\frac{a(1-e^2)}{\mu}} \frac{\cos(\omega+\theta)}{1+e\cos\theta} f_z \tag{7.60}$$

Equation for $\frac{d\Omega}{dt}$

Recall from Section 3.7 that the right ascension of the ascending node satisfies

$$\cos\Omega = \frac{\vec{\mathbf{n}} \cdot \vec{\mathbf{x}}_G}{n}, \tag{7.61}$$

where $\vec{\mathbf{n}} = \vec{\mathbf{z}}_G \times \vec{\mathbf{h}}$. Differentiating this gives

$$-\sin\Omega \frac{d\Omega}{dt} = \frac{\dot{\vec{\mathbf{n}}} \cdot \vec{\mathbf{x}}_G}{n} - \frac{\dot{n}\vec{\mathbf{n}} \cdot \vec{\mathbf{x}}_G}{n^2} = \frac{\dot{\vec{\mathbf{n}}} \cdot \vec{\mathbf{x}}_G}{n} - \frac{\dot{n}\cos\Omega}{n}. \tag{7.62}$$

We will evaluate this expression term by term. First, we have

$$\dot{\vec{\mathbf{n}}} = \vec{\mathbf{z}}_G \times \dot{\vec{\mathbf{h}}},$$

and therefore,

$$\begin{aligned} \dot{\vec{\mathbf{n}}} \cdot \vec{\mathbf{x}}_G &= \vec{\mathbf{x}}_G \cdot (\vec{\mathbf{z}}_G \times \dot{\vec{\mathbf{h}}}), \\ &= \dot{\vec{\mathbf{h}}} \cdot (\vec{\mathbf{x}}_G \times \vec{\mathbf{z}}_G), \\ &= -\dot{\vec{\mathbf{h}}} \cdot \vec{\mathbf{y}}_G. \end{aligned}$$

Using the same arguments that were used to obtain $\vec{\mathbf{z}}_G$ in cylindrical coordinates from the rotation matrix (7.58), we can find that

$$\vec{\mathbf{y}}_G = (s_\Omega c_{\omega+\theta} + c_\Omega c_i s_{\omega+\theta})\vec{\mathbf{x}}_o + (-s_\Omega s_{\omega+\theta} + c_\Omega c_i c_{\omega+\theta})\vec{\mathbf{y}}_o + (-c_\Omega s_i)\vec{\mathbf{z}}_o. \tag{7.63}$$

Making use of (7.54) and (7.63), we have

$$\begin{aligned} \dot{\vec{\mathbf{n}}} \cdot \vec{\mathbf{x}}_G = &-\left[rf_\theta\vec{\mathbf{z}}_o - rf_z\vec{\mathbf{y}}_o\right] \cdot \\ &\left[(s_\Omega c_{\omega+\theta} + c_\Omega c_i s_{\omega+\theta})\vec{\mathbf{x}}_o + (-s_\Omega s_{\omega+\theta} + c_\Omega c_i c_{\omega+\theta})\vec{\mathbf{y}}_o + (-c_\Omega s_i)\vec{\mathbf{z}}_o\right], \\ = &\, rf_\theta \cos\Omega \sin i - rf_z\left[\sin\Omega\sin(\omega+\theta) - \cos\Omega\cos i\cos(\omega+\theta)\right]. \end{aligned} \tag{7.64}$$

Since the inclination i is the angle between $\vec{\mathbf{h}}$ and $\vec{\mathbf{z}}_G$, we immediately have that

$$n = |\vec{\mathbf{h}}||\vec{\mathbf{z}}_G|\sin i = h\sin i. \tag{7.65}$$

Therefore, we have

$$\frac{\vec{\mathbf{n}}\cdot\vec{\mathbf{x}}_G}{n} = \frac{rf_\theta}{h}\cos\Omega - \frac{rf_z}{h\sin i}\left[\sin\Omega\sin(\omega+\theta) - \cos\Omega\cos i\cos(\omega+\theta)\right]. \tag{7.66}$$

The next term we need in (7.62) is $\dot{n}$. Now, this is given by

$$\dot{n} = \frac{\vec{\mathbf{n}}\cdot\dot{\vec{\mathbf{n}}}}{n}. \tag{7.67}$$

We therefore need expressions for $\vec{\mathbf{n}}$ and $\dot{\vec{\mathbf{n}}}$. Let us start with $\vec{\mathbf{n}}$. From its definition, and (7.59) and (7.12), we have

$$\begin{aligned}\vec{\mathbf{n}} = \vec{\mathbf{z}}_g\times\vec{\mathbf{h}} &= h\left[\sin i\sin(\omega+\theta)\vec{\mathbf{x}}_o + \sin i\cos(\omega+\theta)\vec{\mathbf{y}}_o + \cos i\vec{\mathbf{z}}_o\right]\times\vec{\mathbf{z}}_o,\\ &= -h\sin i\sin(\omega+\theta)\vec{\mathbf{y}}_o + h\sin i\cos(\omega+\theta)\vec{\mathbf{x}}_o.\end{aligned} \tag{7.68}$$

Using (7.54) we find,

$$\begin{aligned}\dot{\vec{\mathbf{n}}} = \vec{\mathbf{z}}_G\times\dot{\vec{\mathbf{h}}} &= \vec{\mathbf{z}}_G\times\left[rf_\theta\vec{\mathbf{z}}_o - rf_z\vec{\mathbf{y}}_o\right],\\ &= rf_\theta\vec{\mathbf{z}}_G\times\vec{\mathbf{z}}_o - rf_z\vec{\mathbf{z}}_G\times\vec{\mathbf{y}}_o.\end{aligned} \tag{7.69}$$

Making use of (7.59), we have

$$\begin{aligned}\vec{\mathbf{z}}_G\times\vec{\mathbf{z}}_o &= \left[\sin i\sin(\omega+\theta)\vec{\mathbf{x}}_o + \sin i\cos(\omega+\theta)\vec{\mathbf{y}}_o + \cos i\vec{\mathbf{z}}_o\right]\times\vec{\mathbf{z}}_o,\\ &= -\sin i\sin(\omega+\theta)\vec{\mathbf{y}}_o + \sin i\cos(\omega+\theta)\vec{\mathbf{x}}_o,\end{aligned}$$

and

$$\begin{aligned}\vec{\mathbf{z}}_G\times\vec{\mathbf{y}}_o &= \left[\sin i\sin(\omega+\theta)\vec{\mathbf{x}}_o + \sin i\cos(\omega+\theta)\vec{\mathbf{y}}_o + \cos i\vec{\mathbf{z}}_o\right]\times\vec{\mathbf{y}}_o,\\ &= \sin i\sin(\omega+\theta)\vec{\mathbf{z}}_o - \cos i\vec{\mathbf{x}}_o.\end{aligned}$$

Substituting these expressions into (7.69) leads to

$$\begin{aligned}\dot{\vec{\mathbf{n}}} &= rf_\theta\left[-\sin i\sin(\omega+\theta)\vec{\mathbf{y}}_o + \sin i\cos(\omega+\theta)\vec{\mathbf{x}}_o\right] - rf_z\left[\sin i\sin(\omega+\theta)\vec{\mathbf{z}}_o - \cos i\vec{\mathbf{x}}_o\right],\\ &= r\left[f_\theta\sin i\cos(\omega+\theta) + f_z\cos i\right]\vec{\mathbf{x}}_o - rf_\theta\sin i\sin(\omega+\theta)\vec{\mathbf{y}}_o - rf_z\sin i\sin(\omega+\theta)\vec{\mathbf{z}}_o.\end{aligned} \tag{7.70}$$

Taking the dot product of (7.68) and (7.70), gives

$$\begin{aligned}\vec{\mathbf{n}}\cdot\dot{\vec{\mathbf{n}}} &= rh\sin i\left[f_\theta\sin i\cos^2(\omega+\theta) + f_z\cos i\cos(\omega+\theta)\right] + rh\sin i f_\theta\sin i\sin^2(\omega+\theta),\\ &= rh\sin i\left[\sin i f_\theta + \cos i\cos(\omega+\theta)f_z\right].\end{aligned}$$

Substituting this and (7.65) into (7.67) gives

$$\dot{n} = r\left[\sin i f_\theta + \cos i \cos(\omega + \theta) f_z\right]. \tag{7.71}$$

Finally, making use of (7.65), we can obtain the second term in (7.62) as

$$\frac{\dot{n}\cos\Omega}{n} = \frac{r f_\theta}{h}\cos\Omega + \frac{r f_z}{h\sin i}\cos\Omega\cos i\cos(\omega + \theta). \tag{7.72}$$

Substituting (7.66) and (7.72) into (7.62) gives

$$-\sin\Omega\frac{d\Omega}{dt} = -\frac{r f_z}{h\sin i}\sin\Omega\sin(\omega + \theta),$$

which leads to

$$\frac{d\Omega}{dt} = \frac{r f_z}{h\sin i}\sin(\omega + \theta).$$

Finally, substituting the polar equation for the instantaneous two-body orbit, we have the desired result

$$\frac{d\Omega}{dt} = \sqrt{\frac{a(1 - e^2)}{\mu}}\frac{\sin(\omega + \theta)}{\sin i(1 + e\cos\theta)} f_z \tag{7.73}$$

Equation for $\frac{d\omega}{dt}$

Recall from Section 3.7 that the argument of periapsis satisfies

$$\cos\omega = \frac{\vec{\mathbf{n}}\cdot\vec{\mathbf{e}}}{ne}. \tag{7.74}$$

Differentiating this leads to

$$\begin{aligned} -\sin\omega\frac{d\omega}{dt} &= \frac{\dot{\vec{\mathbf{n}}}\cdot\vec{\mathbf{e}}}{ne} + \frac{\vec{\mathbf{n}}\cdot\dot{\vec{\mathbf{e}}}}{ne} - \frac{\dot{n}\vec{\mathbf{n}}\cdot\vec{\mathbf{e}}}{n^2 e} - \frac{\dot{e}\vec{\mathbf{n}}\cdot\vec{\mathbf{e}}}{ne^2}, \\ &= \frac{\dot{\vec{\mathbf{n}}}\cdot\vec{\mathbf{e}}}{ne} + \frac{\vec{\mathbf{n}}\cdot\dot{\vec{\mathbf{e}}}}{ne} - \frac{\dot{n}\cos\omega}{n} - \frac{\dot{e}\cos\omega}{e}. \end{aligned} \tag{7.75}$$

We will treat this term by term. We see that we need expressions for $\vec{\mathbf{e}}$ and $\dot{\vec{\mathbf{e}}}$. Recall that the eccentricity vector $\vec{\mathbf{e}}$ points towards periapsis (see Section 3.3). Therefore, in perifocal coordinates ($\mathcal{F}_p$), the eccentricity vector is given by

$$\vec{\mathbf{e}} = \vec{\mathcal{F}}_p^T\begin{bmatrix} e \\ 0 \\ 0 \end{bmatrix}.$$

Now, the cylindrical frame is obtained from the perifocal frame by a rotation about $\vec{\mathbf{z}}_o$ through angle θ (the instantaneous true-anomaly). The associated rotation matrix is

$$\mathbf{C}_{op} = \begin{bmatrix} \cos\theta & \sin\theta & 0 \\ -\sin\theta & \cos\theta & 0 \\ 0 & 0 & 1 \end{bmatrix},$$

such that the eccentricity vector is given by

$$\begin{aligned} \vec{\mathbf{e}} &= \vec{\mathcal{F}}_p^T \mathbf{C}_{op} \begin{bmatrix} e \\ 0 \\ 0 \end{bmatrix}, \\ &= e\cos\theta \vec{\mathbf{x}}_o - e\sin\theta \vec{\mathbf{y}}_o. \end{aligned} \tag{7.76}$$

Now, recall the definition of the eccentricity vector

$$\vec{\mathbf{e}} = \frac{\vec{\mathbf{v}} \times \vec{\mathbf{h}}}{\mu} - \frac{\vec{\mathbf{r}}}{r}. \tag{7.77}$$

Differentiating this gives

$$\dot{\vec{\mathbf{e}}} = \frac{\dot{\vec{\mathbf{v}}} \times \vec{\mathbf{h}}}{\mu} - \frac{\dot{\vec{\mathbf{r}}}}{r} + \frac{\dot{r}\vec{\mathbf{r}}}{r^2} + \frac{\vec{\mathbf{v}} \times \dot{\vec{\mathbf{h}}}}{\mu}.$$

Substituting the equation of motion (7.10) for $\dot{\vec{\mathbf{v}}}$, this becomes

$$\begin{aligned} \dot{\vec{\mathbf{e}}} &= \frac{\left(-\frac{\mu}{r^3}\vec{\mathbf{r}}\right) \times \vec{\mathbf{h}}}{\mu} - \frac{\dot{\vec{\mathbf{r}}}}{r} + \frac{\dot{r}\vec{\mathbf{r}}}{r^2} + \frac{\vec{\mathbf{f}}_p \times \vec{\mathbf{h}}}{\mu} + \frac{\vec{\mathbf{v}} \times \dot{\vec{\mathbf{h}}}}{\mu}, \\ &= \frac{\vec{\mathbf{f}}_p \times \vec{\mathbf{h}}}{\mu} + \frac{\vec{\mathbf{v}} \times \dot{\vec{\mathbf{h}}}}{\mu}. \end{aligned} \tag{7.78}$$

Note that we showed in Section 3.2.3 that $-(\mu/r^3)\vec{\mathbf{r}} \times \vec{\mathbf{h}}/\mu - \dot{\vec{\mathbf{r}}}/r + \dot{r}\vec{\mathbf{r}}/r^2 = \vec{\mathbf{0}}$. Let us now evaluate (7.78) term by term. Using (7.13) and (7.12), we have

$$\begin{aligned} \frac{\vec{\mathbf{f}}_p \times \vec{\mathbf{h}}}{\mu} &= \frac{h}{\mu}\left[f_r\vec{\mathbf{x}}_o + f_\theta\vec{\mathbf{y}}_o + f_z\vec{\mathbf{z}}_o\right] \times \vec{\mathbf{z}}_o, \\ &= \frac{1}{\mu}\left[-hf_r\vec{\mathbf{y}}_o + hf_\theta\vec{\mathbf{x}}_o\right]. \end{aligned}$$

Using (7.11) and (7.54), we have

$$\begin{aligned}\frac{\vec{\mathbf{v}} \times \dot{\vec{\mathbf{h}}}}{\mu} &= \frac{1}{\mu}\left[\dot{r}\vec{\mathbf{x}}_o + r\dot{\theta}\vec{\mathbf{y}}_o\right] \times \left[rf_\theta\vec{\mathbf{z}}_o - rf_z\vec{\mathbf{y}}_o\right],\\ &= \frac{1}{\mu}\left[-r\dot{r}f_\theta\vec{\mathbf{y}}_o - r\dot{r}f_z\vec{\mathbf{z}}_o + r^2\dot{\theta}f_\theta\vec{\mathbf{x}}_o\right],\\ &= \frac{1}{\mu}\left[-r\dot{r}f_\theta\vec{\mathbf{y}}_o - r\dot{r}f_z\vec{\mathbf{z}}_o + hf_\theta\vec{\mathbf{x}}_o\right],\end{aligned}$$

where the fact that $h = r^2\dot{\theta}$ has been used. Using these expressions in (7.78), we have

$$\dot{\vec{\mathbf{e}}} = \frac{2hf_\theta}{\mu}\vec{\mathbf{x}}_o - \frac{[hf_r + r\dot{r}f_\theta]}{\mu}\vec{\mathbf{y}}_o - \frac{r\dot{r}f_z}{\mu}\vec{\mathbf{z}}_o. \tag{7.79}$$

Let us now start evaluating the terms in (7.75). Using (7.70), (7.76) and (7.65), the first term is

$$\begin{aligned}\frac{\dot{\vec{\mathbf{n}}}\cdot\vec{\mathbf{e}}}{ne} &= \frac{1}{eh\sin i}\left[e\cos\theta r\left(f_\theta\sin i\cos(\omega+\theta) + f_z\cos i\right)\right.\\ &\quad \left. + e\sin\theta rf_\theta\sin i\sin(\omega+\theta)\right]\\ &= \frac{rf_\theta}{h}\left[\cos\theta\cos(\omega+\theta) + \sin\theta\sin(\omega+\theta)\right] + \frac{rf_z\cos i\cos\theta}{h\sin i}.\end{aligned}$$

Using the trigonometric double angle formula

$$\cos(a-b) = \cos a\cos b + \sin a\sin b,$$

this reduces to

$$\frac{\dot{\vec{\mathbf{n}}}\cdot\vec{\mathbf{e}}}{ne} = \frac{rf_\theta\cos\omega}{h} + \frac{rf_z\cos i\cos\theta}{h\sin i}. \tag{7.80}$$

Next, using (7.68), (7.79) and (7.65), we find that the second term in (7.75) is

$$\begin{aligned}\frac{\vec{\mathbf{n}}\cdot\dot{\vec{\mathbf{e}}}}{ne} &= \frac{1}{eh\sin i}\left[h\sin i\cos(\omega+\theta)\frac{2hf_\theta}{\mu} + h\sin i\sin(\omega+\theta)\frac{[hf_r + r\dot{r}f_\theta]}{\mu}\right],\\ &= \frac{2hf_\theta}{\mu e}\cos(\omega+\theta) + \frac{hf_r}{\mu e}\sin(\omega+\theta) + \frac{r\dot{r}f_\theta}{\mu e}\sin(\omega+\theta).\end{aligned}$$

From Section 3.8, we find that

$$\frac{r\dot{r}}{\mu e} = \frac{a(1-e^2)}{\mu e(1+e\cos\theta)}\sqrt{\frac{\mu}{a(1-e^2)}}e\sin\theta = \sqrt{\frac{a(1-e^2)}{\mu}}\frac{\sin\theta}{1+e\cos\theta} = \frac{h}{\mu}\frac{\sin\theta}{1+e\cos\theta}.$$

Therefore, we have

$$\begin{aligned}
\frac{\vec{\mathbf{n}}\cdot\dot{\vec{\mathbf{e}}}}{ne} &= \frac{2hf_\theta}{\mu e}\cos(\omega+\theta) + \frac{hf_r}{\mu e}\sin(\omega+\theta) + \frac{hf_\theta}{\mu}\frac{\sin\theta\sin(\omega+\theta)}{1+e\cos\theta},\\
&= \frac{hf_\theta}{\mu}\left[\frac{2\cos(\omega+\theta)}{e} + \frac{\sin\theta\sin(\omega+\theta)}{1+e\cos\theta}\right] + \frac{hf_r}{\mu e}\sin(\omega+\theta),\\
&= \frac{hf_\theta}{\mu}\left[\frac{2\cos(\omega+\theta)+2e\cos\theta\cos(\omega+\theta)+e\sin\theta\sin(\omega+\theta)}{e(1+e\cos\theta)}\right]\\
&\quad + \frac{hf_r}{\mu e}\sin(\omega+\theta),\\
&= \frac{hf_\theta}{\mu}\left[\frac{\cos(\omega+\theta)(2+e\cos\theta)+e\left(\cos\theta\cos(\omega+\theta)+\sin\theta\sin(\omega+\theta)\right)}{e(1+e\cos\theta)}\right]\\
&\quad + \frac{hf_r}{\mu e}\sin(\omega+\theta),\\
&= \frac{hf_\theta}{\mu}\left[\frac{\cos(\omega+\theta)(2+e\cos\theta)+e\cos\omega}{e(1+e\cos\theta)}\right] + \frac{hf_r}{\mu e}\sin(\omega+\theta).
\end{aligned} \tag{7.81}$$

Next, using (7.71) and (7.65), we obtain the third term in (7.75)

$$\begin{aligned}
\frac{\dot{n}\cos\omega}{n} &= \frac{\cos\omega}{h\sin i}\left[r\sin i f_\theta + r\cos i\cos(\omega+\theta)f_z\right],\\
&= \frac{rf_\theta}{h}\cos\omega + \frac{rf_z}{h}\frac{\cos i}{\sin i}\cos\omega\cos(\omega+\theta).
\end{aligned} \tag{7.82}$$

Using (7.56), we obtain the fourth term in (7.75) as

$$\begin{aligned}
\frac{\dot{e}\cos\omega}{e} &= \frac{hf_r}{\mu e}\sin\theta\cos\omega + \frac{hf_\theta}{\mu}\frac{2\cos\theta+e(1+\cos^2\theta)}{e(1+e\cos\theta)}\cos\omega,\\
&= \frac{hf_r}{\mu e}\sin\theta\cos\omega + \frac{hf_\theta}{\mu}\frac{\cos\theta\cos\omega(2+e\cos\theta)+e\cos\omega}{e(1+e\cos\theta)}.
\end{aligned} \tag{7.83}$$

Now, substituting the four terms (7.80), (7.81), (7.82) and (7.83) into (7.75), we obtain

$$\begin{aligned}
-\sin\omega\frac{d\omega}{dt} &= \frac{rf_\theta\cos\omega}{h} + \frac{rf_z\cos i\cos\theta}{h\sin i}\\
&\quad + \frac{hf_\theta}{\mu}\left[\frac{\cos(\omega+\theta)(2+e\cos\theta)+e\cos\omega}{e(1+e\cos\theta)}\right] + \frac{hf_r}{\mu e}\sin(\omega+\theta)\\
&\quad - \frac{rf_\theta}{h}\cos\omega - \frac{rf_z}{h}\frac{\cos i}{\sin i}\cos\omega\cos(\omega+\theta)\\
&\quad - \frac{hf_r}{\mu e}\sin\theta\cos\omega - \frac{hf_\theta}{\mu}\frac{\cos\theta\cos\omega(2+e\cos\theta)+e\cos\omega}{e(1+e\cos\theta)},
\end{aligned}$$

$$
\begin{aligned}
&= \frac{hf_r}{\mu e}\left[\sin(\omega+\theta) - \sin\theta\cos\omega\right] \\
&+ \frac{hf_\theta}{\mu}\frac{(2+e\cos\theta)}{e(1+e\cos\theta)}\left[\cos(\omega+\theta) - \cos\theta\cos\omega\right] \\
&+ \frac{rf_z\cos i}{h\sin i}\left[\cos\theta - \cos\omega\cos(\omega+\theta)\right].
\end{aligned}
$$

These terms can be reduced using trigonometric double angle formulas. Specifically,

$$
\begin{aligned}
\cos(\omega+\theta) &= \cos\omega\cos\theta - \sin\omega\sin\theta, \\
\sin(\omega+\theta) &= \sin\omega\cos\theta + \cos\omega\sin\theta,
\end{aligned}
$$

and

$$
\cos\theta = \cos\left((\omega+\theta) - \omega\right) = \cos(\omega+\theta)\cos\omega + \sin(\omega+\theta)\sin\omega.
$$

Making use of these, we obtain

$$
-\sin\omega\frac{d\omega}{dt} = \frac{hf_r}{\mu e}\sin\omega\cos\theta - \frac{hf_\theta}{\mu}\frac{(2+e\cos\theta)}{e(1+e\cos\theta)}\sin\omega\sin\theta + \frac{rf_z\cos i}{h\sin i}\sin(\omega+\theta)\sin\omega.
$$

Noting that

$$
\frac{r}{h} = \frac{h}{\mu}\frac{1}{1+e\cos\theta},
$$

and that

$$
\frac{h}{\mu} = \sqrt{\frac{a(1-e^2)}{\mu}},
$$

we finally obtain the desired result

$$
\frac{d\omega}{dt} = \sqrt{\frac{a(1-e^2)}{\mu}}\left[-\frac{\cos\theta}{e}f_r + \frac{(2+e\cos\theta)\sin\theta}{e(1+e\cos\theta)}f_\theta - \frac{\sin(\omega+\theta)}{\tan i(1+e\cos\theta)}f_z\right]. \quad (7.84)
$$

Notes

As explained in this chapter, true spacecraft orbital motion differs from Keplerian two-body motion (presented in Chapter 3) due to additional perturbing forces acting on the spacecraft. This chapter has examined various methods of determining the perturbed motion. In particular, we have presented two special perturbation techniques (Cowell's method and Enke's method), and one general perturbation technique (Gauss' variational equations). There are several other techniques in existence, and the reader is referred to Vallado (2007) for a general overview of these methods.

In order to use these techniques for dealing with orbital perturbations, mathematical models of the perturbing forces are needed. We have presented a detailed force model for the gravitational perturbations due to a non-spherical primary. However, we have not considered other perturbing forces such as gravitational forces from third and fourth bodies, detailed atmospheric drag models, solar radiation pressure among others. The reader is referred to Montenbruck and Gill (2000) and Vallado (2007) for additional perturbing force models.

References

Montenbruck O and Gill E 2000 *Satellite Orbits: Models, Methods, Applications*. Springer, Berlin Heidelberg.

Vallado DA 2007 *Fundamentals of Astrodynamics and Applications* 3rd edn. Microcosm Press, Hawthorne, CA.

8

Low Thrust Trajectory Analysis and Design

8.1 Problem Formulation

In Chapter 5, we considered transfers between orbits that were accomplished by the application of large thrusts over short periods of time. By assuming the transfer time was short relative to the orbital period, these thrusts could be modeled as impulsive. In this chapter, the other extreme will be considered: the application of low thrusts continually over long periods of time. The methods developed here were pioneered by T. N. Edelbaum in the 1960s. Although approximate, they are easily derived and yield a great deal of insight into low thrust trajectories.

Our major tool will be Gauss's Variational Equations. In particular we will need the equations for the semimajor axis a, the eccentricity e, and the inclination i. From the previous chapter:

$$\frac{da}{dt} = \frac{2a^2}{\sqrt{\mu a(1-e^2)}}[e \sin\theta f_r + (1 + e\cos\theta) f_\theta], \tag{8.1}$$

$$\frac{de}{dt} = \sqrt{\frac{a(1-e^2)}{\mu}}\left[\sin\theta f_r + \frac{2\cos\theta + e(1+\cos^2\theta)}{(1+e\cos\theta)} f_\theta\right], \tag{8.2}$$

$$\frac{di}{dt} = \sqrt{\frac{a(1-e^2)}{\mu}}\frac{\cos(\omega+\theta)}{1+e\cos\theta} f_z, \tag{8.3}$$

where f_r is the radial thrust, f_θ is the tangential thrust, and f_z is the out-of-plane thrust.

The following assumptions will be in effect for the rest of the chapter:

(i) The magnitude of the applied thrust is small and constant.
(ii) The eccentricity of the orbit throughout the maneuver remains small.
(iii) The transfer time (T_m) is long relative to the orbital period (T).

Our objective is to minimize the maneuver time subject to the desired terminal orbit constraints.

Spacecraft Dynamics and Control: An Introduction, First Edition.
Anton H.J. de Ruiter, Christopher J. Damaren and James R. Forbes.

Given assumption (ii), we can expand the right-hand sides of equations (8.1)-(8.3) in a Taylor series in e and retain only the first (zeroth-order) terms. Since the thrusts are also assumed to be small, terms of the order $e^i f_r$, $e^i f_\theta$, and $e^i f_z$, $i = 1, 2, 3, \ldots$, are assumed to be of second or higher order and neglected. The resulting equations are:

$$\frac{da}{dt} = 2\sqrt{\frac{a^3}{\mu}} f_\theta, \tag{8.4}$$

$$\frac{de}{dt} = \sqrt{\frac{a}{\mu}} \left[\sin\theta f_r + 2\cos\theta f_\theta\right], \tag{8.5}$$

$$\frac{di}{dt} = \sqrt{\frac{a}{\mu}} \cos(\omega + \theta) f_z. \tag{8.6}$$

Since the magnitude of the thrust $f = \sqrt{f_r^2 + f_\theta^2 + f_z^2}$ is taken to be constant, it is helpful to express the thrust components in terms of the steering angles α (the angle between the velocity vector and the component of the thrust vector in the orbital plane) and β (the angle between the thrust vector and the orbital plane): $f_r = f\cos\beta\sin\alpha$, $f_\theta = f\cos\beta\cos\alpha$, and $f_z = f\cos\beta$. Equations (8.4)-(8.6) become

$$\frac{da}{dt} = 2\sqrt{\frac{a^3}{\mu}} f\cos\alpha\cos\beta, \tag{8.7}$$

$$\frac{de}{dt} = \sqrt{\frac{a}{\mu}} f\left[\sin\theta\sin\alpha\cos\beta + 2\cos\theta\cos\alpha\cos\beta\right], \tag{8.8}$$

$$\frac{di}{dt} = \sqrt{\frac{a}{\mu}} f\cos\theta\sin\beta, \tag{8.9}$$

where we have taken $\omega = 0$ given assumption (ii) and assumed that θ is measured from the ascending node.

8.2 Coplanar Circle to Circle Transfers

Letting T_m denote the transfer time, we seek to determine the steering angles that minimize T_m subject to the following boundary conditions on $a(t)$: $a(0) = r_1$ and $a(T_m) = r_2$. Here, the notation of Chapter 5 has been adopted where r_1 is the radius of the initial circular orbit and r_2 is the radius of the final circular orbit. In order to minimize T_m, it follows that da/dt should be maximized. From equation (8.7), we take $\alpha = \beta = 0$. This leads to no change in inclination using equation (8.9).

Rearranging (8.7), we obtain

$$\frac{1}{2}\sqrt{\frac{\mu}{a^3}}\frac{da}{dt} = f.$$

The velocity change (Δv) for the maneuver can then be calculated via integration of both sides:

$$\frac{1}{2}\int_{r_1}^{r_2}\sqrt{\frac{\mu}{a^3}}\,da = \int_0^{T_m} f\,dt = \Delta v = fT_m. \tag{8.10}$$

Hence,

$$\Delta v = \sqrt{\frac{\mu}{r_1}} - \sqrt{\frac{\mu}{r_2}}.$$

Denoting the initial circular velocity by $v_1 = \sqrt{\mu/r_1}$, the expression for the velocity change can be nondimensionalized to give

$$\frac{\Delta v}{v_1} = 1 - \sqrt{\frac{r_1}{r_2}}. \tag{8.11}$$

This can be compared with the corresponding expression for the Hohmann transfer developed in Chapter 5:

$$\frac{\Delta v}{v_1} = \sqrt{\frac{r_1}{r_2}} - 1 - \sqrt{\left(\frac{r_1}{r_2}\right)\frac{2}{1+(r_2/r_1)}} + \sqrt{\left(\frac{r_2}{r_1}\right)\frac{2}{1+(r_2/r_1)}}.$$

Both expressions are plotted in Figure 8.1.

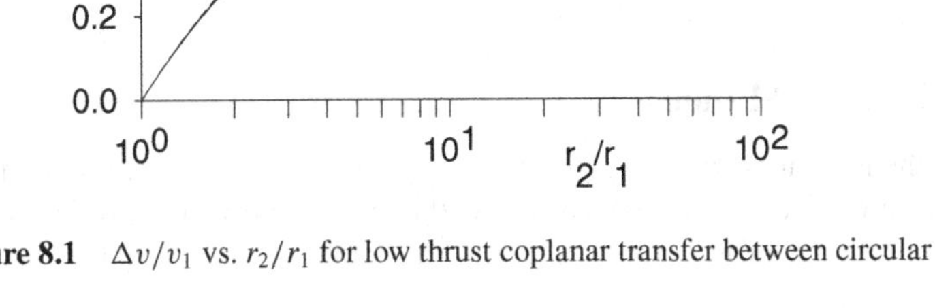

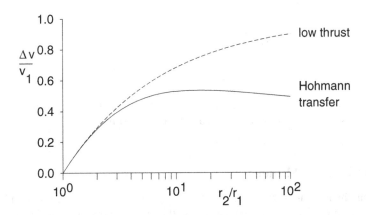

Figure 8.1 $\Delta v/v_1$ vs. r_2/r_1 for low thrust coplanar transfer between circular orbits

An expression for the evolution of the semimajor axis during the maneuver, $a(t)$, can be determined by replacing the upper limit of integration in equation (8.10) by $a(t)$ on the left-hand side and t on the right-hand side. This gives

$$\sqrt{\frac{\mu}{r_1}} - \sqrt{\frac{\mu}{a(t)}} = ft$$

or $a(t) = \mu/(v_1 - ft)^2$. It is noted that the expression for the eccentricity in equation (8.8) for $\alpha = \beta = 0$ becomes

$$\frac{de}{dt} = 2\sqrt{\frac{a}{\mu}} f \cos\theta.$$

For large T_m and small f, the eccentricity remains small.

The spiral nature of the maneuver can be visualized by making the substitution $dt = \sqrt{a^3/\mu}\, d\theta$ in equation (8.10) which gives

$$\frac{1}{2}\int_{r_1}^{a(t)} \frac{\mu}{a^3}\, da = \int_0^{\theta(t)} f\, d\theta.$$

Performing the integrations gives

$$f\theta(t) = \frac{\mu}{4}\left(\frac{1}{r_1^2} - \frac{1}{a^2(t)}\right). \tag{8.12}$$

The total number of revolutions $[\theta(T_m)/(2\pi)]$ can be determined by setting $t = T_m$ and $a(t) = a(T_m) = r_2$. Combining this result with equation (8.12) yields the following expression for $a(\theta)$:

$$\frac{a(\theta)}{r_1} = \frac{1}{\sqrt{1 - \frac{\theta(t)}{\theta(T_m)}\left[1 - \left(\frac{r_1}{r_2}\right)^2\right]}}. \tag{8.13}$$

Neglecting the change in the eccentricity, the Cartesian trajectory y vs. x with $y = a(\theta)\sin\theta$ and $x = a(\theta)\cos\theta$ can be determined such as that given in Figure 8.2.

8.3 Plane Change Maneuver

In this section, the low thrust analog of the problem solved by impulsive thrust in Section 5.3 will be tackled. In particular, we look for a low thrust solution which produces a change in inclination for a circular orbit without affecting the radius or eccentricity. With a view to

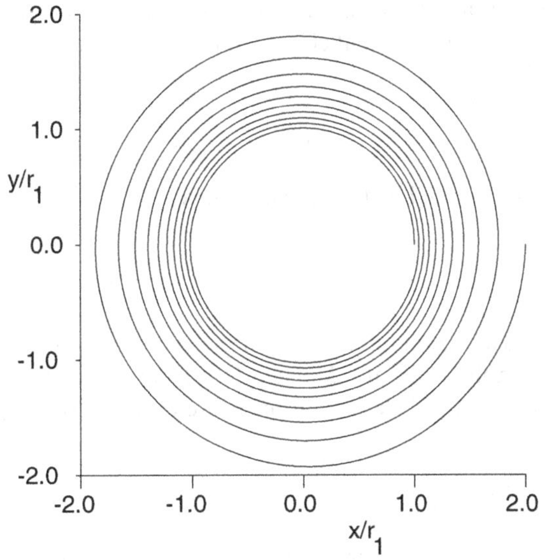

Figure 8.2 Circle-to-circle low thrust transfer for $r_2/r_1 = 2$ and a 10 revolution transfer

minimizing the transfer time T_m we shall maximize di/dt while setting $da/dt = de/dt = 0$. Examining equations (8.7)-(8.9), we set

$$\sin\beta = \text{sgn}(\cos\theta), \quad \text{sgn}(x) = \begin{cases} 1, & x > 0 \\ 0, & x = 0 \\ -1, & x < 0 \end{cases} \tag{8.14}$$

where sgn(x) is the signum function.

Setting $a = r_1$, $v_1 = \sqrt{\mu/r_1}$, and β from equation (8.14), equation (8.9) becomes

$$\frac{di}{dt} = \frac{f}{v_1}|\cos\theta|.$$

Let us now relate the Δv for one orbit ($T_m = T$) to the corresponding change in inclination, Δi. Setting $dt = T/(2\pi)\,d\theta$ and integrating both sides from $t = 0$ to $t = T$ yields

$$\Delta i = \frac{fT}{2\pi v_1}\int_0^{2\pi} |\cos\theta|\,d\theta = \frac{2}{\pi}\frac{\Delta v}{v_1}. \tag{8.15}$$

Note that we have assumed that the desired inclination change Δi is positive. A similar strategy holds for Δi negative. It turns out that the presented strategy is not a minimum time transfer, which requires a varying β from orbit to orbit. The underlying philosophy is that the orbit should be enlarged to lower the velocity, most of the plane change should occur, and then the orbit is reduced to its original value. The strategy developed in equation (8.15) is close to optimal for small changes in inclination.

From Chapter 5, the Δv for an impulsive change in plane with plane change angle Δi is given by $\Delta v = 2v\sin(\Delta i/2)$. For a small change in inclination this yields

$$\frac{\Delta v}{v_1} = \Delta i$$

which should be compared to the low thrust result above.

Notes

This chapter has provided a brief introduction to the subject of low-thrust spacecraft trajectory design, which is based on the work of Edelbaum (1961). For further examples of low-thrust spacecraft trajectory design, the reader is referred to Kemble (2006).

References

Edelbaum TN 1961 Propulsion Requirements for Controllable Satellites. *ARS Journal* **31**, 1079–1089.
Kemble S 2006 *Interplanetary Mission Analysis and Design*. Springer, Chichester, UK.

9

Spacecraft Formation Flying

A topic of significant current interest in the space community is that of spacecraft formation flying. First, we must define what is meant by a spacecraft formation. *By a spacecraft formation, we mean a collection of coordinated spacecraft flying in close proximity.* If they are coordinated, but not in close proximity, we shall not consider it a formation. This distinguishes formations from constellations (see Figure 9.1). There are a number of reasons spacecraft formations are of interest:

1. Cost: a formation of smaller cheaper spacecraft may be less expensive than a single large expensive spacecraft.
   - a formation of smaller spacecraft may be able to achieve the same objectives as a single large spacecraft by distributing instruments among them.
2. Graceful degradation
   - failure of a single spacecraft in a formation does not mean the end of the mission. The mission can continue in reduced capacity with the remaining satellites. Failure of a single large spacecraft on the other hand, means the end of the mission.
3. New possibilities not achievable with a single satellite, for example:
   - sparse aperture radar, interferometry
   - search for planets around distant stars

There are currently two classes of spacecraft formations under investigation:

1. Planet orbiting formations
2. Deep space formations (at Lagrange points (see Chapter 10 for an explanation of the Lagrange points))

We shall restrict our attention to planet orbiting formations.

Spacecraft Dynamics and Control: An Introduction, First Edition.
Anton H.J. de Ruiter, Christopher J. Damaren and James R. Forbes.

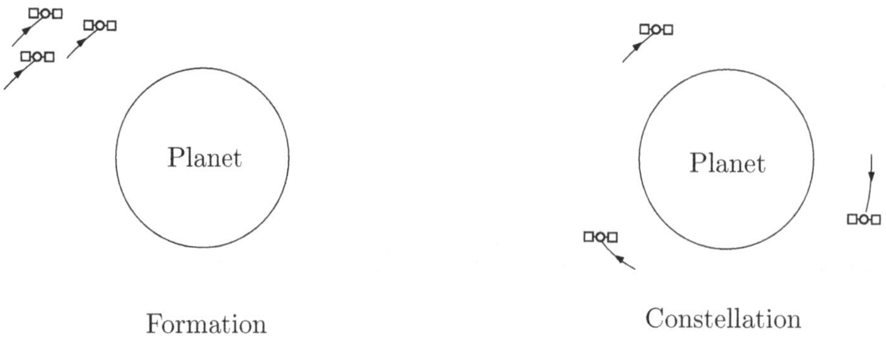

Figure 9.1 Satellite formation and constellation

9.1 Mathematical Description

For nearly circular orbits, the relative motion between spacecraft is described by the Hill or Clohessy-Wiltshire equations. These equations were originally developed for orbital rendezvous and docking operations.

Consider two spacecraft in formation, as shown in Figure 9.2. We shall call the first spacecraft the leader, and the second the follower. The follower is assumed to have thrusters capable of producing a force per unit mass $\vec{\mathbf{f}}_f$ on the follower. This is needed for formation control. For the purposes of our formulation, we shall assume that the leader is unforced.

As shown in Figure 9.2, the follower position relative to the leader is defined as $\vec{\boldsymbol{\rho}} = \vec{\mathbf{r}}_f - \vec{\mathbf{r}}_l$ where $\vec{\mathbf{r}}_f$ and $\vec{\mathbf{r}}_l$ are the leader and follower positions relative to the center of the Earth, respectively. The equations of motion for the follower and leader (assuming the Earth is spherical) are respectively

$$\ddot{\vec{\mathbf{r}}}_f = -\mu \frac{\vec{\mathbf{r}}_f}{|\vec{\mathbf{r}}_f|^3} + \vec{\mathbf{f}}_f, \tag{9.1}$$

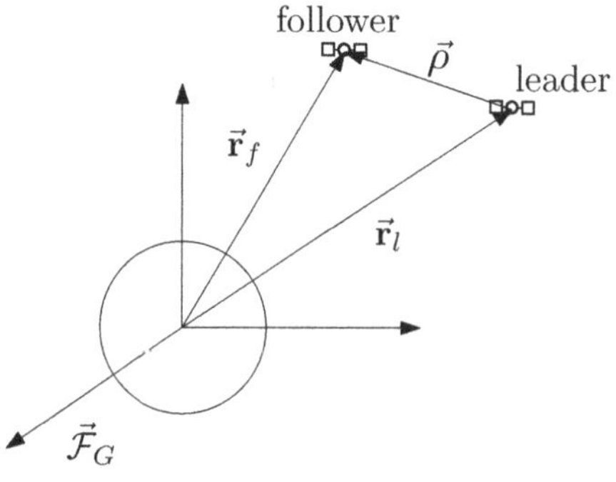

Figure 9.2 Leader-follower formation

and

$$\ddot{\vec{\mathbf{r}}}_l = -\mu \frac{\vec{\mathbf{r}}_l}{|\vec{\mathbf{r}}_l|^3}. \tag{9.2}$$

The equations of relative motion are then obtained by subtracting (9.2) from (9.1)

$$\begin{aligned} \ddot{\vec{\rho}} = \ddot{\vec{\mathbf{r}}}_f - \ddot{\vec{\mathbf{r}}}_l &= -\mu \frac{\vec{\mathbf{r}}_f}{|\vec{\mathbf{r}}_f|^3} + \vec{\mathbf{f}}_f + \mu \frac{\vec{\mathbf{r}}_l}{|\vec{\mathbf{r}}_l|^3}, \\ &= -\mu \frac{(\vec{\mathbf{r}}_l + \vec{\rho})}{|\vec{\mathbf{r}}_l + \vec{\rho}|^3} + \vec{\mathbf{f}}_f + \mu \frac{\vec{\mathbf{r}}_l}{|\vec{\mathbf{r}}_l|^3}. \end{aligned} \tag{9.3}$$

Let us now find a linear approximation to (9.3) (which will be useful later for studying relative motion solutions). First, we note that

$$\begin{aligned} \frac{1}{|\vec{\mathbf{r}}_l + \vec{\rho}|^3} &= \frac{1}{\left((\vec{\mathbf{r}}_l + \vec{\rho}) \cdot (\vec{\mathbf{r}}_l + \vec{\rho})\right)^{3/2}}, \\ &= \frac{1}{\left(\vec{\mathbf{r}}_l \cdot \vec{\mathbf{r}}_l + \vec{\rho} \cdot \vec{\rho} + 2\vec{\mathbf{r}}_l \cdot \vec{\rho}\right)^{3/2}}, \\ &= \frac{1}{\left(r_l^2 + \rho^2 + 2r_l\rho \cos\phi\right)^{3/2}}, \\ &= \frac{1}{r_l^3\left(1 + (\rho/r_l)^2 + 2(\rho/r_l)\cos\phi\right)^{3/2}} \end{aligned} \tag{9.4}$$

where $\rho = |\vec{\rho}|$, $r_l = |\vec{\mathbf{r}}_l|$ and ϕ is the angle between the vectors $\vec{\mathbf{r}}_l$ and $\vec{\rho}$.

Since the spacecraft are in close proximity, we assume $\epsilon \triangleq \rho/r_l \ll 1$. A first-order Taylor series expansion gives

$$\frac{1}{|\vec{\mathbf{r}}_l + \vec{\rho}|^3} = \frac{1}{r_l^3}\left(1 + 2\epsilon\cos\phi + \epsilon^2\right)^{-3/2} \approx \frac{1}{r_l^3}\left(1 + \frac{d}{d\epsilon}\left\{\left(1 + 2\epsilon\cos\phi + \epsilon^2\right)^{-3/2}\right\}\Big|_{\epsilon=0} \epsilon\right).$$

Now,

$$\frac{d}{d\epsilon}\left\{\left(1 + 2\epsilon\cos\phi + \epsilon^2\right)^{-3/2}\right\} = \frac{-3(\cos\phi + \epsilon)}{\left(1 + 2\epsilon\cos\phi + \epsilon^2\right)^{5/2}},$$

such that

$$\left(1 + 2\epsilon\cos\phi + \epsilon^2\right)^{-3/2} \approx 1 - 3\epsilon\cos\phi.$$

Making use of this approximation in (9.4) leads to

$$\frac{1}{|\vec{\mathbf{r}}_l + \vec{\rho}|^3} \approx \frac{1}{r_l^3} - \frac{3(\rho/r_l)\cos\phi}{r_l^3}. \tag{9.5}$$

However, by definition of the dot product we have $\vec{\rho} \cdot \vec{\mathbf{r}}_l = r_l \rho \cos\phi$, which leads to

$$\frac{\rho}{r_l}\cos\phi = \frac{\vec{\rho} \cdot \vec{\mathbf{r}}_l}{r_l^2}.$$

Substituting this into (9.5) gives

$$\frac{1}{|\vec{\mathbf{r}}_l + \vec{\rho}|^3} \approx \frac{1}{r_l^3} - \frac{3\vec{\rho} \cdot \vec{\mathbf{r}}_l}{r_l^5}. \tag{9.6}$$

Now, we can make use of (9.6) in (9.3) to obtain

$$\ddot{\vec{\rho}} \approx -\mu \frac{(\vec{\mathbf{r}}_l + \vec{\rho})}{r_l^3}\left(1 - \frac{3\vec{\rho} \cdot \vec{\mathbf{r}}_l}{r_l^2}\right) + \vec{\mathbf{f}}_f + \mu \frac{\vec{\mathbf{r}}_l}{r_l^3}.$$

Expanding this and neglecting $(3\mu/r_l^3)\left(\vec{\rho} \cdot \vec{\mathbf{r}}_l\right)\vec{\rho}/r_l^2$ (since it is of second order in ρ/r_l), we obtain a first order approximation of the relative dynamics

$$\ddot{\vec{\rho}} \approx -\frac{\mu}{r_l^3}\vec{\rho} + \frac{3\mu}{r_l^3}\frac{\left(\vec{\rho} \cdot \vec{\mathbf{r}}_l\right)\vec{\mathbf{r}}_l}{r_l^2} + \vec{\mathbf{f}}_f. \tag{9.7}$$

Equation (9.7) describes the relative dynamics in physical vector form. As always, we want to express them in a reference frame. We shall assume that the leader is in a circular orbit of radius r_l. A useful frame for expressing the relative dynamics is a reference frame that orbits with the leader spacecraft, $\mathcal{F}_H$ called the Hill frame (see Figure 9.3). It is defined as

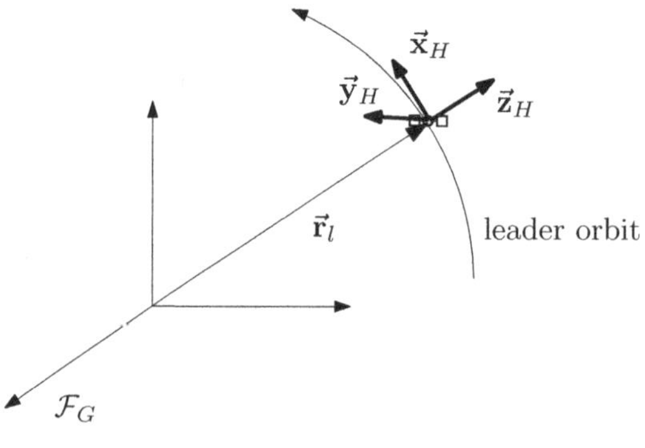

Figure 9.3 Hill-frame definition

follows: $\vec{\mathbf{x}}_H = \vec{\mathbf{r}}_l/|\vec{\mathbf{r}}_l|$, $\vec{\mathbf{z}}_H = \vec{\mathbf{r}}_l/|\vec{\mathbf{r}}_l|$ and $\vec{\mathbf{y}}_H = \vec{\mathbf{z}}_H \times \vec{\mathbf{x}}_H$. The angular velocity of this frame is given by

$$\vec{\boldsymbol{\omega}}_H = \vec{\mathcal{F}}_H^T \begin{bmatrix} 0 \\ \omega_o \\ 0 \end{bmatrix}.$$

where $\omega_o = \sqrt{\mu/r_l^3}$ is the orbital angular velocity of the leader satellite, which is constant (since the leader orbit is circular). In this frame, the leader position vector is given by

$$\vec{\mathbf{r}}_l = \vec{\mathcal{F}}_H^T \begin{bmatrix} 0 \\ 0 \\ r_l \end{bmatrix}.$$

Let $(\mathring{\ })$ denote time-differentiation as seen in frame $\mathcal{F}_H$. Then, as shown in Section 1.5,

$$\ddot{\vec{\rho}} = \mathring{\mathring{\vec{\rho}}} + 2\vec{\boldsymbol{\omega}}_H \times \mathring{\vec{\rho}} + \mathring{\vec{\boldsymbol{\omega}}}_H \times \vec{\rho} + \vec{\boldsymbol{\omega}}_H \times \vec{\boldsymbol{\omega}}_H \times \vec{\rho}.$$

Substituting this into (9.7), and noting that $\mathring{\vec{\boldsymbol{\omega}}}_H = \vec{\mathbf{0}}$ (since the leader orbit is circular), the equation of relative motion becomes

$$\mathring{\mathring{\vec{\rho}}} + 2\vec{\boldsymbol{\omega}}_H \times \mathring{\vec{\rho}} + \vec{\boldsymbol{\omega}}_H \times \vec{\boldsymbol{\omega}}_H \times \vec{\rho} + \omega_o^2 \vec{\rho} - 3\omega_o^2 \frac{(\vec{\rho} \cdot \vec{\mathbf{r}}_l)\,\vec{\mathbf{r}}_l}{r_l^2} = \vec{\mathbf{f}}_f. \tag{9.8}$$

Note that we have also replaced μ/r_l^3 by ω_o^2. Expressing the relative position vector in $\mathcal{F}_H$ coordinates as

$$\vec{\rho} = \vec{\mathcal{F}}_H^T \begin{bmatrix} x \\ y \\ z \end{bmatrix}, \tag{9.9}$$

we have

$$\mathring{\vec{\rho}} = \vec{\mathcal{F}}_H^T \begin{bmatrix} \dot{x} \\ \dot{y} \\ \dot{z} \end{bmatrix} \text{ and } \mathring{\mathring{\vec{\rho}}} = \vec{\mathcal{F}}_H^T \begin{bmatrix} \ddot{x} \\ \ddot{y} \\ \ddot{z} \end{bmatrix}.$$

We can now evaluate the individual terms in (9.8). We have

$$\vec{\boldsymbol{\omega}}_H \times \mathring{\vec{\rho}} = \vec{\mathcal{F}}_H^T \begin{bmatrix} 0 & 0 & \omega_o \\ 0 & 0 & 0 \\ -\omega_o & 0 & 0 \end{bmatrix} \begin{bmatrix} \dot{x} \\ \dot{y} \\ \dot{z} \end{bmatrix} = \vec{\mathcal{F}}_H^T \begin{bmatrix} \omega_o \dot{z} \\ 0 \\ -\omega_o \dot{x} \end{bmatrix},$$

$$\vec{\boldsymbol{\omega}}_H \times \vec{\boldsymbol{\omega}}_H \times \vec{\boldsymbol{\rho}} = \vec{\mathcal{F}}_H^T \begin{bmatrix} 0 & 0 & \omega_o \\ 0 & 0 & 0 \\ -\omega_o & 0 & 0 \end{bmatrix} \begin{bmatrix} \omega_o z \\ 0 \\ -\omega_o x \end{bmatrix} = \begin{bmatrix} -\omega_o^2 x \\ 0 \\ -\omega_o^2 z \end{bmatrix},$$

$$\frac{(\vec{\boldsymbol{\rho}} \cdot \vec{\mathbf{r}}_l)\, \vec{\mathbf{r}}_l}{r_l^2} = \vec{\mathcal{F}}_H^T \left(\begin{bmatrix} x & y & z \end{bmatrix} \begin{bmatrix} 0 \\ 0 \\ r_l \end{bmatrix} \right) \begin{bmatrix} 0 \\ 0 \\ r_l \end{bmatrix} \frac{1}{r_l^2} = \vec{\mathcal{F}}_H^T \begin{bmatrix} 0 \\ 0 \\ z \end{bmatrix}.$$

Let us also express the follower force/unit mass in Hill coordinates as

$$\vec{\mathbf{f}}_f = \vec{\mathcal{F}}_H^T \begin{bmatrix} f_x \\ f_y \\ f_z \end{bmatrix}.$$

Finally, we can use all of the above expressions in (9.7) to obtain the relative equations of motion in Hill coordinates as

$$\begin{bmatrix} \ddot{x} + 2\omega_o \dot{z} \\ \ddot{y} + \omega_o^2 y \\ \ddot{z} - 2\omega_o \dot{x} - 3\omega_o^2 z \end{bmatrix} = \begin{bmatrix} f_x \\ f_y \\ f_z \end{bmatrix}. \tag{9.10}$$

Equations (9.10) are known as the Hill or Clohessy-Wiltshire equations.

By the definition of the Hill frame, the x and z coordinates represent relative spacecraft motion within the orbital plane of the leader. This relative motion is due to differences in the semi-major axis and eccentricity of the two spacecraft. The y coordinate represents relative motion out of the orbital plane of the leader. This motion is due to differences in inclination and right ascension of the ascending node. It can be seen from (9.10) that the in-plane and out of plane relative motions are decoupled. Thus, we have the *in-plane* equations

$$\begin{aligned} \ddot{x} + 2\omega_o \dot{z} &= f_x, \\ \ddot{z} - 2\omega_o \dot{x} - 3\omega_o^2 z &= f_z, \end{aligned} \tag{9.11}$$

and the *out-of-plane* equation

$$\ddot{y} + \omega_o^2 y = f_y, \tag{9.12}$$

which can be treated separately.

9.2 Relative Motion Solutions

Any formation that would be practically useful requires minimal fuel (propellant) to maintain the formation. Ideally, we would seek natural formations that do not require any fuel to maintain, that is formations with $f_x = f_y = f_z = 0$. Of course, due to orbital perturbations this is not possible, and some fuel will always be required for formation maintenance. However, it is still of interest to examine unforced solutions of (9.11) and (9.12), which represent an ideal case.

9.2.1 Out-of-Plane Motion

We first examine out-of-plane motion. From (9.12), the equation to be solved is given by

$$\ddot{y} + \omega_o^2 y = 0, \text{ with initial conditions } y(0) = y_0, \ \dot{y}(0) = \dot{y}_0. \tag{9.13}$$

This is a simple second order ordinary differential equation which may be solved in a variety of ways. Letting $Y(s) = \mathcal{L}(y(t))$ be the Laplace transform of $y(t)$, and taking the Laplace transform of (9.13) gives

$$s^2 Y - s y_0 - \dot{y}_0 + \omega_o^2 Y = 0,$$

which rearranges to give

$$Y(s) = \frac{s}{s^2 + \omega_o^2} y_0 + \frac{\dot{y}_o}{s^2 + \omega_o^2}.$$

Taking the inverse Laplace transform gives the solution

$$y(t) = y_0 \cos \omega_o t + \frac{\dot{y}_0}{\omega_o} \sin \omega_o t. \tag{9.14}$$

We see that unforced out-of-plane motion is sinusoidal, which means that it is always bounded. Physically this makes sense, since out-of-plane motion is due to a difference in orbital plane of the leader and follower satellite (which as mentioned previously is due to a difference in inclination and/or right ascension of the ascending node). See Figure 9.4.

9.2.2 In-Plane Motion

We now turn our attention to relative motion within the orbital plane of the leader. From (9.11), the equations to be solved are

$$\begin{aligned} &\ddot{x} + 2\omega_o \dot{z} = 0, \text{ with initial conditions } x(0) = x_0, \ \dot{x}(0) = \dot{x}_0 \\ &\ddot{z} - 2\omega_o \dot{x} - 3\omega_o^2 z = 0, \text{ with initial conditions } z(0) = z_0, \ \dot{z}(0) = \dot{z}_0. \end{aligned} \tag{9.15}$$

These are a pair of coupled second order ordinary differential equations. As for the out-of-plane equations, there are a number of methods that could be used to solve them. Integrating the first equation in (9.15) gives

$$\int_0^t \ddot{x} d\tau + 2\omega_o \int_0^t \dot{z} d\tau = 0,$$

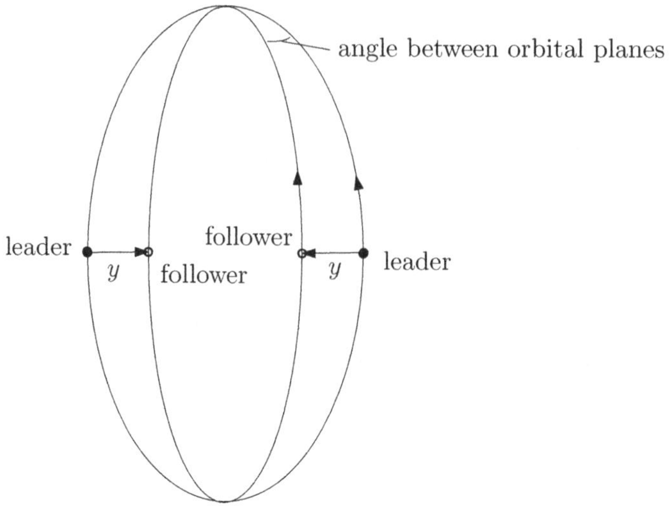

Figure 9.4 Out-of-plane relative motion

which leads to

$$\dot{x} = -2\omega_o z + \dot{x}_0 + 2\omega_o z_0. \tag{9.16}$$

Substituting this into the second equation in (9.15) gives

$$\ddot{z} - 3\omega_o^2 z - 2\omega_o(-2\omega_o z + \dot{x}_0 + 2\omega_o z_0) = 0,$$

which can be simplified to

$$\ddot{z} + \omega_o^2 z = 2\omega_o \dot{x}_0 + 4\omega_o^2 z_0.$$

Taking Laplace transforms of this gives

$$s^2 Z - s z_0 - \dot{z}_0 + \omega_o^2 Z = \frac{2\omega_o \dot{x}_0}{s} + \frac{4\omega_o^2 z_0}{s},$$

which can be rearranged to give

$$Z = \frac{s}{s^2 + \omega_o^2} z_0 + \frac{4\omega_o^2}{s(s^2 + \omega_o^2)} z_0 + \frac{\dot{z}_0}{s^2 + \omega_o^2} + \frac{2\omega_o}{s(s^2 + \omega_o^2)} \dot{x}_0. \tag{9.17}$$

Now, the following partial fraction expansion can be readily verified, and will be needed:

$$\frac{1}{s(s^2 + \omega_o^2)} = \frac{1}{\omega_o^2 s} - \frac{s}{\omega_o^2 (s^2 + \omega_o^2)}. \tag{9.18}$$

Substituting (9.18) into (9.17) leads to

$$Z = \left(\frac{s}{s^2+\omega_o^2} + \frac{4}{s} - \frac{4s}{s^2+\omega_o^2}\right) z_0 + \frac{\dot{z}_0}{s^2+\omega_o^2} + \frac{2}{\omega_o}\left(\frac{1}{s} - \frac{s}{s^2+\omega_o^2}\right)\dot{x}_0.$$

Taking the inverse Laplace transform and rearranging leads to the solution for $z(t)$

$$z(t) = \left(4z_0 + \frac{2\dot{x}_0}{\omega_o}\right) + \frac{\dot{z}_0}{\omega_o}\sin\omega_o t - \left(3z_0 + \frac{2\dot{x}_0}{\omega_o}\right)\cos\omega_o t. \tag{9.19}$$

Returning now to (9.16), we have upon further integration

$$x(t) = x_0 - 2\omega_o \int_0^t z(\tau)d\tau + (\dot{x}_0 + 2\omega_o z_0)\,t. \tag{9.20}$$

Integrating (9.19), we now have

$$\int_0^t z(\tau)d\tau = \left(4t - \frac{3\sin\omega_o t}{\omega_o}\right) z_0 + \frac{\dot{z}_0}{\omega_o^2}(1 - \cos\omega_o t) + \frac{2}{\omega_o}\left(t - \frac{\sin\omega_o t}{\omega_o}\right)\dot{x}_0.$$

Substituting this into (9.20) and rearranging gives

$$x(t) = \left(x_0 - \frac{2\dot{z}_0}{\omega_o}\right) - (3\dot{x}_0 + 6\omega_o z_0)\,t + \left(6z_0 + \frac{4\dot{x}_0}{\omega_o}\right)\sin\omega_o t + \frac{2\dot{z}_0}{\omega_0}\cos\omega_o t. \tag{9.21}$$

In summary, in-plane motion is described by (9.21) and (9.19). The x component is called the *along-track* component since the Hill frame x-axis points in the direction of the leader satellite orbital motion, while the z component is called the *radial* component since the Hill frame z axis points radially outwards from the leader satellite orbit. It can be seen from (9.21) and (9.19) that the in-plane relative motion is sinusoidal plus both an along-track and radial offset and an along-track secular drift.

In order to achieve bounded relative motion (and hence a formation), the secular drift must be zero. To achieve this, the coefficient of t in (9.21) must be zero. That is, the initial conditions must satisfy

$$\dot{x}_0 = -2\omega_0 z_0. \tag{9.22}$$

9.2.3 Alternative Description for In-Plane Relative Motion

As previously mentioned, the in-plane motion can be separated into secular and periodic parts. From equations (9.21) and (9.19), it is seen that the secular part is given by

$$\begin{aligned}\bar{x}(t) &= \left(x_0 - \frac{2\dot{z}_0}{\omega_o}\right) - (3\dot{x}_0 + 6\omega_o z_0)\,t,\\ \bar{z}(t) &= \left(4z_0 + \frac{2\dot{x}_0}{\omega_o}\right).\end{aligned} \tag{9.23}$$

Likewise, the periodic part is given by

$$\begin{aligned}x_p(t) &= 2\left[\left(3z_0 + \frac{2\dot{x}_0}{\omega_o}\right)\sin\omega_o t + \frac{\dot{z}_0}{\omega_0}\cos\omega_o t\right],\\ z_p(t) &= \frac{\dot{z}_0}{\omega_o}\sin\omega_o t - \left(3z_0 + \frac{2\dot{x}_0}{\omega_o}\right)\cos\omega_o t.\end{aligned} \tag{9.24}$$

Therefore, making use of (9.23) and (9.24), the in-plane motion (see (9.21) and (9.19)) can be written as

$$\begin{aligned}x(t) &= \bar{x}(t) + x_p(t),\\ z(t) &= \bar{z}(t) + z_p(t).\end{aligned} \tag{9.25}$$

9.2.3.1 Periodic Part

Let us now turn our attention to the periodic part of the in-plane relative motion. We will make use of the following facts from trigonometry:

$$A\sin\omega_o t + B\cos\omega_o t = (A^2 + B^2)^{\frac{1}{2}}\cos(\omega_o t + \phi),$$

and

$$B\sin\omega_o t - A\cos\omega_o t = (A^2 + B^2)^{\frac{1}{2}}\sin(\omega_o t + \phi),$$

where

$$\sin\phi = \frac{-A}{(A^2 + B^2)^{\frac{1}{2}}}, \quad \cos\phi = \frac{B}{(A^2 + B^2)^{\frac{1}{2}}}.$$

Applying these identities to (9.24), the periodic part of the relative motion can be written as

$$\begin{aligned}x_p(t) &= 2P\cos(\omega_o t + \phi_0),\\ z_p(t) &= P\sin(\omega_o t + \phi_0),\end{aligned} \tag{9.26}$$

where

$$P = \left(\left(3z_0 + \frac{2\dot{x}_0}{\omega_o} \right)^2 + \left(\frac{\dot{z}_0}{\omega_0} \right)^2 \right)^{\frac{1}{2}}. \tag{9.27}$$

and

$$\sin \phi_0 = \frac{-\left(3z_0 + \frac{2\dot{x}_0}{\omega_o}\right)}{P}, \quad \cos \phi_0 = \frac{\dot{z}_0}{\omega_o P}. \tag{9.28}$$

Using (9.26), a simple calculation shows that

$$\frac{x_p(t)^2}{(2P)^2} + \frac{z_p(t)^2}{P^2} = \cos^2(\omega_o t + \phi_0) + \sin^2(\omega_o t + \phi_0) = 1.$$

That is, the periodic part of the in-plane motion describes an ellipse with semi-major axis $2P$ and semi-minor axis P.

Including the secular part of the relative motion, the in-plane relative motion of the follower relative to the leader describes an ellipse centered at $\bar{x}(t)$, $\bar{z}(t)$, as shown in Figure 9.5. An interesting point to note is that the shape of the ellipse is fixed by the ratio of semi-major to semi-minor axes, which is always 2. This ellipse is sometimes called a *football orbit*.

9.2.3.2 Secular Part

Let us now examine the secular part of the in-plane relative motion. Taking the time-derivative of (9.23) gives

$$\frac{d\bar{x}}{dt} = -(3\dot{x}_0 + 6\omega_o z_0) = -\frac{3\omega_o}{2}\bar{z}, \quad \frac{d\bar{z}}{dt} = 0.$$

This shows that the secular radial offset $\bar{z}$ is constant. It also shows that since $\frac{d\bar{x}}{dt} = -\frac{3\omega_o}{2}\bar{z}$, the only way to eliminate the secular drift is if the secular radial offset $\bar{z} = 0$. This makes sense

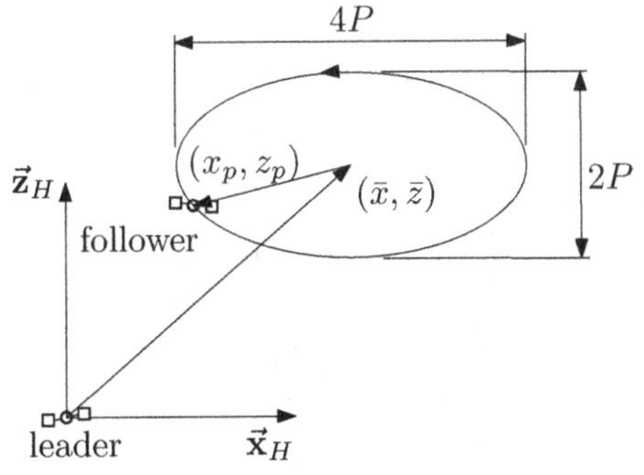

Figure 9.5 In-plane relative motion

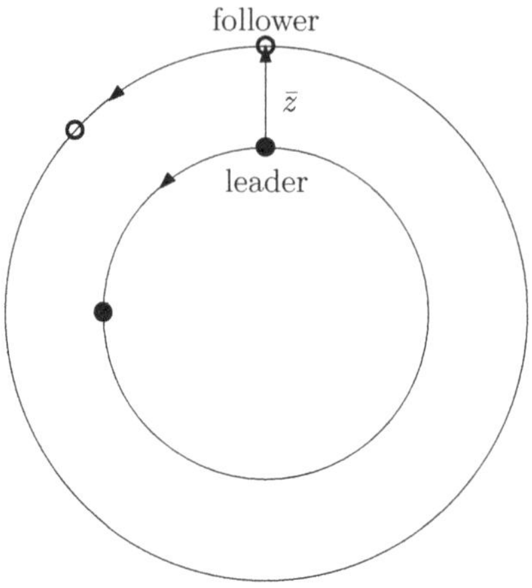

Figure 9.6 Secular part of relative motion due to different semi-major axes

if you consider the fact that a non-zero $\bar{z}$ means a difference in semi-major axis of the leader and follower orbits (see Figure 9.6). The difference in semi-major axes means a difference in orbital periods. For bounded relative motion, both the leader and the follower spacecraft must have the same orbital period.

9.2.4 *Further Examination of In-Plane Motion*

In Sections 3.6 to 3.8, we saw that an alternative to describing orbital motion in terms of position and velocity is to use the orbital elements, which are much more useful for visualizing an orbit. Expressions were derived for transforming from position and velocity to the orbital elements and vice-versa. We can do likewise for in-plane relative motion. As we have just seen, an alternative to using x, z, $\dot{x}$ and $\dot{z}$ could be the secular motion $\bar{x}$, $\bar{z}$ and the size and phase of the football orbit given by P and $\phi = \omega_o t + \phi_0$ respectively. So far, we have seen these quantities defined in terms of the initial conditions. However, since the equations of motion are time-invariant (do not explicitly depend on time), the datum $t = 0$ may be selected arbitrarily. In particular, this means that we can obtain $\bar{x}$, $\bar{z}$, P and $\phi(t) = \omega_o t + \phi_0$ by using the current values for $x(t)$, $z(t)$, $\dot{x}(t)$ and $\dot{z}(t)$ in place of the initial conditions x_0, z_0, $\dot{x}_0$ and $\dot{z}_0$ in (9.23), (9.27) and (9.28) and setting $t = 0$ in the right hand side expression for $\bar{x}(t)$. That is,

$$\begin{aligned}
\bar{x}(t) &= x(t) - \frac{2\dot{z}(t)}{\omega_o}, \quad \bar{z}(t) = 4z(t) + \frac{2\dot{x}(t)}{\omega_o}, \\
P &= \left(\left(3z(t) + \frac{2\dot{x}(t)}{\omega_o} \right)^2 + \left(\frac{\dot{z}(t)}{\omega_0} \right)^2 \right)^{\frac{1}{2}}, \\
\sin\phi(t) &= \frac{-\left(3z(t) + \frac{2\dot{x}(t)}{\omega_o}\right)}{P}, \\
\cos\phi(t) &= \frac{\dot{z}(t)}{\omega_o P}.
\end{aligned} \tag{9.29}$$

We will now demonstrate that the equation for $\bar{z}(t)$ agrees with the previous definition in terms of the initial conditions.

Consider the time derivative

$$\frac{d}{dt}\left(4z + \frac{2\dot{x}}{\omega_o}\right) = 4\dot{z} + \frac{2\ddot{x}}{\omega_o}.$$

Substituting from the equation of motion for $\ddot{x}$ (9.15) gives

$$\frac{d}{dt}\left(4z + \frac{2\dot{x}}{\omega_o}\right) = 0.$$

Integrating gives

$$\int_0^t \frac{d}{d\tau}\left(4z + \frac{2\dot{x}}{\omega_o}\right) d\tau = 0,$$

which leads to

$$4z(t) + \frac{2\dot{x}(t)}{\omega_o} = 4z_0 + \frac{2\dot{x}_0}{\omega_o},$$

which shows that the expression for $\bar{z}$ in terms of the current position and velocity in (9.29) agrees with the expression given in terms of the initial conditions in (9.23). The other expressions in (9.29) can be demonstrated similarly (try it as an exercise).

Summary

To summarize, the relative in-plane motion can be described as

$$\begin{aligned} x(t) &= \bar{x}(t) + 2P\cos\phi(t), \\ z(t) &= \bar{z}(t) + P\sin\phi(t), \end{aligned} \tag{9.30}$$

where

$$\begin{aligned} \bar{x}(t) &= x(t) - \frac{2\dot{z}(t)}{\omega_o}, \\ \bar{z}(t) &= 4z(t) + \frac{2\dot{x}(t)}{\omega_o}, \\ P &= \left(\left(3z(t) + \frac{2\dot{x}(t)}{\omega_o}\right)^2 + \left(\frac{\dot{z}(t)}{\omega_0}\right)^2\right)^{1/2}, \\ \sin\phi(t) &= \frac{-\left(3z(t) + \frac{2\dot{x}(t)}{\omega_o}\right)}{P}, \\ \cos\phi(t) &= \frac{\dot{z}(t)}{\omega_o P}. \end{aligned}$$

9.2.5 Out-of-Plane Motion - Revisited

We have seen that the out-of-plane relative motion is given by

$$y(t) = y_0 \cos \omega_o t + \frac{\dot{y}_0}{\omega_o} \sin \omega_o t.$$

Similar to the in-plane periodic motion, this can be rewritten as

$$y(t) = Q \sin(\omega_o t + \alpha_0) \tag{9.31}$$

where

$$Q = \left(y_0^2 + \left(\frac{\dot{y}_0}{\omega_0} \right)^2 \right)^{1/2}, \quad \sin \alpha_0 = \frac{y_0}{Q}, \quad \cos \alpha_0 = \frac{\dot{y}_0}{\omega_o Q}.$$

Also, as for in-plane periodic motion, the amplitude Q can be computed from the relative position and velocity at any time by

$$Q = \left(y(t)^2 + \left(\frac{\dot{y}(t)}{\omega_0} \right)^2 \right)^{1/2}. \tag{9.32}$$

Similarly, the phase $\alpha(t) = \omega_o t + \alpha_0$ can be computed from

$$\sin \alpha(t) = \frac{y(t)}{Q}, \quad \cos \alpha(t) = \frac{\dot{y}(t)}{\omega_o Q}.$$

We will now demonstrate that the above equations for $\alpha(t)$ agree with the previous definition in terms of the initial conditions. Define an angle $\beta(t)$ by the relationships $\sin \beta(t) = \frac{y(t)}{Q}$ and $\cos \beta(t) = \frac{\dot{y}(t)}{\omega_o Q}$ where $Q = (y(t)^2 + (\frac{\dot{y}(t)}{\omega_0})^2)^{1/2}$. Note that we have not yet established equality with $\alpha(t)$, so we call it $\beta(t)$. At time $t = 0$, we have

$$\sin \beta(0) = \frac{y(0)}{Q} \text{ and } \cos \beta(0) = \frac{\dot{y}(0)}{\omega_o Q},$$

which shows that $\beta(0) = \alpha_0$ by the definition of α_0. Now, take the time-derivative

$$\frac{d}{dt}(\cos \beta(t)) = -\dot{\beta} \sin \beta = \frac{\ddot{y}(t)}{\omega_o Q},$$

Note that we have used the fact that Q is constant (although we have not demonstrated it). Rearranging gives

$$\dot{\beta} = -\frac{\ddot{y}(t)}{\omega_o Q \sin \beta}.$$

Substituting for $\sin\beta$ from its definition, and for $\ddot{y}$ from the equation of motion (9.13) gives

$$\dot{\beta} = -\frac{(-\omega_o^2 y)}{\omega_o Q \frac{y(t)}{Q}} = \omega_o.$$

Integrating this gives

$$\int_0^t \dot{\beta} d\tau = \int_0^t \omega_o d\tau,$$

which leads to

$$\beta(t) = \beta(0) + \omega_o t.$$

Since we have already established that $\beta(0) = \alpha_0$, we have

$$\beta(t) = \alpha(t) = \alpha_0 + \omega_o t,$$

as required. The relationship for Q can be demonstrated similarly (try it as an exercise).

9.3 Special Types of Relative Orbits

9.3.1 Along-Track Orbits

We note from the solutions for in-plane and out-of-plane motion (equations (9.30) and (9.31)) that if we do not want any periodic relative motion, then the amplitudes P and Q must be equal to zero. From (9.29), the requirement that $P = 0$ means that

$$3z(t) + \frac{2\dot{x}(t)}{\omega_o} = 0, \quad \frac{\dot{z}(t)}{\omega_0} = 0.$$

Additionally, we require that there be no secular drift. As shown in Section 9.2.3.2, this means that we must have $\bar{z} = 0$, which from (9.29) means that

$$4z(t) + \frac{2\dot{x}(t)}{\omega_o} = 0.$$

The only way that the above three equations can be satisfied simultaneously is if

$$\dot{x} = \dot{z} = z = 0.$$

That is, the only possible bounded non-periodic in-plane motion is

$$x(t) = constant,$$

$$z(t) = 0,$$

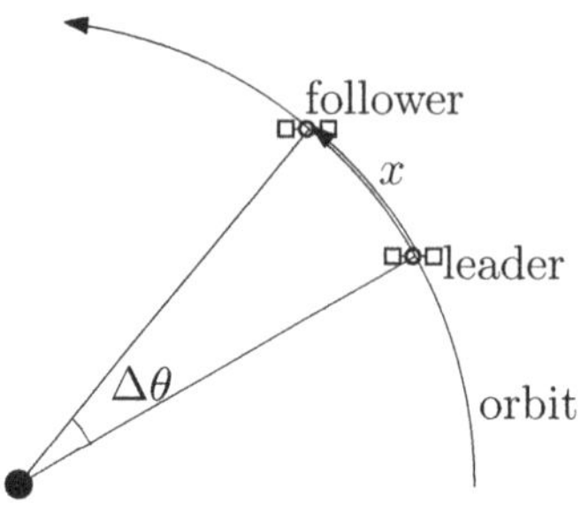

Figure 9.7 Along-track relative orbit

What this means physically is that both the leader and follower orbits are circular with the same radius. The only difference is in the true anomalies, which results in the constant $x(t)$ (along-track) offset.

The requirement that $Q = 0$ means from (9.32) that

$$y(t) = 0.$$

That is, there is no out of plane motion. This means physically that the leader and follower orbits have the same inclination and right ascension of the ascending node. Such a relative orbit is called an *along-track* relative *orbit* (ATO). See Figure 9.7.

9.3.2 Projected Elliptical Orbits

The projection of the relative motion on the $\vec{\mathbf{x}}_H$-$\vec{\mathbf{y}}_H$ plane is of interest, since this plane is facing the Earth directly below (which is often to be observed). As shown in Section 9.2, the $\vec{\mathbf{x}}_H$ and $\vec{\mathbf{y}}_H$ components of the natural relative motion can be written as

$$x(t) = \bar{x}(0) - \frac{3\omega_o}{2}\bar{z}(0)t + 2P\cos(\omega_o t + \phi_0), \tag{9.33}$$

and

$$y(t) = Q\sin(\omega_o t + \alpha_0) \tag{9.34}$$

respectively, where

$$\bar{x}(0) = x_0 - \frac{2\dot{z}_0}{\omega_o},$$

$$\bar{z}(0) = 4z_0 + \frac{2\dot{x}_0}{\omega_o},$$

$$P = \left(\left(3z_0 + \frac{2\dot{x}_0}{\omega_o}\right)^2 + \left(\frac{\dot{z}_0}{\omega_0}\right)^2\right)^{\frac{1}{2}},$$

$$\sin\phi_0 = \frac{-\left(3z_0 + \frac{2\dot{x}_0}{\omega_o}\right)}{P},$$

$$\cos\phi_0 = \frac{\dot{z}_0}{\omega_o P},$$

$$Q = \left(y_0^2 + \left(\frac{\dot{y}_0}{\omega_0}\right)^2\right)^{\frac{1}{2}},$$

$$\sin\alpha_0 = \frac{y_0}{Q},$$

$$\cos\alpha_0 = \frac{\dot{y}_0}{\omega_o Q}.$$

By appropriate selection of the initial conditions, we can obtain a relative motion whose projection on the $\vec{\mathbf{x}}_H$-$\vec{\mathbf{y}}_H$ is an ellipse of a given size specified by P and Q. This is called a *Projected Elliptical Orbit* (PEO). Let us see how to do this.

First of all, the relative motion must be bounded (no secular drift). Therefore, we must have

$$\bar{z}(0) = 4z_0 + \frac{2\dot{x}_0}{\omega_o} = 0,$$

which implies that

$$\dot{x}_0 = -2\omega_o z_0. \tag{9.35}$$

Next, the center of the ellipse should be at the origin. That is,

$$\bar{x}(0) = x_0 - \frac{2\dot{z}_0}{\omega_o} = 0,$$

which implies that

$$\frac{\dot{z}_0}{\omega_o} = \frac{x_0}{2}. \tag{9.36}$$

Substituting these into the expressions for the amplitude P and initial phase ϕ_0, we have

$$P = \left(\left(3z_0 + \frac{2}{\omega_o}(-2\omega_o z_0)\right)^2 + \left(\frac{x_0}{2}\right)^2\right)^{\frac{1}{2}},$$

$$= \left(z_0^2 + \left(\frac{x_0}{2}\right)^2\right)^{\frac{1}{2}},$$

and

$$\sin\phi_0 = \frac{z_0}{P}, \quad \cos\phi_0 = \frac{x_0}{2P}. \tag{9.37}$$

Therefore, by appropriately selecting x_0 and z_0, we can determine the size P and the initial phase ϕ_0.

We can see from (9.33) and (9.34) that for the projected relative motion to describe an ellipse, the initial phases must satisfy

$$\alpha_0 = \phi_0, \text{ or } \alpha_0 = \phi_0 + \pi.$$

This means that either

$$\sin\alpha_0 = \sin\phi_0 \text{ and } \cos\alpha_0 = \cos\phi_0,$$

or

$$\sin\alpha_0 = -\sin\phi_0 \text{ and } \cos\alpha_0 = -\cos\phi_0.$$

In terms of the initial conditions, this means that

$$\frac{y_0}{Q} = \frac{z_0}{P} \text{ and } \frac{\dot{y}_0}{\omega_0 Q} = \frac{x_0}{2P}, \tag{9.38}$$

or

$$\frac{y_0}{Q} = -\frac{z_0}{P} \text{ and } \frac{\dot{y}_0}{\omega_o Q} = -\frac{x_0}{2P}. \tag{9.39}$$

We can now show that the projected motion with these initial conditions is an ellipse. From (9.33) and (9.34),

$$\frac{x(t)^2}{4P^2} + \frac{y(t)^2}{Q^2} = \cos^2(\omega_o t + \phi_0) + \sin^2(\omega_o t + \alpha_0) = 1,$$

when $\alpha_0 = \phi_0$ or $\alpha_0 = \phi_0 + \pi$. It is clear that when $2P > Q$, then $2P$ is the semi-major axis and Q is the semi-minor axis. When $2P < Q$, the roles are reversed.

Summary

To summarize, given a desired PEO with specified semi-major and semi-minor axes $2P$ and Q respectively (or vice versa), and a specified initial phase ϕ_0, the initial conditions may be obtained as follows:

1. From (9.37) $x_0 = 2P\cos\phi_0$ and $z_0 = P\sin\phi_0$.
2. From (9.35) and (9.36), $\dot{x}_0 = -2\omega_o z_0$ and $\dot{z}_0 = \omega_o x_0/2$.
3. From (9.38), $y_0 = Qz_0/P$ and $\dot{y}_0 = \omega_0 Q x_0/2P$.

9.3.3 Projected Circular Orbits

A special case of the projected elliptical orbit is when the projected motion is circular. This is called a *Projected Circular Orbit* (PCO). In this case, the projected semi-major and semi-minor axes must be equal, that is

$$2P = Q,$$

which means that the radius of the projected circular orbit is $R = 2P$.

The phase matching conditions in (9.38) and (9.39) become

$$y_0 = \frac{z_0 Q}{P} = 2z_0 \text{ and } \dot{y}_0 = \frac{\omega_o x_0 Q}{2P} = \omega_o x_o, \tag{9.40}$$

or

$$y_0 = -2z_0 \text{ and } \dot{y}_0 = -\omega_o x_o. \tag{9.41}$$

To an observer viewing the formation along the $\vec{\mathbf{z}}_H$ axis, the follower appears to orbit around the leader in a circle with radius R.

Summary
To summarize, given a desired PCO with specified radius R, and a specified initial phase ϕ_0, the initial conditions may be obtained as follows:

1. From (9.37) $x_0 = R\cos\phi_0$ and $z_0 = \frac{R}{2}\sin\phi_0$.
2. From (9.35) and (9.36), $\dot{x}_0 = -2\omega_o z_0$ and $\dot{z}_0 = \omega_o x_0/2$.
3. From (9.40), $y_0 = 2z_0$ and $\dot{y}_0 = \omega_0 x_0$.

Notes

This chapter has provided a brief introduction to the subject of spacecraft formation flying. In doing so, we have made some significant simplifying assumptions, in particular that there are no perturbing forces acting on the spacecraft, and that the leader spacecraft is in a circular orbit. We have not considered the spacecraft formation control problem at all. Spacecraft formation flying has been a topic of significant interest in recent years, and there are extensive results in the publicly available literature. A detailed overview of the most significant developments may be found in Alfriend et al. (2010), which is the first comprehensive book published on the subject.

Reference

Alfriend KT, Vadali SR, Gurfil P, How JP and Breger LS 2010 *Spacecraft Formation Flying: Dynamics, control and navigation*. Elsevier, Burlington, MA.

10

The Restricted Three-Body Problem

10.1 Formulation

Consider the three masses that interact gravitationally in Figure 10.1.

Assumptions

(i) The gravitational effect of m_3 on m_2 and m_1 is negligible. This permits a two-body solution for the motion of m_1 and m_2.
(ii) The two-body motion of m_1 and m_2 is circular about their mutual center of mass.
(iii) The initial position and velocity of m_3 are in the plane of the two-body motion of m_1 and m_2. Therefore m_3 remains in this plane.

Examples are sun-planet-moon systems and Earth-Moon-spacecraft systems. We shall treat the restricted problem in the context of the Earth-Moon-spacecraft system. In this case, we identify m_1 with the mass of the Earth and m_2 with the Lunar mass. The origin of $\mathcal{F}_I$ will now be placed at their mass center.

This form of the Restricted Three-Body Problem is more correctly called the Circular Restricted Three-Body Problem but we will drop 'Circular' from the title since it is the only version of the problem that we consider. Now consider Figure 10.2. Here, X-Y is an inertial frame centered at the Earth-Moon center of mass. The coordinates x-y refer to the position of m_3 in a frame rotating with an angular velocity equivalent to the mean motion of the Earth-Moon system:

$$\omega = \sqrt{G(m_1 + m_2)/r_{12}^3}, \quad r_{12} = r_1 + r_2.$$

Spacecraft Dynamics and Control: An Introduction, First Edition.
Anton H.J. de Ruiter, Christopher J. Damaren and James R. Forbes.

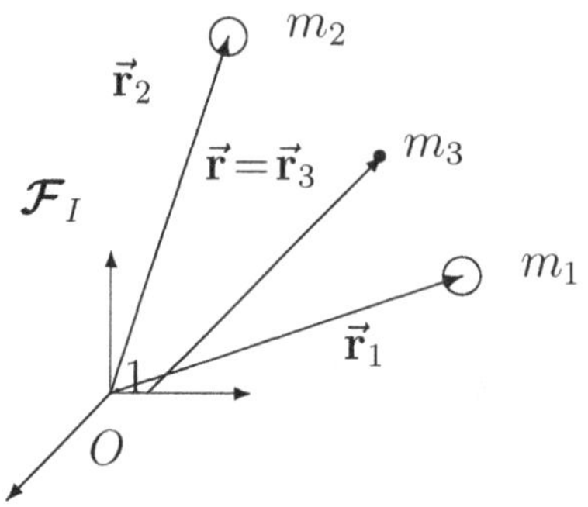

Figure 10.1 The three-body problem

Here, we have introduced r_{12} for the distance between the Earth and the Moon. Note that

$$
\begin{aligned}
m_1 r_1 &= m_2 r_2 \\
r_1 &= \frac{m_2}{m_1 + m_2} r_{12} \\
r_2 &= \frac{m_1}{m_1 + m_2} r_{12}
\end{aligned}
$$

since the origin is the center of mass.

Given the position of m_3 and the angular velocity of the x-y frame,

$$
\begin{aligned}
\vec{\mathbf{r}} &= x\vec{\mathbf{x}}_1 + y\vec{\mathbf{y}}_1 \\
\vec{\boldsymbol{\omega}} &= \omega\vec{\mathbf{z}}_1,
\end{aligned}
$$

let us determine the acceleration of m_3, expressed in the rotating frame:

$$
\ddot{\vec{\mathbf{r}}} = \overset{\circ\circ}{\vec{\mathbf{r}}} + 2\vec{\boldsymbol{\omega}} \times \overset{\circ}{\vec{\mathbf{r}}} + \vec{\boldsymbol{\omega}} \times (\vec{\boldsymbol{\omega}} \times \vec{\mathbf{r}}).
$$

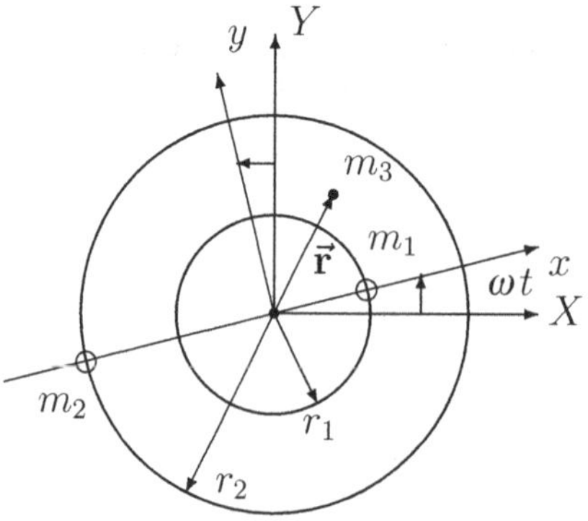

Figure 10.2 The restricted three-body problem

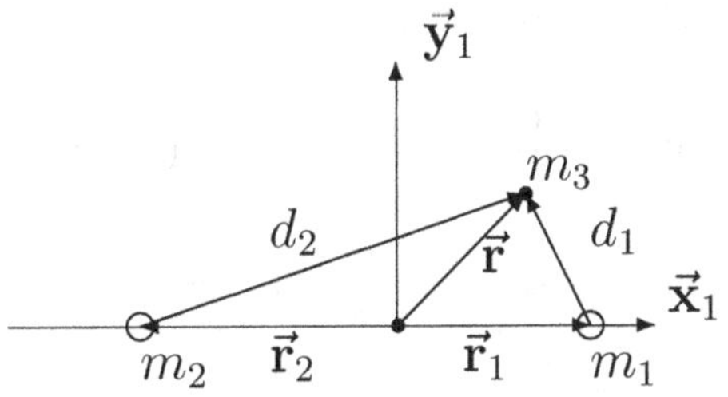

Figure 10.3 The three-body problem geometry

Here,

$$\overset{\circ\circ}{\vec{\mathbf{r}}} = \ddot{x}\vec{\mathbf{x}}_1 + \ddot{y}\vec{\mathbf{y}}_1,$$

$$\vec{\boldsymbol{\omega}}\times \overset{\circ}{\vec{\mathbf{r}}} = -\dot{y}\omega\vec{\mathbf{x}}_1 + \dot{x}\omega\vec{\mathbf{y}}_1,$$

$$\vec{\boldsymbol{\omega}} \times (\vec{\boldsymbol{\omega}} \times \vec{\mathbf{r}}) = -\omega^2 x\vec{\mathbf{x}}_1 - \omega^2 y\vec{\mathbf{y}}_1.$$

Hence, the acceleration of m_3 is

$$\ddot{\vec{\mathbf{r}}} = (\ddot{x} - 2\dot{y}\omega - \omega^2 x)\vec{\mathbf{x}}_1 + (\ddot{y} + 2\dot{x}\omega - \omega^2 y)\vec{\mathbf{y}}_1. \tag{10.1}$$

The gravitational force (per unit mass) acting on m_3 due to m_1 is

$$\begin{aligned}\vec{\mathbf{f}}_1 &= -\frac{Gm_1}{d_1^3}(\vec{\mathbf{r}} - \vec{\mathbf{r}}_1) \\ &= -\frac{Gm_1}{d_1^3}[(x - r_1)\vec{\mathbf{x}}_1 + y\vec{\mathbf{y}}_1]\end{aligned} \tag{10.2}$$

where $d_1 = \sqrt{(x - r_1)^2 + y^2}$. The gravitational force (per unit mass) acting on m_3 due to m_2 is

$$\begin{aligned}\vec{\mathbf{f}}_2 &= -\frac{Gm_2}{d_2^3}(\vec{\mathbf{r}} - \vec{\mathbf{r}}_2) \\ &= -\frac{Gm_2}{d_2^3}[(x + r_2)\vec{\mathbf{x}}_1 + y\vec{\mathbf{y}}_1]\end{aligned} \tag{10.3}$$

where $d_2 = \sqrt{(x + r_2)^2 + y^2}$.

10.1.1 Equations of Motion

Newton's second law for m_3 is:

$$\ddot{\vec{\mathbf{r}}} = \vec{\mathbf{f}}_1 + \vec{\mathbf{f}}_2.$$

Using equations (10.1)–(10.3) we arrive at

$$\ddot{x} - 2\dot{y}\omega - \omega^2 x = -\frac{Gm_1}{d_1^3}(x - r_1) - \frac{Gm_2}{d_2^3}(x + r_2) = f(x, y), \tag{10.4}$$

$$\ddot{y} + 2\dot{x}\omega - \omega^2 y = -G\left[\frac{m_1}{d_1^3} + \frac{m_2}{d_2^3}\right] y = g(x, y). \tag{10.5}$$

These are the differential equations governing the motion of a spacecraft in the rotating system. Given the initial conditions $x(0)$, $y(0)$, $\dot{x}(0)$, and $\dot{y}(0)$, a unique solution for $x(t)$ and $y(t)$ can be determined numerically. It was proven by Poincaré that no analytical solution exists.

10.2 The Lagrangian Points

Let us look for equilibrium solutions of equations (10.4) and (10.5), that is constant solutions for x and y and hence $\dot{x} = \dot{y} = \ddot{x} = \ddot{y} = 0$. Making these substitutions in equations (10.4) and (10.5) gives:

$$-\omega^2 x = -G\left[\frac{m_1}{d_1^3}(x - r_1) + \frac{m_2}{d_2^3}(x + r_2)\right] = f(x, y) \tag{10.6}$$

$$-\omega^2 y = -G\left[\frac{m_1}{d_1^3} + \frac{m_2}{d_2^3}\right] y = g(x, y). \tag{10.7}$$

These represent two nonlinear equations in the two unknowns x and y. From equation (10.7), we have

$$\left[G\frac{m_1}{d_1^3} + G\frac{m_2}{d_2^3} - \omega^2\right] y = 0.$$

Hence, we must have:

$$\text{case (i)} \Rightarrow y = 0 \tag{10.8}$$

or

$$\text{case (ii)} \Rightarrow G\left[\frac{m_1}{d_1^3} + \frac{m_2}{d_2^3}\right] = \omega^2. \tag{10.9}$$

10.2.1 *Case (i)*

Since $y = 0$, $d_1 = |x - r_1|$ and $d_2 = |x + r_2|$. Substituting this into (10.6) gives

$$\omega^2 x = \frac{Gm_1(x - r_1)}{|x - r_1|^3} + \frac{Gm_2(x + r_2)}{|x + r_2|^3}.$$

This equation has three real roots which in thc casc of the Earth-Moon system ($m_2/(m_1+m_2)=0.01213$) are given by:

$$L_1 \Rightarrow x = -0.838 r_{12},$$
$$L_2 \Rightarrow x = -1.156 r_{12},$$
$$L_3 \Rightarrow x = 1.005 r_{12}.$$

The symbol L denotes Lagrange (the discoverer of these points).

10.2.2 Case (ii)

From (10.9),

$$\frac{Gm_1}{d_1^3} + \frac{Gm_2}{d_2^3} = \omega^2 = \frac{G(m_1+m_2)}{r_{12}^3}.$$

Suppose that $d_1 = d_2$. Then the above implies that

$$d_1^3 = d_2^3 = \frac{G(m_1+m_2)}{\omega^2} = r_{12}^3.$$

Therefore,

$$d_1 = d_2 = r_1 + r_2.$$

Notice that this choice of d_1 and d_2 reduces (10.6) to $m_1 r_1 - m_2 r_2 = 0$ which is an identity. This solution corresponds to L_4 and L_5 which are the triangle or equilateral libration points. They are shown below with L_1, L_2, and L_3 for the Earth-Moon system:

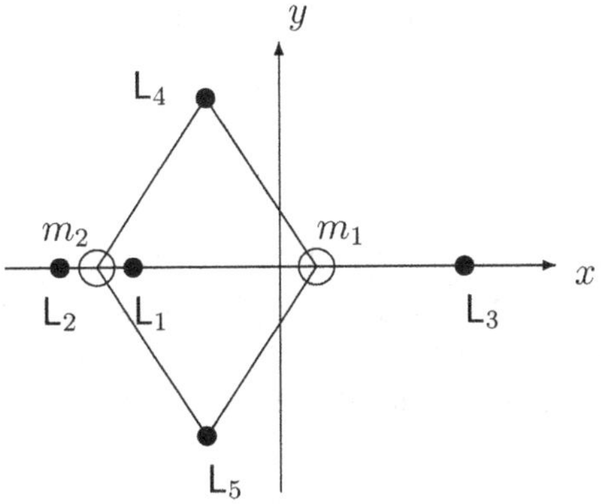

Figure 10.4 The lagrangian points

$L_1, \cdots, L_5$ correspond to points with zero relative velocity and acceleration. A small mass placed at these points remains motionless with respect to the Earth and Moon.

10.3 Stability of the Lagrangian Points

Let (x_e, y_e) denote the components of a Lagrangian point and expand

$$\begin{bmatrix} x(t) \\ y(t) \end{bmatrix} = \begin{bmatrix} x_e \\ y_e \end{bmatrix} + \begin{bmatrix} \delta x(t) \\ \delta y(t) \end{bmatrix} \tag{10.10}$$

where δx and δy denote small perturbations. Linearizing leads to linear differential equations whose stablity can be tested via eigenvalue analysis. This path will be followed here.

Substituting equation (10.10) into equations (10.4) and (10.5) leads to

$$\delta\ddot{x} - 2\omega\delta\dot{y} - \omega^2(x_e + \delta x) = f(x_e + \delta x, y_e + \delta y), \tag{10.11}$$

$$\delta\ddot{y} + 2\omega\delta\dot{x} - \omega^2(y_e + \delta y) = g(x_e + \delta x, y_e + \delta y). \tag{10.12}$$

Linearizing the right-hand sides, we have

$$f(x_e + \delta x, y_e + \delta y) \doteq f(x_e, y_e) + \underbrace{\frac{\partial f}{\partial x}(x_e, y_e)}_{A}\delta x + \underbrace{\frac{\partial f}{\partial y}(x_e, y_e)}_{B}\delta y,$$

$$g(x_e + \delta x, y_e + \delta y) \doteq g(x_e, y_e) + \underbrace{\frac{\partial g}{\partial x}(x_e, y_e)}_{C}\delta x + \underbrace{\frac{\partial g}{\partial y}(x_e, y_e)}_{D}\delta y.$$

Substituting these expressions into equations (10.11) and (10.12) while noting that since (x_e, y_e) satisfy equations (10.6) and (10.7), i.e.,

$$-\omega^2 x_e = f(x_e, y_e)$$

$$-\omega^2 y_e = g(x_e, y_e)$$

we arrive at

$$\delta\ddot{x} - 2\omega\delta\dot{y} - (\omega^2 + A)\delta x - B\delta y = 0, \tag{10.13}$$

$$\delta\ddot{y} + 2\omega\delta\dot{x} - C\delta x - (\omega^2 + D)\delta y = 0. \tag{10.14}$$

Since these are linear constant-coefficient differential equations, we shall look for solutions of the form

$$\begin{bmatrix} \delta x(t) \\ \delta y(t) \end{bmatrix} = \begin{bmatrix} \bar{x} \\ \bar{y} \end{bmatrix} e^{\lambda t}. \tag{10.15}$$

Substituting this into equations (10.13) and (10.14) yields

$$e^{\lambda t}\underbrace{\begin{bmatrix}\lambda^2-\omega^2-A & -2\omega\lambda-B\\ 2\omega\lambda-C & \lambda^2-\omega^2-D\end{bmatrix}}_{\mathbf{T}(\lambda)}\begin{bmatrix}\bar{x}\\ \bar{y}\end{bmatrix}=\begin{bmatrix}0\\0\end{bmatrix}.$$

For nontrivial solutions $[\bar{x}\ \ \bar{y}] \neq [0\ \ 0]$, we must have $\det \mathbf{T}(\lambda) = 0$. This leads to a quartic equation in λ:

$$\lambda^4 + (2\omega^2 - A - D)\lambda^2 + 2\omega(B - C)\lambda + (\omega^2 + A)(\omega^2 + D) - BC = 0.$$

Designating the roots of this equation by $\{\lambda_1, \lambda_2, \lambda_3, \lambda_4\}$, stability is equivalent to boundedness of the temporal solutions $e^{\lambda_i t}$ or $Re\{\lambda_i\} \leq 0, i = 1, \cdots, 4$ (in the event that $Re\{\lambda_i\} = 0$ we also require that the roots be distinct). If any of the roots satisfy $Re\{\lambda_i\} > 0$, then the equilibrium point is unstable.

10.3.1 Comments

1. It can be shown that L_1, L_2, L_3 are unstable.
2. The triangle points L_4 and L_5 are stable. That is, small motions in the vicinity of L_4 and L_5 tend to remain in the vicinity. The stability property of L_4 and L_5 for the restricted problem is only true if

$$\frac{m_2}{m_1 + m_2} < 0.0385 \text{ or } \frac{m_2}{m_1 + m_2} > 0.9615.$$

   For the Earth-Moon system, $m_2/(m_1 + m_2) = 0.01213$ (this is the largest value of this mass ratio in the solar system).
3. The L_4 and L_5 points of the Sun-Jupiter system contain the Trojan asteroids. Voyager discovered several objects in the triangle points of moons within Saturn's system.
4. The Earth-Moon triangle points are actually unstable when the effects of the sun are accounted for.

10.4 Jacobi's Integral

For the two-body problem, we are able to establish some very useful constants of the motion:

$$\begin{aligned}\mathcal{E} &= \frac{1}{2}v^2 - \mu/r\\ \vec{\mathbf{h}} &= \vec{\mathbf{r}} \times \vec{\mathbf{v}}\\ \vec{\mathbf{p}} &= m_1\vec{\mathbf{v}}_1 + m_2\vec{\mathbf{v}}_2.\end{aligned}$$

For the restricted three-body problem, there is only one exact conservation law. Consider equations (10.4) and (10.5). Multiply equation (10.4) by $\dot{x}$ and equation (10.5) by $\dot{y}$ and add the results:

$$\dot{x}\ddot{x} + \dot{y}\ddot{y} - \omega^2(x\dot{x} + y\dot{y}) = -G\left[\frac{m_1}{d_1^3}(x - r_1)\dot{x} + \frac{m_2}{d_2^3}(x + r_2)\dot{x} + \left(\frac{m_1}{d_1^3} + \frac{m_2}{d_2^3}\right)y\dot{y}\right].$$

This can be written as

$$\frac{1}{2}\frac{d}{dt}\left[\dot{x}^2 + \dot{y}^2 - \omega^2(x^2 + y^2)\right] = G\frac{d}{dt}\left[\frac{m_1}{d_1} + \frac{m_2}{d_2}\right]$$

or

$$\frac{d}{dt}\left[\frac{1}{2}(\dot{x}^2 + \dot{y}^2 - \omega^2(x^2 + y^2)) - G\left(\frac{m_1}{d_1} + \frac{m_2}{d_2}\right)\right] = 0. \tag{10.16}$$

Defining the potential energy-like function

$$\Upsilon(x, y) = G\left(\frac{m_1}{d_1} + \frac{m_2}{d_2}\right) + \frac{\omega^2}{2}(x^2 + y^2),$$

equation (10.16) implies that

$$\frac{1}{2}(\dot{x}^2 + \dot{y}^2) - \Upsilon(x, y) = C_J = \text{constant} \tag{10.17}$$

which is known as Jacobi's integral. This is analogous to the energy $\mathcal{E}$ for the two-body problem. However $\frac{1}{2}(\dot{x}^2 + \dot{y}^2)$ is not the true kinetic energy since x and y are measured in the rotating frame. Also, $\Upsilon(x, y)$ contains the true potential energy and a contribution stemming from "centrifugal forces" $[\frac{1}{2}\omega^2(x^2 + y^2)]$.

The value of C_J (the Jacobi constant) is determined by the initial conditions $x(0)$, $y(0)$, $\dot{x}(0)$, and $\dot{y}(0)$.

10.4.1 Hill's Curves

Jacobi's integral, equation (10.17), can be rewritten as

$$\Upsilon(x, y) + C_J = \frac{1}{2}(\dot{x}^2 + \dot{y}^2) \geq 0. \tag{10.18}$$

Hence, $\Upsilon(x, y) \geq -C_J$ which defines admissible values of (x, y) in the x-y plane for a given value of C_J. Values of (x, y) for which

$$\Upsilon(x, y) < -C_J$$

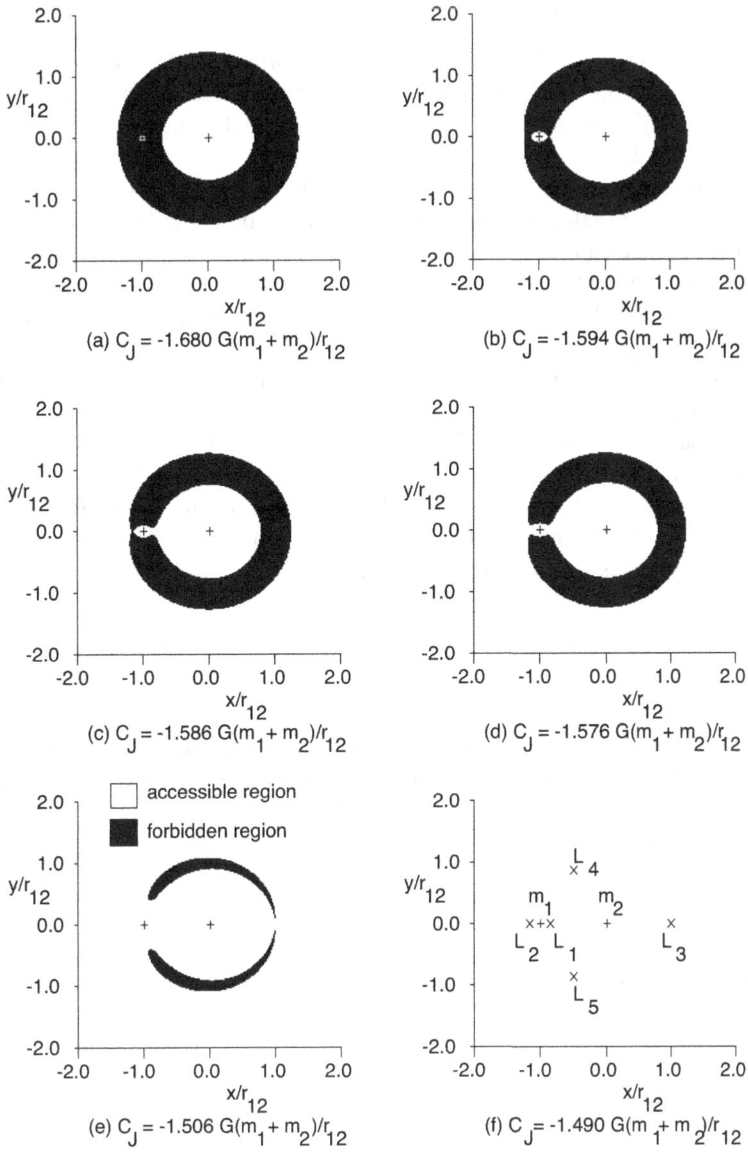

Figure 10.5 Hill's curves

are not physically obtainable (forbidden regions). The boundary between admissible and forbidden regions satisfies

$$\Upsilon(x, y) + C_J = 0 \tag{10.19}$$

These curves for values of C_J are called Hill's curves. From equations (10.18) and (10.19), we see that they are also curves of zero relative velocity ($\dot{x} = \dot{y} = 0$).

10.4.2 *Comments on Figure 10.5*

(a) Motion is physically possible in the vicinity of the Earth or the Moon. This is essentially a two-body orbit of either the Earth or the Moon. Motion outside the larger circle is possible which is two-body motion far from either primary. Note that flight between the Earth and Moon is not possible.
(b) For this value of C_J, motion between the Earth and Moon is just possible. If the spacecraft reaches L_1 with zero relative velocity, it will continue on to the Moon. This corresponds to minimum energy transfer to the Moon.
(c) Now that inner and outer accessible regions touch at L_2, escape from the Earth-Moon system is possible. This is the minimum energy escape trajectory and requires a close fly-by of the Moon.
(d) As C_J is increased, the opening at L_2 widens further and escape is possible through a wider corridor.
(e) For this value of C_J, the inner and outer curves touch at L_3. The forbidden region still encloses L_4 and L_5.
(f) As C_J is increased further, the teardrop-shaped regions enclosing the forbidden regions (and L_4 and L_5) shrink further until they disappear at L_4 and L_5. The triangle points are the most costly points in the x-y plane to reach.

Notes

This chapter has provided a brief introduction to the circular restricted three-body problem. In particular, we have restricted ourselves to motion within the orbital plane of m_1 and m_2, whose orbits are assumed to be circular. Further treatment of the restricted three-body problem may be found in books such as Battin (1999), Schaub and Junkins (2003) and Wie (2008).

References

Battin RH 1999 *An Introduction to the Mathematics and Methods of Astrodynamics, Revised Edition*. American Institute of Aeronautics and Astronautics, Reston, VA.

Schaub H and Junkins JL 2003 *Analytical Mechanics of Space Systems*. American Institute of Aeronautics and Astronautics, Reston, VA.

Wie B 2008 *Space Vehicle Dynamics and Control* 2nd edn. American Institute of Aeronautics and Astronautics, Reston, VA.

11

Introduction to Spacecraft Attitude Stabilization

Typically, the purpose of putting a satellite in orbit is to point an instrument (called the payload) at something. For example, for an astronomy mission, the objective is to point a telescope at a distant star or some other astronomical feature. For an Earth observation satellite, the purpose is to point a camera, radar or other instrument toward a desired location on the Earth below. For a communications satellite, the transmitting and receiving antennas need to point toward the Earth. These are but a few examples. What it means is that the attitude (orientation) of the spacecraft needs to be stabilized to some desired attitude. To achieve this, some means of controlling the attitude is needed. This can be done both actively and passively, depending on the desired attitude, and how accurately it must be maintained. Passive methods of attitude stabilization make use of the spacecraft's natural dynamics to ensure a stable equilibrium at the desired attitude, and are useful when accuracy requirements are coarse. For active control, actuators capable of affecting the attitude are installed on the spacecraft. We shall discuss different types of actuators in Chapter 26. For the present time, it suffices to consider the actuators as a means to apply a torque to the spacecraft. In many cases, passive control methods are augmented by active control methods. We shall examine some passive methods of attitude stabilization in Chapters 13 to 16, but we shall place the majority of our focus on active control methods in Chapters 17 to 25. Now, the question arises as to what torque the actuator should apply to the spacecraft. This is the purpose of the control system. The control system typically consists of three parts:

- **Navigation** - where am I?
- **Guidance** - where do I want to be?
- **Control** - how do I get there?

Spacecraft Dynamics and Control: An Introduction, First Edition.
Anton H.J. de Ruiter, Christopher J. Damaren and James R. Forbes.

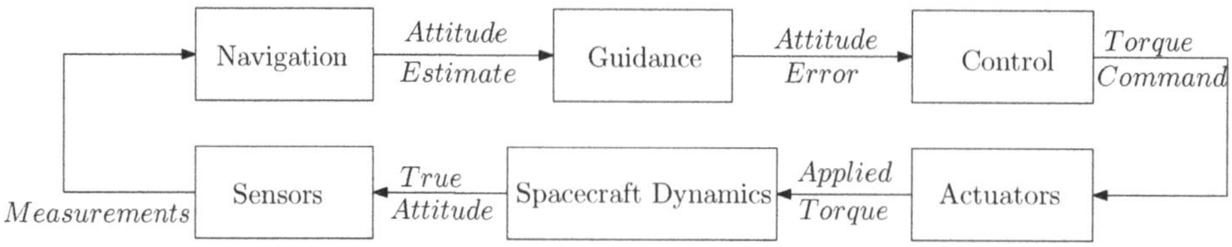

Figure 11.1 Typical spacecraft control system

In a bit more detail, the Navigation part determines the current spacecraft attitude from attitude sensor measurements (this is called attitude determination, and will be covered in more detail in Chapter 25). The Guidance part determines what the desired spacecraft attitude is, and what the error in attitude is (by using information from the Navigation part). The Control part determines what torque should be applied to the spacecraft to correct the attitude error, and sends this as a command to the attitude actuators. In order to determine the torque required to correct the attitude error, it is important to understand how the spacecraft attitude will respond to the applied torque. This response is governed by the spacecraft attitude dynamics. This is why we will study both the spacecraft attitude dynamics, as well as its control. A spacecraft control system can be represented in block-diagram form as shown in Figure 11.1. As we know from Chapter 1, spacecraft motion has both a translational and rotational component. The translational part is the subject of orbital mechanics, and we have studied these in detail in Chapters 3 to 10. Often times, a means of orbit control is needed as well. However, this typically consists of corrections to the orbit applied occasionally (at discrete instances separated by long periods of time). We have already seen how corrections may be made to the orbit by making small impulsive velocity changes. This type of control is quite different from the control that is required for the spacecraft attitude, which is a continuous process (all the time).

11.1 Introduction to Control Systems

A control system is the brain of any system (aerospace, mechanical, electrical, mechatronic, chemical, etc.). The purpose of a control system is to make the system perform a desired task. In the case of a spacecraft, the purpose of the control system is to make the spacecraft achieve and maintain a desired attitude.

Any system to be controlled has parameters the control system can change to make the system achieve the desired behavior. These parameters are called the *control inputs* or *plant input*. **In the case of a spacecraft, the actuator torques are the control inputs**. How the control system determines the inputs is the subject of control theory. There are two types of control: open-loop control and closed-loop control.

11.1.1 Open-loop versus Closed-loop

You might ask the question: if we understand the spacecraft attitude dynamics, and we know what attitude profile we want the spacecraft to follow, why can't we just determine the required torque for that attitude profile beforehand by inverting the dynamics and then apply it? This

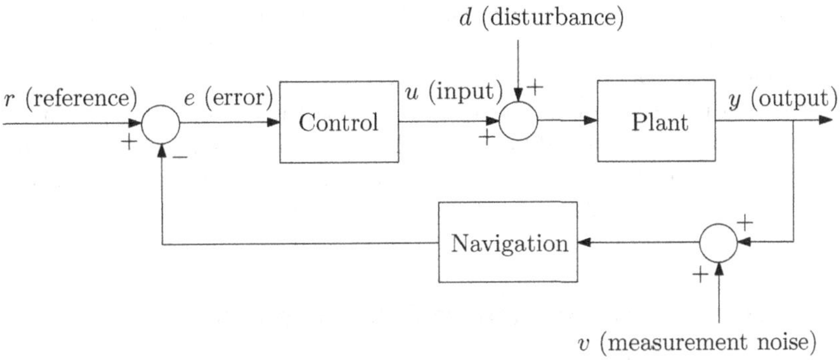

Figure 11.2 Typical feedback control structure

is an example of *open-loop control*, which means that there is no feedback of information to the control system. The problem with this approach is that it is like driving your car with your eyes closed. It would work if you know the curves in the road exactly, and you know the exact initial location and orientation of your car. However, if there are any errors in your knowledge, you will eventually veer off-course, and you won't have the ability to react to any unexpected obstacles that might arise. For a spacecraft (and any system in general), the dynamics are not known exactly, and the control system requires feedback to continuously correct for these errors. This is called *closed-loop control*, also known as *feedback control*. For the above-mentioned reasons, feedback control is the preferred approach.

11.1.2 Typical Feedback Control Structure

Figure 11.2 shows a typical feedback control structure. The system being controlled is called the *plant*, which in our case is the spacecraft attitude kinematics and dynamics. Its behavior is governed by the plant dynamics, which are typically described by a set of differential equations. We need to be able to measure the current state of the plant to be able to provide feedback information to the controller. The quantity that is measured is called the *plant output* (labeled y). In the case of spacecraft attitude control, the output depends on what the available sensors can measure (discussed in more detail in Chapter 26), and could be the attitude (in the case of a star-tracker), the direction of the sun in spacecraft body coordinates (in the case of a sun-sensor), the Earth's magnetic field vector in spacecraft body coordinates (in the case of a magnetometer), etc. In all of these cases, the output is a function of the quantity we are trying to control, which is the attitude.

The sensors themselves, make measurements of the output. However, since no sensor is perfect, the measurement is corrupted by sensor noise (labeled v).

Because the sensors cannot always directly measure the quantity we wish to control, the purpose of the navigation block is to determine that quantity from the available measurements. In the case of spacecraft attitude control, the function of the navigation block is to determine the spacecraft attitude from the available measurements (such as the sun and Earth magnetic field vectors).

A reference (or desired state), r, is defined (in our case, a desired attitude). The control block typically takes as input the error (in our case the error in attitude) and determines what the control input (labeled u) should be and sends this as a command to the actuator (in our case it issues a torque command). The actuator then creates this as an input to the plant. Unfortunately, there may also be additional disturbances acting on the plant (labeled d). In our case, these are environmental disturbance torques acting on the spacecraft (these are treated in detail in Chapter 12).

11.2 Overview of Attitude Representation and Kinematics

This section contains a brief summary of the relevant details from Chapter 1 for the study of spacecraft attitude kinematics.

When talking about the attitude of a spacecraft, it only makes sense if it is relative to something. For example, the inertial attitude is the attitude relative to an inertial frame $\mathcal{F}_G$. The attitude can be fully described by the orientation of a reference frame embedded in the spacecraft $\mathcal{F}_b$ (called a body-fixed frame) relative to the other frame.

As we have seen in Section 1.3, the orientation of $\mathcal{F}_b$ relative to $\mathcal{F}_G$ is fully described by the rotation matrix $\mathbf{C}_{bG}$. Therefore, $\mathbf{C}_{bG}$ may be used as an attitude representation. Alternatively, we could use any parameterization of $\mathbf{C}_{bG}$, such as an Euler rotation sequence or quaternions (or other parameterizations). For example, a common Euler rotation sequence is a 3-2-1 rotation sequence given by:

1. A rotation ψ about the original z-axis (called a "yaw" rotation).
2. A rotation θ about the intermediate y-axis (called a "pitch" rotation).
3. A rotation ϕ about the transformed x-axis (called a "roll" rotation).

The angles ψ, θ and ψ equivalently represent the attitude. Indeed, they fully parameterize the rotation matrix $\mathbf{C}_{bG}$ as

$$\mathbf{C}_{bG}(\phi, \theta, \psi) = \begin{bmatrix} c_\theta c_\psi & c_\theta s_\psi & -s_\theta \\ s_\phi s_\theta c_\psi - c_\phi s_\psi & s_\phi s_\theta s_\psi + c_\phi c_\psi & s_\phi c_\theta \\ c_\phi s_\theta c_\psi + s_\phi s_\psi & c_\phi s_\theta s_\psi - s_\phi c_\psi & c_\phi c_\theta \end{bmatrix}, \tag{11.1}$$

where $s_b = \sin b$ and $c_b = \cos b$. Conversely, given any rotation matrix $\mathbf{C}_{bG}$, the corresponding Euler angles may be uniquely determined, unless the rotation matrix corresponds to the singularity of the Euler sequence (which occurs when $\theta = \pm 90^\circ$ for the 3-2-1 Euler sequence). The relevant expressions can be found in Section 1.3.3. Note that every Euler rotation sequence has a singularity.

The quaternion is a four-parameter representation of the attitude. Unlike an Euler sequence, it does not possess a singularity. The quaternion consists of a vector part $\boldsymbol{\epsilon}$ and a scalar part η, satisfying a unit norm constraint $\boldsymbol{\epsilon}^T\boldsymbol{\epsilon} + \eta^2 = 1$. The rotation matrix in terms of the quaternion is given by

$$\mathbf{C}_{bG} = \left(2\eta^2 - 1\right)\mathbf{1} + 2\boldsymbol{\epsilon}\boldsymbol{\epsilon}^T - 2\eta\boldsymbol{\epsilon}^\times. \tag{11.2}$$

Conversely, given a rotation matrix $\mathbf{C}_{bG}$, the corresponding quaternion may readily be obtained. The relevant expressions can be found in Section 1.3.4.

Since $\mathcal{F}_b$ is embedded in the spacecraft, the spacecraft has angular velocity $\vec{\omega}_{bG}$ relative to $\mathcal{F}_G$ (recall that by $\vec{\omega}_{bG}$, we mean the angular velocity of frame $\mathcal{F}_b$ relative to frame $\mathcal{F}_G$). If we express $\vec{\omega}_{bG}$ in the body frame $\mathcal{F}_b$ as

$$\vec{\omega}_{bG} = \vec{\mathcal{F}}_b^T \boldsymbol{\omega}_{bG},$$

the attitude kinematics in terms of the rotation matrix are

$$\dot{\mathbf{C}}_{bG} = -\boldsymbol{\omega}_{bG}^{\times}\mathbf{C}_{bG}. \tag{11.3}$$

Equivalently, the attitude kinematics in terms of a 3-2-1 Euler rotation sequence are

$$\begin{bmatrix} \dot{\phi} \\ \dot{\theta} \\ \dot{\psi} \end{bmatrix} = \begin{bmatrix} 1 & \sin\phi\tan\theta & \cos\phi\tan\theta \\ 0 & \cos\phi & -\sin\phi \\ 0 & \sin\phi\sec\theta & \cos\phi\sec\theta \end{bmatrix} \boldsymbol{\omega}_{bG}. \tag{11.4}$$

In terms of the quaternion, the attitude kinematics are

$$\begin{aligned} \dot{\boldsymbol{\epsilon}} &= \frac{1}{2}\left(\eta\mathbf{1} + \boldsymbol{\epsilon}^{\times}\right)\boldsymbol{\omega}_{bG}, \\ \dot{\eta} &= -\frac{1}{2}\boldsymbol{\epsilon}^T\boldsymbol{\omega}_{bG}. \end{aligned} \tag{11.5}$$

11.3 Overview of Spacecraft Attitude Dynamics

This section contains a brief summary of the relevant details from Chapter 2 for the study of spacecraft attitude dynamics.

As we have seen in Section 2.3, the spacecraft angular momentum about the center of mass satisfies

$$\dot{\vec{\mathbf{h}}}_c = \vec{\mathbf{T}}_c, \tag{11.6}$$

where $(\dot{\ })$ denotes the inertial time-derivative (as seen in $\mathcal{F}_G$), $\vec{\mathbf{h}}_c$ is the angular momentum vector about the center of mass and $\vec{\mathbf{T}}_c$ is the total external torque about the center of mass. Using the relation $\dot{\vec{\mathbf{r}}} = \mathring{\vec{\mathbf{r}}} + \vec{\omega} \times \vec{\mathbf{r}}$, taking $(\mathring{\ })$ to mean time differentiation as seen in the spacecraft body frame $\mathcal{F}_b$, we can rewrite the attitude dynamics (11.6) as

$$\mathring{\vec{\mathbf{h}}}_c + \vec{\omega}_{bG} \times \vec{\mathbf{h}}_c = \vec{\mathbf{T}}_c. \tag{11.7}$$

As seen in Section 2.3, for a rigid body, the angular momentum vector expressed in $\mathcal{F}_b$ is given by

$$\vec{\mathbf{h}}_c = \vec{\mathcal{F}}_b^T \mathbf{I}\boldsymbol{\omega}_{bG},$$

where $\mathbf{I}$ is the moment of inertia matrix about the center of mass expressed in body coordinates. Since $\mathcal{F}_b$ is embedded in the spacecraft, $\dot{\mathbf{I}} = \mathbf{0}$. From this, we have that in body coordinates,

$$\overset{\circ}{\vec{\mathbf{h}}}_c = \vec{\mathcal{F}}_b^T \mathbf{I}\dot{\boldsymbol{\omega}}_{bG},$$

so that in body coordinates the attitude dynamics become

$$\mathbf{I}\dot{\boldsymbol{\omega}}_{bG} + \boldsymbol{\omega}_{bG}^{\times}\mathbf{I}\boldsymbol{\omega}_{bG} = \mathbf{T}_c, \tag{11.8}$$

where $\vec{\mathbf{T}}_c = \vec{\mathcal{F}}_b^T \mathbf{T}_c$. These are called *Euler's Equations.* Equations (11.3) and (11.8) fully describe the spacecraft attitude motion relative to an inertial frame $\mathcal{F}_G$ (when the spacecraft consists of a single rigid body).

11.3.1 Properties of the Inertia Matrix - A Summary

Here is a summary of some important facts from Section 2.4.

1. The inertia matrix is real, positive definite and symmetric, i.e. $\mathbf{I} = \mathbf{I}^T$ and $\mathbf{x}^T\mathbf{I}\mathbf{x} > 0$ for all non-zero vectors $\mathbf{x} \neq \mathbf{0}$.
2. The inertia matrix is dependent on the orientation of $\mathcal{F}_b$ in the body (we can embed the frame $\mathcal{F}_b$ with arbitrary orientation). Let us consider two body-fixed frames $\mathcal{F}_{b1}$ and $\mathcal{F}_{b2}$. Let $\mathbf{I}_1$ be the inertia matrix as computed in $\mathcal{F}_{b1}$, and $\mathbf{I}_2$ be the inertia matrix as computed in $\mathcal{F}_{b2}$. Then, $\mathbf{I}_2$ and $\mathbf{I}_1$ are related by

   $$\mathbf{I}_2 = \mathbf{C}_{21}\mathbf{I}_1\mathbf{C}_{12},$$

   where $\mathbf{C}_{21}$ is the rotation matrix representing the rotation from frame $\mathcal{F}_{b1}$ to frame $\mathcal{F}_{b2}$.
3. It is always possible to find a body-fixed frame $\mathcal{F}_{b,p}$ such that the inertia matrix as computed in $\mathcal{F}_{b,p}$ is diagonal

   $$\mathbf{I}_p = \begin{bmatrix} I_x & 0 & 0 \\ 0 & I_y & 0 \\ 0 & 0 & I_z \end{bmatrix}.$$

   The frame $\mathcal{F}_{b,p}$ is called a principal axes frame, and is a useful frame to work with. The attitude dynamics are greatly simplified when expressed in this frame. Since $\mathbf{I}_p > 0$, it must be that $I_x > 0$, $I_y > 0$, and $I_z > 0$. I_x, I_y and I_z are called the principal moments of inertia.

Returning to the attitude dynamics (11.8), let us assume that $\mathcal{F}_b$ is a principal axes frame, and let

$$\boldsymbol{\omega}_{bG} = \begin{bmatrix} \omega_x \\ \omega_y \\ \omega_z \end{bmatrix}, \quad \mathbf{T}_c = \begin{bmatrix} T_x \\ T_y \\ T_z \end{bmatrix}.$$

Then, Euler's equations (11.8) become

$$\begin{aligned} I_x\dot{\omega}_x + (I_z - I_y)\omega_y\omega_z &= T_x, \\ I_y\dot{\omega}_y + (I_x - I_z)\omega_x\omega_z &= T_y, \\ I_z\dot{\omega}_z + (I_y - I_x)\omega_x\omega_y &= T_z. \end{aligned} \tag{11.9}$$

12

Disturbance Torques on a Spacecraft

There are a number of external torques acting on a spacecraft which disturb the attitude motion. For a spacecraft in the vicinity of the Earth, the major disturbance torques are:

1. Magnetic torque,
2. Solar radiation pressure torque,
3. Aerodynamic torque, and
4. Gravity-gradient torque.

We shall now look at each one in some detail.

12.1 Magnetic Torque

The cause of the magnetic torque is the interaction between the Earth's magnetic field and any magnetization of the spacecraft (electronic components create an equivalent current loop, which results in a magnetic dipole).

The torque on the spacecraft due to the interaction of the Earth's magnetic field $\vec{\mathbf{b}}$ and the spacecraft residual magnetic dipole moment $\vec{\mathbf{m}}$ is given by

$$\vec{\mathbf{T}}_m = \vec{\mathbf{m}} \times \vec{\mathbf{b}}. \tag{12.1}$$

Spacecraft Dynamics and Control: An Introduction, First Edition.
Anton H.J. de Ruiter, Christopher J. Damaren and James R. Forbes.

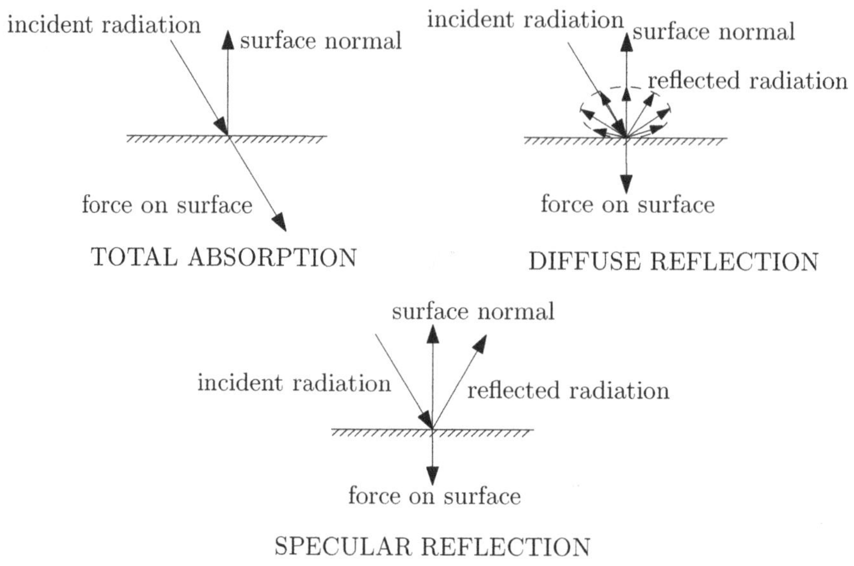

Figure 12.1 Interaction of radiation with surface

12.2 Solar Radiation Pressure Torque

The output from the sun (photons) contain momentum, which produce an effective pressure on spacecraft surfaces (due to momentum transfer), with magnitude near Earth of

$$p = 4.5 \times 10^{-6}\text{N/m}^2.$$

There are different modes of interaction between the solar radiation and the spacecraft surface, which depends on the spacecraft surface properties, as shown in Figure 12.1. In practise, all three types of interaction are present in different proportions. For our purposes, we shall assume total absorption. The projected area of a surface element dS with unit outward normal $\vec{\mathbf{n}}$ (refer to Figure 12.2) in the sun's direction (along $\vec{\mathbf{s}}$) is

$$dA = \cos\alpha_s dS = \vec{\mathbf{n}} \cdot \vec{\mathbf{s}}\, dS.$$

The solar pressure force on dS is given by

$$d\vec{\mathbf{F}}_s = -p\, dA\vec{\mathbf{s}} = -p\vec{\mathbf{n}} \cdot \vec{\mathbf{s}}\, dS\vec{\mathbf{s}}.$$

The torque about the spacecraft center of mass due to the force on dS is given by

$$d\vec{\mathbf{T}}_s = \vec{\rho} \times d\vec{\mathbf{F}}_s = -p\vec{\rho} \times \vec{\mathbf{s}}\left(\vec{\mathbf{n}} \cdot \vec{\mathbf{s}}\right) dS,$$

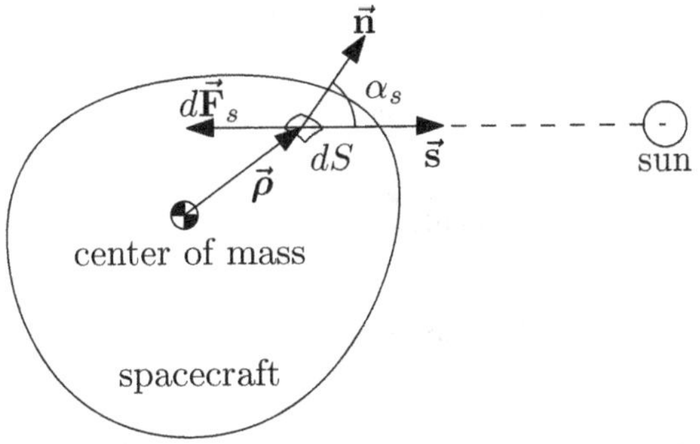

Figure 12.2 Solar radiation pressure force on a surface element

where $\vec{\rho}$ is the location of dS from the spacecraft center of mass. The total torque about the spacecraft center of mass due to the solar pressure force is then given by the integral

$$\vec{\mathbf{T}}_s = \int_{S_{ws}} d\vec{\mathbf{T}}_s = -p\int_{S_{ws}} \vec{\rho} \times \vec{\mathbf{s}}\left(\vec{\mathbf{n}} \cdot \vec{\mathbf{s}}\right) dS,$$

where S_{ws} is the wetted (lit) portion of the spacecraft surface (the portion for which $\vec{\mathbf{n}} \cdot \vec{\mathbf{s}} \geq 0$). This is because there is no solar pressure force on the shaded surfaces of the spacecraft.

The total force due to the solar pressure is given by the integral

$$\vec{\mathbf{F}}_s = \int_{S_{ws}} d\vec{\mathbf{F}}_s = -p\vec{\mathbf{s}}\int_{S_{ws}} \vec{\mathbf{n}} \cdot \vec{\mathbf{s}} dS.$$

Let us define the center of solar pressure (the effective location of solar pressure application) as

$$\vec{\mathbf{c}}_{ps} = \frac{\int_{S_{ws}} \vec{\rho}\left(\vec{\mathbf{n}} \cdot \vec{\mathbf{s}}\right) dS}{\int_{S_{ws}} \left(\vec{\mathbf{n}} \cdot \vec{\mathbf{s}}\right) dS}.$$

Then, the total solar pressure torque is given by

$$\vec{\mathbf{T}}_s = -p\left[\int_{S_{ws}} \vec{\rho}\left(\vec{\mathbf{n}} \cdot \vec{\mathbf{s}}\right) dS\right] \times \vec{\mathbf{s}} = -p\left[\vec{\mathbf{c}}_{ps}\int_{S_{ws}} \left(\vec{\mathbf{n}} \cdot \vec{\mathbf{s}}\right) dS\right] \times \vec{\mathbf{s}},$$

which simplifies to

$$\vec{\mathbf{T}}_s = \vec{\mathbf{c}}_{ps} \times \vec{\mathbf{F}}_s. \qquad (12.2)$$

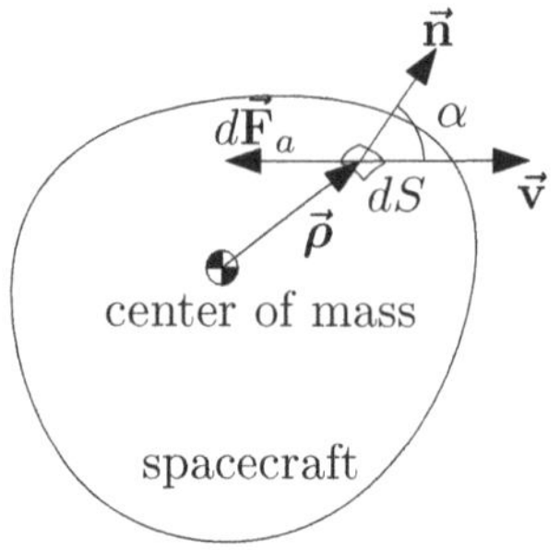

Figure 12.3 Aerodynamic force on a surface element

12.3 Aerodynamic Torque

In low Earth orbits, there is still some residual atmosphere. However, the density is so low that conventional fluid mechanics theories based on a continuum model of the atmosphere do not apply. Instead, the interaction between the atmosphere and the spacecraft must be treated at the molecular level. We make the following assumptions:

1. The momentum of molecules arriving at the surface of the spacecraft is totally lost to the surface (the molecule sticks to it).
2. The thermal motion of the atmosphere is much smaller than the spacecraft speed.
3. The spacecraft is nominally non-spinning.

Assumptions 1 and 3 imply that the molecules have velocity $\vec{\mathbf{v}}$ (the orbital velocity of the spacecraft) upon interaction with the spacecraft surface. Assumption 2 implies that the molecules are taken to have zero velocity prior to interaction with the spacecraft surface. Let us now consider the torque on a spacecraft due to molecules arriving at surface element dS with unit outward normal $\vec{\mathbf{n}}$, located at position $\vec{\rho}$ from the spacecraft center of mass (see Figure 12.3). In particular, consider the mass of molecules arriving at dS during a short time interval δt, as shown in Figure 12.4. The mass in the streamtube arriving at dS over the time interval δt is

$$\delta m = \rho_a v \delta t dA = \rho_a v \delta t \cos \alpha dS,$$

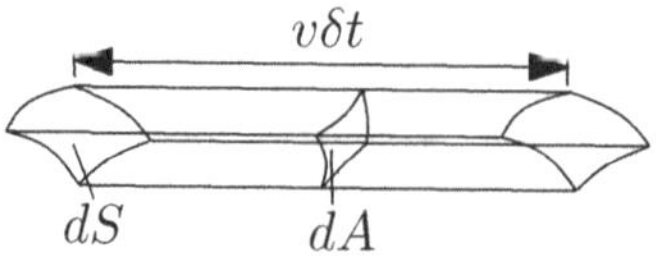

Figure 12.4 Infinitessimal stream-tube

where ρ_a is the atmospheric density, α is the angle between $\vec{\mathbf{n}}$ and $\vec{\mathbf{v}}$, and $v = |\vec{\mathbf{v}}|$. The momentum of the molecules in the stream-tube at time t (before interaction with the surface) is

$$\vec{\mathbf{p}}(t) = \delta m\vec{\mathbf{0}} = \vec{\mathbf{0}}.$$

At time $t + \delta t$, the molecules are stuck to the surface which has velocity $\vec{\mathbf{v}}$. Therefore, the momentum of the molecules in the streamtube at time $t + \delta t$ is

$$\begin{aligned}\vec{\mathbf{p}}(t + \delta t) &= \delta m\vec{\mathbf{v}}, \\ &= \rho_a v \cos\alpha dS\delta t\vec{\mathbf{v}}, \\ &= \rho_a v^2 \cos\alpha dS\delta t\hat{\mathbf{v}},\end{aligned}$$

where $\hat{\mathbf{v}} = \frac{\vec{\mathbf{v}}}{v}$. The rate of change of momentum of the molecules is

$$\frac{d\vec{\mathbf{p}}}{dt} = \lim_{\delta t \to 0} \frac{\vec{\mathbf{p}}(t + \delta t) - \vec{\mathbf{p}}(t)}{\delta t} = \rho_a v^2 \cos\alpha dS\hat{\mathbf{v}}.$$

By Newton's second and third laws, the force acting on the surface element dS due to the arriving molecules is

$$d\vec{\mathbf{F}}_a = -\rho_a v^2 \cos\alpha dS\hat{\mathbf{v}} = -\rho_a v^2 (\vec{\mathbf{n}} \cdot \hat{\mathbf{v}}) dS\hat{\mathbf{v}}.$$

The total aerodynamic force acting on the spacecraft is then given by the integral

$$\vec{\mathbf{F}}_a = \int_{S_{wa}} d\vec{\mathbf{F}}_a = -\rho_a v^2 \hat{\mathbf{v}} \int_{S_{wa}} (\vec{\mathbf{n}} \cdot \hat{\mathbf{v}}) dS,$$

where S_{wa} is the wetted area (area facing the flow, given by $\vec{\mathbf{n}} \cdot \hat{\mathbf{v}} \geq 0$). As for the solar pressure torque, we define the center of aerodynamic pressure (effective location of aerodynamic force application) to be

$$\vec{\mathbf{c}}_{pa} = \frac{\int_{S_{wa}} \vec{\rho}(\vec{\mathbf{n}} \cdot \hat{\mathbf{v}}) dS}{\int_{S_{wa}} (\vec{\mathbf{n}} \cdot \hat{\mathbf{v}}) dS}.$$

In a similar manner to the solar pressure torque, the aerodynamic torque becomes

$$\vec{\mathbf{T}}_a = \vec{\mathbf{c}}_{pa} \times \vec{\mathbf{F}}_a. \tag{12.3}$$

12.4 Gravity-Gradient Torque

The gravity-gradient torque is due to the fact that the Earth's gravitational force is not constant with distance from the Earth's center, but decreases quadratically. That is, the gravitational force on a mass further from the Earth is smaller than the force on a mass that is closer. This gravity-gradient produces a torque.

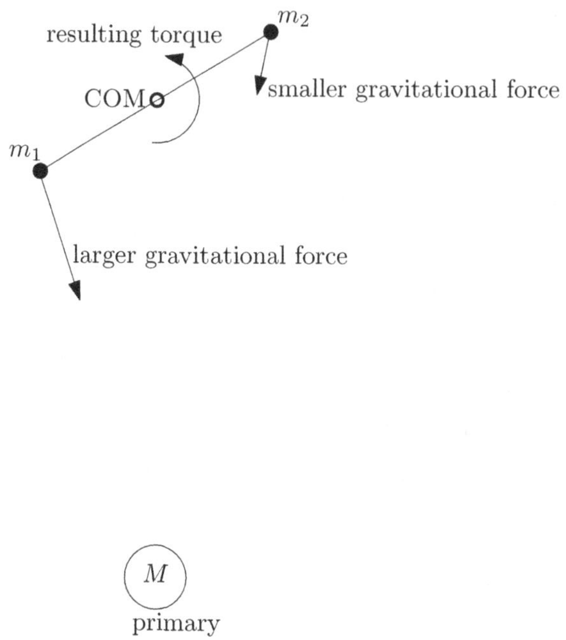

Figure 12.5 Gravity-gradient torque concept

To understand how this works, consider two equal point masses connected by a massless rod, as shown in Figure 12.5. The gravitational force on mass m_2 is smaller than the gravitational force on mass m_1, which results in a net torque about the center of mass. It is clear that the configuration shown in Figure 12.6 is stable (convince yourself of this fact). This torque is made use of for gravity-gradient stabilization (a passive stabilization method), which we shall study in detail in Chapter 16.

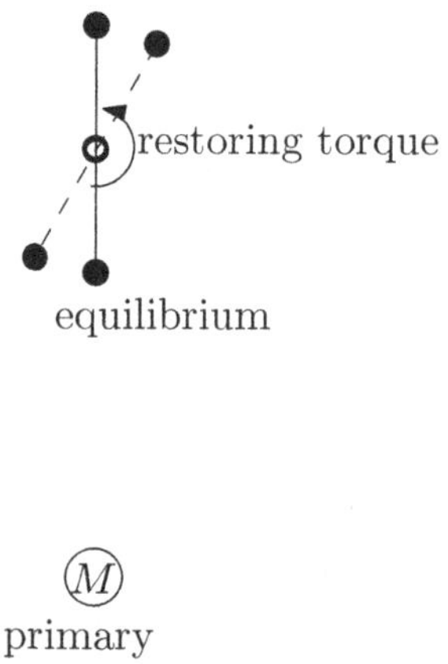

Figure 12.6 Stable gravity-gradient equilibrium

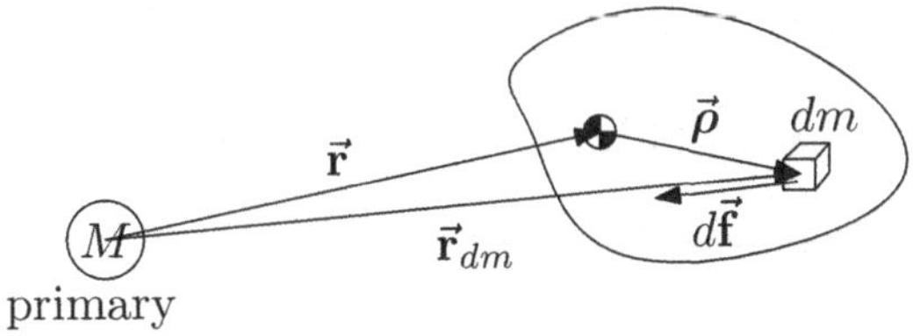

Figure 12.7 Modeling gravity-gradient torque for a continuous body

We shall now obtain an expression for the gravity-gradient torque on a continuous body. Consider Figure 12.7. The gravitational force on a mass element dm located at position $\vec{\rho}$ from the spacecraft center of mass is given by

$$d\vec{\mathbf{f}} = -\frac{\mu\left(\vec{\mathbf{r}}+\vec{\rho}\right)}{\left|\vec{\mathbf{r}}+\vec{\rho}\right|^{3}}dm, \tag{12.4}$$

where $\vec{\mathbf{r}}$ is the orbital position of the spacecraft center of mass. Now, we are going to look for an approximation to this. Typically, we will have $\rho \ll r = |\vec{\mathbf{r}}|$ which means $\rho/r \ll 1$ (the largest dimension of the spacecraft is much smaller than the orbital radius). Therefore, we can approximate $|\vec{\mathbf{r}}+\vec{\rho}|^{-3}$ very well by a first order Taylor series expansion. We re-use the result in equation (9.6) in Section 9.1, which is

$$\left|\vec{\mathbf{r}}+\vec{\rho}\right|^{-3} \approx \frac{1}{r^{3}}\left(1-\frac{3\vec{\mathbf{r}}\cdot\vec{\rho}}{r^{2}}\right),$$

where $r = |\vec{\mathbf{r}}|$. Substituting this approximation into the expression for the force on dm (12.4), we have

$$d\vec{\mathbf{f}} = -\frac{\mu\left(\vec{\mathbf{r}}+\vec{\rho}\right)}{r^{3}}\left(1-\frac{3\vec{\mathbf{r}}\cdot\vec{\rho}}{r^{2}}\right)dm.$$

The total torque about the center of mass is given by

$$\begin{aligned}\vec{\mathbf{T}}_{g} &= \int_{V}\vec{\rho}\times d\vec{\mathbf{f}},\\ &= \int_{V}\vec{\rho}\times\left(-\frac{\mu\left(\vec{\mathbf{r}}+\vec{\rho}\right)}{r^{3}}\right)dm + \int_{V}\vec{\rho}\times\left(\frac{3\mu\vec{\mathbf{r}}\cdot\vec{\rho}}{r^{5}}\left(\vec{\mathbf{r}}+\vec{\rho}\right)\right)dm,\end{aligned}$$

where the integral is taken over the entire spacecraft body. Now, $\int_V \vec{\rho}dm = 0$, since the origin is at the center of mass. Also, recall that $\vec{\rho}\times\vec{\rho} = \vec{\mathbf{0}}$. Therefore, the total torque simplifies to

$$\vec{\mathbf{T}}_{g} = \frac{3\mu}{r^{5}}\int_{V}\vec{\rho}\times\vec{\mathbf{r}}\left(\vec{\mathbf{r}}\cdot\vec{\rho}\right)dm.$$

To evaluate this expression, let us now express all quantities in the spacecraft body frame $\mathcal{F}_b$,

$$\vec{\mathbf{T}}_g = \vec{\mathcal{F}}_b^T \mathbf{T}_g, \quad \vec{\rho} = \vec{\mathcal{F}}_b^T \boldsymbol{\rho}, \quad \vec{\mathbf{r}} = \vec{\mathcal{F}}_b^T \mathbf{r}_b.$$

Then, in spacecraft body coordinates, we have

$$\mathbf{T}_g = \frac{3\mu}{r^5} \int_V \boldsymbol{\rho}^{\times} \mathbf{r}_b \boldsymbol{\rho}^T \mathbf{r}_b dm.$$

Recall the identity (see Section 1.2.4) that $\boldsymbol{\rho}^{\times}\mathbf{r}_b = -\mathbf{r}_b^{\times}\boldsymbol{\rho}$. This leads to

$$\mathbf{T}_g = -\frac{3\mu}{r^5}\mathbf{r}_b^{\times} \int_V \boldsymbol{\rho}\boldsymbol{\rho}^T dm \mathbf{r}_b.$$

Note that we are able to pull $\mathbf{r}_b$ outside the integral since it is independent of $\boldsymbol{\rho}$. Now, we make use of the identity (see Section 1.2.4) $\boldsymbol{\rho}\boldsymbol{\rho}^T = \boldsymbol{\rho}^{\times}\boldsymbol{\rho}^{\times} + \boldsymbol{\rho}^T\boldsymbol{\rho}\mathbf{1}$, to give

$$\mathbf{T}_g = -\frac{3\mu}{r^5}\mathbf{r}_b^{\times} \int_V \boldsymbol{\rho}^{\times}\boldsymbol{\rho}^{\times} dm \mathbf{r}_b - \frac{3\mu}{r^5}\mathbf{r}_b^{\times} \int_V \boldsymbol{\rho}^T\boldsymbol{\rho}\mathbf{1} dm \mathbf{r}_b.$$

Noting that $\boldsymbol{\rho}^T\boldsymbol{\rho}$ is a scalar and that $\mathbf{r}_b^{\times}\boldsymbol{\rho}^T\boldsymbol{\rho}\mathbf{1}\mathbf{r}_b = \boldsymbol{\rho}^T\boldsymbol{\rho}\mathbf{r}_b^{\times}\mathbf{r}_b = \mathbf{0}$, we have

$$\mathbf{T}_g = \frac{3\mu}{r^5}\mathbf{r}_b^{\times} \int_V \left(-\boldsymbol{\rho}^{\times}\boldsymbol{\rho}^{\times}\right) dm \mathbf{r}_b.$$

But, the moment of inertia matrix is (see Section 2.3.2)

$$\mathbf{I} = \int_V \left(-\boldsymbol{\rho}^{\times}\boldsymbol{\rho}^{\times}\right) dm.$$

Therefore, we finally obtain the gravity-gradient torque as

$$\mathbf{T}_g = \frac{3\mu}{r^5}\mathbf{r}_b^{\times}\mathbf{I}\mathbf{r}_b.$$

Note that $\mathbf{r}_b$ is the spacecraft orbital position in spacecraft body coordinates.

Notes

This chapter has provided mathematical models for various disturbance torques acting on a spacecraft. More detailed mathematical models may be found in Hughes (2004).

Reference

Hughes PC 2004 *Spacecraft Attitude Dynamics*. Dover Publications, Mineola, NY.

13

Torque-Free Attitude Motion

The disturbance torques that we derived in Chapter 12 are typically very small, so torque-free motion is a good approximation over short periods of time, and can provide us with an understanding of the dominant behavior of the natural attitude motion of a spacecraft. Note that the disturbance torques cannot be neglected if we are considering periods of time consisting of several orbits.

13.1 Solution for an Axisymmetric Body

We now consider a special case of an axisymmetric spacecraft, since this will allow us to obtain a closed-form solution of the attitude motion. An axisymmetric body is one in which two of the principal inertias are equal, for example, a cylinder of uniform density as shown in Figure 13.1 is axisymmetric. We consider an axisymmetric spacecraft with principal inertias

$$I_x = I_y = I_t, \quad I_z = I_a. \tag{13.1}$$

Note that the subscript 't' means *transverse*, and the subscript 'a' means *axial*. The z-axis is the axis of symmetry (since the moments of inertia about the x- and y-axes are equal). Recall the governing equations of motion from Section 11.3 (Euler's equations)

$$\begin{aligned}
I_x\dot{\omega}_x + (I_z - I_y)\omega_y\omega_z &= T_x, \\
I_y\dot{\omega}_y + (I_x - I_z)\omega_x\omega_z &= T_y, \\
I_z\dot{\omega}_z + (I_y - I_x)\omega_x\omega_y &= T_z.
\end{aligned}$$

Spacecraft Dynamics and Control: An Introduction, First Edition.
Anton H.J. de Ruiter, Christopher J. Damaren and James R. Forbes.

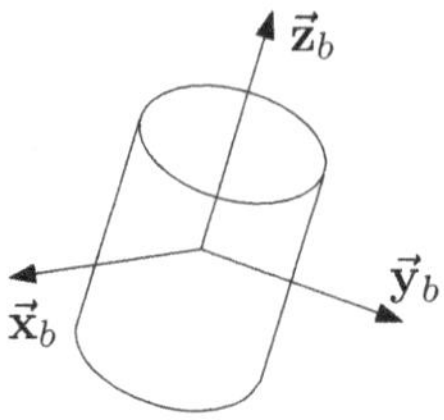

Figure 13.1 Axisymmetric body

For our axisymmetric spacecraft with principal inertias given by (13.1), and setting the disturbance torques to zero ($T_x = T_y = T_z = 0$), the equations of motion become

$$I_t\dot{\omega}_x + (I_a - I_t)\omega_y\omega_z = 0, \tag{13.2}$$

$$I_t\dot{\omega}_y + (I_t - I_a)\omega_x\omega_z = 0, \tag{13.3}$$

$$I_a\dot{\omega}_z = 0. \tag{13.4}$$

Let us denote the initial conditions by

$$\omega_x(0) = \omega_{x0}, \quad \omega_y(0) = \omega_{y0}, \quad \omega_z(0) = \omega_{z0}.$$

We can immediately see from the equation for ω_z (13.4), that

$$\dot{\omega}_z = 0,$$

which means that the angular velocity about the z-axis (the axis of symmetry) remains constant

$$\omega_z(t) = \omega_{z0}.$$

Substituting this into equations (13.2) and (13.3) leads to the equations for the x and y components of the angular velocity, which are

$$\begin{aligned}\dot{\omega}_x - \frac{(I_t - I_a)}{I_t}\omega_{z0}\omega_y &= 0,\\ \dot{\omega}_y + \frac{(I_t - I_a)}{I_t}\omega_{z0}\omega_x &= 0.\end{aligned} \tag{13.5}$$

Let us define the quantity

$$\Omega = \frac{(I_t - I_a)}{I_t}\omega_{z0},$$

which is called the *relative spin rate*. Then, the equations for x and y become

$$\begin{aligned} \dot{\omega}_x - \Omega\omega_y &= 0, \\ \dot{\omega}_y + \Omega\omega_x &= 0. \end{aligned} \tag{13.6}$$

These are two coupled first order linear differential equations. There are a number of methods by which they can be solved. We shall solve them using Laplace transforms. Taking the Laplace transform of (13.6) gives

$$\begin{aligned} s\hat{\omega}_x - \omega_{x0} - \Omega\hat{\omega}_y &= 0, \\ s\hat{\omega}_y - \omega_{y0} + \Omega\hat{\omega}_x &= 0, \end{aligned} \tag{13.7}$$

where $\hat{\omega}_x = \mathcal{L}(\omega_x)$ and $\hat{\omega}_y = \mathcal{L}(\omega_y)$. Equation (13.7) can be rewritten in matrix form as

$$\begin{bmatrix} s & -\Omega \\ \Omega & s \end{bmatrix} \begin{bmatrix} \hat{\omega}_x \\ \hat{\omega}_y \end{bmatrix} = \begin{bmatrix} \omega_{x0} \\ \omega_{y0} \end{bmatrix}, \tag{13.8}$$

which can be solved as

$$\begin{bmatrix} \hat{\omega}_x \\ \hat{\omega}_y \end{bmatrix} = \begin{bmatrix} \dfrac{s}{s^2+\Omega^2} & \dfrac{\Omega}{s^2+\Omega^2} \\ -\dfrac{\Omega}{s^2+\Omega^2} & \dfrac{s}{s^2+\Omega^2} \end{bmatrix} \begin{bmatrix} \omega_{x0} \\ \omega_{y0} \end{bmatrix}. \tag{13.9}$$

Taking inverse Laplace transforms, we have

$$\begin{bmatrix} \omega_x(t) \\ \omega_y(t) \end{bmatrix} = \begin{bmatrix} \cos\Omega t & \sin\Omega t \\ -\sin\Omega t & \cos\Omega t \end{bmatrix} \begin{bmatrix} \omega_{x0} \\ \omega_{y0} \end{bmatrix}. \tag{13.10}$$

Let us define the *transverse angular velocity* by

$$\omega_t = \left(\omega_x^2 + \omega_y^2\right)^{\frac{1}{2}}.$$

Clearly, $\omega_t \geq 0$ by its definition.

We can rewrite (13.10) as

$$\begin{aligned} \omega_x(t) &= \omega_t(0)\left[\frac{\omega_{x0}}{\omega_t(0)}\cos\Omega t + \frac{\omega_{y0}}{\omega_t(0)}\sin\Omega t\right], \\ \omega_y(t) &= \omega_t(0)\left[\frac{\omega_{y0}}{\omega_t(0)}\cos\Omega t - \frac{\omega_{x0}}{\omega_t(0)}\sin\Omega t\right]. \end{aligned} \tag{13.11}$$

Now, we can make use of the trigonometric identities

$$\cos(a+b) = \cos a\cos b - \sin a\sin b, \quad \sin(a+b) = \sin a\cos b + \cos a\sin b,$$

to rewrite (13.11) as

$$\begin{aligned}\omega_x(t) &= \omega_t(0)\sin(\Omega t+\phi),\\ \omega_y(t) &= \omega_t(0)\cos(\Omega t+\phi),\end{aligned}\tag{13.12}$$

by identifying

$$\sin\phi = \frac{\omega_{x0}}{\omega_t(0)} \text{ and } \cos\phi = \frac{\omega_{y0}}{\omega_t(0)}.$$

From (13.12), we immediately see that the transverse angular velocity ω_t is constant, that is,

$$\omega_t(t) = \omega_t(0)\left(\sin^2(\Omega t+\phi)+\cos^2(\Omega t+\phi)\right)^{\frac{1}{2}} = \omega_t(0).$$

Therefore, we have

$$\begin{aligned}\omega_x(t) &= \omega_t\sin(\Omega t+\phi),\\ \omega_y(t) &= \omega_t\cos(\Omega t+\phi),\end{aligned}\tag{13.13}$$

We see that the magnitude of the angular velocity is constant also, that is,

$$\omega = \left(\omega_x^2+\omega_y^2+\omega_z^2\right)^{\frac{1}{2}} = \left(\omega_t^2+\omega_z^2\right)^{\frac{1}{2}} = \mathit{constant}.$$

Let us define the angle $\mu = \Omega t + \phi$. Then, from the previous development, we can summarize the solution for the angular velocity (in spacecraft body coordinates) as

$$\begin{aligned}\omega_x(t) &= \omega_t\sin\mu,\\ \omega_y(t) &= \omega_t\cos\mu,\\ \omega_z(t) &= \omega_{z0}.\end{aligned}\tag{13.14}$$

Pictorially, this is shown in Figure 13.2. What this means is that the angular velocity vector rotates about the spacecraft body z-axis at a constant rate $\dot{\mu} = \Omega$. Let us now examine the angular momentum vector, given in spacecraft body coordinates by $\mathbf{h} = \mathbf{I}\boldsymbol{\omega}$, that is,

$$\begin{aligned}h_x &= I_t\omega_x = I_t\omega_t\sin\mu,\\ h_y &= I_t\omega_y = I_t\omega_t\cos\mu,\\ h_z &= I_a\omega_z = I_a\omega_{z0}.\end{aligned}\tag{13.15}$$

As for the transverse angular velocity, we can define the *transverse angular momentum* as

$$h_t = \left(h_x^2+h_y^2\right)^{\frac{1}{2}} = I_t\left(\omega_x^2+\omega_y^2\right)^{\frac{1}{2}} = I_t\omega_t,$$

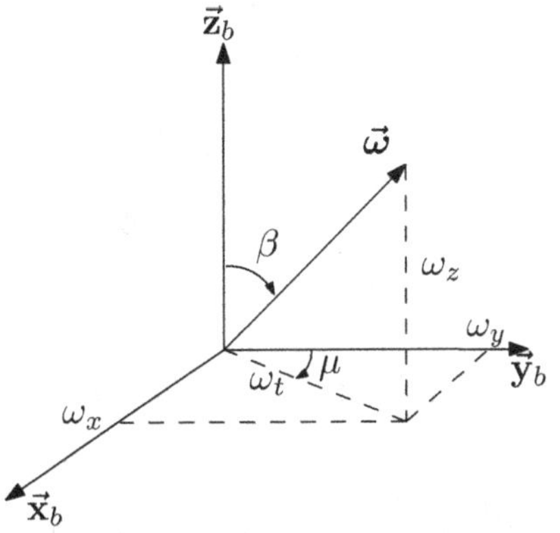

Figure 13.2 Angular velocity in body coordinates

which also is constant. Likewise, the total angular momentum $h = (h_x^2 + h_y^2 + h_z^2)^{\frac{1}{2}} = (h_t^2 + h_z^2)^{\frac{1}{2}}$ is constant also. Then, we can rewrite the solution for the angular momentum in (13.15) as

$$\begin{aligned} h_x &= h_t \sin\mu, \\ h_y &= h_t \cos\mu, \\ h_z &= I_a \omega_{z0}. \end{aligned} \tag{13.16}$$

This shows that just like the angular velocity vector, the angular momentum vector rotates about the spacecraft body z-axis at a rate Ω. The angle that the angular momentum vector makes with the spacecraft body z-axis, γ is given by

$$\sin\gamma = \frac{h_t}{h}, \quad \cos\gamma = \frac{h_z}{h},$$

and is called the *nutation angle* (see Figure 13.3), and is easily seen to be constant.

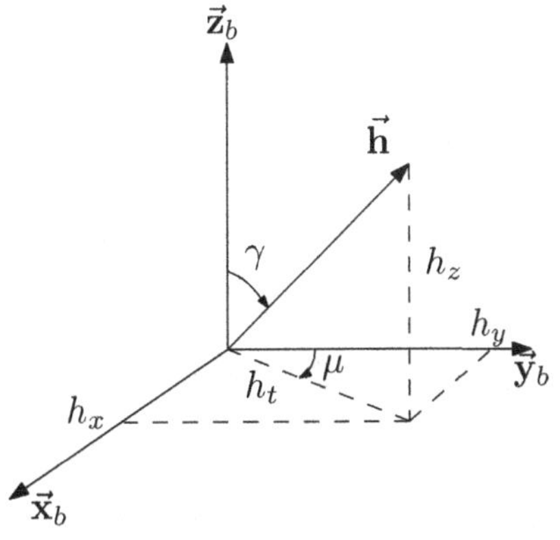

Figure 13.3 Angular momentum in body coordinates

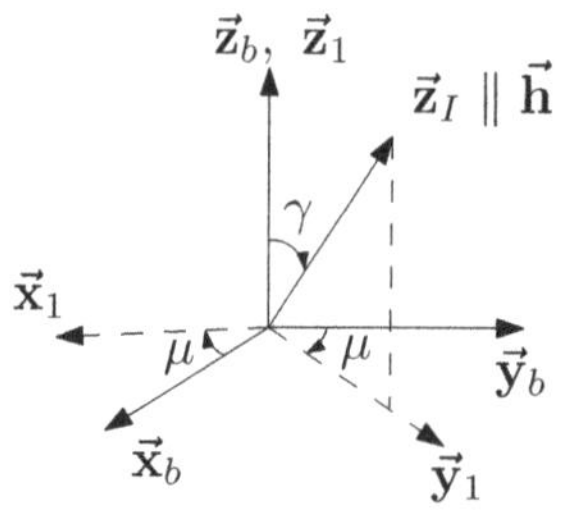

Figure 13.4 Rotation about $\vec{\mathbf{z}}_b$

Now, the physical vector form of the equation of motion for the angular momentum for torque-free motion is given by (see Section 11.3)

$$\dot{\vec{\mathbf{h}}} = \vec{\mathbf{0}},$$

where ($\dot{}$) denotes the time derivative with respect to an inertial frame. Therefore, the angular momentum vector is inertially fixed (is constant in an inertial frame of reference). We can make use of this fact to derive the attitude motion of the body.

Since the angular momentum vector $\vec{\mathbf{h}}$ is inertially fixed, let us choose an inertial frame $\mathcal{F}_I$ with z-axis aligned with the angular momentum vector, that is $\vec{\mathbf{z}}_I = \vec{\mathbf{h}}/h$. Then, we can map the spacecraft body fixed frame $\mathcal{F}_b$ onto the inertial frame as follows.

First, if we rotate $\mathcal{F}_b$ through $-\mu$ about the $\vec{\mathbf{z}}_b$ axis as shown in Figure 13.4, the $\vec{\mathbf{z}}_I$ will lie in the resulting intermediate frame $\mathcal{F}_1$ y-z plane, with angle γ from $\vec{\mathbf{z}}_1$. Next, if we rotate the intermediate frame $\mathcal{F}_1$ through an angle $-\gamma$ about $\vec{\mathbf{x}}_1$, then $\vec{\mathbf{z}}_I$ will be equal to the z-axis of the transformed frame $\mathcal{F}_2$, as shown in Figure 13.5. Finally, since $\vec{\mathbf{z}}_I = \vec{\mathbf{z}}_2$, the frame $\mathcal{F}_I$ can be obtained by a rotation about $\vec{\mathbf{z}}_2$ through some angle $-\theta$, as shown in Figure 13.6. The angle θ is called the *precession angle*. What we have just accomplished, is to find a 3-1-3 Euler rotation sequence (see Section 1.3.3) to map $\mathcal{F}_b$ onto $\mathcal{F}_I$. The associated rotation matrix describing this transformation is

$$\mathbf{C}_{Ib} = \mathbf{C}_z(-\theta)\mathbf{C}_x(-\gamma)\mathbf{C}_z(-\mu).$$

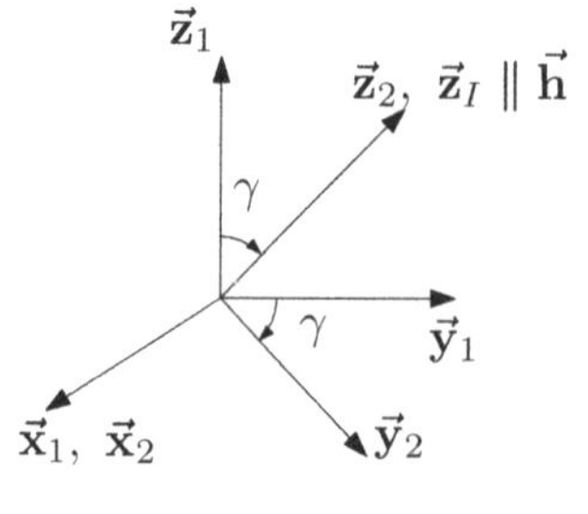

Figure 13.5 Rotation about $\vec{\mathbf{x}}_1$

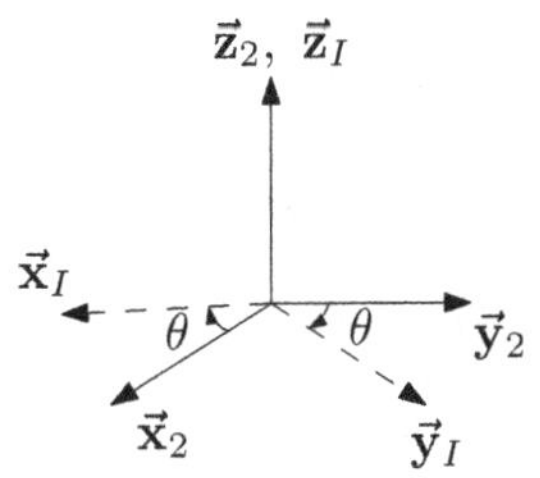

Figure 13.6 Rotation about $\vec{\mathbf{z}}_2$

Finally, we can obtain the rotation matrix describing the attitude of $\mathcal{F}_b$ relative to $\mathcal{F}_I$ as the reverse transformation

$$\mathbf{C}_{bI} = \mathbf{C}_z(\mu)\mathbf{C}_x(\gamma)\mathbf{C}_z(\theta). \tag{13.17}$$

The angular velocity in spacecraft body coordinates in terms of the three Euler angles θ, γ and μ is given by (see Section 1.4.2)

$$\begin{bmatrix} \omega_x \\ \omega_y \\ \omega_z \end{bmatrix} = \begin{bmatrix} 0 \\ 0 \\ \dot{\mu} \end{bmatrix} + \mathbf{C}_z(\mu) \begin{bmatrix} \dot{\gamma} \\ 0 \\ 0 \end{bmatrix} + \mathbf{C}_z(\mu)\mathbf{C}_x(\gamma) \begin{bmatrix} 0 \\ 0 \\ \dot{\theta} \end{bmatrix}.$$

Now, since the angle γ is constant, we have $\dot{\gamma} = 0$. We have also seen previously that $\dot{\mu} = \Omega$. Recall Section 1.3.1 that

$$\mathbf{C}_z(\mu) = \begin{bmatrix} \cos\mu & \sin\mu & 0 \\ -\sin\mu & \cos\mu & 0 \\ 0 & 0 & 1 \end{bmatrix}, \quad \mathbf{C}_x(\gamma) = \begin{bmatrix} 1 & 0 & 0 \\ 0 & \cos\gamma & \sin\gamma \\ 0 & -\sin\gamma & \cos\gamma \end{bmatrix}.$$

Therefore, we find that the angular velocity is given by

$$\begin{bmatrix} \omega_x \\ \omega_y \\ \omega_z \end{bmatrix} = \begin{bmatrix} \dot{\theta}\sin\gamma\sin\mu \\ \dot{\theta}\sin\gamma\cos\mu \\ \Omega + \dot{\theta}\cos\gamma \end{bmatrix}. \tag{13.18}$$

Comparing (13.18) with our previously derived solution for the angular velocity (13.14), we have

$$\begin{bmatrix} \dot{\theta}\sin\gamma\sin\mu \\ \dot{\theta}\sin\gamma\cos\mu \\ \Omega + \dot{\theta}\cos\gamma \end{bmatrix} = \begin{bmatrix} \omega_t\sin\mu \\ \omega_t\cos\mu \\ \omega_{z0} \end{bmatrix}.$$

From this, we obtain

$$\dot{\theta}\sin\gamma = \omega_t,$$

which leads to

$$\dot{\theta} = \frac{\omega_t}{\sin\gamma}.$$

Using the fact that $\sin\gamma = h_t/h$, this leads to

$$\dot{\theta} = \frac{h}{I_t}, \tag{13.19}$$

which is constant. Let define the quantity

$$\Omega_p = \dot{\theta} = \frac{h}{I_t}, \tag{13.20}$$

which is known as the *precession rate*. Physically, the precession rate is the rate at which the spacecraft body z-axis ($\vec{\mathbf{z}}_b$) rotates about the inertial z-axis ($\vec{\mathbf{z}}_I$) (since $\vec{\mathbf{z}}_b$ is fixed in the transformed frame $\mathcal{F}_2$).

13.2 Physical Interpretation of the Motion

In this section we shall obtain a nice physical interpretation of the spacecraft attitude motion. First, let us once again consider the spacecraft angular velocity vector $\vec{\omega}$ and angular momentum vectors $\vec{\mathbf{h}}$ as seen in the spacecraft body frame $\mathcal{F}_b$, as shown in Figures 13.2 and 13.3 respectively.

As before, let γ be the angle between $\vec{\mathbf{z}}_b$ and $\vec{\mathbf{h}}$. We also let β be the angle between $\vec{\mathbf{z}}_b$ and $\vec{\omega}$. Clearly, $\vec{\mathbf{h}}$, $\vec{\omega}$ and $\vec{\mathbf{z}}_b$ all lie in the same plane. From Figures 13.2 and 13.3 we obtain

$$\tan\gamma = \frac{h_t}{h_z}, \quad \tan\beta = \frac{\omega_t}{\omega_z}.$$

From before, we have

$$h_t = I_t\omega_t, \quad h_z = I_a\omega_z.$$

Therefore, we find that

$$\tan\gamma = \frac{I_t}{I_a}\tan\beta.$$

From this, we see that

$$\begin{aligned} &\text{if} \quad I_t > I_a \quad \text{then} \quad \gamma > \beta, \\ &\text{if} \quad I_t < I_a \quad \text{then} \quad \gamma < \beta. \end{aligned}$$

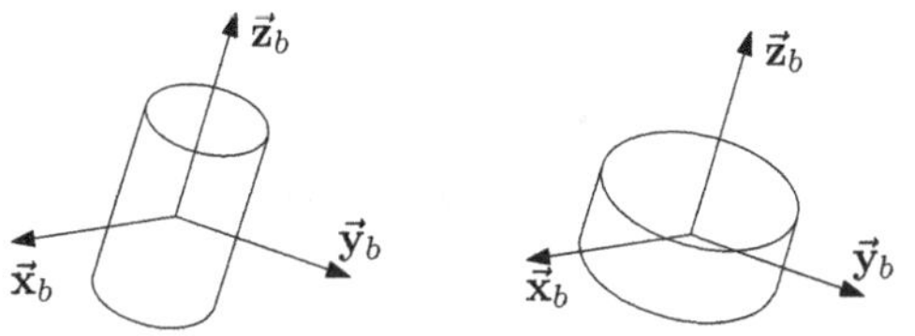

Prolate Spinner
$(I_t > I_a)$

Oblate Spinner
$(I_a > I_t)$

Figure 13.7 Prolate vs. oblate spinner

When $I_t > I_a$, the spacecraft is called a *prolate spinner*, and when $I_t < I_a$, the spacecraft is called an *oblate spinner*. These cases are shown in Figure 13.7.

Case 1: Prolate Spinner ($I_t > I_a$)
From before, we have $\gamma > \beta$. The relationship between vectors $\vec{\mathbf{h}}$, $\vec{\boldsymbol{\omega}}$ and $\vec{\mathbf{z}}_b$ is shown in Figure 13.8. As shown, the angle between $\vec{\mathbf{h}}$ and $\vec{\boldsymbol{\omega}}$ is given by $\alpha = \gamma - \beta$. Noting that the angular momentum vector $\vec{\mathbf{h}}$ is parallel to the z-axis of the inertial frame $\mathcal{F}_I$, let us define a cone $\mathcal{C}_s$ with axis $\vec{\mathbf{z}}_I$ and half angle α. This is called the *space cone*. We also define a cone $\mathcal{C}_b$ with axis $\vec{\mathbf{z}}_b$ and half angle β. This is called the *body cone*. The cones are shown in Figure 13.9. Clearly, the cones touch along the angular velocity vector $\vec{\boldsymbol{\omega}}$. Now, as the axis of the body cone $\vec{\mathbf{z}}_b$ (the body z-axis of the spacecraft) rotates about $\vec{\mathbf{z}}_I$, let the body cone $\mathcal{C}_b$ roll without slipping on the space cone $\mathcal{C}_s$ (which is fixed in $\mathcal{F}_I$). Since $\mathcal{C}_b$ rolls without slipping on $\mathcal{C}_s$, the instantaneous axis of rotation of $\mathcal{C}_b$ must lie along the line of contact between $\mathcal{C}_b$ and $\mathcal{C}_s$. Therefore, the angular velocity vector of $\mathcal{C}_b$, denoted by $\vec{\boldsymbol{\omega}}'$ must be parallel to $\vec{\boldsymbol{\omega}}$, as shown in Figure 13.9. That is, $\vec{\boldsymbol{\omega}}' = a\vec{\boldsymbol{\omega}}$ for some scalar a.

Now, since $\vec{\mathbf{z}}_b$ is the axis of $\mathcal{C}_b$, it is fixed in $\mathcal{C}_b$. Therefore, we must have

$$\dot{\vec{\mathbf{z}}}_b = \vec{\boldsymbol{\omega}}' \times \vec{\mathbf{z}}_b = a\vec{\boldsymbol{\omega}} \times \vec{\mathbf{z}}_b.$$

However, since $\vec{\mathbf{z}}_b$ is also fixed in $\mathcal{F}_b$ we also have

$$\dot{\vec{\mathbf{z}}}_b = \vec{\boldsymbol{\omega}} \times \vec{\mathbf{z}}_b.$$

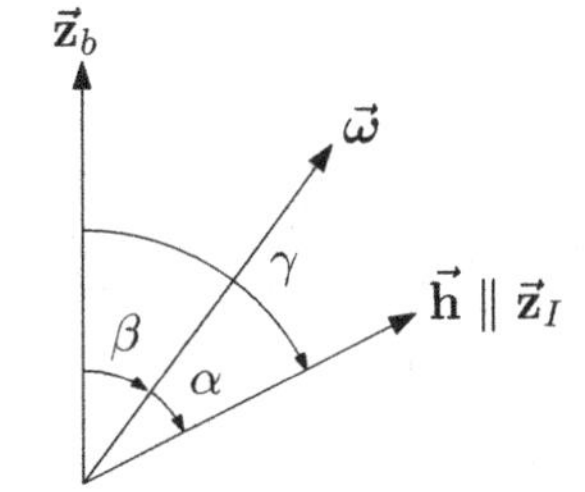

Figure 13.8 $\vec{\mathbf{h}}$, $\vec{\boldsymbol{\omega}}$ and $\vec{\mathbf{z}}_b$ for a prolate spinner

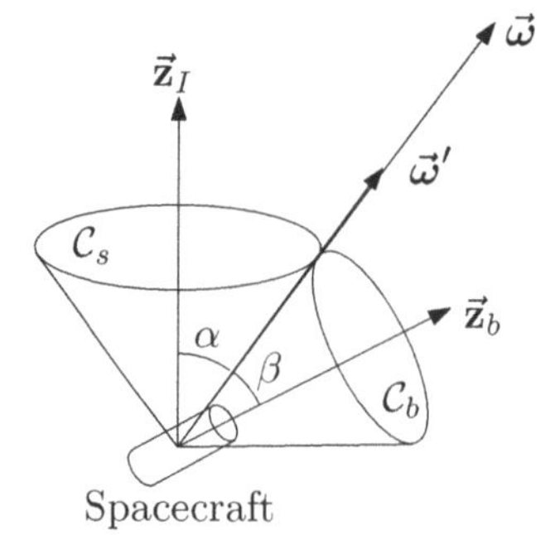

Figure 13.9 Space and body cones in the prolate case

That is,

$$a\vec{\omega} \times \vec{\mathbf{z}}_b = \vec{\omega} \times \vec{\mathbf{z}}_b.$$

Clearly, this is possible only if $a = 1$. Therefore, we must have $\vec{\omega}' = \vec{\omega}$. That is, the body cone C_b has the same angular velocity as the spacecraft body frame $\mathcal{F}_b$. The body cone can therefore be considered attached to the spacecraft.

In conclusion, the spacecraft attitude motion is the same as the motion of the body cone C_b rolling without slipping on the space cone C_s.

Case 2: Oblate Spinner $I_a > I_t$

From before, we have $\beta > \gamma$. The relationship between vectors $\vec{\mathbf{h}}$, $\vec{\omega}$ and $\vec{\mathbf{z}}_b$ is shown in Figure 13.10. In this case, we see that the angle between $\vec{\mathbf{h}}$ and $\vec{\omega}$ is given by $\alpha = \beta - \gamma$. As for the prolate case, we define a space cone C_s with axis $\vec{\mathbf{z}}_I$ and half angle α. We also define a body cone C_b with axis $\vec{\mathbf{z}}_b$ and half angle β. The cones are shown in Figure 13.11.

Using the exact same analysis as for the prolate case, we obtain the conclusion that the spacecraft attitude motion is the same as the motion of the body cone C_b rolling without slipping on the space cone C_s.

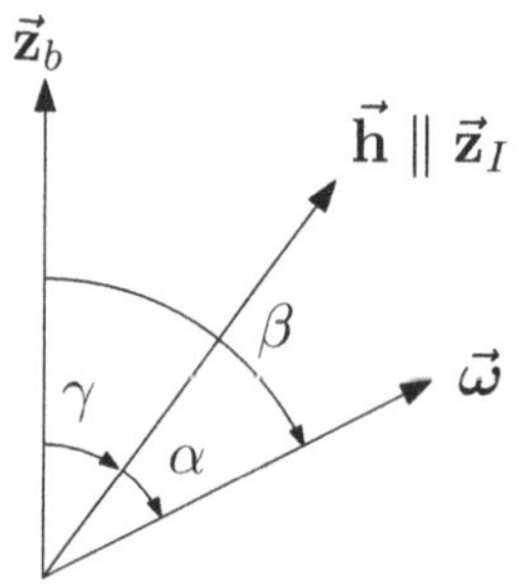

Figure 13.10 $\vec{\mathbf{h}}$, $\vec{\omega}$ and $\vec{\mathbf{z}}_b$ for an oblate spinner

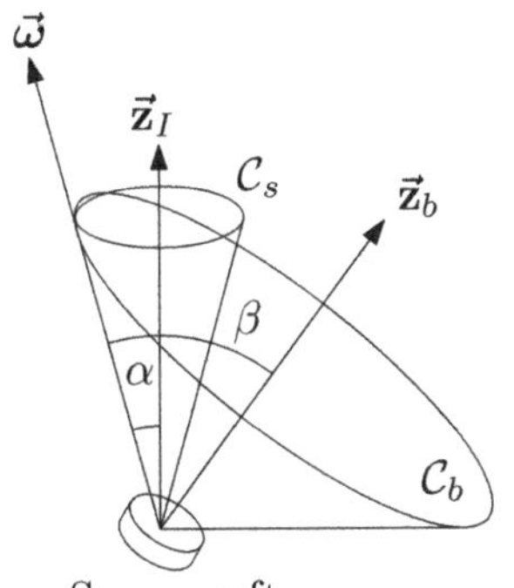

Figure 13.11 Space and body cones in the oblate case

Notes

This chapter has presented an analytical solution to torque-free spacecraft attitude motion as well as a geometric interpretation of the motion in the case of an axisymmetric spacecraft. An analytical solution to torque-free spacecraft attitude motion for the tri-inertial case (all principal moments of inertia being unequal) and its geometric interpretation may be found in Thomson (1986) and Hughes (2004). A derivation of the analytical solution in the tri-inertial case may be found in Wittenburg (1977).

References

Hughes PC 2004 *Spacecraft Attitude Dynamics*. Dover Publications, Mineola, NY.
Thomson WT 1986 *Introduction to Space Dynamics*. Dover Publications, Mineola, NY.
Wittenburg J 1977 *Dynamics of Systems of Rigid Bodies*. Teubner, Stuttgart.

14

Spin Stabilization

In the next three chapters, we shall examine various passive stabilization approaches. Passive stabilization involves putting the spacecraft into a favorable naturally stable equilibrium. Spin stabilization is one of the oldest forms of passive stabilization. If the objective of a mission is to point a single instrument onboard the spacecraft in an inertially fixed direction, then oftentimes only that direction matters (depending upon the instrument), and the orientation about that direction is irrelevant. As such, we can spin the spacecraft about that axis (see Figure 14.1). We will shortly examine under what conditions such a spin is stable. First, we have to define what we mean by *stability*.

14.1 Stability

The stability of a system refers to the asymptotic behavior of the system (as $t \to \infty$). Does the motion remain bounded (stable)? Does the motion grow unbounded (unstable)? Does it approach an equilibrium (asymptotically stable)?

A rigorous definition of stability for general nonlinear systems will be presented in Chapter 24. For the moment however, we shall keep our focus specifically on linear time-invariant systems.

Consider the linear differential equation with constant real coefficients

$$\frac{d^n x}{dt} + b_1 \frac{d^{n-1} x}{dt^{n-1}} + \ldots + b_{n-1} \frac{dx}{dt} + b_n x = 0.$$

Clearly, $x(t) \equiv 0$ is an equilibrium solution of this differential equation. To determine stability of this equilibrium, we wish to see what happens when it is displaced from equilibrium, that is,

Spacecraft Dynamics and Control: An Introduction, First Edition.
Anton H.J. de Ruiter, Christopher J. Damaren and James R. Forbes.

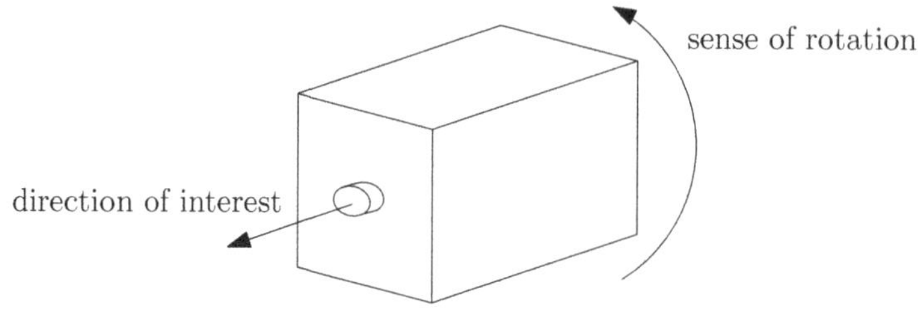

Figure 14.1 Spinning spacecraft

when the initial conditions non-zero. Taking the Laplace transform of the differential equation gives

$$\begin{aligned}&\left[s^n X(s) - s^{n-1}x(0) - s^{n-2}\frac{dx}{dt}(0) - \ldots - \frac{d^{n-1}x}{dt^{n-1}}(0)\right]\\&+ b_1\left[s^{n-1}X(s) - s^{n-2}x(0) - s^{n-3}\frac{dx}{dt}(0) - \ldots - \frac{d^{n-2}x}{dt^{n-2}}(0)\right]\\&+ \ldots + b_{n-1}\left[sX(s) - x(0)\right] + b_n X(s) = 0,\end{aligned}$$

which can be rewritten as

$$\left[s^n + b_1 s^{n-1} + b_2 s^{n-2} + \ldots + b_n\right] X(s) = a_1 s^{n-1} + a_2 s^{n-2} + \ldots + a_n,$$

where the coefficients

$$a_1 = x(0), \quad a_2 = \frac{dx}{dt}(0) + b_1 x(0), \ldots, a_n = \frac{d^{n-1}x}{dt^{n-1}}(0) + b_1\frac{d^{n-2}x}{dt^{n-2}}(0) + \ldots + b_{n-1}x(0),$$

are functions of the initial conditions. We can now solve for $X(s)$

$$X(s) = \frac{a_1 s^{n-1} + a_2 s^{n-2} + \ldots + a_n}{s^n + b_1 s^{n-1} + b_2 s^{n-2} + \ldots + b_n}.$$

Setting the denominator to zero yields the *characteristic equation* of the differential equation

$$s^n + b_1 s^{n-1} + b_2 s^{n-2} + \ldots + b_n = 0. \tag{14.1}$$

Let $p_i, i = 1, \ldots, n$, be the solutions of the characteristic equation (14.1). As is explained in Appendix A, p_i are called the *poles* of $X(s)$. It turns out that they completely determine the asymptotic behavior. For simplicity, we shall assume that the poles are distinct, that is $p_i \neq p_j$ if $i \neq j$. In this case (as shown in Appendix A), $X(s)$ can be expanded in a partial fraction expansion as

$$X(s) = \frac{c_1}{s - p_1} + \frac{c_2}{s - p_2} + \ldots + \frac{c_n}{s - p_n},$$

where c_i are the residues of $X(s)$. Taking the inverse Laplace transform gives the solution for $x(t)$,

$$x(t) = c_1 e^{p_1 t} + c_2 e^{p_2 t} + \ldots + c_n e^{p_n t}. \tag{14.2}$$

Now, since all coefficients a_i and b_i are real, it must be that complex poles of $X(s)$ occur in complex conjugate pairs. That is, if $p = \sigma + j\omega$ is a pole with $\omega \neq 0$, then $\bar{p} = \sigma - j\omega$ must be a pole also. It can readily be shown then that if p_i is a complex pole with corresponding residue c_i, then the residue corresponding to the pole $\bar{p}_i$ is $\bar{c}_i$. Therefore, the part of the solution of $x(t)$ corresponding to a complex conjugate pair of poles $p = \sigma + j\omega$ and $\bar{p} = \sigma - j\omega$ is given by

$$ce^{pt} + \bar{c}e^{\bar{p}t}.$$

Letting $c = a + jb$, this becomes

$$\begin{aligned} ce^{pt} + \bar{c}e^{\bar{p}t} &= (a + jb)\, e^{\sigma t} (\cos \omega t + j \sin \omega t) + (a - jb)\, e^{\sigma t} (\cos \omega t - j \sin \omega t), \\ &= e^{\sigma t} (2a \cos \omega t - 2b \sin \omega t). \end{aligned}$$

Instability Result
From (14.2), we see that if any of the poles p_i have positive real part $Re(p_i) > 0$, then $x(t)$ grows without bound and the system is unstable.

Asymptotic Stability Result
If all of the poles have negative real part, $Re(p_i) < 0$, then $x(t)$ converges to zero and the system is asymptotically stable.

Stability Result
If none of the poles have positive real part, but there are distinct poles on the imaginary axis, then $x(t)$ exhibits bounded oscillatory behavior, and the system is stable. Note that any repeated poles on the imaginary axis make the system unstable.

14.2 Spin Stability of Torque-Free Motion

We shall now examine the stability of spin equilibrium conditions. First of all, only spins about principal axes of inertia are equilibrium conditions. Let us therefore consider the principal axes torque free Euler equations (see Section 11.3)

$$\begin{aligned} I_x\dot{\omega}_x + (I_z - I_y)\omega_y\omega_z &= 0, \\ I_y\dot{\omega}_y + (I_x - I_z)\omega_x\omega_z &= 0, \\ I_z\dot{\omega}_z + (I_y - I_x)\omega_x\omega_y &= 0. \end{aligned}$$

It is readily seen that a principal axis spin is an equilibrium condition. Let us now demonstrate this for a z-axis spin

$$\omega_x(t) = \omega_y(t) = 0, \quad \omega_z(t) = \nu \text{ (constant)}.$$

Substituting this into Euler's equations, we have

$$\begin{aligned} I_x\dot{\omega}_x &= 0, \\ I_y\dot{\omega}_y &= 0, \\ I_z\dot{\omega}_z &= 0, \end{aligned}$$

which means that a principal z-axis spin is indeed an equilibrium condition (the angular velocities do not change).

Let us now see what happens if we perturb the equilibrium. Let

$$\omega_x = \epsilon_x, \quad \omega_y = \epsilon_y, \quad \omega_z = \epsilon_z + \nu,$$

where ϵ_x, ϵ_y and ϵ_z are small angular velocities. Euler's equations now become

$$\begin{aligned} I_x\dot{\epsilon}_x + (I_z - I_y)\epsilon_y(\epsilon_z + \nu) &= 0, \\ I_y\dot{\epsilon}_y + (I_x - I_z)\epsilon_x(\epsilon_z + \nu) &= 0, \\ I_z\dot{\epsilon}_z + (I_y - I_x)\epsilon_x\epsilon_y &= 0. \end{aligned}$$

Since all ϵ are small, we neglect products of ϵ. That is, we linearize Euler's equations about the equilibrium to get

$$\begin{aligned} I_x\dot{\epsilon}_x + (I_z - I_y)\nu\epsilon_y &= 0, \\ I_y\dot{\epsilon}_y + (I_x - I_z)\nu\epsilon_x &= 0, \\ I_z\dot{\epsilon}_z &= 0. \end{aligned}$$

The equations for ϵ_x and ϵ_y are decoupled from the equation for ϵ_z. The equation for ϵ_z implies that

$$\epsilon_z = constant,$$

which means that ϵ_z is stable. Let us now perform a stability analysis on the equations for ϵ_x and ϵ_y, which we rewrite for convenience as

$$\begin{aligned} \dot{\epsilon}_x + \frac{(I_z - I_y)\nu}{I_x}\epsilon_y &= 0, \\ \dot{\epsilon}_y + \frac{(I_x - I_z)\nu}{I_y}\epsilon_x &= 0. \end{aligned}$$

Taking the Laplace transform gives

$$\begin{aligned} s\hat{\epsilon}_x(s) - \epsilon_x(0) + \frac{(I_z - I_y)\nu}{I_x}\hat{\epsilon}_y(s) &= 0, \\ s\hat{\epsilon}_y(s) - \epsilon_y(0) + \frac{(I_x - I_z)\nu}{I_y}\hat{\epsilon}_x(s) &= 0. \end{aligned}$$

This can be rewritten in matrix form as

$$\begin{bmatrix} s & \dfrac{(I_z - I_y)\nu}{I_x} \\ \dfrac{(I_x - I_z)\nu}{I_y} & s \end{bmatrix} \begin{bmatrix} \hat{\epsilon}_x(s) \\ \hat{\epsilon}_y(s) \end{bmatrix} = \begin{bmatrix} \epsilon_x(0) \\ \epsilon_y(0) \end{bmatrix}.$$

Solving this gives

$$\begin{bmatrix} \hat{\epsilon}_x(s) \\ \hat{\epsilon}_y(s) \end{bmatrix} = \frac{1}{s^2 + \alpha} \begin{bmatrix} s & -\dfrac{(I_z - I_y)\nu}{I_x} \\ -\dfrac{(I_x - I_z)\nu}{I_y} & s \end{bmatrix} \begin{bmatrix} \epsilon_x(0) \\ \epsilon_y(0) \end{bmatrix},$$

where

$$\alpha = \frac{(I_z - I_y)(I_z - I_x)\nu^2}{I_x I_y}.$$

That is,

$$\hat{\epsilon}_x(s) = \frac{s\epsilon_x(0) - \frac{(I_z - I_y)\nu}{I_x}\epsilon_y(0)}{s^2 + \alpha},$$

$$\hat{\epsilon}_y(s) = \frac{s\epsilon_y(0) - \frac{(I_x - I_z)\nu}{I_y}\epsilon_x(0)}{s^2 + \alpha}.$$

We see immediately, that both $\hat{\epsilon}_x(s)$ and $\hat{\epsilon}_y(s)$ have the same poles, satisfying

$$s^2 + \alpha = 0.$$

That is, the poles are $\pm\sqrt{-\alpha}$. There are now three possible cases.

1. $\alpha > 0$
   In this case, the poles are $\pm j\sqrt{\alpha}$, which are purely imaginary. This means that the solutions for $\epsilon_x(t)$ and $\epsilon_y(t)$ are oscillatory. Therefore, the motion is stable.
2. $\alpha = 0$
   In this case, there is a repeated pole at the origin. This means that both $\epsilon_x(t)$ and $\epsilon_y(t)$ have a component that grows linearly with time (show this as an exercise, by taking the inverse Laplace transform). Therefore, the motion is unstable.
3. $\alpha < 0$
   In this case, the poles are $\pm\sqrt{-\alpha}$, which are real. One of them is positive, the other is negative. This means that both $\epsilon_x(t)$ and $\epsilon_y(t)$ have a component that grows exponentially with time (show this as an exercise, by taking the inverse Laplace transform). Therefore, the motion is unstable.

In conclusion, for stability, we must have

$$\frac{(I_z - I_y)(I_z - I_x)\nu^2}{I_x I_z} > 0.$$

which means that

$$(I_z - I_y)(I_z - I_x) > 0.$$

This is only possible if either

$$I_z > I_y \text{ and } I_z > I_x,$$

or

$$I_z < I_y \text{ and } I_z < I_x.$$

Therefore, for spin stability, the spacecraft must be spinning either about the major (maximum) or minor (minimum) axis of inertia. A spin about the intermediate axis is unstable. Also, note that asymptotic stability is impossible, since we cannot obtain two poles with negative real parts.

14.3 Effect of Internal Energy Dissipation

The preceding result relies on the fact that the body is **rigid**, and that the motion is **torque free**. In practise, spacecraft are not entirely rigid, and the motion is not torque free (there are disturbances as we have seen in Chapter 12). Spacecraft have flexible elements, which are caused to deform due to the external torques as well as the attitude motion itself. These structural deformations result in a loss of kinetic energy in the form of heat. Because of this energy dissipation, the kinetic energy is non-increasing. This general effect is demonstrated by the following heuristic argument:

14.3.1 Energy Sink Hypothesis

A quasi-rigid body will dissipate energy until a state of minimum kinetic energy is reached. For torque-free motion, the angular momentum is conserved.

From the previous section, we know that only major or minor axis spins are stable. For a major axis spin, the kinetic energy is given by

$$T_{maj} = \frac{1}{2} I_{maj} \omega_{maj}^2 = \frac{h^2}{2I_{maj}}, \qquad h = I_{maj}\omega_{maj}.$$

For a minor axis spin, the kinetic energy is given by

$$T_{min} = \frac{1}{2} I_{min} \omega_{min}^2 = \frac{h^2}{2I_{min}}, \qquad h = I_{min}\omega_{min}.$$

Therefore, since $I_{min} < I_{maj}$, for a given angular momentum h,

$$T_{maj} < T_{min}.$$

That is, kinetic energy is minimized for a major axis spin. A minor axis spin maximizes kinetic energy. Since the kinetic energy reduces due to energy dissipation resulting from spacecraft flexibility, minor axis spins are also unstable. This leads to the following rule.

14.3.2 Major Axis Rule

Spins about the major axis are asymptotically stable. Spins about any other axis are unstable.

Example 14.1 (Explorer 1) *This was the first U.S. satellite. It was axisymmetric, and spin-stabilized about the minor axis of inertia. Explorer 1 had flexible antennas, causing energy dissipation. The satellite began tumbling hours after deployment. This lead to the discovery of the major axis rule.*

Notes

This chapter has presented the elements of spin stabilization for spacecraft. For a more detailed treatment of spin stability, the reader is referred to Hughes (2004). For a discussion of practical issues associated with spin stabilization, the reader is referred to Sidi (1997).

References

Hughes PC 2004 *Spacecraft Attitude Dynamics*. Dover Publications, Mineola, NY.

Sidi MJ 1997 *Spacecraft Dynamics and Control: A Practical Engineering Approach*. Cambridge University Press, New York, NY.

15

Dual-Spin Stabilization

Spin stabilization is a useful passive stabilization technique. By spinning the spacecraft about the major axis, gyroscopic stiffness is provided to the spin axis. This means that the spin axis tries to stay inertially fixed. However, spin stabilization is not always practical. First of all, the major axis rule requires that the spacecraft be spinning about the major axis, which may not be consistent with the shape of the spacecraft. Second of all, it means that the entire spacecraft is spinning. This becomes problematic in instances when it is necessary to point instruments in inertially changing directions, such as in Earth observation missions when an instrument is required to point toward the Earth, which is not an inertially fixed direction as seen from an Earth orbiting satellite, but rotates once per orbit.

It would be useful to have the benefits of spin stabilization (the gyroscopic stability), without having to spin the entire spacecraft, or at least without having to spin about a major axis. To accomplish this, a spinning wheel is mounted in the spacecraft with its spin axis aligned with the desired axis we wish to keep inertially fixed. The spinning wheel adds gyroscopic stability. This is called *Dual-Spin Stabilization*.

15.1 Equations of Motion

For a dual-spin spacecraft, we can consider the spacecraft to be made up of two rigid bodies: the wheel, and the platform (everything else) (see Figure 15.1). The wheel spin axis is fixed within the platform, and the wheel spins with angular velocity ω_s relative to the platform. We consider a body fixed frame $\mathcal{F}_b$ attached to the platform with the origin at the center of mass of the spacecraft (the combined platform and wheel system). It can be shown that the total angular momentum of the spacecraft (platform and wheel) is given in body coordinates by

$$\vec{\mathbf{h}} = \vec{\mathcal{F}}_b^T \left[\mathbf{I}\boldsymbol{\omega} + h_s\mathbf{a}\right], \tag{15.1}$$

Spacecraft Dynamics and Control: An Introduction, First Edition.
Anton H.J. de Ruiter, Christopher J. Damaren and James R. Forbes.

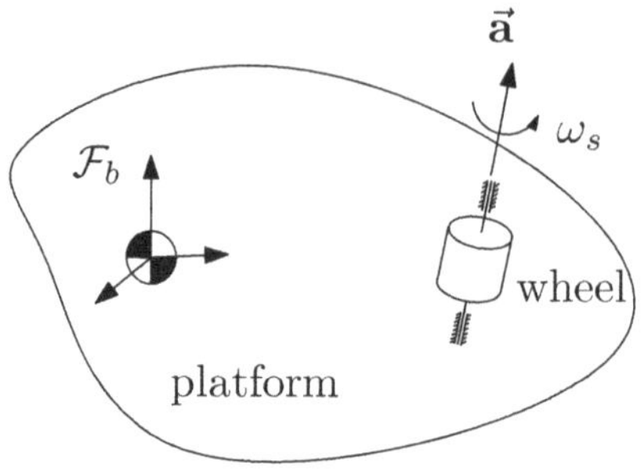

Figure 15.1 Dual-spin spacecraft

where

$\mathbf{I}$ = moment of inertia matrix of the spacecraft (platform and wheel)
$\vec{\boldsymbol{\omega}} = \vec{\mathcal{F}}_b^T \boldsymbol{\omega}$ = angular velocity of platform relative to an inertial frame
$h_s = I_s \omega_s$ = wheel spin angular momentum relative to platform
I_s = wheel moment of inertia about the spin axis
ω_s = wheel angular velocity relative to platform
$\vec{\mathbf{a}} = \vec{\mathcal{F}}_b^T \mathbf{a}$ = wheel spin axis.

The equations of motion are now given by Euler's equations (see Section 11.3)

$$\overset{\circ}{\vec{\mathbf{h}}} + \vec{\boldsymbol{\omega}} \times \vec{\mathbf{h}} = \vec{\mathbf{T}}, \tag{15.2}$$

where $(^{\circ})$ denotes the time-derivative as seen in the body frame $\mathcal{F}_b$, and $\vec{\mathbf{T}} = \vec{\mathcal{F}}_b^T \mathbf{T}$ are the external torques acting on the spacecraft.

From equation (15.1), we find that

$$\overset{\circ}{\vec{\mathbf{h}}} = \vec{\mathcal{F}}_b^T \left[\mathbf{I}\dot{\boldsymbol{\omega}} + \dot{h}_s \mathbf{a} \right], \tag{15.3}$$

since the inertia matrix $\mathbf{I}$ and the wheel spin axis $\mathbf{a}$ are constant in body coordinates. Substituting equations (15.3) and (15.1) into equation (15.2), we finally obtain the equations of motion in body coordinates

$$\mathbf{I}\dot{\boldsymbol{\omega}} + \dot{h}_s \mathbf{a} + \boldsymbol{\omega}^{\times} \left[\mathbf{I}\boldsymbol{\omega} + h_s \mathbf{a} \right] = \mathbf{T}. \tag{15.4}$$

Typically, for a dual spin system, the wheel spin angular velocity is kept constant, such that $\dot{h}_s = I_s \dot{\omega}_s = 0$. This leads to

$$\mathbf{I}\dot{\boldsymbol{\omega}} + \boldsymbol{\omega}^{\times} \left[\mathbf{I}\boldsymbol{\omega} + h_s \mathbf{a} \right] = \mathbf{T}. \tag{15.5}$$

15.2 Stability of Dual-Spin Torque-Free Motion

As we mentioned in the introduction, the purpose of adding a spinning wheel is to provide gyroscopic stability, and remove some of the drawbacks of spin-stabilizing the entire spacecraft. We therefore want to analyze the stability of a dual-spin spacecraft. We will make a direct comparison with the spinning spacecraft in Section 14.2. As in Section 14.2, let us consider the body frame $\mathcal{F}_b$ to be a principal axes frame, such that the inertia matrix is given by

$$\mathbf{I} = \begin{bmatrix} I_x & 0 & 0 \\ 0 & I_y & 0 \\ 0 & 0 & I_z \end{bmatrix}.$$

Since in Section 14.2 we consider a z-axis spin, we let the wheel spin axis coincide with the body z-axis, that is

$$\mathbf{a} = \begin{bmatrix} 0 \\ 0 \\ 1 \end{bmatrix}.$$

With the angular velocity given in body coordinates by

$$\boldsymbol{\omega} = \begin{bmatrix} \omega_x \\ \omega_y \\ \omega_z \end{bmatrix},$$

the torque-free ($\mathbf{T} = \mathbf{0}$) equations of motion in (15.5) become

$$\begin{aligned} I_x\dot{\omega}_x + \left(I_z - I_y\right)\omega_y\omega_z + h_s\omega_y &= 0, \\ I_y\dot{\omega}_y + \left(I_x - I_z\right)\omega_x\omega_z - h_s\omega_x &= 0, \\ I_z\dot{\omega}_z + \left(I_y - I_x\right)\omega_x\omega_y &= 0. \end{aligned} \tag{15.6}$$

As in Section 14.2, we are interested in the stability of equilibrium solutions of (15.6). There are a number of equilibrium solutions of (15.6). However, the most useful solution for comparison with Section 14.2 is a principal z-axis spin with

$$\omega_x(t) = \omega_y(t) = 0, \quad \omega_z(t) = \nu \text{ (constant)}.$$

Substituting this into the equations of motion (15.6), we have

$$\begin{aligned} I_x\dot{\omega}_x &= 0, \\ I_y\dot{\omega}_y &= 0, \\ I_z\dot{\omega}_z &= 0, \end{aligned}$$

which means that a principal z-axis spin is indeed an equilibrium condition (the angular velocities do not change).

As in Section 14.2, to examine the stability of this equilibrium condition, we make a small perturbation

$$\omega_x = \epsilon_x, \quad \omega_y = \epsilon_y, \quad \omega_z = \epsilon_z + \nu,$$

where ϵ_x, ϵ_y and ϵ_z are small angular velocities. The equations of motion now become

$$\begin{aligned} I_x\dot{\epsilon}_x + (I_z - I_y)\epsilon_y(\epsilon_z + \nu) + h_s\epsilon_y &= 0, \\ I_y\dot{\epsilon}_y + (I_x - I_z)\epsilon_x(\epsilon_z + \nu) - h_s\epsilon_x &= 0, \\ I_z\dot{\epsilon}_z + (I_y - I_x)\epsilon_x\epsilon_y &= 0. \end{aligned}$$

Since all ϵ are small, we neglect products of ϵ. That is, we linearize Euler's equations about the equilibrium to get

$$\begin{aligned} I_x\dot{\epsilon}_x + (I_z - I_y)\nu\epsilon_y + h_s\epsilon_y &= 0, \\ I_y\dot{\epsilon}_y + (I_x - I_z)\nu\epsilon_x - h_s\epsilon_x &= 0, \\ I_z\dot{\epsilon}_z &= 0. \end{aligned}$$

As in Section 14.2 for a spinning satellite, the equations for ϵ_x and ϵ_y are decoupled from the equation for ϵ_z. The equation for ϵ_z implies that

$$\epsilon_z = constant,$$

which means that ϵ_z is stable. Let us now perform a stability analysis on the equations for ϵ_x and ϵ_y, which we rewrite for convenience as

$$\begin{aligned} \dot{\epsilon}_x + \frac{(I_z - I_y)\nu + h_s}{I_x}\epsilon_y &= 0, \\ \dot{\epsilon}_y + \frac{(I_x - I_z)\nu - h_s}{I_y}\epsilon_x &= 0. \end{aligned}$$

For a spinning platform ($\nu \neq 0$), it will be convenient to define the quantity

$$\lambda = I_z + \frac{h_s}{\nu}. \tag{15.7}$$

In this case, we can rewrite the equations of motion as

$$\begin{aligned} \dot{\epsilon}_x + \frac{(\lambda - I_y)\nu}{I_x}\epsilon_y &= 0, \\ \dot{\epsilon}_y + \frac{(I_x - \lambda)\nu}{I_y}\epsilon_x &= 0. \end{aligned} \tag{15.8}$$

The related equations of motion from Section 14.2 for a spinning spacecraft are repeated here for convenience

$$\begin{aligned} \dot{\epsilon}_x + \frac{(I_z - I_y)\nu}{I_x}\epsilon_y &= 0, \\ \dot{\epsilon}_y + \frac{(I_x - I_z)\nu}{I_y}\epsilon_x &= 0. \end{aligned} \tag{15.9}$$

Comparing the equations of motion for the dual-spin satellite (15.8) with those of a spinning spacecraft (15.9), it can be seen that the stored wheel angular momentum h_s augments the platform moment of inertia about the spin axis I_z. That is, to complete the stability analysis, we can re-use the results from Section 14.2 by replacing

$$I_z \rightarrow \lambda = I_z + \frac{h_s}{\nu}.$$

Therefore, we can conclude that stability is only possible if either

$$I_z + \frac{h_s}{\nu} = \lambda > I_y \quad \text{and} \quad I_z + \frac{h_s}{\nu} = \lambda > I_x, \tag{15.10}$$

or

$$I_z + \frac{h_s}{\nu} = \lambda < I_y \quad \text{and} \quad I_z + \frac{h_s}{\nu} = \lambda < I_x. \tag{15.11}$$

In this case, even intermediate axis spins can also be made stable by making the wheel stored angular momentum h_s large enough such that either of the above pairs of inequalities are satisfied.

15.3 Effect of Internal Energy Dissipation

The preceding analysis was for a rigid body. However, there is no such thing as a perfectly rigid body. There is always some flexibility in the system, which means that there is always some internal energy dissipation. We would now like to see what effect internal energy dissipation has on the stability of a dual spin spacecraft, just as we did for a spin-stabilized spacecraft in Section 14.3. To be consistent with Section 15.2, we will consider the dual-spin spacecraft with the wheel spin axis aligned with the principal z-axis. In this case, from (15.1), the total angular momentum of the spacecraft is given by

$$\mathbf{h} = \begin{bmatrix} I_x\omega_x \\ I_y\omega_y \\ I_z\omega_z + h_s \end{bmatrix}. \tag{15.12}$$

It can readily be shown that the magnitude of the angular velocity $h = |\mathbf{h}|$ is constant for torque-free motion. Indeed, consider

$$\frac{d}{dt}\left(h^2\right) = 2\mathbf{h}^T\dot{\mathbf{h}}.$$

For torque-free motion, Euler's equation (15.2) gives us

$$\dot{\mathbf{h}} = -\boldsymbol{\omega}^{\times}\mathbf{h},$$

which leads to

$$\frac{d}{dt}\left(h^2\right) = -2\mathbf{h}^T\boldsymbol{\omega}^{\times}\mathbf{h} = 0,$$

since $\boldsymbol{\omega}^{\times}\mathbf{h}$ is perpendicular to $\mathbf{h}$. This shows that h^2 is constant, and therefore, so is h.

Now, for the energy sink analysis, we consider the energy-like function

$$T_o = \frac{1}{2}\left[I_x\omega_x^2 + I_y\omega_y^2 + I_z\omega_z^2\right]. \tag{15.13}$$

Note that this is not the total rotational kinetic energy of the dual spin spacecraft, rather it is a portion of it which is due to the motion of the platform. It can readily be shown that for a rigid dual-spin spacecraft, T_o is constant for torque-free motion. We assume that the momentum wheel is rigid. The energy sink hypothesis then postulates that for a quasi-rigid body, the "energy" T_o decreases to a minimum value, while the angular momentum h remains constant. That is, $\dot{T}_o < 0$ until the energy is a minimum. This minimum energy configuration is therefore stable.

Let us now find out under what conditions our simple spins in Section 15.2 are stable when energy dissipation is accounted for. To do this, we need to solve the constrained minimization problem

$$\text{Find } \boldsymbol{\omega} \text{ to minimize } T_o \text{ subject to } h = constant,$$

where

$$h^2 = \mathbf{h}^T\mathbf{h} = (I_x\omega_x)^2 + \left(I_y\omega_y\right)^2 + (I_z\omega_z + h_s)^2.$$

We shall now transform the problem into an unconstrained minimization problem through a suitable change of variables. To do this, we recognize that the angular momentum vector $\mathbf{h}$ must lie on the sphere of radius h. This means that we can readily describe the angular momentum vector in terms of spherical coordinates, as shown in Figure 15.2. We have,

$$\mathbf{h} = \begin{bmatrix} h\cos\phi\sin\theta \\ h\sin\phi \\ h\cos\phi\cos\theta \end{bmatrix}. \tag{15.14}$$

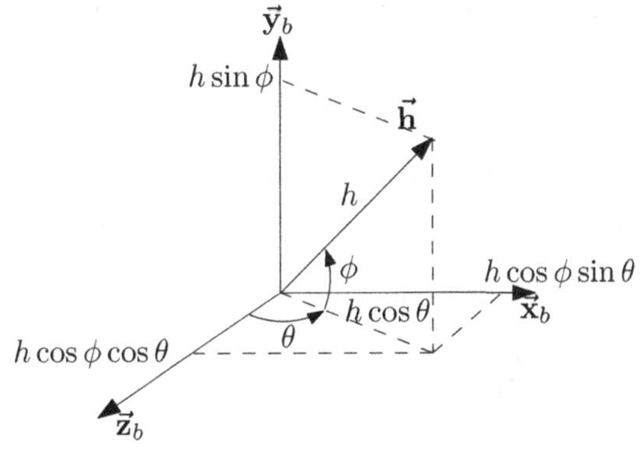

Figure 15.2 Angular momentum vector in spherical coordinates

Note that we have intentionally chosen the spherical coordinate system with origin along the body z-axis so that when $\phi = \theta = 0$, the angular momentum vector is given by

$$\mathbf{h} = \begin{bmatrix} 0 \\ 0 \\ h \end{bmatrix},$$

which corresponds to a simple spin about the z-axis as in Section 15.2. Now, equating the expressions for $\mathbf{h}$ in (15.12) and (15.14), we obtain

$$\begin{aligned} I_x\omega_x &= h\cos\phi\sin\theta, \\ I_y\omega_y &= h\sin\phi, \\ I_z\omega_z + h_s &= h\cos\phi\cos\theta, \end{aligned} \tag{15.15}$$

which give equations for the angular velocities

$$\begin{aligned} \omega_x &= \frac{h}{I_x}\cos\phi\sin\theta, \\ \omega_y &= \frac{h}{I_y}\sin\phi, \\ \omega_z &= \frac{h}{I_z}\cos\phi\cos\theta - \frac{h_s}{I_z}. \end{aligned} \tag{15.16}$$

Equations (15.16) now provide the required transformation from ω_x, ω_y and ω_z (which are constrained by the constant angular momentum h requirement), to the unconstrained spherical coordinates ϕ and θ. Since T_o is differentiable with respect to ϕ and θ, necessary conditions for a minimum to occur are that the partial derivatives are zero at that point,

$$\frac{\partial T_o}{\partial \phi} = 0, \quad \frac{\partial T_o}{\partial \theta} = 0.$$

This means that the function T_o is stationary at a minimum. By the chain-rule, we find that

$$\frac{\partial T_o}{\partial \phi} = \frac{\partial T_o}{\partial \omega_x}\frac{\partial \omega_x}{\partial \phi} + \frac{\partial T_o}{\partial \omega_y}\frac{\partial \omega_y}{\partial \phi} + \frac{\partial T_o}{\partial \omega_z}\frac{\partial \omega_z}{\partial \phi},$$
$$\frac{\partial T_o}{\partial \theta} = \frac{\partial T_o}{\partial \omega_x}\frac{\partial \omega_x}{\partial \theta} + \frac{\partial T_o}{\partial \omega_y}\frac{\partial \omega_y}{\partial \theta} + \frac{\partial T_o}{\partial \omega_z}\frac{\partial \omega_z}{\partial \theta}. \tag{15.17}$$

From equations (15.13) and (15.16), we find that

$$\frac{\partial T_o}{\partial \omega_x} = I_x\omega_x = h\cos\phi\sin\theta,$$
$$\frac{\partial T_o}{\partial \omega_y} = I_y\omega_y = h\sin\phi,$$
$$\frac{\partial T_o}{\partial \omega_z} = I_z\omega_z = h\cos\phi\cos\theta - h_s.$$

From equations (15.16), we obtain

$$\frac{\partial \omega_x}{\partial \phi} = -\frac{h}{I_x}\sin\phi\sin\theta, \quad \frac{\partial \omega_y}{\partial \phi} = \frac{h}{I_y}\cos\phi, \quad \frac{\partial \omega_z}{\partial \phi} = -\frac{h}{I_z}\sin\phi\cos\theta,$$
$$\frac{\partial \omega_x}{\partial \theta} = \frac{h}{I_x}\cos\phi\cos\theta, \quad \frac{\partial \omega_y}{\partial \theta} = 0, \quad \frac{\partial \omega_z}{\partial \theta} = -\frac{h}{I_z}\cos\phi\sin\theta.$$

Substituting the obtained partial derivatives into (15.17), we obtain the required partial derivatives,

$$\frac{\partial T_o}{\partial \phi} = -\frac{h^2}{I_x}\sin\phi\cos\phi\sin^2\theta + \frac{h^2}{I_y}\sin\phi\cos\phi - \frac{h^2}{I_z}\sin\phi\cos\phi\cos^2\theta + \frac{hh_s}{I_z}\sin\phi\cos\theta, \tag{15.18}$$

and

$$\frac{\partial T_o}{\partial \theta} = \frac{h^2}{I_x}\cos^2\phi\sin\theta\cos\theta - \frac{h^2}{I_z}\cos^2\phi\sin\theta\cos\theta + \frac{hh_s}{I_z}\cos\phi\sin\theta. \tag{15.19}$$

It will be useful to rearrange these as

$$\frac{\partial T_o}{\partial \phi} = h\sin\phi\left[-\frac{h}{I_x}\cos\phi\sin^2\theta + \frac{h}{I_y}\cos\phi - \frac{h}{I_z}\cos\phi\cos^2\theta + \frac{h_s}{I_z}\cos\theta\right], \tag{15.20}$$

and

$$\frac{\partial T_o}{\partial \theta} = h\sin\theta\left[\frac{h}{I_x}\cos^2\phi\cos\theta - \frac{h}{I_z}\cos^2\phi\cos\theta + \frac{h_s}{I_z}\cos\phi\right]. \tag{15.21}$$

We immediately see from (15.20) and (15.21) that T_o has stationary values when $\sin\theta = 0$ and $\sin\phi = 0$. This means that $\cos\phi = \pm 1$ and $\cos\theta = \pm 1$. From the equation for the angular momentum vector (15.14), we see that these stationary values correspond to simple spins about the body z-axis with angular momentum vector

$$\mathbf{h} = \begin{bmatrix} 0 \\ 0 \\ \pm h \end{bmatrix},$$

since the product $\cos\theta\cos\phi = \pm 1$. It is clear that we can cover both cases by considering $\phi = 0^\circ$ and $\theta = 0^\circ,\ 180^\circ$.

We have now established that simple spins about the principal z-axis could be minimum T_o configurations. Now we need to establish conditions that in fact make them minima. To establish this, we need to consider the second derivatives. Specifically, a sufficient condition for T_o to be a minimum at a point is if the Hessian matrix

$$\begin{bmatrix} \dfrac{\partial^2 T_o}{\partial\phi^2} & \dfrac{\partial^2 T_o}{\partial\phi\partial\theta} \\ \dfrac{\partial^2 T_o}{\partial\theta\partial\phi} & \dfrac{\partial^2 T_o}{\partial\theta^2} \end{bmatrix}$$

is positive definite at that point. This means that for simple spins about the principal z-axis, the Hessian matrices

$$\left.\begin{bmatrix} \dfrac{\partial^2 T_o}{\partial\phi^2} & \dfrac{\partial^2 T_o}{\partial\phi\partial\theta} \\ \dfrac{\partial^2 T_o}{\partial\theta\partial\phi} & \dfrac{\partial^2 T_o}{\partial\theta^2} \end{bmatrix}\right|_{\phi=0^\circ,\theta=0^\circ} \quad \text{or} \quad \left.\begin{bmatrix} \dfrac{\partial^2 T_o}{\partial\phi^2} & \dfrac{\partial^2 T_o}{\partial\phi\partial\theta} \\ \dfrac{\partial^2 T_o}{\partial\theta\partial\phi} & \dfrac{\partial^2 T_o}{\partial\theta^2} \end{bmatrix}\right|_{\phi=0^\circ,\theta=180^\circ},$$

must be positive definite.

Let us now obtain expressions for the second partial derivatives of T_o. It will be useful to rewrite the partial derivatives (15.18) and (15.19) in a different form by grouping some terms and making use of the trigonometric identity $2\sin a\cos a = \sin 2a$. We obtain

$$\frac{\partial T_o}{\partial\phi} = h\left[\frac{\sin 2\phi}{2}\left(\frac{h}{I_y} - \frac{h}{I_x}\sin^2\theta - \frac{h}{I_z}\cos^2\theta\right) + \frac{h_s}{I_z}\sin\phi\cos\theta\right], \tag{15.22}$$

and

$$\frac{\partial T_o}{\partial\theta} = h\left[\left(\frac{h}{I_x} - \frac{h}{I_z}\right)\frac{\cos^2\phi\sin 2\theta}{2} + \frac{h_s}{I_z}\cos\phi\sin\theta\right]. \tag{15.23}$$

From these, we can now readily obtain the second partial derivatives as

$$\frac{\partial^2 T_o}{\partial\phi^2} = h\left[\cos 2\phi\left(\frac{h}{I_y} - \frac{h}{I_x}\sin^2\theta - \frac{h}{I_z}\cos^2\theta\right) + \frac{h_s}{I_z}\cos\phi\cos\theta\right], \tag{15.24}$$

$$\frac{\partial^2 T_o}{\partial \theta^2} = h\left[\left(\frac{h}{I_x} - \frac{h}{I_z}\right)\cos^2\phi\cos 2\theta + \frac{h_s}{I_z}\cos\phi\cos\theta\right], \tag{15.25}$$

and

$$\frac{\partial^2 T_o}{\partial \theta \partial \phi} = \frac{\partial^2 T_o}{\partial \phi \partial \theta} = -h\left[\left(\frac{h}{I_x} - \frac{h}{I_z}\right)\sin\phi\cos\phi\sin 2\theta + \frac{h_s}{I_z}\sin\phi\sin\theta\right]. \tag{15.26}$$

With the expressions for the partial derivatives in hand, we can now establish conditions which would make a principal z-axis spin a minimum energy configuration, and hence a stable configuration in the presence of energy dissipation.

We first examine the case $\phi = \theta = 0^\circ$. We find that the second partial derivatives in equations (15.24) to (15.26) become

$$\left.\frac{\partial^2 T_o}{\partial \phi^2}\right|_{\phi=0^\circ,\theta=0^\circ} = h\left[\frac{h}{I_y} - \frac{h}{I_z} + \frac{h_s}{I_z}\right],$$

$$\left.\frac{\partial^2 T_o}{\partial \theta^2}\right|_{\phi=0^\circ,\theta=0^\circ} = h\left[\frac{h}{I_x} - \frac{h}{I_z} + \frac{h_s}{I_z}\right],$$

$$\left.\frac{\partial^2 T_o}{\partial \theta \partial \phi}\right|_{\phi=0^\circ,\theta=0^\circ} = 0.$$

Therefore, to guarantee that $\phi = 0^\circ, \theta = 0^\circ$ corresponds to a minimum, positive definiteness of the Hessian matrix reduces to the inequalities

$$h\left[\frac{h}{I_y} - \frac{h}{I_z} + \frac{h_s}{I_z}\right] > 0, \quad h\left[\frac{h}{I_x} - \frac{h}{I_z} + \frac{h_s}{I_z}\right] > 0.$$

In general, $h \geq 0$, but we consider only the case $h > 0$. For consistency with Section 15.2, we also restrict $\omega_z \neq 0$ (a spinning platform). From the third equation in (15.15), we find that for $\phi = 0^\circ, \theta = 0^\circ$,

$$h = I_z\omega_z + h_s = \lambda\omega_z, \tag{15.27}$$

where

$$\lambda = I_z + \frac{h_s}{\omega_z},$$

which is the same as the quantity λ defined in equation (15.7) in Section 15.2 (which played a critical role in the rigid body stability analysis). Now, since $h > 0$, the conditions for a minimum become

$$\frac{h}{I_y} - \frac{h - h_s}{I_z} > 0, \quad \frac{h}{I_x} - \frac{h - h_s}{I_z} > 0.$$

Now, from (15.27), we note that $h - h_s = I_z\omega_z$. Making this substitution, the conditions for a minimum become

$$h - I_y\omega_z > 0, \quad h - I_x\omega_z > 0.$$

Substituting $h = \lambda\omega_z$ (from (15.27)), the conditions become

$$\left(\lambda - I_y\right)\omega_z > 0, \quad \left(\lambda - I_x\right)\omega_z > 0.$$

Finally, we note from (15.27) that since $h > 0$, both ω_z and λ must have the same sign. Making use of this fact, the conditions for a minimum when $\phi = 0°, \theta = 0°$ become

$$\left(\lambda - I_y\right)\lambda > 0, \quad \left(\lambda - I_x\right)\lambda > 0.$$

Next, we consider the case when $\phi = 0°$ and $\theta = 180°$. In this case, the second partial derivatives in equations (15.24) to (15.26) become

$$\left.\frac{\partial^2 T_o}{\partial\phi^2}\right|_{\phi=0°,\theta=180°} = h\left[\frac{h}{I_y} - \frac{h}{I_z} - \frac{h_s}{I_z}\right],$$

$$\left.\frac{\partial^2 T_o}{\partial\theta^2}\right|_{\phi=0°,\theta=180°} = h\left[\frac{h}{I_x} - \frac{h}{I_z} - \frac{h_s}{I_z}\right],$$

$$\left.\frac{\partial^2 T_o}{\partial\theta\,\partial\phi}\right|_{\phi=0°,\theta=180°} = 0.$$

Therefore, to guarantee that $\phi = 0°, \theta = 180°$ corresponds to a minimum, positive definiteness of the Hessian matrix reduces to the inequalities

$$\frac{h}{I_y} - \frac{h}{I_z} - \frac{h_s}{I_z} > 0, \quad \frac{h}{I_x} - \frac{h}{I_z} - \frac{h_s}{I_z} > 0,$$

which rearrange to give

$$\frac{h}{I_y} - \frac{h + h_s}{I_z} > 0, \quad \frac{h}{I_x} - \frac{h + h_s}{I_z} > 0.$$

As before, from the third equation in (15.15), we find that for $\phi = 0°, \theta = 180°$,

$$-h = I_z\omega_z + h_s = \lambda\omega_z, \tag{15.28}$$

Making use of (15.28) as before, we find that $h + h_s = -I_z\omega_z$, and the conditions for a minimum become

$$-\left(\lambda - I_y\right)\omega_z > 0, \quad -\left(\lambda - I_x\right)\omega_z > 0.$$

Finally, we note that since $h > 0$ and $h = -\lambda\omega_z$, it must be that λ and ω_z have the opposite sign. Therefore, the conditions for a minimum when $\phi = 0°, \theta = 180°$ become

$$(\lambda - I_y)\,\lambda > 0, \quad (\lambda - I_x)\lambda > 0,$$

which are exactly the same as for the case $\phi = 0°, \theta = 0°$. Therefore, we conclude that a spin about the principal z-axis is a minimum energy configuration, and therefore stable in the presence of energy dissipation, if the quantity $\lambda = I_z + \frac{h_s}{\omega_z}$ satisfies the inequalities

$$(\lambda - I_y)\,\lambda > 0, \quad (\lambda - I_x)\lambda > 0,$$

It can be seen that these conditions are satisfied only if either

$$\lambda > I_x \quad \text{and} \quad \lambda > I_y, \tag{15.29}$$

or

$$\lambda < 0. \tag{15.30}$$

We can now compare these conditions for stability with those we obtained in Section 15.2. It can be seen that condition (15.29) is exactly the same as the condition (15.10). The condition (15.30) is consistent with condition (15.11), but is more restrictive.

As in Section 15.2, we can conclude that since h_s can be adjusted to make λ as large as we wish, it is possible to stabilize even minor or intermediate axis spins for a dual-spin spacecraft. In fact, this analysis shows that some energy dissipation is advantageous, since the energy decay means that

$$\boldsymbol{\omega} \to \begin{bmatrix} 0 \\ 0 \\ \omega_z \end{bmatrix}.$$

Notes

This chapter has presented the elements of dual-spin stabilization for spacecraft. In our energy-sink analysis, we have assumed that all energy is dissipated in the spacecraft platform, and none in the wheel (we assume that the wheel is perfectly rigid). For a more detailed treatment of dual spin stability where this assumption is not made, the reader is referred to Hughes (2004). For a discussion of practical issues associated with dual spin stabilization, the reader is referred to Sidi (1997).

References

Hughes PC 2004 *Spacecraft Attitude Dynamics*. Dover Publications, Mineola, NY.

Sidi MJ 1997 *Spacecraft Dynamics and Control: A Practical Engineering Approach*. Cambridge University Press, New York, NY.

16

Gravity-Gradient Stabilization

There are a number of satellite applications that require the pointing of an instrument toward the Earth. For these types of spacecraft, it is useful to have one of the spacecraft body axes pointing down (called the nadir direction). Let us define an orbiting reference frame $\mathcal{F}_o$ with basis vectors $\vec{\mathbf{x}}_o$, $\vec{\mathbf{y}}_o$ and $\vec{\mathbf{z}}_o$ as follows.

$$\vec{\mathbf{x}}_o = \vec{\mathbf{y}}_o \times \vec{\mathbf{z}}_o, \quad \vec{\mathbf{y}}_o = -\frac{\vec{\mathbf{r}} \times \vec{\mathbf{v}}}{|\vec{\mathbf{r}} \times \vec{\mathbf{v}}|}, \quad \vec{\mathbf{z}}_o = -\frac{\vec{\mathbf{r}}}{|\vec{\mathbf{r}}|},$$

where $\vec{\mathbf{r}}$ and $\vec{\mathbf{v}}$ are the inertial spacecraft position and velocity relative to the center of the Earth respectively. Figure 16.1 illustrates their definitions. In spacecraft dynamics, the three basis vectors defining the orbiting frame $\mathcal{F}_o$ are given special names.

$$\begin{aligned} \vec{\mathbf{x}}_o &= \text{roll axis} \\ \vec{\mathbf{y}}_o &= \text{pitch axis} \\ \vec{\mathbf{z}}_o &= \text{yaw axis} \end{aligned}$$

Associated with these axes are three rotations: roll rotation, pitch rotation, and yaw rotation respectively.

It should be clear that if the spacecraft body frame $\mathcal{F}_b$ is aligned with the orbiting frame $\mathcal{F}_o$, then the body z-axis points directly to nadir as desired for an Earth pointing satellite. The orbiting frame $\mathcal{F}_o$ defines the desired attitude for the spacecraft. It is therefore of interest to study the attitude motion of the spacecraft relative to $\mathcal{F}_o$.

Spacecraft Dynamics and Control: An Introduction, First Edition.
Anton H.J. de Ruiter, Christopher J. Damaren and James R. Forbes.

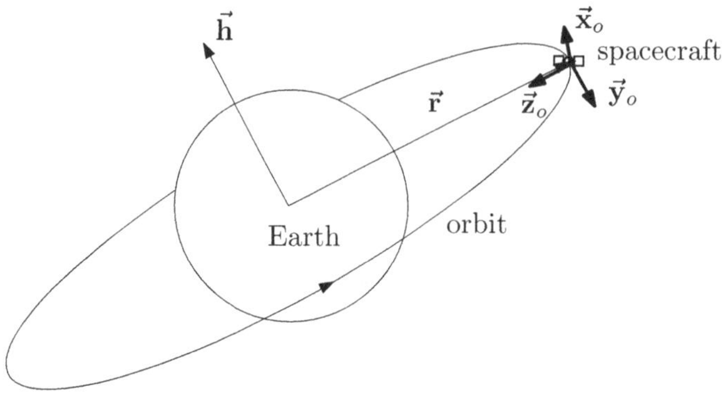

Figure 16.1 Orbiting reference frame definition

16.1 Equations of Motion

Let us describe the spacecraft attitude relative to the orbiting frame by a 3-2-1 attitude sequence (see Sections 1.3.3 and 1.4.2). Namely,

1. A rotation ψ about the original z-axis (called the yaw angle).
2. A rotation θ about the intermediate y-axis (called a pitch angle).
3. A rotation ϕ about the transformed x-axis (called a roll angle).

In this case, the rotation matrix from frame $\mathcal{F}_o$ to frame $\mathcal{F}_b$ is given by

$$\begin{aligned}\mathbf{C}_{bo}(\phi, \theta, \psi) &= \mathbf{C}_x(\phi)\mathbf{C}_y(\theta)\mathbf{C}_z(\psi) \\ &= \begin{bmatrix} c_\theta c_\psi & c_\theta s_\psi & -s_\theta \\ s_\phi s_\theta c_\psi - c_\phi s_\psi & s_\phi s_\theta s_\psi + c_\phi c_\psi & s_\phi c_\theta \\ c_\phi s_\theta c_\psi + s_\phi s_\psi & c_\phi s_\theta s_\psi - s_\phi c_\psi & c_\phi c_\theta \end{bmatrix}\end{aligned} \tag{16.1}$$

where $s_b = \sin b$ and $c_b = \cos b$.

The attitude kinematics relative to the orbiting frame are then given by (see Section 1.4.2)

$$\boldsymbol{\omega}_{bo} = \begin{bmatrix} 1 & 0 & -\sin\theta \\ 0 & \cos\phi & \sin\phi\cos\theta \\ 0 & -\sin\phi & \cos\phi\cos\theta \end{bmatrix} \begin{bmatrix} \dot{\phi} \\ \dot{\theta} \\ \dot{\psi} \end{bmatrix}, \tag{16.2}$$

where $\vec{\boldsymbol{\omega}}_{bo} = \vec{\mathcal{F}}_b^T \boldsymbol{\omega}_{bo}$ is the angular velocity of the body frame $\mathcal{F}_b$ relative to the orbiting frame $\mathcal{F}_o$ expressed in body coordinates.

The attitude dynamics of the spacecraft are given by Euler's equations (see Section 11.3), which describe the dynamics of the spacecraft body frame $\mathcal{F}_b$ relative to an inertial frame $\mathcal{F}_I$. We therefore need to know the inertial angular velocity of $\mathcal{F}_o$, namely $\vec{\boldsymbol{\omega}}_{oI}$. To simplify matters, we shall restrict ourselves to the case of a circular Keplerian orbit. In this case, $\vec{\mathbf{y}}_o$ is inertially fixed since it is parallel to the orbital angular momentum vector, and the orbiting frame $\mathcal{F}_o$ rotates once per orbit about $\vec{\mathbf{y}}_o$ with constant angular velocity given by the orbital rate $\omega_o = \sqrt{\frac{\mu}{r^3}}$ (see Chapter 3). Note that μ is the gravitational constant of the Earth, and $r = |\vec{\mathbf{r}}|$. The angular velocity of $\mathcal{F}_o$ relative to $\mathcal{F}_I$ is therefore given by

$$\vec{\boldsymbol{\omega}}_{oI} = \vec{\mathcal{F}}_o^T \begin{bmatrix} 0 \\ -\omega_o \\ 0 \end{bmatrix}. \tag{16.3}$$

The negative sign is due to the fact that the direction of $\vec{\mathbf{y}}_o$ is the negative of the direction of the orbital angular momentum.

The inertial angular velocity of the spacecraft is now given by

$$\vec{\boldsymbol{\omega}}_{bI} = \vec{\boldsymbol{\omega}}_{bo} + \vec{\boldsymbol{\omega}}_{oI}. \tag{16.4}$$

Including the gravity-gradient torque (see Section 12.4), the attitude dynamics of the spacecraft are given in body coordinates by

$$\mathbf{I}\frac{d}{dt}(\boldsymbol{\omega}_{bo} + \mathbf{C}_{bo}\boldsymbol{\omega}_{oI}) + (\boldsymbol{\omega}_{bo} + \mathbf{C}_{bo}\boldsymbol{\omega}_{oI})^{\times}\,\mathbf{I}\,(\boldsymbol{\omega}_{bo} + \mathbf{C}_{bo}\boldsymbol{\omega}_{oI}) = \mathbf{T}_g, \tag{16.5}$$

where

$$\mathbf{T}_g = \frac{3\mu}{r^5}\mathbf{r}_b^{\times}\mathbf{I}\mathbf{r}_b,$$

and $\vec{\mathbf{r}} = \vec{\mathcal{F}}_b\mathbf{r}_b$ is the spacecraft orbital position vector expressed in body coordinates. Equations (16.2) and (16.5) completely describe the attitude motion of the spacecraft relative to the orbiting frame $\mathcal{F}_o$. To make matters simpler, which will allow us to perform a stability analysis, we shall linearize the equations of motion. That is, we consider only small angles and angular rates, i.e. ϕ, θ, ψ, $\dot{\phi}$, $\dot{\theta}$ and $\dot{\psi}$ are considered small. This allows for the following simplifications

$$\sin\phi \approx \phi, \quad \cos\phi \approx 1, \quad \sin\theta \approx \phi, \quad \cos\theta \approx 1, \quad \sin\psi \approx \phi, \quad \cos\psi \approx 1.$$

Note that these approximations are only valid when the angles are in radians. The small angle and rate assumption also allow us to neglect products between angles and angular rates (for example, we can neglect $\phi\theta$ and $\phi\dot{\phi}$). We shall now simplify Euler's equation (16.5) term by term.

First of all, under these simplifications, the rotation matrix $\mathbf{C}_{bo}$ in (16.1) becomes

$$\mathbf{C}_{bo}(\phi, \theta, \psi) \approx \begin{bmatrix} 1 & \psi & -\theta \\ -\psi & 1 & \phi \\ \theta & -\phi & 1 \end{bmatrix}. \tag{16.6}$$

Next, the angular velocity in (16.2) becomes

$$\boldsymbol{\omega}_{bo} = \begin{bmatrix} \dot{\phi} \\ \dot{\theta} \\ \dot{\psi} \end{bmatrix} \tag{16.7}$$

Therefore, we obtain the inertial angular velocity in body coordinates as

$$\boldsymbol{\omega}_{bo} + \mathbf{C}_{bo}\boldsymbol{\omega}_{oI} = \begin{bmatrix} \dot{\phi} - \psi\omega_o \\ \dot{\theta} - \omega_o \\ \dot{\psi} + \phi\omega_o \end{bmatrix}, \tag{16.8}$$

and the inertial angular acceleration in body coordinates as

$$\frac{d}{dt}\left(\boldsymbol{\omega}_{bo} + \mathbf{C}_{bo}\boldsymbol{\omega}_{oI}\right) = \begin{bmatrix} \ddot{\phi} - \dot{\psi}\omega_o \\ \ddot{\theta} \\ \ddot{\psi} + \dot{\phi}\omega_o \end{bmatrix}, \tag{16.9}$$

since ω_o is constant.

We assume that the body frame $\mathcal{F}_b$ is a principal axes frame, such that the inertia matrix is

$$\mathbf{I} = \begin{bmatrix} I_x & 0 & 0 \\ 0 & I_y & 0 \\ 0 & 0 & I_z \end{bmatrix},$$

which means that

$$\mathbf{I}\frac{d}{dt}\left(\boldsymbol{\omega}_{bo} + \mathbf{C}_{bo}\boldsymbol{\omega}_{oI}\right) = \begin{bmatrix} I_x\left(\ddot{\phi} - \dot{\psi}\omega_o\right) \\ I_y\ddot{\theta} \\ I_z\left(\ddot{\psi} + \dot{\phi}\omega_o\right) \end{bmatrix}. \tag{16.10}$$

We now deal with the second term on the left hand side of Euler's equation (16.5). With the above simplifications, we have

$$\begin{aligned}
&\left(\boldsymbol{\omega}_{bo} + \mathbf{C}_{bo}\boldsymbol{\omega}_{oI}\right)^{\times}\mathbf{I}\left(\boldsymbol{\omega}_{bo} + \mathbf{C}_{bo}\boldsymbol{\omega}_{oI}\right) \\
&= \begin{bmatrix} 0 & -\left(\dot{\psi} + \phi\omega_o\right) & \dot{\theta} - \omega_o \\ \dot{\psi} + \phi\omega_o & 0 & -\left(\dot{\phi} - \psi\omega_o\right) \\ -\left(\dot{\theta} - \omega_o\right) & \dot{\phi} - \psi\omega_o & 0 \end{bmatrix} \begin{bmatrix} I_x & 0 & 0 \\ 0 & I_y & 0 \\ 0 & 0 & I_z \end{bmatrix} \begin{bmatrix} \dot{\phi} - \psi\omega_o \\ \dot{\theta} - \omega_o \\ \dot{\psi} + \phi\omega_o \end{bmatrix} \\
&= \begin{bmatrix} \left(I_z - I_y\right)\left(\dot{\psi} + \phi\omega_o\right)\left(\dot{\theta} - \omega_o\right) \\ \left(I_x - I_z\right)\left(\dot{\psi} + \phi\omega_o\right)\left(\dot{\phi} - \psi\omega_o\right) \\ \left(I_y - I_x\right)\left(\dot{\theta} - \omega_o\right)\left(\dot{\phi} - \psi\omega_o\right) \end{bmatrix}.
\end{aligned}$$

Neglecting all products of angles and angular rates, this simplifies to

$$(\boldsymbol{\omega}_{bo} + \mathbf{C}_{bo}\boldsymbol{\omega}_{oI})^{\times} \mathbf{I} (\boldsymbol{\omega}_{bo} + \mathbf{C}_{bo}\boldsymbol{\omega}_{oI}) = \begin{bmatrix} (I_y - I_z)\left[\omega_o^2\phi + \omega_o\dot{\psi}\right] \\ 0 \\ (I_y - I_x)\left[\omega_o^2\psi - \omega_o\dot{\phi}\right] \end{bmatrix}. \tag{16.11}$$

The only remaining term in the equation of motion (16.5) is the gravity-gradient term. To evaluate this, we first recognize that by the definition of the orbiting frame $\mathcal{F}_o$, the spacecraft position vector is given by

$$\vec{\mathbf{r}} = \vec{\mathcal{F}}_o^T \begin{bmatrix} 0 \\ 0 \\ -r \end{bmatrix}. \tag{16.12}$$

Therefore, in body coordinates, the spacecraft position vector becomes

$$\mathbf{r}_b = \mathbf{C}_{bo} \begin{bmatrix} 0 \\ 0 \\ -r \end{bmatrix} \approx \begin{bmatrix} 1 & \psi & -\theta \\ -\psi & 1 & \phi \\ \theta & -\phi & 1 \end{bmatrix} \begin{bmatrix} 0 \\ 0 \\ -r \end{bmatrix} = \begin{bmatrix} r\theta \\ -r\phi \\ -r \end{bmatrix}.$$

Making use of this approximation, we evaluate the gravity-gradient torque to be approximately

$$\mathbf{T}_g = \frac{3\mu}{r^5}\mathbf{r}_b^{\times}\mathbf{I}\mathbf{r}_b = \frac{3\mu}{r^5} \begin{bmatrix} 0 & r & -r\phi \\ -r & 0 & -r\theta \\ r\phi & r\theta & 0 \end{bmatrix} \begin{bmatrix} I_x & 0 & 0 \\ 0 & I_y & 0 \\ 0 & 0 & I_z \end{bmatrix} \begin{bmatrix} r\theta \\ -r\phi \\ -r \end{bmatrix},$$

which becomes

$$\mathbf{T}_g = \frac{3\mu}{r^5} \begin{bmatrix} (I_z - I_y)\, r^2\phi \\ (I_z - I_x) r^2\theta \\ (I_x - I_y)\, r^2\phi\theta \end{bmatrix}.$$

Neglecting the product $\phi\theta$, and recognizing the orbital angular rate $\omega_o = \sqrt{\frac{\mu}{r^3}}$, the gravity-gradient torque finally becomes

$$\mathbf{T}_g = 3\omega_o^2 \begin{bmatrix} (I_z - I_y)\,\phi \\ (I_z - I_x)\theta \\ 0 \end{bmatrix}. \tag{16.13}$$

Note that there is no torque about the yaw axis. This makes sense, since we are considering small angles, which means that the spacecraft body-frame is close to the orbiting frame. The yaw axis is the line of action of the gravitational force, and the moment vector generated by any force vector is always perpendicular to the direction of that force. Also note that the torque about the roll axis is proportional to the roll angle, and the torque about the pitch axis is proportional to the pitch angle.

We now have all the approximations required to simplify the equations of motion (16.5). Substituting the approximations (16.10), (16.11) and (16.13) into (16.5) gives

$$\begin{bmatrix} I_x\left(\ddot{\phi}-\dot{\psi}\omega_o\right)+\left(I_y-I_z\right)\left[\omega_o^2\phi+\omega_o\dot{\psi}\right] \\ I_y\ddot{\theta} \\ I_z\left(\ddot{\psi}+\dot{\phi}\omega_o\right)+\left(I_y-I_x\right)\left[\omega_o^2\psi-\omega_o\dot{\phi}\right] \end{bmatrix} = \begin{bmatrix} 3\omega_o^2\left(I_z-I_y\right)\phi \\ 3\omega_o^2(I_z-I_x)\theta \\ 0 \end{bmatrix},$$

which can be rearranged to finally give

$$I_x\ddot{\phi}-\left[I_x-I_y+I_z\right]\omega_o\dot{\psi}+4\omega_o^2\left(I_y-I_z\right)\phi=0, \tag{16.14}$$

$$I_y\ddot{\theta}+3\omega_o^2(I_x-I_z)\theta=0 \tag{16.15}$$

$$I_z\ddot{\psi}+\left[I_x-I_y+I_z\right]\omega_o\dot{\phi}+\omega_o^2\left(I_y-I_x\right)\psi=0. \tag{16.16}$$

It is important to note that the above equations describing the attitude motion of the spacecraft body relative to the orbiting frame are approximations assuming small angles and small angular rates.

From equations (16.14), (16.15) and (16.16), it can be seen that the roll and yaw motions of the spacecraft are coupled, but the pitch motion is independent.

16.2 Stability Analysis

In this section we analyze the stability of the attitude motion described by the approximate motion equations (16.14), (16.15) and (16.16).

16.2.1 Pitch Motion

The pitch equation (16.15) can be rewritten as

$$\ddot{\theta}+\alpha\theta=0,$$

where

$$\alpha=3\omega_o^2\frac{(I_x-I_z)}{I_y}.$$

Taking the Laplace transform gives

$$s^2\hat{\theta}(s)-s\theta(0)-\dot{\theta}(0)+\alpha\hat{\theta}(s)=0,$$

which can be rearranged for $\hat{\theta}(s)$

$$\hat{\theta}(s) = \frac{s\theta(0) + \dot{\theta}(0)}{s^2 + \alpha}.$$

It is clear that the poles of $\hat{\theta}(s)$ satisfy

$$s^2 + \alpha = 0.$$

That is, the poles are $\pm\sqrt{-\alpha}$. There are now three possible cases.

1. $\alpha > 0$
   In this case, the poles are $\pm j\sqrt{\alpha}$, which are purely imaginary. This means that the solution for $\theta(t)$ is oscillatory. Therefore, the motion is stable.
2. $\alpha = 0$
   In this case, there is a repeated pole at the origin. This means that $\theta(t)$ has a component that grows linearly with time. Therefore, the motion is unstable.
3. $\alpha < 0$
   In this case, the poles are $\pm\sqrt{-\alpha}$, which are real. One of them is positive, the other is negative. This means that $\theta(t)$ has a component that grows exponentially with time. Therefore, the motion is unstable.

Therefore, we conclude that for the pitch motion to be stable, we must have $\alpha > 0$, which means that

$$I_x > I_z. \tag{16.17}$$

Equation (16.15) shows that the pitch motion behaves like an undamped mass-spring system, with the gravity-gradient torque acting as the spring. The requirement for stability (16.17) means that the effective spring constant of the gravity-gradient torque is positive. That is, it provides a restoring torque.

16.2.2 Roll-Yaw Motion

Let us define the quantities

$$A = \left[I_x - I_y + I_z\right], \quad B = 4\left(I_y - I_z\right), \quad C = \left(I_y - I_x\right). \tag{16.18}$$

Then, we can rewrite the roll and yaw equations (16.14) and (16.16) as

$$\ddot{\phi} - \omega_o \frac{A}{I_x}\dot{\psi} + \omega_o^2 \frac{B}{I_x}\phi = 0, \tag{16.19}$$

$$\ddot{\psi} + \omega_o \frac{A}{I_z}\dot{\phi} + \omega_o^2 \frac{C}{I_z}\psi = 0. \tag{16.20}$$

Let us now take Laplace transforms to get

$$s^2\hat{\phi} - s\omega_o\frac{A}{I_x}\hat{\psi} + \omega_o^2\frac{B}{I_x}\hat{\phi} = s\phi(0) + \dot{\phi}(0) - \omega_o\frac{A}{I_x}\psi(0),$$
$$s^2\hat{\psi} + s\omega_o\frac{A}{I_z}\hat{\phi} + \omega_o^2\frac{C}{I_z}\hat{\psi} = s\psi(0) + \dot{\psi}(0) + \omega_o\frac{A}{I_z}\phi(0).$$

We can write this in matrix form as

$$\begin{bmatrix} s^2 + \omega_o^2\frac{B}{I_x} & -s\omega_o\frac{A}{I_x} \\ s\omega_o\frac{A}{I_z} & s^2 + \omega_o^2\frac{C}{I_z} \end{bmatrix}\begin{bmatrix} \hat{\phi} \\ \hat{\psi} \end{bmatrix} = \begin{bmatrix} s\phi(0) + \dot{\phi}(0) - \omega_o\frac{A}{I_x}\psi(0) \\ s\psi(0) + \dot{\psi}(0) + \omega_o\frac{A}{I_z}\phi(0) \end{bmatrix},$$

which can be inverted to get

$$\begin{bmatrix} \hat{\phi} \\ \hat{\psi} \end{bmatrix} = \frac{1}{\left(s^2 + \omega_o^2\frac{B}{I_x}\right)\left(s^2 + \omega_o^2\frac{C}{I_z}\right) + s^2\omega_o^2\frac{A^2}{I_xI_z}}\begin{bmatrix} s^2 + \omega_o^2\dfrac{C}{I_z} & s\omega_o\dfrac{A}{I_x} \\ -s\omega_o\dfrac{A}{I_z} & s^2 + \omega_o^2\dfrac{B}{I_x} \end{bmatrix}$$
$$\times \begin{bmatrix} s\phi(0) + \dot{\phi}(0) - \omega_o\frac{A}{I_x}\psi(0) \\ s\psi(0) + \dot{\psi}(0) + \omega_o\frac{A}{I_z}\phi(0) \end{bmatrix}.$$

From this, it can be seen that the poles of both $\hat{\phi}(s)$ and $\hat{\psi}(s)$ satisfy the characteristic equation

$$\left(s^2 + \omega_o^2\frac{B}{I_x}\right)\left(s^2 + \omega_o^2\frac{C}{I_z}\right) + s^2\omega_o^2\frac{A^2}{I_xI_z} = 0,$$

which can be expanded as

$$s^4 + \left(\frac{A^2}{I_xI_z} + \frac{B}{I_x} + \frac{C}{I_z}\right)\omega_o^2 s^2 + \frac{BC}{I_xI_z}\omega_o^4 = 0. \tag{16.21}$$

Let us define the quantities

$$p = \frac{A^2}{I_xI_z} + \frac{B}{I_x} + \frac{C}{I_z}, \quad q = \frac{BC}{I_xI_z}. \tag{16.22}$$

Let us also define

$$\lambda = \frac{s^2}{\omega_o^2}.$$

Then, the characteristic equation (16.21) becomes a quadratic in λ

$$\lambda^2 + p\lambda + q = 0. \tag{16.23}$$

There are two solutions λ of (16.23), given by

$$\lambda = \frac{-p \pm \sqrt{p^2 - 4q}}{2}. \tag{16.24}$$

Corresponding to each solution λ are two poles of (16.21) given by

$$s = \pm\omega_o\sqrt{\lambda}.$$

Since there are two solutions to (16.23) as shown in (16.24), say λ_1 and λ_2, there are four poles for (16.21), namely $s = \pm\omega_o\sqrt{\lambda_1},\ \pm\omega_o\sqrt{\lambda_2}$.

There are now four possibilities for each solution λ of (16.23).

1. $\lambda < 0$
   In this case, the corresponding poles are purely imaginary and are given by $s = \pm j\omega_o\sqrt{-\lambda}$. This leads to oscillatory behavior of $\phi(t)$ and $\psi(t)$.
2. $\lambda = 0$
   In this case, there is a pair of poles at $s = 0$. This leads to unbounded linear growth of $\phi(t)$ and $\psi(t)$, and hence the motion is unstable.
3. $\lambda > 0$
   In this case, the corresponding poles are given by $s = \pm\omega_o\sqrt{\lambda}$, one of which is positive. This leads to unbounded exponential growth of $\phi(t)$ and $\psi(t)$, and hence the motion is unstable.
4. $\lambda = a + jb$ with $b \neq 0$
   That is, λ has an imaginary component. It turns out that $\sqrt{\lambda}$ must lie in the right-half complex plane (it must have positive real part). To see why, let us represent λ in polar form $\lambda = re^{j\theta}$, where $r > 0$, and $-\pi < \theta < \pi$. Then, we have $\sqrt{\lambda} = \sqrt{r}e^{j\frac{\theta}{2}}$. Since $-\frac{\pi}{2} < \frac{\theta}{2} < \frac{\pi}{2}$, it must be that $\sqrt{\lambda}$ has a positive real part, as shown in Figure 16.2.

   This means that one of the pair of poles $s = \pm\omega_o\sqrt{\lambda}$ is an unstable pole and leads to unbounded exponential growth of $\phi(t)$ and $\psi(t)$.

From the above four cases, it can be seen that we must have $\lambda < 0$ for the motion to be stable, and that we can at best have bounded oscillatory motion. However, there is an

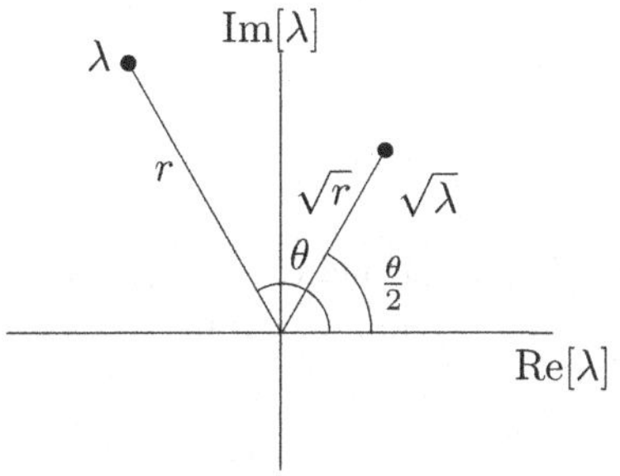

Figure 16.2 Square-root of a complex number

additional restriction. If the two solutions of λ_1 and λ_2 for (16.23) are negative and the same, that is, $\lambda_1 = \lambda_2 < 0$ then the result will be a pair of purely imaginary repeated poles $s = \pm j\omega_o\sqrt{-\lambda_1}$, $\pm j\omega_o\sqrt{-\lambda_1}$. This leads to oscillatory behavior with unbounded linear growth of $\phi(t)$ and $\psi(t)$, which is unstable.

Therefore, we conclude that for stability, the solutions λ_1 and λ_2 of (16.23) must be negative, and distinct. That is, $\lambda_1 < 0$, $\lambda_2 < 0$, $\lambda_1 \neq \lambda_2$. Turning to the expression for the solution (16.24), if $p = 0$, then

$$\lambda = \pm\sqrt{-q}.$$

Then, if $q = 0$, $\lambda = 0, 0$. If $q < 0$, then the solution $\lambda = \sqrt{-q}$ is positive. If $q > 0$, then $\lambda = \pm j\sqrt{q}$ is imaginary. None of these cases are allowed, therefore we must have $p \neq 0$.

Next, we see that to ensure distinct real solutions for λ, we must have

$$p^2 - 4q > 0. \tag{16.25}$$

Given (16.25), if $p < 0$, then the solution $\lambda = \frac{-p+\sqrt{p^2-4q}}{2}$ will be positive. Therefore, we must have

$$p > 0. \tag{16.26}$$

Finally, if $q \leq 0$, then $\sqrt{p^2 - 4q} \geq p$, which means that the solution $\lambda = \frac{-p+\sqrt{p^2-4q}}{2}$ will either be zero or positive. Therefore, we must have

$$q > 0. \tag{16.27}$$

To summarize, for stability, we require the following three conditions:

$$\begin{aligned} p &> 0, \\ q &> 0, \\ p^2\text{-}4q &> 0. \end{aligned}$$

Now, the condition $p > 0$ means by the definition of p in (16.22) that we must have

$$\frac{A^2}{I_x I_z} + \frac{B}{I_x} + \frac{C}{I_z} > 0.$$

This condition is readily satisfied if $B > 0$ and $C > 0$, since the quantity $A^2 \geq 0$.

The condition $q > 0$ means by the definition of q in (16.22) that we must have

$$\frac{BC}{I_x I_z} > 0.$$

Again, this is readily satisfied if $B > 0$ and $C > 0$.

Finally, let us examine the condition $p^2 - 4q > 0$. From the definitions of p and q in (16.22), we have

$$
\begin{aligned}
p^2 - 4q &= \left(\frac{A^2}{I_x I_z} + \frac{B}{I_x} + \frac{C}{I_z}\right)^2 - 4\frac{BC}{I_x I_z}, \\
&= \left(\frac{A^2}{I_x I_z}\right)^2 + 2\frac{A^2}{I_x I_z}\left(\frac{B}{I_x} + \frac{C}{I_z}\right) + \left(\frac{B}{I_x} + \frac{C}{I_z}\right)^2 - 4\frac{BC}{I_x I_z}, \\
&= \left(\frac{A^2}{I_x I_z}\right)^2 + 2\frac{A^2}{I_x I_z}\left(\frac{B}{I_x} + \frac{C}{I_z}\right) + \left(\frac{B}{I_x}\right)^2 + 2\frac{BC}{I_x I_z} + \left(\frac{C}{I_z}\right)^2 - 4\frac{BC}{I_x I_z}, \\
&= \left(\frac{A^2}{I_x I_z}\right)^2 + 2\frac{A^2}{I_x I_z}\left(\frac{B}{I_x} + \frac{C}{I_z}\right) + \left(\frac{B}{I_x} - \frac{C}{I_z}\right)^2.
\end{aligned}
$$

Now, it can readily be shown from the definition of the moments of inertia, that for any three-dimensional body, $A = I_x - I_y + I_z > 0$ (try this as an exercise). Therefore, if $B > 0$ and $C > 0$, we see from the above expression that the condition $p^2 - 4q > 0$ is automatically satisfied. Finally, making use of the definitions of B and C in (16.18), the conditions $B > 0$ and $C > 0$ translate into $I_y > I_z$ and $I_y > I_x$.

To conclude, we have seen that if

$$
I_y > I_z, \quad I_y > I_x, \tag{16.28}
$$

then the roll-yaw motion is stable in the presence of the gravity-gradient torque.

16.2.3 Combined Pitch and Roll/Yaw

Combining the roll/yaw stability condition in (16.28) with the pitch stability requirement in (16.17), we obtain the following result.

The spacecraft attitude motion relative to the orbiting frame $\mathcal{F}_o$ is stable for small attitude deviations from $\mathcal{F}_o$ if $\mathcal{F}_b$ is a principal axes frame, and the principal inertias satisfy

$$
I_y > I_x > I_z. \tag{16.29}
$$

Notes

This chapter has presented the elements of gravity-gradient stabilization for spacecraft. For an in-depth treatment of gravity-gradient stabilization, the reader is referred to Hughes (2004). For a treatment of gravity-gradient stabilization from a practical point of view, the reader is referred to Sidi (1997).

References

Hughes PC 2004 *Spacecraft Attitude Dynamics*. Dover Publications, Mineola, NY.

Sidi MJ 1997 *Spacecraft Dynamics and Control: A Practical Engineering Approach*. Cambridge University Press, New York, NY.

17

Active Spacecraft Attitude Control

In Chapters 13 to 16, we have examined passive means of stabilization. That is, making use of the natural spacecraft dynamics to obtain stability. Unfortunately, the attitude accuracy that can be obtained by this method is not very high, and the disturbance torques on the spacecraft can cause it to deteriorate over time. For example, flight experience with the Radio Astronomy Explorer (RAE) satellite, which is gravity-gradient stabilized, found that the spacecraft pitch was able to stay within ± 20 degrees. For certain types of missions (such as RAE), this kind of accuracy is acceptable. However, in many other applications (for example space astronomy missions) the attitude accuracy requirements are very stringent. For example, the MOST microsatellite (a Canadian space telescope) is required to be able to maintain a given attitude to within 25 arcseconds, so that it can observe a star for long periods of time without blurring the image. To be able to achieve such accuracy, an active attitude control system is needed. This is not to say that what we have studied in terms of passive stabilization is not useful. On the contrary, it is very useful to design a spacecraft that has passive stability (if possible), and then augment this with an active control scheme. Since the spacecraft attitude has natural stability, the control system does not need to work as hard to maintain the required attitude. It also means that the attitude remains stable if the control system fails.

An active spacecraft attitude control system consists of attitude sensors, attitude actuators and typically a processor. The attitude sensors take measurements which are used to compute the current spacecraft attitude and/or angular velocity. The attitude actuators then supply torques to correct the difference between the measured and desired attitude. The mathematical relationship between the measured attitude and the corrective torques is called a *control law*, and is implemented as a program on the processor.

For the next several chapters, we are going to develop some of the tools required to design an active control system. The theory presented will be applicable to a much broader class of problems than just spacecraft attitude control. However, we shall try to illustrate as much of it as possible in terms of spacecraft attitude control. We shall not worry about the sensor and actuator details. Rather, we shall assume that the attitude and/or angular velocity are available for measurement, and that there are actuators capable of supplying the control torques we want.

Spacecraft Dynamics and Control: An Introduction, First Edition.
Anton H.J. de Ruiter, Christopher J. Damaren and James R. Forbes.

We shall discuss spacecraft attitude determination methods and sensor and actuator types and their limitations in Chapters 25 and 26 respectively.

Even though the spacecraft attitude dynamics are nonlinear, in the next several chapters (Chapters 17 to 23) we are going to restrict ourselves to the control design of linear systems. We shall discuss nonlinear spacecraft attitude control design in Chapter 24. The lessons learned from control design for linear systems are very important, even in the control design for nonlinear systems. In fact in practise, spacecraft attitude control designs typically begin with a linear analysis and design of the control system. Only once this has been completed is the full nonlinear system considered. To be able to do this, the spacecraft attitude dynamics and kinematics are linearized about a desired attitude configuration. We have already performed such linearizations several times in the stability analyses of equilibria in spin stabilization (Chapter 14), dual-spin stabilization (Chapter 15) and gravity-gradient stabilization (Chapter 16). In each of these cases when we said we would consider small deviations in attitude and angular velocity from some nominal case, we were performing a linearization.

17.1 Attitude Control for a Nominally Inertially Fixed Spacecraft

In this section, we are going to formulate the attitude control problem in the case where we wish to control the attitude relative to some inertially fixed reference frame $\mathcal{F}_{ref}$. Let us represent the attitude of the body-fixed frame $\mathcal{F}_b$ relative to $\mathcal{F}_{ref}$ by a 3-2-1 Euler sequence (see Section 11.2):

1. A rotation ψ about the original z-axis (called the yaw angle).
2. A rotation θ about the intermediate y-axis (called a pitch angle).
3. A rotation ϕ about the transformed x-axis (called a roll angle).

We further assume that $\mathcal{F}_b$ is a principal axis frame. The spacecraft attitude motion is therefore completely described by the kinematic and dynamic equations (see Sections 11.2 and 11.3)

$$\begin{bmatrix} \omega_x \\ \omega_y \\ \omega_z \end{bmatrix} = \begin{bmatrix} 1 & 0 & -\sin\theta \\ 0 & \cos\phi & \sin\phi\cos\theta \\ 0 & -\sin\phi & \cos\phi\cos\theta \end{bmatrix} \begin{bmatrix} \dot{\phi} \\ \dot{\theta} \\ \dot{\psi} \end{bmatrix}, \tag{17.1}$$

and

$$\begin{aligned} I_x\dot{\omega}_x + (I_z - I_y)\omega_y\omega_z &= T_{xc} + T_{xd}, \\ I_y\dot{\omega}_y + (I_x - I_z)\omega_x\omega_z &= T_{yc} + T_{yd}, \\ I_z\dot{\omega}_z + (I_y - I_x)\omega_x\omega_y &= T_{zc} + T_{zd}, \end{aligned} \tag{17.2}$$

respectively. Note that we have split the external torque into two components, the control torque $T_{\cdot c}$, which can be applied by the actuator, and the disturbance torque $T_{\cdot d}$ which is present but unwanted.

It is clear that the equations (17.1) and (17.2) are nonlinear. However, we shall assume small angles and rates, just as we did in Section 16.1. In this case, equation (17.1) becomes

$$\begin{bmatrix} \omega_x \\ \omega_y \\ \omega_z \end{bmatrix} = \begin{bmatrix} \dot{\phi} \\ \dot{\theta} \\ \dot{\psi} \end{bmatrix},$$

which when substituted into (17.2) leads to (after neglecting second order terms)

$$\begin{aligned} I_x\ddot{\phi} &= T_{xc} + T_{xd}, \\ I_y\ddot{\theta} &= T_{yc} + T_{yd}, \\ I_z\ddot{\psi} &= T_{zc} + T_{zd}. \end{aligned} \tag{17.3}$$

We see that each of the equations in (17.3) are decoupled from each other, and have the same form. Therefore, we only need to consider one axis at a time, which we will generically write as

$$I\ddot{\theta} = T_c + T_d. \tag{17.4}$$

To further introduce some control system terminology, the system we are trying to regulate is called the *plant*. In our case, the plant is described by the spacecraft attitude dynamics (17.4). The quantity that we are interested to control is the attitude angle θ, which is generically called the *plant output*, and is generally labeled $y(t)$, such that on our case,

$$y(t) = \theta(t).$$

The quantity that can be used to affect the attitude is the control torque, which is generically called the *control input* or *plant input*, and is generally labeled $u(t)$, such that in our case,

$$u(t) = T_c(t).$$

17.2 Transfer Function Representation of a System

In many cases in control system design, we are not so much interested in the response of the output to non-zero initial conditions. These will disappear to zero anyway provided the control system is designed to be asymptotically stabilizing (which it should be). Rather, we are interested in how the system responds to an input, which will affect how the control system should be designed. Therefore, we set the initial conditions to zero. Taking the Laplace transform of (17.4) with zero initial conditions, we get

$$Is^2\hat{\theta}(s) = \hat{T}_c(s) + \hat{T}_d(s),$$

and with $Y(s) = \hat{\theta}(s)$ and $U(s) = \hat{T}_c(s)$, we have the relationship between input and output

$$Y(s) = \frac{1}{Is^2}\left(U(s) + \hat{T}_d(s)\right). \tag{17.5}$$

The ratio

$$G(s) = \left.\frac{Y(s)}{U(s)}\right|_{\hat{T}_d(s)=0} = \left.\frac{Y(s)}{\hat{T}_d(s)}\right|_{U(s)=0} = \frac{1}{Is^2}, \tag{17.6}$$

is known as the *transfer function* from plant input to plant output, and it determines how the output behaves in response to a given input. Note that since we have two plant inputs (the control input and the disturbance) we actually have two plant transfer functions, one from control input to output, the other from disturbance to output. As seen in equation (17.5), in this case they are both the same and are given by (17.6). In particular, we see that (17.5) becomes

$$Y(s) = G(s)\left(U(s) + \hat{T}_d(s)\right). \tag{17.7}$$

In general, for a linear time-invariant system with one input and one output, the transfer function will have the form

$$G(s) = \frac{a_0 s^m + a_1 s^{m-1} + \ldots + a_m}{s^n + b_1 s^{n-1} + \ldots b_n},$$

where $m \leq n$, and a_i, b_i are real coefficients. As was detailed in Appendix A, $G(s)$ can be factored into

$$G(s) = K\frac{\prod_{i=1}^{m}(s - z_i)}{\prod_{i=1}^{n}(s - p_i)},$$

where z_i are the roots of $a_0 s^m + a_1 s^{m-1} + \ldots + a_m = 0$, called the *zeros* of $G(s)$, and p_i are the roots of $s^n + b_1 s^{n-1} + \ldots b_n = 0$, called the *poles* of $G(s)$.

17.3 System Response to an Impulsive Input

To find the response of a system with transfer function $G(s)$, subject to an input $u(t)$, we only need to find the Laplace transform of $u(t)$, $U(s) = \mathcal{L}(u(t))$, and then referring to (17.7) with $\hat{T}_d(s) = 0$, the response is given by

$$y(t) = \mathcal{L}^{-1}(Y(s)) = \mathcal{L}^{-1}(G(s)U(s)).$$

A special type of input is the unit impulse, $\delta(t)$. To motivate what an impulse response physically means, consider a short constant input (unit pulse) at time $t = 0$, with magnitude A and duration ΔT, such that the area under the curve is one, as shown in Figure 17.1. That is,

$$f(t) = \begin{cases} 0, & t > \Delta T \\ A, & 0 \leq t \leq \Delta T \\ 0, & t < 0 \end{cases}$$

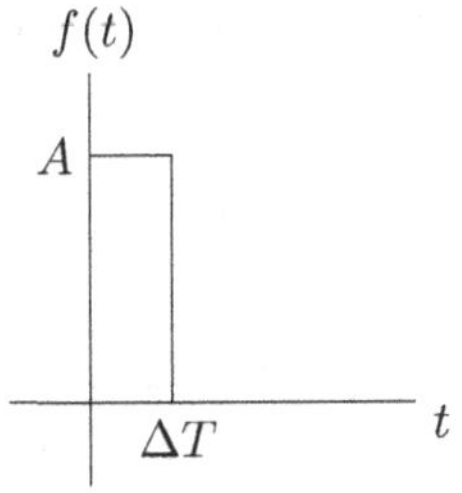

Figure 17.1 Unit pulse

and

$$\int_{-\infty}^{\infty} f(t)dt = A\Delta T = 1.$$

Let us now examine the integral

$$\int_{-\infty}^{\infty} f(t)g(t)dt,$$

for any other function $g(t)$ which is assumed to be continuous on the interval $t \in [0, \Delta T]$. Then,

$$\int_{-\infty}^{\infty} f(t)g(t)dt = \int_{0}^{\Delta T} Ag(t)dt.$$

By the mean-value theorem, the integral is given by

$$\int_{-\infty}^{\infty} f(t)g(t)dt = Ag(t')\Delta t = g(t'),$$

for some $0 \leq t' \leq \Delta T$. Letting $\Delta T \to 0$, we see that

$$\lim_{\Delta T \to 0} \int_{-\infty}^{\infty} f(t)g(t)dt = g(0).$$

This motivates the definition of an impulse function, denoted by $\delta(t)$. It is defined by the integral

$$\int_{-\infty}^{\infty} g(t)\delta(t)dt = g(0). \tag{17.8}$$

Physically, a unit impulse is a unit pulse applied over an infinitely short period of time. It does not exist in practise, but it is a good approximation.

Let us now evaluate a system response to a unit impulse ($u(t) = \delta(t)$). The Laplace transform of an impulse is given by

$$\mathcal{L}\left(\delta(t)\right) = \int_{0^-}^{\infty} e^{-st}\delta(t)dt = e^{-s0} = 1.$$

Therefore, if the input to a system with transfer function $G(s)$ is a unit impulse, the *impulse response* is given by the inverse Laplace transform of the transfer function itself

$$y(t) = \mathcal{L}^{-1}\left(G(s)\right) = g(t),$$

when $u(t) = \delta(t)$. *Therefore, the inverse Laplace transform of the transfer function $g(t) = \mathcal{L}^{-1}\left(G(s)\right)$ is called the impulse response of the system.*

As a result of the convolution theorem (see Appendix A), the system response to any input $u(t)$ is given by the convolution between the system unit impulse response $g(t)$, and the input $u(t)$, which is given by

$$y(t) = \int_0^t g(t-\tau)u(\tau)d\tau.$$

17.4 Block Diagrams

We have seen that a system can be represented by a transfer function $G(s)$, relating the inputs $U(s)$ to the outputs $Y(s)$ by

$$Y(s) = G(s)U(s).$$

This can be represented pictorially using a block diagram, as shown in Figure 17.2. Now, suppose the input $U(s)$ is actually the output of another system with transfer function $H(s)$ with input $W(s)$ ($U(s) = H(s)W(s)$). Then, we have

$$Y(s) = G(s)H(s)W(s),$$

and the transfer function for the combined system is $G(s)H(s)$. This is called a *series connection* of transfer functions. Pictorially, this is represented as shown in Figure 17.3. Therefore, you can see that a block diagram represents the flow of information, with the transfer function blocks operating on the inputs using multiplication to yield the outputs.

Let us look at some other typical block diagram operations.

The *parallel connection* between $G(s)$ and $H(s)$ is

$$Y(s) = G(s)U(s) + H(s)U(s) = \left(G(s) + H(s)\right)U(s).$$

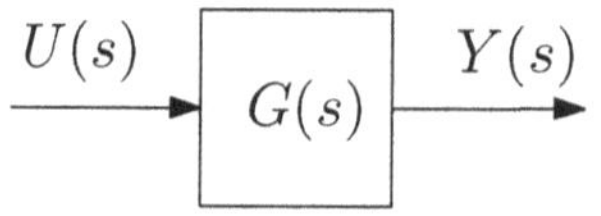

Figure 17.2 Block diagram

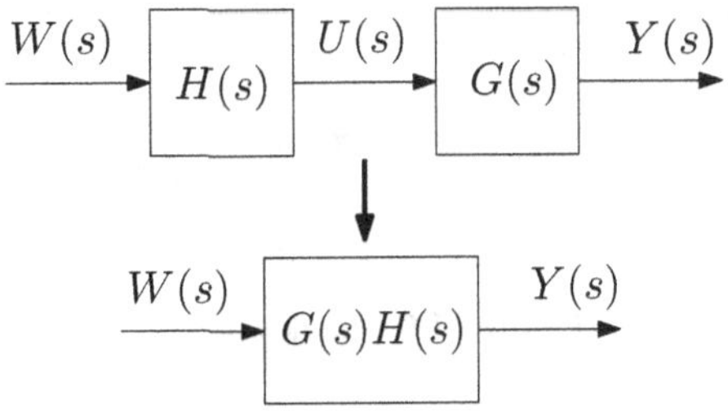

Figure 17.3 Series connection

Pictorially, this is represented as shown in Figure 17.4. The point shown in Figure 17.5 is called a *summing point* because the incoming signals are added. The point shown in Figure 17.6 is called a *branch point*, and the same signal is sent in multiple directions.

The final block diagram operation we will look at is the *feedback interconnection*, and is represented in block diagram form in Figure 17.7. Let us look at how the mathematics of the feedback connection works. Let $A(s) = H(s)Y(s)$ be the output of $H(s)$, and let $B(s) = U(s) - A(s)$. Then,

$$Y(s) = G(s)B(s) = G(s)\,[U(s) - A(s)] = G(s)U(s) - G(s)H(s)Y(s),$$

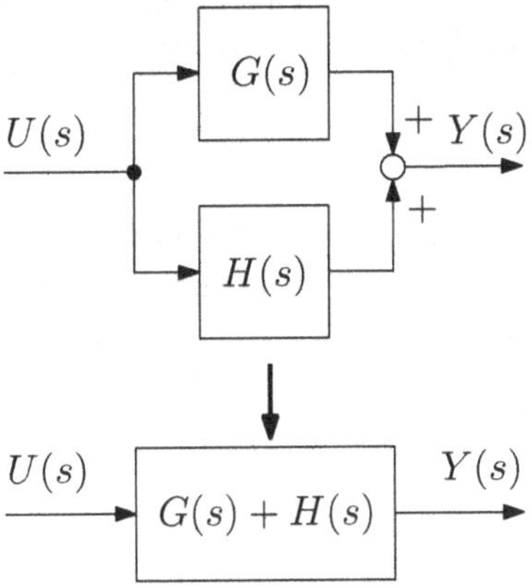

Figure 17.4 Parallel connection

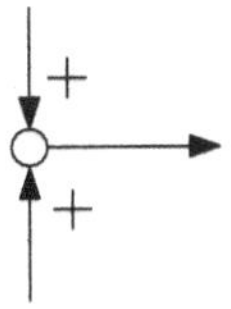

Figure 17.5 Summing point

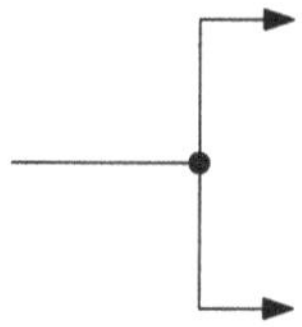

Figure 17.6 Branch point

which can be rearranged to give

$$Y(s) = \frac{G(s)}{1 + G(s)H(s)} U(s).$$

Hence, the transfer function for the feedback interconnection is given by

$$\frac{G(s)}{1 + G(s)H(s)}.$$

Using the block diagram reductions given in Figures 17.3 to 17.7, any complicated block diagram can be reduced to a single block.

17.5 The Feedback Control Problem

Let us now return to our attitude control design problem in Section 17.1. From (17.7), we see that the plant (spacecraft attitude dynamics) are represented by

$$Y(s) = G_p(s) \left(U(s) + \hat{T}_d(s) \right),$$

where $G_p(s) = \frac{1}{Is^2}$ and the subscript p has been added to signify that the transfer function is for the plant. The output $Y(s) = \hat{\theta}(s)$ is the attitude, the input $U(s) = \hat{T}_c(s)$ is the control

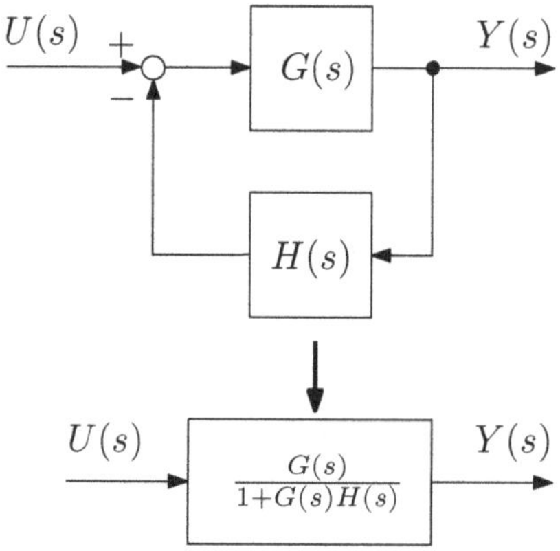

Figure 17.7 Feedback interconnection

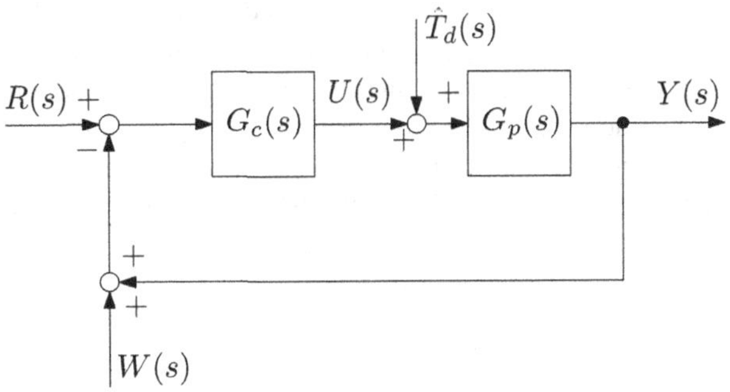

Figure 17.8 Feedback control system

torque, and $\hat{T}_d(s)$ is the disturbance torque. Our control objective is to make the output $y(t)$ (and hence the attitude $\theta(t)$) follow a reference signal $r(t)$ (a desired attitude $\theta_d(t)$), which has Laplace transform $R(s) = \hat{\theta}_d(s)$. Our control input must therefore correct the output error (the error in attitude $e(t) = \theta_d(t) - \theta(t) = r(t) - y(t)$). That is, our control objective is to drive the error $e(t)$ to zero.

Let us therefore represent the control law by a transfer function $G_c(s)$ such that

$$U(s) = G_c(s)E(s).$$

where $E(s) = \mathcal{L}(e(t))$. Unfortunately for any real sensor, the measurement is not perfect, but is corrupted by noise $w(t)$ (which has Laplace transform $W(s)$). That is, our sensor does not provide $y(t)$, but it provides $y(t) + w(t)$. Therefore, we can only provide our control law with the measured error

$$e(t) = r(t) - (y(t) + w(t)),$$

and not the true error. We can now represent the full feedback control system in block diagram form as shown in Figure 17.8. Note that a more general feedback control system includes a sensor transfer function $H(s)$, as shown in Figure 17.9. For now, we shall restrict ourselves to

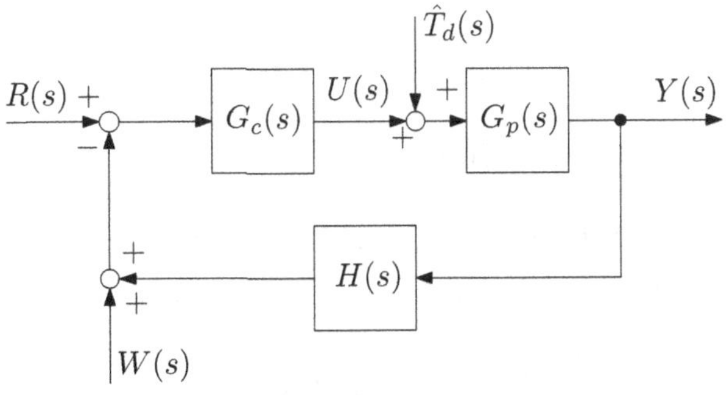

Figure 17.9 General feedback control system

the case with $H(s) = 1$ as in Figure 17.8, which is called a *unity feedback* system. The theory to be presented is still valid for the more general case, provided a little care is taken in its application. Returning to our problem, we can see from Figure 17.8, that our system has three inputs. Namely, the reference signal $R(s)$, the disturbance $\hat{T}_d(s)$ and the measurement noise $W(s)$. We can determine the effect of each input on the output separately. Since the system is linear, the combined effect on the output is just the sum of the individual effects.

Let $Y_r(s)$ be the response to the reference signal $R(s)$ with $\hat{T}_d(s) = W(s) = 0$. Then,

$$Y_r(s) = G_p(s)G_c(s)E(s).$$

But, $E(s) = R(s) - Y_r(s)$, so that

$$Y_r(s) = G_p(s)G_c(s)\left(R(s) - Y_r(s)\right),$$

which rearranges to give

$$Y_r(s) = \frac{G_p(s)G_c(s)}{1 + G_p(s)G_c(s)}R(s).$$

Next, let $Y_d(s)$ be the response to the disturbance $\hat{T}_d(s)$ with $R(s) = W(s) = 0$. Then,

$$Y_d(s) = G_p(s)\left(U(s) + \hat{T}_d(s)\right) = G_p(s)G_c(s)E(s) + G_p(s)\hat{T}_d(s).$$

With $E(s) = -Y_d(s)$, we have

$$Y_d(s) = G_p(s)\left(U(s) + \hat{T}_d(s)\right) = -G_p(s)G_c(s)Y_d(s) + G_p(s)T_d(s),$$

which rearranges to give

$$Y_d(s) = \frac{G_p(s)}{1 + G_p(s)G_c(s)}\hat{T}_d(s).$$

Finally, let $Y_w(s)$ be the response to the measurement noise $W(s)$ with $R(s) = \hat{T}_d(s) = 0$. Then,

$$Y_w(s) = G_p(s)G_c(s)E(s).$$

With $E(s) = -\left(Y_w(s) + W(s)\right)$, we have

$$Y_w(s) = -G_p(s)G_c(s)\left(Y_w(s) + W(s)\right),$$

which rearranges to give

$$Y_w(s) = -\frac{G_p(s)G_c(s)}{1 + G_p(s)G_c(s)}W(s).$$

The combined response to $R(s)$, $\hat{T}_d(s)$ and $W(s)$ is the sum

$$\begin{aligned} Y(s) &= Y_r(s) + Y_d(s) + Y_w(s), \\ &= \frac{G_p(s)G_c(s)}{1+G_p(s)G_c(s)}R(s) + \frac{G_p(s)}{1+G_p(s)G_c(s)}\hat{T}_d(s) - \frac{G_p(s)G_c(s)}{1+G_p(s)G_c(s)}W(s). \end{aligned} \tag{17.9}$$

Ideally our control system design would be such that the output $Y(s)$ matches the reference signal $R(s)$ exactly (the actual attitude follows the desired attitude perfectly), that is $Y(s) = R(s)$. Now, $G_p(s)$ is the plant, and we cannot change that. However, $G_c(s)$ is the controller, and this is up to our discretion. From equation (17.9), we see that if $|G_c(s)| \to \infty$ (let the control be large), then

$$\frac{G_p(s)G_c(s)}{1+G_p(s)G_c(s)} \to 1 \text{ and } \frac{G_p(s)}{1+G_p(s)G_c(s)} \to 0,$$

that is, we get perfect tracking ($Y_r(s) = R(s)$), and the response to disturbances becomes zero ($Y_d(s) = 0$). Unfortunately, the side-effect is that the response to measurement noise has also become large, i.e. $Y_w(s) = -W(s)$. This is the compromise we face as control system designers. We cannot simultaneously reject disturbances and measurement noise. However, all is not lost, as we shall see later when we study frequency response based control design methods in Chapter 23.

17.6 Typical Control Laws

Let us now look at some of the more common types of control. We shall neglect the measurement noise in this discussion (we set $W(s) = 0$). For reference, the equation of motion of our spacecraft attitude is (see (17.4))

$$I\ddot{y} = u(t) + T_d(t).$$

17.6.1 Proportional "P" Control

In proportional control, the control input is just a scaling of the error signal,

$$u(t) = K_p e(t),$$

where $K_p > 0$ is known as the proportional gain. The associated control transfer function is

$$G_c(s) = K_p.$$

For second order systems, such as our attitude control problem, proportional control acts as a spring with constant K_p, as shown in Figure 17.10. Assuming the reference attitude $r(t)$ is

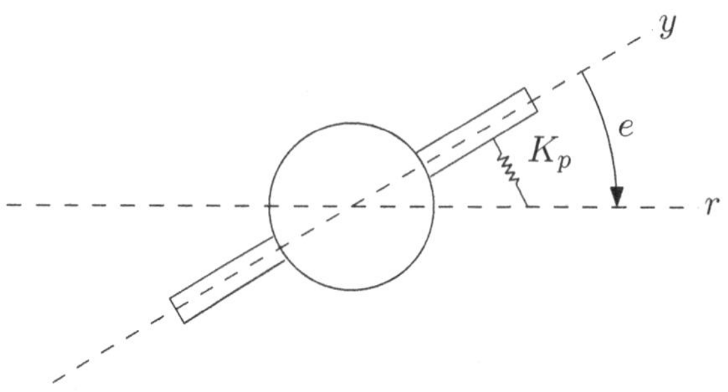

Figure 17.10 Proportional control

constant, and neglecting the disturbance T_d, the equation of motion for the spacecraft becomes

$$I\ddot{y} = K_p e(t) = K_p (r - y(t)).$$

This is just a spring-mass system. It is clear that if the reference attitude $r(t)$ is constant, the resulting closed-loop behavior is undamped oscillatory attitude motion.

17.6.2 Proportional Derivative "PD" Control

Pure undamped oscillatory motion as results from the use of proportional control is not desirable. Therefore, by adding a derivative term to proportional control, we can add some damping. The proportional derivative control law is given by

$$u(t) = K_p e(t) + K_d \dot{e}(t),$$

where $K_d > 0$ is the derivative gain. The associated control transfer function is

$$G_c(s) = K_p + sK_d.$$

For a second order system, such as our attitude control problem, proportional-derivative control acts like a spring and viscous damper, with spring-constant K_p, and damping constant K_d, as shown in Figure 17.11. Assuming the reference attitude $r(t)$ is constant, and neglecting the disturbance T_d, the equation of motion for the spacecraft becomes

$$I\ddot{y} = K_p(r - y(t)) - K_d\dot{y}.$$

This is just a spring-mass-damper system. It is clear that if the reference attitude $r(t)$ is constant, any oscillatory behavior due to the spring dies out due to the damping term.

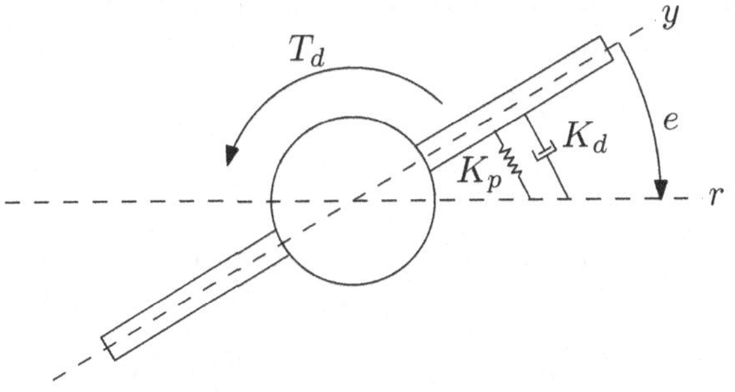

Figure 17.11 Proportional-derivative control

17.6.3 Proportional Integral Derivative "PID" Control

Continuing our assumption that the reference attitude $r(t)$ is constant, let us now add a constant external disturbance torque T_d to our spacecraft with proportional-derivative control, as shown in Figure 17.11. The equation of motion is now

$$I\ddot{y} = K_p(r - y(t)) - K_d\dot{y} + T_d.$$

The effective spring-damper system of the PD control law will cause all oscillations to die out. However, there will be a steady-state error in the attitude. This can readily be seen by a torque balance on the spacecraft. Assuming that our reference attitude $r(t)$ is constant, at steady-state the derivatives of the attitude $\dot{y} = \ddot{y} = 0$. We must therefore have from above

$$K_p(r - y) + T_d = 0,$$

That is, the steady-state error is

$$e = r - y = -\frac{T_d}{K_p},$$

which is clearly non-zero. Graphically, the error at steady-state with PD control is shown in Figure 17.12. The problem is that the proportional-derivative control has no memory. To

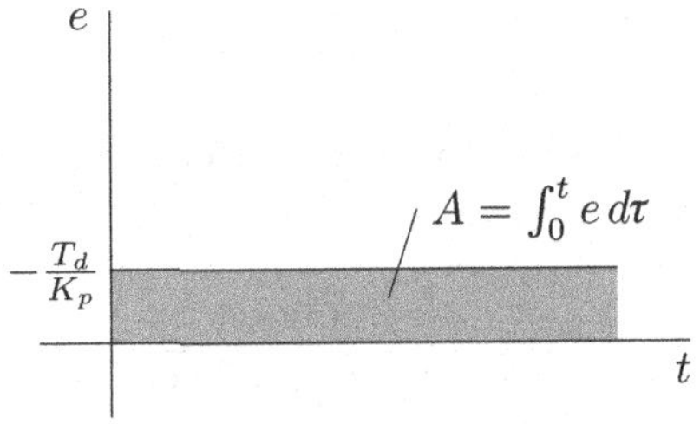

Figure 17.12 Steady-state error with PD control

provide the controller with the capability of driving the error to zero in the presence of a constant disturbance, an integral term is added to the control law:

$$K_i \int_0^t e(\tau)d\tau,$$

where $K_i > 0$ is the integral gain. Therefore the PID control law is given by

$$u(t) = K_p e(t) + K_d \dot{e}(t) + K_i \int_0^t e(\tau)d\tau,$$

and the corresponding transfer function is

$$G_c(s) = K_p + sK_d + \frac{K_i}{s}.$$

To understand why the integral term drives the error to zero, let us again consider Figure 17.12. The integral term has memory, that is, it is proportional to the area under the error curve, A. If the error is non-zero, the integral term grows, providing a larger and larger correcting torque, until the disturbance torque has been compensated and the error is zero. Effectively, the integral term learns what the disturbance torque is.

17.7 Time-Domain Specifications

We have now seen what a control law looks like, how the closed-loop system is formed, and some typical types of control. Now, we need some guidelines as to how to design the controller. That is, how do we want the closed-loop system to perform.

First and foremost, the control law must provide asymptotic stability to the closed-loop system. To illustrate, imagine that we excite the closed-loop system

$$Y(s) = \frac{G_p(s)G_c(s)}{1 + G_p(s)G_c(s)} R(s),$$

with a unit impulse ($r(t) = \delta(t)$). Asymptotic stability means that the impulse response

$$y(t) = \mathcal{L}^{-1}\left(\frac{G_p(s)G_c(s)}{1 + G_p(s)G_c(s)}\right),$$

must asymptotically go to zero. As we have seen in Section 14.1, this means that the poles of the closed-loop transfer function

$$T(s) = \frac{G_p(s)G_c(s)}{1 + G_p(s)G_c(s)},$$

must have negative real parts.

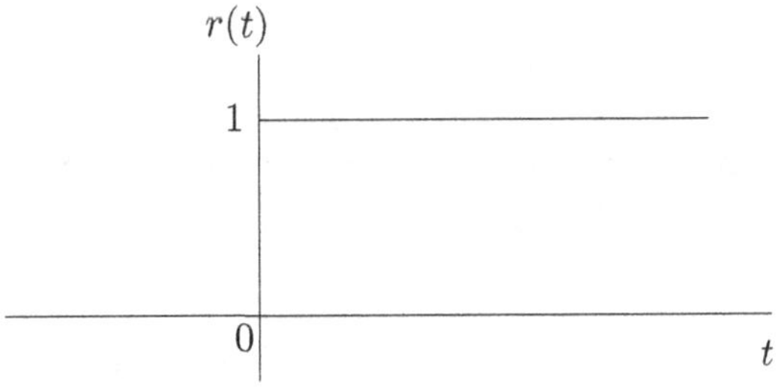

Figure 17.13 Unit step input

Provided that the closed-loop system is asymptotically stable, the time-domain response to a reference input and/or disturbance consists of two parts: transient behavior (behavior that dies out), and steady-state behavior (behavior that persists). Both components are important in specifying the desired time-domain behavior.

17.7.1 Transient Specifications

Transient specifications are often given in terms of the response of the closed-loop system to a unit step command. In our case, this means the attitude response to a unit step command in the desired attitude (see Figure 17.13). That is for a reference attitude satisfying

$$r(t) = \begin{cases} 1, & t \geq 0 \\ 0, & t < 0 \end{cases}$$

Let us examine the unit step response for a modified proportional-derivative control of our spacecraft attitude (the most common control for this type of system). The standard proportional-derivative control law is

$$u(t) = K_p e(t) + K_d \dot{e}(t).$$

This control law becomes problematic to implement when $\dot{e}(t)$ does not exist, such as occurs at step changes in the reference signal $r(t)$. Practically, step changes in $r(t)$, cause a large spike in $\dot{e}(t)$, which effectively results in an impulse within the control $u(t)$, at the time of the step change. To overcome this problem, the proportional-derivative control law can be modified to the form

$$u(t) = K_p e(t) - K_d \dot{y}(t),$$

with $K_p > 0$ and $K_d > 0$. Since $y(t)$ is always differentiable, this avoids the problem. More generally, it avoids large control inputs (possibly outside the actuator capabilities) resulting from rapid changes in reference signal $r(t)$. The block diagram of this implementation is shown in Figure 17.14. Note that measurement noise is not included. To keep the framework the same as in Figure 17.8, we have included the derivative term in the plant $G_p(s)$, rather than the control $G_c(s)$. So, as seen in Figure 17.14, the modified proportional-derivative control is equivalent to proportional control of the plant which is augmented with the derivative term.

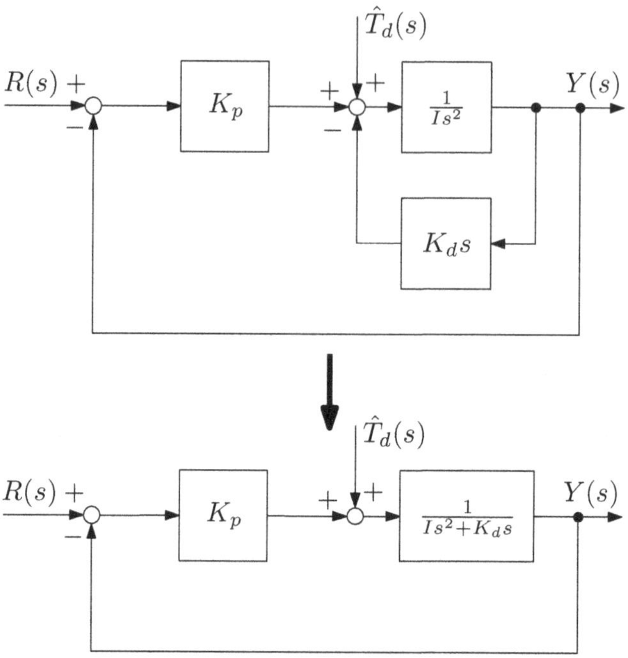

Figure 17.14 Modified Proportional-Derivative Control

Only the response to a step command in attitude $r(t)$ is relevant at this stage, so we set the disturbance and measurement noise to zero.

Referring to Figure 17.14, we have the augmented plant transfer function,

$$G_p(s) = \frac{1}{Is^2 + K_d s},$$

and the effective control transfer function

$$G_c(s) = K_p.$$

The closed-loop response to a reference command $R(s)$ with $\hat{T}_d(s) = 0$ (no disturbance since we are interested in response to a command) is then given by

$$Y(s) = T(s)R(s),$$

where the closed-loop transfer function is

$$\begin{aligned} T(s) &= \frac{G_p(s)G_c(s)}{1 + G_p(s)G_c(s)} = \frac{K_p/(Is^2 + K_d s)}{1 + K_p/(Is^2 + K_d s)}, \\ &= \frac{(K_p/I)}{s^2 + (K_d/I)s + (K_p/I)}. \end{aligned}$$

It will be useful to define

$$\omega_n^2 = \frac{K_p}{I}, \quad 2\zeta\omega_n = \frac{K_d}{I}, \tag{17.10}$$

The quantity ω_n is known as the *undamped natural frequency*, and ζ is known as the *damping ratio*. In terms of these quantities, the closed-loop transfer function becomes

$$T(s) = \frac{\omega_n^2}{s^2 + 2\zeta\omega_n s + \omega_n^2}. \tag{17.11}$$

Equation (17.11) generically describes a second order system, so we see that with PD control, the closed-loop spacecraft attitude system is second order. The closed-loop poles can be obtained using the quadratic equation solution and are given by

$$s = -\zeta\omega_n \pm \omega_n\sqrt{\zeta^2 - 1}.$$

There are now three cases:

Case 1: $0 < \zeta < 1$. The system is called *underdamped*. In this case, $\sqrt{\zeta^2 - 1}$ is imaginary, and there are two complex conjugate poles

$$s = -\zeta\omega_n \pm j\omega_n\sqrt{1 - \zeta^2},$$

as shown in Figure 17.15. We call $\omega_d = \omega_n\sqrt{1 - \zeta^2}$ the *damped natural frequency*. The poles can be rewritten as

$$s = -\zeta\omega_n \pm j\omega_d.$$

These poles lead to decaying oscillatory behavior at frequency ω_d.

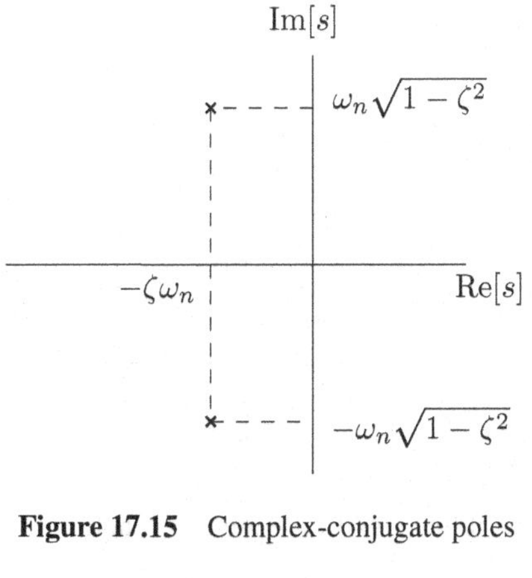

Figure 17.15 Complex-conjugate poles

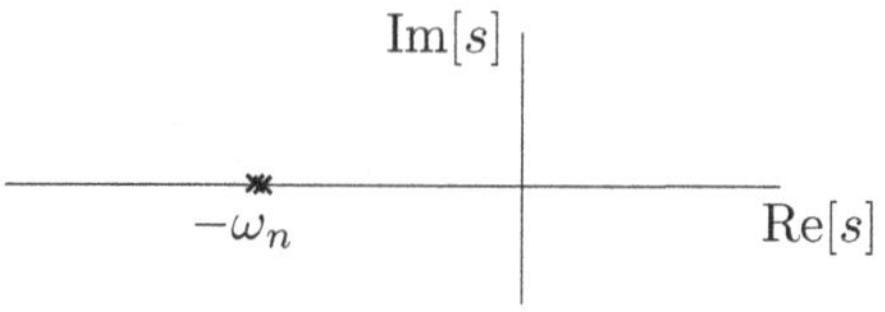

Figure 17.16 Double poles

Case 2: $\zeta = 1$. The system is called *critically damped*. In this case, the poles are

$$s = -\omega_n, -\omega_n,$$

as shown in Figure 17.16.

Case 3: $\zeta > 1$. The system is called *overdamped*. In this case, $\sqrt{\zeta^2 - 1}$ is real, and there are two distinct negative real poles

$$s = -\zeta\omega_n \pm \omega_n\sqrt{\zeta^2 - 1},$$

as shown in Figure 17.17.

Figure 17.18 shows the step response for different values of ζ. It can be seen that in Case 1, the transient behavior has an oscillatory component, which results in some overshoot of steady-state value. It can be seen that in both Cases 2 and 3, the transient response decays exponentially, but does not have any oscillatory component, and hence no overshoot. It can be seen in Figure 17.18 that this lack of overshoot comes at the expense of slower response to the step command. In some applications, it is desirable to have no overshoot. However, for spacecraft attitude control, this is typically not the case, and some overshoot allows us to have faster response. Hence, we shall restrict ourselves to Case 1 (the underdamped case).

Let us now examine the step response in the underdamped case. We have reference command

$$R(s) = \frac{1}{s},$$

so that the step response is given by

$$Y(s) = \frac{\omega_n^2}{s^2 + 2\zeta\omega_n s + \omega_n^2}\frac{1}{s}.$$

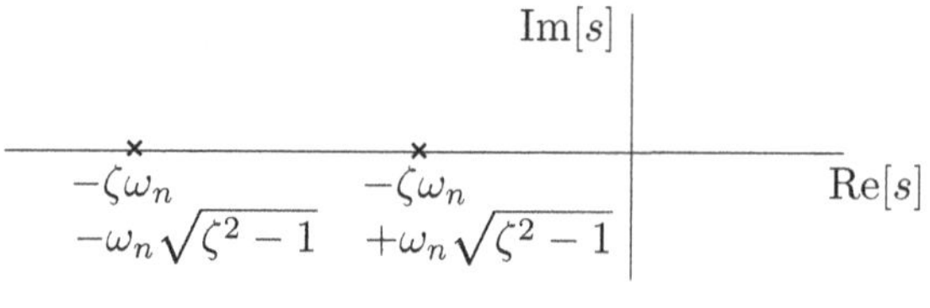

Figure 17.17 Distinct real poles

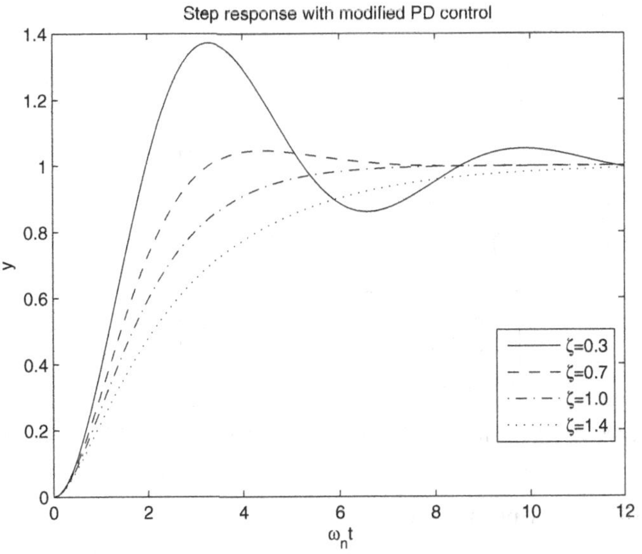

Figure 17.18 Unit step response

We can expand $Y(s)$ in a partial fraction expansion (check it for yourself)

$$Y(s) = \frac{1}{s} - \frac{s + 2\zeta\omega_n}{s^2 + 2\zeta\omega_n s + \omega_n^2} = \frac{1}{s} - \frac{s + \zeta\omega_n}{s^2 + 2\zeta\omega_n s + \omega_n^2} - \frac{\zeta\omega_n}{s^2 + 2\zeta\omega_n s + \omega_n^2}.$$

We now complete the square in the denominator to get

$$\begin{aligned} s^2 + 2\zeta\omega_n s + \omega_n^2 &= \left(s^2 + 2\zeta\omega_n s + \zeta^2\omega_n^2\right) + \omega_n^2 - \zeta^2\omega_n^2, \\ &= (s + \zeta\omega_n)^2 + \omega_n^2\left(1 - \zeta^2\right), \\ &= (s + \zeta\omega_n)^2 + \omega_d^2. \end{aligned}$$

Therefore, we can rewrite the step response as

$$Y(s) = \frac{1}{s} - \frac{s + \zeta\omega_n}{(s + \zeta\omega_n)^2 + \omega_d^2} - \frac{\zeta\omega_n}{\omega_d}\frac{\omega_d}{(s + \zeta\omega_n)^2 + \omega_d^2}.$$

Taking the inverse Laplace transform using the table in Section A.5 (try it for yourself), we obtain the step response in the time-domain

$$y(t) = 1 - e^{-\zeta\omega_n t}\left[\cos\omega_d t + \frac{\zeta\omega_n}{\omega_d}\sin\omega_d t\right]. \tag{17.12}$$

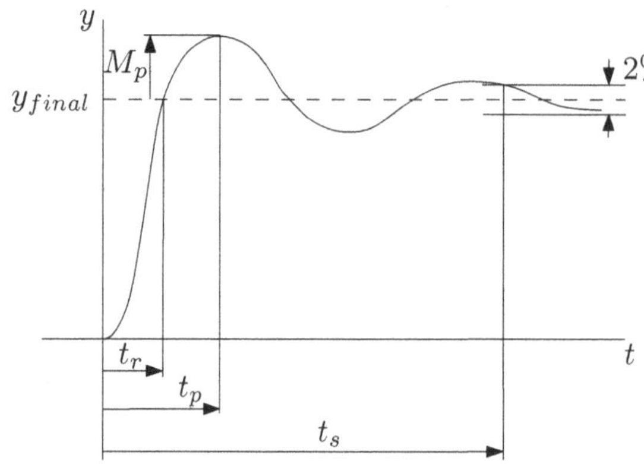

Figure 17.19 Time-domain specifications based on a unit step response

17.7.1.1 Time-Domain Specifications

Based on the unit step response, a number of different time-domain specifications may be given.

- **Rise time** t_r - the time taken for the step response to first reach the final value.
- **Peak time** t_p - the time taken to first achieve the peak response.
- **Maximum overshoot** M_p - the maximum percentage overshoot from the final value.
- **Settling time** t_s - the time taken for the output to get to within 2% of the final value and stay there.

The meanings of each of these specifications are illustrated in Figure 17.19.

Note that the above definition of rise-time is only useful if the damping ratio, ζ is not close to 1. If ζ is close to 1, the system almost behaves like a critically damped system. It may well be that the rise-time under the definition given here is very long, but if the poles are very fast, it could be that the settling time is less than the rise-time, as shown in Figure 17.20. *That is, the response is close to the final value in much less time than is indicated by the rise-time as defined above. In this case, a different definition of rise-time is more useful. The definition used most often is the time taken to rise from 10% to 90% of the final value. This concept is illustrated in* Figure 17.20.

Making use of our knowledge of the unit step response, we can now obtain expressions for each of the specifications in terms of the damping ratio ζ and the undamped natural frequency ω_n.

17.7.1.2 Rise Time

The rise time is the time taken to first reach the final value. From (17.12), we see that at $t = t_r$, the output is $y(t_r) = 1$. Therefore, from (17.12) we must solve

$$\cos \omega_d t_r + \frac{\zeta \omega_n}{\omega_d} \sin \omega_d t_r = 0.$$

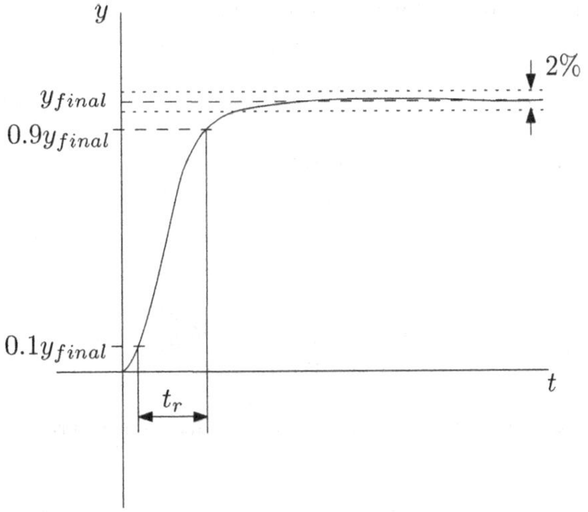

Figure 17.20 Alternative rise-time definition for damping ratios close to 1

This rearranges to give

$$\tan \omega_d t_r = \frac{\omega_d}{-\zeta \omega_n} = \frac{\omega_n \sqrt{1-\zeta^2}}{-\omega_n \zeta}.$$

But, this can readily be obtained from the pole location, as shown in Figure 17.21. Note that $-\omega_n \zeta$ is the real part of the pole, and ω_d is the imaginary part of the pole. From the plot of the pole, we have

$$\omega_d t_r = \pi - \beta,$$

which leads to the expression for the rise time

$$t_r = \frac{\pi - \beta}{\omega_d}, \tag{17.13}$$

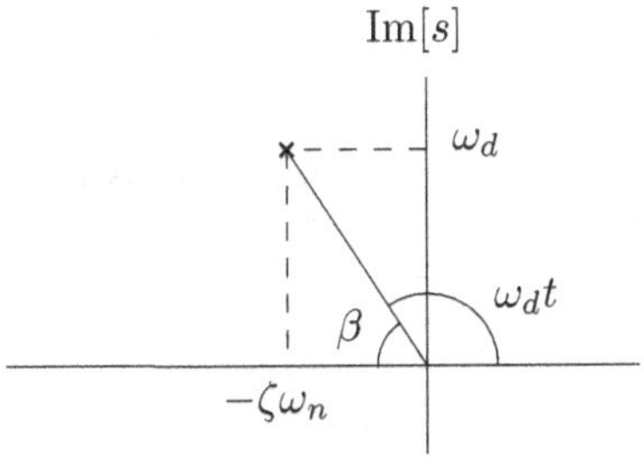

Figure 17.21 Relationship between rise-time and pole location

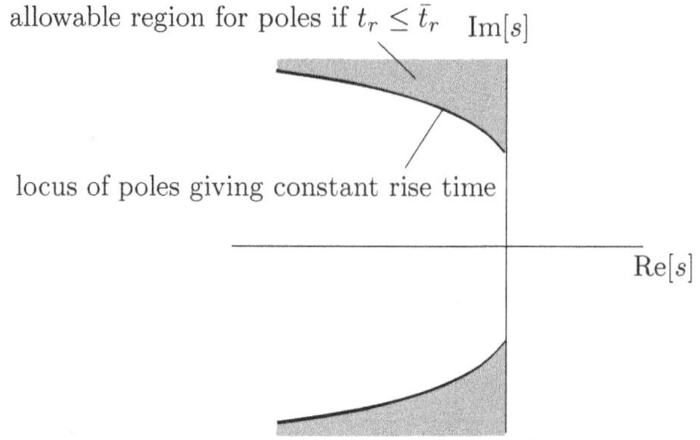

Figure 17.22 Relationship between pole locations and rise-time

where

$$\beta = \tan^{-1}\left(\frac{\sqrt{1-\zeta^2}}{\zeta}\right). \tag{17.14}$$

Note that we have made use of the fact that $\omega_d/(\zeta\omega_n) = \sqrt{1-\zeta^2}/\zeta$. See Figure 17.22 for the relationship between pole locations and rise-time. In particular, the allowable pole locations are shown for a given rise-time constraint $t_r \leq \bar{t}_r$, where $\bar{t}_r > 0$ is the given maximum allowable rise-time.

17.7.1.3 Peak Time

The peak time occurs at the first time when $dy/dt = 0$. Taking the derivative of (17.12), we obtain

$$\begin{aligned}\frac{dy}{dt} &= \zeta\omega_n e^{-\zeta\omega_n t}\left[\cos\omega_d t + \frac{\zeta\omega_n}{\omega_d}\sin\omega_d t\right] - e^{-\zeta\omega_n t}\left[-\omega_d\sin\omega_d t + \frac{\zeta\omega_n}{\omega_d}\omega_d\cos\omega_d t\right],\\ &= e^{-\zeta\omega_n t}\sin\omega_d t\left[\frac{\zeta^2\omega_n^2}{\omega_d} + \omega_d\right].\end{aligned}$$

From this, we find that at the peak time $t = t_p$, we must have

$$\sin\omega_d t_p = 0.$$

That means, $\omega_d t = 0, \pi, 2\pi, 3\pi, \ldots$. Clearly, the first peak corresponds to $\omega_d t_p = \pi$. Therefore, the peak time is

$$t_p = \frac{\pi}{\omega_d}. \tag{17.15}$$

From (17.15) we see that the peak time depends only in the imaginary part of the pole $s = \zeta\omega_n \pm j\omega_d$. See Figure 17.23 for the relationship between pole locations and peak-time.

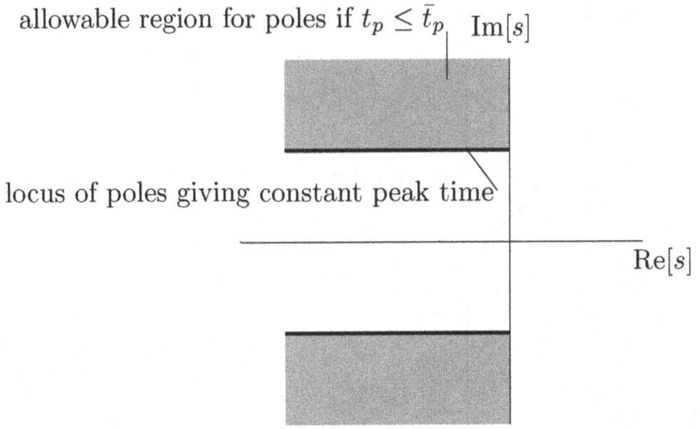

Figure 17.23 Relationship between pole locations and peak-time

In particular, the allowable pole locations are shown for a given peak-time constraint $t_p \leq \bar{t}_p$, where $\bar{t}_p > 0$ is the given maximum allowable peak-time.

17.7.1.4 Maximum Overshoot

The maximum overshoot occurs at the peak time $t = t_p$, with the maximum response being

$$y_p = 1 - e^{-\zeta\omega_n t}\left[\cos\omega_d t_p + \frac{\zeta\omega_n}{\omega_d}\sin\omega_d t_p\right].$$

But, we know that $\omega_d t_p = \pi$, such that $\cos\omega_d t_p = -1$ and $\sin\omega_d t_p = 0$. Therefore, the maximum response is

$$y_p = 1 + e^{-\pi\zeta\omega_n/\omega_d} = 1 + e^{-\pi\zeta/\sqrt{1-\zeta^2}}.$$

The maximum percentage overshoot of the final value is then

$$M_p = \frac{y_p - 1}{1} \times 100\% = e^{-\pi\zeta/\sqrt{1-\zeta^2}} \times 100\%. \tag{17.16}$$

It is clear from (17.16) that the maximum overshoot depends only on the damping ratio, ζ. Likewise, the ratio between the real and imaginary part of the closed-loop poles (given by $s = -\zeta\omega_n \pm j\omega_n\sqrt{1-\zeta^2}$) also depends only on the damping ratio.

$$\frac{Re(s)}{Im(s)} = \mp\frac{\zeta}{\sqrt{1-\zeta^2}}.$$

See Figure 17.24 for the relationship between pole locations and percent overshoot. In particular, the allowable pole locations are shown for a given maximum percentage overshoot constraint $M_p \leq \bar{M}_p$, where $\bar{M}_p > 0$ is the given maximum allowable percent overshoot.

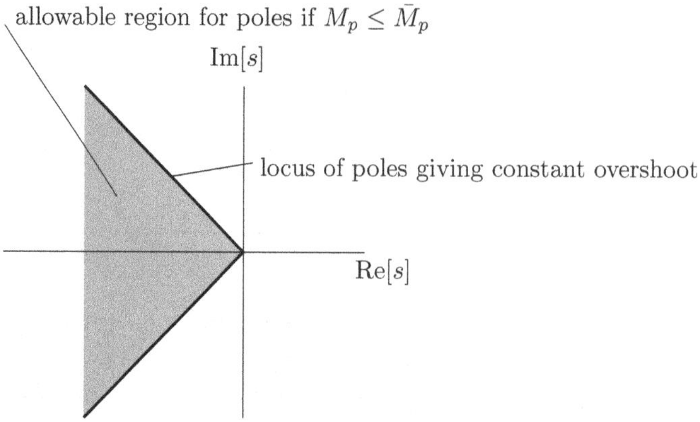

Figure 17.24 Relationship between pole locations and overshoot

17.7.1.5 Settling Time

As seen in (17.12), the difference between the current and final value of $y(t)$ is

$$y(t) - 1 = -e^{-\zeta\omega_n t}\left[\cos\omega_d t + \frac{\zeta\omega_n}{\omega_d}\sin\omega_d t\right].$$

We can rewrite

$$\cos\omega_d t + \frac{\zeta\omega_n}{\omega_d}\sin\omega_d t = \left(\frac{\zeta^2}{1-\zeta^2} + 1\right)^{\frac{1}{2}}\sin\left(\omega_d t + \phi\right),$$

where

$$\phi = \tan^{-1}\left(\frac{1}{\frac{\zeta}{\sqrt{1-\zeta^2}}}\right).$$

Noting that

$$\frac{\zeta^2}{1-\zeta^2} + 1 = \frac{1}{1-\zeta^2},$$

we have

$$y(t) - 1 = -\frac{e^{-\zeta\omega_n t}}{\sqrt{1-\zeta^2}}\sin\left(\omega_d t + \phi\right).$$

Since we know that $|\sin(\omega_d t + \phi)| \leq 1$, we obtain the bound

$$|y(t) - 1| \leq \frac{e^{-\zeta\omega_n t}}{\sqrt{1-\zeta^2}}.$$

We are therefore guaranteed that the percentage deviation from the final value satisfies

$$\frac{|y(t)-1|}{1} \times 100\% \leq 2\%,$$

when

$$\frac{e^{-\zeta\omega_n t}}{\sqrt{1-\zeta^2}} \leq 0.02.$$

We therefore take the settling time to satisfy

$$\frac{e^{-\zeta\omega_n t_s}}{\sqrt{1-\zeta^2}} = 0.02,$$

which leads to the expression for the settling time

$$t_s = \frac{\ln\left(0.02\sqrt{1-\zeta^2}\right)}{-\zeta\omega_n}. \tag{17.17}$$

For damping ratios between 0.1 and 0.9, the quantity $-\ln(0.02\sqrt{1-\zeta^2})$, varies between 3.9 and 4.8. We may therefore approximate the expression for the settling time (17.17) by

$$t_s \approx \frac{4.4}{\zeta\omega_n}. \tag{17.18}$$

This shows that the settling time depends primarily on the real part of the poles. See Figure 17.25 for the relationship between pole locations and settling time. In particular, the allowable pole locations are shown for a given settling-time constraint $t_s \leq \bar{t}_s$, where $\bar{t}_s > 0$ is the given maximum allowable settling-time.

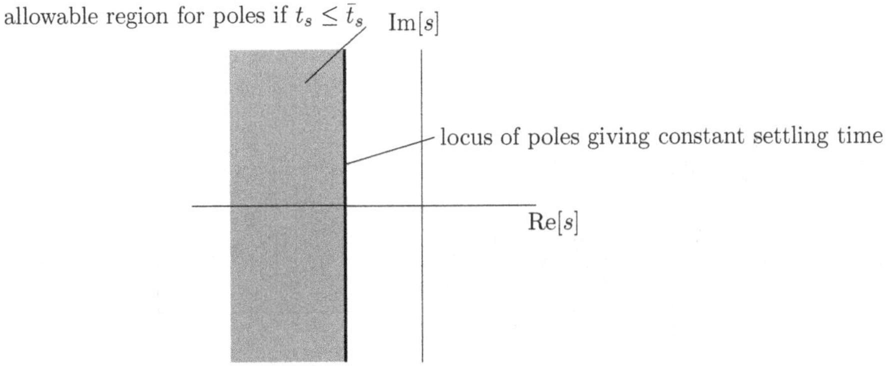

Figure 17.25 Relationship between pole locations and settling time

Example 17.1 (PD controller design for spacecraft attitude control) *Suppose that our spacecraft has moment of inertia $I = 1$ kg·m^2, and that the following time-domain design specifications are given.*

1. *Rise-time constraint, $t_r \leq 30$ seconds.*
2. *Maximum overshoot constraint, $M_p \leq 30\%$.*
3. *Settling-time constraint, $t_s \leq 100$ seconds.*

From equation (17.13), *we see that the rise-time constraint leads to the requirement*

$$\frac{\pi - \beta}{\omega_d} \leq 30,$$

which can be rearranged to give

$$\beta \geq \pi - 30\omega_d.$$

The allowable region for closed-loop poles to satisfy the rise-time constraint is shown in Figure 17.26. From (17.16), *the maximum overshoot constraint leads to*

$$e^{-\frac{\pi\zeta}{\sqrt{1-\zeta^2}}} \leq 0.3,$$

which rearranges to give

$$\frac{\zeta}{\sqrt{1-\zeta^2}} \geq -\frac{\ln 0.3}{\pi}.$$

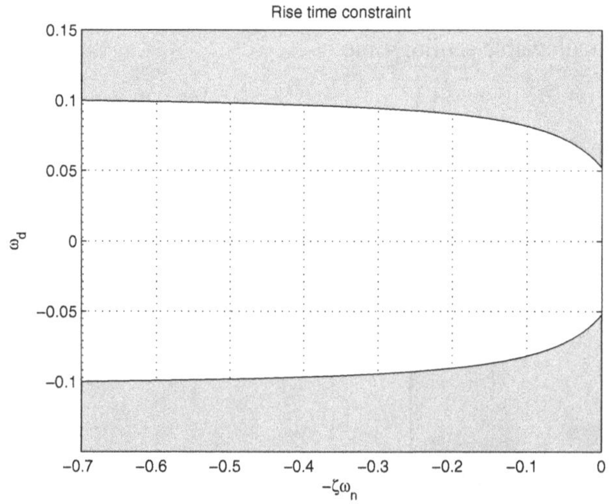

Figure 17.26 Allowable region for rise time constraint

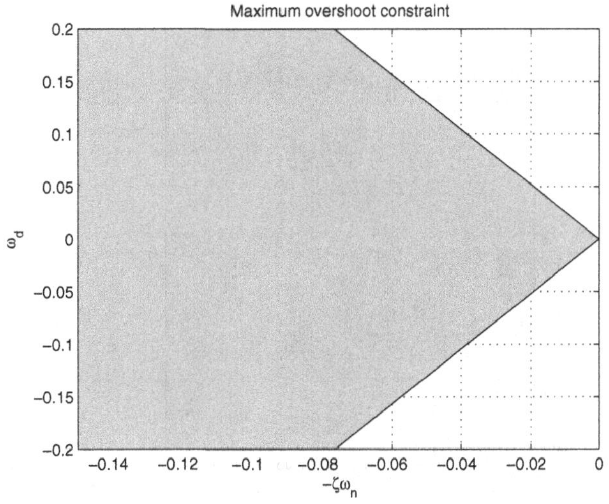

Figure 17.27 Allowable region for overshoot constraint

This can be further rearranged to give

$$\frac{\sqrt{1-\zeta^2}}{\zeta} \leq -\frac{\pi}{\ln 0.3}.$$

Recognizing that the angle $\beta = \tan^{-1}(\frac{\sqrt{1-\zeta^2}}{\zeta})$ *(see (17.14)), we have*

$$\beta \leq \tan^{-1}\left(-\frac{\pi}{\ln 0.3}\right) = 1.205\,(69^\circ).$$

Therefore, to satisfy the overshoot requirement, the closed-loop poles must lie in the shaded region shown in Figure 17.27. From equation (17.18), we see that the settling time constraint leads to the requirement

$$\frac{4.4}{\zeta\omega_n} \leq 100,$$

which can be rearranged to give

$$\zeta\omega_n \geq \frac{4.4}{100} = 0.044. \tag{17.19}$$

Therefore, to satisfy the settling time constraint, the closed-loop poles must lie in the shaded region shown in Figure 17.28. Now that we have found the allowable closed-loop pole regions for each of the individual constraints, we can combine them to find the allowable closed-loop pole region for the combined constraints. This is just the intersection as shown in Figure 17.29.

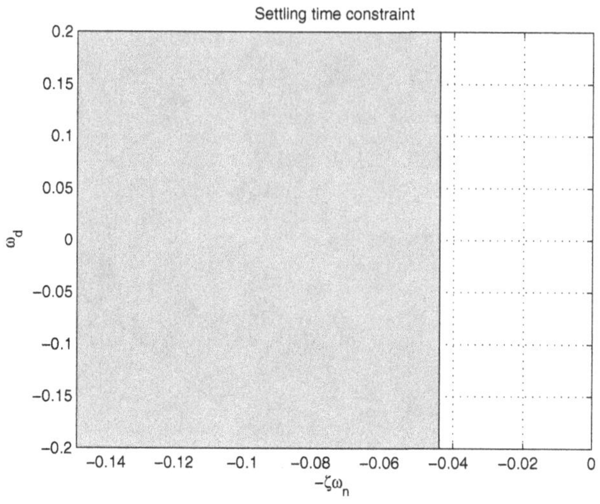

Figure 17.28 Allowable region for settling-time constraint

Unless there are any other constraints, we are free to choose the closed-loop poles anywhere in the shaded region. In particular, we can see that the closed-loop poles

$$s = -0.05 \pm j0.1$$

satisfy all constraints.

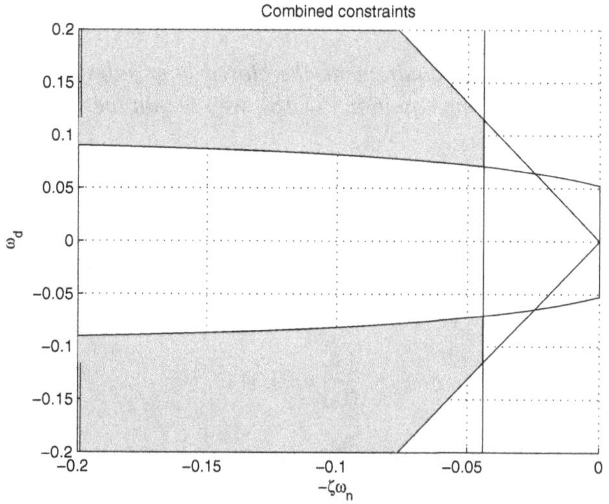

Figure 17.29 Allowable region for combined constraints

From the general expression for the closed-loop poles

$$s = -\zeta\omega_n \pm j\omega_n\sqrt{1-\zeta^2},$$

we see that

$$\omega_n = |s|.$$

We can therefore compute ω_n^2 *as*

$$\omega_n^2 = |s|^2 = 0.05^2 + 0.1^2 = 0.0125.$$

From (17.10), we can now finally compute our PD control gains, namely

$$K_p = \omega_n^2 I = 0.0125, \quad K_d = 2\zeta\omega_n I = 2 \times 0.05 = 0.1.$$

A control law that satisfies the time-domain specifications is then given by

$$u(t) = 0.0125e(t) - 0.1\dot{y}(t).$$

Figure 17.30 shows the unit step response of the spacecraft attitude with the designed PD controller. It can be seen that all time-domain specifications are met.

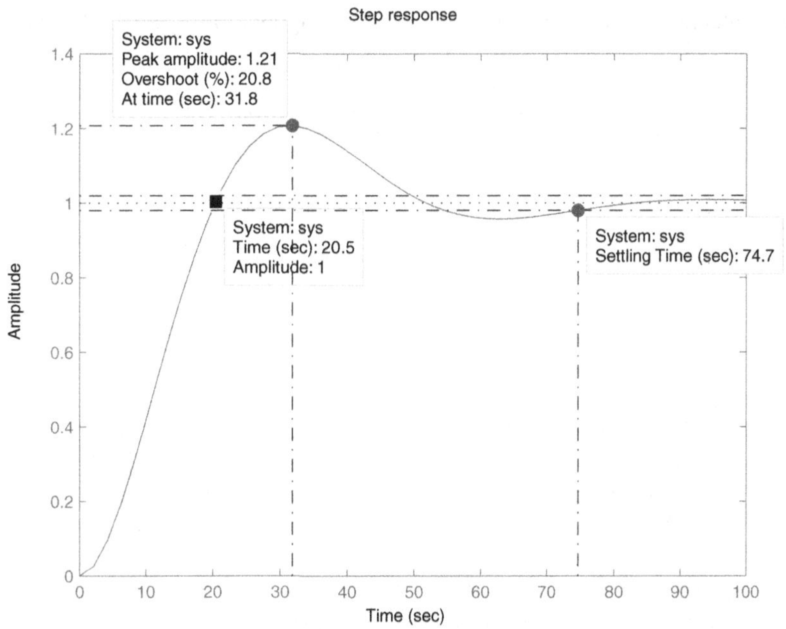

Figure 17.30 Step response with our PD design

17.8 Factors that Modify the Transient Behavior

There are two factors that can modify the transient behavior that we have seen in the previous section (and in general). The first is the presence of zeros in the closed-loop transfer function. The second is the presence of additional poles. We shall deal with each of these separately.

17.8.1 Effect of Zeros

The step response we have looked at is for a system with no closed-loop zeros (see (17.11)). If we had instead used standard proportional-derivative control as in Section 17.6.2 ($u(t) = K_p e(t) + K_d \dot{e}(t)$), the closed-loop transfer function would have been (try this as an exercise)

$$T(s) = \frac{2\zeta\omega_n s + \omega_n^2}{s^2 + 2\zeta\omega_n s + \omega_n^2},$$

The closed-loop poles are exactly the same as with modified PD control (namely $s = \zeta\omega_n \pm \omega_n\sqrt{\zeta^2 - 1}$). However, there is now a zero at $s = -\omega_n/(2\zeta)$. In the underdamped case, the associated step response is given by (try this as an exercise)

$$y(t) = 1 - e^{-\zeta\omega_n t}\left[\cos\omega_d t - \frac{\zeta\omega_n}{\omega_d}\sin\omega_d t\right],$$

where $\omega_d = \omega_n\sqrt{1-\zeta^2}$. Comparing this with (17.12), it can be seen that the frequency of oscillation ω_d and rate of decay $\zeta\omega_n$ are the same. These are determined by the poles. However, the coefficients of the terms are different, which is due to the presence of the zero. This results in changed transient characteristics. Figures 17.31 and 17.32 show the step

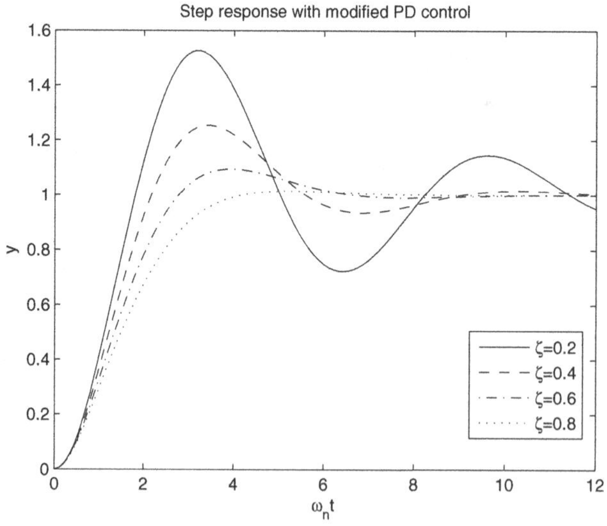

Figure 17.31 Unit step response with modified PD control

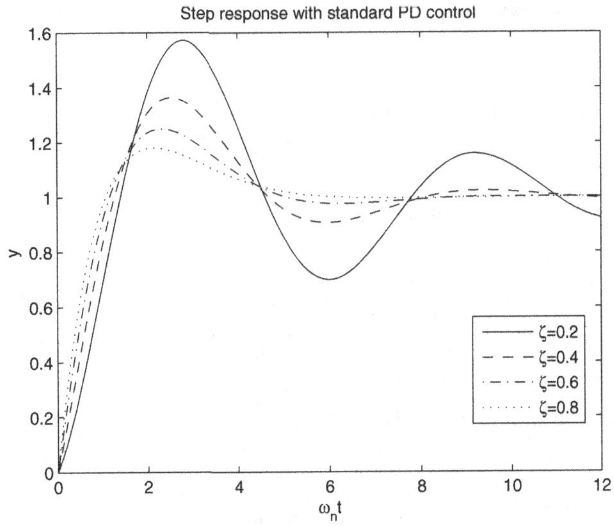

Figure 17.32 Unit step response with standard PD control

responses for the modified PD control and standard PD control. It can be seen that the settling times and rates of oscillation are similar. However, the rise and peak times with the standard PD control are faster than with the modified PD control. Also, the responses with standard PD control exhibit significantly more overshoot.

In general, the rule of thumb is that zeros far away from the poles do not have significant effects on the transient behavior. Zeros close to the poles do have significant effect on the transient behavior. A zero very close to a pole tends to reduce the contribution of that pole since a pole-zero cancelation almost occurs in the transfer function.

17.8.2 Effect of Additional Poles

In our formulation of the attitude control problem with PD control, we have a second order system. That is, there are two closed-loop poles. However, in reality the actuators that are used to control the attitude also have dynamics (which we have neglected). The attitude determination scheme may also have some dynamics. These can all contribute poles to the system. The control law itself could also be more elaborate, and contribute additional poles itself (for example, PID control contributes a pole). Therefore, it is important to understand what the effects of the additional poles are, and when we may still approximate our system by a second order system.

Generally speaking, the closed-loop transfer function will have the form

$$T(s) = K \frac{\prod_{m}^{i=1} (s - z_i)}{\prod_{n}^{i=1} (s - p_i)},$$

where $m \leq n$, and z_i are the closed-loop zeros and p_i are the closed-loop poles. Since we assume that the control law has asymptotically stabilized the system, we have $Re(p_i) < 0$.

The step response of this system is given by

$$Y(s) = T(s)\frac{1}{s} = K\frac{\prod_m^{i=1}(s - z_i)}{\prod_n^{i=1}(s - p_i)}\frac{1}{s}.$$

Assuming distinct poles, the partial fraction expansion has the form

$$Y(s) = \frac{a}{s} + \sum_{i=1}^{n} \frac{b_i}{s - p_i}.$$

In the time-domain, the step response is given by

$$y(t) = a + \sum_{i=1}^{n} b_i e^{p_i t}.$$

The contribution of each pole to the response depends on the associated coefficient b_i and how quickly it decays (how negative $Re(p_i)$ is).

If the ratios of the real parts of the poles near to the imaginary axis and poles further away are greater than 5, then the poles closest to the imaginary axis dominate the response (see Figure 17.33), since their contribution decays the slowest (lasts the longest). If there is a complex conjugate pair of dominant poles (which results in decaying oscillatory behavior), then the system approximately behaves as an underdamped second order system with those poles only (recall from Section 14.1 that a pair of stable complex conjugate poles results in decaying oscillatory behavior).

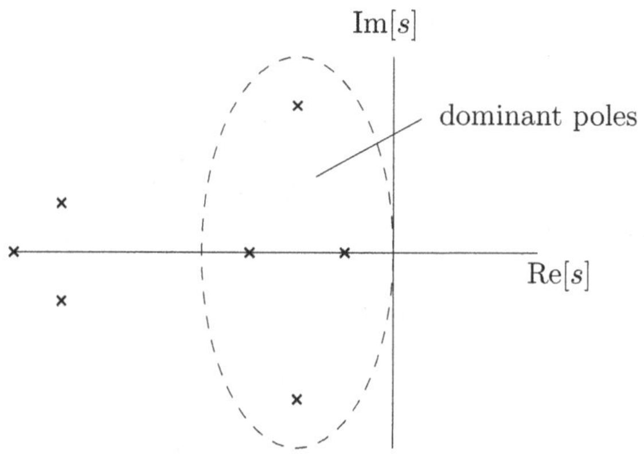

Figure 17.33 Dominant poles

17.9 Steady-State Specifications and System Type

As has been mentioned previously, the response of the output to a reference signal can be broken into two parts. The transient part which vanishes with time, and the steady-state part, which persists. We have seen how specifications for the transient part may be defined. Now we shall look at the steady-state part.

Consider the feedback control system shown in Figure 17.8 with $W(s) = 0$ (we do not consider measurement noise at this time). Note that every control system we have looked at can fit within this framework. Typically, the long-term (steady-state) objective is for the output $y(t)$ to match the reference signal $r(t)$. That is, once the transients have died out, we would like to have $e(t) = r(t) - y(t) \approx 0$. In our attitude control case, it means that we would like the actual attitude to eventually match the desired attitude. In practise, there are factors which prevent the error from going to zero. First of all, any non-vanishing disturbances $T_d(t)$ may cause a non-zero steady-state error. Secondly, the closed-loop system itself may not be capable of tracking the reference signal $r(t)$.

The closed-loop system corresponding to Figure 17.8 (with $W(s) = 0$) is given by

$$Y(s) = \frac{G_p(s)G_c(s)}{1 + G_p(s)G_c(s)} R(s) + \frac{G_p(s)}{1 + G_p(s)G_c(s)} T_d(s).$$

Since $E(s) = R(s) - Y(s)$, we find that

$$E(s) = \frac{1}{1 + G_p(s)G_c(s)} R(s) - \frac{G_p(s)}{1 + G_p(s)G_c(s)} T_d(s). \tag{17.20}$$

17.9.1 System Type

When we discussed transient specifications, they were given in terms of the response to a step input ($r(t) = constant$), which is a common type of reference signal. However in practise, the reference signal we wish to track may not be constant. We may or may not know in advance what $r(t)$ is going to be, and even if we do know it, it may not have a simple form that is useful for control design. We therefore examine the closed-loop response to test signals, which are similar to the expected reference signal $r(t)$. Common test signals include:

Step input $r(t) = 1$, $R(s) = 1/s$ (see Figure 17.34) – appropriate when the reference signal $r(t)$ is approximately constant for long periods of time. For example, we may want to occasionally change the attitude from one fixed reference attitude to another.
Ramp input $r(t) = t$, $R(s) = 1/s^2$ (see Figure 17.35) – appropriate when the reference signal $r(t)$ is slowly varying. For example, we may want the satellite to perform a slow attitude maneuver (such as tracking a target on the ground).
Acceleration input $r(t) = t^2/2$, $R(s) = 1/s^3$ (see Figure 17.36) – this is not commonly used.

Usually, if the closed-loop system behaves well for the test signals, it will behave well for the real signals.

Now, let us determine what types of signals the closed-loop system in equation (17.20) can track perfectly, what types of signals it can track but with some steady-state error, and what

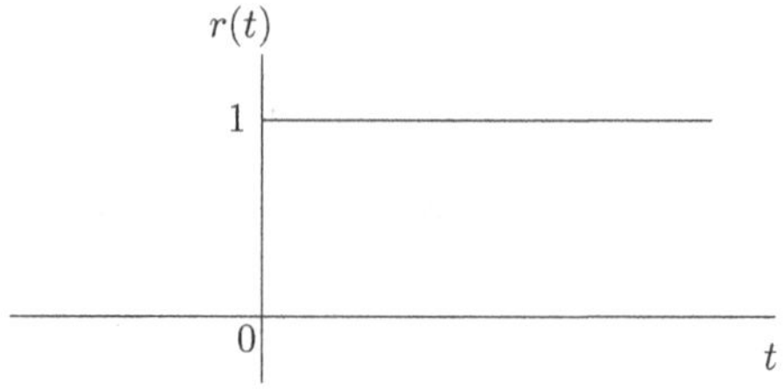

Figure 17.34 Unit step input

types of signals it cannot track at all. This introduces the notion of *system type*. To determine what kind of signal the closed-loop system can track, we must consider the transfer function from reference signal $R(s)$ to error $E(s)$ (we ignore the disturbance $T_d(s)$), and from (17.20) we have

$$E(s) = \frac{1}{1 + G_o(s)} R(s).$$

where we have defined the *open-loop transfer function* to be the product of all transfer functions around the loop (see Figure 17.8)

$$G_o(s) = G_p(s) G_c(s). \tag{17.21}$$

The open-loop transfer function will play a major role in a lot of the control system analysis we will perform in the next several chapters.

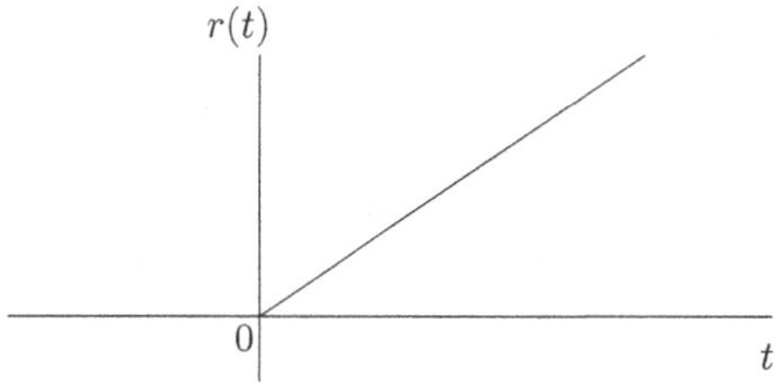

Figure 17.35 Ramp input

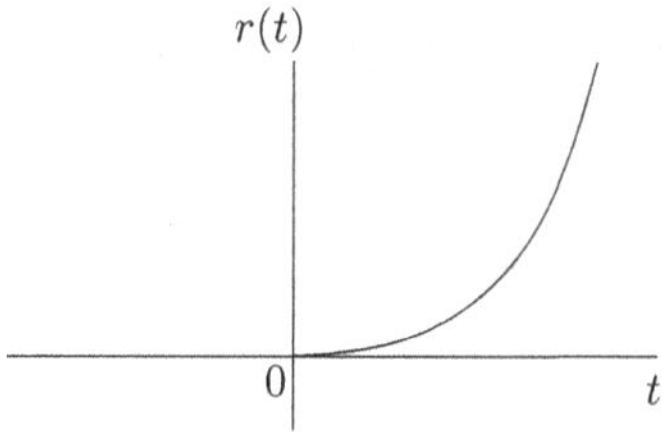

Figure 17.36 Acceleration input

In general, the open-loop transfer function can be written as

$$G_o(s) = \bar{K}\frac{(s - z_1)\dots(s - z_m)}{s^N(s - p_1)\dots(s - p_n)}. \tag{17.22}$$

where z_i are the open-loop zeros, and $p_i \neq 0$ are the non-zero open loop poles. There are also N open-loop poles at $s = 0$.

The type of the system is now classified as the value of N. For example,

- $N = 0$ - type 0 system
- $N = 1$ - type 1 system
- $N = 2$ - type 2 system

As we shall see, this is directly related to the type of reference signal the closed-loop system is able to track.

It will be useful to rewrite the open-loop transfer function (17.22) as

$$G_o(s) = K\frac{(T_{a1}s + 1)\dots(T_{am}s + 1)}{s^N(T_{b1}s + 1)\dots(T_{bn}s + 1)}, \tag{17.23}$$

where

$$K = \bar{K}\frac{(-z_1)\dots(-z_m)}{(-p_1)\dots(-p_n)},$$

and

$$T_{a1} = \frac{-1}{z_1}, \dots, T_{b1} = \frac{-1}{p_1}, \dots$$

Assuming that the closed-loop system is asymptotically stable (that is $1 + G_o(s) = 0$ only has roots with negative real parts), as it should be, we can examine the steady-state error to different types of reference signals using the final value theorem (see Appendix A.3). That is, the steady-state error is given by

$$e_{ss} = \lim_{t\to\infty} e(t) = \lim_{s\to 0} sE(s) = \lim_{s\to 0}\frac{s}{1 + G_o(s)}R(s). \tag{17.24}$$

Let us now examine the steady-state error for different types of inputs.

17.9.2 Step Input $R(s) = \frac{1}{s}$ $(r(t) = 1)$.

Applying the final value theorem, we have

$$e_{ss} = \lim_{s\to 0}\frac{s}{1 + G_o(s)}\frac{1}{s} = \frac{1}{1 + \lim_{s\to 0} G_o(s)}.$$

The *static position error constant* is defined as

$$K_{sp} = \lim_{s \to 0} G_o(s).$$

Then, the steady-state error to a step input is given by

$$e_{ss} = \frac{1}{1 + K_{sp}}.$$

For a type 0 system ($N = 0$),

$$K_{sp} = \lim_{s \to 0} K \frac{(T_{a1}s + 1) \ldots (T_{am}s + 1)}{(T_{b1}s + 1) \ldots (T_{bn}s + 1)} = K.$$

For type 1 or higher ($N \geq 1$),

$$K_{sp} = \lim_{s \to 0} K \frac{(T_{a1}s + 1) \ldots (T_{am}s + 1)}{s^N (T_{b1}s + 1) \ldots (T_{bn}s + 1)} = \infty.$$

Therefore, for type 0 systems, the steady-state error to a step input is

$$e_{ss} = \frac{1}{1 + K},$$

which is non-zero but finite. Type 1 or higher systems can track step inputs perfectly, and the steady-state error is

$$e_{ss} = \frac{1}{1 + \infty} = 0.$$

17.9.3 Ramp Input $R(s) = \frac{1}{s^2}$ $(r(t) = t)$

Applying the final value theorem, we have

$$e_{ss} = \lim_{s \to 0} \frac{s}{1 + G_o(s)} \frac{1}{s^2} = \lim_{s \to 0} \frac{1}{s + sG_o(s)} = \frac{1}{\lim_{s \to 0} sG_o(s)}.$$

The *static velocity error constant* is defined as

$$K_{sv} = \lim_{s \to 0} sG_o(s).$$

Then, the steady-state error to a ramp input is given by

$$e_{ss} = \frac{1}{K_{sv}}.$$

For a type 0 system ($N = 0$),

$$K_{sv} = \lim_{s\to 0} K \frac{s(T_{a1}s+1)\dots(T_{am}s+1)}{(T_{b1}s+1)\dots(T_{bn}s+1)} = 0.$$

For a type 1 system ($N = 1$),

$$K_{sv} = \lim_{s\to 0} K \frac{(T_{a1}s+1)\dots(T_{am}s+1)}{(T_{b1}s+1)\dots(T_{bn}s+1)} = K.$$

For type 2 or higher ($N \geq 2$),

$$K_{sp} = \lim_{s\to 0} K \frac{(T_{a1}s+1)\dots(T_{am}s+1)}{s^{N-1}(T_{b1}s+1)\dots(T_{bn}s+1)} = \infty.$$

Therefore, for type 0 systems, the steady-state error to a ramp input is

$$e_{ss} = \frac{1}{0} = \infty,$$

which shows that a type 0 system cannot track a ramp input. For a type 1 system, the steady-state error is

$$e_{ss} = \frac{1}{K},$$

which is non-zero but finite. Type 2 or higher systems can track step inputs perfectly, and the steady-state error is

$$e_{ss} = \frac{1}{\infty} = 0.$$

We can continue the argument for a type 2 system (with acceleration input) and higher.

The type of a system, N, is the number of integrators in the loop, which can either exist in the plant or controller (it doesn't matter where). In general, the more integrators in the loop (the higher the type), the higher the order of signal that can be tracked. To decrease the steady-state error to a reference input, we can either increase the open-loop gain, K, or increase the type of the system by adding integrators ($1/s$).

Example 17.2 (PD spacecraft attitude control) *Let us examine our proportional-derivative attitude control problem in Figure 17.14. The plant transfer function is*

$$G_p(s) = \frac{1}{Is^2 + K_d s} = \frac{1}{K_d s\left(\frac{I}{K_d}s + 1\right)}.$$

The control transfer function is

$$G_c(s) = K_p.$$

Therefore, the open-loop transfer function is

$$G_o(s) = G_p(s)G_c(s) = \frac{K_p}{K_d s\left(\frac{I}{K_d}s + 1\right)}.$$

Clearly, we have a type 1 system. This means that we can track step attitude commands with zero steady-state error.

The static velocity error constant is

$$K_{sv} = \lim_{s\to 0} sG_o(s) = \frac{K_p}{K_d}.$$

Therefore, we can track ramp inputs with steady-state error

$$e_{ss} = \frac{1}{K_{sv}} = \frac{K_d}{K_p}.$$

This is not very surprising, since a ramp input corresponds to a constant angular velocity command. However, the derivative term in the control law $-K_p\dot{y}$ *corresponds to a torque directly opposing any angular velocity. If we want to track ramp inputs with zero steady-state error, we can add an integral term to our controller, which increases the system type to 2. Alternatively, we could use standard proportional derivative control as in Section 17.6.2, which also increases the system type to 2. Hence, standard PD control is more appropriate if the control objective is to track a time-varying attitude command.*

17.10 Effect of Disturbances

We shall now look at the effect disturbances have on the steady-state error. Therefore, we shall ignore the contribution of the reference signal $R(s)$, and consider from (17.20)

$$E(s) = -\frac{G_p(s)}{1 + G_p(s)G_c(s)}T_d(s). \tag{17.25}$$

In practise, we do not know the nature of the disturbances (if we did, we could directly compensate for them). However, for spacecraft problems they are slowly varying with time, and a constant disturbance $T_d(t) = \hat{T}_d$ is a good approximation for the purposes of initial controller design.

As we did for the open-loop transfer function in equation (17.23), we can generally write the plant and control transfer functions in the form

$$G_p(s) = K_{pl}\frac{(T_{a1p}s + 1)\ldots(T_{amp}s + 1)}{s^{N_p}(T_{b1p}s + 1)\ldots(T_{bnp}s + 1)} = K_{pl}\frac{\mathcal{N}_p(s)}{s^{N_p}\mathcal{D}_p(s)},$$

and

$$G_c(s) = K_c\frac{(T_{a1c}s + 1)\ldots(T_{amc}s + 1)}{s^{N_c}(T_{b1c}s + 1)\ldots(T_{bnc}s + 1)} = K_c\frac{\mathcal{N}_c(s)}{s^{N_c}\mathcal{D}_c(s)}.$$

It is easy to see that

$$K_{pl} = \lim_{s\to 0} s^{N_p} G_p(s), \qquad K_c = \lim_{s\to 0} s^{N_c} G_c(s).$$

We can now form the transfer function from disturbance to error as

$$\frac{G_p(s)}{1+G_p(s)G_c(s)} = \frac{K_{pl}s^{N_c}\mathcal{N}_p(s)\mathcal{D}_c(s)}{s^{N_p+N_c}\mathcal{D}_p(s)\mathcal{D}_c(s)+K_{pl}K_c\mathcal{N}_p(s)\mathcal{N}_c(s)}.$$

We can now apply the final value theorem (assuming the closed-loop system is asymptotically stable) to find the steady-state error to a step disturbance $T_d(s) = \hat{T}_d/s$. That is, the steady-state error is given by

$$e_{ss} = \lim_{s\to 0} sE(s) = -\lim_{s\to 0} s\frac{K_{pl}s^{N_c}\mathcal{N}_p(s)\mathcal{D}_c(s)}{s^{N_p+N_c}\mathcal{D}_p(s)\mathcal{D}_c(s)+K_{pl}K_c\mathcal{N}_p(s)\mathcal{N}_c(s)}\frac{\hat{T}_d}{s},$$

which leads to

$$e_{ss} = -\lim_{s\to 0}\frac{K_{pl}s^{N_c}\mathcal{N}_p(s)\mathcal{D}_c(s)}{s^{N_p+N_c}\mathcal{D}_p(s)\mathcal{D}_c(s)+K_{pl}K_c\mathcal{N}_p(s)\mathcal{N}_c(s)}\hat{T}_d.$$

It is now clear that to be able to completely reject a constant disturbance (get $e_{ss} = 0$), we must have $N_c \geq 1$. That is, there must be an integrator $(1/s)$ in the controller to reject a constant disturbance.

Let us now look at the value for the steady-state error if there is no integrator in the controller ($N_c = 0$). Since,

$$\mathcal{N}_p(0) = \mathcal{N}_c(0) = \mathcal{D}_p(0) = \mathcal{D}_c(0) = 1,$$

we have

$$e_{ss} = \begin{cases} -\dfrac{K_{pl}\hat{T}_d}{1+K_{pl}K_c}, & N_p = 0, \\ -\dfrac{\hat{T}_d}{K_c}, & N_p \geq 1. \end{cases}$$

We see in both cases above, that if there is no integrator in the controller ($N_c = 0$), then we must increase the controller gain K_c in order to reduce the steady-state error to a disturbance.

Example 17.3 (PD spacecraft attitude control) *Let us again revisit our proportional-derivative attitude control problem in Figure 17.14. Let us assume that we have a constant disturbance torque $T_d(t) = \hat{T}_d$. Again, the plant transfer function is*

$$G_p(s) = \frac{1}{Is^2 + K_d s} = \frac{1}{K_d s\left(\frac{I}{K_d}s + 1\right)}.$$

and the control transfer function is

$$G_c(s) = K_p,$$

from which we get the plant and controller gains $K_{pl} = 1/K_d$ and $K_c = K_p$ respectively. It is clear that there is no integrator in the controller ($N_c = 0$). However, the plant does have one integrator $N_p = 1$. Therefore, the steady-state error to a constant disturbance torque is

$$e_{ss} = -\frac{\hat{T}_d}{K_p}.$$

If we are given a specification on the steady-state attitude error that $|e_{ss}| \leq e_{\max,ss}$, then the proportional gain K_p must be chosen such that

$$K_p \geq \frac{\hat{T}_d}{e_{max,ss}}.$$

Since $K_p/I = \omega_n^2$ (see (17.10)), this provides an additional restriction on the closed-loop pole locations. We have seen that $|s|^2 = \omega_n^2$ for underdamped poles. This means that the closed-loop poles must satisfy

$$|s|^2 \geq \frac{\hat{T}_d}{Ie_{max,ss}}.$$

Therefore, the allowable region for closed-loop poles is as shown in Figure 17.37.

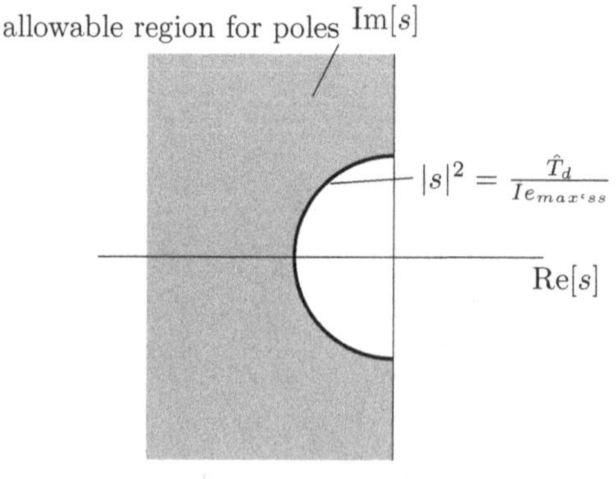

Figure 17.37 Allowable pole locations with disturbance rejection constraints

17.11 Actuator Limitations

The torque on the spacecraft demanded by our control law must be applied by an actuator. Any real actuator has limitations on the amount of torque that it can apply. That is, $|u(t)| = |T_c(t)| \leq T_{max}$. We should therefore ensure that our control law will not demand more torque than the actuator is capable of.

Let e_{max} be the largest attitude error that we expect to encounter. The corresponding restoring torque from our proportional-derivative control law (assuming zero angular velocity) will have magnitude

$$u = K_p e_{max}.$$

Therefore, to ensure that the control torque stays within the actuator capabilities, we must have

$$K_p e_{max} \leq T_{max},$$

which leads to a constraint that

$$K_p \leq \frac{T_{max}}{e_{max}}.$$

That is, the proportional gain should not be too large. Since $K_p/I = \omega_n^2$ (see (17.10)), this again provides an additional restriction on the closed-loop pole locations. As we have seen, $|s|^2 = \omega_n^2$ for underdamped poles. This means that the closed-loop poles must satisfy

$$|s|^2 \leq \frac{T_{max}}{I e_{max}}.$$

Therefore, the allowable region for closed-loop poles is as shown in Figure 17.38. We could similarly form a constraint on the derivative gain K_d considering the maximum expected angular velocity $\dot{y}_{max}$. We could also form a constraint on both gains K_p and K_d considering

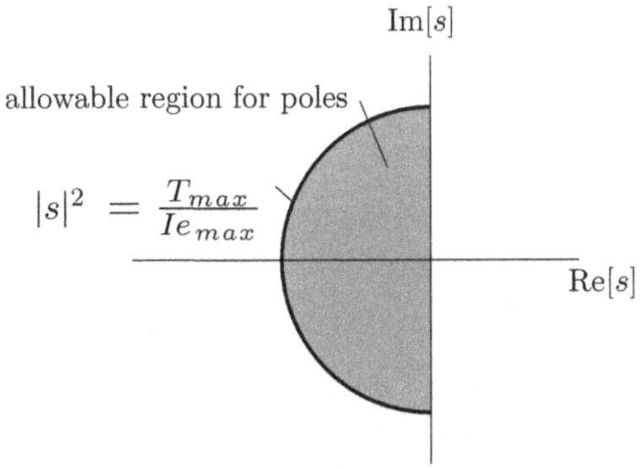

Figure 17.38 Allowable pole locations with actuator limitation constraints

e_{max} and $\dot{y}_{max}$ simultaneously. This is not an exact science; there is art involved in control design.

Notes

This chapter has presented an introduction to control systems, and in particular second-order systems such as spacecraft attitude dynamics. There are several excellent texts on this subject (for example Qiu and Zhou (2010), Franklin et al. (2010) and Ogata (2010)), and the reader is encouraged to consult them for further details.

References

Franklin GF, Powell JD and Emami-Naeini A 2010 *Feedback Control of Dynamic Systems* 6th edn. Prentice Hall, Upper Saddle River, NJ.

Ogata K 2010 *Modern Control Engineering* 5th edn. Prentice Hall, Upper Saddle River, NJ.

Qiu L and Zhou K 2010 *Introduction to Feedback Control*. Prentice Hall, Upper Saddle River, NJ.

18

Routh's Stability Criterion

Given a general transfer function of the form

$$G(s) = \frac{a_0 s^n + a_1 s^{n-1} + \ldots + a_n}{b_0 s^n + b_1 s^{n-1} + \ldots + b_n},$$

where $b_0 > 0$ (usually we have $b_0 = 1$), we would like to be able to determine whether the system is asymptotically stable. We could just compute the poles numerically, and be done with it, but it is not very insightful. The poles obtained would be specific to the coefficients b_i, and no further insight on what range of coefficients b_i provide stability.

There is a means to determine stability without computing the poles. This is called *Routh's stability criterion*. Recall that the poles are the roots of the polynomial

$$b_0 s^n + b_1 s^{n-1} + \ldots + b_n = 0.$$

The Routh procedure is as follows.

1. First of all, the coefficients $b_1, b_2, \ldots, b_n$ must all be *positive*. If any of them are negative or zero, a pole exists that either lies on the imaginary axis, or has a positive real part. That is, the system is not asymptotically stable.
2. If all coefficients are positive, form the following array (called the Routh array).

$$\begin{array}{c|lllll}
s^n & b_0 & b_2 & b_4 & b_6 & \cdots \\
s^{n-1} & b_1 & b_3 & b_5 & b_7 & \cdots \\
s^{n-2} & c_1 & c_2 & c_3 & c_4 & \cdots \\
s^{n-2} & d_1 & d_2 & d_3 & d_4 & \cdots \\
\vdots & \vdots & \vdots & & & \\
s^2 & f_1 & f_2 & & & \\
s^1 & g_1 & & & & \\
s^0 & h_1 & & & &
\end{array}$$

Spacecraft Dynamics and Control: An Introduction, First Edition.
Anton H.J. de Ruiter, Christopher J. Damaren and James R. Forbes.

Note that the table has powers of s in descending order. The coefficients $c_1, c_2, c_3, \ldots$ are computed as

$$c_1 = \frac{b_1 b_2 - b_3 b_0}{b_1}, \quad c_2 = \frac{b_1 b_4 - b_5 b_0}{b_1}, \quad c_3 = \frac{b_1 b_6 - b_7 b_0}{b_1}, \text{ etc.}$$

The coefficients $d_1, d_2, d_3, \ldots$ are computed as

$$d_1 = \frac{c_1 b_3 - c_2 b_1}{c_1}, \quad d_2 = \frac{c_1 b_5 - c_3 b_1}{c_1}, \quad d_3 = \frac{c_1 b_7 - c_4 b_1}{c_1}, \text{ etc.}$$

Coefficients $e_1, e_2, e_3, \ldots$ are computed similarly using the rows containing $c_1, c_2, c_3, \ldots$ and $d_1, d_2, d_3, \ldots$ In general, coefficients are computed using the previous two rows.

Once the Routh array has been constructed, asymptotic stability can be determined by the first column of coefficients. The transfer function $G(s)$ is asymptotically stable if and only if *all coefficients in the first column are positive.* That is, $b_1, c_1, d_1, e_1, \ldots$ must all be positive.

18.1 Proportional-Derivative Control with Actuator Dynamics

Let us return to our modified proportional derivative design for spacecraft attitude control in Section 17.7.1. Let us assume that the proportional and derivative gains K_p and K_d have been designed to give a desired pair of underdamped poles, $s = -\zeta\omega_n \pm j\omega_n\sqrt{1-\zeta^2}$. In the design in Section 17.7.1, it has been assumed that there is an actuator capable of providing the control torque $u(t) = K_p e(t) - K_d \dot{y}(t)$. In reality, actuators themselves have some dynamics as well, and there is a delay between the demanded control torque, and the actual torque delivered by the actuator. For reaction wheels, a reasonable actuator model is the first order transfer function

$$U(s) = \frac{1}{Ts+1} U_c(s),$$

where $T > 0$ is a time-constant of the actuator, $U(s)$ is the actual torque delivered by the actuator, and $U_c(s)$ is the torque commanded by the control law. Let us look at what the actual torque looks like for a step torque command $u_c(t) = \hat{u}$. We have, $U_c(s) = \hat{u}/s$, such that

$$U(s) = \frac{1/T}{s + 1/T} \frac{\hat{u}}{s}.$$

Taking the partial fraction expansion, we have

$$U(s) = \frac{\hat{u}}{s} - \frac{\hat{u}}{s + 1/T}. \tag{18.1}$$

Taking inverse Laplace transforms, we have

$$u(t) = \hat{u}\left(1 - e^{-t/T}\right). \tag{18.2}$$

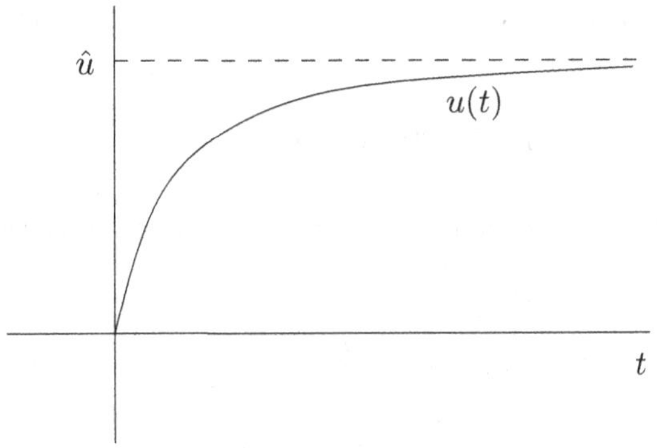

Figure 18.1 Actuator step response

This is shown graphically in Figure 18.1. We see from equation (18.2) that the time constant T determines the speed of the actuator response to a torque command, and the smaller T is, the faster the response. We expect that the faster the response of the actuator is, the better the closed-loop performance. Let us now use the Routh analysis to determine requirements on the actuator time constant T to ensure asymptotic stability. The torque commanded by the control law is

$$u_c(t) = K_p e(t) - K_d \dot{y}(t).$$

With the actuator dynamics given by (18.1), the block diagram in Figure 17.14 becomes as shown in Figure 18.2. The closed-loop transfer function from commanded attitude $R(s)$ to actual attitude $Y(s)$ can be found to be

$$Y(s) = \frac{K_p}{ITs^3 + Is^2 + K_d s + K_p} R(s).$$

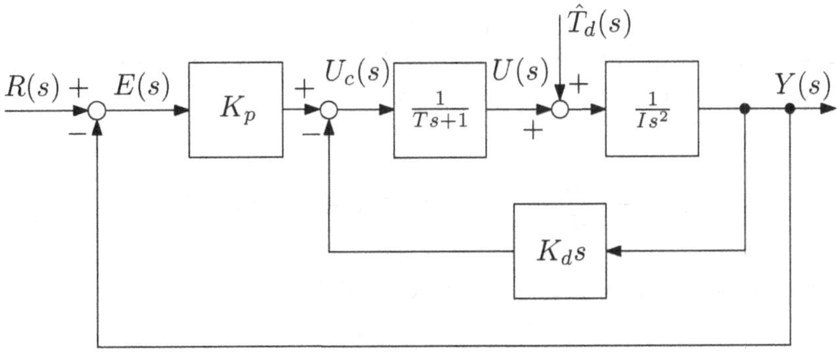

Figure 18.2 Modified proportional-derivative control with actuator dynamics

The poles when actuator dynamics are included are therefore the roots of the characteristic equation

$$ITs^3 + Is^2 + K_d s + K_p = 0. \tag{18.3}$$

Clearly, all coefficients in (18.3) are positive. Therefore, we can proceed with constructing a Routh table.

s^3	IT	K_d
s^2	I	K_p
s^1	c_1	
s^0	d_1	

We have

$$c_1 = \frac{IK_d - ITK_p}{I} = K_d - K_p T,$$

and

$$d_1 = K_p.$$

For asymptotic stability, we must have $c_1 > 0$ and $d_1 > 0$. Clearly, $d_1 > 0$ is automatically satisfied since $K_p > 0$. It is the coefficient c_1 that places a constraint on the time constant T. Namely, we must have

$$K_d - K_p T > 0,$$

equivalently

$$T < \frac{K_d}{K_p}.$$

If $T > K_d/K_p$, the closed-loop system is unstable. Since a smaller T means a faster actuator, this confirms our intuition that the actuator must be sufficiently fast.

If $T = K_d/K_p$, we expect there to be a closed-loop pole on the imaginary axis (since this is the boundary of stability). Let us check this by setting $T = K_d/K_p$ and $s = j\omega$ in (18.3). We have

$$-j\frac{IK_d}{K_p}\omega^3 - I\omega^2 + jK_d\omega + K_p = 0.$$

Separating this into real and imaginary components, we have

$$j\frac{IK_d}{K_p}\omega^3 = jK_d\omega,$$

and

$$I\omega^2 = K_p.$$

Both of these equations yield the same solution $\omega = \pm\sqrt{K_p/I}$. This confirms that $s = \pm j\sqrt{K_p/I}$ are closed-loop poles when $T = K_d/K_p$.

18.2 Active Dual-Spin Stabilization

We shall now apply the Routh stability analysis to a more complicated example, namely the control design for a dual-spin stabilized satellite augmented with active control. We recall the equations of motion from Section 15.1. Namely,

$$\mathbf{I}\dot{\boldsymbol{\omega}} + \boldsymbol{\omega}^{\times}\left[\mathbf{I}\boldsymbol{\omega} + h_s\mathbf{a}\right] = \mathbf{T}_c. \tag{18.4}$$

where $\mathbf{T}_c$ is the control torque. Note that we are neglecting any disturbance torques. We shall also neglect measurement noise. As in Section 15.2, we shall assume that the spacecraft body frame $\mathcal{F}_b$ is a principal axis frame, that is

$$\mathbf{I} = \begin{bmatrix} I_x & 0 & 0 \\ 0 & I_y & 0 \\ 0 & 0 & I_z \end{bmatrix}.$$

Unlike Section 15.2, we shall assume that the momentum wheel spin axis is aligned with the body-frame y-axis, that is

$$\mathbf{a} = \begin{bmatrix} 0 \\ 1 \\ 0 \end{bmatrix}.$$

The control objective here is to control the spacecraft attitude relative to an inertially fixed reference frame $\mathcal{F}_I$. As in Section 17.1, we shall represent the attitude of the $\mathcal{F}_b$ relative to $\mathcal{F}_I$ by a 3-2-1 Euler sequence

1. A rotation ψ about the original z-axis (called the yaw angle).
2. A rotation θ about the intermediate y-axis (called a pitch angle).
3. A rotation ϕ about the transformed x-axis (called a roll angle).

The attitude kinematics relative to the inertial frame are then given by (see Section 11.2)

$$\boldsymbol{\omega}_{bI} = \begin{bmatrix} 1 & 0 & -\sin\theta \\ 0 & \cos\phi & \sin\phi\cos\theta \\ 0 & -\sin\phi & \cos\phi\cos\theta \end{bmatrix}\begin{bmatrix} \dot{\phi} \\ \dot{\theta} \\ \dot{\psi} \end{bmatrix}, \tag{18.5}$$

where $\vec{\omega}_{bI} = \vec{\mathcal{F}}_b^T \omega_{bI}$ is the angular velocity of the body frame $\mathcal{F}_b$ relative to the inertial frame $\mathcal{F}_I$ expressed in body coordinates. We shall assume that the angles and angular velocity are small, that is, ϕ, θ, ψ, $\dot{\phi}$, $\dot{\theta}$ and $\dot{\psi}$ are considered small. This allows the following simplifications

$$\sin\phi \approx \phi, \quad \cos\phi \approx 1, \quad \sin\theta \approx \phi, \quad \cos\theta \approx 1, \quad \sin\psi \approx \phi, \quad \cos\psi \approx 1.$$

Under this approximation, the attitude kinematics in (18.5) become

$$\boldsymbol{\omega}_{bI} = \begin{bmatrix} \dot{\phi} \\ \dot{\theta} \\ \dot{\psi} \end{bmatrix}, \tag{18.6}$$

and the angular acceleration becomes

$$\dot{\boldsymbol{\omega}}_{bI} = \begin{bmatrix} \ddot{\phi} \\ \ddot{\theta} \\ \ddot{\psi} \end{bmatrix}. \tag{18.7}$$

Substituting approximations (18.6) and (18.7) into the dynamical equation (18.4) and neglecting products of angular rates gives

$$I_x\ddot{\phi} - h_s\dot{\psi} = T_x, \tag{18.8}$$

$$I_y\ddot{\theta} = T_y, \tag{18.9}$$

$$I_z\ddot{\psi} + h_s\dot{\phi} = T_z. \tag{18.10}$$

We note that the pitch equation (18.9) is decoupled from the roll and yaw equations ((18.8) and (18.10) respectively). The pitch equation is just the same as the simple attitude equation in Section 17.1, for which a PD controller can be designed, as shown in detail in Chapter 17.

The roll and yaw equations on the other hand are coupled, and it is not immediately obvious if a simple PD control law is appropriate. Let us examine a different possible control law design for the roll-yaw equations. First, let us make the approximations $|h_s\dot{\psi}| \gg |I_x\ddot{\phi}|$, and $|h_s\dot{\phi}| \gg |I_z\ddot{\psi}|$. The roll and yaw equations (18.8) and (18.10) respectively, become

$$\dot{\psi} = -\frac{T_x}{h_s}, \tag{18.11}$$

$$\dot{\phi} = \frac{T_z}{h_s}. \tag{18.12}$$

Let us choose the gyroscopic control law

$$T_x = K_G\psi, \quad T_z = -K_G\phi. \tag{18.13}$$

Substituting these into (18.11) and (18.12), we have

$$\dot{\psi} = -\frac{K_G}{h_s}\psi,$$

$$\dot{\phi} = -\frac{K_G}{h_s}\phi,$$

both of which have the same form and solution

$$\psi(t) = e^{-t/T}\psi(0), \quad \phi(t) = e^{-t/T}\phi(0),$$

where

$$T = \frac{h_s}{K_G}.$$

Clearly for asymptotic stability, we must have $T > 0$, which means that the gyroscopic gain

$$K_G = \frac{h_s}{T},$$

must have the same sign as h_s. The quantity T is then a time constant related to the rate of decay, and can be used as a control design parameter.

Let us now examine the stability of the closed-loop system when the control law (18.13) is applied to (18.8) and (18.10). We have

$$I_x\ddot{\phi} - h_s\dot{\psi} = K_G\psi,$$

$$I_x\ddot{\psi} + h_s\dot{\phi} = -K_G\phi.$$

Taking the Laplace transform gives

$$I_x s^2\phi - sI_x\phi(0) - I_x\dot{\phi}(0) - sh_s\psi + h_s\psi(0) = K_G\psi,$$

$$I_z s^2\psi - sI_z\psi(0) - I_z\dot{\psi}(0) + sh_s\phi - h_s\phi(0) = -K_G\phi,$$

which can be put in matrix form

$$\begin{bmatrix} s^2 I_x & -(sh_s + K_G) \\ sh_s + K_G & s^2 I_z \end{bmatrix}\begin{bmatrix} \phi \\ \psi \end{bmatrix} = \begin{bmatrix} sI_x\phi(0) + I_x\dot{\phi}(0) - h_s\psi(0) \\ sI_z\psi(0) - I_z\dot{\psi}(0) + h_s\phi(0) \end{bmatrix}.$$

The closed-loop characteristic equation is the determinant of the matrix on the left, namely

$$\det\begin{bmatrix} s^2 I_x & -(sh_s + K_G) \\ sh_s + K_G & s^2 I_z \end{bmatrix} = 0.$$

Evaluating the determinant, we see that the closed-loop poles satisfy

$$I_x I_z s^4 + h_s^2 s^2 + 2h_s K_G s + K_G^2 = 0.$$

We can immediately see by Step 1 of the Routh analysis that since the coefficient of s^3 is zero, the closed-loop system is not asymptotically stable. This suggests the need to add some damping. Let us try the control law

$$T_x = K_G\psi - I_x K_d\dot{\phi}, \quad T_z = -K_G\phi - I_z K_d\dot{\psi}, \tag{18.14}$$

where K_d is a derivative gain. Substituting (18.14) into (18.8) and (18.10), the closed-loop equations now become

$$\begin{aligned} I_x\ddot{\phi} - h_s\dot{\psi} &= K_G\psi - I_x\dot{\phi}, \\ I_z\ddot{\psi} + h_s\dot{\phi} &= -K_G\phi - I_z\dot{\psi}. \end{aligned}$$

Taking Laplace transforms leads to

$$\begin{bmatrix} s^2 I_x + sI_x K_d & -(sh_s + K_G) \\ sh_s + K_G & s^2 I_z + sI_z K_d \end{bmatrix} \begin{bmatrix} \phi \\ \psi \end{bmatrix} = \begin{bmatrix} I_x(s+1)\phi(0) + I_x\dot{\phi}(0) - h_s\psi(0) \\ I_z(s+1)\psi(0) - I_z\dot{\psi}(0) + h_s\phi(0) \end{bmatrix}.$$

As we saw previously, the stability analysis only depends on the determinant of the matrix on the left. The right-hand side contains the initial conditions, and is not relevant for stability analysis. The characteristic equation now becomes

$$\det\begin{bmatrix} s^2 I_x + sI_x K_d & -(sh_s + K_G) \\ sh_s + K_G & s^2 I_z + sI_z K_d \end{bmatrix} = 0,$$

which is evaluated to be

$$I_x I_z s^4 + 2K_d I_x I_z s^3 + (I_x I_z K_d^2 + h_s^2)s^2 + 2h_s K_G s + K_G^2 = 0. \tag{18.15}$$

We can now perform a Routh analysis to determine constraints on the derivative gain K_d to make the closed-loop system asymptotically stable.

Step 1: All coefficients in (18.15) must be positive. We see that the coefficients of s^4, s^2, s^1, s^0 are already positive (recall that K_G and h_s have the same sign). Therefore, we obtain our first condition on K_d from the coefficient of s^3. Namely, that

$$K_d > 0 \tag{18.16}$$

Step 2: Create the Routh array.

$$\begin{array}{c|ccc} s^4 & I_x I_z & I_x I_z K_d^2 + h_s^2 & K_G^2 \\ s^3 & 2K_d I_x I_z & 2h_s K_G & \\ s^2 & c_1 & c_2 & \\ s^1 & d_1 & & \\ s^0 & e_1 & & \end{array}$$

We now have

$$c_1 = \frac{2K_d I_x I_z (I_x I_z K_d^2 + h_s^2) - 2h_s K_G I_x I_z}{2K_d I_x I_z} = \frac{K_d (I_x I_z K_d^2 + h_s^2) - h_s K_G}{K_d},$$

$$c_2 = K_G^2,$$

$$d_1 = \frac{\frac{K_d (I_x I_z K_d^2 + h_s^2) - h_s K_G}{K_d} 2h_s K_G - K_G^2 2K_d I_x I_z}{\frac{K_d (I_x I_z K_d^2 + h_s^2) - h_s K_G}{K_d}},$$

$$= \frac{2h_s K_G (K_d (I_x I_z K_d^2 + h_s^2) - h_s K_G) - 2K_d^2 I_x I_z K_G^2}{K_d (I_x I_z K_d^2 + h_s^2) - h_s K_G}.$$

$$e_1 = K_G^2.$$

Now, for asymptotic stability, we require $c_1 > 0$, $d_1 > 0$ and $e_1 > 0$. We automatically have $e_1 = K_G^2 > 0$. Let us now consider c_1 and d_1 sequentially. First of all, we have already seen that we must have $K_d > 0$. Therefore, the requirement that $c_1 > 0$ becomes

$$K_d (I_x I_z K_d^2 + h_s^2) - h_s K_G > 0. \tag{18.17}$$

Now, it can readily be seen that (18.17) can only be satisfied if (18.16) is satisfied (see what happens to (18.17) if $K_d \leq 0$). Therefore, if (18.17) is satisfied, (18.16) is automatically satisfied (condition (18.17) supercedes condition (18.16)). Next, since $c_1 > 0$, the requirement that $d_1 > 0$ becomes

$$h_s K_G (K_d (I_x I_z K_d^2 + h_s^2) - h_s K_G) - K_d^2 I_x I_z K_G^2 > 0. \tag{18.18}$$

Now, it can readily be seen that (18.18) can only be satisfied if (18.17) is satisfied (see what happens to (18.18) if $K_d (I_x I_z K_d^2 + h_s^2) - h_s K_G \leq 0$, remembering that $h_s K_G > 0$). Therefore, if (18.18) is satisfied, (18.17) and (18.16) are automatically satisfied (condition (18.18) supercedes conditions (18.17) and (18.16)). As such, we only need to find what range of K_d satisfy (18.18).

Now, let us factorize the left hand side of (18.18) to obtain

$$h_s K_G \left(K_d - \frac{K_G}{h_s} \right) (I_x I_z K_d^2 + h_s^2) > 0. \tag{18.19}$$

From (18.19) **it is very easily seen that since** $h_s K_G > 0$**, the closed-loop system is asymptotically stable if and only if**

$$K_d > \frac{K_G}{h_s} = \frac{1}{T}.$$

Of course, in designing the control gains K_G and K_d, we also need to account for steady-state error limitations and actuator limitations, just as we did in Sections 17.10 and 17.11.

Notes

This chapter has presented an introduction to Routh's stability criterion. We have then applied it to a spacecraft attitude control design to analyze the closed-loop stability when actuator dynamics are included. For further treatments of spacecraft attitude actuator dynamics models, the reader is referred to Sidi (1997). We have presented the Routh stability criterion from the point of view of whether or not the system is asymptotically stable (since this is the most important aspect for a control system). However, the Routh criterion in its more general form allows for the determination of the number of unstable poles (if there are any). For the more general presentation of the Routh criterion, the reader is referred to any of the excellent texts on classical control, including Qiu and Zhou (2010), Franklin et al. (2010) and Ogata (2010). The reader may find a mathematical proof of the Routh criterion in Qiu and Zhou (2010).

References

Franklin GF, Powell JD and Emami-Naeini A 2010 *Feedback Control of Dynamic Systems* 6th edn. Prentice Hall, Upper Saddle River, NJ.

Ogata K 2010 *Modern Control Engineering* 5th edn. Prentice Hall, Upper Saddle River, NJ.

Qiu L and Zhou K 2010 *Introduction to Feedback Control.* Prentice Hall, Upper Saddle River, NJ.

Sidi MJ 1997 *Spacecraft Dynamics and Control: A Practical Engineering Approach.* Cambridge University Press, New York, NY.

19

The Root Locus

Thus far in Chapters 17 and 18, we have seen three important factors for a closed-loop system. They are stability, transient response and steady-state error. The first two are influenced by the locations of the closed-loop poles. For stability, the poles must be in the left half complex plane. The transient response (as we saw in Chapter 17) is influenced by the locations of the dominant poles. It is therefore very useful to be able to see what closed-loop pole locations are achievable for a given control structure.

Example 19.1 (Root locus for spacecraft PD control) *Consider again the proportional-derivative attitude control problem as in Section 7.7.1. We found that the closed-loop transfer function is given by*

$$\frac{Y(s)}{R(s)} = \frac{(K_p/I)}{s^2 + (K_d/I)s + (K_p/I)}.$$

and the closed-loop poles are given by the roots of the characteristic equation

$$s^2 + \frac{K_d}{I}s + \frac{K_p}{I} = 0. \tag{19.1}$$

Problem Statement: *What closed-loop poles are possible for any proportional gain $K_p > 0$ given a fixed derivative gain K_d?*

For simplicity, let us set $I = K_d = 1$. We can solve (19.1) *to find the closed-loop poles*

$$s = -\frac{1}{2} \pm \frac{\sqrt{1 - 4K_p}}{2}. \tag{19.2}$$

We see from (19.2) *that*

- *when $0 \le K_p < 1/4$, the closed-loop poles are real, and lie between -1 and 0,*
- *when $K_p = 1/4$, there are two repeated poles at $s = -1/2$,*

Spacecraft Dynamics and Control: An Introduction, First Edition.
Anton H.J. de Ruiter, Christopher J. Damaren and James R. Forbes.

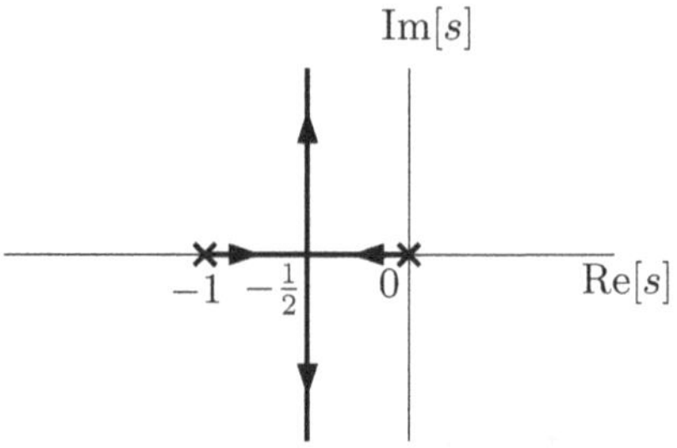

Figure 19.1 Root-locus for PD attitude control

- *when $K_p > 1/4$, the roots are imaginary, with $s = -1/2 \pm j\sqrt{4K_p - 1}/2$, that is, the real part is constant, and the imaginary part grows with increasing K_p.*

We can plot the closed-loop pole locations as shown in Figure 19.1. The plot in Figure 19.1 is called a root locus, *and it shows all possible closed-loop pole locations for $K_p \geq 0$ when $K_d = 1$ is fixed. Note that at the point $s = -1/2$, the root loci on the real axis meet, and then split into the complex plane. This is called a* breakaway point. *Note also that the root locus only shows how the closed-loop pole locations change with a single parameter (in this case K_p).*

For this example, we were able to determine the root locus analytically. In general, this is not possible. It can be generated numerically using a package such as MATLAB®. Historically, however, there is a graphical method of obtaining rough root locus plots. This method is important to understand, since it provides insight into the root locus, and how it changes with different control structures.

19.1 Rules for Constructing the Root Locus

Let us consider the general feedback system as shown in Figure 19.2. This has the closed-loop transfer function

$$\frac{Y(s)}{R(s)} = \frac{G_p(s)G_c(s)}{1 + G_p(s)G_c(s)},$$

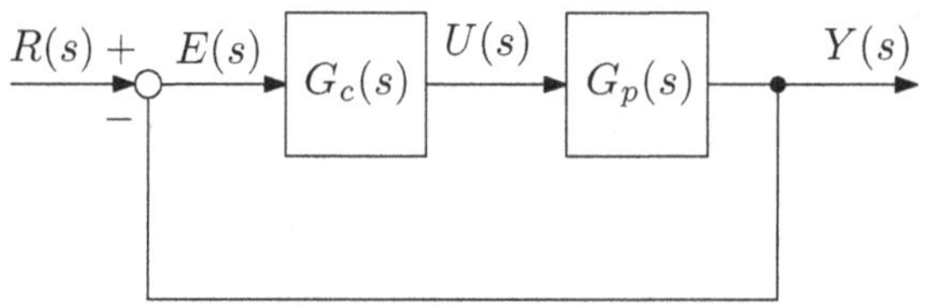

Figure 19.2 General feedback system

and the closed-loop poles satisfy

$$1 + G_p(s)G_c(s) = 0. \tag{19.3}$$

As we have seen in Section 17.9, $G_p(s)G_c(s)$ is the open-loop transfer function,

$$G_o(s) = G_p(s)G_c(s),$$

which may be written as

$$G_o(s) = K\frac{(s - z_1)...(s - z_m)}{(s - p_1)...(s - p_n)},$$

where z_1, ..., z_m are the open-loop zeros and p_1, ..., p_n are the open-loop poles, and $m \leq n$. Let us call the open-loop numerator and denominator

$$a(s) = (s - z_1)...(s - z_m), \quad b(s) = (s - p_1)...(s - p_n),$$

respectively. We can now represent the closed-loop characteristic equation (19.3) in several equivalent ways.

$$1 + K\frac{a(s)}{b(s)} = 0, \tag{19.4}$$

$$b(s) + Ka(s) = 0, \tag{19.5}$$

$$\frac{a(s)}{b(s)} = -\frac{1}{K}, \quad \text{if} \quad K \neq 0, \tag{19.6}$$

$$\frac{1}{K}b(s) + a(s) = 0, \quad \text{if} \quad K \neq 0, \tag{19.7}$$

The root locus is the set of all s satisfying the above equations (19.4) to (19.7) for $K \geq 0$. Recall that a complex number may be represented by a magnitude r and a phase θ (see Appendix A, Figure A.1),

$$s = r(\cos\theta + j\sin\theta) = re^{j\theta}.$$

With this in mind, since we take $K \geq 0$, the equation (19.6) shows that the root locus is the set of all s such that $a(s)/b(s)$ is negative and real, which can be interpreted in terms of the angle condition

$$\angle\frac{a(s)}{b(s)} = 180^\circ + 360^\circ(l - 1), \tag{19.8}$$

where l is an integer. Recall (see Appendix A) that the product of complex numbers $s_1 = r_1e^{j\theta_1}$,...,$s_n = r_ne^{j\theta_n}$ has phase

$$\angle s_1s_2...s_n = \theta_1 + \theta_2 + ... + \theta_n,$$

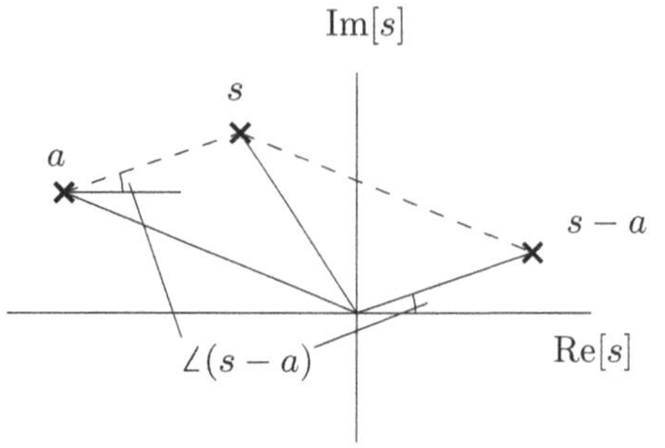

Figure 19.3 Determining argument graphically

and that for any two complex numbers $s_1 = r_1 e^{j\theta_1}$ and $s_2 = r_2 e^{j\theta_2}$, the quotient s_1/s_2 has phase

$$\angle\frac{s_1}{s_2} = \theta_1 - \theta_2.$$

Therefore, we have

$$\begin{aligned}\angle\frac{a(s)}{b(s)} &= \angle\left(\frac{(s-z_1)...(s-z_m)}{(s-p_1)...(s-p_n)}\right),\\ &= (\angle(s-z_1) + \angle(s-z_2) + ... + \angle(s-z_m))\\ &\quad -(\angle(s-p_1) + \angle(s-p_2) + ... + \angle(s-p_n)).\end{aligned}$$

We now need to find a method of determining $\theta = \angle(s-a)$. We could just do this numerically. However, a graphical method may provide more insight. Consider the diagram shown in Figure 19.3. We see that the angle $\angle(s-a)$ is just the angle of a line from point a to point s, measured counterclockwise from the positive real axis.

Let us now apply this knowledge to our root locus. Let s be a test point for the root locus. Let us plot the open-loop poles as $\times$, and the open-loop zeros as $\circ$, as in Figure 19.4. Then,

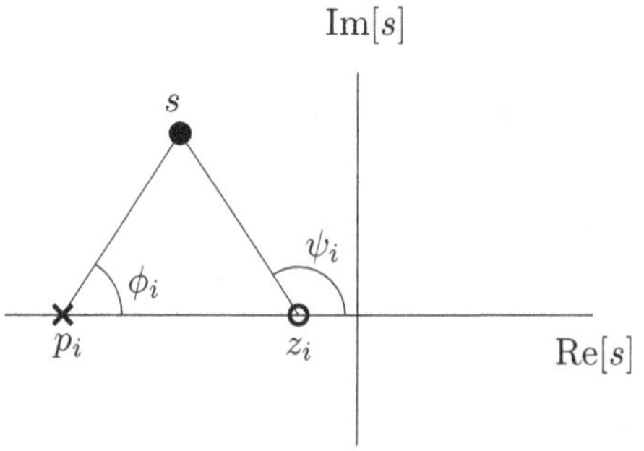

Figure 19.4 Test point for root locus

using the angle condition (19.8), s is on the root locus (an achievable closed-loop pole for some $K \geq 0$) if and only if

$$\sum \psi_i - \sum \phi_i = 180^\circ + 360^\circ(l-1), \tag{19.9}$$

for some integer l, where $\phi_i = \angle(s - p_i)$ are the angles of the lines drawn from the open-loop poles to the test point s, and $\psi_i = \angle(s - z_i)$ are the angles of the lines drawn from the open-loop zeros to the test point s.

19.1.1 Rules for constructing the root locus

Using the angle condition (19.9) as well as conditions (19.4) to (19.7), it is possible to derive a complete set of rules for constructing the root locus. We shall now state these rules. Their derivations may be found in Section 19.3.

Given an open-loop transfer function with m zeros and n poles such that $m \leq n$, the rules for constructing the root locus are:

1. There are n branches of the root locus (since there are n poles). Each branch starts at an open-loop pole, p_i (when $K = 0$). m branches end at an open-loop zero z_i (when $K \to \infty$). The other $n - m$ branches go to infinity along an asymptote (when $K \to \infty$), as seen in Rule 3.
2. The root locus includes all points on the real axis to the left of an odd number of open-loop real poles and zeros, as shown in Figure 19.5.
3. As $K \to \infty$, $n - m$ branches approach $n - m$ asymptotes with angles measured counter-clockwise from the positive real axis given by

$$\phi_l = \frac{180^\circ + 360^\circ(l-1)}{n-m}, \quad l = 1, 2, ..., n - m.$$

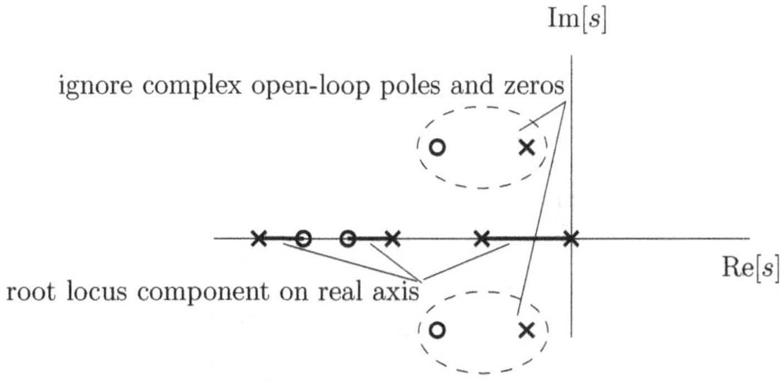

Figure 19.5 Root locus component on real axis

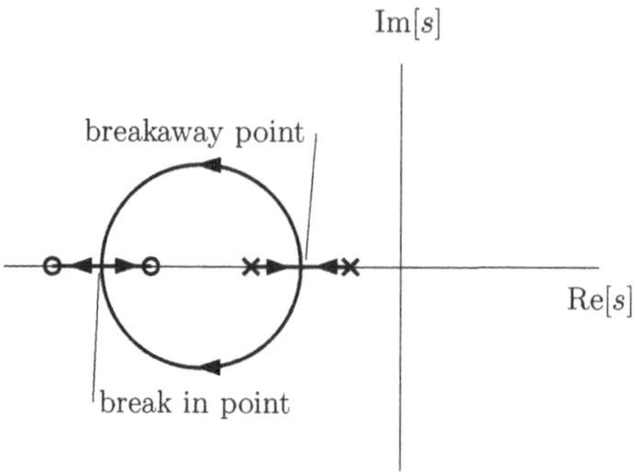

Figure 19.6 Breakaway and break-in points

The asymptotes have origin on the real axis at

$$\alpha = \frac{\sum_{i=1}^{n} p_i - \sum_{i=1}^{m} z_i}{n - m}.$$

4. The *break-away points* (where the root-loci meet and split), and the *break-in points* (where the root-loci meet and join) satisfy

$$a(s)\frac{db(s)}{ds} - b(s)\frac{da(s)}{ds} = 0.$$

Note that only solutions of this equation that lie on the root-locus may be considered as breakaway or break in points (see Figure 19.6).

5. The angle of departure ϕ_k of a branch from a pole p_k of multiplicity n_k (see Figure 19.7) is given by

$$n_k\phi_k = \sum_{i=1}^{m} \angle(p_k - z_i) - \sum_{j=1, j\neq g \text{ if } p_g = p_k}^{n} \angle(p_k - p_j) + 180^\circ + 360^\circ(l - 1).$$

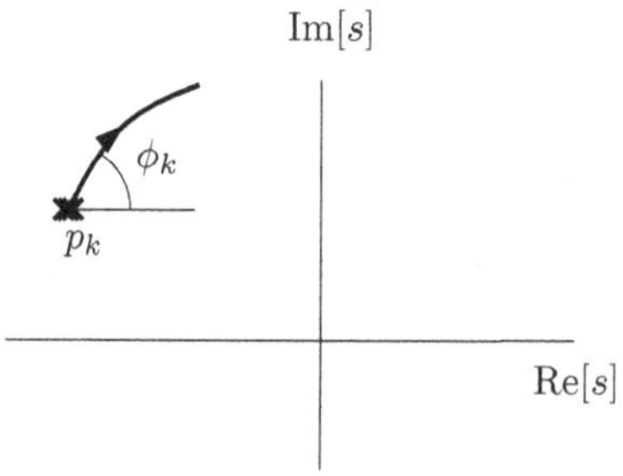

Figure 19.7 Departure angle from a repeated pole

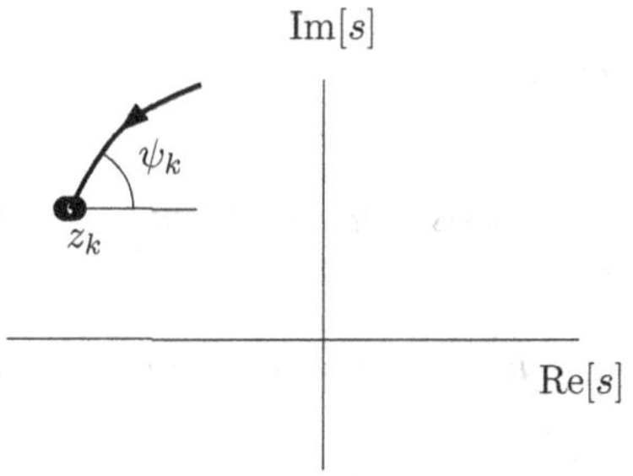

Figure 19.8 Arrival angle at a repeated zero

The angle of arrival ψ_k at a zero z_k of multiplicity n_k (see Figure 19.8) is given by

$$n_k\psi_k = -\sum_{i=1, i\neq g \text{ if } z_g=z_k}^{m} \angle(z_k - z_i) + \sum_{j=1}^{n} \angle(z_k - p_j) + 180^\circ + 360^\circ(l-1).$$

Note that l can be any integer in both cases.

6. The points where the root loci cross the imaginary axis may be determined using the Routh stability analysis method.

Example 19.2 *Consider a system with open-loop transfer function*

$$G_o(s) = \frac{K}{s(s+1)(s+2)}.$$

There are no open-loop zeros, so $m = 0$. The open-loop poles are $s = 0, -1, -2$, so $n = 3$. Let us now construct the root-locus using the applicable rules above. The completed root locus is shown in Figure 19.9.

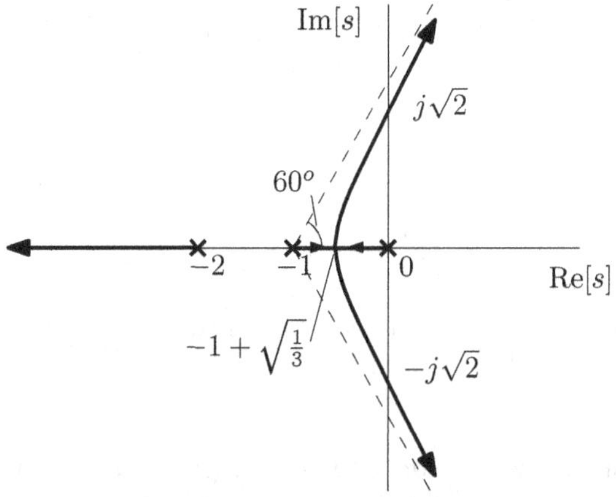

Figure 19.9 Example root locus

Step 1
Plot the open-loop poles and zeros.

Step 2
Determine the root locus on the real axis (between 0 and −1, and to the left of −2).

Step 3
Determine the asymptotes. There are $n - m = 3$ asymptotes. The angles of the asymptotes are

$$\phi_1 = \frac{180^\circ}{n-m} = \frac{180^\circ}{3} = 60^\circ,$$
$$\phi_2 = \frac{180^\circ + 360^\circ}{n-m} = \frac{180^\circ + 360^\circ}{3} = 180^\circ,$$
$$\phi_3 = \frac{180^\circ + 360^\circ \times 2}{n-m} = \frac{180^\circ + 360^\circ \times 2}{3} = 300^\circ \text{ (equivalently } -60^\circ\text{)}.$$

The center of the asymptotes is at

$$\alpha = \frac{\sum_{i=1}^{n} p_i - \sum_{i=1}^{m} z_i}{n-m} = \frac{-1-2}{3} = -1.$$

Step 4
Determine the breakaway point. We have $a(s) = 1$, and $b(s) = s(s+1)(s+2) = s^3 + 3s^2 + 2s$. Therefore,

$$\frac{da}{ds} = 0, \quad \frac{db}{ds} = 3s^2 + 6s + 2.$$

The breakaway point satisfies

$$a(s)\frac{db(s)}{ds} - b(s)\frac{da(s)}{ds} = 3s^2 + 6s + 2 = 0,$$

which has solution

$$s = -1 \pm \sqrt{\frac{1}{3}}.$$

Now, the point $s = -1 - \sqrt{1/3}$ lies between $s = -1$ and $s = -2$, and is therefore not on the root locus. Therefore, the breakaway point is at $s = -1 + \sqrt{1/3}$.

Step 5
Determine the points where the root loci cross the imaginary axis (they must exist, since two of the asymptotes enter the right-half plane). We could use a Routh analysis, however we shall demonstrate another approach.

The closed-loop characteristic equation can be found to be (see (19.5))

$$s^3 + 3s^2 + 2s + K = 0.$$

A pole on the imaginary axis would have the form $s = j\omega$. Substituting this into the above, we have

$$-j\omega^3 - 3\omega^2 + j2\omega + K = 0.$$

Equating the real and imaginary parts to zero, we have two equations for ω and the corresponding gain K.

$$\omega(\omega^2 - 2) = 0,$$

and

$$K - 3\omega^2 = 0.$$

The first of these gives $\omega = 0, \pm\sqrt{2}$. We already know that $\omega = 0$ lies on the root-locus since it is an open-loop pole. The imaginary axis crossing must therefore occur at $s = \pm j\sqrt{2}$. From the second equation, we see that the corresponding gain is $K = 6$.

Step 6
Draw the complete root-locus (shown in Figure 19.9).

19.1.2 Important Points

- If there is an open-loop zero in the right-half plane, the closed-loop system will eventually become unstable for a large enough gain K, as shown in Figure 19.10.
- If there are three or more open-loop zeros than poles ($n - m \geq 3$), the closed-loop system will eventually become unstable for a large enough gain K. To see this, consider that the number of asymptotes is determined by the number of open-loop zeros and poles, and has nothing to do with their location. The first asymptote has angle

$$\phi_1 = \frac{180^\circ}{n - m} < 90^\circ \quad \text{if} \quad n - m \geq 3.$$

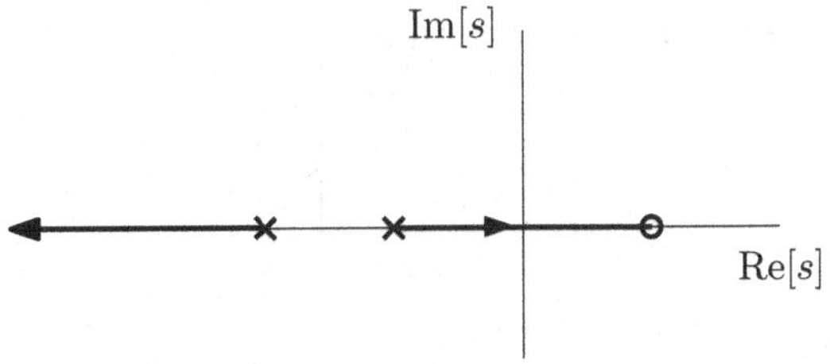

Figure 19.10 Root locus with right-half plane open-loop zero

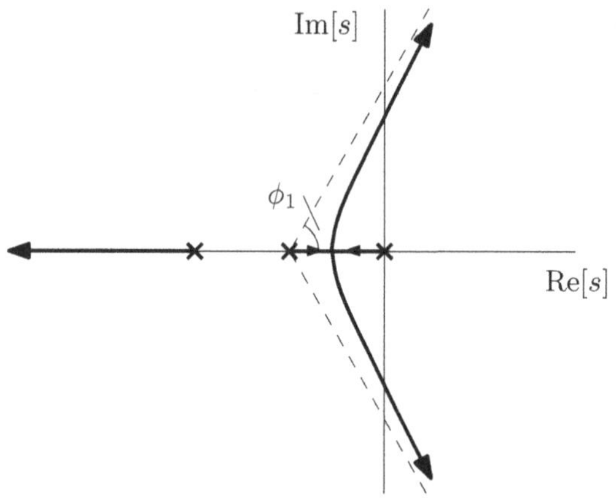

Figure 19.11 Root locus with $n - m = 3$

That is, the first asymptote will always enter the right-half plane if $n - m \geq 3$, as shown in Figure 19.11. This problem can be remedied by adding stable open-loop zeros to the system to make $0 \leq n - m \leq 2$.

- If $n - m = 2$, and the center of the asymptote is in the right-half plane, the closed-loop system will eventually become unstable for a large enough gain K, as illustrated in Figure 19.12. This problem can be remedied by ensuring that the asymptote center satisfies

$$\alpha = \frac{\sum_{i=1}^{n} p_i - \sum_{i=1}^{m} z_i}{n - m} < 0,$$

by choosing the controller poles and zeros appropriately (the poles and zeros of $G_c(s)$).

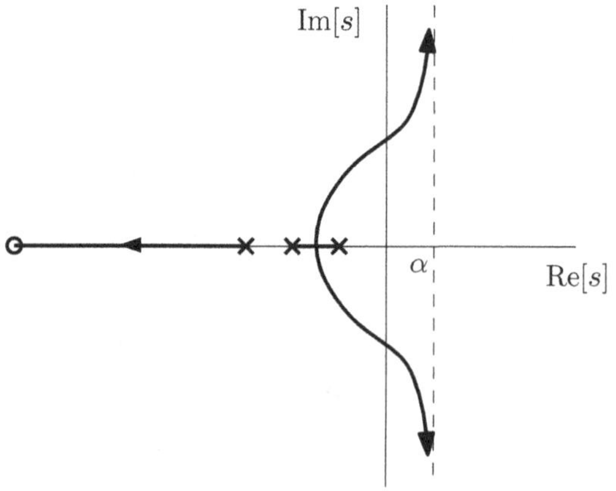

Figure 19.12 Root locus with $n - m = 2$

19.2 PD Attitude Control with Actuator Dynamics - Revisited

Let us revisit the proportional-derivative attitude control problem in Section 18.1, in which we incorporated actuator dynamics. In Section 18.1, we used Routh stability analysis to determine requirements on the actuator time-constant T for asymptotic stability. Let us now take this analysis a bit further, by plotting a root-locus as a function of T, for a given pair of proportional and derivative gains K_p and K_d.

We found in Section 18.1 that the closed-loop poles satisfy the characteristic equation

$$ITs^3 + Is^2 + K_d s + K_p = 0. \tag{19.10}$$

Let us divide this by IT to obtain

$$s^3 + \frac{1}{T}\left(s^2 + \frac{K_d}{I}s + \frac{K_p}{I}\right) = 0.$$

This can be put in equivalent root-locus form

$$1 + \frac{1}{T}\frac{\left(s^2 + (K_d/I)s + (K_p/I)\right)}{s^3} = 0. \tag{19.11}$$

Note that

$$G_o(s) \triangleq \frac{1}{T}\frac{\left(s^2 + (K_d/I)s + (K_p/I)\right)}{s^3}, \tag{19.12}$$

is **not** the open-loop transfer function corresponding to the block diagram in Figure 18.2. However, it can be treated as an open-loop transfer function for the purposes of applying the rules for constructing the root-locus for $T \geq 0$. Since the "open-loop gain" in (19.12) is $1/T$, we shall plot the root-locus for $1/T : 0 \to \infty$ (equivalently $T : \infty \to 0$).

Let us assume that the proportional and derivative gains K_p and K_d were designed to give a nominal underdamped pair of closed-loop poles (neglecting actuator dynamics) given by

$$s = -\zeta\omega_n \pm j\omega_n\sqrt{1-\zeta^2},$$

where $0 < \zeta < 1$, $\omega_n^2 = K_p/I$ and $2\zeta\omega_n = K_d/I$.

From (19.12), we now see that the "open-loop" zeros are the nominal underdamped closed-loop poles

$$z_1 = -\zeta\omega_n + j\omega_n\sqrt{1-\zeta^2}, \quad z_2 = -\zeta\omega_n - j\omega_n\sqrt{1-\zeta^2},$$

and that $m = 2$. The "open-loop" poles corresponding to (19.12) are

$$p_1 = 0, \quad p_2 = 0, \quad p_3 = 0,$$

and $n = 3$. Let us now construct the root-locus.

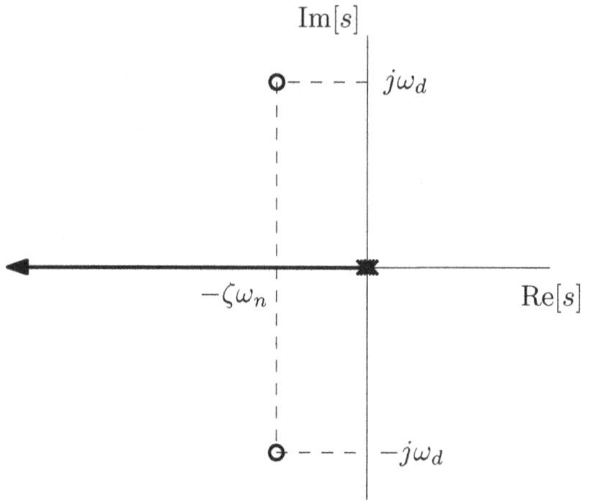

Figure 19.13 Steps 1 and 2

Step 1
Plot the open-loop poles and zeros (shown in Figure 19.13).

Step 2
Determine the root locus on the real axis (shown in Figure 19.13).

Step 3
Determine the asymptotes. There is $n - m = 1$ asymptote. The angle of the asymptote is

$$\phi_1 = \frac{180^\circ}{n-m} = \frac{180^\circ}{1} = 180^\circ,$$

which lies along the negative real axis. This is already covered by Step 2 (it lies on the negative real axis), so we do not need to examine it further.

Step 4
There are no breakaway points, so we do not need to determine them.

Step 5
Determine the angle of departure from the open-loop poles. There are three repeated open-loop poles at $p_1 = p_2 = p_3 = 0$. We therefore have the departure angle

$$\phi = \frac{\angle(p_1 - z_1) + \angle(p_1 - z_2) + 180^\circ + 360^\circ(l-1)}{3}.$$

for any integer l. From the diagram shown in Figure 19.14, we see that

$$\angle(p_1 - z_1) = 180^\circ + \alpha, \quad \angle(p_1 - z_2) = 180^\circ - \alpha.$$

Therefore, we have the departure angles

$$\phi = 60^\circ, \quad 180^\circ, \quad -60^\circ.$$

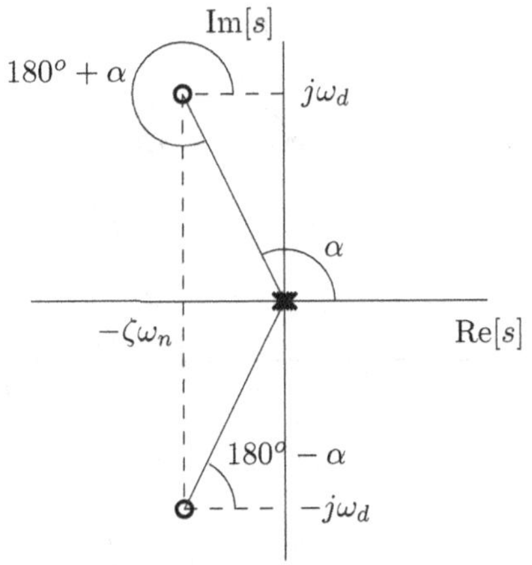

Figure 19.14 Step 5 - departure angles

Determine the arrival angles at the open-loop zeros. Since the root-locus is symmetric about the real axis (poles must occur in complex conjugate pairs), we only need to determine the arrival angle at z_1. Now, z_1 is not a repeated zero, so the arrival angle at z_1 satisfies

$$\psi = -\angle(z_1 - z_2) + \sum_{i=1}^{3} \angle(z_1 - p_i) + 180^\circ + 360^\circ(l - 1),$$

for any integer l. Let us pick $l = 0$. From the diagram shown in Figure 19.15, we see that

$$\angle(z_1 - p_i) = \alpha,$$

and

$$\angle(z_1 - z_2) = 90^\circ.$$

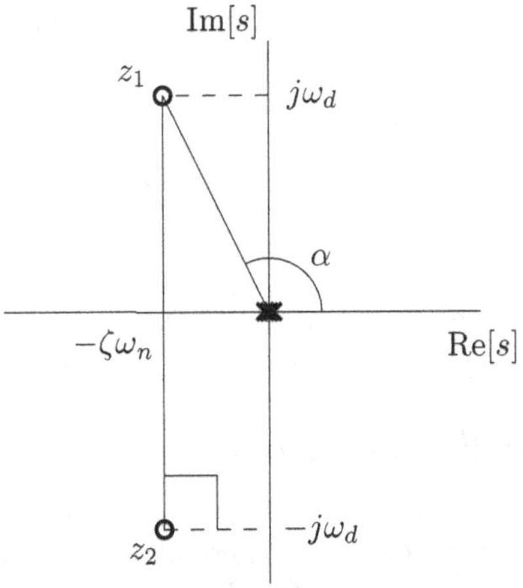

Figure 19.15 Step 5 - arrival angles

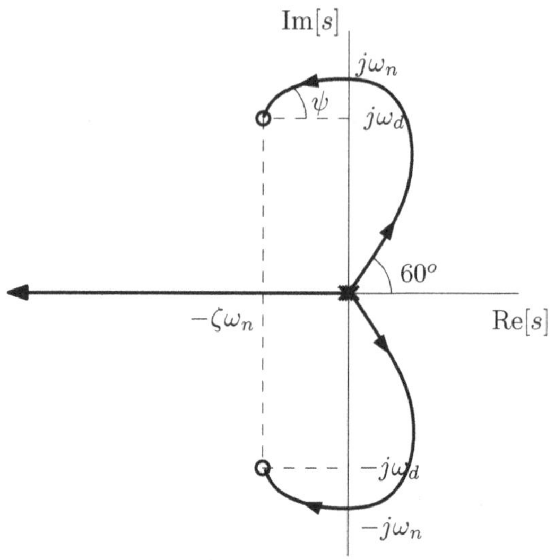

Figure 19.16 Root locus for PD spacecraft attitude control as the actuator time-constant $T : \infty \rightarrow 0$

Therefore, we have the arrival angle

$$\psi = -90^\circ + 3\alpha - 180^\circ = 3(\alpha - 90^\circ).$$

Since $0 < \alpha - 90^\circ < 90^\circ$ for underdamped pair of asymptotically stable poles, the arrival angle must lie in the range $0 < \psi < 270^\circ$.

Step 6
We saw in Step 5 that two of the root-locus branches depart from the open-loop poles into the right-half plane. By Rule 1, these must eventually terminate at the open-loop zeros in the left-half plane. Therefore, there must be imaginary axis crossings. We already determined the imaginary axis crossing in Section 18.1 when we performed the Routh analysis. We found that the root-locus crosses the imaginary axis at

$$s = \pm j\sqrt{\frac{K_p}{I}} = \pm j\omega_n.$$

Step 7
Draw the complete root-locus (shown in Figure 19.16).

19.2.1 Interpretation of the root locus

As $1/T$ increases, that is the actuator time-constant T becomes smaller (and hence the actuator becomes faster), two of the closed-loop poles approach $s = -\zeta\omega_n \pm j\omega_n\sqrt{1-\zeta^2}$, which is what we designed the closed-loop system for. The third closed-loop pole, becomes more negative (and hence will die off faster) as $1/T$ increases. Therefore, for a sufficiently fast actuator, we will have a pair of dominant poles close to the nominal closed-loop poles $s = -\zeta\omega_n \pm j\omega_n\sqrt{1-\zeta^2}$, and the system will behave as if the actuator is ideal. This shows that if the actuator is sufficiently fast, we can neglect actuator dynamics when initially performing the controller design (as we did in Section 17.7.1).

19.3 Derivation of the Rules for Constructing the Root Locus

In this section, we demonstrate where the rules for constructing the root locus come from.

19.3.1 Rule 1

There are n branches of the root locus (since there are n poles). Each branch starts at an open-loop pole, p_i (when $K = 0$). m branches end at an open-loop zero z_i (when $K \to \infty$). The other $n - m$ branches go to infinity along an asymptote (when $K \to \infty$), as seen in Rule 3.

To see this, consider the two forms of the closed-loop characteristic equation given in (19.5) and (19.6), repeated here for convenience

$$b(s) + Ka(s) = 0, \tag{19.13}$$

$$\frac{a(s)}{b(s)} = -\frac{1}{K}, \quad \text{if} \quad K \neq 0, \tag{19.14}$$

where

$$a(s) = (s - z_1)...(s - z_m), \quad b(s) = (s - p_1)...(s - p_n),$$

and z_i, $i = 1, ..., m$ and p_i, $i = 1, ..., m$ are the open-loop zeros and poles respectively. It is assumed that $m \leq n$. We note that the root locus is the set of all closed-loop poles for $K \geq 0$. It is clear that for $n \geq m$, the characteristic equation has n roots (closed-loop poles) for each value of K. Therefore, there are n branches to the root locus, corresponding to the n closed-loop poles. Now, letting $K = 0$ in (19.13), we see that the characteristic equation becomes

$$b(s) = 0,$$

which means that the closed-loop poles with $K = 0$ are equal to the open-loop poles. These are the starting points for each of the n branches of the root locus. To see where the branches finish, let us now consider the form of the characteristic equation given by (19.14), and let $K \to \infty$. We therefore have

$$\frac{a(s)}{b(s)} \to 0.$$

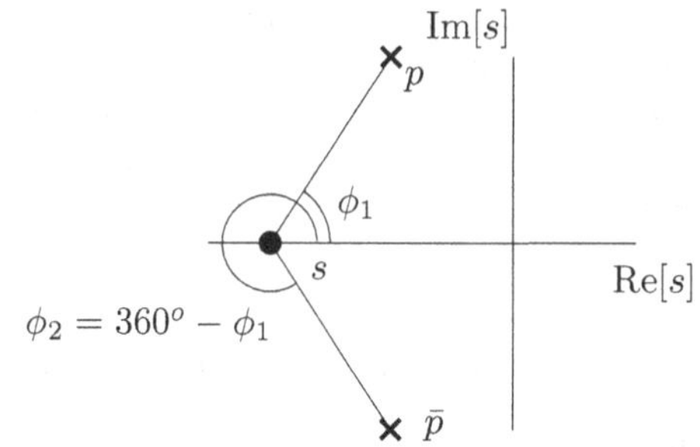

Figure 19.17 Test point on real axis with complex conjugate pair of open-loop poles

There are now two possibilities. The first is that

$$a(s) = 0,$$

which is satisfied by the m open-loop zeros. The second possibility is that $|s| \rightarrow \infty$ in such a way that $\angle(a(s)/b(s)) = -180^\circ + 360^\circ(l-1)$, for some integer l. We shall examine this in detail in the derivation of Rule 3. As we shall see, this is possible for only $n-m$ branches of the root locus. Therefore, we conclude that m of the branches of the root locus terminate at the open-loop zeros (one branch per open-loop zero). The other $n-m$ go to infinity, as we shall see.

19.3.2 Rule 2

The root locus includes all points on the real axis to the left of an odd number of open-loop real poles and zeros.

To see this, we first note that complex-conjugate open-loop poles and zeros do not affect the root locus on the real axis. Consider a test point s on the real axis, and a complex-conjugate pair of open-loop poles p and $\bar{p}$, as shown in Figure 19.17. Then, $\phi_1 + \phi_2 = 360^\circ$. The same argument can be made for complex-conjugate pairs of open-loop zeros. Since the condition for the test point s to lie on the root locus is

$$\sum \psi_i - \sum \phi_i = 180^\circ + 360^\circ(l-1),$$

it is clear that complex-conjugate open-loop poles and zeros only adjust the integer l. Therefore, root locii on the real axis is determined by real open-loop poles and zeros only. Now, as shown in Figure 19.18, real poles and zeros to the right of a test point s contribute an angle 180° with the positive real axis. The real poles and zeros to the left of the test point s contribute and angle 0° with the real axis. Therefore, only open-loop poles and zeros to the right of the test point s determine whether or not s lies on the root locus.

If the number of open-loop poles and zeros to the right of s is even, then $\sum \psi_i - \sum \phi_i = 0^\circ + 360^\circ(l-1)$ for some integer l, and the point s is not on the root locus. If the number of

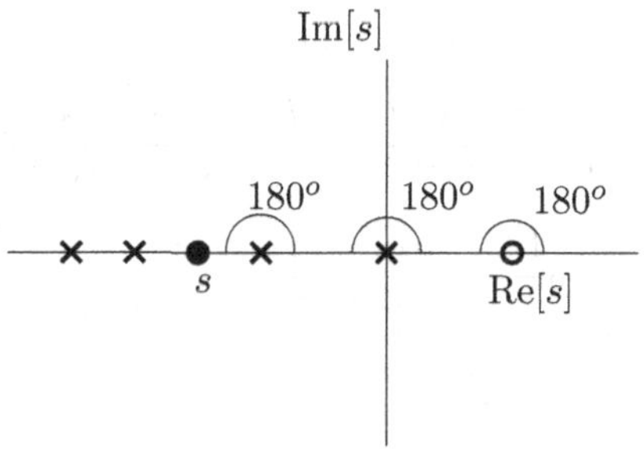

Figure 19.18 Test point on real axis with real open-loop poles and zeros

open-loop poles and zeros to the right of s is odd, then $\sum \psi_i - \sum \phi_i = 180^\circ + 360^\circ(l-1)$ for some integer l, and thc point s is on the root locus.

19.3.3 Rule 3

As $K \to \infty$, $n-m$ branches approach $n-m$ asymptotes with angles measured counter-clockwise from the positive real axis, given by

$$\phi_l = \frac{180^\circ + 360^\circ(l-1)}{n-m}, \quad l = 1, 2, ..., n-m.$$

The asymptotes have origin on the real axis at

$$\alpha = \frac{\sum_{i=1}^{n} p_i - \sum_{i=1}^{m} z_i}{n-m}.$$

As mentioned in the derivation of Rule 1, we need to examine the behavior of $a(s)/b(s)$ for large s. To do this, we are going to approximate it. First, we note that

$$a(s) = (s-z_1)...(s-z_m) = s^m + a_1 s^{m-1} + ... + a_m,$$

with $a_1 = -\sum_{i=1}^{m} z_i$, and

$$b(s) = (s-p_1)...(s-p_n) = s^n + b_1 s^{n-1} + ... + b_n,$$

with $b_1 = -\sum_{i=1}^{n} p_i$. This can be shown by simple expansions of the products.

Now, let us consider

$$\frac{b(s)}{a(s)} = \frac{s^n + b_1 s^{n-1} + ... + b_n}{s^m + a_1 s^{m-1} + ... + a_m}.$$

We can expand this as follows

$$\begin{aligned}\frac{b(s)}{a(s)} &= \frac{s^n + a_1 s^{n-1} + \ldots + a_m s^{n-m}}{s^m + a_1 s^{m-1} + \ldots + a_m} + \frac{(b_1 - a_1)s^{n-1} + (b_2 - a_2)s^{n-2} + \ldots}{s^m + a_1 s^{m-1} + \ldots + a_m},\\ &= s^{n-m} + (b_1 - a_1)\frac{\left(s^{n-1} + a_1 s^{n-2} + \ldots + a_m s^{n-m-1}\right)}{s^m + a_1 s^{m-1} + \ldots + a_m}\\ &\quad + \frac{(b_2 - a_2 - (b_1 - a_1)a_1)s^{n-2} + \ldots}{s^m + a_1 s^{m-1} + \ldots + a_m}\\ &= s^{n-m} + (b_1 - a_1)s^{n-m-1} + \ldots\end{aligned}$$

Continuing this process, we would obtain a power series expansion for $b(s)/a(s)$ with each term having a lower power of s. For large values of s, the power series is dominated by the highest powers of s. Let us choose the first two terms, namely

$$\frac{b(s)}{a(s)} \approx s^{n-m} + (b_1 - a_1)s^{n-m-1}.$$

Now, we note that

$$(s - \alpha)^{n-m} = s^{n-m} - (n - m)\alpha s^{n-m-1} + \ldots$$

Making use of this expansion, we see that for very large s, we can make the approximation

$$\frac{b(s)}{a(s)} \approx \left(s - \frac{(a_1 - b_1)}{n - m}\right)^{n-m}.$$

Therefore, for very large s, we can approximate the closed-loop characteristic equation in (19.6) by

$$\frac{1}{(s - \alpha)^{n-m}} = -\frac{1}{K}, \tag{19.15}$$

with

$$\alpha = \frac{(a_1 - b_1)}{n - m} = \frac{\sum_{i=1}^{n} p_i - \sum_{i=1}^{m} z_i}{n - m}.$$

What this approximation essentially means is that when s is very large, all of the open-loop poles and zeros appear clustered at the same point near the origin, namely $s = \alpha$, as shown in Figure 19.19. The m open-loop zeros cancel with m of the open-loop poles, leaving $n - m$ open-loop poles at $s = \alpha$. Therefore, for very large values of K, $n - m$ branches of the root locus approach the root locus of (19.15). We can find the root locus of (19.15) analytically. We note that we can write

$$-K = Ke^{j(\pi + 2\pi(l-1))},$$

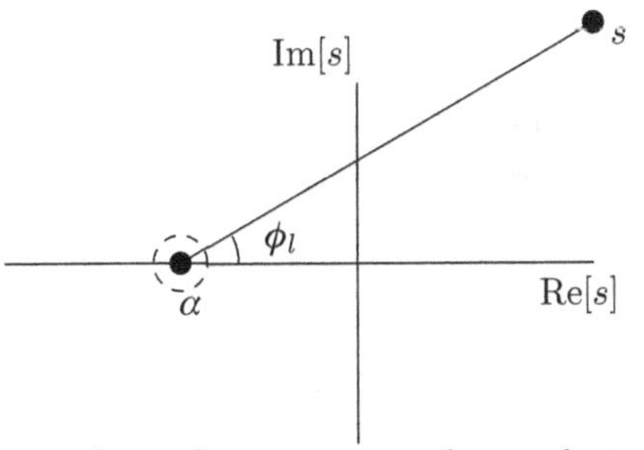

Figure 19.19 Far away test-point

for some integer l. Then, the root locus of (19.15) is given by

$$s = \alpha + K^{1/(n-m)} e^{j(\pi + 2\pi(l-1))/(n-m)}, \quad l = 1, ..., n-m.$$

That is, they are $n - m$ straight lines with origin $s = \alpha$, making angles

$$\phi_l = \frac{180^\circ + 360^\circ(l-1)}{n-m}, \quad l = 1, ..., n-m.$$

with the positive real axis.

19.3.4 Rule 4

The break-away points (where the root-loci meet and split), and the break-in points (where the root-loci meet and join) satisfy

$$a(s)\frac{db(s)}{ds} - b(s)\frac{da(s)}{ds} = 0.$$

To see this, suppose that $s = r_1$ is a break-in or break-away point, with corresponding gain K_1. Then, since $s = r_1$ is a repeated closed-loop pole for the gain K_1, we can write the corresponding characteristic equation as

$$b(s) + K_1 a(s) = (s - r_1)^q f(s) = 0, \tag{19.16}$$

where $q \geq 2$ is the multiplicity of the pole $s = r_1$, and $f(s)$ is a polynomial in s. Differentiating this gives

$$\frac{db(s)}{ds} + K_1\frac{da(s)}{ds} = q\,(s - r_1)^{q-1} f(s) + (s - r_1)^q \frac{df}{ds}.$$

Evaluating this at $s = r_1$, we have

$$\frac{db(r_1)}{ds} + K_1 \frac{da(r_1)}{ds} = 0. \tag{19.17}$$

From (19.16), we have

$$K_1 = -\frac{b(r_1)}{a(r_1)},$$

Substituting this into (19.17) gives

$$\frac{db(r_1)}{ds} - \frac{b(r_1)}{a(r_1)} \frac{da(r_1)}{ds} = 0.$$

Rearranging this gives the desired result,

$$a(r_1)\frac{db(r_1)}{ds} - b(r_1)\frac{da(r_1)}{ds} = 0.$$

19.3.5 Rule 5

The angle of departure ϕ_k of a branch from a pole p_k of multiplicity n_k is given by

$$n_k \phi_k = \sum_{i=1}^{m} \angle(p_k - z_i) - \sum_{j=1, j \neq g \text{ if } p_g = p_k}^{n} \angle(p_k - p_j) + 180^\circ + 360^\circ(l-1).$$

The angle of arrival ψ_k at a zero z_k of multiplicity n_k is given by

$$n_k \psi_k = - \sum_{i=1, i \neq g \text{ if } z_g = z_k}^{m} \angle(z_k - z_i) + \sum_{j=1}^{n} \angle(z_k - p_j) + 180^\circ + 360^\circ(l-1).$$

Note that l can be any integer in both cases.

To see this, consider first an open-loop pole p_k of multiplicity n_k, and a test point s very closed to p_k, as shown in Figure 19.20.

Since s is very close to p_k, we have

$$\angle(p_k - z_i) \approx \angle(s - z_i),$$

and

$$\angle(p_k - p_j) \approx \angle(s - p_j),$$

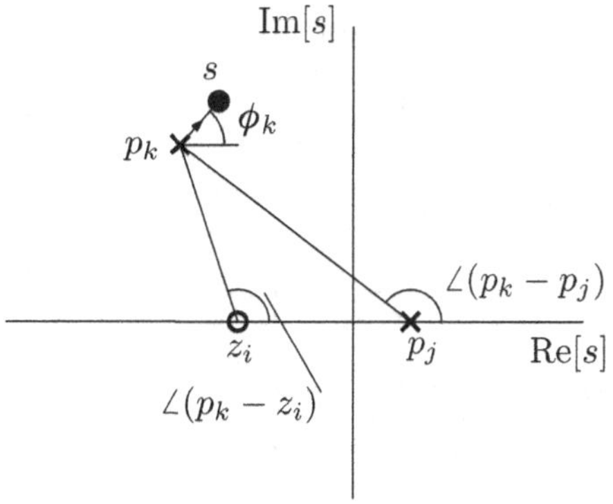

Figure 19.20 Determination of departure angle

provided $p_k \neq p_j$. The departure angle of a branch from p_k (note that there must be n_k branches leaving from p_k), is given by

$$\phi_k = \angle (s - p_k),$$

for points s on the root locus arbitrarily close to p_k. The angle condition for s to lie on the root locus is then

$$\sum_{i=1}^{m} \angle (s - z_i) - \sum_{j=1}^{n} \angle (s - p_j) = 180^\circ + 360^\circ(l - 1),$$

which becomes

$$\sum_{i=1}^{m} \angle (p_k - z_i) - \sum_{j=1, j \neq g \text{ if } p_g = p_k}^{n} \angle (p_k - p_j) - n_k \phi_k = 180^\circ + 360^\circ(l - 1),$$

for any integer l. Rearranging, we get

$$n_k \phi_k = \sum_{i=1}^{m} \angle (p_k - z_i) - \sum_{j=1, j \neq g \text{ if } p_g = p_k}^{n} \angle (p_k - p_j) - 180^\circ - 360^\circ(l - 1).$$

Since the integer l is arbitrary, we can rewrite this as

$$n_k \phi_k = \sum_{i=1}^{m} \angle (p_k - z_i) - \sum_{j=1, j \neq g \text{ if } p_g = p_k}^{n} \angle (p_k - p_j) + 180^\circ + 360^\circ(l - 1).$$

To find the arrival angle of a branch at an open-loop zero z_k of multiplicity n_k, consider a test point s very close to z_k, as shown in Figure 19.21. As for the departure from an open-loop

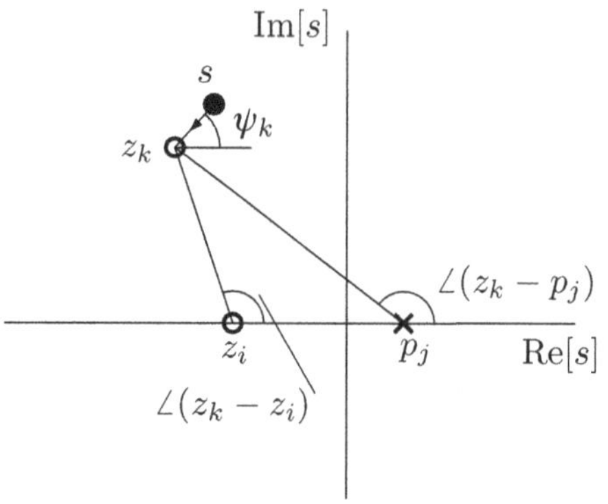

Figure 19.21 Determination of arrival angle

pole, since s is very close to z_k, we have

$$\angle(z_k - z_i) \approx \angle(s - z_i),$$

provided $z_k \neq z_i$, and

$$\angle\left(z_k - p_j\right) \approx \angle\left(s - p_j\right).$$

The arrival angle of a branch at z_k (note that there must be n_k branches arriving at z_k), is given by

$$\psi_k = \angle(s - z_k),$$

for points s on the root locus arbitrarily close to z_k. The angle condition for s to lie on the root locus is then

$$\sum_{i=1}^{m} \angle(s - z_i) - \sum_{j=1}^{n} \angle\left(s - p_j\right) = 180^\circ + 360^\circ(l - 1),$$

which becomes

$$\sum_{i=1, i\neq g \text{ if } z_g = z_k}^{m} \angle(p_k - z_i) + n_k \psi_k - \sum_{j=1}^{n} \angle\left(p_k - p_j\right) = 180^\circ + 360^\circ(l - 1),$$

for any integer l. Rearranging gives

$$n_k \psi_k = - \sum_{i=1, i\neq g \text{ if } z_g = z_k}^{m} \angle(z_k - z_i) + \sum_{j=1}^{n} \angle(z_k - p_j) + 180^\circ + 360^\circ(l - 1).$$

Notes

This chapter has presented the root locus. We have then applied the root locus technique to examine how the closed-loop poles for a spacecraft attitude control design vary with the speed of the actuator response. As intuition would suggest, the faster the actuator, the less effect it has on the location of the originally designed closed-loop poles. Additional examples on how to apply the root locus techniques may be found in the many excellent books on classical control systems, including Qiu and Zhou (2010), Franklin et al. (2010) and Ogata (2010).

References

Franklin GF, Powell JD and Emami-Naeini A 2010 *Feedback Control of Dynamic Systems* 6th edn. Prentice Hall, Upper Saddle River, NJ.

Ogata K 2010 *Modern Control Engineering* 5th edn. Prentice Hall, Upper Saddle River, NJ.

Qiu L and Zhou K 2010 *Introduction to Feedback Control*. Prentice Hall, Upper Saddle River, NJ.

20

Control Design by the Root Locus Method

We have seen how to construct a root locus plot for a given system. This gives the information about all of the possible closed-loop pole locations for a given control structure. However, we typically want to have a pair of dominant poles with specified transient step response characteristics (e.g. rise-time, overshoot, settling time, see Section 17.7.1). As seen in Section 17.7.1, these are dictated by the damping ratio ζ and the undamped natural frequency ω_n, which fully specifies the dominant pole location. We must therefore choose our control structure (transfer function) such that the desired dominant poles lie on the root locus. We consider again the general feedback system as shown in Figure 20.1. What the control law $G_c(s)$ does, is it adds poles and zeros to the open-loop transfer function $G_o(s) = G_c(s)G_p(s)$. As control designers, we can choose the number of poles and zeros we wish to add, as well as their locations. By adding appropriate poles and zeros, the root locus can be modified such that the desired dominant poles are on the root locus.

As control system designers, it is important to understand what the effects of additional poles and zeros are on the root locus.

Effect of Adding Open-Loop Poles

Adding poles increases the difference between open-loop poles and zeros $n - m$, which decreases the angle of the first asymptote, as seen in Figure 20.2. The overall effect is that the root-locus is pushed toward the right-half plane, which results in longer settling times.

The location of the additional pole affects the center of the asymptote location. Recall that the center of the asymptote is given by

$$\alpha = \frac{\sum_{i=1}^{n} p_i - \sum_{i=1}^{m} z_i}{n - m},$$

where p_i are the open-loop poles, and z_i are the open-loop zeros.

Spacecraft Dynamics and Control: An Introduction, First Edition.
Anton H.J. de Ruiter, Christopher J. Damaren and James R. Forbes.

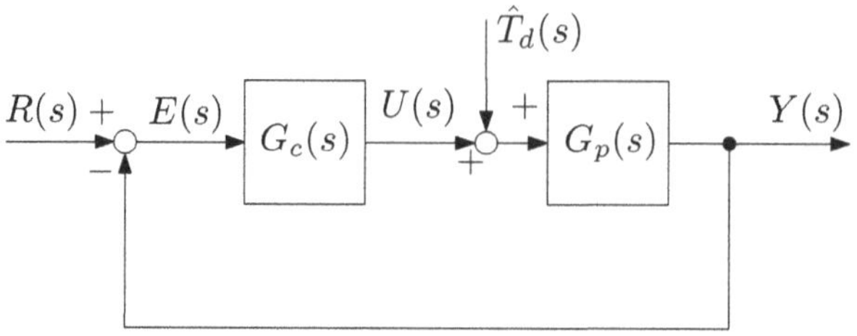

Figure 20.1 General feedback system

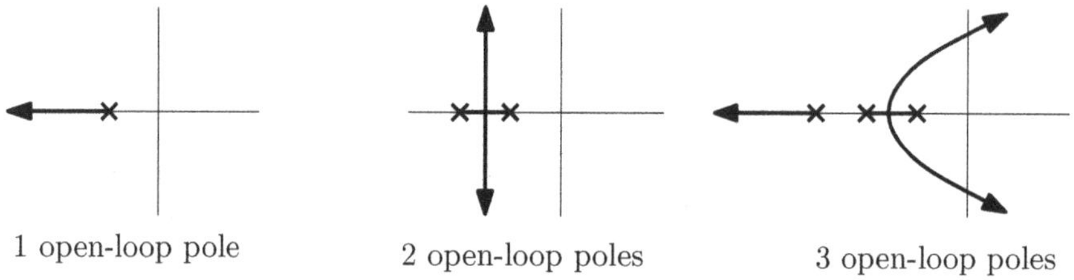

Figure 20.2 Effect of open-loop poles on the root locus

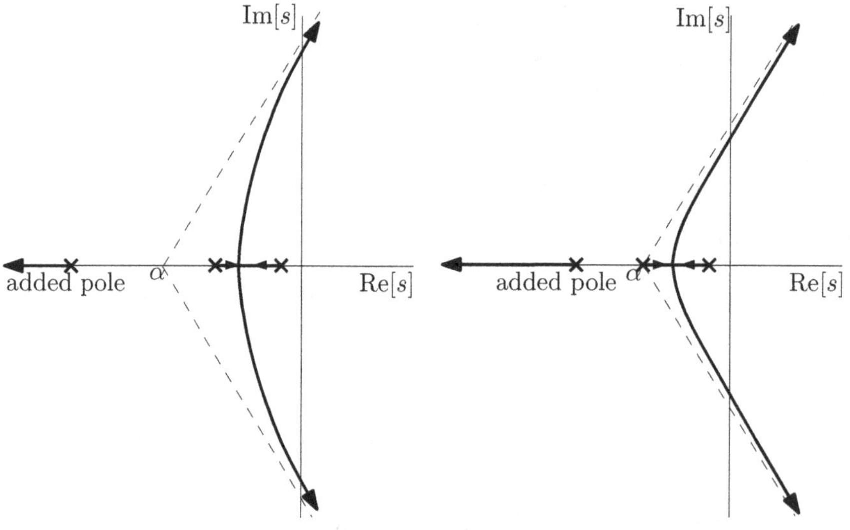

Figure 20.3 Effect of open-loop poles on the root locus

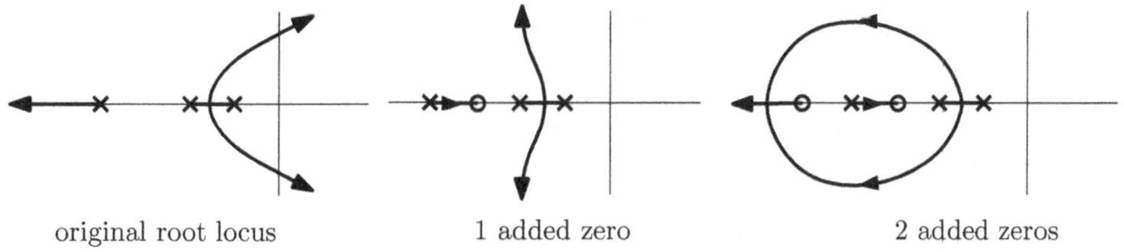

Figure 20.4 Effect of open-loop zeros on the root locus

As shown in Figure 20.3, as the added pole moves to the right, the center of the asymptote moves to the right, and the effect of pushing the root locus to the right becomes more pronounced. **Therefore, if added open-loop poles are needed, try to put them to the far left of the dominant poles.**

Effect of Adding Additional Open-Loop Zeros

Adding zeros decreases $n - m$, which decreases the number of the asymptotes, increasing the angle of the first asymptote, as shown in Figure 20.4. The overall effect is to pull the root locus to the left, which makes the system more stable, and speeds up the transient response.

The location of the added zero affects the center of the asymptote.

$$\alpha = \frac{\sum_{i=1}^{n} p_i - \sum_{i=1}^{m} z_i}{n - m}.$$

As the added zero moves to the right, the center of the asymptote moves to the left, as shown in Figure 20.5. Of course, the zero should still remain in the left-half plane (since poles go to zeros as the open-loop gain $K \to \infty$).

20.1 Typical Types of Controllers

In this section, we will examine some typical controllers, and what their effects are on the root-locus.

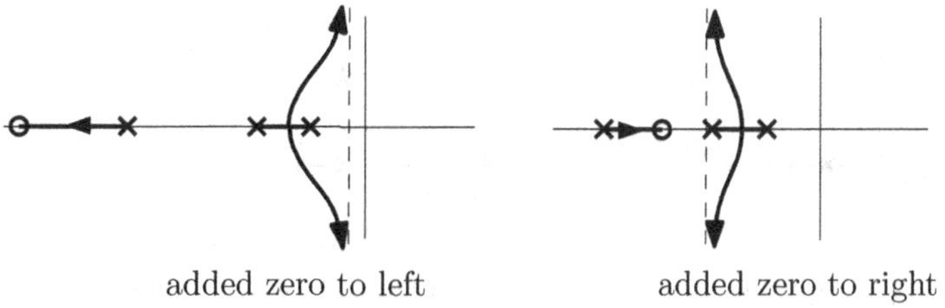

Figure 20.5 Effect of open-loop zero location on the root locus

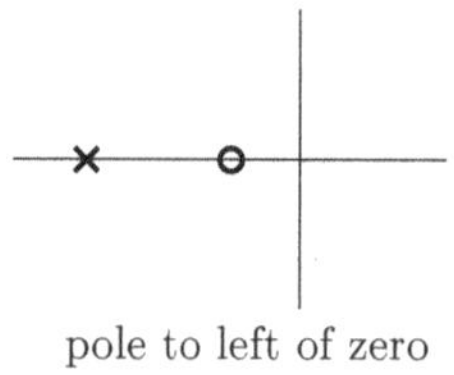

Figure 20.6 Lead compensator pole and zero

20.1.1 Lead

A lead controller has a transfer function of the form

$$G_c(s) = K_{lead}\frac{s + 1/T}{s + 1/(\alpha T)}, \quad 0 < \alpha < 1. \tag{20.1}$$

Clearly, a lead controller adds an open-loop pole and zero. This means that the number of asymptotes does not change. As shown in Figure 20.6, the pole is to the left of the zero, so as shown in Figure 20.7, it shifts the root locus to the left (improves transient response), and we can change the location of the dominant poles.

20.1.2 Lag

A lag controller has a transfer function of the form

$$G_c(s) = K_{lag}\frac{s + 1/T}{s + 1/(\beta T)}, \quad \beta > 1, \quad K_{lag} \approx 1. \tag{20.2}$$

A lag controller therefore adds an open-loop pole and zero, as shown in Figure 20.8. This means that the number of the asymptotes does not change. Lag compensation can be used if the steady-state error is to be reduced without affecting the transient response (dominant pole locations). To illustrate, let us consider the spacecraft attitude control problem as formulated in Section 17.1. That is, the plant is given by

$$G_p(s) = \frac{1}{Is^2}.$$

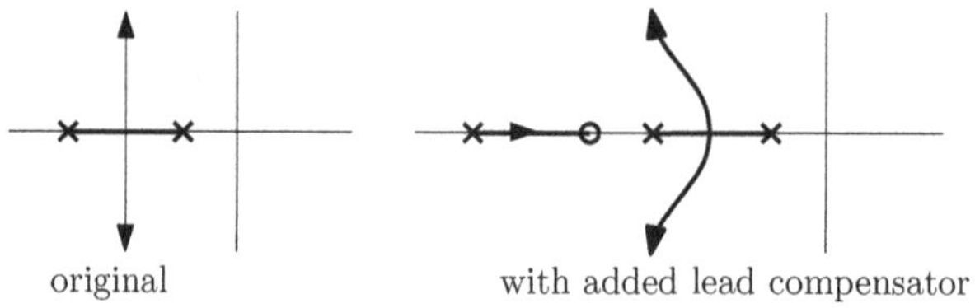

Figure 20.7 Effect of lead compensator on root locus

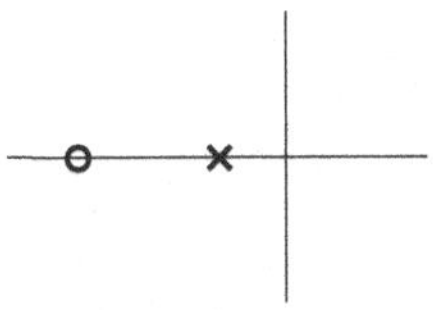

Figure 20.8 Lag compensator pole and zero

Suppose we have designed some controller $G_c(s)$ as in Figure 20.1, which provides the required transient performance, but does not contain an integrator. Then, as we saw in Section 17.10, the steady-state error to a constant disturbance $\hat{T}_d$, is

$$e_{ss} = -\frac{\hat{T}_d}{K_c},$$

where $K_c = G_c(0)$.

Now, suppose we would like to reduce the steady-state error without significantly affecting the transient response (we do not wish to shift the dominant poles). One way of doing this is to append a lag compensator to our controller,

$$\bar{G}_c(s) = \frac{s + 1/T}{s + 1/(\beta T)} G_c(s).$$

With the added lag compensator, the steady-state error to a constant disturbance becomes

$$e_{ss} = -\frac{\hat{T}_d}{\bar{K}_c},$$

where

$$\bar{K}_c = \beta K_c.$$

By choosing an appropriate $\beta > 1$, we can reduce the magnitude of the steady-state error as much as we like. So as not to affect the rest of the root locus (and therefore the dominant poles), we need to place the lag compensator pole and zero close to the origin (by the choice of an appropriately large T). To see why, let s be a point on the original root locus. As shown in Figure 20.9, the angles of the lines from the lag compensator pole and zero to the point s are approximately equal, and by the angle condition for the root-locus (see equation (19.9) in Section 19.1) their effects on the root locus approximately cancel. That is, even though the point s will not be on the root locus, a point very close to s will be. Hence, the shape of the root locus stays the same, and the dominant poles with the original controller $G_c(s)$ will be approximately retained. This is illustrated in Figure 20.10. We can see that the lag compensator has added a very slow pole to the open-loop system. However, the lag compensator has also added a closed-loop zero very close to it (since open-loop zeros are closed-loop zeros of the transfer function from reference input $R(s)$ to output $Y(s)$ (check

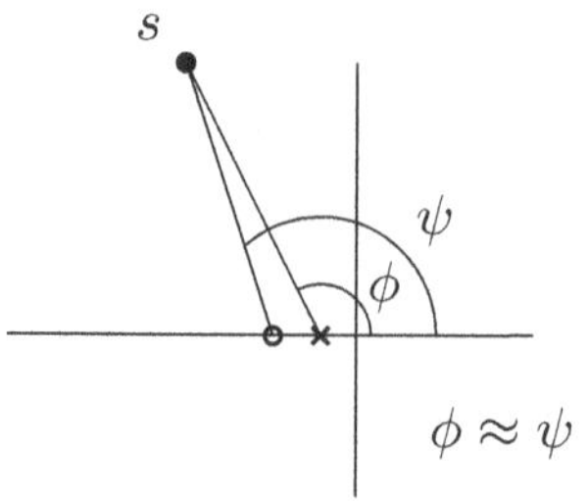

Figure 20.9 Lag compensator pole and zero effect on root locus

this as an exercise)). The added zero has the effect of reducing the amplitude of the transient component due to the slow pole (since they approximately cancel). Therefore, the transient response will be approximately the same as with the original controller $G_c(s)$.

20.1.3 Proportional-Derivative (PD)

A proportional-derivative controller can be written as

$$G_c(s) = K_{pd}\left(s + \frac{1}{T}\right). \tag{20.3}$$

Therefore, a PD controller adds a single zero to the root-locus, which decreases the number of asymptotes on the root-locus by one (since $n - m$ is reduced by one). As such, the angle of the first asymptote increases. Comparing the PD controller (20.3) with the lead controller (20.1), it can be seen that the PD controller is just a limiting case of a lead controller when the lead pole goes to negative infinity ($\alpha \rightarrow 0$), as shown in Figure 20.11. Therefore, a PD controller can be designed similarly to a lead controller, and can be used to improve transient performance.

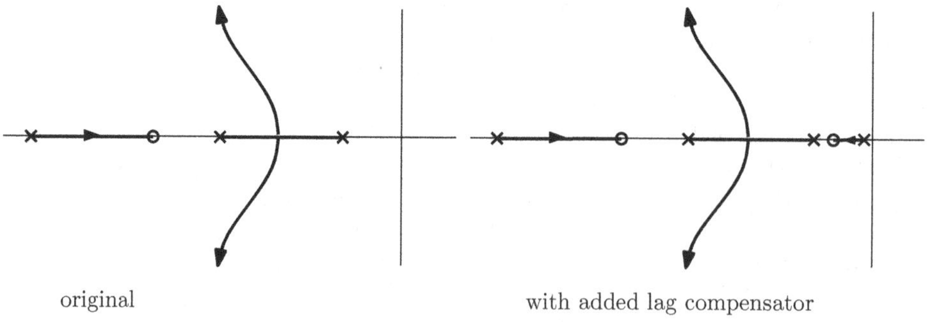

Figure 20.10 Lag compensator pole and zero effect on root locus

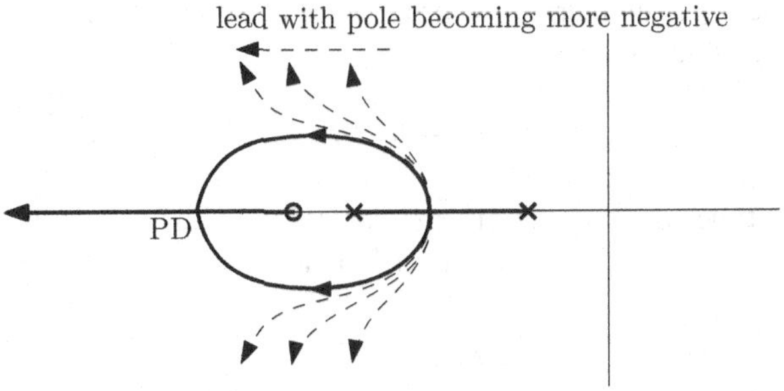

Figure 20.11 Root locus with PD compensation and comparison with lead compensation

20.1.4 Proportional-Integral (PI)

A proportional-integral controller can be written as

$$G_c(s) = K_{pi}\frac{s + 1/T}{s}. \tag{20.4}$$

Therefore, a PI controller adds a small stable open-loop zero, and an open-loop pole at the origin. Comparing the PI controller (20.4) to the lag controller (20.2), it can be seen that a PI controller is just a limiting case of the lag controller when the lag pole goes to zero ($\beta \to \infty$), as shown in Figure 20.12. That is, the PI controller can be used designed like a lag controller to reduce the steady-state error (to zero by increasing the system type) while leaving the transient performance almost unchanged.

We have seen four basic types of controllers, and their effects on the root locus. We can use any combination of these as required to achieve the desired transient and steady-state performance.

20.2 PID Design for Spacecraft Attitude Control

We shall now apply the root-locus design method to the spacecraft attitude control design problem in Example 17.1, with the added requirement that the steady-state error must be zero

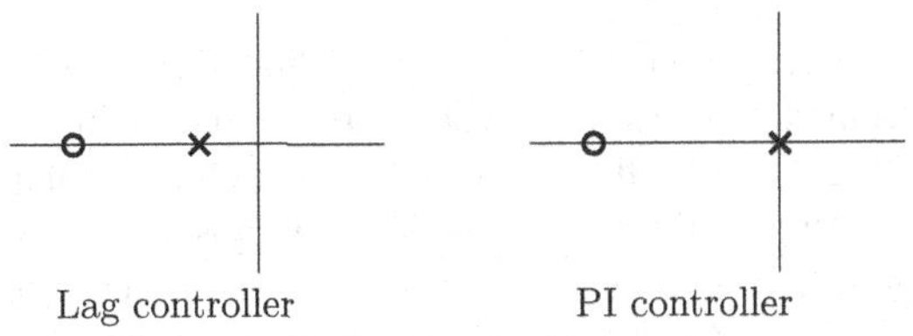

Figure 20.12 Root loci for lag and PI controllers

for a step disturbance. To summarize the problem in Example 17.1, we are considering the control of the spacecraft attitude given by

$$I\ddot{y} = u + T_d,$$

where the spacecraft inertia $I = 1$ kg·m^2, the control torque is u, and T_d is the disturbance torque. Recall that the associated plant transfer function is

$$G_p(s) = \frac{1}{s^2}.$$

We must design a controller so that the closed-loop system satisfies the following specifications.

1. Rise-time constraint, $t_r \leq 30$ seconds.
2. Maximum overshoot constraint, $M_p \leq 30\%$.
3. Settling-time constraint, $t_s \leq 100$ seconds.
4. The steady-state error for a step disturbance $\hat{T}_d$ must be zero.

Now, the first three requirements are transient specifications which are affected by the closed-loop dominant pole locations. We have seen that PD control can be used to obtain the desired dominant poles. As seen in Section 17.10, the fourth constraint means that the controller must contain an integrator. This suggests the use of PID control. Now, the question is how to design the PID gains.

It turns out that a PID controller can be designed as a combined PD and PI controller. To see this, consider the combined PD and PI controller

$$\begin{aligned}
G_c(s) &= K\left(s + \frac{1}{T_{pd}}\right)\left(\frac{s + 1/T_{pi}}{s}\right), \\
&= K\left(s + \frac{1}{T_{pd}} + \frac{1}{T_{pi}} + \frac{1}{T_{pd}T_{pi}s}\right), \\
&= K_p + sK_d + \frac{K_i}{s},
\end{aligned}$$

which is a PID controller with proportional, derivative and integral gains given by

$$K_p = K\left(\frac{1}{T_{pd}} + \frac{1}{T_{pi}}\right), \quad K_d = K, \quad K_i = \frac{K}{T_{pd}T_{pi}}, \tag{20.5}$$

respectively.

This means that we can design our PID controller sequentially with PD and PI controllers. The typical approach would be to design the PD controller first to obtain the desired dominant poles, and then design the PI controller so as to not affect the dominant poles significantly. In our case, this would result in the PI zero being very close to the origin, and as a result, the closed-loop system will have a very slow pole. This is because the root locus will have a branch starting at the PI pole (at the origin) and finishing at the PI zero (very close to the origin).

We shall therefore follow the opposite approach. We shall first design the PI controller, and then we shall design the PD controller to give the desired dominant poles.

Before continuing, we need to specify the desired dominant poles. We have seen in Example 17.1, that a pair of dominant poles that yield a satisfactory transient response are

$$s = -0.05 \pm j0.1.$$

20.2.1 PI Design

The design of the PI part amounts to choosing the PI zero. Let us try $T_{pi} = 20$, which leads to a PI zero $z_{PI} = -1/T_{pi} = -0.05$. That is, the PI part of the controller is given by

$$\left(\frac{s + 0.05}{s}\right).$$

20.2.2 PD Design

The design of the PD part now requires us to find the PD zero to ensure that the desired dominant poles $s = -0.05 \pm j0.1$ are closed-loop poles for some controller gain K. That is, we need to choose the PD zero such that $s = -0.05 \pm j0.1$ lie on the root locus of

$$1 + G_o(s) = 0, \tag{20.6}$$

where the open-loop transfer function is

$$G_o(s) = G_c(s)G_p(s) = K\left(s + \frac{1}{T_{pd}}\right)\left(\frac{s + 1/T_{pi}}{s}\right)\left(\frac{1}{s^2}\right). \tag{20.7}$$

Note that complex poles always occur in complex conjugate pairs, so we only need to ensure that $s = -0.05 + j0.1$ is on the root locus.

Recall from Section 19.1, that for the point s to lie on the root locus, it must satisfy the angle condition

$$\angle G_o(s) = 180^\circ + 360^\circ(l - 1),$$

for some integer l. In terms of the open-loop transfer function (20.7), this becomes

$$\angle\left(s + \frac{1}{T_{pd}}\right) + \angle\left(\frac{s + 1/T_{pi}}{s}\right) + \angle\left(\frac{1}{s^2}\right) = 180^\circ + 360^\circ(l - 1), \tag{20.8}$$

for some integer l.

Let us first determine the phase of the plant, $\angle\left(1/s^2\right)$. We see that the plant has a double pole at the origin.

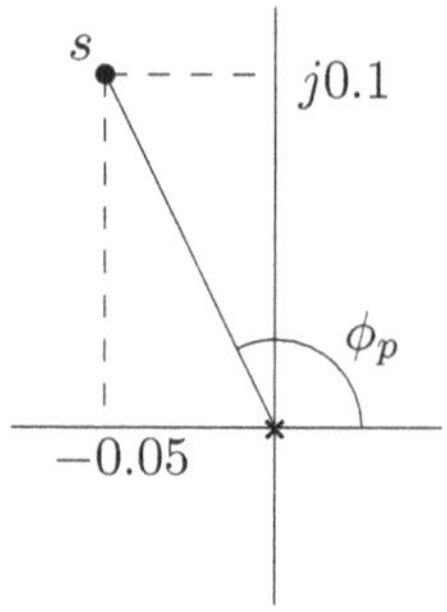

Figure 20.13 Desired dominant pole location

From Figure 20.13, we can see that

$$\phi_p = 180^\circ - \tan^{-1}(0.1/0.05) = 116.57^\circ.$$

Therefore, we have

$$\angle\left(\frac{1}{s^2}\right) = -2\phi_p = -233.13^\circ.$$

Now, let us determine the phase of the PI part, $\angle\left((s + 1/T_{pi})/s\right)$. We see that the PI part has a zero at $-1/T_{pi} = -0.05$ and a pole at the origin.

From Figure 20.14, we can see that

$$\phi_{pi} = \phi_p = 116.57^\circ, \quad \psi_{pi} = 90^\circ.$$

Therefore, we have

$$\angle\left(\frac{s + 1/T_{pi}}{s}\right) = \psi_{pi} - \phi_{pi} = -26.5651^\circ.$$

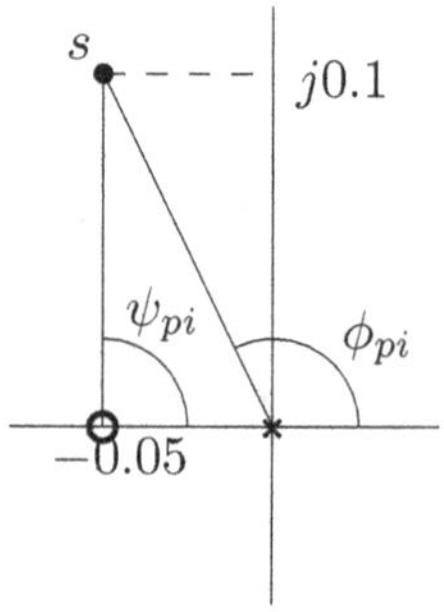

Figure 20.14 PI zero and pole location

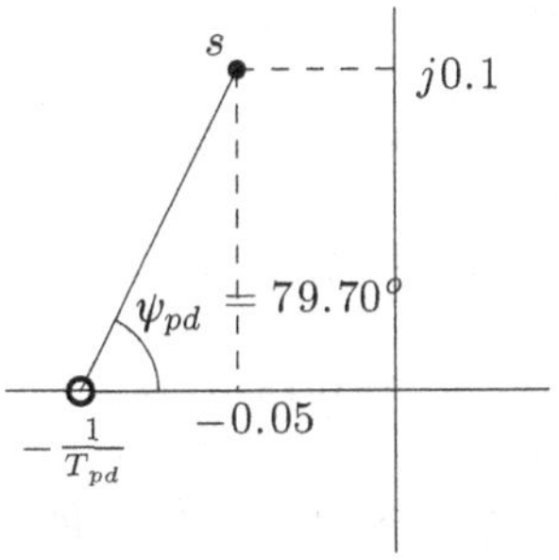

Figure 20.15 PD zero location

From the angle condition (20.8), we now find that the PD part must add the phase

$$\angle\left(s+\frac{1}{T_{pd}}\right) = -180^\circ - \angle\left(\frac{s+1/T_{pi}}{s}\right) - \angle\left(\frac{1}{s^2}\right) = 79.70^\circ.$$

Now, we see that the PD part adds a single zero at $-1/T_{pd}$. We can determine it's location from the above angle condition.

We see from Figure 20.15 that

$$\tan\psi_{pd} = \frac{0.1}{1/T_{pd} - 0.05},$$

which rearranges to give

$$\frac{1}{T_{pd}} = \frac{0.1}{\tan 79.70^\circ} + 0.05 = 0.0682.$$

Note that this means that $T_{pd} = 16.67$.

20.2.3 Computation of the Gain K

With this choice of PD zero, we have ensured that the desired dominant pole $s = -0.05 + j0.1$ is on the root locus. The final part of the controller design is determining the controller gain K such that $s = -0.05 + j0.1$ is a closed-loop pole. From the characteristic equation (20.6), we see that we must have

$$1 + G_o(s) = 1 + K\left(s+\frac{1}{T_{pd}}\right)\left(\frac{s+1/T_{pi}}{s}\right)\left(\frac{1}{s^2}\right) = 0.$$

This can be rearranged to give

$$K = \left|\frac{s^3}{\left(s+1/T_{pd}\right)\left(s+1/T_{pi}\right)}\right|.$$

Evaluating this at $s = -0.05 + j0.1$, we obtain the gain

$$K = 0.1375.$$

The corresponding PID gains are (see (20.5))

$$K_p = 0.0163, \quad K_d = 0.1375, \quad K_i = 4.69 \times 10^{-4}.$$

To summarize, we have designed a PID controller for implementation as in Figure 20.1

$$G_c(s) = K_p + K_d s + \frac{K_i}{s}, \tag{20.9}$$

which means that the control torque on the spacecraft is given by

$$u(t) = K_p e(t) + K_d \dot{e}(t) + K_i \int_0^t e(\tau)d\tau. \tag{20.10}$$

That is, part of the control torque is proportional to the derivative of the tracking error $\dot{e}(t)$. Recall from Section 17.7.1, that there may be situations where we do not wish to do this (likewise, there are situations where we do), and we would rather the control torque has the form

$$u(t) = K_p e(t) - K_d \dot{y}(t) + K_i \int_0^t e(\tau)d\tau, \tag{20.11}$$

similar to the modified proportional derivative control in Section 17.7.1.

Let us divide the control law (20.9) into two parts

$$G_c(s) = G_{ce}(s) + G_{cy}(s),$$

where $G_{ce}(s) = K_p + K_i/s$ and $G_{cy}(s) = K_d s$. The standard control implementation of (20.10) is shown in Figure 20.16 (compare this with Figure 20.1). The modified control implementation of (20.11) is shown in Figure 20.17. We can now find the closed-loop transfer functions from $R(s)$ to $Y(s)$ as

$$\frac{Y(s)}{R(s)} = \frac{(G_{ce}(s) + G_{cy}(s))G_p(s)}{1 + (G_{ce}(s) + G_{cy}(s))G_p(s)},$$

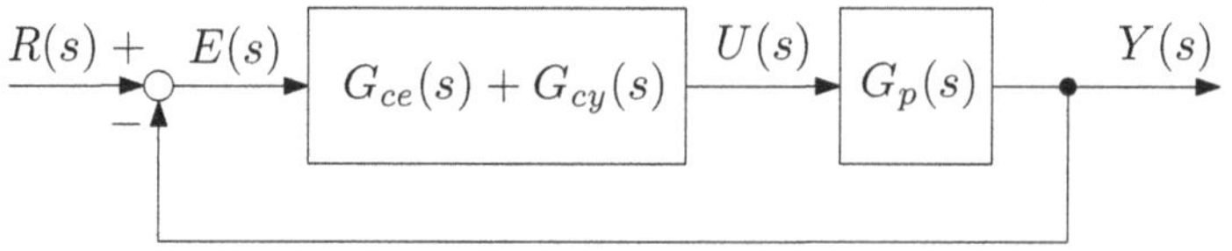

Figure 20.16 Standard control implementation

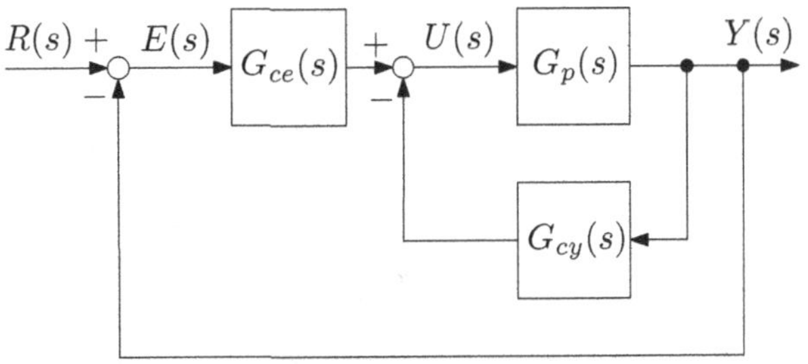

Figure 20.17 Modified control implementation

for the standard control implementation (Figure 20.16), and

$$\frac{Y(s)}{R(s)} = \frac{G_{ce}(s)G_p(s)}{1 + (G_{ce}(s) + G_{cy}(s))G_p(s)},$$

for the modified control implementation (Figure 20.17). Comparing the two closed-loop transfer functions, we can see that both have the same characteristic equation

$$1 + (G_{ce}(s) + G_{cy}(s))G_p(s) = 0.$$

That is, the closed-loop poles for both implementations are the same. In particular, the modified control implementation (20.11) also provides the desired dominant closed-loop poles. Figure 20.18 now shows the attitude response to a unit step command with the designed controller using the modified control implementation (20.11). We can see from Figure 20.18, that both the settling-time and rise-time are within specifications, but that the percent overshoot is too large. This is due to the additional pole arising from the integral term, as well as a closed-loop zero that is present.

To satisfy all closed-loop specifications, we need to tweak the control design a little bit. Since the percentage overshoot is directly related to the damping ratio of the dominant poles (see Section 17.7.1), let us increase the damping ratio of the dominant poles. Since the settling-time is almost on the boundary of the specification, let us also speed up the dominant poles (by making the real part more negative). Let us try the new desired dominant poles

$$s = -0.07 \pm j0.09.$$

Repeating the above PID design procedure with the new desired dominant poles, we obtain the PID gains

$$K_p = 0.0179, \quad K_d = 0.1753, \quad K_i = 4.59 \times 10^{-4}.$$

Figure 20.19 shows the attitude step response with the new controller gains. We can see that now all of the transient closed-loop specifications are satisfied. Figure 20.20 shows that we have also achieved a zero steady-state error to a step disturbance $\hat{T}_d$.

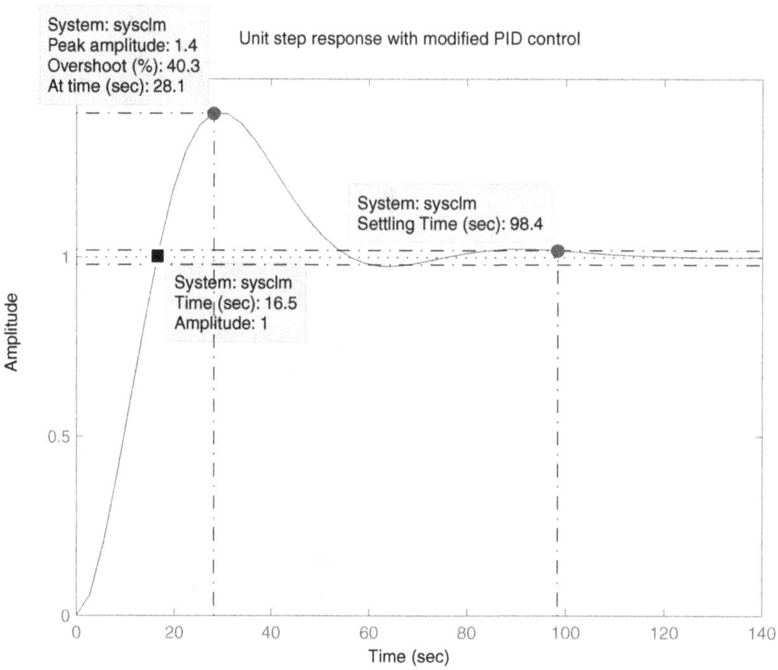

Figure 20.18 Unit step response $K_p = 0.0163$, $K_d = 0.1375$, $K_i = 4.69 \times 10^{-4}$

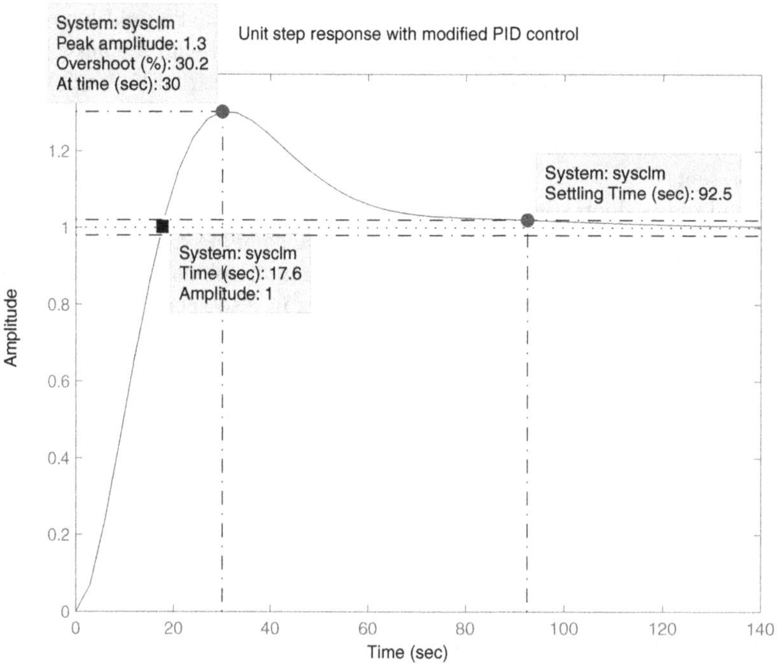

Figure 20.19 Unit step response $K_p = 0.0179$, $K_d = 0.1753$, $K_i = 4.59 \times 10^{-4}$

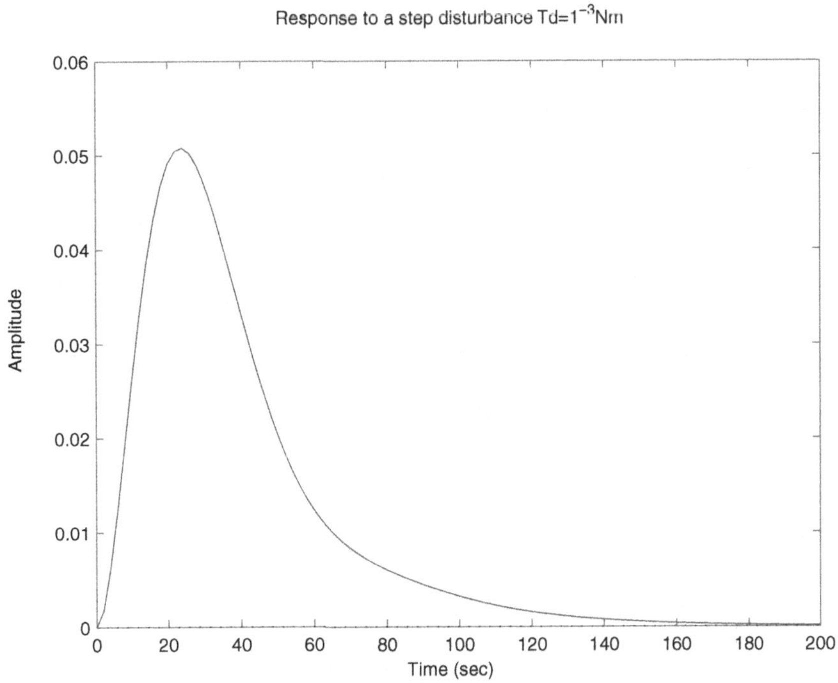

Figure 20.20 Attitude response for a step disturbance $\hat{T}_d = 10^{-3}$Nm

Notes

This chapter has examined how the root locus technique may be applied to the control system design problem. In particular, we have examined how common controller types affect the root locus. We have then applied the root locus technique to the design of a PID controller for spacecraft attitude control. For further examples on how to apply the root locus technique to control system design, the reader is referred to any of the excellent books on classical control systems, including Qiu and Zhou (2010), Franklin et al. (2010) and Ogata (2010).

References

Franklin GF, Powell JD and Emami-Naeini A 2010 *Feedback Control of Dynamic Systems* 6th edn. Prentice Hall, Upper Saddle River, NJ.

Ogata K 2010 *Modern Control Engineering* 5th edn. Prentice Hall, Upper Saddle River, NJ.

Qiu L and Zhou K 2010 *Introduction to Feedback Control*. Prentice Hall, Upper Saddle River, NJ.

21

Frequency Response

Up to now, all control laws we have designed have not taken into consideration measurement noise. As was mentioned in Section 17.5, the sensors will never provide the true system output $y(t)$, but instead will provide a version corrupted by measurement noise $y(t) + w(t)$, as shown in Figure 21.1.

In the case of spacecraft attitude control, this means that the sensor will never provide the true attitude $\theta(t)$, but will provide instead $\theta(t) + w(t)$. Let us see what this implies for proportional-derivative control of spacecraft attitude. The proportional part of the control torque demanded from the actuators becomes

$$K_p \left(r(t) - y(t) - w(t) \right).$$

This means that instead of a nice smooth control torque, the actuator will expend unnecessary effort due to the measurement noise. This will result in a time-varying tracking error. The problem becomes even more pronounced if we examine the derivative control term. If the attitude sensor suite does not contain a rate-gyro which provides a low noise measurement of the angular velocity $\dot{y}$ directly, the angular velocity must be obtained by differentiating the attitude measurement. In this case, the demanded control torque due to the derivative term becomes

$$K_d \frac{d}{dt} \left(y(t) + w(t) \right).$$

It can be seen in Figure 21.2, that directly differentiating a noisy measurement amplifies the noise content of the derivative. This example highlights the need to remove as much of the measurement noise as possible. Fortunately, as Figures 21.1 and 21.2 illustrate, the output of interest (the attitude in our case) $y(t)$ typically varies slowly with time compared with the measurement noise $w(t)$. That is, the output $y(t)$ is generally "low-frequency" in nature, while the measurement noise $w(t)$ has more "high-frequency" content. We can therefore remove a significant portion of the measurement noise $w(t)$ by the application of an appropriate filter $H(s)$, which passes low-frequency content ($y(t)$) but blocks high frequency content ($w(t)$).

Spacecraft Dynamics and Control: An Introduction, First Edition.
Anton H.J. de Ruiter, Christopher J. Damaren and James R. Forbes.

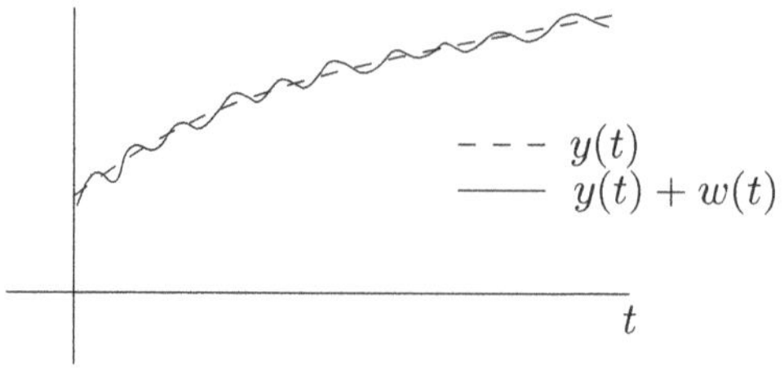

Figure 21.1 Real and ideal measurements

With the application of a low-pass filter $H(s)$, the block diagram for our proportional-derivative spacecraft attitude control problem from Section 17.7.1, becomes as shown in Figure 21.3.

To understand how to design the low-pass filter, we need to study the frequency response.

21.1 Frequency Response and Bode Plots

By the frequency response of a system described by transfer function $y(s)/u(s) = G(s)$, we mean the steady-state response to a sinusoidal input $u(t) = A \sin \omega t$, as shown in Figure 21.4.

Obviously, the concept of frequency response only makes sense if the system $G(s)$ is asymptotically stable (otherwise, the transient response does not vanish). We shall therefore consider only asymptotically stable systems $G(s)$ (which means the poles of $G(s)$ lie in the open left half plane).

It can be shown that the steady-state response of an asymptotically stable system $y(s) = G(s)u(s)$ to a sinusoidal input

$$u(t) = A \sin \omega t,$$

is given by

$$y_{ss}(t) = A|G(j\omega)| \sin(\omega t + \phi), \tag{21.1}$$

where $\phi = \angle G(j\omega)$. Note that ω is the frequency of the input, and A is the amplitude.

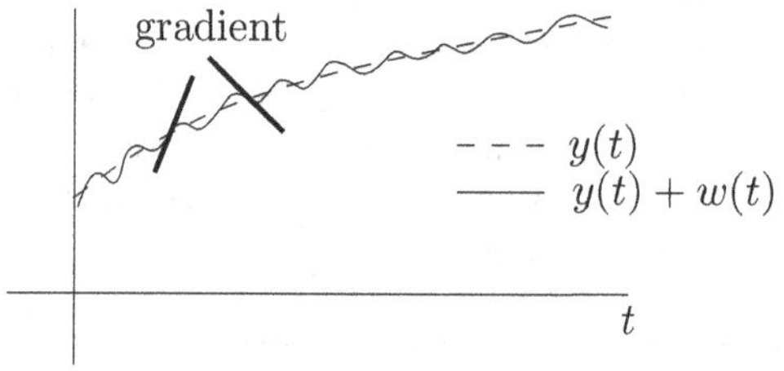

Figure 21.2 Differentiation of noisy measurements

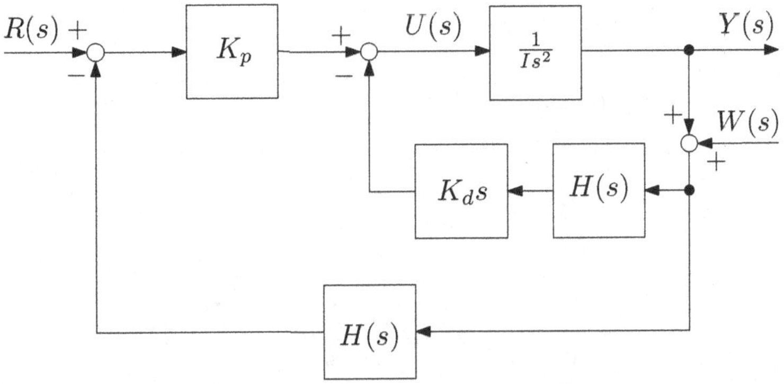

Figure 21.3 Modified proportional-derivative control with low-pass filter

Equation (21.1) shows two key features. First of all, the amplitude of the response is equal to the amplitude of the input scaled by $|G(j\omega)|$ (the magnitude of the transfer function evaluated at $j\omega$). Secondly, the phase of the output is shifted from the phase of the input by $\phi = \angle G(j\omega)$, which is the phase of the transfer function evaluated at $j\omega$.

What this means is that if $|G(j\omega)| \ll 1$ is very small for a given frequency ω, then the response of $G(s)$ is attenuated at that frequency. If $|G(j\omega)| \approx 1$ and $\angle G(j\omega) \approx 0$ for a given frequency ω, then $y_{ss}(t) \approx u(t)$.

Applying this knowledge to our low-pass filter $H(s)$ for measurement noise attenuation, we see that we want $|H(j\omega)| \approx 1$ and $\angle H(j\omega) \approx 0$ for low frequencies and $|H(j\omega)| \ll 1$ for high frequencies. Related to this are two important quantities, namely the filter bandwidth and cutoff rate. These are defined as follows:

Bandwidth, ω_b: the largest frequency such that $|H(j\omega)| \geq (1/\sqrt{2})|H(0)|$ for all frequencies ω in the interval $[0, \omega_b]$, as shown in Figure 21.5. The bandwidth indicates the range of frequencies that the system $H(s)$ passes inputs without significant attenuation.

Cutoff rate: the negative slope of $|H(j\omega)|$ at frequencies $\omega \geq \omega_b$. The larger the cutoff rate, the better the signal attenuation outside of the system bandwidth, and consequently better noise rejection.

For the purposes of filter design then, the filter bandwidth is chosen according to the range of frequencies the filter is required to pass. To be able to design an appropriate filter $H(s)$, we need to understand how $|H(j\omega)|$ and $\angle H(j\omega)$ vary as a function of frequency ω.

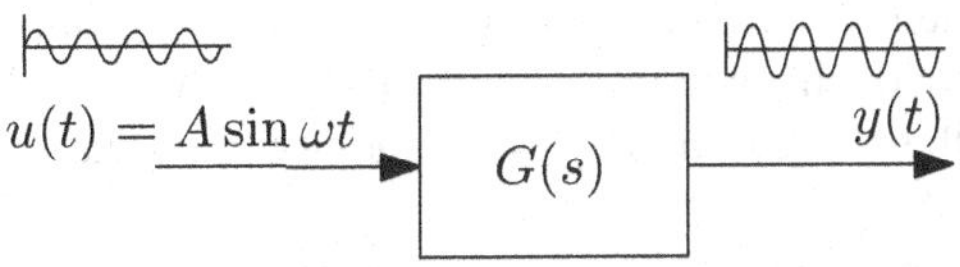

Figure 21.4 Frequency response of a system

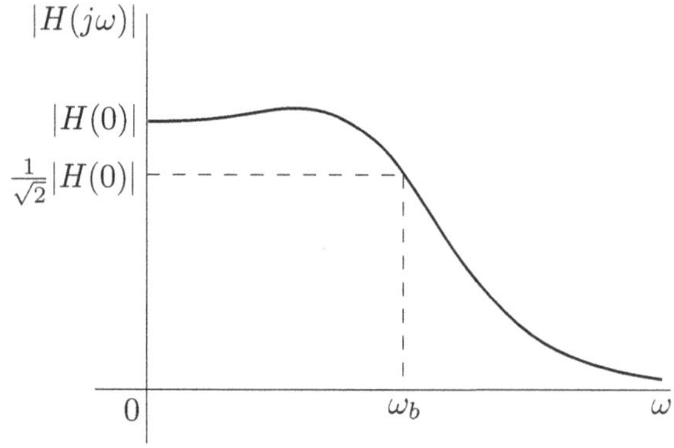

Figure 21.5 System bandwidth

21.1.1 Plotting the Frequency Response as a Function of ω (Bode Plots)

In general, we can write an asymptotically stable transfer function with integrators in the form

$$G(s) = \frac{K \prod_{i=1}^{k} (T_i s + 1) \prod_{j=1}^{p} \left((s/\omega_{nj})^2 + 2\zeta_j (s/\omega_{nj}) + 1\right)}{s^N \prod_{i=1}^{l} (\bar{T}_i s + 1) \prod_{j=1}^{q} \left((s/\bar{\omega}_{nj})^2 + 2\bar{\zeta}_j (s/\bar{\omega}_{nj}) + 1\right)}. \tag{21.2}$$

where $T_i, \bar{T}_i, \omega_{nj}, \bar{\omega}_{nj} > 0$ and $0 < \zeta_j, \bar{\zeta}_j < 1$. Note that the quadratic factors result from complex conjugate pairs of poles and zeros. Now, for complex numbers we have the following magnitude and phase relationships (see Appendix A)

$$\left| \frac{z_1 z_2 \ldots z_m}{s_1 s_2 \ldots s_n} \right| = \frac{|z_1||z_2|\ldots|z_m|}{|s_1||s_2|\ldots|s_n|},$$

and

$$\angle \left(\frac{z_1 z_2 \ldots z_m}{s_1 s_2 \ldots s_n} \right) = \sum_{i=1}^{m} \angle z_i - \sum_{j=1}^{n} \angle s_j.$$

Applying these relationships to the transfer function (21.2), we see that to find $|G(j\omega)|$ and $\angle G(j\omega)$, we just need to find the magnitudes and phases of the individual factors in the transfer function.

It can also be seen that the phases of the individual factors are additive. This is very useful for manual plotting of the phase frequency response ($\angle G(j\omega)$ vs. ω). That is, we can plot the phase of each factor individually, and then graphically sum them. It also provides an intuitive understanding for how the phase of an individual factor affects the phase frequency response of the entire transfer function.

Unfortunately, the magnitudes of the factors are multiplicative, which does not lend itself to manual plotting of the gain frequency response ($|G(j\omega)|$ vs. ω), nor does it provide an intuitive understanding of how the magnitude of an individual factor affects the gain frequency response of the entire function. To make them additive, we instead plot the log magnitude. Specifically, we obtain the gain $|G(j\omega)|$ at each frequency ω in *decibels*, which is defined as

$$20 \log |G(j\omega)|.$$

The reason for this is that

$$\log\left(\frac{|z_1||z_2|\ldots|z_m|}{|s_1||s_2|\ldots|s_n|}\right) = \sum_{i=1}^{m} \log|z_i| - \sum_{j=1}^{n} \log|s_j|,$$

which shows that the log magnitudes are additive. Therefore, for the gain frequency response, we plot $20 \log |G(j\omega)|$ vs ω, for which the contributions from each individual factor in the transfer function can be added graphically.

Examining the transfer function (21.2), we see that there are four basic types of factors we need to deal with in a transfer function.

1. Gain, K.
2. Integrators $1/s$ and derivatives s.
3. First order factors $Ts + 1$ and $(Ts + 1)^{-1}$.
4. Second order factors $(s/\omega_n)^2 + 2\zeta(s/\omega_n) + 1$ and $\left((s/\omega_n)^2 + 2\zeta(s/\omega_n) + 1\right)^{-1}$.

If we are familiar with the shapes of the frequency response for these factors, we will be able to construct frequency response plots for more complex transfer functions by addition of the frequency responses of the individual factors.

1. **Gain,** K.
   The log-magnitude is given by

   $$20 \log K = constant.$$

   The phase is given by

   $$\angle K = 0.$$

   The log-magnitude and phase frequency responses are plotted in Figure 21.6.
2. (a) **Derivative** s.
   The log magnitude is given by

   $$20 \log |j\omega| = 20 \log \omega.$$

   That is, the gain increases 20 dB for every decade of ω. We typically plot the frequency response with a logarithmic scale for frequency, which makes the above gain relationship linear.

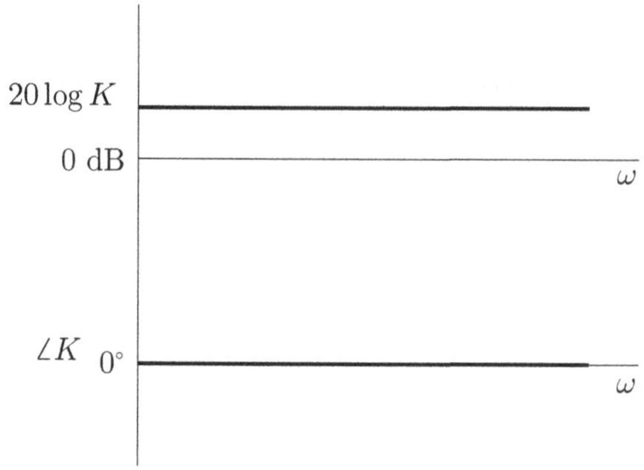

Figure 21.6 Frequency response of a constant gain

The phase is

$$\angle(j\omega) = 90^\circ.$$

The log-magnitude and phase frequency responses are plotted in Figure 21.7.

It is clear from the gain frequency response why it is bad to differentiate a noisy measurement. High frequency content in the noise is amplified (see the log-magnitude plot in Figure 21.7).

(b) **Integrator $1/s$.**

The log magnitude is

$$20 \log \left| \frac{1}{j\omega} \right| = -20 \log \omega,$$

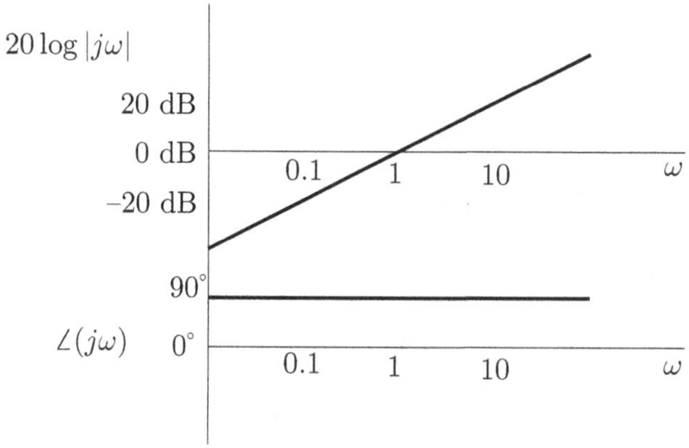

Figure 21.7 Frequency response of a derivative

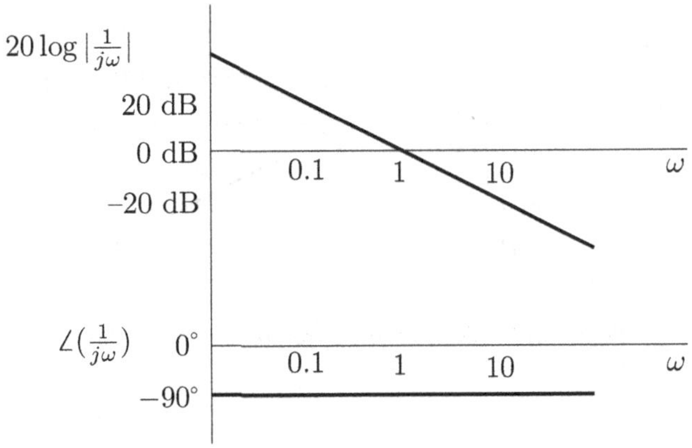

Figure 21.8 Frequency response of an integrator

and the phase is

$$\angle\left(\frac{1}{j\omega}\right) = -90°.$$

It is clear that the integrator frequency response is just the negative of the derivative frequency response. The log-magnitude and phase frequency responses are plotted in Figure 21.8.

3. **First order factors** $(Ts + 1)^{\pm 1}$

   Let us consider

$$\frac{1}{Ts+1}$$

We have

$$\left|\frac{1}{j\omega T+1}\right| = \frac{1}{|j\omega T+1|} = \frac{1}{\sqrt{\omega^2T^2+1}}.$$

Therefore, the log magnitude is

$$20\log\left|\frac{1}{j\omega T+1}\right| = -20\log\sqrt{\omega^2T^2+1}.$$

For very small ωT, this is approximately

$$20\log\left|\frac{1}{j\omega T+1}\right| \approx -20\log 1 = 0, \quad \omega T \ll 1.$$

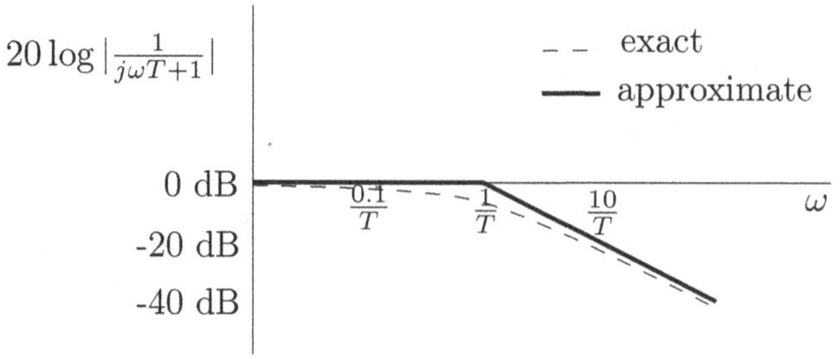

Figure 21.9 Gain frequency response of $1/(Ts + 1)$

That is, the gain frequency response has an asymptote at $20 \log |1/(j\omega T + 1)| = 0$ dB. For large ωT, the log magnitude is approximately

$$20 \log \left| \frac{1}{j\omega T + 1} \right| \approx -20 \log \omega T = -20 \log \omega - 20 \log T, \quad \omega T \gg 1.$$

That is, the gain frequency response has an asymptote with slope -20 dB/decade of ω, which crosses $20 \log |1/(j\omega T + 1)| = 0$ when $\omega = 1/T$. Therefore, the gain frequency response may be approximated by the two asymptotes, which meet at $\omega_c = 1/T$, which is called the *corner frequency*. The log-magnitude frequency response is plotted in Figure 21.9.

The maximum error in this asymptotic approximation occurs at the corner frequency, where the gain is $-20 \log \sqrt{2} = -3.03$ dB.

The phase at $\omega = 0$ is

$$\angle \left(\frac{1}{j\omega T + 1} \right) = 0^\circ, \quad \omega = 0.$$

The phase for very large ωT is

$$\angle \left(\frac{1}{j\omega T + 1} \right) \approx \angle \left(\frac{1}{j\omega T} \right) = -90^\circ, \quad \omega T \gg 1.$$

At the corner frequency, the phase is

$$\angle \left(\frac{1}{j\omega T + 1} \right) = \angle \left(\frac{1}{j1 + 1} \right) = -45^\circ.$$

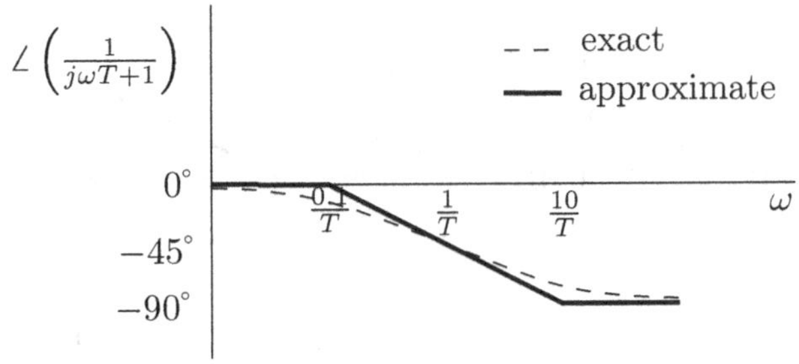

Figure 21.10 Phase frequency response of $1/(Ts+1)$

The phase frequency response may also be represented by a straight line approximation. To see this, let us examine the phases at $\omega = 0.1\omega_c$ and $\omega = 10\omega_c$ (one decade to the left and right of the corner frequency). We have

$$\angle\left(\frac{1}{j\omega T+1}\right) = \angle\left(\frac{1}{j0.1+1}\right) = -5.7^\circ \approx 0^\circ, \quad \omega = 0.1\omega_c = \frac{0.1}{T},$$
$$\angle\left(\frac{1}{j\omega T+1}\right) = \angle\left(\frac{1}{j10+1}\right) = -84.3^\circ \approx -90^\circ, \quad \omega = 10\omega_c = \frac{10}{T},$$

respectively. Due to this, we can approximate the phase frequency response between $\omega = 0.1\omega_c$ and $\omega = 10\omega_c$ by a straight line from 0° to 90°. Figure 21.10 shows the corresponding straight-line approximation to the phase frequency response, and the exact phase frequency response. It can be seen from Figures 21.9 and 21.10 that the approximations capture the main features of the gain and phase frequency responses.

For the first order factor $Ts + 1$, the frequency response plots are just the negative of those for $1/(Ts + 1)$, as shown in Figure 21.11.

4. **Quadratic Factors** $\left((s/\omega_n)^2 + 2\zeta(s/\omega_n) + 1\right)^{\pm 1}, \quad 0 < \zeta < 1$
   Consider the factor

$$\frac{1}{(j\omega/\omega_n)^2 + 2\zeta(j\omega/\omega_n) + 1} = \frac{1}{1 - (\omega/\omega_n)^2 + j2\zeta(\omega/\omega_n)}.$$

We have

$$\left|\frac{1}{(j\omega/\omega_n)^2 + 2\zeta(j\omega/\omega_n) + 1}\right| = \frac{1}{\sqrt{\left(1-(\omega/\omega_n)^2\right)^2 + (2\zeta(\omega/\omega_n))^2}}.$$

Therefore, the log magnitude becomes

$$\begin{aligned} &20\log\left|1/\left((j\omega/\omega_n)^2 + 2\zeta(j\omega/\omega_n) + 1\right)\right| \\ &= -20\log\sqrt{\left(1-(\omega/\omega_n)^2\right)^2 + (2\zeta(\omega/\omega_n))^2}. \end{aligned}$$

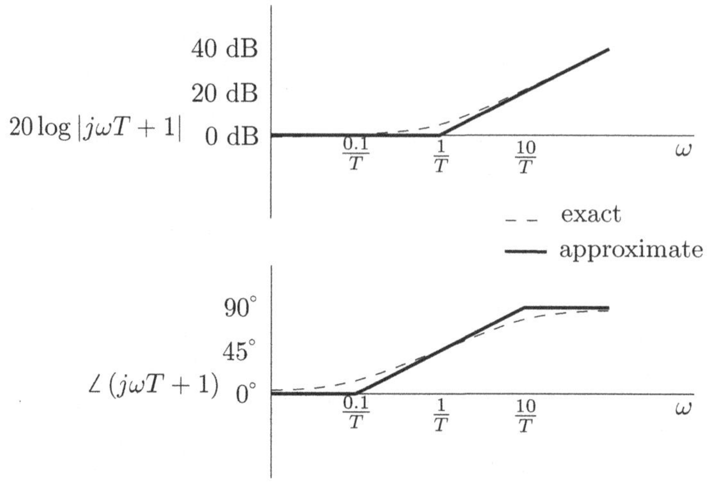

Figure 21.11 Frequency response of $Ts + 1$

For very small ω, the log magnitude is approximately

$$-20\log\sqrt{\left(1-(\omega/\omega_n)^2\right)^2+(2\zeta(\omega/\omega_n))^2} \approx -20\log 1 = 0, \quad \omega \ll \omega_n.$$

Hence, the gain frequency response has an asymptote at 0 dB. For very large ω, this log magnitude becomes approximately

$$\begin{aligned}-20\log\sqrt{\left(1-(\omega/\omega_n)^2\right)^2+(2\zeta(\omega/\omega_n))^2} &\approx -20\log(\omega/\omega_n)^2\\ &= -40\log(\omega/\omega_n),\\ &= -40\log\omega + 40\log\omega_n,\end{aligned}$$

for $\omega \gg \omega_n$. This is an asymptote with slope -40 dB/decade, which crosses 0 dB at $\omega = \omega_n$. That is, ω_n, the undamped natural frequency is the corner frequency for a quadratic factor, and the slope of the log magnitude at high frequencies is double that of a first order factor. The log-magnitude frequency response is plotted in Figure 21.12.

The phase at $\omega = 0$ is

$$\angle\left(\frac{1}{(j\omega/\omega_n)^2+2\zeta(j\omega/\omega_n)+1}\right) = 0^\circ, \quad \omega = 0.$$

At very large frequencies, the phase is approximately

$$\angle\left(\frac{1}{(j\omega/\omega_n)^2+2\zeta(j\omega/\omega_n)+1}\right) \approx \angle\left(\frac{1}{(j\omega/\omega_n)^2}\right) = -180^\circ, \quad \omega \gg \omega_n.$$

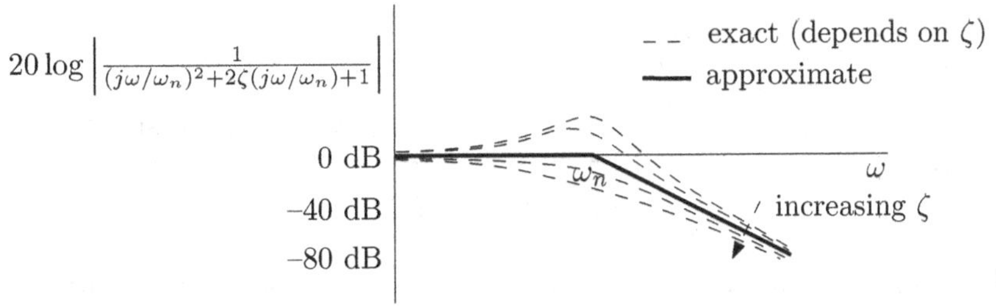

Figure 21.12 Gain frequency response of $\left((s/\omega_n)^2 + 2\zeta(s/\omega_n) + 1\right)^{-1}$

At the corner frequency, the phase is

$$\angle\left(\frac{1}{(j\omega/\omega_n)^2 + 2\zeta(j\omega/\omega_n) + 1}\right) = \angle\left(\frac{1}{2\zeta j}\right) = -90^\circ.$$

Similar to the first order factor, we can use a straight line approximation to the phase frequency response. Figure 21.13 shows the corresponding straight-line approximation to the phase frequency response, and the exact phase frequency response. From Figures 21.12 and 21.13 it can be seen that the approximations capture the main features of the gain and phase frequency responses, although not as well as for a first-order factor (compare to Figures 21.9 and 21.10).

For the quadratic factor $(s/\omega_n)^2 + 2\zeta(s/\omega_n) + 1$, the frequency response plots are just the negative of those for $\left((s/\omega_n)^2 + 2\zeta(s/\omega_n) + 1\right)^{-1}$, as seen in Figure 21.14.

The log magnitude and phase frequency response plots we have presented in Figures 21.6 to 21.14 are called *Bode Diagrams*. Using the responses for the four factors in a transfer function, we may now present a procedure for plotting Bode Diagrams for any transfer function made up of those factors.

Procedure for plotting Bode Diagrams

1. Rewrite $G(s)$ in terms of the basic factors K, s, $Ts + 1$, $(s/\omega_n)^2 + 2\zeta(s/\omega_n) + 1$.

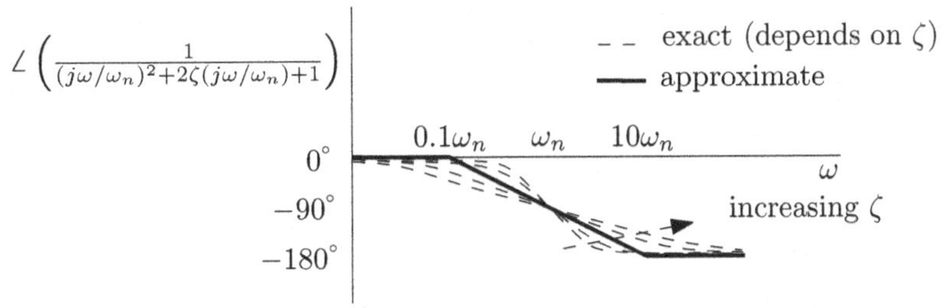

Figure 21.13 Phase frequency response of $\left((s/\omega_n)^2 + 2\zeta(s/\omega_n) + 1\right)^{-1}$

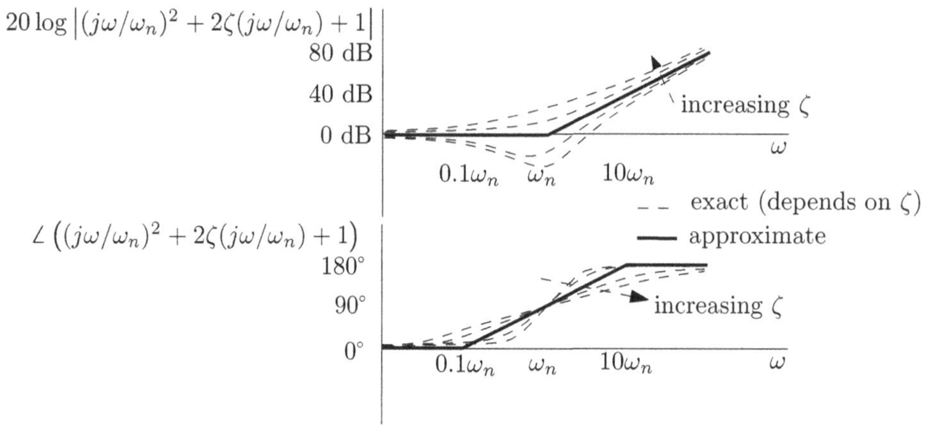

Figure 21.14 Frequency response of $(s/\omega_n)^2 + 2\zeta(s/\omega_n) + 1$

2. Plot the log magnitude frequency responses for each of these factors. Sum these to get the overall response.
3. Plot the phase frequency responses for each of these factors. Sum these to get the overall response.

Example 21.1 *Let us draw the Bode-plot for the transfer function*

$$G(s) = \frac{10(s+3)}{s(s^2+s+2)}.$$

First, we need to rewrite the factors in the standard form for Bode plots. We have

$$s + 3 = 3\left(\frac{1}{3}s + 1\right), \quad s^2 + s + 2 = 2\left(\left(\frac{s}{\sqrt{2}}\right)^2 + \frac{s}{2} + 1\right).$$

For the quadratic factor, we have undamped natural frequency $\omega_n = \sqrt{2}$ *and damping ratio* $\zeta = 1/(2\sqrt{2})$. *Therefore, in terms of these factors, the transfer function becomes*

$$G(s) = 15\frac{((1/3)s+1)}{s\left(\left(s/\sqrt{2}\right)^2 + s/2 + 1\right)}.$$

This transfer function has four factors

1. 15
2. $(1/3)s + 1$
3. $1/s$
4. $\left((s/\sqrt{2})^2 + (s/2) + 1\right)^{-1}$

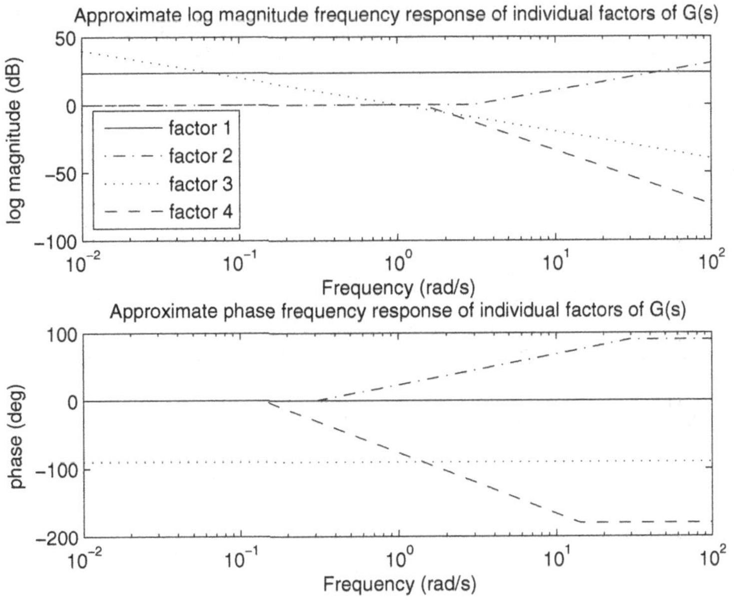

Figure 21.15 Approximate Bode diagrams for the individual factors of $G(s)$

Figure 21.15 shows the approximate frequency responses of each of the individual factors. Figure 21.16 shows the sum of the individual factors. For comparison, Figure 21.17 shows the exact frequency response of $G(s)$. Comparing Figures 21.16 and 21.17, it can be seen that the approximate gain frequency response is close to the true gain frequency response, while the approximate phase frequency response does not match the true phase frequency response very well. This illustrates that the approximate methods of obtaining frequency responses are good for obtaining a qualitative understanding, however any real analysis should be performed with precise numerically obtained frequency responses.

Finally, we note that we can still plot Bode diagrams for an unstable transfer function $G(s)$ (just plot $20\log|G(j\omega)|$ and $\angle G(j\omega)$). They just do not have the physical interpretation of being the steady-state frequency response. Also, they cannot be obtained experimentally (whereas they can for asymptotically stable systems).

21.2 Low-Pass Filter Design

Let us now return to the problem of finding a low-pass filter $H(s)$ for the purposes of measurement noise attenuation as in Figure 21.3. Recall that we wish $H(j\omega) \approx 1$ for small frequencies, and $H(j\omega) \approx 0$ for high frequencies. From the basic factors in Section 21.1.1, we see that a first order factor

$$H(s) = \frac{1}{Ts+1},$$

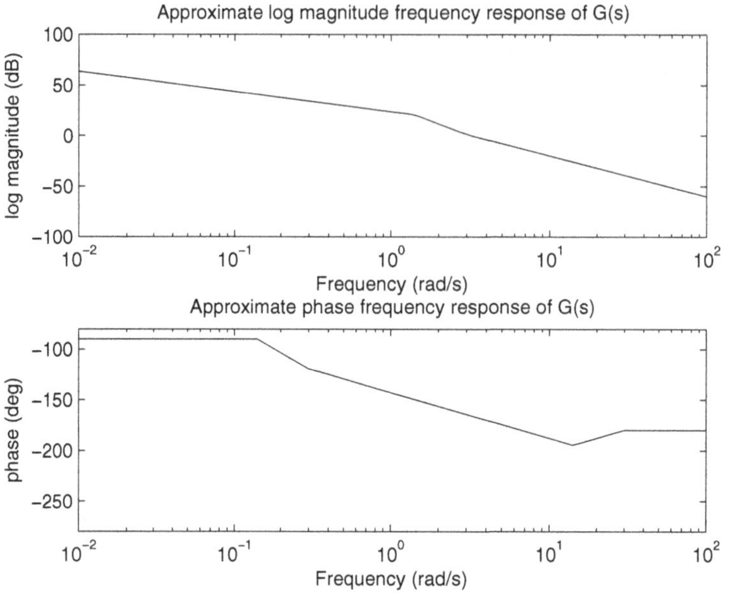

Figure 21.16 Approximate Bode Diagrams for $G(s)$

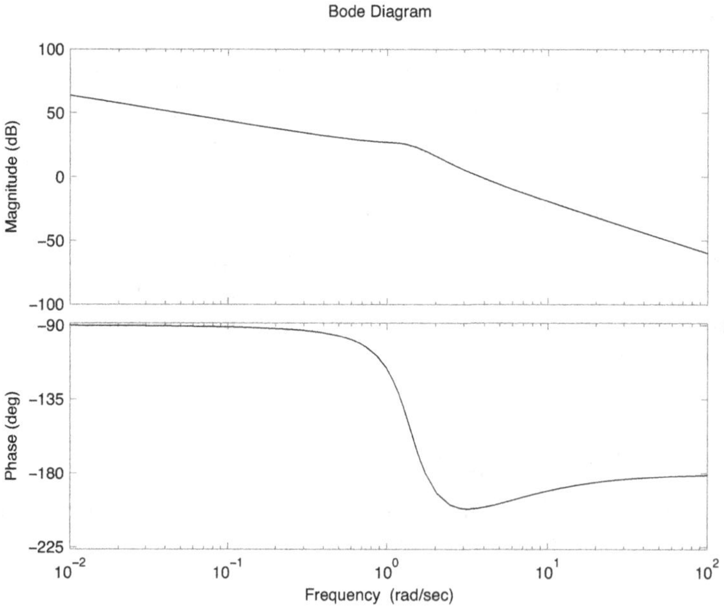

Figure 21.17 Exact Bode Diagrams for $G(s)$

with appropriately chosen corner frequency $\omega_c = (1/T)$ to give the desired bandwidth, is a good candidate for the filter $H(s)$. In fact, depending on the amount of measurement noise attenuation we require, we can use more than one in series, for example,

$$H(s) = \left(\frac{1}{Ts+1}\right)^2.$$

This increases the cut-off rate.

Notes

This chapter has introduced the concept of the frequency response of a system, and how it to relates the filtering of measurement noise in a control system. We have then examined techniques for determining an approximate frequency response. For further examples on how to determine an approximate frequency response, the reader is referred to any of the excellent books on classical control systems, including Franklin et al. (2010) and Ogata (2010).

References

Franklin GF, Powell JD and Emami-Naeini A 2010 *Feedback Control of Dynamic Systems* 6th edn. Prentice Hall, Upper Saddle River, NJ.

Ogata K 2010 *Modern Control Engineering* 5th edn. Prentice Hall, Upper Saddle River, NJ.

22

Relative Stability

We have seen that an important requirement on a closed-loop system is for it to be stable. In this chapter we shall study how close a system is to becoming unstable, which is the concept of *relative stability*. This is important to understand, since our mathematical models of the system being controlled are never perfect. It is very important that a closed-loop system with the controller that we have designed remains stable given the uncertainties in the plant model. It turns out that the system frequency response provides useful tools in this regard.

22.1 Polar Plots

We have seen that the steady-state frequency response of the asymptotically stable system

$$y(s) = G(s)u(s),$$

to the input $u(t) = A \sin \omega t$, depends on $G(j\omega)$ in the sense that

$$y_{ss}(t) = A|G(j\omega)| \sin(\omega t + \angle G(j\omega)).$$

Since,

$$G(j\omega) = |G(j\omega)| [\cos \angle G(j\omega) + j \sin \angle G(j\omega)], \tag{22.1}$$

it fully determines the amplitude and phase of the steady-state response. One method of presenting $G(j\omega)$ for different frequencies is the Bode Diagram. As we have seen, this contains two plots. One for the magnitude $|G(j\omega)|$ and one for the phase $\angle G(j\omega)$.

We could alternatively represent $G(j\omega)$ fully in the complex plane, as shown in Figure 22.1.

Letting ω vary from 0 to ∞, as in the Bode Diagram, we obtain a curve for $G(j\omega)$ in the complex plane. What we are doing is mapping the positive imaginary axis $s = j\omega$ to a curve in the complex plane through the transfer function $G(s)$, as shown in Figure 22.2.

Spacecraft Dynamics and Control: An Introduction, First Edition.
Anton H.J. de Ruiter, Christopher J. Damaren and James R. Forbes.

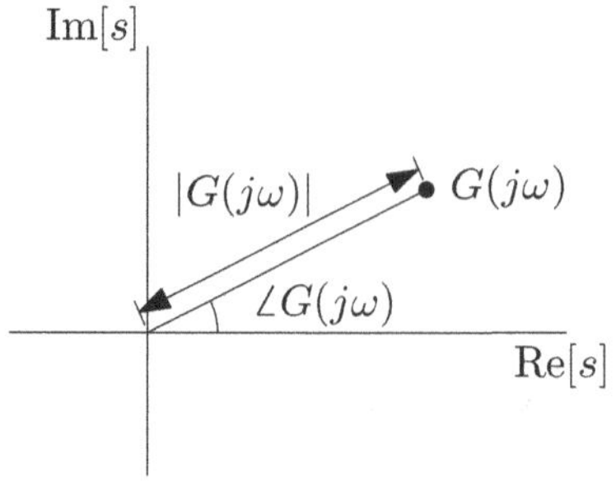

Figure 22.1 Representation of $G(j\omega)$ in the complex plane

The advantage of this is that we can include all information (magnitude and phase) on a single plot (and it can provide insight into system stability as we shall see shortly). The disadvantage is that the contributions of individual factors in the transfer function $G(s)$ cannot be seen clearly (unlike the Bode Diagram).

Let us now look at the general shapes of polar plots. Consider the transfer function

$$G(s) = \frac{a_0 s^m + a_1 s^{m-1} + \ldots + a_m}{s^n + b_1 s^{n-1} + \ldots + b_n}, \tag{22.2}$$

with $n > m$. For simplicity, we restrict $a_i > 0$ and $b_j > 0$ for $i = 1, \ldots, m$ and $j = 1, \ldots, n$.

22.1.1 Type 0 Systems ($N = 0$)

Let us first consider type 0 systems. Recall from Section 17.9 that the type of a system is the number of integrators, N in the transfer function. For a type 0 system, there are therefore no integrators, which means that $b_n \neq 0$.

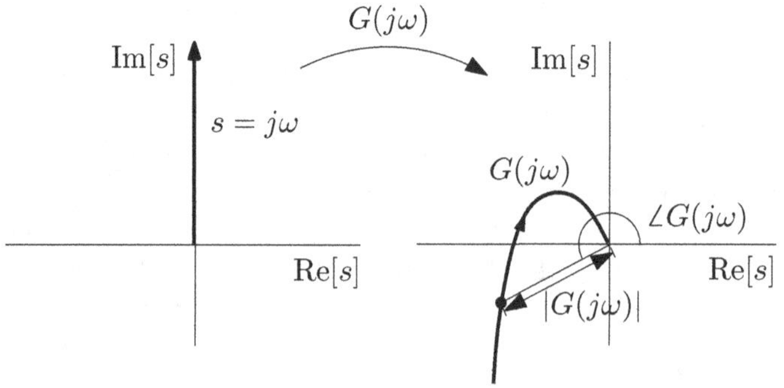

Figure 22.2 Mapping of $s = j\omega$ through $G(s)$

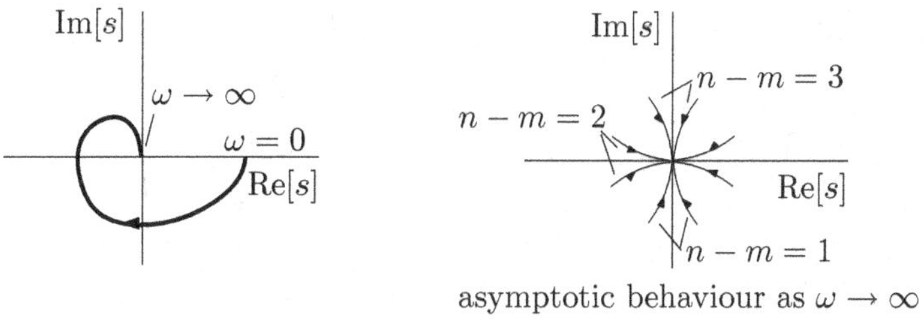

Figure 22.3 Polar plot for a type 0 system

Let us now consider the extreme behaviors of the polar plot of $G(j\omega)$. Referring to (22.2), we have

$$G(j\omega) = \frac{a_m}{b_n}, \text{ when } \omega = 0,$$

and for large ω, the highest powers of $j\omega$ in the numerator and denominator dominate such that we have

$$G(j\omega) \approx \frac{a_0}{(j\omega)^{n-m}} \to 0, \text{ when } \omega \to \infty.$$

Since $\angle a_0/(j\omega)^{n-m}$ is a multiple of -90°, it must be that the polar plot approaches the origin along one of the coordinate axes. Which axis depends on the relative degree, $n - m$ of $G(s)$, as shown in Figure 22.3.

22.1.2 *Type 1 Systems ($N = 1$)*

For a type 1 system we have $a_m \neq 0$, $b_{n-1} \neq 0$ and $b_n = 0$. Therefore, from (22.2), for very small values of ω, the lowest powers of $j\omega$ in the numerator and denominator dominate such that

$$G(j\omega) \approx \frac{a_m}{b_{n-1} j\omega}, \text{ when } \omega \to 0.$$

Therefore, we have

$$|G(j\omega)| \to \infty, \quad \angle G(j\omega) \to -90^\circ \text{ when } \omega \to 0.$$

As for type 0 systems, for large ω, the highest powers of $j\omega$ in the numerator and denominator dominate such that

$$G(j\omega) \approx \frac{a_0}{(j\omega)^{n-m}} \to 0, \text{ when } \omega \to \infty.$$

Figure 22.4 shows a typical polar plot for a type 1 system.

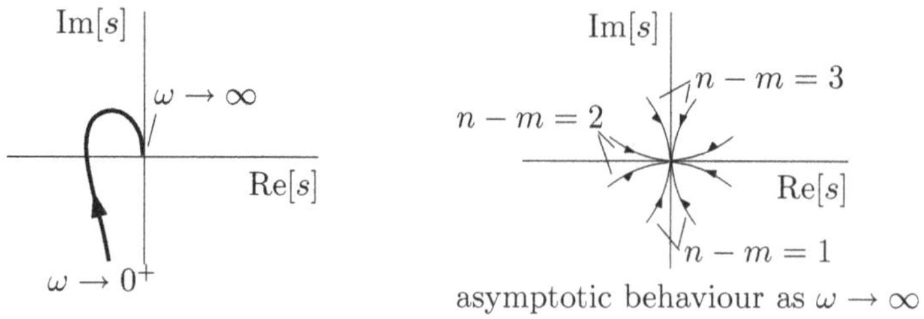

Figure 22.4 Polar plot for a type 1 system

22.2 Nyquist Stability Criterion

We shall now see how the frequency response can be used to determine closed-loop stability. It is best illustrated using the polar plot, and once that is understood, it can be related back to the Bode Diagram.

Consider the general feedback system as shown in Figure 22.5. Note that $G(s)$ may contain the controller as well as the plant. As seen in Section 17.4, the closed-loop transfer function is

$$T(s) = \frac{Y(s)}{R(s)} = \frac{G(s)}{1 + G(s)H(s)}. \tag{22.3}$$

As we have seen in Section 17.9, the open-loop transfer function is

$$G_o(s) = G(s)H(s).$$

The closed-loop stability is determined by the poles of (22.3), and they are therefore the zeros of the characteristic equation

$$1 + G_o(s) = 0. \tag{22.4}$$

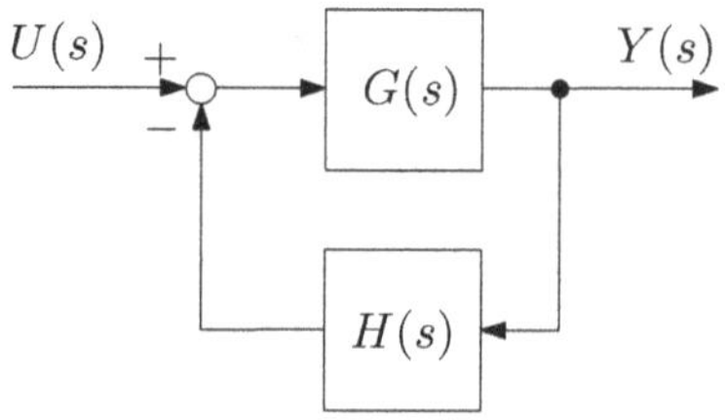

Figure 22.5 General feedback system

Therefore, we would like to know if the zeros of (22.4) lie in the open left-half plane. We do not need to know their exact values. Now, in general, the open-loop transfer function can be written as

$$G_o(s) = \frac{a_0 s^m + a_1 s^{m-1} + \ldots + a_m}{s^n + b_1 s^{n-1} + \ldots + b_n} = \frac{a(s)}{b(s)},$$

where $m \leq n$. Therefore, we can write

$$1 + G_o(s) = \frac{b(s) + a(s)}{b(s)}.$$

As an aside, this shows that the poles of $1 + G_o(s)$ are equal to the poles of the open-loop transfer function $G_o(s)$.

Now, there is a very important result in complex analysis, which will allow us to relate the frequency response of the open-loop transfer function $G_o(s)$ to the stability of the closed-loop transfer function $T(s)$. The result is called the *argument principle*.

22.2.1 Argument Principle

Let

$$F(s) = \frac{a_0 s^m + a_1 s^{m-1} + \ldots + a_m}{s^n + b_1 s^{n-1} + \ldots b_n}, \tag{22.5}$$

be some transfer function. Clearly, for every point s, except at the poles of $F(s)$, there is a corresponding point $F(s)$ in the complex plane. That is, $F(s)$ maps points in the complex plane to points in the complex plane. Likewise, $F(s)$ maps contours in the complex plane to contours in the complex plane. For example, let

$$s(t) = c(t) + jy(t), \quad t \in [t_1, t_2],$$

where $x(t)$ and $y(t)$ are real continuous functions of t, be a contour in the complex plane. Then, $F(s(t))$ is also a contour in the complex plane, as shown in Figure 22.6.

Theorem 22.1 (Argument Principle) *Consider $F(s)$ as in (22.5). Let $s(t)$ be a simple (does not intersect itself) closed clockwise contour in the complex plane that does not pass through any zeros or poles of $F(s)$. Let P be the number of poles of $F(s)$ inside the contour, and Z be the number of zeros of $F(s)$ inside the contour. Then, the number of clockwise encirclements of the origin by the curve $F(s(t))$ is given by*

$$N = Z - P.$$

See Figure 22.7 for some examples.

Note: The exact shape of the contour $s(t)$ does not matter, as long as it is simple, closed and clockwise, and does not pass through any poles or zeros of $F(s)$.

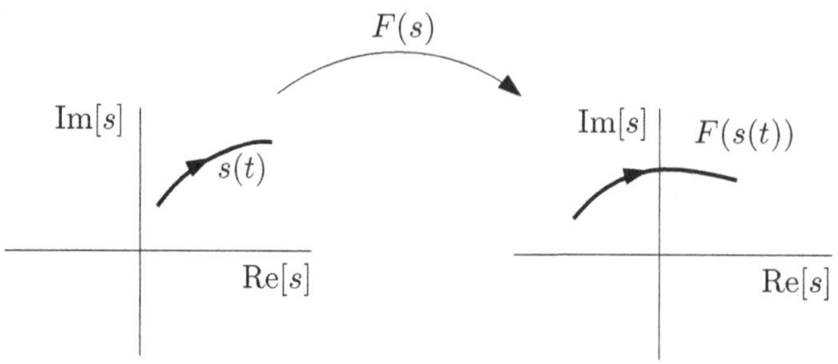

Figure 22.6 Mapping of $s(t)$ through $F(s)$

22.2.2 *Stability Analysis of the Closed-Loop System*

We have already seen that the closed-loop poles of $T(s)$ in (22.3) are given by the zeros of

$$1 + G_o(s).$$

If $T(s)$ is unstable, it must have poles in the right-half plane. That is, $1 + G_o(s)$ must have zeros in the right-half plane. From our knowledge of the open-loop transfer function $G_o(s)$, we already know the number of poles of $1 + G_o(s)$ in the right-half plane (this is equal to the number of poles of $G_o(s)$ in the right-half plane). Let

$$P = \text{number of poles of } G_o(s) \text{ in the right-half plane.}$$

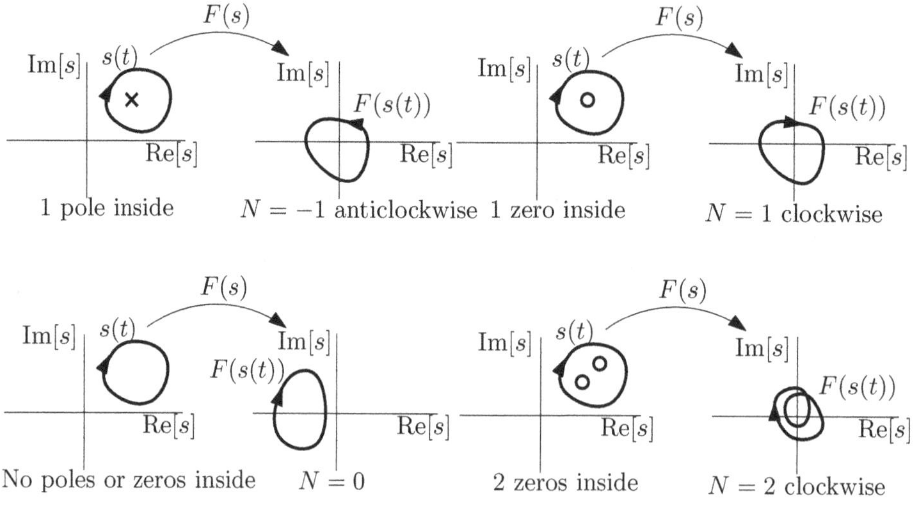

Figure 22.7 Examples of argument principle

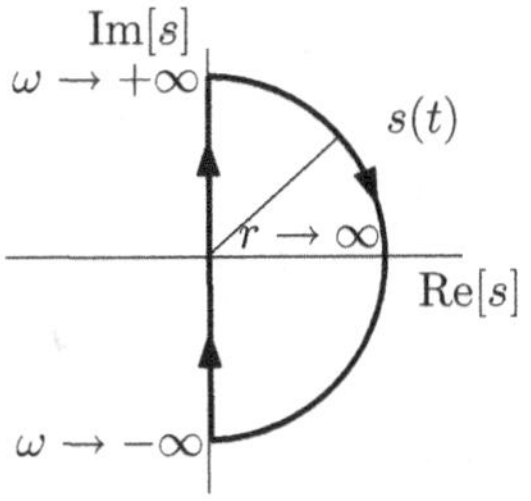

Figure 22.8 Nyquist contour

We can now determine

$$Z = \text{ number of zeros of } 1 + G_o(s) \text{ in the right-half plane,}$$

by using the argument principle. We do this by defining the contour shown in Figure 22.8 (assuming $1 + G_o(s)$ has no poles on the imaginary axis). This is called the Nyquist contour, and it will contain any right-half plane zeros of $1 + G_o(s)$.

Using the argument principle, the closed-loop system stability can be determined by the number of clockwise encirclements N of the origin of the contour $1 + G_o(s(t))$, that is,

$$Z = N + P,$$

where P is known from $G_o(s)$, and N is determined from the mapping of the Nyquist contour.

Now, we note that the contour $1 + G_o(s(t))$ is just the contour of $G_o(s(t))$ shifted to the right by 1. Therefore, encirclements of the origin by $1 + G_o(s(t))$ is equivalent to encirclements of -1 by $G_o(s(t))$, which is the contour of the open-loop system, as shown in Figure 22.9. The mapping of the Nyquist contour through the open-loop transfer function is called the *Nyquist plot*.

The Nyquist contour can be divided into three parts:

$$\begin{aligned} &s_1(\omega) = j\omega, \quad \omega : 0 \to \infty, \\ &s_2(\theta) = re^{j\theta}, \quad \theta : \tfrac{\pi}{2} \to -\tfrac{\pi}{2}, \quad r \to \infty, \\ &s_3(\omega) = -j\omega, \; \omega : \infty \to 0. \end{aligned}$$

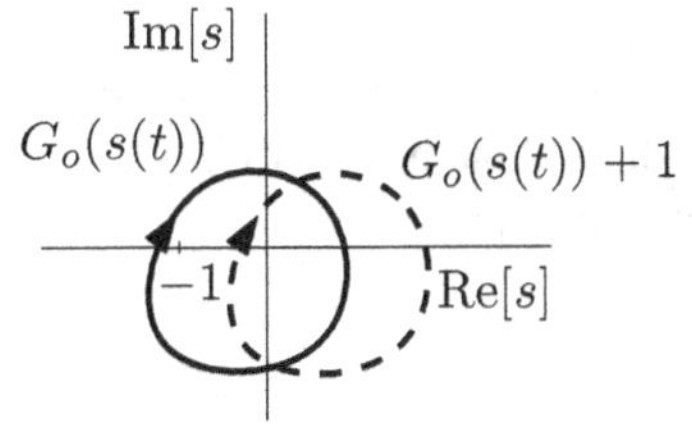

Figure 22.9 Shift of open-loop Nyquist plot

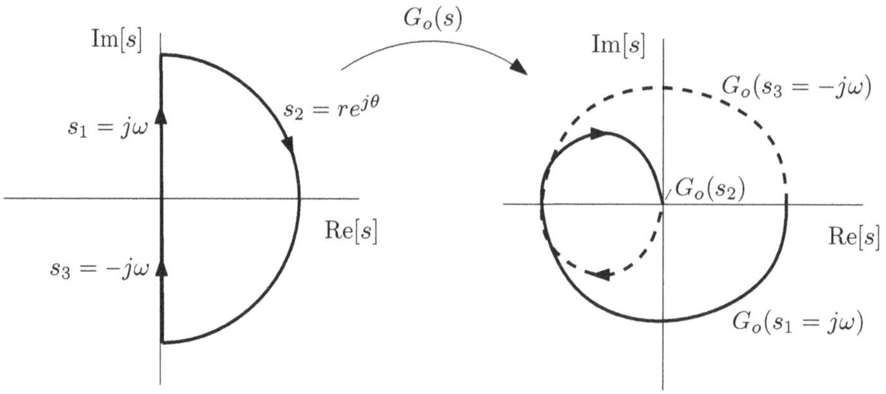

Figure 22.10 Nyquist plot with no open-loop poles on the imaginary axis

Let the open-loop transfer function be given by

$$G_o(s) = \frac{a_0 s^m + a_1 s^{m-1} + \ldots + a_m}{s^n + b_1 s^{n-1} + \ldots b_n}, \quad m \leq n.$$

Then, the Nyquist plot can also be divided into three parts

$$G_o(s_1) = G_o(j\omega), \text{ the open-loop frequency response, as illustrated in Figure 22.10,}$$
$$G_o(s_2) = \begin{cases} 0, & m < n, \\ a_0, & m = n. \end{cases}$$
$$G_o(s_3) = G_o(-j\omega) = \overline{G(j\omega)}, \text{ the mirror image of the open-loop frequency response.}$$

Since $G_o(s_2) = \textit{constant}$, we only need to plot the contours $G_o(j\omega)$ and $G_o(-j\omega)$ to determine encirclements of -1. Based on the Nyquist plot of the open-loop transfer function, we can now apply the argument principle to determine stability of the closed-loop system. The result is called the Nyquist Stability Criterion.

Nyquist Stability Criterion

Let the open-loop transfer function $G_o(s)$ have no poles on the imaginary axis. Let

N_{-1}^{cw} = number of clockwise encirclements of -1 of $G_o(j\omega)$ as $\omega : -\infty \to \infty$,
$P_{G_o}^{+}$ = number of right-half plane poles of $G_o(s)$,
$Z_{1+G_o}^{+}$ = number of right-half plane zeros of $1 + G_o(s)$.

Then,

$$Z_{1+G_o}^{+} = N_{-1}^{cw} + P_{G_o}^{+}. \tag{22.6}$$

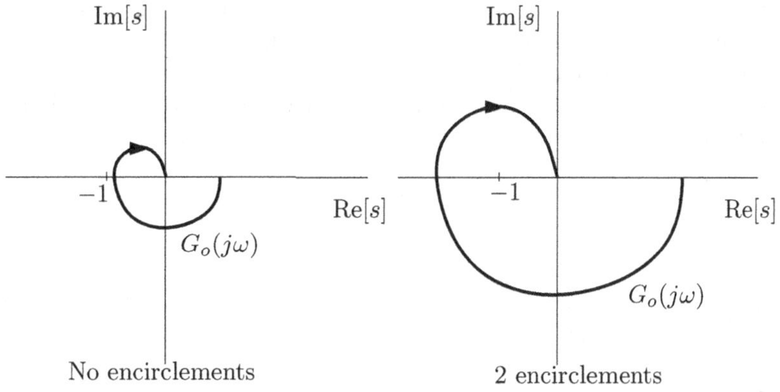

Figure 22.11 Determining number of encirclements from the frequency response

Since $Z^+_{1+G_o}$ is also the number of unstable closed-loop poles, the condition for stability is

$$Z^+_{1+G_o} = 0, \text{ for stability.} \tag{22.7}$$

Note that since $G_o(-j\omega)$ is the mirror image of $G_o(j\omega)$, we only need to plot the frequency response $G_o(j\omega)$ for $\omega : 0 \to \infty$ to obtain enough information to determine encirclements of -1, and hence closed-loop stability. That is, all we need to determine closed-loop stability is the open-loop frequency response (which can also be presented in the Bode Diagram), as shown in Figure 22.11.

Modification for Open-Loop Poles on the Imaginary Axis

If the open-loop transfer function $G_o(s)$ has poles on the imaginary axis, we must modify the Nyquist contour to pass around them, as shown in Figure 22.12.

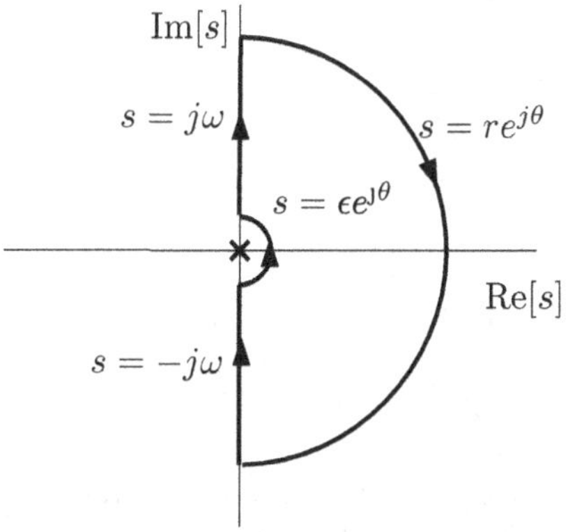

Figure 22.12 Modified Nyquist contour with an open-loop pole at the origin

That is, if jp is a pole of $G_o(s)$, then we modify the Nyquist contour around $s = jp$ to be $jp + \epsilon e^{j\theta}$, where we let $\epsilon \to 0$, and $\theta : -\frac{\pi}{2} \to \frac{\pi}{2}$. Note that when $\omega \to p$, $|G(j\omega)| \to \infty$. Therefore, the Nyquist mapping $G_o(jp + \epsilon e^{j\theta})$ will make an arc about the complex plane at infinity as $\epsilon \to 0$. We must account for any possible encirclements of -1 by this arc. Having accounted for this, the Nyquist stability criterion remains the same, namely

$$Z^+_{1+G_o} = N^{cw}_{-1} + P^+_{G_o}, \tag{22.8}$$

with

$$Z^+_{1+G_o} = 0, \text{ for stability.} \tag{22.9}$$

Example 22.1 (PID spacecraft attitude control) *Let us now apply the Nyquist stability criterion to the PID controller design for spacecraft attitude control in Section 20.2, where the plant is given by*

$$G_p(s) = \frac{1}{Is^2},$$

and the controller is

$$G_c(s) = K_p + K_d s + \frac{K_i}{s}.$$

In Section 20.2, we have taken the spacecraft inertia to be $I = 1$ kg·m^2. The PID control gains were designed as

$$K_p = 0.0179, \quad K_d = 0.1753, \quad K_i = 4.59 \times 10^{-4}.$$

We find that the open-loop transfer function is given by

$$G_o(s) = \frac{K_d s^2 + K_p s + K_i}{s^3}. \tag{22.10}$$

The open-loop transfer function has a triple pole at the origin. Therefore, the Nyquist contour must pass around it, as shown in Figure 22.13.

The Nyquist contour is therefore made up of four parts

$$\begin{aligned}
s_1(\omega) &= j\omega, \quad \omega : \epsilon \to r, \\
s_2(\theta) &= re^{j\theta}, \quad \theta : \tfrac{\pi}{2} \to -\tfrac{\pi}{2}, \\
s_3(\omega) &= -j\omega, \quad \omega : r \to \epsilon, \\
s_4(\theta) &= \epsilon e^{j\theta}, \quad \theta : -\tfrac{\pi}{2} \to \tfrac{\pi}{2}.
\end{aligned}$$

To ensure that the Nyquist contour encircles the entire right-half plane, we let $\epsilon \to 0$ and $r \to \infty$. We have $G_o(s_2) \to 0$ as $r \to \infty$. Since $G_o(s_3)$ is the mirror image of $G_o(s_1)$, we need only examine $G_o(s_1)$ and $G_o(s_4)$ to determine the entire Nyquist plot. Let us start with $G_o(s_1)$.

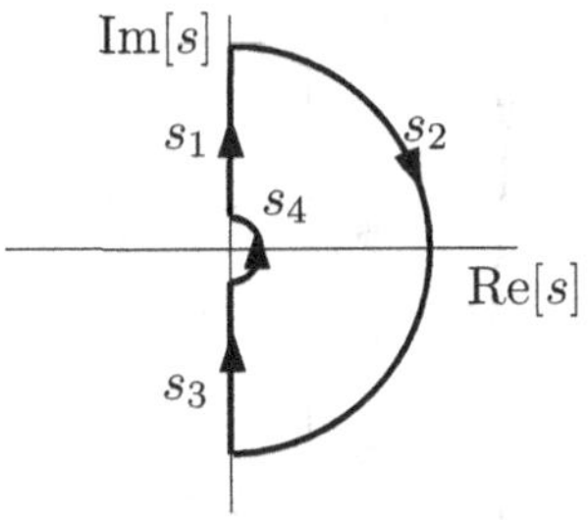

Figure 22.13 Nyquist contour for PID spacecraft attitude control example

When $\omega = \epsilon$, for very small ϵ, lowest powers of $j\omega$ in the numerator and denominator dominate such that

$$G_o(j\omega) \approx \frac{K_i}{(j\epsilon)^3} = \frac{K_i}{-j\epsilon^3}.$$

Therefore, we have

$$|G_o(j\omega)| \to \infty, \ \angle G_o(j\omega) \to -270^\circ, \ \text{as } \omega = \epsilon \to 0^+.$$

That is, the frequency response $G_o(j\omega)$ comes in from infinity parallel to the positive imaginary axis.

When $\omega = r$, for very large r, highest powers of $j\omega$ in the numerator and denominator dominate such that

$$G_o(j\omega) \approx \frac{K_d}{(jr)}.$$

Therefore, we have

$$|G_o(j\omega)| \to 0, \ \angle G_o(j\omega) \to -90^\circ, \ \text{as } \omega = r \to \infty.$$

That is, the frequency response $G_o(j\omega)$ approaches the origin along the negative imaginary axis. This means that the frequency response must cross the real axis at some point. Consider

$$G_o(j\omega) = \frac{K_d(j\omega)^2 + K_p(j\omega) + K_i}{(j\omega)^3} = \frac{(K_i - K_d\omega^2) + jK_p\omega}{-j\omega^3}. \quad (22.11)$$

We see that since the denominator is purely imaginary, $G_o(j\omega)$ can only be real if the numerator is also purely imaginary. That means that at the real axis crossing occurs at the frequency satisfying,

$$K_i - K_d\omega^2 = 0.$$

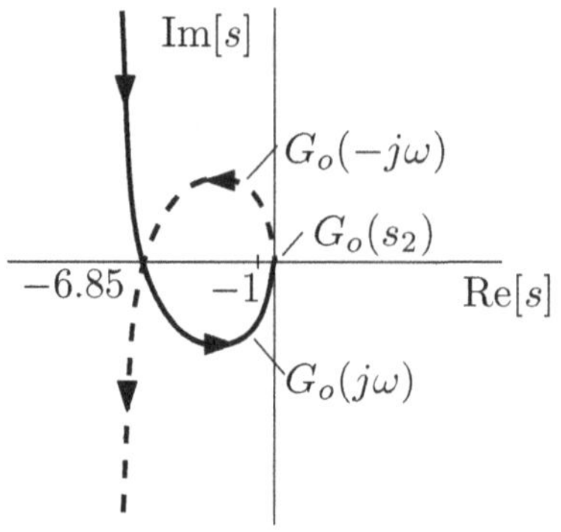

Figure 22.14 Frequency response of $G_o(j\omega)$ and its mirror image

That is, when the frequency is

$$\omega = \sqrt{\frac{K_i}{K_d}},$$

the frequency response $G_o(j\omega)$ crosses the real axis. Substituting this into (22.11), we find that the real axis crossing occurs at

$$G_o(j\sqrt{\frac{K_i}{K_d}}) = -\frac{K_p K_d}{K_i} = -6.85.$$

It is also clear from (22.11) that $G_o(j\omega)$ does not cross the imaginary axis. We can now sketch the frequency response $G_o(j\omega)$ and its mirror image $G_o(-j\omega)$, as shown in Figure 22.14.

The final part needed to complete the Nyquist plot is $G_o(s_4)$. When $s = \epsilon e^{j\theta}$ for very small ϵ, lowest powers of s dominate such that

$$G_o(s) \approx \frac{K_i}{(\epsilon e^{j\theta})^3} = \frac{K_i e^{-3j\theta}}{\epsilon^3}.$$

Therefore, for $s_4 = \epsilon e^{j\theta}$, with $\theta : -\frac{\pi}{2} \to \frac{\pi}{2}$, we have

$$|G_o(s_4)| \to \infty, \text{ as } \epsilon \to 0^+ \text{ and } \angle G_o(s_4) : \frac{3\pi}{2} \to -\frac{3\pi}{2}.$$

That is, $G_o(s_4)$ encircles the entire right-half plane one and a half times, starting at the end of $G_o(-j\omega)$ and finishing at the beginning of $G_o(j\omega)$, as shown in Figure 22.15.

We can now see from the Nyquist plot in Figure 22.15 that there is one clockwise encirclement of −1 and one counterclockwise encirclement of −1. Therefore, we have

$$N_{-1}^{cw} = 1 - 1 = 0.$$

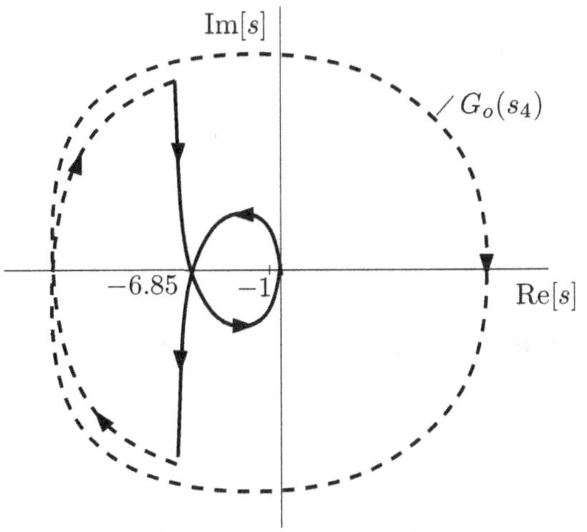

Figure 22.15 Complete Nyquist plot for PID spacecraft attitude control example

From the open-loop transfer function (22.10), we know that there are no open-loop right-half plane poles, so

$$P_{G_o}^{+} = 0.$$

Therefore, we find that the number of closed-loop right-half plane poles is

$$Z_{1+G_o}^{+} = N_{-1}^{cw} + P_{G_o}^{+} = 0,$$

and the closed-loop system is stable.

22.3 Stability Margins

As explained at the beginning of this chapter, the mathematical models we use for designing a control system are never perfect. It is therefore very important that the control system we design is robustly stable given uncertainties in the mathematical model of the system. To fix ideas, let the plant model we use for the design of a control system be given by

$$\frac{Y(s)}{U(s)} = G_p(s),$$

while the true plant has transfer function

$$\frac{Y(s)}{U(s)} = \bar{G}_p(s) = G_p(s)G_u(s),$$

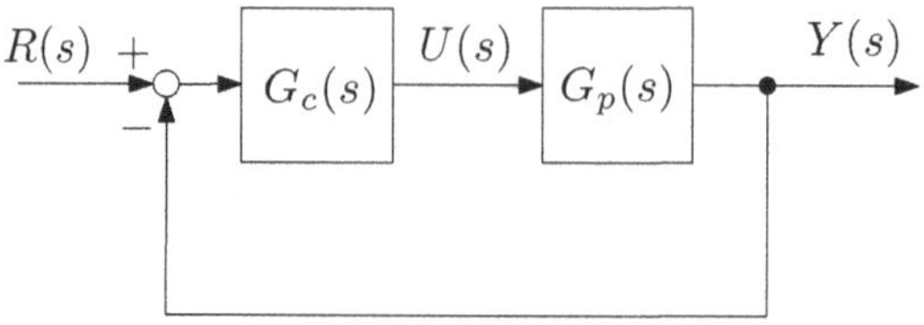

Figure 22.16 Ideal closed-loop system

where $G_u(s)$ is a transfer function representing the uncertainty. Let us assume that we have designed a controller

$$U(s) = G_c(s)E(s),$$

which asymptotically stabilizes the modeled plant $G_p(s)$ as shown in Figure 22.16. When the controller is applied to the true system, the closed-loop system is as shown in Figure 22.17. Comparing Figures 22.16 and 22.17, we see that the ideal open-loop transfer function (the one we designed) is

$$G_o(s) = G_c(s)G_p(s),$$

while the true open-loop transfer function is

$$\bar{G}_o(s) = G_o(s)G_u(s).$$

Therefore, the designed closed-loop poles satisfy the characteristic equation

$$1 + G_o(s) = 0,$$

while the true closed-loop poles satisfy

$$1 + G_o(s)G_u(s) = 0.$$

We therefore need to design our closed-loop system with some stability margin such that the real closed-loop system will still be stable.

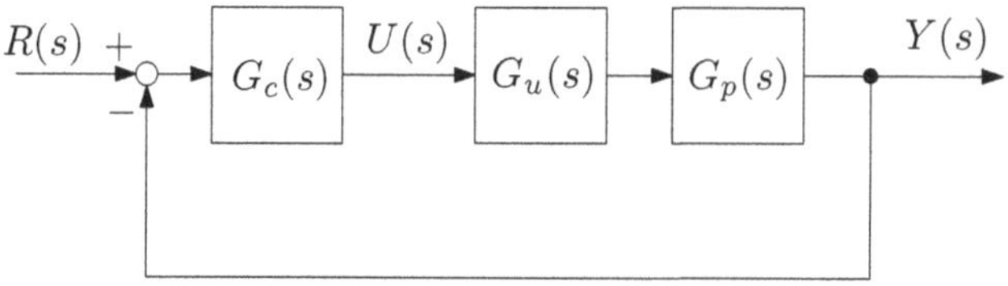

Figure 22.17 True closed-loop system

Modified Implementation

As explained in Sections 17.7.1 and 20.2, there are occasions where it is desirable to divide the control law into two parts

$$G_c(s) = G_{ce}(s) + G_{cy}(s),$$

and then implement the control law with the modified form shown in Figure 22.18. It is readily found that the closed-loop characteristic equations for the standard and modified implementations (Figures 22.17 and 22.18, respectively) are identical. Therefore, for the purposes of a closed-loop robustness analysis, we need only consider the standard implementation shown in Figure 22.17.

22.3.1 Stability Margin Definitions

We shall assume that we have designed a control system $G_c(s)$ which renders the ideal closed-loop system (in Figure 22.16) asymptotically stable.

As we have just seen in Section 22.2.2, the Nyquist stability criterion tells us this means that the Nyquist plot of the ideal open-loop system ($G_o(s)$) has the required number of clockwise encirclements of -1, given by

$$N_{-1}^{cw} = -P_{G_o}^{+},$$

where $P_{G_o}^{+}$ is the number of open-loop poles in the right-half plane. *We shall assume that $G_u(s)$ has no right-half plane poles (if it does, we have done a poor job of modeling the plant).* Therefore, for robust stability, the Nyquist plot of the real open-loop system ($G_o(s)G_u(s)$) must also have the same number of clockwise encirclements of -1 as the Nyquist plot of the ideal open-loop system $G_o(s)$.

As we saw in Chapter 21, the effect of multiplying one transfer function by another is to add gain and phase to the frequency response. That is, $G_u(s)$ adds gain and phase to the open-loop transfer function $G_o(s)$. We can therefore define stability margins on the basis of how much pure gain or phase can be added to the open-loop transfer function by $G_u(s)$ before the real closed-loop system (in Figure 22.17) becomes unstable.

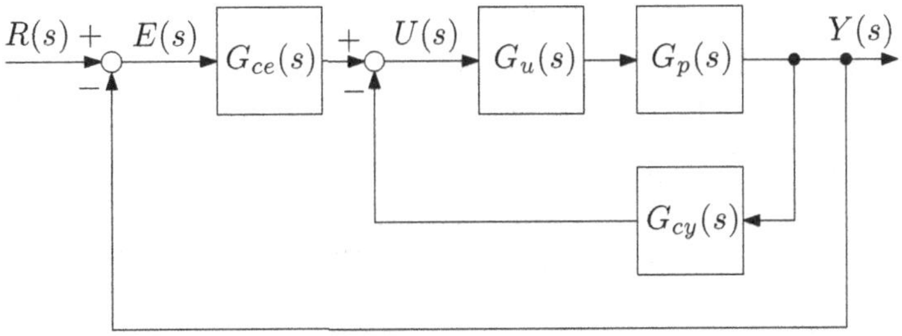

Figure 22.18 True closed-loop system with modified control implementation

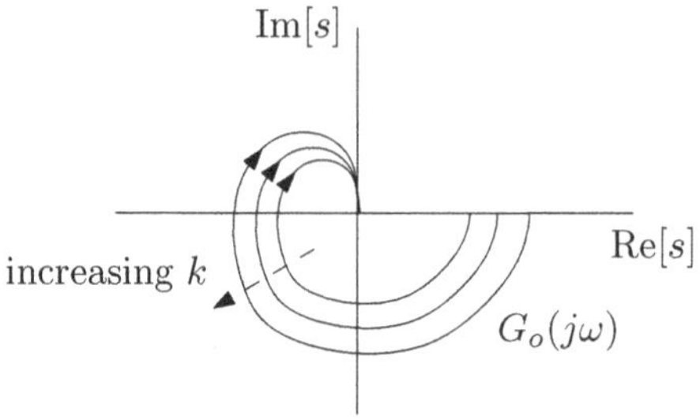

Figure 22.19 Effect of changing gain on the Nyquist plot

22.3.1.1 Gain Margin

To define the gain margin, we let the uncertainty in Figure 22.17 be a constant gain

$$G_u(s) = k > 0.$$

The gain margin is then defined by the range k over which the real closed-loop system is asymptotically stable. To determine this, we need to examine the Nyquist plot of the ideal open-loop system $G_o(s)$. First, we note that the effect of the gain k is simply to scale the Nyquist plot, as shown in Figure 22.19. That is, only the magnitude changes, but the phase stays the same

$$|kG_o(j\omega)| = k|G_o(j\omega)|, \quad \angle(kG_o(j\omega)) = \angle(G_o(j\omega)).$$

Since the ideal closed-loop system is asymptotically stable (by design), we determine the gain margin by varying k from the ideal value ($k = 1$ for the ideal closed-loop system) until the number of encirclements of -1 changes. We may be able to do this both by increasing and/or decreasing k. To do this, we determine all frequencies, ω_p at which

$$\angle G_o(j\omega_p) = -180^\circ + 360^\circ(l - 1),$$

for some integer l. That is, we determine all frequencies at which the Nyquist plot crosses the negative real axis, as shown in Figure 22.20. These frequencies are called *phase cross-over frequencies*.

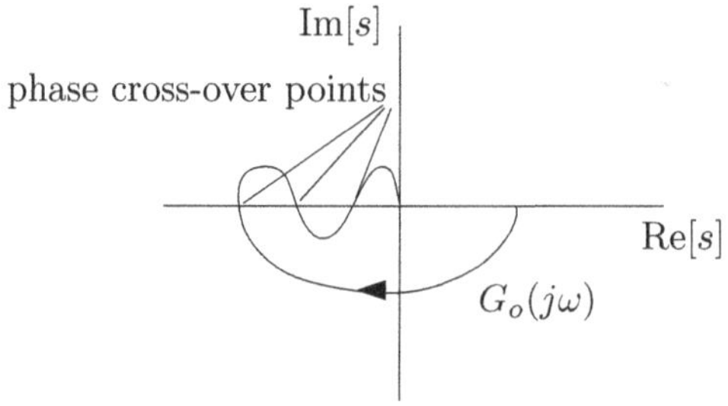

Figure 22.20 Phase cross-over points

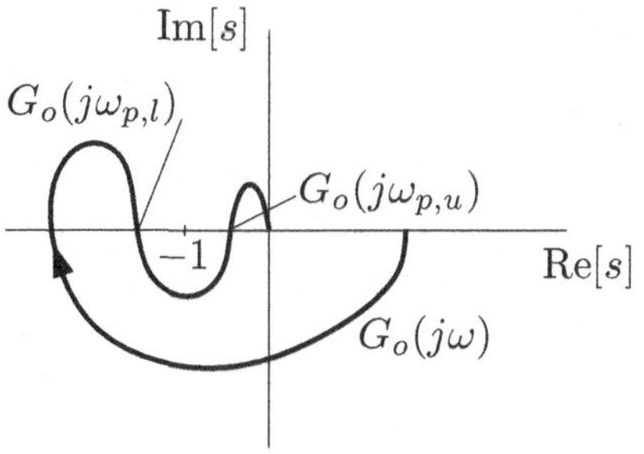

Figure 22.21 Upper and lower phase cross-over points

We then choose the two phase cross-over frequencies for which $G_o(j\omega_p)$ are immediately to the left right of -1, respectively. Let us call these frequencies $\omega_{p,l}$ and $\omega_{p,u}$ respectively, as shown in Figure 22.21.

Clearly, the upper and lower limits on the range of k for closed-loop stability are

$$k_l = \frac{1}{|G_o(j\omega_{p,l})|} \text{ and } k_u = \frac{1}{|G_o(j\omega_{p,u})|},$$

respectively. To see why, note that as we decrease k, the negative-real axis cross-over point of the Nyquist plot immediately to the left of -1 moves to the right until $k = k_l$, at which point

$$k_l G_o(j\omega_{p,l}) = \frac{1}{|G_o(j\omega_{p,l})|} G_o(j\omega_{p,l}) = -1,$$

which indicates a change in the number of encirclements of -1. Similarly increasing k results in the negative-real axis cross-over point of the Nyquist plot immediately to the right of -1 moving to the left until $k = k_u$, at which point

$$k_u G_o(j\omega_{p,u}) = \frac{1}{|G_o(j\omega_{p,u})|} G_o(j\omega_{p,u}) = -1,$$

indicating a change in the number of encirclements of -1. We call k_u and k_l upper and lower gain margins, respectively. It is common practice to express these gain margins in decibels as

$$GM_p = 20 \log k_u = -20 \log |G_o(j\omega_{p,u})|,$$

and

$$GM_n = 20 \log k_l = -20 \log |G_o(j\omega_{p,l})|,$$

respectively. Note that $GM_p > 0$ (since $|G_o(j\omega_{p,u})| < 1$) and is therefore called the *positive gain margin*. Likewise, $GM_n < 0$ (since $|G_o(j\omega_{p,l})| > 1$) and is called the *negative gain margin*.

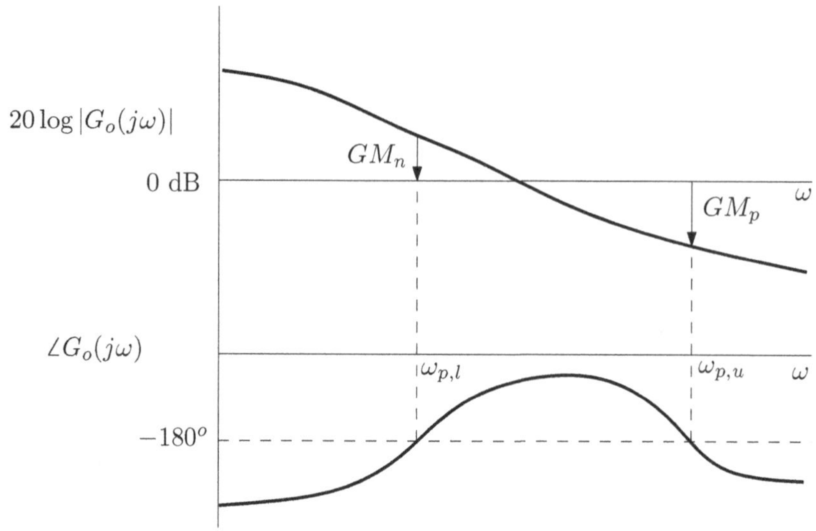

Figure 22.22 Gain margins from Bode plot

It is very useful to note that positive and negative gain margins can also be found from the Bode plot, as shown in Figure 22.22. As illustrated, we identify the phase cross-over frequencies from the phase frequency response plot. The gain margins are then obtained from the log-magnitude frequency response plot as shown.

22.3.1.2 Phase Margin

We can also change the number of encirclements of -1 by rotating the polar plot of $G_o(s)$, as shown in Figure 22.23.

This happens when pure phase is added to the open-loop transfer function. The phase margin then gives how much phase may be added or removed from the open-loop system before it becomes unstable.

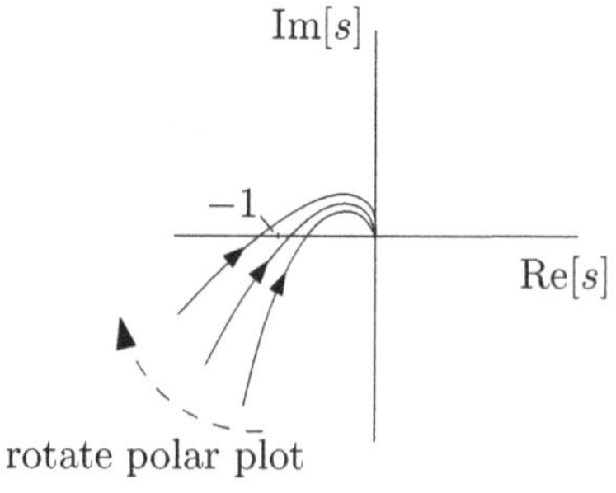

Figure 22.23 Rotation of polar plot

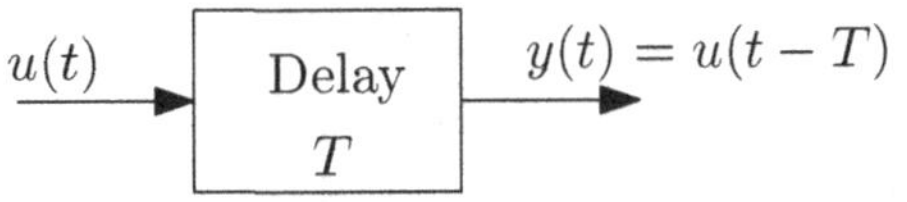

Figure 22.24 Unit delay

A very important case when this occurs is the presence of delays in the feedback control system. Up to now, everything we have formulated assumes that the process between sensor measurements to the execution of control commands occurs instantaneously. In reality, there is some delay in the communication of the measurements to the control system, some delay due to the computation, and delay in communication with the actuators to execute the command. This results in some additional phase lag in the frequency response, and has a destabilizing effect on the closed-loop system. It is therefore very important to build in some phase margin to accommodate delays in the feedback loop (as well as other uncertainties).

Let us now look at the transfer function for a unit delay of time T, represented in block diagram form in Figure 22.24.

We have

$$y(t) = u(t - T).$$

The Laplace transform of $y(t)$ is by definition

$$Y(s) = \mathcal{L}(y(t)) \int_0^\infty y(t)e^{-st}dt = \int_0^\infty u(t - T)e^{-st}dt.$$

Making the change of variables $\tau = t - T$, this becomes

$$Y(s) = \int_{-T}^\infty u(\tau)e^{-s(\tau+T)}dt = e^{-sT}\int_{-T}^\infty u(\tau)e^{-s\tau}dt.$$

In the definition of the one-sided Laplace transform (which we are using), it is assumed that all functions are zero for $t < 0$. In particular, it is assumed that $u(t) = 0$ when $t < 0$. Therefore, we have

$$Y(s) = e^{-sT}\int_0^\infty u(\tau)e^{-s\tau}dt = e^{-sT}\mathcal{L}(u(t)) = e^{-sT}U(s).$$

So, we have the transfer function for a unit delay

$$\frac{Y(s)}{U(s)} = e^{-sT}.$$

We note that when $s = j\omega$, the unit delay has gain

$$\left|e^{-j\omega T}\right| = 1,$$

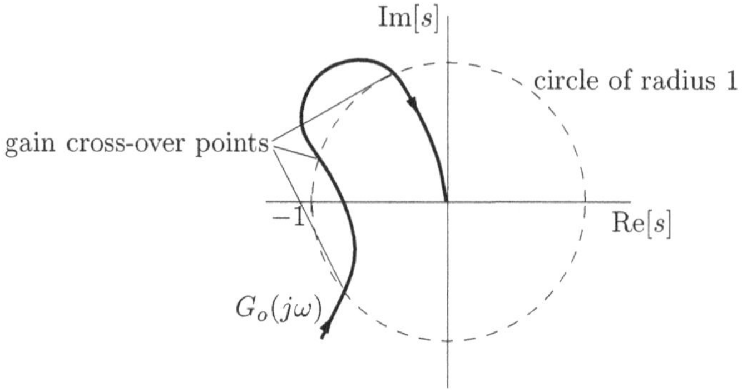

Figure 22.25 Gain cross-over points

and phase

$$\angle e^{-j\omega T} = -\omega T.$$

Therefore, we see that the unit delay only adds phase lag to the open-loop transfer function, but no gain. To define the phase margin, we must examine points on the frequency response that cross the circle of radius 1, centered at the origin. These points are called gain cross-over points, as shown in Figure 22.25.

Adding or removing pure phase rotates the gain cross-over points about the circle. By adding or removing enough pure phase, these points will cross -1, and change the number of encirclements of -1.

We therefore determine all frequencies ω_g corresponding to the gain cross-over points. That is, all frequencies for which

$$|G_o(j\omega_g)| = 1.$$

These frequencies are called *gain cross-over frequencies*. We then choose the two gain cross-over frequencies for which $G_o(j\omega_g)$ are immediately below and above -1, respectively. Let us call these frequencies $\omega_{g,u}$ and $\omega_{g,l}$ respectively, as shown in Figure 22.26.

The phase margins are then defined as how much phase lag (negative phase) can be added to the open-loop system without making it unstable. Clearly, as shown in the diagram in Figure 22.26, the upper and lower limits are PM_p and PM_n, and depend on $\angle G_o(j\omega_{g,u})$ and $\angle G_o(j\omega_{g,l})$ respectively. Note that PM_p is given as a positive angle while PM_n is given as a negative angle (they are measured positive counter-clockwise from the negative real axis). Accordingly, PM_p is called the *positive phase margin* and specifies how much phase lag can be added to the open-loop system, while PM_n is called the *negative phase margin* and specifies how much phase lag can be removed from the open-loop system. Note that since a unit delay adds phase lag to the open-loop system (of amount ωT at each frequency), we can

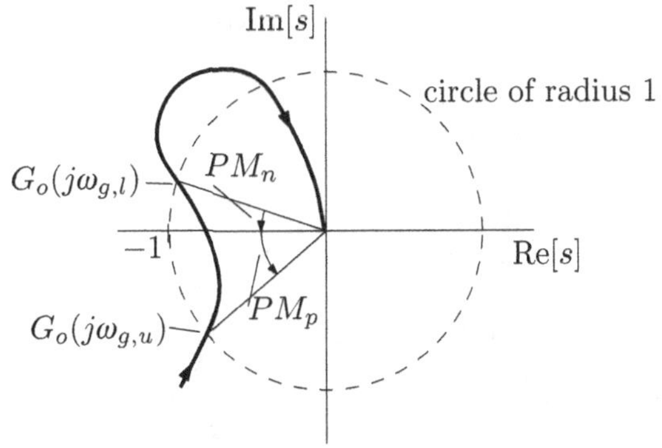

Figure 22.26 Phase margins from Nyquist plot

use the positive phase margin to specify the allowable delay in the feedback loop. Specifically, for closed-loop stability, the unit delay must satisfy

$$T < \frac{PM_p}{\omega_{g,u}}.$$

Since the unit delay adds increasing phase lag with increasing frequency, we must check this condition at every gain cross-over frequency.

As for the gain margin, we can also obtain the phase margins from the Bode plots of the frequency response, as shown in Figure 22.27. As shown, we identify the gain cross-over

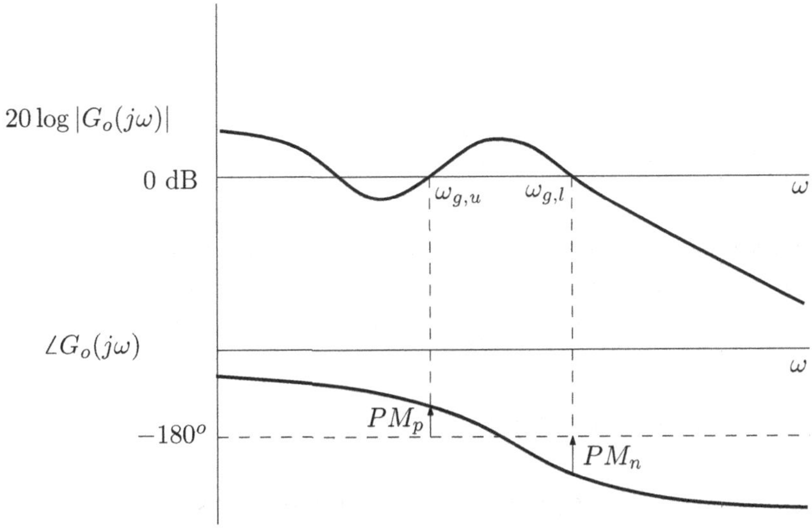

Figure 22.27 Phase margins from Bode plot

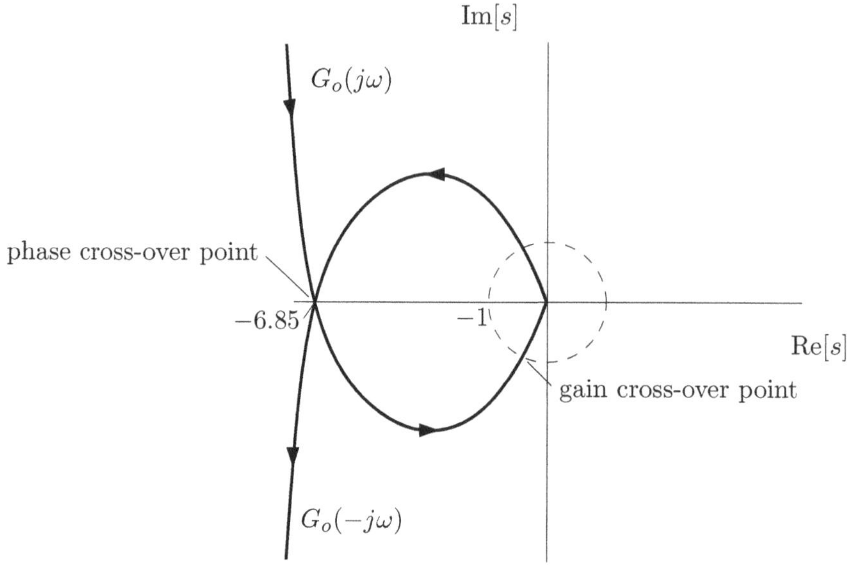

Figure 22.28 Nyquist plot for PID spacecraft attitude control example

frequencies from the log-magnitude frequency response plot. The phase margins are then determined from the phase frequency response plots, as shown.

Example 22.2 (PID spacecraft attitude control) *Let us now return to the PID controller design for spacecraft attitude control from Example 22.1. We have seen the Nyquist plot of the open-loop transfer function in Figure 22.15. Let us now determine the associated stability margins.*

Gain Margins

From the Nyquist plot (reproduced in Figure 22.28), we see that there is one phase cross-over frequency (negative real axis crossing). We already found the phase cross-over frequency, $\omega_{p,l} = \sqrt{K_i/K_d} = 0.0512$ *rad/s. The corresponding negative real axis crossing is* $G_o(j\omega_{p,l}) = -6.85$*, which occurs to the left of −1. Note that there is no crossing of the negative real axis between −1 and 0. Therefore, there is no increase in open-loop gain that will result in a change in encirclements of −1 by the Nyquist plot (which means that the closed-loop system remains stable for any increase in open-loop gain). The upper and lower gain margins are then*

$$k_u = \infty, \quad k_l = \frac{1}{|G_o(j\omega_{p,l})|} = 0.146.$$

In terms of log magnitude, we have positive and negative gain margins

$$GM_p = 20 \log k_u = \infty, \quad GM_n = 20 \log k_l = -16.7dB.$$

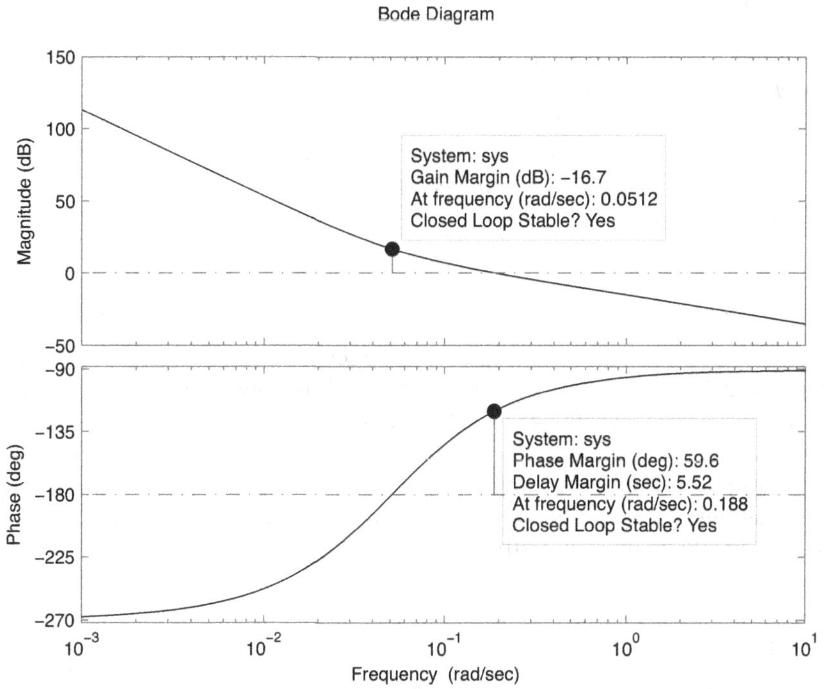

Figure 22.29 Bode plot for $G_o(j\omega)$ for PID spacecraft attitude control example

The Bode plot for the open-loop transfer function is shown in Figure 22.29. It can be seen that the gain margins can be determined from the Bode plot also.

Phase Margins

From the Nyquist diagram in Figure 22.28, we see that there is one gain cross-over frequency, and this will lead to a positive phase margin. Rather than computing the gain cross-over frequency and then the phase, we can determine it graphically from the Bode diagram in Figure 22.29. As shown in the diagram, the gain cross-over frequency is $\omega_{g,u} = 0.188$ *rad/s. The positive phase margin is*

$$PM_p = 59.6^\circ.$$

From the Nyquist diagram, we see that we can also change the number of encirclements of -1 *by rotating the gain cross-over point by* $360^\circ - 59.6^\circ = 300.4^\circ$ *in the counterclockwise direction. This corresponds to a* 300.4° *phase advance at the gain cross-over frequency. Therefore, the negative phase margin is*

$$PM_n = -300.4^\circ.$$

Time Delay Margin

Given the positive phase margin and associated gain cross-over frequency, the total delay in the feedback loop T must satisfy

$$T < \frac{PM_p}{\omega_{g,u}} = \frac{59.6}{0.188} \cdot \frac{\pi}{180} = 5.52 \text{ seconds.}$$

Notes

This chapter has presented the Nyquist plot, and the related Nyquist stability criterion. Based on the Nyquist stability criterion, we have then defined stability margins which indicate how close a stable system is to becoming unstable. The Nyquist stability criterion was then applied to a PID spacecraft attitude control example, and stability margins were determined. For further examples of the application of the Nyquist stability criterion and the determination of stability margins, the reader is referred to any of the excellent books on classical control systems, including Franklin et al. (2010) and Ogata (2010).

References

Franklin GF, Powell JD and Emami-Naeini A 2010 *Feedback Control of Dynamic Systems* 6th edn. Prentice Hall, Upper Saddle River, NJ.

Ogata K 2010 *Modern Control Engineering* 5th edn. Prentice Hall, Upper Saddle River, NJ.

23

Control Design in the Frequency Domain

In Chapter 17, we have examined specifications on the closed-loop performance, namely on the transient response and the steady-state error. It was shown how the transient specifications place constraints on the closed-loop poles, while the steady-state specifications place constraints on the open-loop and controller transfer functions when evaluated at $s = 0$ (namely $G_o(0)$ and $G_c(0)$). This then lead to a control design procedure in Chapter 20 using the root locus to obtain appropriate closed-loop poles.

As we have seen in Chapters 21 and 22, there are other important factors (in addition to the closed-loop performance) for a successful control system design, namely robustness to modeling uncertainty (leading to specifications on stability margins), and attenuation of measurement noise. These factors are best dealt with in the frequency domain. It turns out that the closed-loop performance can also be dealt with in the frequency domain. As such, it is possible to perform the design of a control system exclusively in the frequency domain, based on the open-loop frequency response.

Another benefit of performing control system design in the frequency domain is that unlike in the root locus-based method, an analytical model of the plant is not necessary. All that is required is the open-loop frequency response, which could be obtained experimentally. This is particularly useful when an analytical model for the plant (or parts of it) is not available. For example, for spacecraft attitude control, suppliers may not provide detailed analytical models for the actuators.

Let us now see how the open-loop frequency response provides information on the closed-loop transient behavior in the time-domain. This will be illustrated in the following example.

Example 23.1 (PD Spacecraft Attitude Control) *In Section 17.7.1, we derived expressions for rise-time, settling-time and percentage overshoot for PD spacecraft attitude control, implemented in the modified form shown in Figure 23.1(a). As explained in Section 22.3, for the purpose of evaluation of stability margins, the open-loop frequency response corresponding to the related standard control implementation shown in Figure 23.1(b) should be considered.*

Spacecraft Dynamics and Control: An Introduction, First Edition.
Anton H.J. de Ruiter, Christopher J. Damaren and James R. Forbes.

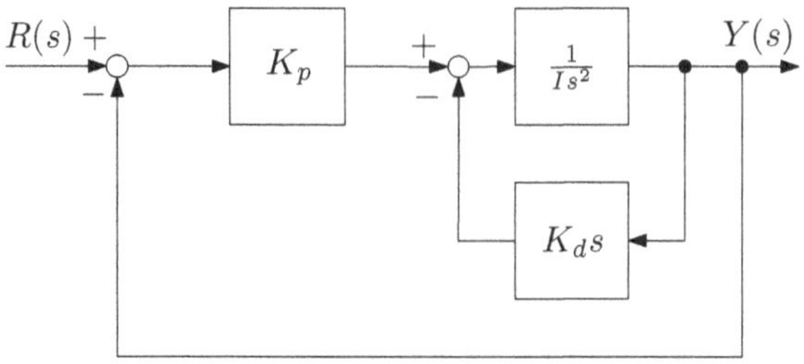

(a) Modified Implementation

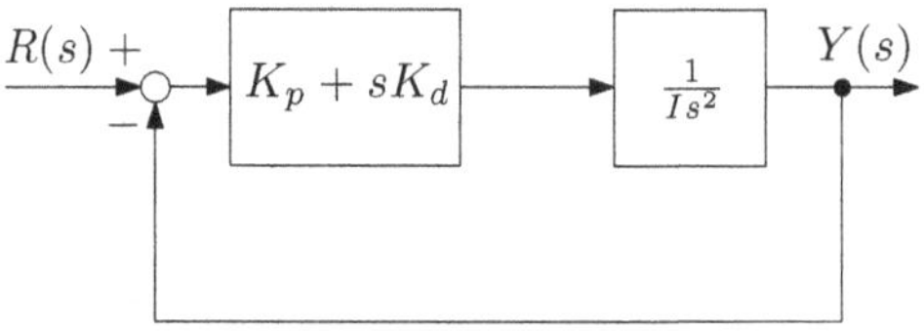

(b) Standard Implementation

Figure 23.1 Modified and standard PD spacecraft attitude control

Let us now proceed by first obtaining the open-loop transfer function corresponding to Figure 23.1(b), which we find to be

$$G_o(s) = \frac{K_p + K_d s}{Is^2}. \tag{23.1}$$

As in Section 17.7.1, we define

$$\omega_n^2 = \frac{K_p}{I}, \quad 2\zeta\omega_n = \frac{K_d}{I},$$

where ω_n is the undamped natural frequency and ζ is the damping ratio. Accordingly, the open-loop transfer function in (23.1) becomes

$$G_o(s) = \frac{2\zeta\omega_n s + \omega_n^2}{s^2}. \tag{23.2}$$

Let us now determine the open-loop frequency response. Setting $s = j\omega$ in (23.2), we obtain

$$G_o(j\omega) = \frac{2\zeta\omega_n j\omega + \omega_n^2}{-\omega^2} = -\left(\frac{\omega_n}{\omega}\right)^2 - j2\zeta\left(\frac{\omega_n}{\omega}\right). \tag{23.3}$$

The open-loop gain frequency response is obtained from (23.3) as

$$|G_o(j\omega)| = \left(\left(\frac{\omega_n}{\omega}\right)^4 + 4\zeta^2\left(\frac{\omega_n}{\omega}\right)^2\right)^{\frac{1}{2}}, \tag{23.4}$$

while the open-loop phase frequency response is found to be

$$\begin{aligned}\angle G_o(j\omega) &= \frac{\angle\left(2\zeta\omega_n j\omega + \omega_n^2\right)}{\angle\left(-\omega^2\right)},\\ &= \angle\left(2\zeta\omega_n j\omega + \omega_n^2\right) - 180^\circ,\\ &= \tan^{-1}\left(\frac{2\zeta\omega}{\omega_n}\right) - 180^\circ.\end{aligned} \tag{23.5}$$

Since the frequency ω is positive, (23.5) shows that the phase frequency response lies in the range

$$-180^\circ < \angle G_o(j\omega) < -90^\circ. \tag{23.6}$$

We now determine an expression for the phase margin. First, we need to find all gain cross-over frequencies, ω_g. We set

$$|G_o(j\omega_g)| = 1. \tag{23.7}$$

Substituting (23.4) into (23.7) and squaring, we obtain

$$\left(\frac{\omega_n}{\omega_g}\right)^4 + 4\zeta^2\left(\frac{\omega_n}{\omega_g}\right)^2 = 1. \tag{23.8}$$

Setting $q = \left(\omega_n/\omega_g\right)^2$, we find that (23.8) becomes a quadratic in q,

$$q^2 + 4\zeta^2 q - 1 = 0, \tag{23.9}$$

which has solution

$$q = -2\zeta^2 \pm \sqrt{4\zeta^4 + 1}. \tag{23.10}$$

Since we seek positive real solutions for ω_g, only the positive solution in (23.10) is valid, and using the definition of q, we find that there is only one gain cross-over frequency, which is given by

$$\omega_g = \frac{\omega_n}{\left(\sqrt{4\zeta^4 + 1} - 2\zeta^2\right)^{\frac{1}{2}}}. \tag{23.11}$$

From the phase frequency response range (23.6), it is clear that the associated phase margin is positive, and is given by

$$PM_p = 180^\circ + \angle G_o(j\omega_g). \tag{23.12}$$

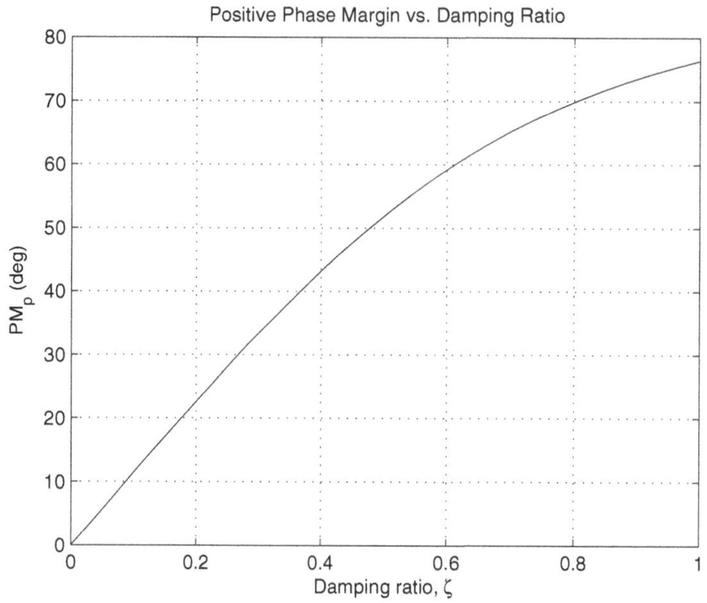

Figure 23.2 Relationship between phase margin and damping ratio

Substituting (23.11) and (23.5) into (23.12), we obtain the positive phase margin

$$PM_p = \tan^{-1}\left(\frac{2\zeta}{\left(\sqrt{4\zeta^4+1}-2\zeta^2\right)^{\frac{1}{2}}}\right). \tag{23.13}$$

From (23.13), it is seen that the positive phase margin depends purely on the damping ratio, ζ. The relationship in Equation (23.12) is shown graphically in Figure 23.2.

Recall from Section 17.1.1, that for an underdamped pair of closed-loop poles (with damping ratio $0 < \zeta < 1$), the maximum overshoot, is given by

$$M_p = e^{-\pi\zeta/\sqrt{1-\zeta^2}} \times 100\%, \tag{23.14}$$

which also depends exclusively on the damping ratio. As a consequence, maximum overshoot is directly related to the phase margin, and Equations (23.13) and (23.14) may be combined to obtain this relationship as shown in Figure 23.3. Therefore, using Figure 23.3, a closed-loop time-domain specification on the maximum overshoot may be recast as an open-loop frequency response specification on the phase margin.

Suppose now that the phase margin has been specified. As a consequence the damping ratio is given. We shall assume that the damping ratio lies in the range $0 < \zeta < 1$ (underdamped). Now consider rise-time and settling time constraints,

$$t_r \le \bar{t}_r, \quad t_s \le \bar{t}_s. \tag{23.15}$$

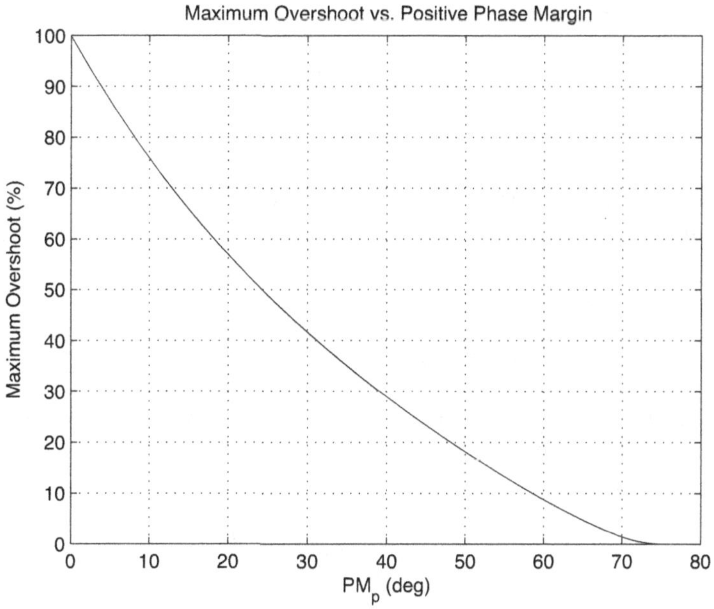

Figure 23.3 Relationship between phase margin and maximum overshoot

Recall the expression for rise-time from Section 17.7.1,

$$t_r = \frac{\pi - \beta}{\omega_n \sqrt{1 - \zeta^2}}, \tag{23.16}$$

where

$$\beta = \tan^{-1}\left(\frac{\sqrt{1-\zeta^2}}{\zeta}\right).$$

Since the damping ratio is given, we can rewrite the rise-time constraint in (23.15) as a constraint on the undamped natural frequency, ω_n as follows

$$\omega_n \geq \frac{\pi - \beta}{\bar{t}_r \sqrt{1 - \zeta^2}}. \tag{23.17}$$

Now, from (23.11), we see that we can transform the constraint on ω_n in (23.17) into a constraint on gain cross-over frequency,

$$\omega_g \geq \frac{\pi - \beta}{\bar{t}_r \sqrt{\left(1 - \zeta^2\right)\left(\sqrt{4\zeta^4 + 1} - 2\zeta^2\right)}}. \tag{23.18}$$

Likewise, recall the expression for the settling time

$$t_s = \frac{-\ln\left(0.02\sqrt{1-\zeta^2}\right)}{\zeta\omega_n}. \tag{23.19}$$

Again, since ζ is given, the settling time constraint in (23.15) may be transformed into a constraint on ω_n,

$$\omega_n \geq \frac{-\ln\left(0.02\sqrt{1-\zeta^2}\right)}{\zeta\bar{t}_s}. \tag{23.20}$$

As for (23.17), using (23.11) the constraint on ω_n in (23.20) may also be transformed into a constraint on the gain cross-over frequency,

$$\omega_g \geq \frac{-\ln\left(0.02\sqrt{1-\zeta^2}\right)}{\zeta\bar{t}_s\left(\sqrt{4\zeta^4+1}-2\zeta^2\right)^{\frac{1}{2}}}. \tag{23.21}$$

Therefore, we have seen that if the phase margin is given, specifications on the speed of closed-loop response, namely the rise-time and settling time may be recast as specifications on the gain cross-over frequency.

Example 23.1 has shown that the transient behavior of the closed-loop response is affected by the cross-over region of the open-loop frequency response. For the PD spacecraft attitude control given in Example 23.1, we were able to derive exact correspondences between the closed-loop transient characteristics and the open-loop frequency response. For more complex systems, we generally cannot find such exact correspondences. However, assuming that the control system design will result in a complex-conjugate pair of dominant poles (see Section 17.8.2), we can use the expressions obtained for the second-order system in Example 23.1 as guidelines to place specifications on the cross-over region of the open-loop frequency response of more complex systems. Generally speaking, the phase margin affects the closed-loop system damping (and hence overshoot), while the gain cross-over frequency affects the closed-loop system speed of response (and hence rise-time and settling time).

23.1 Feedback Control Problem - Revisited

Let us now revisit the standard feedback control problem introduced in Section 17.5, which is reproduced in Figure 23.4. As we have seen, the combined closed-loop response to the reference input $R(s)$, the external disturbance $\hat{T}_d(s)$ and the measurement noise $W(s)$, is given by

$$Y(s) = \frac{G_o(s)}{1+G_o(s)}R(s) + \frac{G_p(s)}{1+G_o(s)}\hat{T}_d(s) - \frac{G_o(s)}{1+G_o(s)}W(s), \tag{23.22}$$

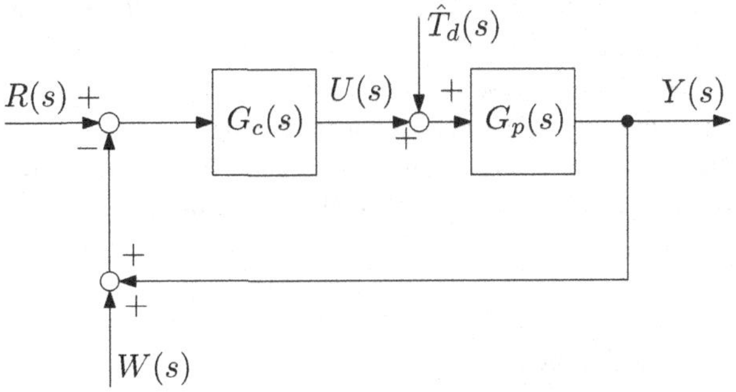

Figure 23.4 Standard feedback control system

where $G_o(s) = G_p(s)G_c(s)$ is the open-loop transfer function. The combined closed-loop response of the error $E(s) = R(s) - Y(s)$ is then readily found to be

$$E(s) = E_r(s) + E_d(s) + E_w(s), \tag{23.23}$$

where

$$E_r(s) = \frac{1}{1+G_o(s)}R(s), \quad E_d(s) = -\frac{G_p(s)}{1+G_o(s)}\hat{T}_d(s), \text{ and } E_w(s) = \frac{G_o(s)}{1+G_o(s)}W(s), \tag{23.24}$$

are the closed-loop error in response to the reference input, external disturbance and measurement noise, respectively. Also of interest is the closed-loop control effort, which is readily found to be

$$U(s) = U_r(s) + U_d(s) + U_w(s), \tag{23.25}$$

where

$$U_r(s) = \frac{G_c(s)}{1+G_o(s)}R(s), \quad U_d(s) = -\frac{G_o(s)}{1+G_o(s)}\hat{T}_d(s), \text{ and } U_w(s) = -\frac{G_c(s)}{1+G_o(s)}W(s), \tag{23.26}$$

are the closed-loop control effort in response to the reference input, external disturbance and measurement noise, respectively.

23.1.1 Closed-Loop Tracking Error

Let us first examine the closed-loop error. As explained in Section 17.5, ideally we would have $E_r(s) = E_d(s) = E_w(s) = 0$, since this represents perfect tracking. However, as demonstrated in Section 17.5, this is not possible. Fortunately, the typical reference signals to be tracked and external disturbances to be rejected vary relatively slowly with time. That is, they have low frequency content. For example, a step input can be thought of as a signal with zero frequency. On the other hand, measurement noise typically varies rapidly with time. That is, it has high frequency content. As such, we only need to consider $E_r(s)$ and $E_d(s)$ over low frequencies to ensure good reference signal tracking and disturbance rejection, while $E_w(s)$ only needs to be considered over high frequencies to ensure good measurement noise attenuation. Let us consider all three separately.

23.1.1.1 Reference Signal Tracking

From equation (23.24), we see that

$$|E_r(j\omega)| = \left|\frac{1}{1+G_o(j\omega)}\right| |R(j\omega)|.$$

Therefore, for good reference signal tracking, we should have

$$\left|\frac{1}{1+G_o(j\omega)}\right| \ll 1,$$

over the appropriate (low) frequency range. To accomplish this, we should have

$$|G_o(j\omega)| \gg 1.$$

That is, the open-loop gain should be large over low frequencies.

Note that this is entirely compatible with the requirement to keep the steady-state error to a step response (which can be thought of as a signal with zero frequency) be small. Indeed, with $R(s) = 1/s$, applying the final value theorem (see Appendix A.3), we find that the steady-state error due to a step response is

$$e_{r,ss} = \lim_{t\to\infty} e_r(t) = \lim_{s\to 0} sE_r(s) = \lim_{s\to 0} \frac{1}{1+G_o(s)}.$$

This shows that $|G_o(0)| \gg 1$ for small steady-state error to a step input.

23.1.1.2 Disturbance Rejection

From equation (23.24), we see that

$$|E_d(j\omega)| = \left|\frac{G_p(j\omega)}{1+G_o(j\omega)}\right| |\hat{T}_d(j\omega)|.$$

Therefore, for good external disturbance rejection, we should have

$$\left|\frac{G_p(j\omega)}{1+G_o(j\omega)}\right| |\hat{T}_d(j\omega)| \ll 1,$$

over the appropriate (low) frequency range. Assuming that we already have $|G_o(j\omega)| \gg 1$ over the desired frequency range to ensure good reference signal tracking, we find that

$$\left|\frac{G_p(j\omega)}{1+G_o(j\omega)}\right| \approx \frac{|G_p(j\omega)|}{|G_o(j\omega)|} = \frac{1}{|G_c(j\omega)|}.$$

Therefore, we can make the error due to an external disturbance small by making $|G_c(j\omega)|$ large enough over the appropriate frequency range.

Note that this is entirely compatible with the steady-state error analysis due to a constant external disturbance in Section 17.10, which placed constraints on $K_c = G_c(0)$.

23.1.1.3 Measurement Noise Attenuation

From equation (23.24), we see that

$$|E_w(j\omega)| = \left|\frac{G_o(j\omega)}{1+G_o(j\omega)}\right| |W(j\omega)|.$$

Therefore, for good measurement noise attenuation, we should have

$$\left|\frac{G_o(j\omega)}{1+G_o(j\omega)}\right| \ll 1,$$

over the appropriate (high) frequency range. To accomplish this, we should have

$$|G_o(j\omega)| \ll 1.$$

That is, the open-loop gain should be very small over high frequencies.

23.1.2 Closed-Loop Control Effort

Let us now look at the closed-loop control effort. As noted in Section 17.11, any real actuator has limitations in the amount of control effort that can be applied. Therefore, the control law should be designed so as to avoid the control signal from becoming too large.

We have seen that for good reference signal tracking and disturbance rejection, the open-loop gain should be large ($|G_o(j\omega)| \gg 1$) for low frequencies, while for good measurement noise attenuation, the open-loop gain should be very small ($|G_o(j\omega)| \ll 1$) for high frequencies. Let us look at some of the implications on the closed-loop control effort.

23.1.2.1 Large Open-Loop Gain

The reference signal and external disturbance typically have primarily low frequency content, and as such we have seen that we try to make the open-loop gain large over those frequencies.

For large open-loop gains ($|G_o(j\omega)| \gg 1$), we find from equation (23.26)

$$|U_r(j\omega)| = \left|\frac{G_c(j\omega)}{1+G_o(j\omega)}\right| |R(j\omega)| \approx \frac{|G_c(j\omega)|}{|G_o(j\omega)|}|R(j\omega)| = \frac{1}{|G_p(j\omega)|}|R(j\omega)|, \tag{23.27}$$

$$|U_w(j\omega)| = \left|\frac{G_c(j\omega)}{1+G_o(j\omega)}\right| |W(j\omega)| \approx \frac{|G_c(j\omega)|}{|G_o(j\omega)|}|W(j\omega)| = \frac{1}{|G_p(j\omega)|}|W(j\omega)|, \tag{23.28}$$

and

$$|U_d(j\omega)| = \left|\frac{G_o(j\omega)}{1+G_o(j\omega)}\right| |\hat{T}_d(j\omega)| \approx |\hat{T}_d(j\omega)|.$$

Equations (23.27) and (23.28) show that the control effort can become very large if the plant gain is very small, while the open-loop gain is large (if $|G_p(j\omega)| \ll 1$ while $|G_o(j\omega)| \gg 1$). Therefore, we should try to avoid the situation where the open-loop bandwidth is significantly larger than the plant bandwidth. In particular, any measurement noise in that frequency range could result in large control effort.

23.1.2.2 Small Open-Loop Gain

The measurement noise typically has primarily high frequency content, and as such, we try to make the open-loop gain very small over those frequencies. For very small open-loop gains ($|G_o(j\omega)| \ll 1$), we find from equation (23.26) that

$$|U_r(j\omega)| = \left|\frac{G_c(j\omega)}{1+G_o(j\omega)}\right| |R(j\omega)| \approx |G_c(j\omega)||R(j\omega)|,$$

$$|U_w(j\omega)| = \left|\frac{G_c(j\omega)}{1+G_o(j\omega)}\right| |W(j\omega)| \approx |G_c(j\omega)||W(j\omega)|, \tag{23.29}$$

and

$$|U_d(j\omega)| = \left|\frac{G_o(j\omega)}{1+G_o(j\omega)}\right| |\hat{T}_d(j\omega)| \approx |G_o(j\omega)||\hat{T}_d(j\omega)|.$$

Equation (23.29) shows that not only should the open-loop gain be very small ($|G_o(j\omega)| \ll 1$) at high frequencies, but the control gain should also be very small ($|G_c(j\omega)| \ll 1$) so as to avoid unnecessary control chatter resulting from measurement noise. Therefore, the control-law itself will often have a specified bandwidth.

23.1.3 Modified Control Implementation

As explained in Sections 17.7.1, 20.2 and 22.3, there are occasions where it is desirable to partition the control law into two parts

$$G_c(s) = G_{ce}(s) + G_{cy}(s), \tag{23.30}$$

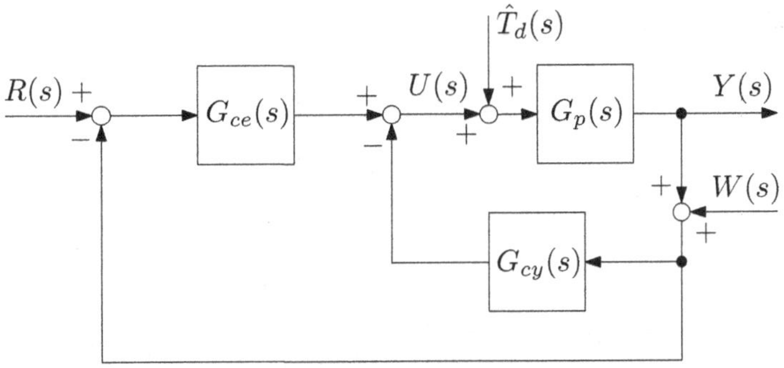

Figure 23.5 Modified feedback control system

and implement it in the modified form shown in Figure 23.5. The closed-loop error response for the modified control implementation is readily found to be

$$E(s) = E_r(s) + E_d(s) + E_w(s), \tag{23.31}$$

where

$$E_r(s) = \frac{1 + G_p(s)G_{cy}(s)}{1 + G_o(s)}R(s), \quad E_d(s) = -\frac{G_p(s)}{1 + G_o(s)}\hat{T}_d(s), \tag{23.32}$$

and

$$E_w(s) = \frac{G_o(s)}{1 + G_o(s)}W(s), \tag{23.33}$$

are the closed-loop errors in response to the reference input, external disturbance and measurement noise, respectively, and $G_o(s) = G_p(s)\left(G_{ce}(s) + G_{cy}(s)\right)$ is the relevant open-loop transfer function. Likewise, the closed-loop control effort for the modified implementation is

$$U(s) = U_r(s) + U_d(s) + U_w(s), \tag{23.34}$$

where

$$U_r(s) = \frac{G_{ce}(s)}{1 + G_o(s)}R(s), \quad U_d(s) = -\frac{G_o(s)}{1 + G_o(s)}\hat{T}_d(s), \text{ and } U_w(s) = -\frac{G_c(s)}{1 + G_o(s)}W(s), \tag{23.35}$$

are the closed-loop control efforts in response to the reference input, external disturbance and measurement noise, respectively.

Repeating the analyses in Sections 23.1.1 and 23.1.2 with equations (23.31) to (23.35), it is readily found that the conclusions of Sections 23.1.1 and 23.1.2 hold also for the modified implementation, with the additional restriction that the partitioning of the control law in equation (23.30) be performed such that $|G_{cy}(j\omega)| \ll |G_{ce}(j\omega)|$ when $|G_o(j\omega)| \gg 1$.

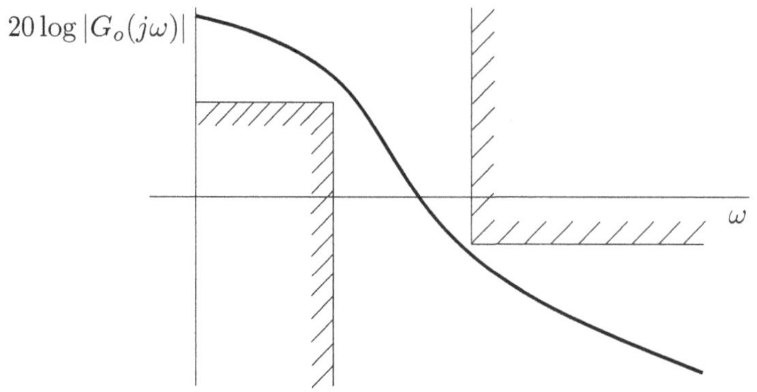

Figure 23.6 Desired open-loop gain frequency response

23.2 Control Design

To summarize the findings of Section 23.1 and Example 23.1, control design can be performed in the frequency domain by appropriately shaping the open-loop gain and phase frequency responses. In particular, closed-loop tracking and disturbance rejection requirements generally place restrictions on the open-loop gain over an appropriately determined low frequency range. Closed-loop measurement noise attenuation requirements place restrictions on the open-loop gain above an appropriate frequency. Figure 23.6 shows an example of these restrictions. The desired transient response and stability margins place restrictions on the cross-over region of the open-loop gain and phase frequency responses. Namely, they place restrictions on the gain and phase margins. So, given a desired open-loop frequency response, how do we go about designing an appropriate controller to achieve this? Recall that log magnitude and phase frequency responses are additive for systems connected in series. The open-loop transfer function is a series connection of the plant and controller transfer functions, $G_o(s) = G_p(s)G_c(s)$. Therefore, given the plant frequency response, $G_p(j\omega)$, we can determine what the controller frequency response $G_c(j\omega)$ should look like to yield an open-loop frequency response with the desired characteristics. We can then construct a controller $G_c(s)$ to achieve this.

Example 23.2 (Adjusting gain and phase margins using proportional control) *Consider the standard feedback control in Figure 23.4 with the proportional control*

$$G_c(s) = K,$$

Suppose that the plant frequency response $G_p(j\omega)$ is given, and that we simply wish to adjust the gain or phase margin. The open-loop transfer function is given by

$$G_o(s) = KG_p(s).$$

As we have seen in Chapter 21, the effect of the gain K on the frequency response as represented in the Bode Diagrams is simply to shift the log magnitude frequency response $20 \log |G_p(j\omega)|$ up by a factor $20 \log K$, while leaving the phase unchanged. As illustrated in Figure 23.7,

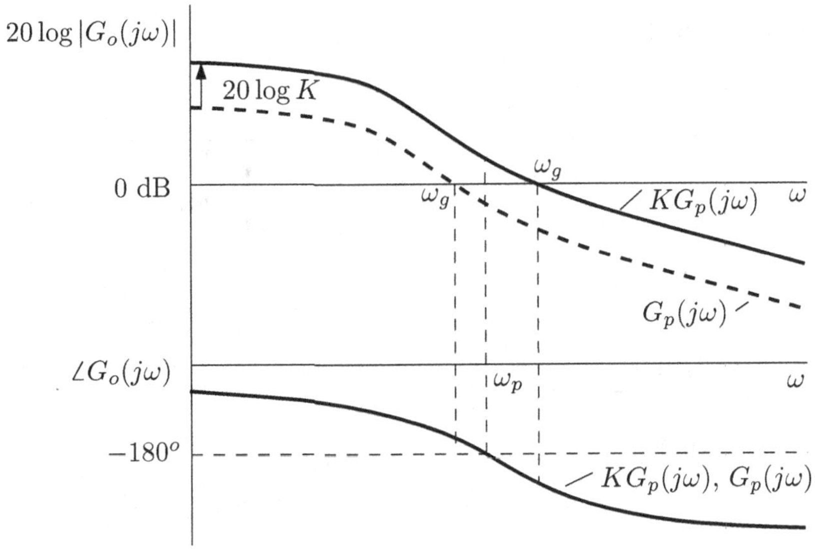

Figure 23.7 Effect of proportional control on the frequency response

since the phase cross-over frequency ω_p is unchanged, the upward shift of the log magnitude frequency response results in a change in gain margin. In addition, the upward shift of the log magnitude frequency response results in a shift in the gain cross-over frequency ω_g. Since the phase is unchanged, this results in a change in phase margin. With this knowledge, we can now find simple procedures to adjust the gain or phase margins.

Adjustment of Gain Margin

Suppose that it is required to find the gain K which yields a specified gain margin GM_{KG_p}. To do this, we first look at the Bode plot of $G_p(j\omega)$, and find the original gain margin at the phase cross-over frequency, ω_p. That is, we find $GM_{G_p} = -20 \log |G_p(j\omega_p)|$. Since the gain K does not chance the phase cross-over frequency, when the gain K is adjusted, the new gain margin is

$$\begin{aligned} GM_{KG_p} &= -20 \log |KG_p(j\omega_p)| \\ &= -20 \log K - 20 \log |G_p(j\omega_p)| \\ &= -20 \log K + GM_{G_p}. \end{aligned}$$

We can then rearrange this to find the required log magnitude

$$-20 \log K = GM_{KG_p} - GM_{G_p},$$

and consequently the required gain, K as

$$K = 10^{(GM_{G_p} - GM_{KG_p})/20}.$$

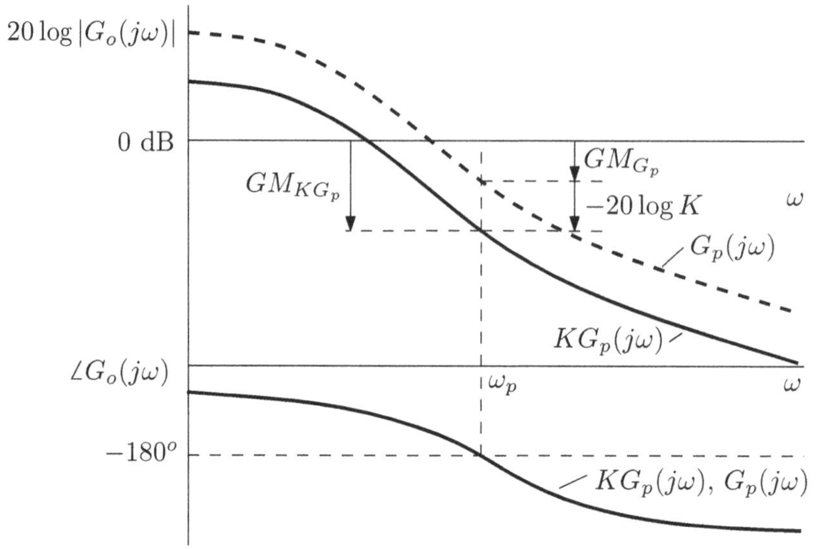

Figure 23.8 Adjusting gain margin using proportional control

The procedure of adjusting the gain margin as described above is demonstrated graphically in Figure 23.8.

Adjustment of Phase Margin

Suppose that it is required to find the gain K which yields a specified phase margin, PM_{KG_p}. Since the phase of $KG_p(j\omega)$ is the same as the phase of $G_p(j\omega)$, we first examine the Bode Diagrams of $G_p(j\omega)$ and find the required gain cross-over frequency ω_g which will give the required phase margin PM_{KG_p}. That is, as illustrated in Figure 23.9, we find ω_g such that

$$\angle G_p(j\omega_g) = PM_{KG_p} - 180^\circ.$$

Now, the gain K is adjusted so that ω_g is the gain cross-over frequency for $KG_p(j\omega)$. That is, such that

$$20\log|KG_p(j\omega_g)| = 0,$$

which rearranges to

$$K = 10^{-(20\log|G_p(j\omega_g)|)/20}.$$

Note that we leave the expression for K in this form, since we can read $20\log|G_p(j\omega_g)|$ directly from the Bode Diagram, as illustrated in Figure 23.9.

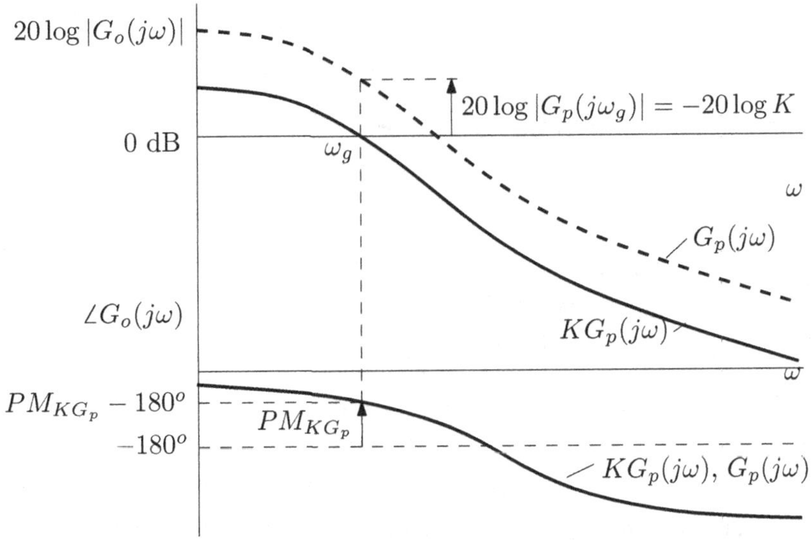

Figure 23.9 Adjusting phase margin using proportional control

23.2.1 Frequency Responses for Common Controllers

Let us now examine the frequency responses for some typical controllers.

23.2.1.1 Lead

A lead controller has the form

$$G_c(s) = K_{lead}\frac{Ts+1}{\alpha Ts+1}, \qquad 0 < \alpha < 1. \tag{23.36}$$

Note that this is equivalent to the transfer function representation given in Section 20.1. As we see in equation (23.36), a lead controller has a gain and two first order factors with corresponding corner frequencies at $\omega = 1/T$ and $\omega = 1/(\alpha T)$. The corresponding frequency response is shown in Figure 23.10 for lead controller gain $K_{lead} = 1$. It can be seen that a lead controller adds positive phase (phase lead, hence the name), which results in an increase the phase margin. As seen in Example 23.1, this results in an increase in damping ratio, or a decrease in overshoot. This is entirely compatible with what we observed in Section 20.1, where it was shown that a lead controller shifts the root locus to the left.

Let us now determine the maximum phase that a lead controller can add. From equation (23.36), we have

$$\angle G_c(j\omega) = \angle(j\omega T + 1) - \angle(j\omega\alpha T + 1). \tag{23.37}$$

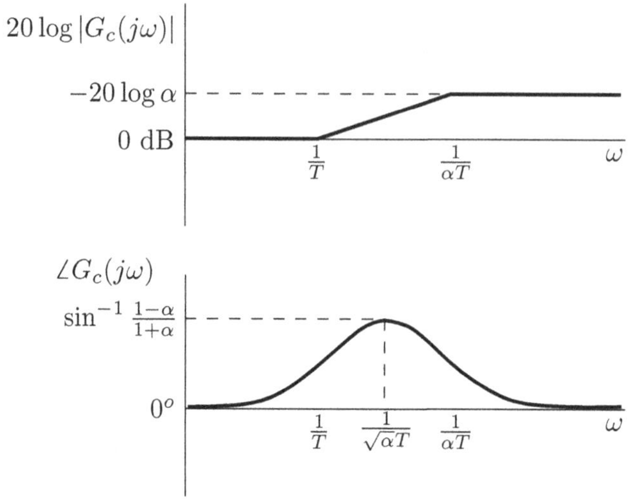

Figure 23.10 Frequency response of a lead controller with $K_{lead} = 1$

Now, $0 \leq \angle(j\omega T + 1), \angle(j\omega\alpha T + 1) < 90^\circ$ for non-negative frequencies ($\omega \geq 0$). Therefore, we can write

$$\angle G_c(j\omega) = \tan^{-1}(\omega T) - \tan^{-1}(\omega\alpha T). \tag{23.38}$$

The maximum phase must occur at the frequency for which

$$\frac{d}{d\omega}(\angle G_c(j\omega)) = 0.$$

Using the fact that

$$\frac{d}{dx}\left(\tan^{-1} x\right) = \frac{1}{1+x^2},$$

by differentiating (23.38) we obtain

$$\frac{d}{d\omega}(\angle G_c(j\omega)) = \frac{T}{1+\omega^2 T^2} - \frac{\alpha T}{1+\omega^2\alpha^2 T^2}. \tag{23.39}$$

Setting (23.39) to zero and solving, we find that the maximum phase occurs at the frequency

$$\omega_{\max} = \frac{1}{\sqrt{\alpha} T}. \tag{23.40}$$

At frequency $\omega_{\max}$, we find that the frequency response is given by

$$\begin{aligned} G_c(j\omega_{\max}) &= K_{lead}\frac{j(1/\sqrt{\alpha})+1}{j\sqrt{\alpha}+1}, \\ &= K_{lead}\frac{(j(1/\sqrt{\alpha})+1)(1-j\sqrt{\alpha})}{1+\alpha}, \\ &= K_{lead}\frac{2+(1-\alpha)/\sqrt{\alpha}j}{1+\alpha}, \\ &= \frac{K_{lead}}{\sqrt{\alpha}}\frac{2\sqrt{\alpha}+(1-\alpha)j}{1+\alpha}. \end{aligned} \tag{23.41}$$

Now, note that

$$\left|2\sqrt{\alpha}+(1-\alpha)j\right| = \sqrt{4\alpha+(1-\alpha)^2} = \sqrt{(1+\alpha)^2} = 1+\alpha.$$

Therefore, from the imaginary part of (23.41), we find the maximum phase

$$\angle G_c(j\omega_{\max}) = \sin^{-1}\frac{1-\alpha}{1+\alpha}. \tag{23.42}$$

Since a lead controller adds phase, it is primarily used to improve the phase margin. Suppose that we know the amount of phase, $\phi_{\max}$, we wish to add at the (given) frequency $\omega_{\max}$. Then, we can solve equations (23.42) and (23.40) to obtain the lead controller parameters

$$\alpha = \frac{1-\sin\phi_{\max}}{1+\sin\phi_{\max}}, \quad T = \frac{1}{\sqrt{\alpha}\omega_{\max}}. \tag{23.43}$$

The controller gain K_{lead} can then be chosen to give the desired gain cross-over frequency.

23.2.1.2 Lag

A lag controller has the form

$$G_c(s) = K_{lag}\frac{Ts+1}{\beta Ts+1}, \quad \beta > 1. \tag{23.44}$$

Note that this is equivalent to the transfer function representation given in Section 20.1. As we see in equation (23.44), a lag controller has a gain and two first order factors with corresponding corner frequencies at $\omega = 1/T$ and $\omega = 1/(\beta T)$. The corresponding frequency response is shown in Figure 23.11. It can be seen that a lag controller adds open-loop gain at low frequencies. As seen in Section 23.1.1, this results in improved steady-state reference signal tracking and external disturbance rejection. A lag controller also adds some negative phase (phase lag). As shown in Figure 23.11 by making the corner frequencies sufficiently small, the lag controller can be designed to primarily affect the low frequency part of the open-loop frequency response, while leaving the cross-over and high frequency regions largely unaffected. This is compatible with what we observed in Section 20.1, where it was shown

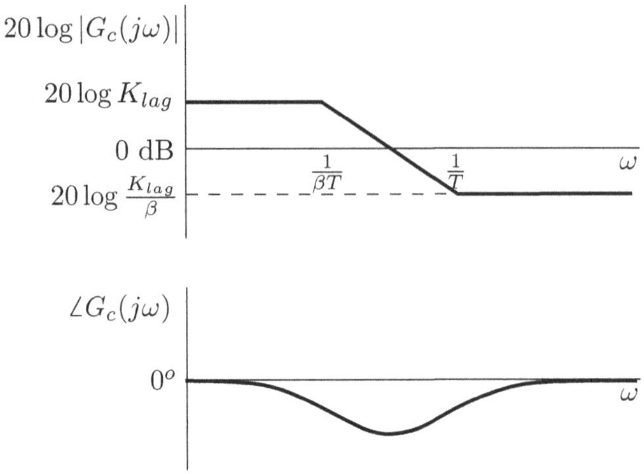

Figure 23.11 Frequency response of a lag controller

that a lag controller is used to increase $G_o(0)$, while leaving the rest of the root locus (and therefore the dominant poles) largely unchanged.

It is readily found from (23.44) that the low frequency gain is given by

$$|G_c(0)| = K_{lag}, \tag{23.45}$$

and the high frequency gain is given by

$$|G_c(j\omega)| \approx \frac{K_{lag}}{\beta}, \text{ for } \omega \gg \frac{1}{T}. \tag{23.46}$$

The lag controller parameters can then be selected as follows: K_{lag} is chosen to yield the appropriate low-frequency open-loop and controller gains. The corner frequency $1/T$ is chosen to be sufficiently small (equivalently, T sufficiently large) such that the phase component of the lag controller frequency response does not affect the cross-over and high frequency components of the open-loop phase frequency response. Finally, β is chosen to yield the appropriate high frequency gain, for example to yield the desired phase or gain margins (in a similar method to Example 23.2).

23.2.1.3 Proportional-Derivative (PD)

A proportional-derivative controller has the form

$$G_c(s) = K_{pd}(Ts + 1). \tag{23.47}$$

Note that this is equivalent to the transfer function representation given in Section 20.1. As we see in equation (23.47), a proportional-derivative controller has a gain and one first order factor with a corner frequency at $\omega = 1/T$. The corresponding frequency response is shown in Figure 23.12 controller gain $K_{pd} = 1$. It can be seen that like a lead controller,

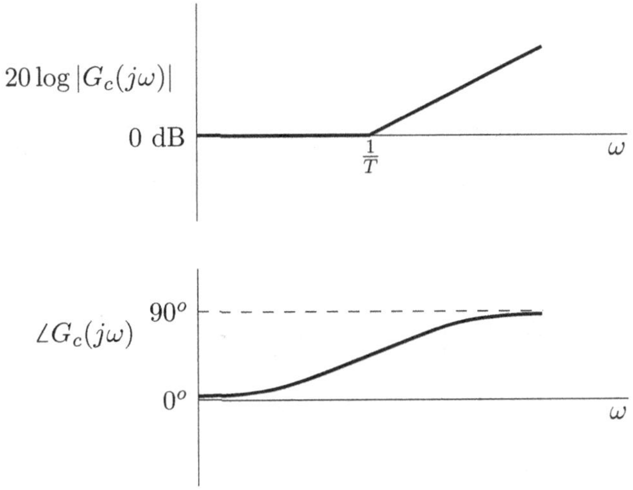

Figure 23.12 Frequency response of a proportional-derivative controller with $K_{pd} = 1$

a proportional-derivative controller adds positive phase (phase lead). Therefore, just as for the lead controller, this results in an increase in positive phase margin and consequently an increase in damping ratio, or equivalently, a decrease in overshoot (as seen in Example 23.1). Again, this is entirely compatible with what we observed in Section 20.1, where it was shown that a proportional-derivative controller shifts the root locus to the left.

Let us now determine the phase that a proportional-derivative controller can add. From equation (23.47), it is readily found that

$$\angle G_c(j\omega) = \tan^{-1}(\omega T), \tag{23.48}$$

and that $0 \leq \angle G_c(j\omega) < 90°$ for $\omega > 0$. Suppose now that it is desired to add the phase ϕ at a specified frequency ω_c. From equation (23.48), we obtain the controller parameter

$$T = \frac{\tan\phi}{\omega_c}. \tag{23.49}$$

The controller gain K_{pd} can then be chosen to give the desired gain cross-over frequency.

23.2.1.4 Proportional-Integral (PI)

A proportional-integral controller has the form

$$G_c(s) = K_{pi}\frac{Ts+1}{s}. \tag{23.50}$$

Note that this is equivalent to the transfer function representation given in Section 20.1. As we see in equation (23.50), a proportional-integral controller has a gain, one first order factor with corner frequency at $\omega = 1/T$ and one integrator. The corresponding frequency response is shown in Figure 23.13. It can be seen that similar to a lag controller, a proportional-integral

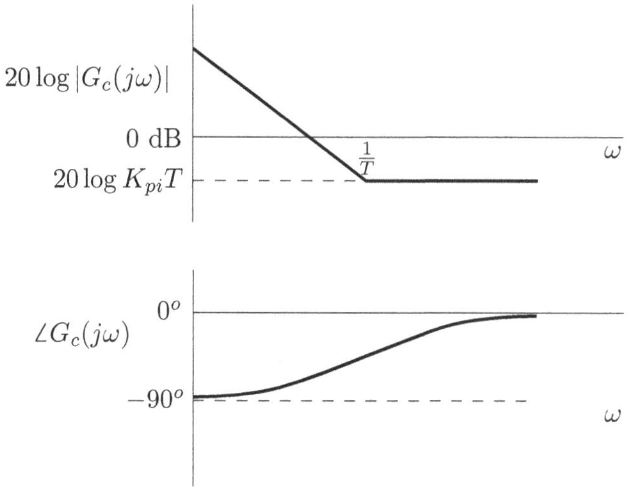

Figure 23.13 Frequency response of a proportional-integral controller

controller adds open-loop gain at low frequencies (in fact, the gain becomes infinite when $\omega = 0$). Just as with a lag controller, this results in improved steady-state reference signal tracking and external disturbance rejection. In fact, it increases the system type by one. A proportional-integral controller also adds some negative phase (phase lag). As shown in Figure 23.13 by making the corner frequency sufficiently small, the proportional-integral controller can be designed to primarily affect the low frequency part of the open-loop frequency response, while leaving the cross-over and high frequency regions largely unaffected.

It is readily found from (23.50) that the high frequency gain is given by

$$|G_c(j\omega)| \approx K_{pi}T, \text{ for } \omega \gg \frac{1}{T}. \tag{23.51}$$

The proportional-integral controller parameters can then be selected as follows: the corner frequency $1/T$ is chosen to be sufficiently small (equivalently, T sufficiently large) such that the phase component of the lag controller frequency response does not affect the cross-over and high frequency components of the open-loop phase frequency response. The gain K_{pi} is then chosen to yield the appropriate high frequency gain, for example to yield the desired phase or gain margins (in a similar method to Example 23.2).

23.3 Example - PID Design for Spacecraft Attitude Control

We shall now redesign the PID control spacecraft attitude control law in Section 20.2 using the frequency-domain design method. Recall from Section 20.2 that we are considering the control of the spacecraft attitude given by

$$I\ddot{y} = u + T_d,$$

where the spacecraft inertia $I = 1$ kg·m^2, the control torque is u, and T_d is the disturbance torque, and that the closed-loop system must satisfy the following specifications.

1. Rise-time constraint, $t_r \leq 30$ seconds.
2. Maximum overshoot constraint, $M_p \leq 30\%$.
3. Settling-time constraint, $t_s \leq 100$ seconds.
4. The steady-state error for a step disturbance $\hat{T}_d$ must be zero.

Recall also that the associated plant transfer function is

$$G_p(s) = \frac{1}{s^2}.$$

As shown in Section 20.2, we can solve this design problem with a PID controller, which can be designed as a combined PD and PI controller. Using the PI and PD forms in this chapter (equations (23.47) and (23.50)), we have

$$\begin{aligned} G_c(s) &= K\left(T_{pd}s + 1\right)\left(\frac{T_{pi}s + 1}{s}\right), \\ &= K_p + sK_d + \frac{K_i}{s}, \end{aligned}$$

which is a PID controller with proportional, derivative and integral gains given by

$$K_p = K\left(T_{pd} + T_{pi}\right), \quad K_d = KT_{pd}T_{pi}, \quad K_i = K, \tag{23.52}$$

respectively. As in Section 20.2, this means that we can design our PID controller sequentially with PD and PI controllers.

First, let us recast the time-domain specifications as frequency-domain specifications. Anticipating a pair of dominant poles, let us use the second order system of Example 23.1 as a guideline. First, we shall handle the overshoot constraint. From Figure 23.3, we find that the overshoot constraint translates to a phase margin constraint $PM_p \geq 38°$. To give ourselves some margin, let us choose

$$PM_p = 45°.$$

From Figure 23.2, this results in a damping ratio of $\zeta = 0.42$. Now we shall deal with the rise-time and settling-time constraints. As shown in Example 23.1, with a given phase margin constraint, these yield constraints on the gain cross-over frequency. Using inequality (23.18), we find that the rise-time constraint translates to $\omega_g \geq 0.0875$ rad/s. Using inequality (23.21), the settling-time constraint translates to $\omega_g \geq 0.1135$ rad/s. To give ourselves some margin, let us choose

$$\omega_g = 0.12 \text{ rad/s}.$$

For convenience, let us write the control law as

$$G_c(s) = KG_{pd}(s)G_{pi}(s),$$

where

$$G_{pd}(s) = T_{pd}s + 1,$$

is the PD part and

$$G_{pi}(s) = \frac{T_{pi}s + 1}{s},$$

is the PI part. The open-loop transfer function then becomes

$$G_o(s) = KG_p(s)G_{pd}(s)G_{pi}(s). \tag{23.53}$$

We shall first design the PD part of the controller to provide the required phase margin. We shall assume that the PI part of the controller will add approximately 5 degrees phase lag at the desired cross-over frequency (that is, $\angle G_{pi}(j\omega_g) \approx -5°$). Now, the positive phase margin is given by

$$PM_p = 180° + \angle G_o(j\omega_g) \approx 180° + \angle G_p(j\omega_g) + \angle G_{pd}(j\omega_g) - 5°.$$

Given that $\angle G_p(j\omega) = -180°$ for all $\omega > 0$, we find that the required phase for the PD part of the controller at the cross-over frequency is

$$\angle G_{pd}(j\omega_g) = PM_p + 5° = 50°.$$

From equation (23.49), we find the PD time constant

$$T_{pd} = \frac{\tan PM_p}{\omega_g} = \frac{\tan 50°}{0.12} = 9.93 \text{ s}.$$

We now design the PI part of the controller. Recall that we choose the PI corner frequency to be sufficiently small so that the resulting PI phase lag does not have significant effect on the cross-over region of the open-loop frequency response. Let us therefore set the PI corner frequency to

$$\frac{1}{T_{pi}} = 0.1\omega_g,$$

which results in

$$T_{pi} = \frac{1}{0.1\omega_g} = 83.3 \text{ s}.$$

The final part of the PID design is to choose the gain K such that $\omega_g = 0.12$ rad/s is indeed the gain cross-over frequency. From (23.53), it can be seen that we must have

$$1 = |G_o(j\omega_g)| = K|G_p(j\omega_g)||G_{pd}(j\omega_g)||G_{pi}(j\omega_g)|,$$

which leads to

$$K = \frac{1}{|G_p(j0.12)||G_{pd}(j0.12)||G_{pi}(j0.12)|} = 1.105 \times 10^{-4}.$$

Using equation (23.52), we finally obtain the PID gains

$$K_p = 0.0103, \quad K_d = 0.0915, \quad K_i = 1.105 \times 10^{-4}.$$

As in Section 20.2, we implement the PID control law in the form

$$u(t) = K_p e(t) - K_d \dot{y}(t) + K_i \int_0^t e(\tau) d\tau.$$

Figure 23.14 shows the unit step response of the corresponding closed-loop system. It can be seen that with the designed control gains, the response has a bit too much overshoot, and the settling time is too long. To reduce the overshoot, let us increase the desired phase margin to

$$PM_p = 50°.$$

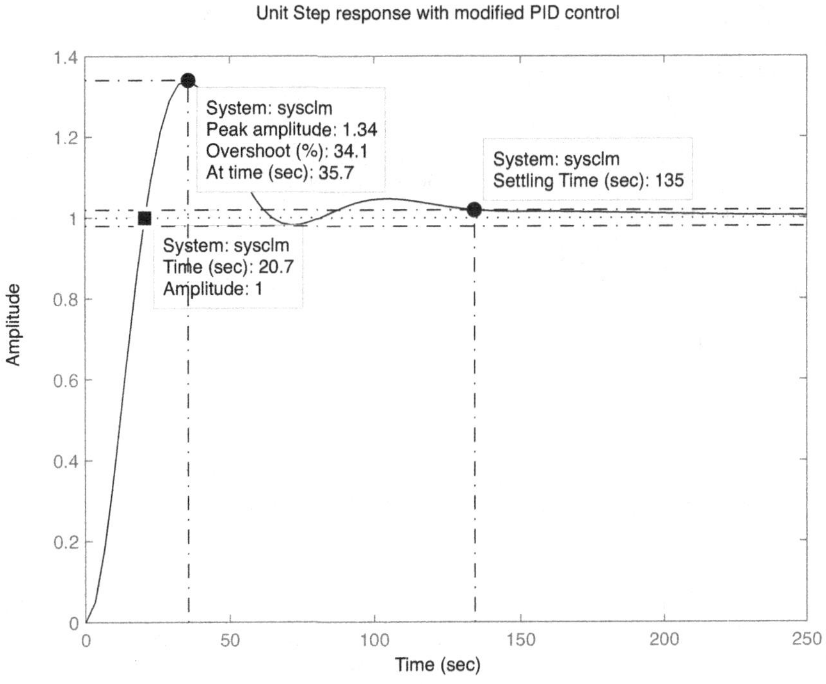

Figure 23.14 Unit step response $K_p = 0.0103$, $K_d = 0.0915$, $K_i = 1.105 \times 10^{-4}$

To reduce the settling time, let us increase the gain cross-over frequency to

$$\omega_g = 0.2 \text{ rad/s}.$$

Repeating the above control design procedure with the new phase margin and gain cross-over frequency, we end up with the gains

$$K_p = 0.0261, \quad K_d = 0.1630, \quad K_i = 4.566 \times 10^{-4}.$$

Figure 23.15 shows the closed-loop system unit step response with the new gains. It can be seen that all transient specifications are satisfied. Figure 23.16 shows the attitude response to a step disturbance. Comparing Figure 23.15 and 23.16 to Figures 20.19 and 20.20 respectively, we see that performing the control design in the frequency domain has resulted in a system with very similar closed-loop performance to the system designed using the root locus. However, unlike the root locus, the frequency domain method also allows the designer to consider issues such as robustness and noise rejection, as well as providing the ability to design a controller for a system for which a plant model is unavailable using an experimentally obtained frequency response.

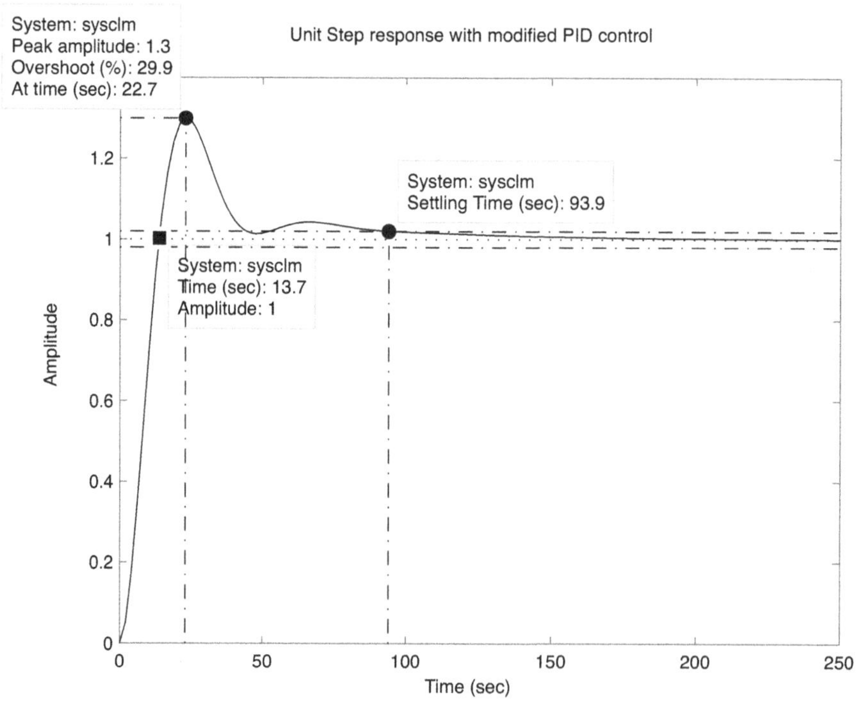

Figure 23.15 Unit step response $K_p = 0.0261$, $K_d = 0.1630$, $K_i = 4.566 \times 10^{-4}$

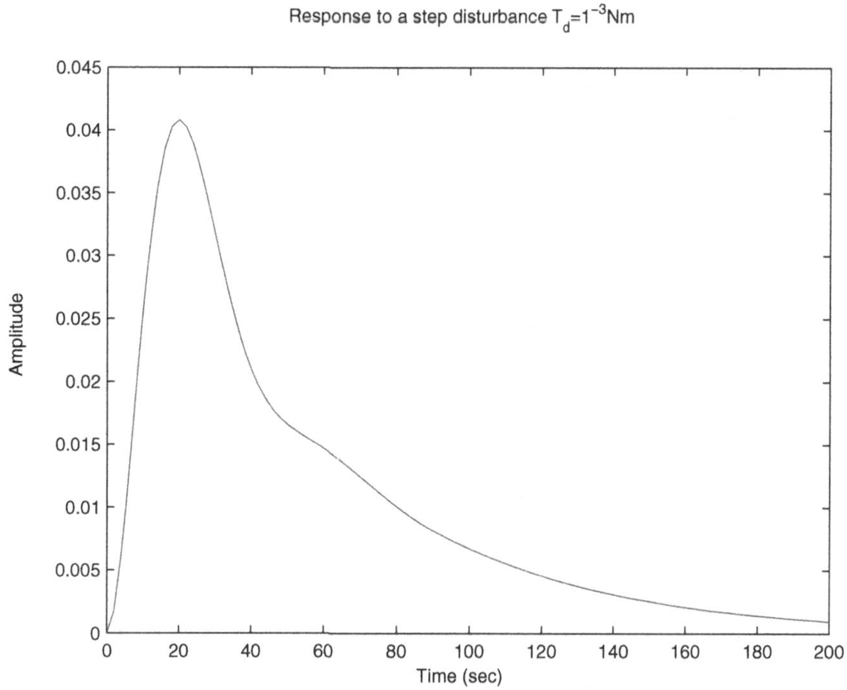

Figure 23.16 Attitude response for a step disturbance $\hat{T}_d = 10^{-3}$Nm

Notes

In this chapter, we have examined how the open-loop frequency response relates to the closed-loop system performance. In particular, we examined how a typical desired open-loop frequency response might be specified. As such, we saw that a control system design may be performed purely in the frequency domain. We then examined how different types of common controllers affect the open-loop frequency response. We finally performed a PID controller design for spacecraft attitude control in the frequency domain, and demonstrated that it met the closed-loop specifications in the time-domain. For further examples on how to perform control system design in the frequency domain, the reader is referred to any of the excellent books on classical control systems, including Qiu and Zhou (2010), Franklin et al. (2010) and Ogata (2010).

References

Franklin GF, Powell JD and Emami-Naeini A 2010 *Feedback Control of Dynamic Systems* 6th edn. Prentice Hall, Upper Saddle River, NJ.

Ogata K 2010 *Modern Control Engineering* 5th edn. Prentice Hall, Upper Saddle River, NJ.

Qiu L and Zhou K 2010 *Introduction to Feedback Control*. Prentice Hall, Upper Saddle River, NJ.

24

Nonlinear Spacecraft Attitude Control

In Chapters 17 to 23, we have examined how to perform spacecraft attitude control using linear control design techniques. This required making the assumption of small angles and angular velocities in order to obtain linear approximations to the nonlinear attitude equations of motion. Since the attitude control design requirements typically include a requirement on the steady-state attitude errors being small, this is certainly a reasonable approach. However, there are occasions when large angle maneuvers are required. For example, when a spacecraft is initially deployed from the launch vehicle, the spacecraft will be tumbling in an uncontrolled manner. When the attitude control system is initially switched on, it must acquire the desired attitude from an arbitrary initial attitude. Therefore, there could be a large initial attitude error.

It is clear then, that in order to perform a complete spacecraft attitude control design for a given mission, there are components of the spacecraft attitude operations sequence for which the small angle and angular velocity assumption cannot be made. In order to guarantee closed-loop stability, we need to analyze the full nonlinear attitude equations. This chapter will present some relevant techniques.

24.1 State-Space Representation of the Spacecraft Attitude Equations

In order to study the full nonlinear spacecraft attitude motion (in particular the stability), we need to put the equations of motion (see Sections 11.2 and 11.3) into a suitable form. For linear differential equations with constant coefficients such as those studied in Chapters 17 to 22, taking Laplace transforms leads to proper transfer functions (the ratio of two polynomials). The resulting transfer functions were easily analyzed, providing a great deal of insight into closed-loop stability and performance. Unfortunately, this approach is not very helpful when dealing with nonlinear differential equations (such as the spacecraft attitude equations of motion). Taking Laplace transforms does not in general lead to nice proper transfer functions that can be readily analyzed. Instead, we must deal directly with the nonlinear differential equations in the time-domain.

Spacecraft Dynamics and Control: An Introduction, First Edition.
Anton H.J. de Ruiter, Christopher J. Damaren and James R. Forbes.

A finite-dimensional nonlinear system, such as we consider in spacecraft dynamics and control can be put in the form

$$\dot{\mathbf{x}} = \mathbf{f}(\mathbf{x}, t, \mathbf{u}), \tag{24.1}$$

$$\mathbf{y} = \mathbf{h}(\mathbf{x}, t), \tag{24.2}$$

where t denotes time, $\mathbf{x}$ is a vector containing the system states, $\mathbf{y}$ is a vector containing the system measurements, $\mathbf{u}$ is a vector containing the system inputs, and $\mathbf{f}$ and $\mathbf{h}$ are vector-valued real functions.

The state vector $\mathbf{x}$ contains all parameters necessary to fully describe the system motion at any time, and the equations (24.1) and (24.2) are collectively referred to as a *state-space realization* of the system. Individually, we call (24.1) the *state equation* of the system, and (24.2) the *measurement equation*. For the purposes of the stability analyses in this chapter, we shall only deal with the state equation for the closed-loop system. Furthermore, for the spacecraft attitude control designs in this chapter, we shall assume that the full state-vector is available for feedback, that is $\mathbf{y} = \mathbf{x}$ (this situation is called *state-feedback*). The measurement equation will become important in Chapter 25 when we examine state estimation in detail, in particular with regards to spacecraft navigation.

Example 24.1 (State-space description of spacecraft attitude motion equations) *Let us find a state-space description for a rigid spacecraft's attitude motion. From Sections 11.2 and 11.3, we see that the attitude motion is fully described by the attitude and angular velocity. These in turn are governed by the attitude kinematics and attitude dynamics. As explained in Section 11.2, there are several different parameterizations that may be used to represent the spacecraft attitude. Let us use the quaternion* $(\boldsymbol{\epsilon}, \eta)$ *to represent the attitude. From Section 11.2, we obtain the attitude kinematics*

$$\begin{aligned} \dot{\boldsymbol{\epsilon}} &= \frac{1}{2}\left(\eta \mathbf{1} + \boldsymbol{\epsilon}^{\times}\right)\boldsymbol{\omega}, \\ \dot{\eta} &= -\frac{1}{2}\boldsymbol{\epsilon}^{T}\boldsymbol{\omega}, \end{aligned} \tag{24.3}$$

where $\boldsymbol{\omega}$ *is the spacecraft angular velocity in body coordinates. Likewise, from Section 11.3, the attitude dynamics are*

$$\mathbf{I}\dot{\boldsymbol{\omega}} + \boldsymbol{\omega}^{\times}\mathbf{I}\boldsymbol{\omega} = \mathbf{T}, \tag{24.4}$$

where $\mathbf{I}$ *is the spacecraft inertia matrix and* $\mathbf{T}$ *is the external torque acting on the spacecraft (represented in body coordinates).*

Equations (24.3) and (24.4) fully describe the spacecraft attitude motion, and can be put in state-space form as follows. Since the quaternion $(\boldsymbol{\epsilon}, \eta)$ *and angular velocity vector* $\boldsymbol{\omega}$ *fully specify the attitude motion, let us define the state-vector as*

$$\mathbf{x} = \begin{bmatrix} \boldsymbol{\epsilon} \\ \eta \\ \boldsymbol{\omega} \end{bmatrix}.$$

Likewise, as seen in (24.4), the system input is the external torque, so we set $\mathbf{u} = \mathbf{T}$. *Now, suppose that the only measurements we have available are the vector part of the quaternion. In this case, we set* $\mathbf{y} = \boldsymbol{\epsilon}$.

With the above definitions in hand, the equations of motion (24.3) and (24.4) may be put into the state-space form of equations (24.1) and (24.2) by defining

$$\mathbf{f}(\mathbf{x}, \mathbf{u}) = \begin{bmatrix} \frac{1}{2}\left(\eta\mathbf{1} + \boldsymbol{\epsilon}^{\times}\right)\boldsymbol{\omega} \\ -\frac{1}{2}\boldsymbol{\epsilon}^{T}\boldsymbol{\omega} \\ \mathbf{I}^{-1}\left[-\boldsymbol{\omega}^{\times}\mathbf{I}\boldsymbol{\omega} + \mathbf{u}\right] \end{bmatrix}, \tag{24.5}$$

$$\mathbf{h}(\mathbf{x}) = \boldsymbol{\epsilon}. \tag{24.6}$$

When the state-equation system in (24.1) does not explicitly contain an input $\mathbf{u}$, it takes the form

$$\dot{\mathbf{x}} = \mathbf{f}(\mathbf{x}, t), \tag{24.7}$$

and we say that it is *unforced*. Note that if we have some feedback control $\mathbf{u} = \boldsymbol{\gamma}(\mathbf{y}, t)$ in the system described by (24.1) and (24.2), we can consider the closed-loop system as unforced. That is, the closed-loop system satisfies

$$\dot{\mathbf{x}} = \bar{\mathbf{f}}(\mathbf{x}, t), \tag{24.8}$$

where

$$\bar{\mathbf{f}}(\mathbf{x}, t) = \mathbf{f}(\mathbf{x}, t, \boldsymbol{\gamma}(\mathbf{h}(\mathbf{x}, t), t)).$$

In the case that the unforced state equation (24.7) is *independent* of time,

$$\dot{\mathbf{x}} = \mathbf{f}(\mathbf{x}), \tag{24.9}$$

we say that the system is *autonomous*. Otherwise, we say that the system is *non-autonomous* (as in equation (24.7)).

Example 24.2 *Let us now find the closed-loop state equation for the spacecraft in Example 24.1 with state-feedback control law*

$$\mathbf{u} = -k_p\boldsymbol{\epsilon} - k_d\boldsymbol{\omega}. \tag{24.10}$$

Substituting (24.10) into (24.5), we readily find that the closed-loop system is given by

$$\dot{\mathbf{x}} = \bar{\mathbf{f}}(\mathbf{x}), \tag{24.11}$$

where

$$\bar{\mathbf{f}}(\mathbf{x}) = \begin{bmatrix} \frac{1}{2}\left(\eta\mathbf{1} + \boldsymbol{\epsilon}^{\times}\right)\boldsymbol{\omega} \\ -\frac{1}{2}\boldsymbol{\epsilon}^{T}\boldsymbol{\omega} \\ \mathbf{I}^{-1}\left[-\boldsymbol{\omega}^{\times}\mathbf{I}\boldsymbol{\omega} - k_p\boldsymbol{\epsilon} - k_d\boldsymbol{\omega}\right]. \end{bmatrix}$$

Clearly the closed-loop state equation (24.11) is independent of time. Therefore, the closed-loop system is autonomous.

Now, suppose that there is a known time-varying disturbance torque $\mathbf{T}_d(t)$ acting on the spacecraft. Then, the attitude dynamics become

$$\mathbf{I}\dot{\boldsymbol{\omega}} + \boldsymbol{\omega}^{\times}\mathbf{I}\boldsymbol{\omega} = \mathbf{u} + \mathbf{T}_d(t). \tag{24.12}$$

We readily find that the closed-loop state equation with the feedback control law (24.10) now becomes

$$\dot{\mathbf{x}} = \bar{\mathbf{f}}(\mathbf{x}, t), \tag{24.13}$$

where

$$\bar{\mathbf{f}}(\mathbf{x}, t) = \begin{bmatrix} \frac{1}{2}\left(\eta\mathbf{1} + \boldsymbol{\epsilon}^{\times}\right)\boldsymbol{\omega} \\ -\frac{1}{2}\boldsymbol{\epsilon}^{T}\boldsymbol{\omega} \\ \mathbf{I}^{-1}\left[-\boldsymbol{\omega}^{\times}\mathbf{I}\boldsymbol{\omega} - k_p\boldsymbol{\epsilon} - k_d\boldsymbol{\omega} + \mathbf{T}_d(t)\right]. \end{bmatrix}$$

In this case, the closed-loop state equation (24.13) is not independent of time. Therefore, the closed-loop system in this case is non-autonomous.

24.2 Stability Definitions

In Section 14.1, we presented some notions of stability for linear time-invariant systems. We made some general arguments as to the meaning of stability in terms of the long-term behavior of the system, but we did not give a precise definition. To determine the stability of a linear time-invariant system, we examined the system poles. This is not appropriate for general nonlinear systems, and we shall need a precise definition of stability and different techniques are needed for its analysis.

24.2.1 Equilibrium Points

A very important concept is that of an equilibrium point. A point $\mathbf{x}^*$ is said to be an equilibrium point if it has the property that whenever the state of the system, $\mathbf{x}$, starts at $\mathbf{x}^*$, it will remain at

$\mathbf{x}^*$ for all future time. Clearly this is a very desirable property when designing and implementing a control system. For example, if the goal of a spacecraft attitude control system is to hold the spacecraft to a desired attitude, then the desired attitude must be an equilibrium point of the closed-loop system.

Let us now consider the system (24.7). Then, the point $\mathbf{x} = \mathbf{x}^*$ is an equilibrium point of (24.7) at time $t = t_0$ if

$$\mathbf{f}\left(\mathbf{x}^*, t\right) = \mathbf{0}, \quad \text{for all } t \geq t_0. \tag{24.14}$$

If the system is autonomous (as in (24.9)), the condition for an equilibrium point becomes independent of the initial time, t_0. That is, the point $\mathbf{x}^*$ is an equilibrium point of (24.9) if

$$\mathbf{f}\left(\mathbf{x}^*\right) = \mathbf{0}. \tag{24.15}$$

We shall only consider systems with equilibrium at the origin $\mathbf{x}^* = \mathbf{0}$. To see why this is reasonable, consider the system with state-equation

$$\dot{\mathbf{z}} = \mathbf{g}\left(\mathbf{z}, t\right),$$

with equilibrium point $\mathbf{z} = \mathbf{z}^* \neq \mathbf{0}$ at time $t = t_0$. Now, consider the change of variables $\mathbf{x} = \mathbf{z} - \mathbf{z}^*$. Then, $\dot{\mathbf{x}} = \dot{\mathbf{z}} = \mathbf{g}\left(\mathbf{x} + \mathbf{z}^*, t\right)$. Defining $\mathbf{f}\left(\mathbf{x}, t\right) = \mathbf{g}\left(\mathbf{x} + \mathbf{z}^*, t\right)$, we see that the system

$$\dot{\mathbf{x}} = \mathbf{f}\left(\mathbf{x}, t\right),$$

has an equilibrium point at the origin ($\mathbf{x} = \mathbf{0}$) at time $t = t_0$. Since we can always make such a change of variables, it is reasonable for us to consider only systems with equilibrium at the origin.

24.2.2 Stability of Equilibria

Just as important as the concept of an equilibrium point is the concept of stability of that equilibrium point. Loosely speaking, if a state $\mathbf{x}$ is close to an equilibrium point $\mathbf{x}^*$, then we want to ensure that $\mathbf{x}$ stays close to $\mathbf{x}^*$ for all future time. This is essentially what stability means. In our spacecraft attitude control example, if the desired attitude is an equilibrium point of the closed-loop system, but is not stable, it is not practically useful, since any disturbance to the system will push the attitude away from the desired attitude, and it will not come back.

We shall now formalize the notion of stability. There are a number of different stability definitions. We shall use a very widely used definition, called *Lyapunov stability*. Once again consider the system (24.7), and suppose that $\mathbf{x} = \mathbf{0}$ is an equilibrium point at time $t = t_0$.

Definition 24.1 *The equilibrium point* $\mathbf{x} = \mathbf{0}$ *of (24.7) is*

1. *stable if for any* $\epsilon > 0$, *a* $\delta > 0$ *can be found such that*

$$\|\mathbf{x}(t_0)\| < \delta \text{ implies that } \|\mathbf{x}(t)\| < \epsilon \text{ for all } t \geq t_0$$

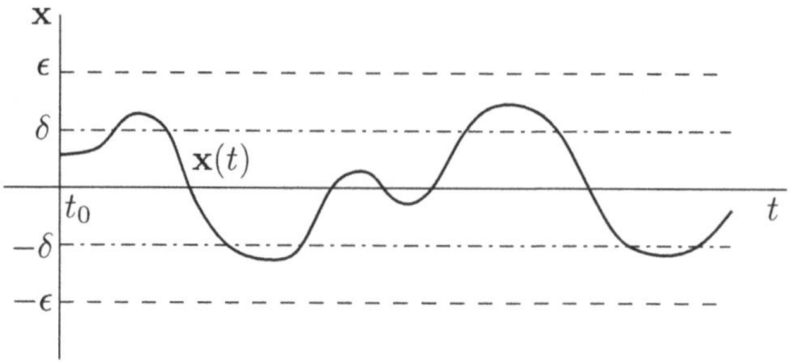

Figure 24.1 Example of Lyapunov stability

2. *unstable if it is not stable*
3. *asymptotically stable if it is stable and* $\delta > 0$ *can be chosen such that*

$$\|\mathbf{x}(t_0)\| < \delta \textit{ implies that } \lim_{t\to\infty} \mathbf{x}(t) = \mathbf{0}.$$

4. *globally asymptotically stable if it is stable and*

$$\lim_{t\to\infty} \mathbf{x}(t) = \mathbf{0}, \textit{ for all } \mathbf{x}(t_0).$$

Definition 24.1 mathematically formalizes our intuitive idea of stability, namely that solutions $\mathbf{x}(t)$ starting close to the equilibrium $\mathbf{x} = \mathbf{0}$ should stay close. This is illustrated graphically in Figure 24.1. We have also formalized the notion of asymptotic stability, which is even more important in control system design, namely, $\mathbf{x}(t)$ will approach the equilibrium point provided $\mathbf{x}(t_0)$ is sufficiently close to it. For our attitude control example, if some disturbance pushes the attitude away from the desired attitude, we want the attitude to return to the desired attitude.

Note the distinction between asymptotic stability and global asymptotic stability. Asymptotic stability is a local definition, which only guarantees that $\mathbf{x}(t)$ will approach the equilibrium point if it starts sufficiently close to it. Global asymptotic stability is a much stronger condition, since it means that $\mathbf{x}(t)$ will approach the equilibrium point regardless of its starting value.

24.3 Stability Analysis

It will be very useful to be able to determine the stability of an equilibrium point without having to solve the associated differential equation (which is typically very difficult or impossible to do analytically). We shall now demonstrate such a procedure by considering the detumbling operation of a rigid spacecraft.

24.3.1 Detumbling of a Rigid Spacecraft

When a spacecraft is deployed from a launch vehicle, it is typically tumbling with some non-zero angular velocity $\boldsymbol{\omega}(t)$. One of the first spacecraft attitude operations will be to detumble the spacecraft, that is to make $\boldsymbol{\omega}(t) \to \mathbf{0}$. Therefore, we want $\boldsymbol{\omega} = \mathbf{0}$ to be an asymptotically stable (globally if possible) equilibrium point of the closed-loop spacecraft attitude dynamics.

Recall from Section 11.3 that the spacecraft attitude dynamics (in the absence of external disturbances) are given by

$$\mathbf{I}\dot{\boldsymbol{\omega}} + \boldsymbol{\omega}^{\times}\mathbf{I}\boldsymbol{\omega} = \mathbf{u}, \tag{24.16}$$

where $\mathbf{I}$ is the spacecraft inertia matrix and $\mathbf{u}$ is the control torque. Since only the angular velocity $\boldsymbol{\omega}$ is of concern to detumbling the spacecraft (which is fully governed by (24.16)), we do not require the spacecraft kinematics. Let us choose the detumbling control law

$$\mathbf{u} = -k\boldsymbol{\omega}, \tag{24.17}$$

where $k > 0$ is a constant gain. Then, the closed-loop system obtained by substituting (24.17) into (24.16) can be written in the form

$$\dot{\boldsymbol{\omega}} = \mathbf{f}(\boldsymbol{\omega}), \tag{24.18}$$

where we set

$$\mathbf{f}(\boldsymbol{\omega}) = \mathbf{I}^{-1}\left[-\boldsymbol{\omega}^{\times}\mathbf{I}\boldsymbol{\omega} - k\boldsymbol{\omega}\right].$$

Clearly, $\mathbf{f}(\mathbf{0}) = \mathbf{0}$, so that $\boldsymbol{\omega} = \mathbf{0}$ is an equilibrium point of the closed-loop system.

To study the stability of the equilibrium, let us consider the rotational kinetic energy (see Section 2.5),

$$T_r(t) = \frac{1}{2}\boldsymbol{\omega}^T(t)\mathbf{I}\boldsymbol{\omega}(t). \tag{24.19}$$

We note that since the inertia matrix, $\mathbf{I}$, is positive definite, the rotational kinetic energy satisfies $T \geq 0$, with equality $T = 0$ only at the equilibrium $\boldsymbol{\omega} = \mathbf{0}$. Now, let I_1, I_2 and I_3 be the principal moments of inertia corresponding to $\mathbf{I}$ (see Section 2.4.3), ordered such that

$$I_1 \leq I_2 \leq I_3. \tag{24.20}$$

Then, we have the following bounds on the rotational kinetic energy

$$\frac{1}{2}I_1\|\boldsymbol{\omega}(t)\|^2 \leq T_r(t) \leq \frac{1}{2}I_3\|\boldsymbol{\omega}(t)\|^2, \tag{24.21}$$

where $\|\boldsymbol{\omega}\| = \sqrt{\boldsymbol{\omega}^T\boldsymbol{\omega}}$. Rearranging the inequality on the left, shows that at any given time,

$$\|\boldsymbol{\omega}(t)\| \leq \sqrt{\frac{2T_r(t)}{I_1}}. \tag{24.22}$$

Thus, inequality (24.22) shows that the angular velocity, $\boldsymbol{\omega}(t)$, is bounded by the rotational kinetic energy. In particular, if we can demonstrate that $T_r(t)$ is bounded, then from (24.22) we find that $\boldsymbol{\omega}(t)$ is bounded also. This fact will be useful for establishing closed-loop stability. Likewise, if we can demonstrate that $T_r(t)$ shrinks to zero, then (24.22) shows that $\boldsymbol{\omega}(t)$ also shrinks to zero. This will then establish asymptotic stability. Therefore, we can study the stability properties of the closed-loop system by examining the qualitative behavior of the rotational kinetic energy.

Before proceeding with the detailed stability analysis, let us first demonstrate the inequalities in (24.21). Consider an arbitrary angular velocity $\boldsymbol{\omega}$. As shown in Section 2.4.3, it is always possible to find a rotation matrix $\mathbf{C}$ such that

$$\mathbf{I}_p \triangleq \begin{bmatrix} I_1 & 0 & 0 \\ 0 & I_2 & 0 \\ 0 & 0 & I_3 \end{bmatrix} = \mathbf{C}\mathbf{I}\mathbf{C}^T. \tag{24.23}$$

Now, consider the vector

$$\mathbf{a} \triangleq \begin{bmatrix} a_1 \\ a_2 \\ a_3 \end{bmatrix} = \mathbf{C}\boldsymbol{\omega}. \tag{24.24}$$

Making use of (24.23) and (24.24), we find that

$$T_r = \frac{1}{2}\boldsymbol{\omega}^T\mathbf{I}\boldsymbol{\omega} = \frac{1}{2}\mathbf{a}^T\mathbf{I}_p\mathbf{a} = \frac{1}{2}\left[I_1a_1^2 + I_2a_2^2 + I_3a_3^2\right]. \tag{24.25}$$

Making use of the first inequality in (24.20), we find from (24.25) that

$$\begin{aligned} T_r &\geq \frac{1}{2}I_1\left[a_1^2 + a_2^2 + a_3^2\right], \\ &= \frac{1}{2}I_1\mathbf{a}^T\mathbf{a}, \\ &= \frac{1}{2}I_1\left(\boldsymbol{\omega}^T\mathbf{C}^T\right)(\mathbf{C}\boldsymbol{\omega}), \\ &= \frac{1}{2}I_1\boldsymbol{\omega}^T\boldsymbol{\omega}, \end{aligned}$$

which establishes the first inequality in (24.21). The second inequality is obtained in similar fashion.

Let us now proceed with our stability analysis, by examining how the rotational kinetic energy behaves as a function of time along a trajectory $\boldsymbol{\omega}(t)$ of the closed-loop system. Differentiating (24.19) with respect to time, and applying the product rule, we have

$$\begin{aligned} \dot{T}_r &= \frac{1}{2}\dot{\boldsymbol{\omega}}^T\mathbf{I}\boldsymbol{\omega} + \frac{1}{2}\boldsymbol{\omega}^T\mathbf{I}\dot{\boldsymbol{\omega}}, \\ &= \boldsymbol{\omega}^T\mathbf{I}\dot{\boldsymbol{\omega}}. \end{aligned} \tag{24.26}$$

Substituting for $\dot{\boldsymbol{\omega}}$ from (24.18), we have

$$\dot{T}_r = \boldsymbol{\omega}^T \mathbf{II}^{-1} \left[-\boldsymbol{\omega}^\times \mathbf{I}\boldsymbol{\omega} - k\boldsymbol{\omega} \right]. \tag{24.27}$$

Noting that, $\boldsymbol{\omega}^T \boldsymbol{\omega}^\times = \mathbf{0}$, this becomes

$$\dot{T}_r = -k\boldsymbol{\omega}^T \boldsymbol{\omega} = -k\|\boldsymbol{\omega}(t)\|^2 \leq 0. \tag{24.28}$$

This shows that time-derivative of the rotational kinetic energy is nonpositive. As a consequence, the rotational kinetic energy cannot increase, and is bounded by its initial value. Indeed, integrating both sides of the inequality (24.28), we have

$$T_r(t) - T_r(0) = \int_0^t \dot{T}_r d\tau \leq 0, \text{ for all } t \geq 0, \tag{24.29}$$

and therefore

$$T_r(t) \leq T_r(0), \text{ for all } t \geq 0. \tag{24.30}$$

Therefore, from inequalities (24.30) and (24.22), we have

$$\|\boldsymbol{\omega}(t)\| \leq \sqrt{\frac{2T_r(0)}{I_1}}, \text{ for all } t \geq 0. \tag{24.31}$$

Let us now use this to establish the closed-loop stability condition. Referring to Definition 24.1, suppose an arbitrary $\epsilon > 0$ is given. Then, provided the initial rotational kinetic energy satisfies

$$\sqrt{\frac{2T_r(0)}{I_1}} < \epsilon, \tag{24.32}$$

inequality (24.31) shows that we will then have

$$\|\boldsymbol{\omega}(t)\| < \epsilon, \text{ for all } t \geq 0, \tag{24.33}$$

which is the condition required in Definition 24.1. We shall now turn condition (24.32) into a bound on $\|\boldsymbol{\omega}(0)\|$. This bound will give us the required $\delta > 0$ in Definition 24.1. Making use of the second inequality in (24.21), we find that

$$\sqrt{\frac{2T_r(0)}{I_1}} \leq \|\boldsymbol{\omega}(0)\| \sqrt{\frac{I_3}{I_1}}. \tag{24.34}$$

Therefore, if we select the initial angular velocity $\boldsymbol{\omega}(0)$ such that

$$\|\boldsymbol{\omega}(0)\|\sqrt{\frac{I_3}{I_1}} < \epsilon, \tag{24.35}$$

inequality (24.32) will guarantee that (24.33) is satisfied. Rearranging (24.35), we obtain the required bound on the initial angular velocity

$$\|\boldsymbol{\omega}(0)\| < \epsilon\sqrt{\frac{I_1}{I_3}}, \tag{24.36}$$

and correspondingly we choose $\delta = \epsilon\sqrt{I_1/I_3}$. To summarize, we have shown that given any $\epsilon > 0$, we can find a corresponding $\delta > 0$ such that

$$\|\boldsymbol{\omega}(0)\| < \delta \text{ implies that } \|\boldsymbol{\omega}(t)\| < \epsilon \text{ for all } t \geq 0.$$

Therefore, the equilibrium $\boldsymbol{\omega} = \mathbf{0}$ is stable.

Now, let us try to demonstrate asymptotic stability of $\boldsymbol{\omega} = \mathbf{0}$. Suppose that the initial condition $\boldsymbol{\omega}(0)$ is given. We have seen that the rotational kinetic energy $T_r(t)$ is non-increasing ($\dot{T}_r \leq 0$ as in (24.28)), and that it is bounded below $T_r(t) \geq 0$ (since it is non-negative by definition). Now, suppose that the rotational kinetic energy does not approach zero. Then there must be a positive scalar $c > 0$ such that $T_r(t) \geq c$ for all $t \geq 0$. We shall now show that this is impossible, and that therefore $T_r(t) \to 0$ as $t \to \infty$. From the second inequality in (24.21), we find that $T_r(t) \geq c$ implies that

$$\|\boldsymbol{\omega}(t)\|^2 \geq \frac{2c}{I_3}. \tag{24.37}$$

From (24.28), we then find that this implies that

$$\dot{T}_r = -k\|\boldsymbol{\omega}(t)\|^2 \leq -\frac{2kc}{I_3}. \tag{24.38}$$

Integrating both sides of (24.38), we find that

$$T_r(t) - T_r(0) = \int_0^t \dot{T}_r d\tau \leq -\int_0^t \frac{2kc}{I_3} d\tau, \tag{24.39}$$

which leads to

$$T_r(t) \leq T_r(0) - \frac{2kc}{I_3} t. \tag{24.40}$$

The right-hand side of (24.40) is a linearly decreasing function of time, which eventually becomes negative. This implies that if $T_r(t) \geq c > 0$ for all $t \geq 0$, then $T_r(t)$ will eventually become negative, which is a contradiction. Therefore, it must be that $T_r(t) \to 0$ as $t \to \infty$.

From (24.22), it then follows that $\boldsymbol{\omega}(t) \to \mathbf{0}$ as $t \to \infty$. Therefore, from Definition 24.1, $\boldsymbol{\omega} = \mathbf{0}$ is an asymptotically stable equilibrium. Since our choice of initial condition $\boldsymbol{\omega}(0)$ was arbitrary (there is not restriction on its size), the result is global, and we can conclude that $\boldsymbol{\omega} = \mathbf{0}$ is in fact globally asymptotically stable.

24.3.2 Lyapunov Stability Theorems

In Section 24.3.1, we made use of a positive definite function (the rotational kinetic energy) to study the stability of the equilibrium without having to solve the equation of motion. The method presented in Section 24.3.1 is a special case of what is known as *Lyapunov stability analysis*. The method presented in Section 24.3.1 is generalized in the following famous theorem.

Theorem 24.1 (Lyapunov's Stability Theorem) *Consider the system with state-equation*

$$\dot{\mathbf{x}} = \mathbf{f}(\mathbf{x}), \tag{24.41}$$

where $\mathbf{x} \in \mathbb{R}^n$ *and assume the existence of solutions* $\mathbf{x}(t)$ *to the state-equation in some open set* $D \subset \mathbb{R}^n$ *containing the origin (that is* $\mathbf{0} \in D$*). Let* $\mathbf{x} = \mathbf{0}$ *be an equilibrium point. Consider a continuously differentiable scalar function* $V(\mathbf{x}) : D \to \mathbb{R}$*, such that* $V(\mathbf{0}) = 0$ *and* $V(\mathbf{x}) > 0$ *for* $\mathbf{x} \in D$ *with* $\mathbf{x} \neq \mathbf{0}$*. Suppose that along solutions* $\mathbf{x}(t) \in D$*,*

$$\dot{V}(\mathbf{x}) \leq 0. \tag{24.42}$$

Then, $\mathbf{x} = \mathbf{0}$ *is stable. Furthermore, if*

$$\dot{V}(\mathbf{x}) < 0 \; for \; \mathbf{x} \neq \mathbf{0}, \tag{24.43}$$

along solutions $\mathbf{x}(t) \in D$*, then* $\mathbf{x} = \mathbf{0}$ *is asymptotically stable.*

Note that the assumption of the existence of solutions to (24.41) is a necessary technical condition that needs to be made, since solutions are not guaranteed to exist for general nonlinear systems. However, for most practical systems (such as closed-loop spacecraft attitude control), this assumption will be valid.

If a function $V(\mathbf{x})$ can be found satisfying the condition (24.42) in Theorem 24.1, it is called a *Lyapunov function*. In the example in Section 24.3.1, the rotational kinetic energy plays the role of the Lyapunov function. It is important to note that $\dot{V}(\mathbf{x})$ is the time-derivative of $V(\mathbf{x})$ along a trajectory $\mathbf{x}(t)$ of (24.41), and is evaluated using the chain-rule as

$$\dot{V}(\mathbf{x}) = \frac{\partial V}{\partial \mathbf{x}^T}\dot{\mathbf{x}} = \frac{\partial V}{\partial \mathbf{x}^T}\mathbf{f}(\mathbf{x}), \tag{24.44}$$

where

$$\frac{\partial V}{\partial \mathbf{x}^T} = \begin{bmatrix} \frac{\partial V}{\partial x_1} & \cdots & \frac{\partial V}{\partial x_n} \end{bmatrix}.$$

Theorem 24.1 gives a local stability result only. To conclude global asymptotic stability, further conditions must be imposed, as outlined in the following theorem.

Theorem 24.2 (Global Version of Lyapunov's Stability Theorem) *Suppose that all conditions of Theorem 24.1 are satisfied with $D = \mathbb{R}^n$ (that is, globally). In addition, suppose that $V(\mathbf{x})$ has the property that*

$$\|\mathbf{x}\| \to \infty \textit{ implies that } V(\mathbf{x}) \to \infty. \tag{24.45}$$

Then, $\mathbf{x} = \mathbf{0}$ is globally asymptotically stable.

The crucial extra condition is (24.45). When $V(\mathbf{x})$ satisfies this condition, it is said to be *radially unbounded.* It is easily seen that the rotational kinetic energy in Section 24.3.1 satisfies this condition.

Important Points

- Theorems 24.1 and 24.2 provide *sufficient* conditions for stability. If we cannot find a Lyapunov function $V(\mathbf{x})$ satisfying the conditions of either Theorems 24.1 and 24.2, it does not mean that the system is unstable.
- A positive definite function $V(\mathbf{x})$ is only called a Lyapunov function if it has been demonstrated to satisfy the conditions in either Theorem 24.1 or 24.2. Until that has been demonstrated, it is referred to as a *Lyapunov function candidate*.
- As demonstrated in Section 24.3.1, a natural choice for Lyapunov function candidate is often the "energy" of the system.

It will be noted that Theorems 24.1 and 24.2 have been presented for autonomous systems. Analogous results exists for non-autonomous systems, and can readily be found in any text on nonlinear control systems.

24.4 LaSalle's Theorem

In Section 24.3, we demonstrated how the closed-loop stability may be determined using Lyapunov stability analysis. Unfortunately, while it is useful for determining stability, it sometimes breaks down when trying to determine asymptotic stability. This problem is particularly prevalent when the systems under consideration are second-order mechanical systems. We shall now illustrate this issue with an example.

Example 24.3 *Consider the simple spring-mass-damper system with equation of motion*

$$m\ddot{x} + d\dot{x} + kx = 0, \tag{24.46}$$

where $m > 0$ is the mass, $d > 0$ is the viscous damping coefficient and $k > 0$ is the spring constant. It is well known that the equilibrium $\dot{x} = x = 0$ is asymptotically stable. Let us try *to demonstrate that using the Lyapunov stability theory from Section 24.3. First, we write the system (24.46) in first-order state-space form by defining the states*

$$x_1 = x \text{ (position)},$$
$$x_2 = \dot{x} \text{ (velocity)}.$$

Form (24.46), we then have

$$\begin{aligned}\dot{x}_1 &= x_2,\\ \dot{x}_2 &= -\tfrac{d}{m}x_2 - \tfrac{k}{m}x_1.\end{aligned} \tag{24.47}$$

As a Lyapunov-function candidate, let us try the total energy of the system

$$V(x_1, x_2) = \frac{1}{2}mx_2^2 + \frac{1}{2}kx_1^2. \tag{24.48}$$

Taking the time-derivative of (24.48) along a trajectory of (24.47), we obtain

$$\begin{aligned}\dot{V}(x_1, x_2) &= mx_2\dot{x}_2 + kmx_1\dot{x}_1,\\ &= mx_2\left[-\frac{d}{m}x_2 - \frac{k}{m}x_1\right] + kx_1x_2,\\ &= -dx_2^2,\\ &\leq 0.\end{aligned}$$

Therefore, condition (24.42) in Theorem 24.1 is satisfied, so we can conclude that the equilibrium $x_1 = x_2 = 0$ is stable. However, since $\dot{V} = 0$ for all $x_1 \neq 0$ provided $x_2 = 0$, condition (24.43) is not satisfied, and we cannot conclude asymptotic stability using Theorem 24.1.

Example 24.1 has demonstrated a potential shortcoming of Lyapunov stability analysis for second order mechanical systems. In particular this will be problematic when we design a spacecraft attitude control law. Fortunately, there is another tool that can be used to tackle this problem. This tool is called LaSalle's Theorem, and will be presented in this section.

First, we must introduce some new concepts.

Definition 24.2 *Consider the system*

$$\dot{\mathbf{x}} = \mathbf{f}(\mathbf{x}), \tag{24.49}$$

where $\mathbf{x} \in \mathbb{R}^n$. Then, we say that a set $M \subset \mathbb{R}^n$ is **invariant** *with respect to (24.49) if*

$$\mathbf{x}(0) \in M \textit{ implies that } \mathbf{x}(t) \in M \textit{ for all } t \in \mathbb{R}. \tag{24.50}$$

That is, if a solution $\mathbf{x}(t)$ belongs to M at some time instant, then it belongs to M for all past and future times.

Definition 24.3 *We say that* $\mathbf{x}(t)$ *approaches a set* M *as* $t \to \infty$ *if for each* $\epsilon > 0$ *there is a* $T > 0$ *such that*

$$dist(\mathbf{x}(t), M) < \epsilon, \text{ for all } t \geq T, \tag{24.51}$$

where $dist(\mathbf{p}, M)$ *denotes the distance from a point* $\mathbf{p}$ *to set* M, *which is defined as*

$$dist(\mathbf{p}, M) = \inf_{\mathbf{q} \in M} \|\mathbf{p} - \mathbf{q}\|.$$

Note that "inf" (infinimum) denotes the highest lower bound on a set of real numbers (it can be thought of as the minimum).

With Definitions 24.2 and 24.3 in hand, we can now state LaSalle's theorem.

Theorem 24.3 (LaSalle's Theorem) *Consider the system with state-equation*

$$\dot{\mathbf{x}} = \mathbf{f}(\mathbf{x}), \tag{24.52}$$

where $\mathbf{x} \in \mathbb{R}^n$ *and assume the existence of solutions* $\mathbf{x}(t)$ *to the state-equation in some open set* $D \subset \mathbb{R}^n$. *Suppose that for some solution* $\mathbf{x}(t)$, *a closed and bounded set* $\Omega \subset D$ *can be found such that* $\mathbf{x}(t) \in \Omega$ *for all* $t \geq 0$. *Let* $V(\mathbf{x}) : D \to \mathbb{R}$ *be a continuously differentiable function such that along the trajectory* $\mathbf{x}(t)$, *the time derivative of* V *satisfies* $\dot{V} \leq 0$ *in* Ω. *Let* E *be the set of all points in* Ω *where* $\dot{V}(\mathbf{x}) = 0$, *that is* $E = \{\mathbf{x} \in \Omega : \dot{V}(\mathbf{x}) = 0\}$. *Let* M *be the largest invariant set of the system (24.52) contained in* E. *Then,* $\mathbf{x}(t) \to M$ *as* $t \to \infty$.

It is very important to note that LaSalle's theorem only holds for autonomous systems. There are some special classes of non-autonomous systems for which similar results can be found. However, in general these results are NOT applicable to non-autonomous systems.

A comparison of LaSalle's theorem (Theorem 24.3) with Lyapunov's theorem (Theorem 24.1) shows that they are quite similar. Some subtle differences that are worth noting are the following: in Lyapunov's theorem, the function $V(\mathbf{x})$ is required to be positive definite, while in LaSalle's theorem it is not. On the other hand, LaSalle's theorem requires the prior knowledge that a trajectory can be contained in a closed and bounded set Ω, while Lyapunov's theorem does not. Often times, a Lyapunov analysis can be used to find such a closed and bounded set Ω.

Example 24.4 *Let us now return to the spring-mass-damper problem in Example 24.3. Consider an arbitrary initial condition* $(x_1(0), x_2(0))$. *From the Lyapunov analysis there, we found that* $\dot{V} \leq 0$, *which implies that*

$$V(x_1(t), x_2(t)) \leq V(x_1(0), x_2(0)).$$

Therefore, we may take the closed and bounded set in Theorem 24.3 to be

$$\Omega = \left\{(x_1, x_2) \in \mathbb{R}^2 : V(x_1(t), x_2(t)) \leq V(x_1(0), x_2(0))\right\}.$$

Next, let us find the set E in Theorem 24.3. It is readily found to be

$$E = \{(x_1, x_2) \in \Omega : x_2 = 0\}$$

Finally, let us find the largest invariant set of the system (24.47) contained in E. Now, for any trajectory contained in the set E, we have $x_2(t) \equiv 0$. Therefore, $\dot{x}_2(t) \equiv 0$ for trajectories contained in E. From the system equations (24.47), *this implies that trajectories contained in E also satisfy*

$$0 \equiv -\frac{d}{m} \cdot 0 - \frac{k}{m} x_1(t),$$

which implies that $x_1(t) \equiv 0$ as well. Therefore, we see that the largest invariant set M is given by

$$M = (0, 0).$$

All conditions in Theorem 24.3 are now satisfied, and upon application of the theorem we conclude that $x_1(t) \to 0$, $x_2(t) \to 0$. As shown in Example 24.3, the point $x_1 = x_2 = 0$ is stable. Therefore, the point is $x_1 = x_2 = 0$ stable in the sense of Lyapunov. Since the initial condition was arbitrary, the result is global.

24.5 Spacecraft Attitude Control with Quaternion and Angular Rate Feedback

Suppose that it is desired to control the spacecraft to some desired constant inertial attitude. Let us represent the attitude error by the quaternion $(\boldsymbol{\epsilon}, \eta)$, as in Example 24.1. Then, as in Example 24.1, the spacecraft attitude motion is governed by the quaternion kinematics and Euler's equations, given in equations (24.3) and (24.4) respectively.

The control objective is then to render the closed-loop system globally convergent to the equilibrium $\boldsymbol{\omega} = \boldsymbol{\epsilon} = \mathbf{0}$, since this corresponds to zero attitude error. Note that due to the unit length constraint in the quaternion, we cannot render the closed-loop system stable in the sense of Definition 24.1, since this requires that $(\boldsymbol{\epsilon}, \eta)$ can be made arbitrarily small.

Let us try the PD-like control law (as in Example 24.2)

$$\mathbf{T} = -k_p \boldsymbol{\epsilon} - k_d \boldsymbol{\omega}, \tag{24.53}$$

where $k_d > 0$ is a derivative gain and $k_p > 0$ is a proportional gain. Thus, we are feeding back the vector part of the attitude error quaternion, as well as the angular velocity vector. The closed-loop system is then given by

$$\begin{aligned}
\dot{\boldsymbol{\epsilon}} &= \frac{1}{2}\left(\eta \mathbf{1} + \boldsymbol{\epsilon}^{\times}\right)\boldsymbol{\omega}, \\
\dot{\eta} &= -\frac{1}{2}\boldsymbol{\epsilon}^T \boldsymbol{\omega}, \\
\dot{\boldsymbol{\omega}} &= \mathbf{I}^{-1}\left[-\boldsymbol{\omega}^{\times}\mathbf{I}\boldsymbol{\omega} - k_p\boldsymbol{\epsilon} - k_d\boldsymbol{\omega}\right].
\end{aligned} \tag{24.54}$$

As noted in Example 24.2, the closed-loop system given by equation (24.54) is autonomous. Next, we recall from Section 1.3.4 that the quaternion representation of the attitude satisfies the unit length constraint $\boldsymbol{\epsilon}^T\boldsymbol{\epsilon} + \eta^2 = 1$. It can readily be shown from the quaternion kinematics in (24.54) that provided the initial quaternion $(\boldsymbol{\epsilon}(0), \eta(0))$ satisfies the unit length constraint, then the quaternion trajectory $(\boldsymbol{\epsilon}(t), \eta(t))$ will satisfy the unit length constraint for all $t \in \mathbb{R}$ (try this as an exercise). In particular, this leads to the inequalities

$$\|\boldsymbol{\epsilon}(t)\| \leq 1, \quad |\eta(t)| \leq 1, \text{ for all } t \in \mathbb{R}. \tag{24.55}$$

Now, consider the function

$$V(\boldsymbol{\epsilon}, \eta, \boldsymbol{\omega}) = \frac{1}{2}\boldsymbol{\omega}^T\mathbf{I}\boldsymbol{\omega} - 2k_p\eta. \tag{24.56}$$

By the second inequality in (24.55), we have

$$\frac{1}{2}\boldsymbol{\omega}^T\mathbf{I}\boldsymbol{\omega} - 2k_p \leq V(\boldsymbol{\epsilon}, \eta, \boldsymbol{\omega}). \tag{24.57}$$

As in Section 24.3.1, let I_1, I_2 and I_3 be the principal moments of inertia corresponding to $\mathbf{I}$, ordered such that

$$I_1 \leq I_2 \leq I_3. \tag{24.58}$$

Then, making use of equation (24.19) and the first inequality in (24.21), the inequality in (24.57) becomes

$$\frac{1}{2}I_1\|\boldsymbol{\omega}(t)\|^2 \leq V(\boldsymbol{\epsilon}(t), \eta(t), \boldsymbol{\omega}(t)) + 2k_p,$$

which leads to

$$\|\boldsymbol{\omega}(t)\| \leq \sqrt{\frac{2V(\boldsymbol{\epsilon}(t), \eta(t), \boldsymbol{\omega}(t)) + 4k_p}{I_1}}, \tag{24.59}$$

for any closed-loop trajectory with initial quaternion satisfying $\boldsymbol{\epsilon}(0)^T\boldsymbol{\epsilon}(0) + \eta(0)^2 = 1$.

Now, let the initial condition $\boldsymbol{\epsilon}(0)$, $\eta(0)$ and $\boldsymbol{\omega}(0)$ be given, such that the initial quaternion satisfies the unit length constraint $\boldsymbol{\epsilon}(0)^T\boldsymbol{\epsilon}(0) + \eta(0)^2 = 1$. Let us take the time-derivative of V in (24.56) along the resulting closed-loop trajectory $\boldsymbol{\epsilon}(t)$, $\eta(t)$ and $\boldsymbol{\omega}(t)$ of (24.54). Making use of (24.54), we find

$$\begin{aligned}
\dot{V} &= \boldsymbol{\omega}^T\mathbf{I}\dot{\boldsymbol{\omega}} - 2k_p\dot{\eta}, \\
&= \boldsymbol{\omega}^T\mathbf{I}\mathbf{I}^{-1}\left[-\boldsymbol{\omega}^{\times}\mathbf{I}\boldsymbol{\omega} - k_p\boldsymbol{\epsilon} - k_d\boldsymbol{\omega}\right] - 2k_p\left[-\frac{1}{2}\boldsymbol{\epsilon}^T\boldsymbol{\omega}\right], \\
&= -k_d\boldsymbol{\omega}^T\boldsymbol{\omega}, \\
&\leq 0.
\end{aligned} \tag{24.60}$$

Integrating both sides of (24.60), this shows that $V(\boldsymbol{\epsilon}(t), \eta(t), \boldsymbol{\omega}(t)) \le V(\boldsymbol{\epsilon}(0), \eta(0), \boldsymbol{\omega}(0))$. Therefore, from (24.59), we find that

$$\|\boldsymbol{\omega}(t)\| \le \sqrt{\frac{2V(\boldsymbol{\epsilon}(0), \eta(0), \boldsymbol{\omega}(0)) + 4k_p}{I_1}}, \quad \text{for all } t \ge 0. \tag{24.61}$$

Combining inequalities (24.55) and (24.61), let us define the set

$$\Omega = \left\{(\boldsymbol{\epsilon}, \eta, \boldsymbol{\omega}) \in \mathbb{R}^3 \times \mathbb{R} \times \mathbb{R}^3 : \|\boldsymbol{\epsilon}\| \le 1, \ |\eta| \le 1, \ \|\boldsymbol{\omega}\| \le a\right\}. \tag{24.62}$$

where

$$a = \sqrt{\frac{2V(\boldsymbol{\epsilon}(0), \eta(0), \boldsymbol{\omega}(0)) + 4k_p}{I_1}}.$$

The set Ω in (24.62) is closed and bounded, and we have shown that the closed-loop trajectory $(\boldsymbol{\epsilon}(t), \eta(t), \boldsymbol{\omega}(t)) \in \Omega$ for all $t \ge 0$. We have also shown that $\dot{V} = -k_d \boldsymbol{\omega}^T \boldsymbol{\omega} \le 0$ on Ω. Therefore, all conditions in Theorem 24.3 are satisfied, and we may apply it. Making use of (24.60), we readily find that the set E in Theorem 24.1 is given by

$$E = \{(\boldsymbol{\epsilon}, \eta, \boldsymbol{\omega}) \in \Omega : \|\boldsymbol{\epsilon}\| \le 1, \ |\eta| \le 1, \ \boldsymbol{\omega} = \mathbf{0}\}. \tag{24.63}$$

We now search for the largest invariant set in E. For any trajectory contained in E, we see from (24.63) that $\boldsymbol{\omega}(t) \equiv \mathbf{0}$, and that consequently $\dot{\boldsymbol{\omega}}(t) \equiv \mathbf{0}$ also. From the closed-loop equations (24.54), we consequently see that any trajectory contained in E satisfies

$$\begin{aligned} \dot{\boldsymbol{\epsilon}} &= \mathbf{0}, \\ \dot{\eta} &= 0, \\ \mathbf{0} &= -k_p \mathbf{I}^{-1} \boldsymbol{\epsilon}, \end{aligned}$$

which implies that $\boldsymbol{\epsilon}(t) \equiv \mathbf{0}$ and that $\eta(t)$ is constant. By the unit length constraint on the quaternion, this implies that $\eta(t) \equiv \pm 1$. Therefore, the largest invariant set in E is given by

$$M = \{(\boldsymbol{\epsilon}, \eta, \boldsymbol{\omega}) \in \Omega : \boldsymbol{\epsilon} = \mathbf{0}, \ \eta = \pm 1, \ \boldsymbol{\omega} = \mathbf{0}\}. \tag{24.64}$$

By application of Theorem 24.3, we conclude that $\boldsymbol{\epsilon}(t) \to \mathbf{0}$ and $\boldsymbol{\omega}(t) \to \mathbf{0}$ as $t \to \infty$. Since the choice of initial condition $(\boldsymbol{\epsilon}(0), \eta(0), \boldsymbol{\omega}(0))$ was arbitrary (subject to the unit length constraint in the initial quaternion), we conclude that this is true for any closed-loop trajectory. Hence, we can conclude global convergence.

Useful features of the closed-loop system

- Our stability analysis has demonstrated that the PD-like quaternion and angular velocity feedback in (24.53) will drive the attitude and angular velocity errors to zero regardless of the initial attitude and angular velocity of the spacecraft. This is a very useful property when large angle maneuvers are required.

- An examination of the convergence analysis shows that no knowledge of the spacecraft inertia matrix was required, other than it being positive definite, which is always true for a real spacecraft. Hence, the global convergence property of the closed-loop system holds for any real spacecraft. Therefore, the control law is inherently very robust. That is, even if we assume inertia values in order to design the control gains based on closed-loop performance requirements, errors in the assumed inertia values will not destabilize the closed-loop system when the control law is applied to the real spacecraft.

24.5.1 Controller Gain Selection

Having demonstrated that the closed-loop system is globally convergent for any positive gains $k_p > 0$ and $k_d > 0$, we now turn to the question of how to select appropriate control gains, given some closed-loop performance requirements. One possible method is to select the gains based on the desired steady-state performance. To facilitate the stability analysis in this chapter, disturbance torques and measurement noise have been neglected. As we have shown, in the absence of these, the steady-state error is zero. However, in reality, there will be disturbance torques and measurement errors, which will lead to non-zero steady-state errors. Typically, closed-loop performance specifications will include requirements that the steady-state errors be small. As such, one approach to selecting the control gains is to examine the linearized closed-loop equations for small angles and rates.

Let us therefore examine the small angle and rate approximations to the closed-loop equations (24.54). First, we note from the definition of the quaternion in Section 1.3.4, that the small angle approximation implies that the vector part of the quaternion, $\boldsymbol{\epsilon}$, is small, and consequently from the unit length constraint, $\eta \approx \pm 1$. Now, assuming small $\boldsymbol{\epsilon}$ and $\boldsymbol{\omega}$, we see that the kinematics in (24.54) become to first order

$$\begin{aligned} \dot{\boldsymbol{\epsilon}} &\approx \pm\frac{1}{2}\boldsymbol{\omega}, \\ \dot{\eta} &\approx 0. \end{aligned} \tag{24.65}$$

Note that the two possible signs for $\dot{\boldsymbol{\epsilon}}$ arise from the fact that we have allowed $\eta = \pm 1$. We shall resolve this issue shortly by selecting the sign of η that leads to a convergent linear approximation (since we already know that the full nonlinear closed-loop equations are convergent). Substituting (24.65) into the closed-loop dynamic equation in (24.54) and neglecting higher order terms, we obtain

$$\eta 2\ddot{\boldsymbol{\epsilon}} \approx \mathbf{I}^{-1}\left[-k_p\boldsymbol{\epsilon} - \eta 2k_d\dot{\boldsymbol{\epsilon}}\right], \tag{24.66}$$

where it is understood that $\eta = \pm 1$. It will be useful to write (24.66) in first order form for the purposes of determining its stability. We have

$$\begin{bmatrix} \dot{\boldsymbol{\epsilon}} \\ \ddot{\boldsymbol{\epsilon}} \end{bmatrix} = \begin{bmatrix} \mathbf{0} & \mathbf{1} \\ -\dfrac{k_p}{2\eta}\mathbf{I}^{-1} & -k_d\mathbf{I}^{-1} \end{bmatrix} \begin{bmatrix} \boldsymbol{\epsilon} \\ \dot{\boldsymbol{\epsilon}} \end{bmatrix}. \tag{24.67}$$

It will be useful to make a pair of coordinate transformations in order to be able to compute the associated poles. First, let I_1, I_2 and I_3 be the principal moments of inertia corresponding to the inertia matrix $\mathbf{I}$. As in Section 24.3.1, let $\mathbf{C}$ be a rotation matrix such that the corresponding principal inertia matrix is given by (24.23). Now, consider the coordinate transformation

$$\boldsymbol{\alpha} = \mathbf{C}\boldsymbol{\epsilon}. \tag{24.68}$$

Then, under this transformation, equations (24.67) become

$$\begin{bmatrix} \dot{\boldsymbol{\alpha}} \\ \ddot{\boldsymbol{\alpha}} \end{bmatrix} = \begin{bmatrix} \mathbf{0} & \mathbf{1} \\ -\frac{k_p}{2\eta}\mathbf{I}_p^{-1} & -k_d\mathbf{I}_p^{-1} \end{bmatrix} \begin{bmatrix} \boldsymbol{\alpha} \\ \dot{\boldsymbol{\alpha}} \end{bmatrix}. \tag{24.69}$$

Let

$$\boldsymbol{\alpha} = \begin{bmatrix} \alpha_1 \\ \alpha_2 \\ \alpha_3 \end{bmatrix}.$$

Then, we can rewrite equations (24.69) as

$$\begin{bmatrix} \dot{\alpha}_1 \\ \ddot{\alpha}_1 \\ \dot{\alpha}_2 \\ \ddot{\alpha}_2 \\ \dot{\alpha}_3 \\ \ddot{\alpha}_3 \end{bmatrix} = \begin{bmatrix} 0 & 1 & 0 & 0 & 0 & 0 \\ -\frac{k_p}{2I_1\eta} & -\frac{k_d}{I_1} & 0 & 0 & 0 & 0 \\ 0 & 0 & 0 & 1 & 0 & 0 \\ 0 & 0 & -\frac{k_p}{2I_2\eta} & -\frac{k_d}{I_2} & 0 & 0 \\ 0 & 0 & 0 & 0 & 0 & 1 \\ 0 & 0 & 0 & 0 & -\frac{k_p}{2I_3\eta} & -\frac{k_d}{I_3} \end{bmatrix} \begin{bmatrix} \alpha_1 \\ \dot{\alpha}_1 \\ \alpha_2 \\ \dot{\alpha}_2 \\ \alpha_3 \\ \dot{\alpha}_3 \end{bmatrix}. \tag{24.70}$$

The stability of equations (24.67) is equivalent to stability of equations (24.70). In fact, their solutions have the same poles. Therefore, we shall analyze the stability of (24.70). Equation (24.70) has the form

$$\dot{\mathbf{x}} = \mathbf{A}\mathbf{x}.$$

Taking Laplace transforms, we obtain

$$(s\mathbf{1} - \mathbf{A})\,\hat{\mathbf{x}} = \mathbf{x}(0),$$

where $\hat{\mathbf{x}} = \mathcal{L}\,(\mathbf{x}(t))$. The poles are now readily found to satisfy the characteristic equation

$$\det\,(s\mathbf{1} - \mathbf{A}) = 0,$$

that is, the poles are the eigenvalues of the matrix **A**. It is straightforward to show that the characteristic equation corresponding to (24.70) is

$$\left(s^2 + \frac{k_d}{I_1}s + \frac{k_p}{2I_1\eta}\right)\left(s^2 + \frac{k_d}{I_2}s + \frac{k_p}{2I_2\eta}\right)\left(s^2 + \frac{k_d}{I_3}s + \frac{k_p}{2I_3\eta}\right) = 0. \tag{24.71}$$

Since k_p, k_d, I_1, I_2 and I_3 are all positive, it can be immediately seen from equation (24.71) that the closed-loop poles are asymptotically stable (have negative real part) if and only if η is positive. Therefore, we set $\eta = 1$ in the linearized equations (24.66), to obtain

$$\ddot{\boldsymbol{\epsilon}} = \mathbf{I}^{-1}\left[-\frac{k_p}{2}\boldsymbol{\epsilon} - k_d\dot{\boldsymbol{\epsilon}}\right], \tag{24.72}$$

Now, suppose that as assumed in Chapter 17, the body frame $\mathcal{F}_b$ is a principal axes frame such that

$$\mathbf{I} = \begin{bmatrix} I_x & 0 & 0 \\ 0 & I_y & 0 \\ 0 & 0 & I_z \end{bmatrix},$$

and let

$$\boldsymbol{\epsilon} = \begin{bmatrix} \epsilon_x \\ \epsilon_y \\ \epsilon_z \end{bmatrix}.$$

Then, the linearized equations (24.72) take the form

$$\begin{aligned} \ddot{\epsilon}_x &= -\frac{k_p}{2I_x}\epsilon_x - \frac{k_d}{I_x}\dot{\epsilon}_x, \\ \ddot{\epsilon}_y &= -\frac{k_p}{2I_y}\epsilon_y - \frac{k_d}{I_y}\dot{\epsilon}_y, \\ \ddot{\epsilon}_z &= -\frac{k_p}{2I_z}\epsilon_z - \frac{k_d}{I_z}\dot{\epsilon}_z. \end{aligned} \tag{24.73}$$

Equations (24.73) are three identical decoupled second order differential equations, of the same form as those considered extensively in Chapter 17 (linearized PD spacecraft attitude control). Measurement noise and disturbance torques can readily be added to equations (24.73) as needed. As such, similar linear control design techniques such as those presented in Chapter 17 (and subsequent chapters) may be used to design the control gains.

Notes

This chapter has presented a brief introduction to nonlinear control systems, with applications to spacecraft attitude control. The fields of nonlinear control and its application to spacecraft attitude control are vast, and there is a significant amount of literature in both books and journals that the reader is encouraged to consult. Excellent texts on the subject of nonlinear control include Khalil (2002) and Vidyasagar (2002). For an overview of many different applications of nonlinear control to the spacecraft attitude control problem, the reader is suggested to consult Wie (2008).

References

Khalil HK 2002 *Nonlinear Systems* 3rd edn. Prentice Hall, Upper Saddle River, NJ.

Vidyasagar M 2002 *Nonlinear Systems Analysis* 2nd edn. Society for Industrial and Applied Mathematics, Philadelphia, PA.

Wie B 2008 *Space Vehicle Dynamics and Control* 2nd edn. American Institute of Aeronautics and Astronautics, Reston, VA.

25

Spacecraft Navigation

In previous chapters we considered how to control the attitude of a spacecraft. However, before attitude control can be attempted, the current attitude of the spacecraft must be *estimated*. We say the attitude must be estimated because it is not possible to determine the exact attitude of a spacecraft given that all sensors, such as sun sensors, star trackers, rate gyros, and others, provide imperfect measurements that are corrupted by *noise*. As such, spacecraft navigation, that is, determining what the actual attitude of a spacecraft is, really comes down to *state estimation*. Strictly speaking, spacecraft navigation involves both orbit and attitude estimation. We shall restrict ourselves to the attitude estimation problem, since it requires some special considerations. However, the filtering techniques developed in this chapter can be adapted for the orbit estimation problem also.

In this chapter we will consider various techniques for spacecraft attitude estimation. In particular, we will consider batch estimation approaches (weighted least squares) and the use of the Kalman filter. As we shall see, because the system we are dealing with is nonlinear, when it comes to Kalman filtering we must use an extended Kalman filter (EKF). However, en route to deriving the EKF we will investigate the standard Kalman filter. Both batch estimation and the EKF have found many uses in other engineering disciplines.

25.1 Review of Probability Theory

As previously mentioned, noise corrupts all measurements. As such, one way of looking at the spacecraft attitude estimation problem is determining how to *filter* noise in some optimal way. Because noise is *random* in nature, we will rely on concepts from probability theory to describe the properties of the noise that we are interested in filtering.

25.1.1 Continuous Random Variables and Probability Density Functions

Consider a *continuous random variable*, x, and an associated *probability density function* (pdf), $p(x)$, where x can lie within the interval $[a, b]$. The random variable x is said to be

Spacecraft Dynamics and Control: An Introduction, First Edition.
Anton H.J. de Ruiter, Christopher J. Damaren and James R. Forbes.

distributed according to the pdf $p(x)$. The pdf must satisfy the axiom of total probability, which means that

$$\int_a^b p(x)dx = 1. \tag{25.1}$$

The probability that x takes on a value between $[\alpha, \beta]$ is

$$Pr(\alpha \le x \le \beta) = \int_\alpha^\beta p(x)dx.$$

Consider a set of N continuous random variables, $\mathbf{x} = [x_1 \ \dots \ x_N]^T$, $x_i \in [a_i, b_i]$, $i = 1, \dots, N$, and an associated *joint probability density function* (joint pdf), $p(\mathbf{x})$ where

$$p(\mathbf{x}) = p(x_1, x_2, \dots, x_N).$$

We call $\mathbf{x}$ a continuous random vector. Even in the N-dimensional case $p(\mathbf{x})$ must satisfy the axiom of total probability, that is

$$\int_{\mathbf{a}}^{\mathbf{b}} p(\mathbf{x})d\mathbf{x} = 1, \tag{25.2}$$

which is short-hand notation for

$$\int_{\mathbf{a}}^{\mathbf{b}} p(\mathbf{x})d\mathbf{x} = \int_{a_N}^{b_N} \cdots \int_{a_2}^{b_2} \int_{a_1}^{b_1} p(x_1, x_2, \dots, x_N)dx_1dx_2 \dots dx_N = 1,$$

where $\mathbf{a} = [a_1 \ \dots a_N \ ]^T$ and $\mathbf{b} = [b_1 \ \dots b_N \ ]^T$.

Example 25.1 *Consider a spacecraft equipped with one bias momentum wheel. According to the manufacturer of the momentum wheel, the probability of failure of the momentum wheel bearing is characterized by the pdf*

$$p(x) = \begin{cases} 0, & 0 \le x < T \\ c\exp(-\lambda x), & T \le x \le \infty \end{cases} \tag{25.3}$$

where $T = 730$ *(days) and* $\lambda = 1/100$ *(1/days).*

1. *What value must c take? What are the units of c? Given* T, λ, *and the c just computed, plot* $p(x)$.
2. *What is the probability that the momentum wheel bearing will fail within* 730 *days? Between* 730 *days and* 1095 *days (i.e., between 2 and 3 years)?*

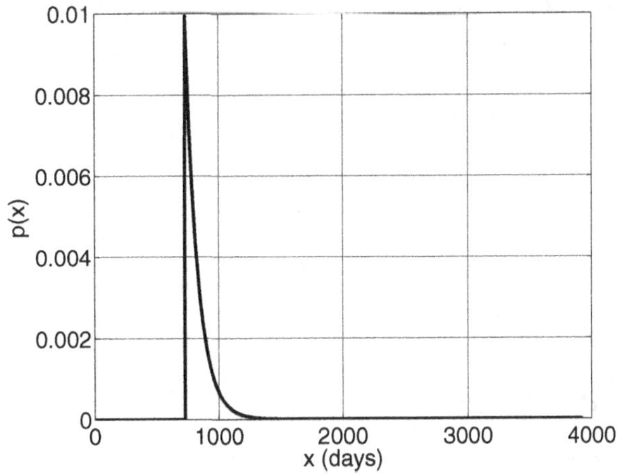

Figure 25.1 Bearing pdf

Solution

1. *The pdf given in* (25.3) *must satisfy the axiom of total probability. From* (25.1) *we have the following:*

$$\begin{aligned}\int_{-\infty}^{\infty} p(x)dx &= \int_0^T 0\,dx + \int_T^{\infty} c\exp(-\lambda x)\,dx \\ &= -\frac{c}{\lambda}\exp(-\lambda x)\Big|_T^{\infty} \\ &= \frac{c}{\lambda}\exp(-\lambda T) \\ &= 1.\end{aligned}$$

Thus, we have that $c = \frac{\lambda}{\exp(-\lambda T)}$. *Given that the units of* λ *are* $1/days$, *the units of c are* $1/days$ *as well. The plot of* $p(x)$ *given* $T = 730$ *days and c as computed is shown in Figure 25.1 .*

2. *The probability the momentum wheel bearing will fail within* 730 *days can be computed using (25.1.1):*

$$\begin{aligned}Pr(0 \le x \le 730) &= \int_0^{730} p(x)dx \\ &= \int_0^{730} 0\,dx \\ &= 0.\end{aligned}$$

Similarly, the probability the momentum wheel bearing will fail between 730 *days and* 1095 *days can be computed using (25.1.1):*

$$\begin{aligned}
Pr(730 \leq x \leq 1095) &= \int_{730}^{1095} c \exp(-\lambda x)\, dx \\
&= -\frac{c}{\lambda} \exp(-\lambda x)\Big|_{730}^{1095} \\
&= 0.9740.
\end{aligned}$$

25.1.2 Mean and Covariance

Consider a continuous random variable, $x \in [a_i, b_i]$, and a pdf, $p(x)$. The mean, or expected value of x is

$$\bar{x} = E[x] = \int_a^b x p(x) dx,$$

where $E[\cdot]$ is the expectation operator. The variance is

$$\sigma^2 = E[(x - \bar{x})^2] = \int_a^b (x - \bar{x})^2 p(x) dx.$$

The standard deviation is simply σ. For a general function of the random variable x, say $f(x)$, the expectation is

$$E[f(x)] = \int_a^b f(x) p(x) dx.$$

Notice that the expectation operator is linear. To see this, say both $f(x)$ and $g(x)$ are functions of x where x is distributed according to $p(x)$, and k is a constant. Then

$$E[kx] = \int_a^b k x p(x) dx = k \int_a^b x p(x) dx = k E[x] = k\bar{x},$$

$$E[f(x) + g(x)] = \int_a^b (f(x) + g(x))\, p(x) dx = E[f(x)] + E[g(x)].$$

Thus, the expectation operator is a linear one.

Now consider the N-dimensional case where we have $\mathbf{x} = [x_1 \ \dots \ x_N]^T$, $x_i \in [a_i, b_i]$, $i = 1, \dots, N$, and a joint pdf, $p(\mathbf{x})$. The mean of the continuous random vector $\mathbf{x}$ is

$$\bar{\mathbf{x}} = E[\mathbf{x}] = \int_{\mathbf{a}}^{\mathbf{b}} \mathbf{x} p(\mathbf{x}) d\mathbf{x},$$

while the covariance is

$$\mathbf{Q} = E[(\mathbf{x} - \bar{\mathbf{x}})(\mathbf{x} - \bar{\mathbf{x}})^T] = \begin{bmatrix} \sigma_1^2 & \cdots & \sigma_{1N} \\ \vdots & \ddots & \vdots \\ \sigma_{N1} & \cdots & \sigma_N^2 \end{bmatrix}$$

where

$$\sigma_{ij} = E[(x_i - \bar{x}_i)(x_j - \bar{x}_j)], \quad i, j = 1, \ldots, N.$$

For a general matrix function of the random vector **x**, $\mathbf{F}(\mathbf{x})$, the expectation is

$$E[\mathbf{F}(\mathbf{x})] = \int_{\mathbf{a}}^{\mathbf{b}} \mathbf{F}(\mathbf{x}) p(\mathbf{x}) d\mathbf{x}.$$

Note that the covariance **Q** is a $N \times N$ matrix. In fact, it is a symmetric positive semi-definite matrix. Recall that a matrix is symmetric if $\mathbf{Q} = \mathbf{Q}^T$. If a matrix is positive semi-definite, usually written $\mathbf{Q} \geq 0$, it means that $\mathbf{x}^T\mathbf{Q}\mathbf{x} \geq 0$ for all $\mathbf{x} \in \mathbb{R}^N$ (written $\forall \mathbf{x} \in \mathbb{R}^N$). There are many ways to check if a matrix, such as **Q**, is positive semi-definite. For instance, if $\underline{\lambda}\{\mathbf{Q}\} \geq 0$ where $\underline{\lambda}\{\cdot\}$ is the minimum eigenvalue of **Q**, then $\mathbf{Q} \geq 0$.

As in the one dimensional case, the expectation operator is still linear. Consider the functions $\mathbf{f}(\mathbf{x})$ and $\mathbf{g}(\mathbf{x})$ where **x** is distributed according to $p(\mathbf{x})$, and the constant matrix **K**. Then

$$E[\mathbf{K}\mathbf{x}] = \mathbf{K}E[\mathbf{x}] = \mathbf{K}\bar{\mathbf{x}},$$
$$E[\mathbf{f}(\mathbf{x}) + \mathbf{g}(\mathbf{x})] = E[\mathbf{f}(\mathbf{x})] + E[\mathbf{g}(\mathbf{x})],$$

indicating the expectation operator is linear.

Example 25.2 *A continuous random variable is called uniform if its pdf is constant between* $[a, b]$.

1. *What is* $p(x)$*?*
2. *If* $a = 1$ *and* $b = 3$ *find the mean,* $\bar{x}$*, and the variance* σ^2*. Draw* $p(x)$.

Solution

1. *Recall that* $p(x)$ *must satisfy the axiom of total probability given in* (25.1). *If* $p(x) = c$ *between* $[a, b]$ *where* c *is a constant number then*

$$1 = \int_a^b p(x)dx = \int_a^b c dx = c(b - a).$$

Solving for c *we have that* $c = \frac{1}{b-a}$ *and*

$$p(x) = \begin{cases} \frac{1}{b-a} & x \in [a, b] \\ 0 & \text{otherwise} \end{cases}.$$

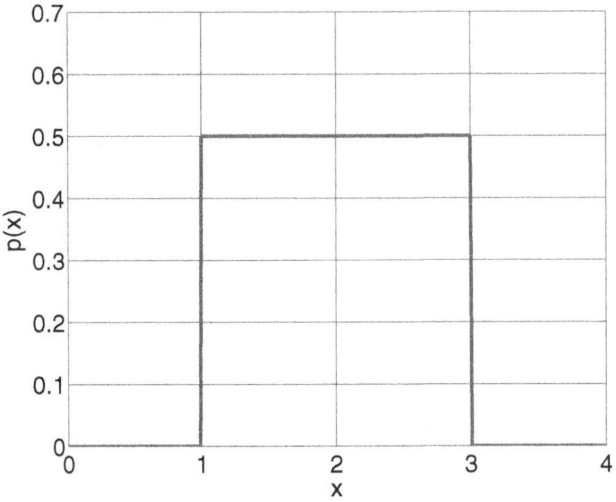

Figure 25.2 Uniform pdf where $a = 1$ and $b = 3$

2. *If* $a = 1$ *and* $b = 3$ *then the mean of* x, $\bar{x}$, *is*

$$\bar{x} = \int_1^3 xp(x)dx = \int_1^3 \frac{x}{2}dx = \left.\frac{x^2}{4}\right|_1^3 = 2,$$

and the variance of x, σ^2, *is*

$$\sigma^2 = \int_1^3 (x - \bar{x})^2 p(x)dx = \int_1^3 \frac{(x-2)^2}{2}dx = \frac{1}{3}.$$

Shown in Figure 25.2 is $p(x)$.

25.1.3 Gaussian Probability Density Functions

A continuous random variable is said to have a *normal* or *Gaussian* distribution if the pdf associated with the random variable x is given by

$$p(x|\bar{x}, \sigma^2) = \frac{1}{\sqrt{2\pi\sigma^2}} \exp\left(-\frac{(x-\bar{x})^2}{2\sigma^2}\right).$$

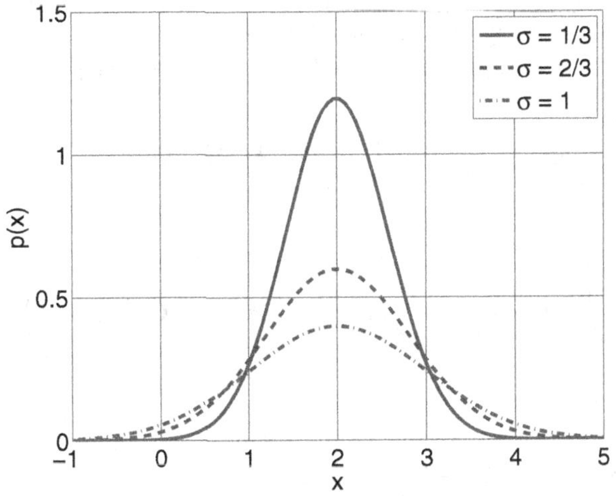

Figure 25.3 Gaussian pdfs where $\bar{x} = 2$ and σ takes on values of $1/3$, $2/3$, and 1

The pdf $p(x|\bar{x}, \sigma^2)$ is called a *conditional probability density function* (conditional pdf) because x is conditioned on $\bar{x}$, the mean, and σ^2, the variance (where σ is the standard deviation). Being a pdf, it can be shown that

$$\int_{-\infty}^{\infty} \frac{1}{\sqrt{2\pi\sigma^2}} \exp\left(-\frac{(x-\bar{x})^2}{2\sigma^2}\right) dx = 1,$$

the mean is

$$\bar{x} = \int_{-\infty}^{\infty} x \frac{1}{\sqrt{2\pi\sigma^2}} \exp\left(-\frac{(x-\bar{x})^2}{2\sigma^2}\right) dx,$$

and the variance is

$$\sigma^2 = \int_{-\infty}^{\infty} (x-\bar{x})^2 \frac{1}{\sqrt{2\pi\sigma^2}} \exp\left(-\frac{(x-\bar{x})^2}{2\sigma^2}\right) dx.$$

Shown in Figure 25.3 are three normal distributions. The mean of each distribution is $\bar{x} = 2$, while the standard deviation of each are $1/3$, $2/3$, and 1, respectively. A short-hand notation for indicating x is normally distributed is $x \sim \mathcal{N}(\bar{x}, \sigma^2)$.

In the N-dimensional case, a continuous random vector $\mathbf{x} \in \mathbb{R}^N$ is said to have a normal or Gaussian distribution if the pdf associated with the random vector $\mathbf{x}$ is given by

$$p(\mathbf{x}|\bar{\mathbf{x}}, \mathbf{Q}) = \frac{1}{\sqrt{(2\pi)^N \det \mathbf{Q}}} \exp\left(-\frac{1}{2}(\mathbf{x}-\bar{\mathbf{x}})^T \mathbf{Q}^{-1} (\mathbf{x}-\bar{\mathbf{x}})\right),$$

where $\bar{\mathbf{x}}$ is the mean and $\mathbf{Q}$ is the covariance matrix. The covariance matrix is symmetric and positive definite (thus ensuring $\mathbf{Q}$ is not singular, and thus $\mathbf{Q}^{-1}$ exists). The pdf $p(\mathbf{x}|\bar{\mathbf{x}}, \mathbf{Q})$ is also a conditional pdf because $\mathbf{x}$ is conditioned on $\bar{\mathbf{x}}$ and $\mathbf{Q}$. Being a pdf, it can be shown that

$$\int_{-\infty}^{\infty} \frac{1}{\sqrt{(2\pi)^N \det \mathbf{Q}}} \exp\left(-\frac{1}{2}(\mathbf{x}-\bar{\mathbf{x}})^T \mathbf{Q}^{-1}(\mathbf{x}-\bar{\mathbf{x}})\right) d\mathbf{x} = 1,$$

the mean is

$$\bar{\mathbf{x}} = \int_{-\infty}^{\infty} \mathbf{x} \frac{1}{\sqrt{(2\pi)^N \det \mathbf{Q}}} \exp\left(-\frac{1}{2}(\mathbf{x}-\bar{\mathbf{x}})^T \mathbf{Q}^{-1}(\mathbf{x}-\bar{\mathbf{x}})\right) d\mathbf{x},$$

and the covariance is

$$\mathbf{Q} = \int_{-\infty}^{\infty} (\mathbf{x}-\bar{\mathbf{x}})(\mathbf{x}-\bar{\mathbf{x}})^T \frac{1}{\sqrt{(2\pi)^N \det \mathbf{Q}}} \exp\left(-\frac{1}{2}(\mathbf{x}-\bar{\mathbf{x}})^T \mathbf{Q}^{-1}(\mathbf{x}-\bar{\mathbf{x}})\right) d\mathbf{x}.$$

A short-hand notation for indicating $\mathbf{x}$ is normally distributed is $\mathbf{x} \sim \mathcal{N}(\bar{\mathbf{x}}, \mathbf{Q})$.

25.1.4 *Discrete-Time White Noise*

Consider a sequence of discrete measurements, $\mathbf{y}_k = \mathbf{x}_k + \mathbf{w}_k$, where $\mathbf{y}_k \in \mathbb{R}^N$ is the measurement at time k, $\mathbf{x}_k \in \mathbb{R}^N$ is the true value of the quantity being measured at k, and $\mathbf{w}_k \in \mathbb{R}^N$ is *noise*. All real systems are corrupted by noise. Noise can represent unknown disturbances, sensor distortion (that is, electrical noise), or model uncertainty. Noise is often assumed be normally distributed with a mean of zero, $\mathbf{w}_k \sim \mathcal{N}(\mathbf{0}, \mathbf{Q}_k)$, and *white*,

$$E[\mathbf{w}_k \mathbf{w}_l^T] = \mathbf{Q}_k \delta_{kl},$$

where $\delta_{kl} = 1$ when $k = l$ and $\delta_{kl} = 0$ when $k \neq l$, which is to say the noise at one instant is completely uncorrelated with the noise at any other instant.

25.1.5 *Simulating Noise*

If we were to simulate a spacecraft mission, to make the simulation realistic we should corrupt our simulated measurements, such as the position of a star, the Earth's magnetic field, and so on, with noise at every time-step within the simulation. As such, given $\mathbf{y}_k = \mathbf{x}_k + \mathbf{w}_k$ where $\mathbf{y}_k \in \mathbb{R}^N$ is a measurement at a particular time-step k, $\mathbf{x}_k \in \mathbb{R}^N$ is the true value of the quantity being measured, and $\mathbf{w}_k \in \mathbb{R}^N$ is zero-mean Gaussian white noise, $\mathbf{w}_k \sim \mathcal{N}(\mathbf{0}, \mathbf{Q}_k)$, how do we actually generate $\mathbf{w}_k$ at each time-step in order to simulate the measurement properly?

Let us assume that the mean and covariance of the noise do not change with time, that is, $\mathbf{w}_k \sim \mathcal{N}(\mathbf{0}, \mathbf{Q})$ where $\mathbf{Q}$ is constant. We start by performing an eigenvalue decomposition on $\mathbf{Q}$:

$$\mathbf{Q} = \mathbf{E}\mathbf{\Lambda}\mathbf{E}^{-1},$$

where the N eigenvectors of $\mathbf{Q}$ are the columns of $\mathbf{E}$, and the eigenvalues of $\mathbf{Q}$, $\lambda_i, i = 1 \ldots N$, are placed along the diagonal of $\mathbf{\Lambda}$. Because $\mathbf{Q}$ is symmetric $\mathbf{E}^{-1} = \mathbf{E}^T$, thus

$$\mathbf{Q} = \mathbf{E}\mathbf{\Lambda}\mathbf{E}^T.$$

Next, we compute $\mathbf{v}_k = [v_{k,1} \ \ldots \ v_{k,N}]^T$ where $v_{k,i} = \sqrt{\lambda_i} r_{k,i}, i = 1 \ldots N$, and $r_{k,i}$ is an independent Gaussian random variable with $\bar{r}_{k,i} = 0$ and $\sigma_{k,i} = 1$. We define $\mathbf{r}_k = [r_{k,1} \ \ldots \ r_{k,N}]^T$ and note that $E[\mathbf{r}_k] = \mathbf{0}$ and $E[\mathbf{r}_k \mathbf{r}_k^T] = \mathbf{1}$, which is to say $\mathbf{r}_k \sim \mathcal{N}(\mathbf{0}, \mathbf{1})$ (which is called a standard normal distribution). Most computer programming languages, such as C, Fortran, or MATLAB®, have built in functions that can generate each $r_{k,i}$. Last, we set $\mathbf{w}_k = \mathbf{E}\mathbf{v}_k$ to generate the noise sample.

To confirm that we have indeed generated a noise sample that has zero mean and covariance $\mathbf{Q}$, we will compute each. We will start with the mean of our sample:

$$E[\mathbf{w}_k] = E[\mathbf{E}\mathbf{v}_k] = \mathbf{E}E[\mathbf{v}_k] = \mathbf{E}\sqrt{\mathbf{\Lambda}}E[\mathbf{r}_k] = \mathbf{0}$$

owing to the fact that $E[\mathbf{r}_k] = \mathbf{0}$. Notice that the matrices $\mathbf{E}$ and $\sqrt{\mathbf{\Lambda}}$ can be brought out of the expectation operator because the expectation operator is a linear operator. Thus, $\mathbf{w}_k$ has zero mean. Next, consider computation of the covariance of $\mathbf{w}_k$:

$$\begin{aligned}
E[\mathbf{w}_k \mathbf{w}_k^T] &= E[\mathbf{E}\mathbf{v}_k \mathbf{v}_k^T \mathbf{E}^T] \\
&= E[\mathbf{E}\sqrt{\mathbf{\Lambda}}\mathbf{r}_k \mathbf{r}_k^T \sqrt{\mathbf{\Lambda}}^T \mathbf{E}^T] \\
&= \mathbf{E}\sqrt{\mathbf{\Lambda}} \underbrace{E[\mathbf{r}_k \mathbf{r}_k^T]}_{=\mathbf{1}} \sqrt{\mathbf{\Lambda}}^T \mathbf{E}^T \\
&= \mathbf{E}\sqrt{\mathbf{\Lambda}}\sqrt{\mathbf{\Lambda}}^T \mathbf{E}^T \\
&= \mathbf{E}\mathbf{\Lambda}\mathbf{E}^T \\
&= \mathbf{Q}.
\end{aligned}$$

Again, the matrices $\mathbf{E}$ and $\sqrt{\mathbf{\Lambda}}$ can be brought out of the expectation operator because of the linear nature of $E[\cdot]$. Thus, the covariance of $\mathbf{w}_k$ is indeed $\mathbf{Q}$.

25.2 Batch Approaches for Spacecraft Attitude Estimation

Consider a spacecraft orbiting the Earth. Embedded within the spacecraft is a reference frame, $\mathcal{F}_b$, while centred in the Earth is an inertial frame, $\mathcal{F}_a$, the ECI frame. The spacecraft is equipped with N sensors, such as a magnetometer, a sun sensor, etc. These sensors provide $k = 1, \ldots, N$ measurements in the body frame, $\vec{s}_k = \vec{\mathcal{F}}_b^T \mathbf{s}_{bk}$. Given knowledge of the orbital position of the spacecraft and various mathematical models (e.g., a model of the Earth's magnetic field), the vectors are assumed to be known in the inertial frame, $\vec{s}_k = \vec{\mathcal{F}}_a^T \mathbf{s}_{ak}$. For convenience, $\vec{s}_k$ vectors are normalized: $\vec{\hat{s}}_k = \vec{\mathcal{F}}_b^T \hat{\mathbf{s}}_{bk} = \vec{\mathcal{F}}_a^T \hat{\mathbf{s}}_{ak}$

Given all these measurements, how do we go about estimating the attitude of the spacecraft? More formally, how do we estimate the rotation matrix $\mathbf{C}_{ba}$ given the $k = 1, \ldots, N$

measurements, $\vec{s}_k = \vec{\mathcal{F}}_b^T \hat{\mathbf{s}}_{bk} = \vec{\mathcal{F}}_a^T \hat{\mathbf{s}}_{ak}$, where $\hat{\mathbf{s}}_{bk} = \mathbf{C}_{ba}\hat{\mathbf{s}}_{ak}$? In this section we will look at two methods that solve this problem by posing a weighted-least squares optimization problem: Davenport's q-Method and the QUEST Algorithm. We will also consider a very simple attitude estimation method that uses only two measurements, the TRIAD algorithm. Although the TRIAD algorithm can be derived within an optimization framework, we will present a deterministic derivation. These attitude estimation methods are called batch methods because the attitude of the spacecraft is estimated given a "batch" of measurements all at one time.

25.2.1 *Wahba's Problem*

Given $\hat{\mathbf{s}}_{bk}$ and $\hat{\mathbf{s}}_{ak}$ our objective is to determine $\mathbf{C}_{ba}$ subject to $\mathbf{C}_{ba}^T\mathbf{C}_{ba} = \mathbf{1}$ such that $\hat{\mathbf{s}}_{bk} = \mathbf{C}_{ba}\hat{\mathbf{s}}_{ak}$. Said another way, we want to minimize $\mathbf{r}_k$ where $\mathbf{r}_k = \hat{\mathbf{s}}_{bk} - \mathbf{C}_{ba}\hat{\mathbf{s}}_{ak}$ subject to $\mathbf{C}_{ba}^T\mathbf{C}_{ba} = \mathbf{1}$. To this end, consider the following constrained optimization problem:

$$\begin{aligned} \text{Minimize} \quad J(\mathbf{C}_{ba}) &= \frac{1}{2}\sum_{k=1}^{N} w_k \mathbf{r}_k^T \mathbf{r}_k \\ &= \frac{1}{2}\sum_{k=1}^{N} w_k \left(\hat{\mathbf{s}}_{bk} - \mathbf{C}_{ba}\hat{\mathbf{s}}_{ak}\right)^T \left(\hat{\mathbf{s}}_{bk} - \mathbf{C}_{ba}\hat{\mathbf{s}}_{ak}\right) \end{aligned} \tag{25.4}$$

such that

$$\mathbf{C}_{ba}^T\mathbf{C}_{ba} = \mathbf{1}$$

where $w_k > 0$ is a weight. The objective function to be minimized is $J(\mathbf{C}_{ba})$, while the constraint that must be satisfied is $\mathbf{C}_{ba}^T\mathbf{C}_{ba} = \mathbf{1}$. This optimization problem was originally posed by Grace Wahba in 1965, and is often called *Wahba's Problem*.

We introduce the weights w_k in order to place more emphasis on certain measurements. For instance, say we have a star tracker and a magnetometer. In general, a magnetometer measurement is very noisy compared to a star tracker measurement. As such, we want to place more emphasis on our star tracker measurement as compared to our magnetometer measurement while solving for $\mathbf{C}_{ba}$ by minimizing (25.4). Often the weights are chosen to be $w_k = 1/\sigma_k^2$ where σ_k^2 is the variance of the measurement. As such, if we are very confident in a measurement (that is, σ_k^2 is very small) then the weight w_k will be very large, placing more emphasis on that measurement in the objective function in (25.4). Conversely, if we are not confident in a measurement (that is, σ_k^2 is very large) then w_k will be very small, emphasizing that measurement less in the objective function in (25.4).

25.2.2 *Davenport's q-Method*

The q-Method, developed by Paul Davenport in a 1968 NASA report, reposes Wahba's problem as a maximization problem with the constraint $\mathbf{C}_{ba}^T\mathbf{C}_{ba} = \mathbf{1}$, and then parameterizes $\mathbf{C}_{ba}$ in terms of quaternions, $\mathbf{q} = \begin{bmatrix}\boldsymbol{\epsilon}^T & \eta\end{bmatrix}^T$ while replacing the constraint $\mathbf{C}_{ba}^T\mathbf{C}_{ba} = \mathbf{1}$ with the unit length quaternion constraint $\mathbf{q}^T\mathbf{q} = 1$.

Consider Wahba's Problem once more, but with the objective function of (25.4) rearranged in the following way:

$$\begin{aligned}
J(\mathbf{C}_{ba}) &= \frac{1}{2}\sum_{k=1}^{N} w_k \left(\hat{\mathbf{s}}_{bk} - \mathbf{C}_{ba}\hat{\mathbf{s}}_{ak}\right)^T \left(\hat{\mathbf{s}}_{bk} - \mathbf{C}_{ba}\hat{\mathbf{s}}_{ak}\right) \\
&= \frac{1}{2}\sum_{k=1}^{N} w_k \left(\hat{\mathbf{s}}_{bk}^T\hat{\mathbf{s}}_{bk} - 2\hat{\mathbf{s}}_{bk}^T\mathbf{C}_{ba}\hat{\mathbf{s}}_{ak} + \hat{\mathbf{s}}_{ak}^T\mathbf{C}_{ba}^T\mathbf{C}_{ba}\hat{\mathbf{s}}_{ak}\right) \\
&= \sum_{k=1}^{N} w_k \left(1 - \hat{\mathbf{s}}_{bk}^T\mathbf{C}_{ba}\hat{\mathbf{s}}_{ak}\right) \\
&= \left(\sum_{k=1}^{N} w_k\right) - \left(\sum_{k=1}^{N} w_k\hat{\mathbf{s}}_{bk}^T\mathbf{C}_{ba}\hat{\mathbf{s}}_{ak}\right) \qquad (25.5)
\end{aligned}$$

where $\hat{\mathbf{s}}_{bk}^T\hat{\mathbf{s}}_{bk} = 1$, $\mathbf{C}_{ba}^T\mathbf{C}_{ba} = \mathbf{1}$, and $\hat{\mathbf{s}}_{ak}^T\hat{\mathbf{s}}_{ak} = 1$ have been used to simplify. The summation of all the weights is a constant number. Hence, minimizing the objective function $J(\mathbf{C}_{ba})$ is equivalent to maximizing the objective function

$$\hat{J}(\mathbf{C}_{ba}) = \sum_{k=1}^{N} w_k\hat{\mathbf{s}}_{bk}^T\mathbf{C}_{ba}\hat{\mathbf{s}}_{ak}. \qquad (25.6)$$

Our problem is to now find $\mathbf{C}_{ba}$ that maximizes $\hat{J}(\mathbf{C}_{ba})$ given in (25.6) subject to the constraint $\mathbf{C}_{ba}^T\mathbf{C}_{ba} = \mathbf{1}$. To solve this problem we will first simplify (25.6) using the trace operation. The trace of a matrix $\mathbf{A} \in \mathbb{R}^{n\times n}$ is

$$\mathrm{tr}\left[\mathbf{A}\right] = \mathrm{tr}\left[\mathbf{A}^T\right] = \sum_{i=1}^{n} a_{ii} = \sum_{i=1}^{n} \lambda_i$$

where $\lambda_i, i = 1\ldots n$ are the eigenvalues of $\mathbf{A}$. Also, if $\mathbf{A} = \mathbf{BCD}$ where $\mathbf{B} \in \mathbb{R}^{n\times m}, \mathbf{C} \in \mathbb{R}^{m\times p}$, and $\mathbf{D} \in \mathbb{R}^{p\times n}$ then

$$\mathrm{tr}\left[\mathbf{A}\right] = \mathrm{tr}\left[\mathbf{BCD}\right] = \mathrm{tr}\left[\mathbf{CDB}\right] = \mathrm{tr}\left[\mathbf{DBC}\right].$$

Using the trace operation we can write the objective function in (25.6) as

$$\begin{aligned}
\hat{J}(\mathbf{C}_{ba}) &= \mathrm{tr}\left[\sum_{k=1}^{N} w_k\hat{\mathbf{s}}_{bk}^T\mathbf{C}_{ba}\hat{\mathbf{s}}_{ak}\right] = \mathrm{tr}\left[\sum_{k=1}^{N} w_k\mathbf{C}_{ba}\hat{\mathbf{s}}_{ak}\hat{\mathbf{s}}_{bk}^T\right] \\
&= \mathrm{tr}\left[\mathbf{C}_{ba}\underbrace{\left(\sum_{k=1}^{N} w_k\hat{\mathbf{s}}_{ak}\hat{\mathbf{s}}_{bk}^T\right)}_{\mathbf{B}^T}\right]. \qquad (25.7)
\end{aligned}$$

Rather than attempting to maximize (25.7) subject to $\mathbf{C}_{ba}^T\mathbf{C}_{ba} = \mathbf{1}$, quaternions will be used to parameterize the rotation matrix, then the quaternion that maximizes (25.7) subject to the constraint $\mathbf{q}^T\mathbf{q} = 1$ will be found. To this end, note that equation (1.33) in Section 1.3.4 may be rewritten as

$$\mathbf{C}_{ba} = (\eta^2 - \boldsymbol{\epsilon}^T\boldsymbol{\epsilon})\mathbf{1} + 2\boldsymbol{\epsilon}\boldsymbol{\epsilon}^T - 2\eta\boldsymbol{\epsilon}^\times. \tag{25.8}$$

Substituting (25.8) into the objective function (25.7) and simplifying gives

$$\begin{aligned}\hat{J}(\mathbf{q}) &= \mathrm{tr}\left[\eta^2\mathbf{B}^T - \boldsymbol{\epsilon}^T\boldsymbol{\epsilon}\mathbf{B}^T + 2\boldsymbol{\epsilon}\boldsymbol{\epsilon}^T\mathbf{B}^T - 2\eta\boldsymbol{\epsilon}^\times\mathbf{B}^T\right] \\ &= \eta^2\mathrm{tr}\left[\mathbf{B}^T\right] - \boldsymbol{\epsilon}^T\boldsymbol{\epsilon}\mathrm{tr}\left[\mathbf{B}^T\right] + 2\boldsymbol{\epsilon}^T\mathbf{B}^T\boldsymbol{\epsilon} - 2\eta\mathrm{tr}\left[\boldsymbol{\epsilon}^\times\mathbf{B}^T\right].\end{aligned}$$

Letting $k_{22} = \mathrm{tr}\left[\mathbf{B}^T\right] = \mathrm{tr}\left[\mathbf{B}\right]$ and noting that $2\boldsymbol{\epsilon}^T\mathbf{B}^T\boldsymbol{\epsilon} = \boldsymbol{\epsilon}^T\left(\mathbf{B} + \mathbf{B}^T\right)\boldsymbol{\epsilon}$ we have

$$\hat{J}(\mathbf{q}) = \boldsymbol{\epsilon}^T\left(\mathbf{B} + \mathbf{B}^T - k_{22}\mathbf{1}\right)\boldsymbol{\epsilon} - 2\eta\mathrm{tr}\left[\boldsymbol{\epsilon}^\times\mathbf{B}^T\right] + \eta^2 k_{22}.$$

The term $-2\eta\mathrm{tr}\left[\boldsymbol{\epsilon}^\times\mathbf{B}^T\right]$ can be simplified in the following way:

$$\begin{aligned}-2\eta\mathrm{tr}\left[\boldsymbol{\epsilon}^\times\mathbf{B}^T\right] &= -2\eta\mathrm{tr}\left[\begin{bmatrix} 0 & -\epsilon_3 & \epsilon_2 \\ \epsilon_3 & 0 & -\epsilon_1 \\ -\epsilon_2 & \epsilon_1 & 0 \end{bmatrix}\begin{bmatrix} B_{11} & B_{12} & B_{13} \\ B_{21} & B_{22} & B_{23} \\ B_{31} & B_{32} & B_{33} \end{bmatrix}^T\right] \\ &= 2\eta\left[\epsilon_1(B_{23} - B_{32}) + \epsilon_2(B_{31} - B_{13}) + \epsilon_3(B_{12} - B_{21})\right].\end{aligned}$$

Letting $\mathbf{k}_{12} = [(B_{23} - B_{32}) \;\; (B_{31} - B_{13}) \;\; (B_{12} - B_{21})]^T$ we can write $\hat{J}(\mathbf{q})$ as

$$\begin{aligned}\hat{J}(\mathbf{q}) &= \boldsymbol{\epsilon}^T\left(\mathbf{B} + \mathbf{B}^T - k_{22}\mathbf{1}\right)\boldsymbol{\epsilon} + \eta\boldsymbol{\epsilon}^T\mathbf{k}_{12} + \eta\mathbf{k}_{12}^T\boldsymbol{\epsilon} + \eta^2 k_{22} \\ &= \begin{bmatrix}\boldsymbol{\epsilon}^T & \eta\end{bmatrix}\underbrace{\begin{bmatrix}\mathbf{K}_{11} & \mathbf{k}_{12} \\ \mathbf{k}_{12}^T & k_{22}\end{bmatrix}}_{\mathbf{K}}\begin{bmatrix}\boldsymbol{\epsilon} \\ \eta\end{bmatrix} \\ &= \mathbf{q}^T\mathbf{K}\mathbf{q}\end{aligned} \tag{25.9}$$

where $\mathbf{K}_{11} = \mathbf{B} + \mathbf{B}^T - k_{22}\mathbf{1} = \mathbf{K}_{11}^T$. Our problem is to now find $\mathbf{q}$ that maximizes (25.9) subject to the constraint $\mathbf{q}^T\mathbf{q} = 1$. To maximize (25.9) subject to the unit length quaternion constraint we will use a Lagrange multiplier to define a new objective function that consists of the objective function $\hat{J}(\mathbf{q})$ augmented with the constraint $1 - \mathbf{q}^T\mathbf{q} = 0$:

$$\hat{\hat{J}}(\mathbf{q}, \lambda) = \mathbf{q}^T\mathbf{K}\mathbf{q} + \lambda\left(1 - \mathbf{q}^T\mathbf{q}\right). \tag{25.10}$$

To find the optimal quaternion, we take the derivative $\hat{J}(\mathbf{q}, \lambda)$ with respect to $\mathbf{q}$ and λ and set the results to zero:

$$\frac{\partial \hat{J}(\mathbf{q}, \lambda)}{\partial \mathbf{q}} = \mathbf{q}^T \mathbf{K} - \lambda \mathbf{q}^T = \mathbf{0},$$

which can be written as

$$\mathbf{K}\mathbf{q} = \lambda \mathbf{q}. \tag{25.11}$$

Also, we have

$$\frac{\partial \hat{J}(\mathbf{q}, \lambda)}{\partial \lambda} = 1 - \mathbf{q}^T \mathbf{q} = 0,$$

which is the quaternion constraint. Hence, maximizing (25.10), which is equivalent to maximizing (25.9) subject to the constraint $\mathbf{q}^T \mathbf{q} = 1$, reduces to the eigenvalue problem shown in equation (25.11). However, there are four eigenvalues and four eigenvectors associated with the eigenproblem (25.11) owing to the fact $\mathbf{K}$ is a 4×4 matrix. Which eigenpair maximizes (25.10)? To find the $\mathbf{q}$ that maximizes the objective function $\hat{J}(\mathbf{q})$ we must find the largest eigenvalue and its corresponding eigenvector associated with (25.11). To see that the largest eigenvalue $\bar{\lambda}$ and corresponding eigenvector (i.e., quaternion) $\bar{\mathbf{q}}$ maximize (25.9), consider any eigenvalue λ with corresponding unit eigenvector $\mathbf{q}$ satisfying (25.11). The associated value of (25.9) is

$$\hat{J}(\mathbf{q}) = \mathbf{q}^T \underbrace{\mathbf{K}\mathbf{q}}_{=\lambda \mathbf{q}} = \lambda \underbrace{\mathbf{q}^T \mathbf{q}}_{=1} = \lambda, \tag{25.12}$$

which is clearly maximized when $\lambda = \bar{\lambda}$ and $\mathbf{q} = \bar{\mathbf{q}}$. Thus, $\bar{\mathbf{q}}$ is the optimal quaternion that along with (25.8) provides the optimal estimate of $\mathbf{C}_{ba}$.

25.2.3 The QUEST Algorithm

Davenport's q-Method reduces Wahba's problem to an eigenvalue problem where optimal attitude estimate corresponds to the eigenvector (that is, the quaternion) associated with the optimal eigenvalue. Solving an eigenvalue problem on a desktop computer is trivial, but solving the same eigenvalue problem on a spacecraft flight computer with limited memory may not be. To circumvent the need to solve an eigenvalue problem Malcolm Shuster developed the QUEST algorithm, which stands for "QUaternion ESTimator", in 1978.

Consider the eigenvalue problem of (25.11) once more:

$$\begin{bmatrix} \mathbf{K}_{11} & \mathbf{k}_{12} \\ \mathbf{k}_{12}^T & k_{22} \end{bmatrix} \begin{bmatrix} \boldsymbol{\epsilon} \\ \eta \end{bmatrix} = \begin{bmatrix} \mathbf{K}_{11}\boldsymbol{\epsilon} + \mathbf{k}_{12}\eta \\ \mathbf{k}_{12}^T \boldsymbol{\epsilon} + k_{22}\eta \end{bmatrix} = \lambda \begin{bmatrix} \boldsymbol{\epsilon} \\ \eta \end{bmatrix}$$

which can be written as two separate equations,

$$\mathbf{K}_{11}\boldsymbol{\epsilon} + \mathbf{k}_{12}\eta = \lambda\boldsymbol{\epsilon}, \tag{25.13}$$

$$\mathbf{k}_{12}^T\boldsymbol{\epsilon} + k_{22}\eta = \lambda\eta. \tag{25.14}$$

Recalling that $\mathbf{K}_{11} = \mathbf{S} - k_{22}\mathbf{1}$ where $\mathbf{S} = \mathbf{B} + \mathbf{B}^T$ we can write (25.13) as

$$((\lambda + k_{22})\mathbf{1} - \mathbf{S}))\boldsymbol{\epsilon} = \mathbf{k}_{12}\eta.$$

Dividing the above equation by η gives

$$((\lambda + k_{22})\mathbf{1} - \mathbf{S}))\,\mathbf{p} = \mathbf{k}_{12}$$

where $\mathbf{p} = \boldsymbol{\epsilon}/\eta = \mathbf{a}\tan(\phi/2)$ is the Gibbs vector. By taking the inverse of $((\lambda + k_{22})\mathbf{1} - \mathbf{S}))$ we can solve for $\mathbf{p}$:

$$\mathbf{p} = ((\lambda + k_{22})\mathbf{1} - \mathbf{S}))^{-1}\,\mathbf{k}_{12}. \tag{25.15}$$

The associated quaternion can be computed from the Gibbs vector in the following way:

$$\mathbf{q} = \frac{1}{\sqrt{1 + \mathbf{p}^T\mathbf{p}}}\begin{bmatrix}\mathbf{p}\\ 1\end{bmatrix}.$$

In solving for $\mathbf{q}$ we have taken the inverse of $((\lambda + k_{22})\mathbf{1} - \mathbf{S}))$. Although this matrix is only a 3×3 matrix, it is best if we can avoid taking its inverse. It can be shown that

$$((\lambda + k_{22})\mathbf{1} - \mathbf{S}))^{-1} = \frac{1}{\gamma}\left(\alpha\mathbf{1} + \beta\mathbf{S} + \mathbf{S}^2\right) \tag{25.16}$$

where

$$\alpha = \lambda^2 - k_{22}^2 + \mathrm{tr}\left[\mathrm{adj}(\mathbf{S})\right], \qquad \beta = \lambda - k_{22}, \qquad \gamma = (\lambda + k_{22})\alpha - \det\mathbf{S},$$

and, as before, $k_{22} = \mathrm{tr}[\mathbf{B}] = \frac{1}{2}\mathrm{tr}[\mathbf{S}]$. Note that adj($\mathbf{S}$) is the classical adjoint of the matrix $\mathbf{S}$. By substituting (25.16) into (25.15) and using the best estimate of λ we have

$$\mathbf{x} = \left(\alpha\mathbf{1} + \beta\mathbf{S} + \mathbf{S}^2\right)\mathbf{k}_{12}$$

where $\mathbf{p} = \mathbf{x}/\gamma$ and

$$\mathbf{q} = \frac{1}{\sqrt{\gamma^2 + \mathbf{x}^T\mathbf{x}}}\begin{bmatrix}\mathbf{x}\\ \gamma\end{bmatrix}, \tag{25.17}$$

the quaternion associated with the spacecraft attitude.

The preceding calculations provide $\mathbf{q}$ given λ. The estimate of the attitude will not be the correct estimate unless λ is equal to, or at least very close to, the optimal eigenvalue associated with the eigenproblem of (25.11). Next we will consider how to find $\bar{\lambda}$ without actually solving the eigenvalue problem of (25.11). Having found $\bar{\lambda}$, the optimal estimate of $\mathbf{q}$, $\bar{\mathbf{q}}$, can be found from (25.17). To start, by rearranging (25.14) and dividing by η we have

$$\mathbf{k}_{12}^T\mathbf{p} = \lambda - k_{22}. \tag{25.18}$$

Next, by premultiplying (25.15) by $\mathbf{k}_{12}^T$ and using (25.16) we have

$$\mathbf{k}_{12}^T\mathbf{p} = \frac{1}{\gamma}\mathbf{k}_{12}^T\left(\alpha\mathbf{1} + \beta\mathbf{S} + \mathbf{S}^2\right)\mathbf{k}_{12}. \tag{25.19}$$

Together (25.18) and (25.19) yield

$$\lambda = k_{22} + \frac{1}{\gamma}\mathbf{k}_{12}^T\left(\alpha\mathbf{1} + \beta\mathbf{S} + \mathbf{S}^2\right)\mathbf{k}_{12}.$$

After some algebra it can be shown that

$$\lambda^4 - (a+b)\lambda^2 - c\lambda + (ab + ck_{22} - d) = 0 \tag{25.20}$$

where

$$a = k_{22}^2 - \mathrm{tr}\left[\mathrm{adj}(\mathbf{S})\right],\ b = k_{22}^2 + \mathbf{k}_{12}^T\mathbf{k}_{12},\ c = \det[\mathbf{S}] + \mathbf{k}_{12}^T\mathbf{S}\mathbf{k}_{12},\ d = \mathbf{k}_{12}^T\mathbf{S}^2\mathbf{k}_{12}.$$

The roots of (25.20) can be found by using Newton's method: let $f(\lambda_j) = \lambda_j^4 - (a+b)\lambda_j^2 - c\lambda_j + (ab + ck_{22} - d)$, then with an initial guess λ_1, solve for λ_j in an iterative manner via

$$\lambda_{j+1} = \lambda_j - \frac{f(\lambda_j)}{f'(\lambda_j)}$$

until $|\lambda_{j+1} - \lambda_j| < \epsilon_{\mathrm{tol}}$ where ϵ_{tol} is a tolerance. Recall that we are interested in the root that corresponds to the optimal estimate attitude estimate, $\bar{\lambda}$. In order to find $\bar{\lambda}$, Newton's method must be given an initial guess that is close to $\bar{\lambda}$. It turns out that the optimal eigenvalue associated with optimal attitude estimate is approximately equal to the sum of the weights w_k. This can be shown in the following way. To start, assume we have an optimal estimate of the attitude that can be expressed in terms of a quaternion, $\bar{\mathbf{q}}$, or a rotation matrix, $\bar{\mathbf{C}}_{ba}$. When the attitude estimate is very close to the true estimate $\hat{\mathbf{s}}_{bk} \approx \bar{\mathbf{C}}_{ba}\hat{\mathbf{s}}_{ak}$. Therefore, our original objective function given in (25.5) is approximately

$$J(\bar{\mathbf{C}}_{ba}) \approx 0.$$

Recognizing the final term in (25.5) as the alternative objective function $\hat{J}(\mathbf{C}_{ba})$ from (25.6), we have

$$\hat{J}(\mathbf{C}_{ba}) \approx \left(\sum_{k=1}^{N} w_k\right).$$

Since (25.6) and (25.9) are equivalent, we have using (25.12)

$$\bar{\lambda} \approx \sum_{k=1}^{N} w_k.$$

Therefore, by setting $\lambda_1 = \sum_{k=1}^{N} w_k$ when solving for λ we can expect Newton's method to converges to a value that is very close to $\bar{\lambda}$.

25.2.4 The TRIAD Algorithm

The TRIAD algorithm is a simple attitude estimation method that uses only two (nonparallel) vector measurements. The TRIAD algorithm was developed before the q-Method or QUEST, and was originally known as the Algebraic Method. Because the TRIAD algorithm is simple and computationally less demanding than the q-Method and QUEST, it is worth mentioning.

Consider a spacecraft equipped with two sensors, for example, a sun sensor that provides $\vec{\hat{\mathbf{s}}}$ and a magnetometer that provides $\vec{\hat{\mathbf{b}}}$. (The vector measurements are normalized.) As with the q-Method and QUEST, it is assumed that the components of $\vec{\hat{\mathbf{s}}}$ and $\vec{\hat{\mathbf{b}}}$ in both $\mathcal{F}_b$ and $\vec{\mathcal{F}}_a$, the body frame and an inertial frame, are known. Additionally, it is assumed throughout that $\vec{\hat{\mathbf{s}}}$ and $\vec{\hat{\mathbf{b}}}$ are not parallel. The rotation matrix $\mathbf{C}_{ba}$ defines the relationship between components of $\vec{\hat{\mathbf{s}}}$ and $\vec{\hat{\mathbf{b}}}$ in $\mathcal{F}_b$ and $\mathcal{F}_a$: $\hat{\mathbf{s}}_b = \mathbf{C}_{ba}\hat{\mathbf{s}}_a$, $\hat{\mathbf{b}}_b = \mathbf{C}_{ba}\hat{\mathbf{b}}_a$. The TRIAD algorithm is composed of three steps.

Step 1: Create an intermediate reference frame, $\vec{\mathcal{F}}_t^T = [\vec{\mathbf{x}}_t \ \vec{\mathbf{y}}_t \ \vec{\mathbf{z}}_t]$, where

$$\vec{\mathbf{x}}_t = \vec{\hat{\mathbf{s}}}, \quad \vec{\mathbf{y}}_t = \frac{\vec{\hat{\mathbf{s}}} \times \vec{\hat{\mathbf{b}}}}{|\vec{\hat{\mathbf{s}}} \times \vec{\hat{\mathbf{b}}}|}, \quad \vec{\mathbf{z}}_t = \vec{\mathbf{x}}_t \times \vec{\mathbf{y}}_t.$$

Step 2: Express the components of $\mathcal{F}_t$ in both $\mathcal{F}_a$ and $\mathcal{F}_b$, that is

$$\vec{\mathbf{x}}_t = \vec{\mathcal{F}}_a^T \mathbf{x}_{t,a} = \vec{\mathcal{F}}_a^T \hat{\mathbf{s}}_a, \quad \vec{\mathbf{y}}_t = \vec{\mathcal{F}}_a^T \mathbf{y}_{t,a} = \vec{\mathcal{F}}_a^T \frac{\hat{\mathbf{s}}_a^{\times} \hat{\mathbf{b}}_a}{|\hat{\mathbf{s}}_a^{\times} \hat{\mathbf{b}}_a|}, \quad \vec{\mathbf{z}}_t = \vec{\mathcal{F}}_a^T \mathbf{z}_{t,a} = \vec{\mathcal{F}}_a^T \mathbf{x}_{t,a}^{\times} \mathbf{y}_{t,a}$$

and

$$\vec{\mathbf{x}}_t = \vec{\mathcal{F}}_b^T \mathbf{x}_{t,b} = \vec{\mathcal{F}}_b^T \hat{\mathbf{s}}_b, \quad \vec{\mathbf{y}}_t = \vec{\mathcal{F}}_b^T \mathbf{y}_{t,b} = \vec{\mathcal{F}}_b^T \frac{\hat{\mathbf{s}}_b^{\times} \hat{\mathbf{b}}_b}{|\hat{\mathbf{s}}_b^{\times} \hat{\mathbf{b}}_b|}, \quad \vec{\mathbf{z}}_t = \vec{\mathcal{F}}_b^T \mathbf{z}_{t,b} = \vec{\mathcal{F}}_b^T \mathbf{x}_{t,b}^{\times} \mathbf{y}_{t,b}.$$

Step 3: Solve for $\mathbf{C}_{ba}$. To do so, recall from Section 1.3 that

$$\mathbf{C}_{at} = \begin{bmatrix}\mathbf{x}_{t,a} & \mathbf{y}_{t,a} & \mathbf{z}_{t,a}\end{bmatrix}, \quad \mathbf{C}_{bt} = \begin{bmatrix}\mathbf{x}_{t,b} & \mathbf{y}_{t,b} & \mathbf{z}_{t,b}\end{bmatrix}.$$

Furthermore, recall from Section 1.3 that

$$\mathbf{C}_{ba} = \mathbf{C}_{bt}\mathbf{C}_{ta},$$

and that $\mathbf{C}_{ta} = \mathbf{C}_{at}^T$. Therefore, it follows that

$$\mathbf{C}_{ba} = \mathbf{C}_{bt}\mathbf{C}_{at}^T.$$

Notice that to compute $\mathbf{C}_{ba}$ we have simply arranged the components of various vectors in a matrix form, and then multiplied two matrices together.

25.2.5 Example

We will now compute the attitude of a spacecraft given vector measurements using the q-Method, QUEST, and the TRIAD algorithm. Say the true rotation matrix describing a spacecraft's attitude at a particular instant is $\mathbf{C}_{ba}$ and we have $k = 1 \ldots N$ vector measurements provided by star trackers, sun sensors, a magnetometer, and other sensors. The measurements are not perfect, and we assume that are they are characterized by

$$\hat{\mathbf{s}}_{bk} = \mathbf{C}_{ba}\hat{\mathbf{s}}_{ak} + \mathbf{v}_k \tag{25.21}$$

where $\mathbf{v}_k$ is noise that is constrained to satisfy $\mathbf{v}_k^T\mathbf{C}_{ba}\hat{\mathbf{s}}_{ak} = 0$. The constraint $\mathbf{v}_k^T\mathbf{C}_{ba}\hat{\mathbf{s}}_{ak} = 0$ ensures the noise is perpendicular to the $\mathbf{C}_{ba}\hat{\mathbf{s}}_{ak}$, which ensures that the lengths of $\hat{\mathbf{s}}_{bk}$ and $\hat{\mathbf{s}}_{ak}$ are the same to first order. This measurement model is often referred to as the QUEST measurement model. The noise is assumed to be Gaussian with

$$E[\mathbf{v}_k] = \mathbf{0}, \quad E[\mathbf{v}_k\mathbf{v}_k^T] = \sigma_k^2\left[\mathbf{1} - (\mathbf{C}_{ba}\hat{\mathbf{s}}_{ak})(\mathbf{C}_{ba}\hat{\mathbf{s}}_{ak})^T\right].$$

On a real mission we would be given $\hat{\mathbf{s}}_{ak}$ (based on various models, such as a star catalog, a geomagnetic field model, etc.), the measurements $\hat{\mathbf{s}}_{bk}$, the statistics σ_k, and then we would compute $\mathbf{C}_{ba}$ using the q-Method, QUEST, or the TRIAD algorithm. In this example we are given $\hat{\mathbf{s}}_{ak}$ and σ_k, and using the QUEST measurement model we will generate the measurements $\hat{\mathbf{s}}_{bk}$. To do so, we begin by drawing noise samples, $\mathbf{v}_k'$, constrain the samples to be perpendicular to $\mathbf{C}_{ba}\hat{\mathbf{s}}_{ak}$ via $\mathbf{v}_k = -(\mathbf{C}_{ba}\hat{\mathbf{s}}_{ak})^\times(\mathbf{C}_{ba}\hat{\mathbf{s}}_{ak})^\times\mathbf{v}_k'$, and then generate $\hat{\mathbf{s}}_{bk}$ via equation (25.21) which is further normalized. For this numerical example we are given

$$\mathbf{C}_{ba} = \mathbf{C}_z(60°)\mathbf{C}_y(-30°)\mathbf{C}_x(45°) = \begin{bmatrix} 0.4330 & 0.4356 & 0.7891 \\ -0.7500 & 0.6597 & 0.0474 \\ -0.5000 & -0.6124 & 0.6124 \end{bmatrix},$$

$$\mathbf{s}_{a1} = \begin{bmatrix} 0 \\ 1 \\ 2 \end{bmatrix}, \quad \mathbf{s}_{a2} = \begin{bmatrix} 1 \\ 3 \\ 0 \end{bmatrix}, \quad \mathbf{s}_{a3} = \begin{bmatrix} -5 \\ 0 \\ 1 \end{bmatrix}, \quad \mathbf{s}_{a4} = \begin{bmatrix} 1 \\ -1 \\ 4 \end{bmatrix}, \quad \mathbf{s}_{a5} = \begin{bmatrix} 1 \\ 1 \\ 1 \end{bmatrix},$$

which are all normalized to obtain $\hat{\mathbf{s}}_{ak}$. For the measurement noise, we have the statistics

$$\sigma_1 = 0.0100, \quad \sigma_2 = 0.0325, \quad \sigma_3 = 0.0550, \quad \sigma_4 = 0.0775, \quad \sigma_5 = 0.1000.$$

Correspondingly, we obtain the measurements

$$\hat{\mathbf{s}}_{b1} = \begin{bmatrix} 0.9082 \\ 0.3185 \\ 0.2715 \end{bmatrix}, \quad \hat{\mathbf{s}}_{b2} = \begin{bmatrix} 0.5670 \\ 0.3732 \\ -0.7343 \end{bmatrix}, \quad \hat{\mathbf{s}}_{b3} = \begin{bmatrix} -0.2821 \\ 0.7163 \\ 0.6382 \end{bmatrix},$$

$$\hat{\mathbf{s}}_{b4} = \begin{bmatrix} 0.7510 \\ -0.3303 \\ 0.5718 \end{bmatrix}, \quad \hat{\mathbf{s}}_{b5} = \begin{bmatrix} 0.9261 \\ -0.2053 \\ -0.3166 \end{bmatrix},$$

We take as weights $w_k = 1/\sigma_k^2$. For the TRIAD algorithm, we use only the first two measurements ($k = 1, 2$). The q-Method and QUEST provide identical results, with the estimated rotation matrix being

$$\bar{\mathbf{C}}_{ba} = \begin{bmatrix} 0.4153 & 0.4472 & 0.7921 \\ -0.7562 & 0.6537 & 0.0274 \\ -0.5056 & -0.6104 & 0.6097 \end{bmatrix}.$$

Using the TRIAD algorithm, the estimated rotation matrix is

$$\bar{\mathbf{C}}_{ba} = \begin{bmatrix} 0.4156 & 0.4504 & 0.7902 \\ -0.7630 & 0.6456 & 0.0333 \\ -0.4952 & -0.6167 & 0.6119 \end{bmatrix}$$

is the estimated attitude. The performance index for each estimate of the rotation matrix, $J(\bar{\mathbf{C}}_{ba})$, is tabulated in Table 25.1. We can also compare the performance of each method by computing the rotation angle, ϕ_e associated with rotation matrix error. The relationship between the estimated attitude, $\bar{\mathbf{C}}_{ba}$, the true attitude, $\mathbf{C}_{ba}$, and the attitude error is $\bar{\mathbf{C}}_{ba} = \mathbf{C}_{ba,e}\mathbf{C}_{ba}$, and therefore $\mathbf{C}_{ba,e} = \bar{\mathbf{C}}_{ba}\mathbf{C}_{ba}^T$ where $\mathbf{C}_{ba,e}$ is the attitude error. The rotation angle associated with $\mathbf{C}_{ba,e}$ is $\cos\phi_e = \frac{1}{2}(\text{tr}[\mathbf{C}_{ba,e}] - \mathbf{1})$. The rotation angle associated with the error of each method is also presented in Table 25.1.

Table 25.1 Performance and error of the attitude estimate provided by the q-Method, QUEST, and the TRIAD algorithm

	TRIAD	q-Method	QUEST
$J(\bar{\mathbf{C}}_{ba})$	4.2449	4.0333	4.0333
ϕ_e	1.3622°	1.2644°	1.2644°

25.3 The Kalman Filter

In the previous section we considered the problem of estimating the attitude of a spacecraft given many vector measurements. Unfortunately, as new measurements become available the previous attitude estimate is not used to compute the new attitude estimate. One would think that the previous attitude estimate could be quite useful in determining the new attitude estimate. Additionally, only vector measurements were used; it would ideal if we could use rate measurements or even knowledge of the spacecraft kinematics and dynamics to somehow help estimate the spacecraft attitude. In this section we will look at the Kalman filter, a sequential estimator that has a predictor-corrector structure. First the system state (which is related to the spacecraft attitude) is predicted given the previous estimate of the state along with a model of the system kinematics and dynamics. Then, the state estimate is corrected using the measurements that become available. Additionally, using the Kalman filter we can directly compute the covariance associated with the state estimate, providing us with a measure of uncertainty (or confidence) in the state estimate.

25.3.1 The Discrete-Time Kalman Filter

Consider the following linear discrete-time system:

$$\mathbf{x}_k = \mathbf{F}_{k-1}\mathbf{x}_{k-1} + \mathbf{G}_{k-1}\mathbf{u}_{k-1} + \mathbf{L}_{k-1}\mathbf{w}_{k-1} \tag{25.22}$$

$$\mathbf{y}_k = \mathbf{H}_k\mathbf{x}_k + \mathbf{M}_k\mathbf{v}_k. \tag{25.23}$$

Equation (25.22) is called the process or plant model, where the n dimensional system state is $\mathbf{x}_k$, $\mathbf{u}_k$ is the known control input, and $\mathbf{w}_k \sim \mathcal{N}(\mathbf{0}, \mathbf{Q}_k)$ is the process noise. The measurement model in (25.23) describes the relationship between a measurement, $\mathbf{y}_k$, and the state, $\mathbf{x}_k$. The measurement noise is $\mathbf{v}_k \sim \mathcal{N}(\mathbf{0}, \mathbf{R}_k)$. Not only are the process and measurement noise zero mean and Gaussian, but both $\mathbf{w}_k$ and $\mathbf{v}_k$ are white and uncorrelated with each other meaning $E[\mathbf{v}_k\mathbf{w}_k^T] = \mathbf{0}$ for all k.

At this point, we must discuss the concept of observability. A discrete-time system is *observable* if for any initial state $\mathbf{x}_0$ and some final time k the initial state $\mathbf{x}_0$ can be uniquely determined from knowledge of the input $\mathbf{u}_j$ and the output $\mathbf{y}_j$ for all $j \in [0, k]$. If a system is observable then not only can the initial state be determined, but all states between the initial and final time can be determined from the measurements and knowledge of the control input. Note that the definition of observability assumes no process or measurement noise. If the system is time-invariant (so that $\mathbf{F}_k = \mathbf{F}$, $\mathbf{G}_k = \mathbf{G}$, $\mathbf{H}_k = \mathbf{H}$ do not depend on time), then the system is observable if and only if

$$\mathcal{O} = \begin{bmatrix} \mathbf{H}^T & (\mathbf{HF})^T & \cdots & (\mathbf{HF}^{n-1})^T \end{bmatrix}^T$$

has rank n.

Before deriving the Kalman filter, let us define various quantities. The Kalman filter has a predictor-corrector structure. Let $\hat{\mathbf{x}}_k^-$ be the estimate of the state before a measurement has been processed; this is the predicted state estimate, or the *a priori* state estimate. The error between the true state and the predicted state estimate is $\mathbf{e}_k^- = \mathbf{x}_k - \hat{\mathbf{x}}_k^-$. Let $\hat{\mathbf{x}}_k$ be the estimate of the state after a measurement has been processed; this is the corrected state estimate, or the *a posteriori* state estimate. Likewise, the error between the true state and the corrected state estimate is $\mathbf{e}_k = \mathbf{x}_k - \hat{\mathbf{x}}_k$. We will assume that the process and measurement noise are also uncorrelated with both the a priori and a posteriori state estimates, which means

$$E[\mathbf{e}_k^- \mathbf{w}_k^T] = \mathbf{0}, \quad E[\mathbf{e}_k \mathbf{w}_k^T] = \mathbf{0}, \quad E[\mathbf{e}_k^- \mathbf{v}_k^T] = \mathbf{0}, \quad E[\mathbf{e}_k \mathbf{v}_k^T] = \mathbf{0},$$

for any k.

Using the definitions of $\mathbf{e}_k^-$ and $\mathbf{e}_k$ we can write out the a priori covariance, $\mathbf{P}_k^-$, and the a posteriori covariance, $\mathbf{P}_k$:

$$\mathbf{P}_k^- = E\left[\mathbf{e}_k^- \mathbf{e}_k^{-T}\right] = E\left[(\mathbf{x}_k - \hat{\mathbf{x}}_k^-)(\mathbf{x}_k - \hat{\mathbf{x}}_k^-)^T\right],$$

$$\mathbf{P}_k = E\left[\mathbf{e}_k \mathbf{e}_k^T\right] = E\left[(\mathbf{x}_k - \hat{\mathbf{x}}_k)(\mathbf{x}_k - \hat{\mathbf{x}}_k)^T\right].$$

These covariance matrices represent the confidence (or uncertainty) of our a priori and a posteriori estimates, $\hat{\mathbf{x}}_k^-$ and $\hat{\mathbf{x}}_k$. For instance, if $\mathbf{P}_k$ is "small" we can be confident $\hat{\mathbf{x}}_k$ is a good estimate of the true state $\mathbf{x}_k$. If $\mathbf{P}_k$ is "large", we cannot be confident that our estimate closely resembles the true state.

25.3.1.1 The Prediction Step

As mentioned previously, the Kalman filter has a predictor-corrector structure. We first predict $\hat{\mathbf{x}}_k^-$, the a priori state estimate, and then correct the a priori estimate using a measurement giving us $\hat{\mathbf{x}}_k$, the a posteriori state estimate. The prediction step is nothing but a prediction of the state given the process model in (25.22) and the best previous estimate $\hat{\mathbf{x}}_{k-1}$:

$$\hat{\mathbf{x}}_k^- = \mathbf{F}_{k-1}\hat{\mathbf{x}}_{k-1} + \mathbf{G}_{k-1}\mathbf{u}_{k-1}.$$

Notice that the control input, $\mathbf{u}_{k-1}$, is known. Also notice that there is no noise term in the prediction step because we don't know, and never will know, the noise. Given this prediction and the process model in (25.22), the a priori estimation error becomes

$$\begin{aligned}
\mathbf{e}_k^- &= \mathbf{x}_k - \hat{\mathbf{x}}_k^- \\
&= \mathbf{F}_{k-1}(\mathbf{x}_{k-1} - \hat{\mathbf{x}}_{k-1}^+) + \mathbf{L}_{k-1}\mathbf{w}_{k-1} \\
&= \mathbf{F}_{k-1}\mathbf{e}_{k-1} + \mathbf{L}_{k-1}\mathbf{w}_{k-1}.
\end{aligned}$$

This expression for $\mathbf{e}_k^-$ can be used to find an expression for $\mathbf{P}_k^-$:

$$\begin{aligned}
\mathbf{P}_k^- &= E\left[\mathbf{e}_k^- \mathbf{e}_k^{-T}\right] \\
&= E\left[(\mathbf{F}_{k-1}\mathbf{e}_{k-1} + \mathbf{L}_{k-1}\mathbf{w}_{k-1})(\mathbf{F}_{k-1}\mathbf{e}_{k-1} + \mathbf{L}_{k-1}\mathbf{w}_{k-1})^T\right] \\
&= E\left[\mathbf{F}_{k-1}\mathbf{e}_{k-1}\mathbf{e}_{k-1}^T\mathbf{F}_{k-1}^T + 2\mathbf{L}_{k-1}\mathbf{w}_{k-1}\mathbf{e}_{k-1}^T\mathbf{F}_{k-1}^T + \mathbf{L}_{k-1}\mathbf{w}_{k-1}\mathbf{w}_{k-1}^T\mathbf{L}_{k-1}^T\right] \\
&= \mathbf{F}_{k-1}E\left[\mathbf{e}_{k-1}\mathbf{e}_{k-1}^T\right]\mathbf{F}_{k-1}^T + 2\mathbf{L}_{k-1}E\left[\mathbf{w}_{k-1}\mathbf{e}_{k-1}^T\right]\mathbf{F}_{k-1}^T \\
&\qquad + \mathbf{L}_{k-1}E\left[\mathbf{w}_{k-1}\mathbf{w}_{k-1}^T\right]\mathbf{L}_{k-1}^T.
\end{aligned}$$

Recall that $\mathbf{w}_{k-1}$ is uncorrelated with $\mathbf{e}_{k-1}$, and thus $E\left[\mathbf{w}_{k-1}\mathbf{e}_{k-1}^T\right] = \mathbf{0}$. It follows that the a priori covariance update is then

$$\mathbf{P}_k^- = \mathbf{F}_{k-1}\mathbf{P}_{k-1}\mathbf{F}_{k-1}^T + \mathbf{L}_{k-1}\mathbf{Q}_{k-1}\mathbf{L}_{k-1}^T$$

where $\mathbf{P}_{k-1} = E\left[\mathbf{e}_{k-1}\mathbf{e}_{k-1}^T\right]$ and $\mathbf{Q}_{k-1} = E[\mathbf{w}_{k-1}\mathbf{w}_{k-1}^T]$.

25.3.1.2 The Correction Step

Given our a priori estimate of the state, we would like to refine the estimate of the state by incorporating the measurement $\mathbf{y}_k$. To do so, we define the linear correction equation, also called the linear update equation:

$$\hat{\mathbf{x}}_k = \hat{\mathbf{x}}_k^- + \mathbf{K}_k(\mathbf{y}_k - \hat{\mathbf{y}}_k^-) \tag{25.24}$$

where $\hat{\mathbf{y}}_k^- = \mathbf{H}_k\hat{\mathbf{x}}_k^-$ is the predicted measurement given the a priori estimate, and $\mathbf{K}_k$ is a gain to be computed. Using (25.23) we can write (25.24) as

$$\hat{\mathbf{x}}_k = (\mathbf{1} - \mathbf{K}_k\mathbf{H}_k)\hat{\mathbf{x}}_k^- + \mathbf{K}_k\mathbf{H}_k\mathbf{x}_k + \mathbf{K}_k\mathbf{M}_k\mathbf{v}_k.$$

This expression can be used to compute the a posterior error:

$$\begin{aligned}
\mathbf{e}_k &= \mathbf{x}_k - \hat{\mathbf{x}}_k \\
&= \mathbf{x}_k - (\mathbf{1} - \mathbf{K}_k\mathbf{H}_k)\hat{\mathbf{x}}_k^- - \mathbf{K}_k\mathbf{H}_k\mathbf{x}_k - \mathbf{K}_k\mathbf{M}_k\mathbf{v}_k \\
&= (\mathbf{1} - \mathbf{K}_k\mathbf{H}_k)(\mathbf{x}_k - \hat{\mathbf{x}}_k^-) - \mathbf{K}_k\mathbf{M}_k\mathbf{v}_k \\
&= (\mathbf{1} - \mathbf{K}_k\mathbf{H}_k)\mathbf{e}_k^- - \mathbf{K}_k\mathbf{M}_k\mathbf{v}_k.
\end{aligned}$$

We can now use this expression to find $\mathbf{P}_k$, the a posterior covariance matrix:

$$\begin{aligned}
\mathbf{P}_k &= E\left[\mathbf{e}_k\mathbf{e}_k{}^T\right] \\
&= E\left[\left((\mathbf{1} - \mathbf{K}_k\mathbf{H}_k)\mathbf{e}_k^- - \mathbf{K}_k\mathbf{M}_k\mathbf{v}_k\right)\left((\mathbf{1} - \mathbf{K}_k\mathbf{H}_k)\mathbf{e}_k^- - \mathbf{K}_k\mathbf{M}_k\mathbf{v}_k\right)^T\right] \\
&= E\Big[(\mathbf{1} - \mathbf{K}_k\mathbf{H}_k)\mathbf{e}_k^-\mathbf{e}_k^{-T}(\mathbf{1} - \mathbf{K}_k\mathbf{H}_k)^T - 2(\mathbf{1} - \mathbf{K}_k\mathbf{H}_k)\mathbf{e}_k^-\mathbf{v}_k^T\mathbf{M}_k^T\mathbf{K}_k^T \\
&\qquad + \mathbf{K}_k\mathbf{M}_k\mathbf{v}_k\mathbf{v}_k^T\mathbf{M}_k^T\mathbf{K}_k^T\Big] \\
&= (\mathbf{1} - \mathbf{K}_k\mathbf{H}_k)E\left[\mathbf{e}_k^-\mathbf{e}_k^{-T}\right](\mathbf{1} - \mathbf{K}_k\mathbf{H}_k)^T \\
&\qquad - 2(\mathbf{1} - \mathbf{K}_k\mathbf{H}_k)E\left[\mathbf{e}_k^-\mathbf{v}_k^T\right]\mathbf{M}_k^T\mathbf{K}_k^T + \mathbf{K}_k\mathbf{M}_kE\left[\mathbf{v}_k\mathbf{v}_k^T\right]\mathbf{M}_k^T\mathbf{K}_k^T.
\end{aligned}$$

The noise $\mathbf{v}_k$ is uncorrelated with $\mathbf{e}_k^-$, thus $E\left[\mathbf{e}_k^-\mathbf{v}_k^T\right] = \mathbf{0}$. Recalling that $\mathbf{P}_k^- = E\left[\mathbf{e}_k^-\mathbf{e}_k^{-T}\right]$ and $\mathbf{R}_k = E\left[\mathbf{v}_k\mathbf{v}_k^T\right]$ the a posterior covariance is

$$\mathbf{P}_k = (\mathbf{1} - \mathbf{K}_k\mathbf{H}_k)\mathbf{P}_k^-(\mathbf{1} - \mathbf{K}_k\mathbf{H}_k)^T + \mathbf{K}_k\mathbf{M}_k\mathbf{R}_k\mathbf{M}_k^T\mathbf{K}_k^T,$$

which is called the Joseph formula.

25.3.1.3 The Optimal Gain Matrix

How do we go about picking the gain $\mathbf{K}_k$ optimally? To find an optimal $\mathbf{K}_k$ we will pose an optimization problem. Consider the following optimization problem:

$$\begin{aligned}\text{Minimize} \quad J_k(\mathbf{K}_k) &= \operatorname{tr}[\mathbf{P}_k] \\ &= \operatorname{tr}\left[(\mathbf{1}-\mathbf{K}_k\mathbf{H}_k)\mathbf{P}_k^-(\mathbf{1}-\mathbf{H}_k^T\mathbf{K}_k^T)+\mathbf{K}_k\mathbf{M}_k\mathbf{R}_k\mathbf{M}_k^T\mathbf{K}_k^T\right] \\ &= \operatorname{tr}\left[\mathbf{P}_k^- - \mathbf{K}_k\mathbf{H}_k\mathbf{P}_k^- - \mathbf{P}_k^-\mathbf{H}_k^T\mathbf{K}_k^T + \mathbf{K}_k\mathbf{W}_k\mathbf{K}_k^T\right]\end{aligned}$$

where

$$\mathbf{W}_k = \mathbf{H}_k\mathbf{P}_k^-\mathbf{H}_k^T + \mathbf{M}_k\mathbf{R}_k\mathbf{M}_k^T.$$

Why are we minimizing $\operatorname{tr}[\mathbf{P}_k]$? To motivate the minimization of $\operatorname{tr}[\mathbf{P}_k]$, consider $\mathbf{P}_k$ written in terms of its eigendecomposition, and then simplified using the properties of the trace operation:

$$\operatorname{tr}[\mathbf{P}_k] = \operatorname{tr}\left[\mathbf{E}_{\mathbf{P}_k}\boldsymbol{\Lambda}_{\mathbf{P}_k}\mathbf{E}_{\mathbf{P}_k}^{-1}\right] = \operatorname{tr}\left[\boldsymbol{\Lambda}_{\mathbf{P}_k}\mathbf{E}_{\mathbf{P}_k}^{-1}\mathbf{E}_{\mathbf{P}_k}\right] = \operatorname{tr}\left[\boldsymbol{\Lambda}_{\mathbf{P}_k}\right] = \sum_{i=1}^{n}\lambda_{\mathbf{P}_k,i}$$

where $\lambda_{\mathbf{P}_k,i}$, $i = 1, 2, \ldots, n$ are the eigenvalues of $\mathbf{P}_k$. Minimizing $\operatorname{tr}[\mathbf{P}_k]$ is equivalent to minimizing the sum of the eigenvalues of $\mathbf{P}_k$, which is equivalent minimizing the uncertainty of our a posteriori estimate. Minimizing the a posteriori estimate uncertainty is reasonable in that we want to be very confident that our estimate of the state, $\hat{\mathbf{x}}_k$, is a good estimate.

To minimize the objective function $J_k(\mathbf{K}_k)$ we will take the derivative of $J_k(\mathbf{K}_k)$ with respect to $\mathbf{K}_k$ and set it equal to zero:

$$\frac{\partial J_k(\mathbf{K}_k)}{\partial \mathbf{K}_k} = 2\left(-\mathbf{P}_k^-\mathbf{H}_k^T + \mathbf{K}_k\mathbf{W}_k\right) = \mathbf{0}.$$

Thus, the optimal gain is

$$\mathbf{K}_k = \mathbf{P}_k^-\mathbf{H}_k^T\mathbf{W}_k^{-1}.$$

This optimal $\mathbf{K}_k$ is called the Kalman gain.

25.3.1.4 Summary of the Kalman Filter

$$\begin{aligned}\text{System:} \qquad \mathbf{x}_k &= \mathbf{F}_{k-1}\mathbf{x}_{k-1} + \mathbf{G}_{k-1}\mathbf{u}_{k-1} + \mathbf{L}_{k-1}\mathbf{w}_{k-1} \\ \mathbf{y}_k &= \mathbf{H}_k\mathbf{x}_k + \mathbf{M}_k\mathbf{v}_k \\ \mathbf{w}_k &\sim \mathcal{N}(\mathbf{0}, \mathbf{Q}_k) \\ \mathbf{v}_k &\sim \mathcal{N}(\mathbf{0}, \mathbf{R}_k) \\ E[\mathbf{w}_k\mathbf{v}_k^T] &= \mathbf{0}\end{aligned}$$

$$\begin{aligned}
\text{Initialization:} \quad & \hat{\mathbf{x}}_0 = E\left[\mathbf{x}_0\right] \\
& \mathbf{P}_0 = E\left[(\mathbf{x}_0 - \hat{\mathbf{x}}_0)(\mathbf{x}_0 - \hat{\mathbf{x}}_0)^T\right] \\
\text{Prediction:} \quad & \hat{\mathbf{x}}_k^- = \mathbf{F}_{k-1}\hat{\mathbf{x}}_{k-1} + \mathbf{G}_{k-1}\mathbf{u}_{k-1} \\
& \mathbf{P}_k^- = \mathbf{F}_{k-1}\mathbf{P}_{k-1}\mathbf{F}_{k-1}^T + \mathbf{L}_{k-1}\mathbf{Q}_{k-1}\mathbf{L}_{k-1}^T \\
\text{Correction:} \quad & \mathbf{W}_k = \mathbf{H}_k\mathbf{P}_k^-\mathbf{H}_k^T + \mathbf{M}_k\mathbf{R}_k\mathbf{M}_k^T \\
& \mathbf{K}_k = \mathbf{P}_k^-\mathbf{H}_k^T\mathbf{W}_k^{-1} \\
& \hat{\mathbf{x}}_k = \hat{\mathbf{x}}_k^- + \mathbf{K}_k(\mathbf{y}_k - \hat{\mathbf{y}}_k^-) \\
& \mathbf{P}_k = (\mathbf{1} - \mathbf{K}_k\mathbf{H}_k)\mathbf{P}_k^-(\mathbf{1} - \mathbf{K}_k\mathbf{H}_k)^T + \mathbf{K}_k\mathbf{M}_k\mathbf{R}_k\mathbf{M}_k^T\mathbf{K}_k^T \\
& \quad = \mathbf{P}_k^- - \mathbf{K}_k\mathbf{H}_k\mathbf{P}_k^- - \mathbf{P}_k^-\mathbf{H}_k^T\mathbf{K}_k^T + \mathbf{K}_k\mathbf{W}_k\mathbf{K}_k^T
\end{aligned}$$

25.3.2 The Norm-Constrained Kalman Filter

In the previous derivation of the Kalman filter it was assumed that the state was unconstrained. We will now consider the case where the state can be partitioned into two components

$$\mathbf{x}_k = \begin{bmatrix} \mathbf{z}_k \\ \mathbf{q}_k \end{bmatrix}$$

where $\mathbf{z}_k$ are the unconstrained states and $\mathbf{q}_k$ are the constrained states. In particular, the constrained states must satisfy the following norm constraint:

$$\|\mathbf{q}_k\| = \sqrt{\ell}$$

which can also be written as

$$\mathbf{q}_k^T\mathbf{q}_k = \ell.$$

We are addressing the issue of Kalman filtering in the presence of a norm constraint because we will soon consider spacecraft attitude estimation. In particular, the states we will be estimating will be the spacecraft angular velocity and a set of states related to the spacecraft attitude. If we choose to parameterize the states in terms of a quaternion, then we will have to ensure our quaternion estimate has unit length; said another way, we must ensure our quaternion estimate satisfies a unit norm constraint.

25.3.2.1 The Prediction Step

The prediction step of the norm-constrained Kalman filter remains the same as the standard Kalman filter. As such,

$$\hat{\mathbf{x}}_k^- = \mathbf{F}_{k-1}\hat{\mathbf{x}}_{k-1} + \mathbf{G}_{k-1}\mathbf{u}_{k-1}$$

and

$$\mathbf{P}_k^- = \mathbf{F}_{k-1}\mathbf{P}_{k-1}\mathbf{F}_{k-1}^T + \mathbf{L}_{k-1}\mathbf{Q}_{k-1}\mathbf{L}_{k-1}^T$$

where $\hat{\mathbf{x}}_k^-$ is the a priori state estimate and $\mathbf{P}_k^-$ is the covariance associated with $\hat{\mathbf{x}}_k^-$. Notice that we have predicted the whole state without any concern for the constraint $\hat{\mathbf{q}}_k^{-\,T}\hat{\mathbf{q}}_k^- = \ell$. We will enforce the norm constraint during the correction step. At this point it will be helpful to break up $\mathbf{P}_k^-$ into various submatrices:

$$\mathbf{P}_k^- = \begin{bmatrix}\mathbf{P}_{1,k}^- & \mathbf{P}_{2,k}^-\end{bmatrix} = \begin{bmatrix}\mathbf{P}_{zz,k}^- & \mathbf{P}_{zq,k}^- \\ \mathbf{P}_{qz,k}^- & \mathbf{P}_{qq,k}^-\end{bmatrix}. \tag{25.25}$$

25.3.2.2 The Correction Step

As in the unconstrained case, we will assume a linear update:

$$\begin{bmatrix}\hat{\mathbf{z}}_k \\ \hat{\mathbf{q}}_k\end{bmatrix} = \begin{bmatrix}\hat{\mathbf{z}}_k^- \\ \hat{\mathbf{q}}_k^-\end{bmatrix} + \begin{bmatrix}\mathbf{K}_{z,k} \\ \mathbf{K}_{q,k}\end{bmatrix}(\mathbf{y}_k - \hat{\mathbf{y}}_k^-) \tag{25.26}$$

where $\mathbf{K}_k$ has been partitioned into $\mathbf{K}_{z,k}$ and $\mathbf{K}_{q,k}$. We can equivalently write (25.26) as two equations,

$$\begin{aligned}\hat{\mathbf{z}}_k &= \hat{\mathbf{z}}_k^- + \mathbf{K}_{z,k}(\mathbf{y}_k - \hat{\mathbf{y}}_k^-),\\ \hat{\mathbf{q}}_k &= \hat{\mathbf{q}}_k^- + \mathbf{K}_{q,k}(\mathbf{y}_k - \hat{\mathbf{y}}_k^-).\end{aligned} \tag{25.27}$$

As before, the a posteriori covariance is given by the Joseph formula

$$\begin{aligned}\mathbf{P}_k &= (\mathbf{1} - \mathbf{K}_k\mathbf{H}_k)\mathbf{P}_k^-(\mathbf{1} - \mathbf{H}_k^T\mathbf{K}_k^T) + \mathbf{K}_k\mathbf{M}_k\mathbf{R}_k\mathbf{M}_k^T\mathbf{K}_k^T\\ &= \mathbf{P}_k^- - \mathbf{K}_k\mathbf{H}_k\mathbf{P}_k^- - \mathbf{P}_k^-\mathbf{H}_k^T\mathbf{K}_k^T + \mathbf{K}_k\mathbf{W}_k\mathbf{K}_k^T\end{aligned}$$

where

$$\mathbf{W}_k = \mathbf{H}_k\mathbf{P}_k^-\mathbf{H}_k^T + \mathbf{M}_k\mathbf{R}_k\mathbf{M}_k^T.$$

Strictly speaking, we are assuming the Joseph formula is valid, and it can be argued that it is indeed appropriate. Given the partitioning of $\mathbf{P}_k^-$ in (25.25) we can write

$$\begin{aligned}\mathbf{K}_k\mathbf{H}_k\mathbf{P}_k^- &= \begin{bmatrix}\mathbf{K}_{z,k}\mathbf{H}_k\mathbf{P}_{1,k}^- & \mathbf{K}_{z,k}\mathbf{H}_k\mathbf{P}_{2,k}^- \\ \mathbf{K}_{q,k}\mathbf{H}_k\mathbf{P}_{1,k}^- & \mathbf{K}_{q,k}\mathbf{H}_k\mathbf{P}_{2,k}^-\end{bmatrix},\\ \mathbf{K}_k\mathbf{W}_k\mathbf{K}_k^T &= \begin{bmatrix}\mathbf{K}_{z,k}\mathbf{W}_k\mathbf{K}_{z,k}^T & \mathbf{K}_{z,k}\mathbf{W}_{z,k}\mathbf{K}_{q,k}^T \\ \mathbf{K}_{q,k}\mathbf{W}_k\mathbf{K}_{z,k}^T & \mathbf{K}_{q,k}\mathbf{W}_{q,k}\mathbf{K}_{q,k}^T\end{bmatrix},\end{aligned}$$

and now write $\mathbf{P}_k$ as

$$\mathbf{P}_k = \begin{bmatrix} \mathbf{P}_{zz,k} & \mathbf{P}_{zq,k} \\ \mathbf{P}_{qz,k} & \mathbf{P}_{qq,k} \end{bmatrix} \tag{25.28}$$

where

$$\mathbf{P}_{zz,k} = \mathbf{P}^-_{zz,k} - \mathbf{K}_{z,k}\mathbf{H}_k\mathbf{P}^-_{1,k} - \mathbf{P}^{-\,T}_{1,k}\mathbf{H}_k^T\mathbf{K}_{z,k}^T + \mathbf{K}_{z,k}\mathbf{W}_k\mathbf{K}_{z,k}^T, \tag{25.29}$$

$$\begin{aligned}
\mathbf{P}_{zq,k} &= \mathbf{P}^-_{zq,k} - \mathbf{K}_{z,k}\mathbf{H}_k\mathbf{P}^-_{2,k} - \mathbf{P}^{-\,T}_{1,k}\mathbf{H}_k^T\mathbf{K}_{q,k}^T + \mathbf{K}_{z,k}\mathbf{W}_k\mathbf{K}_{q,k}^T, \\
\mathbf{P}_{qz,k} &= \mathbf{P}^-_{qz,k} - \mathbf{K}_{q,k}\mathbf{H}_k\mathbf{P}^-_{1,k} - \mathbf{P}^{-\,T}_{2,k}\mathbf{H}_k^T\mathbf{K}_{z,k}^T + \mathbf{K}_{q,k}\mathbf{W}_k\mathbf{K}_{z,k}^T, \\
\mathbf{P}_{qq,k} &= \mathbf{P}^-_{qq,k} - \mathbf{K}_{q,k}\mathbf{H}_k\mathbf{P}^-_{2,k} - \mathbf{P}^{-\,T}_{2,k}\mathbf{H}_k^T\mathbf{K}_{q,k}^T + \mathbf{K}_{q,k}\mathbf{W}_k\mathbf{K}_{q,k}^T.
\end{aligned} \tag{25.30}$$

25.3.2.3 The Optimal Gain Matrix Considering the Constraint

We now seek to find the optimal gains $\mathbf{K}_{z,k}$ and $\mathbf{K}_{q,k}$. Consider the following constrained optimization problem:

$$\text{Minimize} \quad J_k(\mathbf{K}_{z,k}, \mathbf{K}_{q,k}) = \text{tr}\,[\mathbf{P}_k]$$

such that

$$\hat{\mathbf{q}}_k^T\hat{\mathbf{q}}_k - \ell = 0.$$

Although the objective function is the same objective function used to derive the standard Kalman filter, we now have the norm constraint. Using a Lagrange multiplier we can augment the constraint with the objective function to create the following unconstrained optimization problem:

$$\text{Minimize} \quad \hat{J}_k(\mathbf{K}_{z,k}, \mathbf{K}_{q,k}, \lambda_k) = \text{tr}\,[\mathbf{P}_k] + \lambda_k\left(\hat{\mathbf{q}}_k^T\hat{\mathbf{q}}_k - \ell\right).$$

Given the partitioning of $\mathbf{P}_k$ in (25.28), the objective function can be written

$$\begin{aligned}
\text{Minimize} \quad \hat{J}_k(\mathbf{K}_{z,k}, \mathbf{K}_{q,k}, \lambda_k) &= \text{tr}\,[\mathbf{P}_k] + \lambda_k\left(\hat{\mathbf{q}}_k^T\hat{\mathbf{q}}_k - \ell\right) \\
&= \text{tr}\left(\begin{bmatrix} \mathbf{P}_{zz,k} & \mathbf{P}_{zq,k} \\ \mathbf{P}_{qz,k} & \mathbf{P}_{qq,k} \end{bmatrix}\right) + \lambda_k\left(\hat{\mathbf{q}}_k^T\hat{\mathbf{q}}_k - \ell\right) \\
&= \text{tr}\left[\mathbf{P}_{zz,k}\right] + \text{tr}\left[\mathbf{P}_{qq,k}\right] + \lambda_k\left(\hat{\mathbf{q}}_k^T\hat{\mathbf{q}}_k - \ell\right) \\
&= \hat{J}_k(\mathbf{K}_{z,k}) + \hat{J}_k(\mathbf{K}_{q,k}, \lambda_k)
\end{aligned}$$

where $\hat{J}_k(\mathbf{K}_{z,k}) = \text{tr}\left[\mathbf{P}_{zz,k}\right]$ and $\hat{J}_k(\mathbf{K}_{q,k}, \lambda_k) = \text{tr}\left[\mathbf{P}_{qq,k}\right] + \lambda_k\left(\hat{\mathbf{q}}_k^T\hat{\mathbf{q}}_k - \ell\right)$. By minimizing the sum of $\hat{J}_k(\mathbf{K}_{z,k})$ and $\hat{J}_k(\mathbf{K}_{q,k}, \lambda_k)$ the posteriori covariance is minimized. Notice from equations (25.29) and (25.30) that $\text{tr}\left[\mathbf{P}_{zz,k}\right]$ is only a function of $\mathbf{K}_{z,k}$ and $\text{tr}\left[\mathbf{P}_{qq,k}\right]$ is only a function of

$\mathbf{K}_{q,k}$. As such, minimizing $\hat{J}_k(\mathbf{K}_{z,k})$ to find $\mathbf{K}_{z,k}$ and minimizing $\hat{J}_k(\mathbf{K}_{q,k}, \lambda_k)$ to find $\mathbf{K}_{q,k}$ can be done independently.

Because the states $\mathbf{z}_k$ are not constrained, $\mathbf{K}_{z,k}$ can be found by minimizing $\hat{J}_k(\mathbf{K}_{z,k}) = \operatorname{tr}\left[\mathbf{P}_{zz,k}\right]$ as outlined in Section (25.1), the solution being

$$\mathbf{K}_{z,k} = \mathbf{P}_{1,k}^{-\,T}\mathbf{H}_k^T\mathbf{W}_k^{-1}.$$

On the other hand, to find $\mathbf{K}_{q,k}$ we must solve the following optimization problem:

$$\text{Minimize} \quad \hat{J}_k(\mathbf{K}_{q,k}, \lambda_k) = \operatorname{tr}\left[\mathbf{P}_{qq,k}\right] + \lambda_k\left(\hat{\mathbf{q}}_k^T\hat{\mathbf{q}}_k - \ell\right).$$

Using equation (25.26) the constraint $\hat{\mathbf{q}}_k^T\hat{\mathbf{q}}_k - \ell = 0$ can be written

$$\hat{\mathbf{q}}_k^{-\,T}\hat{\mathbf{q}}_k^- + 2\mathbf{r}_k^T\mathbf{K}_{q,k}^T\hat{\mathbf{q}}_k^- + \mathbf{r}_k^T\mathbf{K}_{q,k}^T\mathbf{K}_{q,k}\mathbf{r}_k - \ell = 0 \tag{25.31}$$

where $\mathbf{r}_k = \mathbf{y}_k - \hat{\mathbf{y}}_k^- = \mathbf{y}_k - \mathbf{H}_k\hat{\mathbf{x}}_k^-$ is the measurement residual. Using the trace operation, equation (25.31) can alternatively be written

$$\operatorname{tr}\left[\hat{\mathbf{q}}_k^-\hat{\mathbf{q}}_k^{-\,T} + 2\hat{\mathbf{q}}_k^-\mathbf{r}_k^T\mathbf{K}_{q,k}^T + \mathbf{K}_{q,k}\mathbf{r}_k\mathbf{r}_k^T\mathbf{K}_{q,k}^T\right] - \ell = 0.$$

Using this form of the constraint, the optimization problem to be solved is

$$\begin{aligned}\text{Minimize} \quad \hat{J}_k(\mathbf{K}_{q,k}, \lambda_k) &= \operatorname{tr}\left[\mathbf{P}_{qq,k}\right] \\ &\quad +\lambda_k\left(\operatorname{tr}\left[\hat{\mathbf{q}}_k^-\hat{\mathbf{q}}_k^{-\,T} + 2\hat{\mathbf{q}}_k^-\mathbf{r}_k^T\mathbf{K}_{q,k}^T + \mathbf{K}_{q,k}\mathbf{r}_k\mathbf{r}_k^T\mathbf{K}_{q,k}^T\right] - \ell\right).\end{aligned}$$

To find the optimal gain $\mathbf{K}_{q,k}$ we take the derivative of $\hat{J}_k(\mathbf{K}_{q,k}, \lambda_k)$ with respect to $\mathbf{K}_{q,k}$ and λ_k and set the results to zero:

$$\frac{\partial \hat{J}_k(\mathbf{K}_{q,k}, \lambda_k)}{\partial \mathbf{K}_{q,k}} = 2\left(-\mathbf{P}_{2,k}^{-\,T}\mathbf{H}_k^T + \mathbf{K}_{q,k}\mathbf{W}_k\right) + 2\lambda_k\left(\hat{\mathbf{q}}_k^-\mathbf{r}_k^T + \mathbf{K}_{q,k}\mathbf{r}_k\mathbf{r}_k^T\right) = \mathbf{0},$$

$$\frac{\partial \hat{J}_k(\mathbf{K}_{q,k}, \lambda_k)}{\partial \lambda_k} = \operatorname{tr}\left[\hat{\mathbf{q}}_k^-\hat{\mathbf{q}}_k^{-\,T} + 2\hat{\mathbf{q}}_k^-\mathbf{r}_k^T\mathbf{K}_{q,k}^T + \mathbf{K}_{q,k}\mathbf{r}_k\mathbf{r}_k^T\mathbf{K}_{q,k}^T\right] - \ell = 0.$$

Therefore, the optimal gain $\mathbf{K}_{q,k}$ is

$$\mathbf{K}_{q,k} = \left(\mathbf{P}_{2,k}^{-\,T}\mathbf{H}_k^T - \lambda_k\hat{\mathbf{q}}_k^-\mathbf{r}_k^T\right)\left(\mathbf{W}_k + \lambda_k\mathbf{r}_k\mathbf{r}_k^T\right)^{-1}. \tag{25.32}$$

In order to find a closed-form expression for $\mathbf{K}_{q,k}$ we need to find the optimal Lagrange multiplier, λ_k. To do so, we will simplify our expression for $\mathbf{K}_{q,k}$ and then substitute it into (25.31). To simplify $\mathbf{K}_{q,k}$ we will use the matrix inversion lemma: if $\det(\mathbf{C} + \mathbf{VAU}) \neq 0$ then

$$\left(\mathbf{A}^{-1} + \mathbf{UC}^{-1}\mathbf{V}\right)^{-1} = \mathbf{A} - \mathbf{AU}\left(\mathbf{C} + \mathbf{VAU}\right)^{-1}\mathbf{VA}.$$

Letting $\mathbf{A}^{-1} = \mathbf{W}_k$, $\mathbf{C}^{-1} = \lambda_k \mathbf{1}$, $\mathbf{U} = \mathbf{r}_k$, and $\mathbf{V} = \mathbf{r}_k^T$ we can write $\mathbf{K}_{q,k}$ as

$$\mathbf{K}_{q,k} = \mathbf{P}_{2,k}^{-\,T}\mathbf{H}_k^T\mathbf{W}_k^{-1} - \frac{\lambda_k \mathbf{P}_{2,k}^{-\,T}\mathbf{H}_k^T\mathbf{W}_k^{-1}\mathbf{r}_k\mathbf{r}_k^T\mathbf{W}_k^{-1}}{\left(1+\lambda_k\mathbf{r}_k^T\mathbf{W}_k^{-1}\mathbf{r}_k\right)}$$
$$-\lambda_k\hat{\mathbf{q}}_k^{-}\mathbf{r}_k^T\mathbf{W}_k^{-1} + \frac{\lambda_k^2\hat{\mathbf{q}}_k^{-}\mathbf{r}_k^T\mathbf{W}_k^{-1}\mathbf{r}_k\mathbf{r}_k^T\mathbf{W}_k^{-1}}{\left(1+\lambda_k\mathbf{r}_k^T\mathbf{W}_k^{-1}\mathbf{r}_k\right)} \tag{25.33}$$

Substituting the above expression for $\mathbf{K}_{q,k}$ into the constraint equation given in (25.31) and simplifying gives a quadratic equation in λ_k:

$$a\lambda_k^2 + b\lambda_k + c = 0$$

where $a = -r_k^2\ell$, $b = -2r_k\ell$,

$$c = \left(\mathbf{r}_k^T\mathbf{W}_k^{-1}\mathbf{H}_k\mathbf{P}_{2,k}^{-}\mathbf{P}_{2,k}^{-\,T}\mathbf{H}_k^T\mathbf{W}_k^{-1}\mathbf{r}_k + 2\hat{\mathbf{q}}_k^{-\,T}\mathbf{P}_{2,k}^{-\,T}\mathbf{H}_k^T\mathbf{W}_k^{-1}\mathbf{r}_k + \hat{\mathbf{q}}_k^{-\,T}\hat{\mathbf{q}}_k^{-} - \ell\right),$$

and $r_k = \mathbf{r}_k^T\mathbf{W}_k^{-1}\mathbf{r}_k$. The roots of this quadratic equation are

$$\lambda_k = \frac{-b \pm \sqrt{b^2-4ac}}{2a} = \frac{2r_k\ell \pm \sqrt{4r_k^2\ell^2 + 4r_k^2\ell c}}{-2r_k^2\ell} = \frac{1 \pm \sqrt{1+c/\ell}}{-r_k}. \tag{25.34}$$

One root will minimize the objective function, while the other will maximize the objective function. The quantity $1 + c/\ell$ can be written

$$\begin{aligned}1 + c/\ell &= \left(\mathbf{r}_k^T\mathbf{W}_k^{-1}\mathbf{H}_k\mathbf{P}_{2,k}^{-}\mathbf{P}_{2,k}^{-\,T}\mathbf{H}_k^T\mathbf{W}_k^{-1}\mathbf{r}_k + 2\hat{\mathbf{q}}_k^{-\,T}\mathbf{P}_{2,k}^{-\,T}\mathbf{H}_k^T\mathbf{W}_k^{-1}\mathbf{r}_k + \hat{\mathbf{q}}_k^{-\,T}\hat{\mathbf{q}}_k^{-}\right)/\ell \\ &= \left(\hat{\mathbf{q}}_k^{-} + \mathbf{P}_{2,k}^{-\,T}\mathbf{H}_k^T\mathbf{W}_k^{-1}\mathbf{r}_k\right)^T\left(\hat{\mathbf{q}}_k^{-} + \mathbf{P}_{2,k}^{-\,T}\mathbf{H}_k^T\mathbf{W}_k^{-1}\mathbf{r}_k\right)/\ell\end{aligned}$$

which is greater than or equal to zero. Hence, the quantity $\sqrt{1+c/\ell}$ will always be real, and therefore λ_k will be a real number. Substituting the expression for $1 + c/\ell$ into (25.34) gives

$$\lambda_k = \frac{-1}{r_k} \pm \left[-\frac{\left\|\hat{\mathbf{q}}_k^{-} + \mathbf{P}_{2,k}^{-\,T}\mathbf{H}_k^T\mathbf{W}_k^{-1}\mathbf{r}_k\right\|}{r_k\sqrt{\ell}}\right].$$

To minimize the objective function $\hat{J}(\mathbf{K}_{q,k}, \lambda_k)$ we pick the negative root:

$$\begin{aligned}\lambda_k &= \frac{-1}{r_k} - \left[-\frac{\left\|\hat{\mathbf{q}}_k^{-} + \mathbf{P}_{2,k}^{-\,T}\mathbf{H}_k^T\mathbf{W}_k^{-1}\mathbf{r}_k\right\|}{r_k\sqrt{\ell}}\right] \\ &= \frac{-1}{r_k} + \frac{\left\|\hat{\mathbf{q}}_k^{-} + \mathbf{P}_{2,k}^{-\,T}\mathbf{H}_k^T\mathbf{W}_k^{-1}\mathbf{r}_k\right\|}{r_k\sqrt{\ell}}.\end{aligned} \tag{25.35}$$

It can be shown by more rigorous means that the negative root indeed minimizes the objective function, while the positive root maximizes the objective function.

The solution to our optimization problem is given by $\mathbf{K}_{q,k}$ in equation (25.33) and λ_k in equation (25.35). Let us now discuss what the norm-constrained Kalman filter is doing differently than the nominal (or unconstrained) Kalman filter. First, let $\tilde{\mathbf{K}}_{q,k} = \mathbf{P}_{2,k}^{-\,T}\mathbf{H}_k^T\mathbf{W}_k^{-1}$ be the unconstrained gain. Also, let

$$\tilde{\mathbf{q}}_k = \hat{\mathbf{q}}_k^- + \tilde{\mathbf{K}}_{q,k}\mathbf{r}_k \tag{25.36}$$

be the state update that does not enforce the norm constraint, where $\tilde{\mathbf{q}}_k$ is the estimate of the state that should satisfy $\tilde{\mathbf{q}}_k^T\tilde{\mathbf{q}}_k - \ell = 0$, but given our use of $\tilde{\mathbf{K}}_{q,k}$ may not. Now, let us return to our expression for λ_k given in equation (25.35). Using $\tilde{\mathbf{K}}_{q,k}$ and the expression for $\tilde{\mathbf{q}}_k$ in (25.36) we can write λ_k as

$$\lambda_k = \frac{-1}{r_k} + \frac{\|\hat{\mathbf{q}}_k^- + \tilde{\mathbf{K}}_{q,k}\mathbf{r}_k\|}{r_k\sqrt{\ell}} = \frac{-1}{r_k} + \frac{\|\tilde{\mathbf{q}}_k\|}{r_k\sqrt{\ell}}.$$

Using this alternate form of λ_k we can write $\mathbf{K}_{q,k}$ in equation (25.33) as

$$\mathbf{K}_{q,k} = \tilde{\mathbf{K}}_{q,k} + \frac{1}{r_k}\left(\frac{\sqrt{\ell}}{\|\tilde{\mathbf{q}}_k\|} - 1\right)\tilde{\mathbf{q}}_k\mathbf{r}_k^T\mathbf{W}_k^{-1}. \tag{25.37}$$

Now, returning to our update expression for $\hat{\mathbf{q}}_k$ given in equation (25.27), notice that

$$\begin{aligned}
\hat{\mathbf{q}}_k &= \hat{\mathbf{q}}_k^- + \mathbf{K}_{q,k}\mathbf{r}_k \\
&= \hat{\mathbf{q}}_k^- + \tilde{\mathbf{K}}_{q,k}\mathbf{r}_k + \frac{1}{r_k}\left(\frac{\sqrt{\ell}}{\|\tilde{\mathbf{q}}_k\|} - 1\right)\tilde{\mathbf{q}}_k\mathbf{r}_k^T\mathbf{W}_k^{-1}\mathbf{r}_k \\
&= \frac{\sqrt{\ell}}{\|\tilde{\mathbf{q}}_k\|}\tilde{\mathbf{q}}_k
\end{aligned}$$

where, as before, $r_k = \mathbf{r}_k^T\mathbf{W}_k^{-1}\mathbf{r}_k$. This shows that the norm-constrained state estimate $\hat{\mathbf{q}}$ is the brute-force normalization of the estimate $\tilde{\mathbf{q}}_k$. As such, it turns out that brute-force normalization is in fact optimal.

Now let us investigate the covariance of our estimate, $\mathbf{P}_{qq,k}$. By substituting the expression for $\mathbf{K}_{q,k}$ given in equation (25.37) into equation (25.30) and simplifying we have the following

$$\begin{aligned}
\mathbf{P}_{qq,k} &= \underbrace{\mathbf{P}_{qq,k}^- - \tilde{\mathbf{K}}_{q,k}\mathbf{H}_k\mathbf{P}_{2,k}^- - \mathbf{P}_{2,k}^{-\,T}\mathbf{H}_k^T\tilde{\mathbf{K}}_{q,k}^T + \tilde{\mathbf{K}}_{q,k}\mathbf{W}_k\tilde{\mathbf{K}}_{q,k}^T}_{\tilde{\mathbf{P}}_{qq,k}} + \frac{1}{r_k}\left(\frac{\sqrt{\ell}}{\|\tilde{\mathbf{q}}_k\|} - 1\right)^2\tilde{\mathbf{q}}_k\tilde{\mathbf{q}}_k^T \\
&= \tilde{\mathbf{P}}_{qq,k} + \frac{1}{r_k}\left(\frac{\sqrt{\ell}}{\|\tilde{\mathbf{q}}_k\|} - 1\right)^2\tilde{\mathbf{q}}_k\tilde{\mathbf{q}}_k^T
\end{aligned}$$

where $\tilde{\mathbf{P}}_{qq,k}$ is the covariance associated with $\tilde{\mathbf{q}}_k$. Notice the covariance $\mathbf{P}_{qq,k}$ associated with the norm-constrained estimate $\mathbf{q}_k$ is not equivalent to the covariance $\tilde{\mathbf{P}}_{qq,k}$ associated with $\tilde{\mathbf{q}}_k$. The a posterior covariance $\mathbf{P}_{qq,k}$ equals $\tilde{\mathbf{P}}_{qq,k}$ plus a second order effect.

25.3.2.4 Summary of the Norm-Constrained Kalman Filter

$$
\begin{aligned}
\text{System:} \quad \mathbf{x}_k &= \mathbf{F}_{k-1}\mathbf{x}_{k-1} + \mathbf{G}_{k-1}\mathbf{u}_{k-1} + \mathbf{L}_{k-1}\mathbf{w}_{k-1} \\
\mathbf{y}_k &= \mathbf{H}_k\mathbf{x}_k + \mathbf{M}_k\mathbf{v}_k \\
\mathbf{x}_k &= \begin{bmatrix}\mathbf{z}_k^T & \mathbf{q}_k^T\end{bmatrix}^T \\
\mathbf{q}_k^T\mathbf{q}_k &= \ell \\
\mathbf{w}_k &\sim \mathcal{N}(\mathbf{0}, \mathbf{Q}_k) \\
\mathbf{v}_k &\sim \mathcal{N}(\mathbf{0}, \mathbf{R}_k) \\
E[\mathbf{w}_k\mathbf{v}_k^T] &= \mathbf{0} \\
\text{Initialization:} \quad \hat{\mathbf{x}}_0 &= E[\mathbf{x}_0] \\
\mathbf{P}_0 &= E\left[(\mathbf{x}_0 - \hat{\mathbf{x}}_0)(\mathbf{x}_0 - \hat{\mathbf{x}}_0)^T\right] \\
\text{Prediction:} \quad \hat{\mathbf{x}}_k^- &= \mathbf{F}_{k-1}\hat{\mathbf{x}}_{k-1} + \mathbf{G}_{k-1}\mathbf{u}_{k-1} \\
\mathbf{P}_k^- &= \mathbf{F}_{k-1}\mathbf{P}_{k-1}\mathbf{F}_{k-1}^T + \mathbf{L}_{k-1}\mathbf{Q}_{k-1}\mathbf{L}_{k-1}^T \\
&= \begin{bmatrix}\mathbf{P}_{1,k}^- & \mathbf{P}_{2,k}^-\end{bmatrix} = \begin{bmatrix}\mathbf{P}_{zz,k}^- & \mathbf{P}_{zq,k}^- \\ \mathbf{P}_{qz,k}^- & \mathbf{P}_{qq,k}^-\end{bmatrix} \\
\text{Correction:} \quad \mathbf{W}_k &= \mathbf{H}_k\mathbf{P}_k^-\mathbf{H}_k^T + \mathbf{M}_k\mathbf{R}_k\mathbf{M}_k^T \\
\mathbf{K}_{z,k} &= \mathbf{P}_{1,k}^{-\,T}\mathbf{H}_k^T\mathbf{W}_k^{-1} \\
\hat{\mathbf{z}}_k &= \hat{\mathbf{z}}_k^- + \mathbf{K}_{z,k}(\mathbf{y}_k - \hat{\mathbf{y}}_k^-) \\
\tilde{\mathbf{K}}_{q,k} &= \mathbf{P}_{2,k}^{-\,T}\mathbf{H}_k^T\mathbf{W}_k^{-1} \\
\mathbf{r}_k &= \mathbf{y}_k - \hat{\mathbf{y}}_k^- \\
r_k &= \mathbf{r}_k^T\mathbf{W}_k^{-1}\mathbf{r}_k \\
\tilde{\mathbf{q}}_k &= \hat{\mathbf{q}}_k^- + \tilde{\mathbf{K}}_{q,k}\mathbf{r}_k \\
\mathbf{K}_{q,k} &= \tilde{\mathbf{K}}_{q,k} + \frac{1}{r_k}\left(\frac{\sqrt{\ell}}{\|\tilde{\mathbf{q}}_k\|} - 1\right)\tilde{\mathbf{q}}_k\mathbf{r}_k^T\mathbf{W}_k^{-1} \\
\hat{\mathbf{q}}_k &= \hat{\mathbf{q}}_k^- + \mathbf{K}_{q,k}\mathbf{r}_k = \frac{\sqrt{\ell}}{\|\tilde{\mathbf{q}}_k\|}\tilde{\mathbf{q}}_k \\
\mathbf{P}_{zz,k} &= \mathbf{P}_{zz,k}^- - \mathbf{K}_{z,k}\mathbf{H}_k\mathbf{P}_{1,k}^- - \mathbf{P}_{1,k}^{-\,T}\mathbf{H}_k^T\mathbf{K}_{z,k}^T + \mathbf{K}_{z,k}\mathbf{W}_k\mathbf{K}_{z,k}^T \\
\mathbf{P}_{zq,k} &= \mathbf{P}_{zq,k}^- - \mathbf{K}_{z,k}\mathbf{H}_k\mathbf{P}_{2,k}^- - \mathbf{P}_{1,k}^{-\,T}\mathbf{H}_k^T\mathbf{K}_{q,k}^T + \mathbf{K}_{z,k}\mathbf{W}_k\mathbf{K}_{q,k}^T
\end{aligned}
$$

$$\mathbf{P}_{qz,k} = \mathbf{P}^-_{qz,k} - \mathbf{K}_{q,k}\mathbf{H}_k\mathbf{P}^-_{1,k} - \mathbf{P}^{-\,T}_{2,k}\mathbf{H}^T_k\mathbf{K}^T_{z,k} + \mathbf{K}_{q,k}\mathbf{W}_k\mathbf{K}^T_{z,k}$$

$$\mathbf{P}_{qq,k} = \mathbf{P}^-_{qq,k} - \mathbf{K}_{q,k}\mathbf{H}_k\mathbf{P}^-_{2,k} - \mathbf{P}^{-\,T}_{2,k}\mathbf{H}^T_k\mathbf{K}^T_{q,k} + \mathbf{K}_{q,k}\mathbf{W}_k\mathbf{K}^T_{q,k}$$

$$\mathbf{P}_k = \begin{bmatrix} \mathbf{P}_{zz,k} & \mathbf{P}_{zq,k} \\ \mathbf{P}_{qz,k} & \mathbf{P}_{qq,k} \end{bmatrix}$$

25.3.3 Spacecraft Attitude Estimation Using the Norm-Constrained Extended Kalman Filter

We will now consider spacecraft attitude estimation using the norm-constrained extended Kalman filter. We cannot use the norm-constrained Kalman filter directly because the spacecraft attitude estimation problem is nonlinear, and the norm-constrained Kalman filter is for linear systems. As such, we linearize the nonlinearities associated with the spacecraft dynamics, kinematics, and measurements, then use the norm-constrained Kalman filter on the linearized system. The estimation algorithm is referred to the norm-constrained extended Kalman filter (EKF), where the word extended refers to the linearization procedure we use.

Recall (see Section 11.3) that the dynamics of a rigid body spacecraft are described by

$$\mathbf{I}\dot{\boldsymbol{\omega}} + \boldsymbol{\omega}^\times\mathbf{I}\boldsymbol{\omega} = \mathbf{u} + \mathbf{w} \tag{25.38}$$

where $\mathbf{u}$ is the control input and $\mathbf{w}$ are disturbances. Often the kinematics are parameterized in terms of the unit-length quaternion, $\mathbf{q}^T = [\boldsymbol{\epsilon}^T \; \eta]$, which must satisfy $\mathbf{q}^T\mathbf{q} = 1$. The relationship between $\boldsymbol{\omega}$ and $\dot{\boldsymbol{\epsilon}}$, and $\dot{\eta}$ is

$$\begin{bmatrix} \dot{\boldsymbol{\epsilon}} \\ \dot{\eta} \end{bmatrix} = \frac{1}{2}\begin{bmatrix} \eta\mathbf{1} + \boldsymbol{\epsilon}^\times \\ -\boldsymbol{\epsilon}^T \end{bmatrix}\boldsymbol{\omega} \tag{25.39}$$

As discussed in Chapter 24, the spacecraft dynamics and kinematics in (25.38) and (25.39) can be expressed in first order state-space form:

$$\underbrace{\begin{bmatrix} \dot{\boldsymbol{\omega}} \\ \dot{\boldsymbol{\epsilon}} \\ \dot{\eta} \end{bmatrix}}_{\dot{\mathbf{x}}} = \underbrace{\begin{bmatrix} \mathbf{I}^{-1}[-\boldsymbol{\omega}^\times\mathbf{I}\boldsymbol{\omega} + \mathbf{u} + \mathbf{w}] \\ \frac{1}{2}(\eta\mathbf{1} + \boldsymbol{\epsilon}^\times)\boldsymbol{\omega} \\ -\frac{1}{2}\boldsymbol{\epsilon}^T\boldsymbol{\omega} \end{bmatrix}}_{\mathbf{f}(\mathbf{x},\mathbf{u},\mathbf{w})}. \tag{25.40}$$

We assume that the spacecraft is equipped with a magnetometer that provides $\hat{\mathbf{b}}_b$, the Earth magnetic field vector in the spacecraft body frame (normalized), and a sun sensor that provides $\hat{\mathbf{s}}_b$, the sun position vector in the spacecraft body frame (also normalized). It is assumed the Earth's magnetic field vector and the sun position vector expressed in the inertial frame, $\hat{\mathbf{b}}_a$ and $\hat{\mathbf{s}}_a$ respectively, are known. The relationship between $\hat{\mathbf{b}}_b$, $\hat{\mathbf{b}}_a$, $\hat{\mathbf{s}}_b$, and $\hat{\mathbf{s}}_a$ is $\hat{\mathbf{b}}_b = \mathbf{C}_{ba}\hat{\mathbf{b}}_a$ and $\hat{\mathbf{b}}_s = \mathbf{C}_{ba}\hat{\mathbf{s}}_a$ where

$$\mathbf{C}_{ba} = (\eta^2 - \boldsymbol{\epsilon}^T\boldsymbol{\epsilon})\mathbf{1} + 2\boldsymbol{\epsilon}\boldsymbol{\epsilon}^T - 2\eta\boldsymbol{\epsilon}^\times. \tag{25.41}$$

The measurement equation is then

$$\mathbf{y} = \begin{bmatrix} \hat{\mathbf{b}}_b \\ \hat{\mathbf{s}}_b \end{bmatrix} + \mathbf{v} = \begin{bmatrix} \mathbf{C}_{ba}\hat{\mathbf{b}}_a \\ \mathbf{C}_{ba}\hat{\mathbf{s}}_a \end{bmatrix} + \mathbf{v}.$$

Using equation (25.41) and the identity $-\boldsymbol{\epsilon}^{\times}\boldsymbol{\epsilon}^{\times} = \boldsymbol{\epsilon}^T\boldsymbol{\epsilon}\mathbf{1} - \boldsymbol{\epsilon}\boldsymbol{\epsilon}^T$ we can write $\mathbf{C}_{ba}\hat{\mathbf{a}}_a$ where $\hat{\mathbf{a}}_a$ replaces $\hat{\mathbf{b}}_a$ or $\hat{\mathbf{s}}_a$ as

$$\underbrace{\begin{bmatrix} \eta\mathbf{1} - \boldsymbol{\epsilon}^{\times} & -\boldsymbol{\epsilon} \end{bmatrix} \begin{bmatrix} \hat{\mathbf{a}}_a^{\times} & \hat{\mathbf{a}}_a \\ -\hat{\mathbf{a}}_a^T & 0 \end{bmatrix}}_{\mathbf{Y}(\hat{\mathbf{a}}_a,\mathbf{q})} \begin{bmatrix} \boldsymbol{\epsilon} \\ \eta \end{bmatrix} = \mathbf{Y}(\hat{\mathbf{a}}_a, \mathbf{q})\mathbf{q} \tag{25.42}$$

so that

$$\mathbf{y} = \begin{bmatrix} \mathbf{0} & \mathbf{Y}(\hat{\mathbf{b}}_a, \mathbf{q}) \\ \mathbf{0} & \mathbf{Y}(\hat{\mathbf{s}}_a, \mathbf{q}) \end{bmatrix} \mathbf{x} + \mathbf{v}. \tag{25.43}$$

25.3.3.1 Linearization of the Discrete Time Process and Measurement Models

In order to use the norm-constrained EKF we need a discrete time process model and a discrete time measurement model. Also, in order to adapt the norm-constrained Kalman filter to the nonlinear estimation problem at hand, we will have to linearize the discrete time process and measurement models. We will start with developing the discrete time process model. Although there are many ways to discretize a system, we will discretize (25.40) by simply making the approximation

$$\begin{aligned} \frac{\mathbf{x}_k - \mathbf{x}_{k-1}}{T} &= \mathbf{f}(\mathbf{x}_{k-1}, \mathbf{u}_{k-1}, \mathbf{w}_{k-1}), \\ \mathbf{x}_k &= \underbrace{\mathbf{x}_{k-1} + T\mathbf{f}(\mathbf{x}_{k-1}, \mathbf{u}_{k-1}, \mathbf{w}_{k-1})}_{\mathbf{f}_{k-1}(\mathbf{x}_{k-1},\mathbf{u}_{k-1},\mathbf{w}_{k-1})} \end{aligned} \tag{25.44}$$

where $T = t_k - t_{k-1}$ is the sample period and $\mathbf{u}_k$ is constant between t_{k-1} and t_k. We assume that $\mathbf{w}_k \sim \mathcal{N}(\mathbf{0}, \mathbf{Q}_k)$.

We will now linearize our discrete time system. To do so we perform a Taylor series expansion in $\mathbf{x}_k$ and $\mathbf{w}_k$. In $\mathbf{x}_k$, $\bar{\mathbf{x}}_k$ is the nominal solution and $\delta\mathbf{x}_k$ is a perturbation such that $\mathbf{x}_k = \bar{\mathbf{x}}_k + \delta\mathbf{x}_k$. In $\mathbf{w}_k$, we assume $\bar{\mathbf{w}}_k = \mathbf{0}$ is the nominal disturbance and $\delta\mathbf{w}_k$ a perturbation. Substituting $\mathbf{x}_k = \bar{\mathbf{x}}_k + \delta\mathbf{x}_k$ and $\mathbf{w}_{k-1} = \mathbf{0} + \delta\mathbf{w}_{k-1}$ into (25.44) yields

$$\begin{aligned} \bar{\mathbf{x}}_k + \delta\mathbf{x}_k &= \bar{\mathbf{x}}_{k-1} + \delta\mathbf{x}_{k-1} + T\mathbf{f}(\bar{\mathbf{x}}_{k-1} + \delta\mathbf{x}_{k-1}, \mathbf{u}_{k-1}, \mathbf{0} + \delta\mathbf{w}_{k-1}) \\ &= \underbrace{\bar{\mathbf{x}}_{k-1} + T\mathbf{f}(\bar{\mathbf{x}}_{k-1}, \mathbf{u}_{k-1}, \mathbf{0})}_{\mathbf{f}_{k-1}(\bar{\mathbf{x}}_{k-1},\mathbf{u}_{k-1},\mathbf{0})} \end{aligned}$$

$$+\underbrace{\left[\mathbf{1}+T\begin{bmatrix}\mathbf{I}^{-1}[-\boldsymbol{\omega}_{k-1}^{\times}\mathbf{I}+(\mathbf{I}\boldsymbol{\omega}_{k-1})^{\times}] & \mathbf{0} & \mathbf{0}\\ \frac{1}{2}(\bar{\eta}_{k-1}\mathbf{1}+\bar{\boldsymbol{\epsilon}}_{k-1}^{\times}) & -\frac{1}{2}\bar{\boldsymbol{\omega}}_{k-1}^{\times} & \frac{1}{2}\bar{\boldsymbol{\omega}}_{k-1}\\ -\frac{1}{2}\bar{\boldsymbol{\epsilon}}_{k-1}^{T} & -\frac{1}{2}\bar{\boldsymbol{\omega}}_{k-1}^{T} & 0\end{bmatrix}\right]}_{\mathbf{F}_{k-1}}$$

$$\times\begin{bmatrix}\delta\boldsymbol{\omega}_{k-1}\\ \delta\boldsymbol{\epsilon}_{k-1}\\ \delta\eta_{k-1}\end{bmatrix}+\underbrace{\begin{bmatrix}T\mathbf{I}^{-1}\\ \mathbf{0}\\ \mathbf{0}\end{bmatrix}}_{\mathbf{L}_{k-1}}\delta\mathbf{w}_{k-1}$$

$$=\mathbf{f}_{k-1}(\bar{\mathbf{x}}_{k-1},\mathbf{u}_{k-1},\mathbf{0})+\mathbf{F}_{k-1}\delta\mathbf{x}_{k-1}+\mathbf{L}_{k-1}\delta\mathbf{w}_{k-1} \tag{25.45}$$

where we have neglected products of $\delta\boldsymbol{\epsilon}_{k-1}$, $\delta\eta_{k-1}$, and $\delta\boldsymbol{\omega}_{k-1}$. We can rearrange (25.45) and write

$$\begin{aligned}\mathbf{x}_k &= \mathbf{f}_{k-1}(\bar{\mathbf{x}}_{k-1},\mathbf{u}_{k-1},\mathbf{0})+\mathbf{F}_{k-1}(\mathbf{x}_{k-1}-\bar{\mathbf{x}}_{k-1})+\mathbf{L}_{k-1}\mathbf{w}_{k-1}\\ &= \mathbf{F}_{k-1}\mathbf{x}_{k-1}+\underbrace{(\mathbf{f}_{k-1}(\bar{\mathbf{x}}_{k-1},\mathbf{u}_{k-1},\mathbf{0})-\mathbf{F}_{k-1}\bar{\mathbf{x}}_{k-1})}_{\boldsymbol{u}_{k-1}}+\mathbf{L}_{k-1}\mathbf{w}_{k-1}\\ &= \mathbf{F}_{k-1}\mathbf{x}_{k-1}+\boldsymbol{u}_{k-1}+\mathbf{L}_{k-1}\mathbf{w}_{k-1}\end{aligned} \tag{25.46}$$

where $\delta\mathbf{w}_{k-1}=\mathbf{w}_{k-1}$ and $\boldsymbol{u}_{k-1}$ can be thought of as a fictitious input that will be useful when we are defining the prediction step of the norm-constrained EKF.

We will now consider the discrete time measurement model. Recall that the spacecraft is equipped with a magnetometer and sun sensor. The measurement equation is given by (25.43). The discrete time measurement model is

$$\mathbf{y}_k=\underbrace{\begin{bmatrix}\mathbf{0} & \mathbf{Y}(\hat{\mathbf{b}}_a,\mathbf{q}_k)\\ \mathbf{0} & \mathbf{Y}(\hat{\mathbf{s}}_a,\mathbf{q}_k)\end{bmatrix}\mathbf{x}_k+\mathbf{v}_k}_{\mathbf{h}_k(\mathbf{x}_k,\mathbf{v}_k)} \tag{25.47}$$

where $\mathbf{v}_k\sim\mathcal{N}(\mathbf{0},\mathbf{R}_k)$.

To linearize the discrete time measurement model we must linearize $\mathbf{Y}(\hat{\mathbf{a}}_a,\mathbf{q}_k)$ given in (25.42). Again we perform a Taylor series expansion in $\mathbf{x}_k$ and $\mathbf{v}_k$ about $\bar{\mathbf{x}}_k$ and $\mathbf{0}$, respectively, giving $\mathbf{x}_k=\bar{\mathbf{x}}_k+\delta\mathbf{x}_k$ (which $\mathbf{q}_k=\bar{\mathbf{q}}_k+\delta\mathbf{q}_k$ is part of) and $\mathbf{v}_k=\mathbf{0}+\delta\mathbf{v}_k$. Substituting these expressions into (25.42) gives

$$\begin{aligned}&\mathbf{Y}(\hat{\mathbf{a}}_a,\bar{\mathbf{q}}_k+\delta\mathbf{q}_k)(\bar{\mathbf{q}}_k+\delta\mathbf{q}_k)\\ &=\mathbf{Y}(\hat{\mathbf{a}}_a,\bar{\mathbf{q}}_k)\bar{\mathbf{q}}_k+\underbrace{\left(\mathbf{Y}(\hat{\mathbf{a}}_a,\bar{\mathbf{q}}_k)+\left[\left((\hat{\mathbf{a}}_a^{\times}\bar{\boldsymbol{\epsilon}}_k+\hat{\mathbf{a}}_a\bar{\eta}_k)^{\times}+\hat{\mathbf{a}}_a^{T}\bar{\boldsymbol{\epsilon}}_k\mathbf{1}\right)\quad(\hat{\mathbf{a}}_a^{\times}\bar{\boldsymbol{\epsilon}}_k+\hat{\mathbf{a}}_a\bar{\eta}_k)\right]\right)}_{\bar{\mathbf{Y}}(\hat{\mathbf{a}}_a,\bar{\mathbf{q}}_k)}\delta\mathbf{q}_k\end{aligned}$$

where products of $\delta\epsilon_k$ and $\delta\eta_k$ have been neglected. Letting $\mathbf{y}_k = \bar{\mathbf{y}}_k + \delta\mathbf{y}_k$, equation (25.47) is

$$
\begin{aligned}
\bar{\mathbf{y}}_k + \delta\mathbf{y}_k &= \begin{bmatrix} \mathbf{0} & \mathbf{Y}(\hat{\mathbf{b}}_a, \bar{\mathbf{q}}_k + \delta\mathbf{q}_k) \\ \mathbf{0} & \mathbf{Y}(\hat{\mathbf{s}}_a, \bar{\mathbf{q}}_k + \delta\mathbf{q}_k) \end{bmatrix} (\bar{\mathbf{x}}_k + \delta\mathbf{x}_k) + \delta\mathbf{v}_k \\
&= \underbrace{\begin{bmatrix} \mathbf{0} & \mathbf{Y}(\hat{\mathbf{b}}_a, \bar{\mathbf{q}}_k) \\ \mathbf{0} & \mathbf{Y}(\hat{\mathbf{s}}_a, \bar{\mathbf{q}}_k) \end{bmatrix} \bar{\mathbf{x}}_k}_{\mathbf{h}_k(\bar{\mathbf{x}}_k, \mathbf{0})} + \underbrace{\begin{bmatrix} \mathbf{0} & \bar{\mathbf{Y}}(\hat{\mathbf{b}}_a, \bar{\mathbf{q}}_k) \\ \mathbf{0} & \bar{\mathbf{Y}}(\hat{\mathbf{s}}_a, \bar{\mathbf{q}}_k) \end{bmatrix}}_{\mathbf{H}_k} \delta\mathbf{x}_k + \underbrace{\mathbf{1}}_{\mathbf{M}_k} \delta\mathbf{v}_k
\end{aligned} \tag{25.48}
$$

Equation (25.48) can also be written as

$$
\begin{aligned}
\mathbf{y}_k &= \mathbf{h}_k(\bar{\mathbf{x}}_k, \mathbf{0}) + \mathbf{H}_k(\mathbf{x}_k - \bar{\mathbf{x}}_k) + \mathbf{M}_k\mathbf{v}_k \\
&= \mathbf{H}_k\mathbf{x}_k + \underbrace{(\mathbf{h}_k(\bar{\mathbf{x}}_k, \mathbf{0}) - \mathbf{H}_k\bar{\mathbf{x}}_k)}_{y_k} + \mathbf{M}_k\mathbf{v}_k
\end{aligned} \tag{25.49}
$$

where $\delta\mathbf{v}_k = \mathbf{v}_k$ and $\boldsymbol{y}_y$ is defined to simplify the derivation of the correction step of the norm-constrained EKF.

Notice that equation (25.46) is linear in $\mathbf{x}_{k-1}$, $\boldsymbol{u}_{k-1}$, and $\mathbf{w}_{k-1}$, and equation (25.49) is linear in $\mathbf{x}_k$, $\boldsymbol{y}_k$, and $\mathbf{v}_k$. These equations will be applied to the norm-constrained Kalman filter that, after some additional manipulation, will yield the norm-constrained EKF.

25.3.3.2 The Prediction Step

As with the Kalman filter and the norm-constrained Kalman filter, we start with the prediction step where we predict the state $\hat{\mathbf{x}}_k^-$ given the process model. In our case, the process model is described by equation (25.46)

$$\hat{\mathbf{x}}_k^- = \mathbf{F}_{k-1}\hat{\mathbf{x}}_{k-1} + \boldsymbol{u}_{k-1}$$

where $\mathbf{F}_{k-1}$ and $\boldsymbol{u}_{k-1}$ are evaluated at the best prior estimate of the state, $\hat{\mathbf{x}}_{k-1}$ (i.e., $\hat{\mathbf{x}}_{k-1}$ replaces $\bar{\mathbf{x}}_{k-1}$ in equation (25.46)). Notice that the above prediction is equivalent to

$$
\begin{aligned}
\hat{\mathbf{x}}_k^- &= \mathbf{F}_{k-1}\hat{\mathbf{x}}_{k-1} + \boldsymbol{u}_{k-1} \\
&= \mathbf{F}_{k-1}\hat{\mathbf{x}}_{k-1} + (\mathbf{f}_{k-1}(\hat{\mathbf{x}}_{k-1}, \mathbf{u}_{k-1}, \mathbf{0}) - \mathbf{F}_{k-1}\hat{\mathbf{x}}_{k-1}) \\
&= \mathbf{f}_{k-1}(\hat{\mathbf{x}}_{k-1}, \mathbf{u}_{k-1}, \mathbf{0})
\end{aligned}
$$

which is just the nonlinear discrete time process model evaluated at $\hat{\mathbf{x}}_{k-1}$, $\mathbf{u}_{k-1}$, and $\mathbf{w}_k = \mathbf{0}$. As with the Kalman filter, we do not perform a prediction step that includes disturbances because the disturbances are unknown.

In practice, often a more accurate numerical integration is used to predict $\hat{\mathbf{x}}_k$. For example, we can predict $\hat{\mathbf{x}}_k$ by using a fourth-order Runge-Kutta numerical integration method (see Appendix B):

$$\hat{\mathbf{x}}_k^- = \hat{\mathbf{x}}_{k-1} + \int_{t_{k-1}}^{t_k} \mathbf{f}(\hat{\mathbf{x}}, \mathbf{u}, \mathbf{0})d\tau$$

where $\mathbf{f}(\mathbf{x}, \mathbf{u}, \mathbf{w})$ is from equation (25.40). (Note, $\mathbf{f}(\mathbf{x}, \mathbf{u}, \mathbf{w})$ in (25.40) is not the same as $\mathbf{f}_{k-1}(\mathbf{x}_{k-1}, \mathbf{u}_{k-1}, \mathbf{w}_{k-1})$ in (25.44); the former is the continuous time model while the latter is the discrete time model.)

The a priori covariance is computed in the same way as in the Kalman filter and the norm-constrained Kalman filter, except the linearized process model is used. Doing so gives

$$\begin{aligned}\mathbf{P}_k^- &= \mathbf{F}_{k-1}\mathbf{P}_{k-1}\mathbf{F}_{k-1}^T + \mathbf{L}_{k-1}\mathbf{Q}_{k-1}\mathbf{L}_{k-1}^T \\ &= \begin{bmatrix}\mathbf{P}_{1,k}^- & \mathbf{P}_{2,k}^-\end{bmatrix} \\ &= \begin{bmatrix}\mathbf{P}_{\omega\omega,k}^- & \mathbf{P}_{\omega q,k}^- \\ \mathbf{P}_{q\omega,k}^- & \mathbf{P}_{qq,k}^-\end{bmatrix}\end{aligned}$$

where, in preparation to use the norm-constrained Kalman filter structure, we have partitioned $\mathbf{P}_k^-$ appropriately.

25.3.3.3 The Correction Step

Again we will assume a linear update of the form

$$\begin{bmatrix}\hat{\boldsymbol{\omega}}_k \\ \hat{\mathbf{q}}_k\end{bmatrix} = \begin{bmatrix}\hat{\boldsymbol{\omega}}_k^- \\ \hat{\mathbf{q}}_k^-\end{bmatrix} + \begin{bmatrix}\mathbf{K}_{\omega,k} \\ \mathbf{K}_{q,k}\end{bmatrix}(\mathbf{y}_k - \hat{\mathbf{y}}_k^-) \tag{25.50}$$

where $\mathbf{K}_k$ has been partitioned into $\mathbf{K}_{\omega,k}$ and $\mathbf{K}_{q,k}$. The predicted measurement $\hat{\mathbf{y}}_k^-$ is calculated using (25.49):

$$\hat{\mathbf{y}}_k^- = \mathbf{H}_k\hat{\mathbf{x}}_k^- + y_k^-$$

where $\mathbf{H}_k$ and y_y are evaluated at the best a priori estimate of the state, $\hat{\mathbf{x}}_k^-$ (i.e., $\hat{\mathbf{x}}_k^-$ replaces $\bar{\mathbf{x}}_k$ in equation (25.49)). The predicted measurement can also be written

$$\begin{aligned}\hat{\mathbf{y}}_k^- &= \mathbf{H}_k\hat{\mathbf{x}}_k^- + y_k^- \\ &= \mathbf{H}_k\hat{\mathbf{x}}_k^- + \left(\mathbf{h}_k(\hat{\mathbf{x}}_k^-, \mathbf{0}) - \mathbf{H}_k\hat{\mathbf{x}}_k^-\right) \\ &= \mathbf{h}_k(\hat{\mathbf{x}}_k^-, \mathbf{0})\end{aligned}$$

which is the nonlinear discrete time measurement model evaluated at $\hat{\mathbf{x}}_k^-$, the a priori state estimate. Equation (25.50) can now be written as

$$\hat{\boldsymbol{\omega}}_k = \hat{\boldsymbol{\omega}}_k^- + \mathbf{K}_{\omega,k}\left(\mathbf{y}_k - \mathbf{h}_k(\hat{\mathbf{x}}_k^-, \mathbf{0})\right), \tag{25.51}$$

$$\hat{\mathbf{q}}_k = \hat{\mathbf{q}}_k^- + \mathbf{K}_{q,k}\left(\mathbf{y}_k - \mathbf{h}_k(\hat{\mathbf{x}}_k^-, \mathbf{0})\right). \tag{25.52}$$

The spacecraft angular velocity is not constrained in any way. As such, $\hat{\boldsymbol{\omega}}_k$ is computed from (25.51) where $\mathbf{K}_{\omega,k} = \mathbf{P}_{1,k}^{-\,T}\mathbf{H}_k^T\mathbf{W}_k^{-1}$, $\mathbf{W}_k = \mathbf{H}_k\mathbf{P}_k^-\mathbf{H}_k^T + \mathbf{M}_k\mathbf{R}_k\mathbf{M}_k^T$, and $\mathbf{M}_k = \mathbf{1}$. On the other hand, the quaternion is constrained to have unit length: $\mathbf{q}_k^T\mathbf{q}_k = 1$. Following the norm-constrained Kalman filter formulation with $\ell = 1$ the correction step is given by (25.52) where

$$\mathbf{K}_{q,k} = \tilde{\mathbf{K}}_{q,k} + \frac{1}{r_k}\left(\frac{1}{\|\tilde{\mathbf{q}}_k\|} - 1\right)\tilde{\mathbf{q}}_k\mathbf{r}_k^T\mathbf{W}_k^{-1},$$

$\tilde{\mathbf{K}}_{q,k} = \mathbf{P}_{2,k}^{-\,T}\mathbf{H}_k^T\mathbf{W}_k^{-1}$, $\mathbf{r}_k = \mathbf{y}_k - \mathbf{h}_k(\hat{\mathbf{x}}_k^-, \mathbf{0})$ is the measurement residual, $r_k = \mathbf{r}_k^T\mathbf{W}_k^{-1}\mathbf{r}_k$, and

$$\tilde{\mathbf{q}}_k = \hat{\mathbf{q}}_k^- + \tilde{\mathbf{K}}_{q,k}\mathbf{r}_k$$

is an update that is not assured to satisfy the unit-quaternion constraint. The update expression for $\hat{\mathbf{q}}_k$ given in equation (25.52) can also be written

$$\begin{aligned}
\hat{\mathbf{q}}_k &= \hat{\mathbf{q}}_k^- + \mathbf{K}_{q,k}\mathbf{r}_k \\
&= \hat{\mathbf{q}}_k^- + \tilde{\mathbf{K}}_{q,k}\mathbf{r}_k + \frac{1}{r_k}\left(\frac{1}{\|\tilde{\mathbf{q}}_k\|} - 1\right)\tilde{\mathbf{q}}_k\mathbf{r}_k^T\mathbf{W}_k^{-1}\mathbf{r}_k \\
&= \tilde{\mathbf{q}}_k + \left(\frac{1}{\|\tilde{\mathbf{q}}_k\|} - 1\right)\tilde{\mathbf{q}}_k \\
&= \frac{1}{\|\tilde{\mathbf{q}}_k\|}\tilde{\mathbf{q}}_k
\end{aligned}$$

Following the norm-constrained EKF formulation the matrix $\mathbf{P}_k$ is

$$\mathbf{P}_k = \begin{bmatrix} \mathbf{P}_{\omega\omega,k} & \mathbf{P}_{\omega q,k} \\ \mathbf{P}_{q\omega,k} & \mathbf{P}_{qq,k} \end{bmatrix}$$

where

$$\begin{aligned}
\mathbf{P}_{\omega\omega,k} &= \mathbf{P}_{\omega\omega,k}^- - \mathbf{K}_{\omega,k}\mathbf{H}_k\mathbf{P}_{1,k}^- - \mathbf{P}_{1,k}^{-\,T}\mathbf{H}_k^T\mathbf{K}_{\omega,k}^T + \mathbf{K}_{\omega,k}\mathbf{W}_k\mathbf{K}_{\omega,k}^T, \\
\mathbf{P}_{\omega q,k} &= \mathbf{P}_{\omega q,k}^- - \mathbf{K}_{\omega,k}\mathbf{H}_k\mathbf{P}_{2,k}^- - \mathbf{P}_{1,k}^{-\,T}\mathbf{H}_k^T\mathbf{K}_{q,k}^T + \mathbf{K}_{\omega,k}\mathbf{W}_k\mathbf{K}_{q,k}^T, \\
\mathbf{P}_{q\omega,k} &= \mathbf{P}_{q\omega,k}^- - \mathbf{K}_{q,k}\mathbf{H}_k\mathbf{P}_{1,k}^- - \mathbf{P}_{2,k}^{-\,T}\mathbf{H}_k^T\mathbf{K}_{\omega,k}^T + \mathbf{K}_{q,k}\mathbf{W}_k\mathbf{K}_{\omega,k}^T, \\
\mathbf{P}_{qq,k} &= \mathbf{P}_{qq,k}^- - \mathbf{K}_{q,k}\mathbf{H}_k\mathbf{P}_{2,k}^- - \mathbf{P}_{2,k}^{-\,T}\mathbf{H}_k^T\mathbf{K}_{q,k}^T + \mathbf{K}_{q,k}\mathbf{W}_k\mathbf{K}_{q,k}^T.
\end{aligned}$$

25.3.3.4 Summary of the Norm-Constrained Extended Kalman Filter

System:
$$\mathbf{x}_k = \mathbf{f}_{k-1}(\mathbf{x}_{k-1}, \mathbf{u}_{k-1}, \mathbf{w}_{k-1})$$
$$\mathbf{y}_k = \mathbf{h}_k(\mathbf{x}_k, \mathbf{v}_k)$$
$$\mathbf{x}_k = \begin{bmatrix}\boldsymbol{\omega}_k^T & \mathbf{q}_k^T\end{bmatrix}^T$$
$$\mathbf{q}_k^T \mathbf{q}_k = 1$$
$$\mathbf{w}_k \sim \mathcal{N}(\mathbf{0}, \mathbf{Q}_k)$$
$$\mathbf{v}_k \sim \mathcal{N}(\mathbf{0}, \mathbf{R}_k)$$
$$E[\mathbf{w}_k \mathbf{v}_k^T] = \mathbf{0}$$

Initialization:
$$\hat{\mathbf{x}}_0 = E[\mathbf{x}_0]$$
$$\mathbf{P}_0 = E\left[(\mathbf{x}_0 - \hat{\mathbf{x}}_0)(\mathbf{x}_0 - \hat{\mathbf{x}}_0)^T\right]$$

Prediction:
$$\hat{\mathbf{x}}_k^- = \mathbf{f}_{k-1}(\hat{\mathbf{x}}_{k-1}, \mathbf{u}_{k-1}, \mathbf{0}) \quad \text{or} \quad \hat{\mathbf{x}}_k^- = \hat{\mathbf{x}}_{k-1} + \int_{t_{k-1}}^{t_k} \mathbf{f}(\hat{\mathbf{x}}, \mathbf{u}, \mathbf{0}) d\tau$$
$$\mathbf{P}_k^- = \mathbf{F}_{k-1}\mathbf{P}_{k-1}\mathbf{F}_{k-1}^T + \mathbf{L}_{k-1}\mathbf{Q}_{k-1}\mathbf{L}_{k-1}^T$$
$$= \begin{bmatrix}\mathbf{P}_{1,k}^- & \mathbf{P}_{2,k}^-\end{bmatrix} = \begin{bmatrix}\mathbf{P}_{\omega\omega,k}^- & \mathbf{P}_{\omega q,k}^- \\ \mathbf{P}_{q\omega,k}^- & \mathbf{P}_{qq,k}^-\end{bmatrix}$$

Correction:
$$\mathbf{W}_k = \mathbf{H}_k\mathbf{P}_k^-\mathbf{H}_k^T + \mathbf{M}_k\mathbf{R}_k\mathbf{M}_k^T$$
$$\mathbf{K}_{\omega,k} = \mathbf{P}_{1,k}^{-\,T}\mathbf{H}_k^T\mathbf{W}_k^{-1}$$
$$\hat{\boldsymbol{\omega}}_k = \hat{\boldsymbol{\omega}}_k^- + \mathbf{K}_{\omega,k}\left(\mathbf{y}_k - \mathbf{h}_k(\hat{\mathbf{x}}_k^-, \mathbf{0})\right)$$
$$\tilde{\mathbf{K}}_{q,k} = \mathbf{P}_{2,k}^{-\,T}\mathbf{H}_k^T\mathbf{W}_k^{-1}$$
$$\mathbf{r}_k = \mathbf{y}_k - \mathbf{h}_k(\hat{\mathbf{x}}_k^-, \mathbf{0})$$
$$r_k = \mathbf{r}_k^T\mathbf{W}_k^{-1}\mathbf{r}_k$$
$$\tilde{\mathbf{q}}_k = \hat{\mathbf{q}}_k^- + \tilde{\mathbf{K}}_{q,k}\mathbf{r}_k$$
$$\mathbf{K}_{q,k} = \tilde{\mathbf{K}}_{q,k} + \frac{1}{r_k}\left(\frac{1}{\|\tilde{\mathbf{q}}_k\|} - 1\right)\tilde{\mathbf{q}}_k\mathbf{r}_k^T\mathbf{W}_k^{-1}$$
$$\hat{\mathbf{q}}_k = \hat{\mathbf{q}}_k^- + \mathbf{K}_{q,k}\mathbf{r}_k = \frac{1}{\|\tilde{\mathbf{q}}_k\|}\tilde{\mathbf{q}}_k$$
$$\mathbf{P}_{\omega\omega,k} = \mathbf{P}_{\omega\omega,k}^- - \mathbf{K}_{\omega,k}\mathbf{H}_k\mathbf{P}_{1,k}^- - \mathbf{P}_{1,k}^{-\,T}\mathbf{H}_k^T\mathbf{K}_{\omega,k}^T + \mathbf{K}_{\omega,k}\mathbf{W}_k\mathbf{K}_{\omega,k}^T$$
$$\mathbf{P}_{\omega q,k} = \mathbf{P}_{\omega q,k}^- - \mathbf{K}_{\omega,k}\mathbf{H}_k\mathbf{P}_{2,k}^- - \mathbf{P}_{1,k}^{-\,T}\mathbf{H}_k^T\mathbf{K}_{q,k}^T + \mathbf{K}_{\omega,k}\mathbf{W}_k\mathbf{K}_{q,k}^T$$
$$\mathbf{P}_{q\omega,k} = \mathbf{P}_{q\omega,k}^- - \mathbf{K}_{q,k}\mathbf{H}_k\mathbf{P}_{1,k}^- - \mathbf{P}_{2,k}^{-\,T}\mathbf{H}_k^T\mathbf{K}_{\omega,k}^T + \mathbf{K}_{q,k}\mathbf{W}_k\mathbf{K}_{\omega,k}^T$$
$$\mathbf{P}_{qq,k} = \mathbf{P}_{qq,k}^- - \mathbf{K}_{q,k}\mathbf{H}_k\mathbf{P}_{2,k}^- - \mathbf{P}_{2,k}^{-\,T}\mathbf{H}_k^T\mathbf{K}_{q,k}^T + \mathbf{K}_{q,k}\mathbf{W}_k\mathbf{K}_{q,k}^T$$
$$\mathbf{P}_k = \begin{bmatrix}\mathbf{P}_{\omega\omega,k} & \mathbf{P}_{\omega q,k} \\ \mathbf{P}_{q\omega,k} & \mathbf{P}_{qq,k}\end{bmatrix}$$

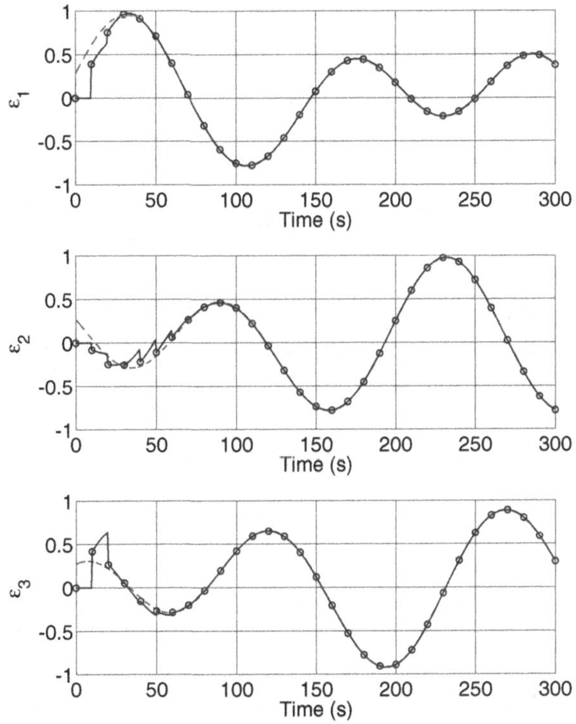

Figure 25.4 Estimate of $\boldsymbol{\epsilon}_k$ (solid line) and the actual value of $\boldsymbol{\epsilon}_k$ (dashed line) versus time. Circles indicate when measurements become available

Example 25.3 *We will consider an uncontrolled spacecraft (that is,* $\mathbf{u}_k = \mathbf{0}$ *for all k) subject to a magnetic disturbance (caused by a residual dipole moment) and a gravity gradient disturbance. The residual dipole moment is equal to* $\mathbf{m}_r = [0.1\ 0.1\ 0.1]^T\ A \cdot m^2$. *The spacecraft mass moment of inertia is* $\mathbf{I} = diag\,\{27, 17, 25\}\ kg \cdot m^2$. *The spacecraft is in a circular orbit at an altitude of* 450 *km and inclination of* 87°. *The initial angular velocity and attitude is* $\boldsymbol{\omega}(0) = [0.05\ \ -0.05\ 0.05]^T\ rad/s$ *and* $\boldsymbol{\epsilon}(0) = (\sin(1/2)/\sqrt{3})[1\ 1\ 1]^T$, $\eta(0) = \cos(1/2)$. *In simulation this truth model is numerically integrated using a fourth-order Runge-Kutta integrator with a time-step of* 0.05 *s. The truth model incorporates the disturbances.*

For the norm-constrained EKF, the process model assumes there are no disturbances. The initial state is $\hat{\boldsymbol{\omega}}(0) = \mathbf{0}\ rad/s$, $\hat{\boldsymbol{\epsilon}}(0) = \mathbf{0}$, $\hat{\eta}(0) = 1$. *The initial covariance is* $\mathbf{P}_0 = diag\,\{0.1\ 0.1\ 0.1\ 0.1\ 0.1\ 0.1\ 0.1\}$. *The process noise covariance is* $\mathbf{Q}_k = \sigma_q^2\mathbf{1}$ *where* $\sigma_q = 0.001$. *The measurement noise covariance associated with the sun sensor and a magnetometer is* $\mathbf{R}_k = diag\,\left\{\sigma_m^2, \sigma_m^2, \sigma_m^2, \sigma_s^2, \sigma_s^2, \sigma_s^2\right\}$ *where* $\sigma_m = 0.01$ *and* $\sigma_s = 0.005$. *Note that we are assuming isotropic measurement noise for each vector measurement, which differs from the QUEST measurement model considered in Section 25.2.5. This assumption has the benefit that* $\mathbf{R}_k$ *is positive definite. It is reasonable provided the vector measurements are not normalized prior to use in the correction step of the EKF. Measurements are acquired every* 10 *seconds* ($T = 10$ *s). During the prediction step, a fourth-order Runge-Kutta is used to predict the state every* 0.5 *s.*

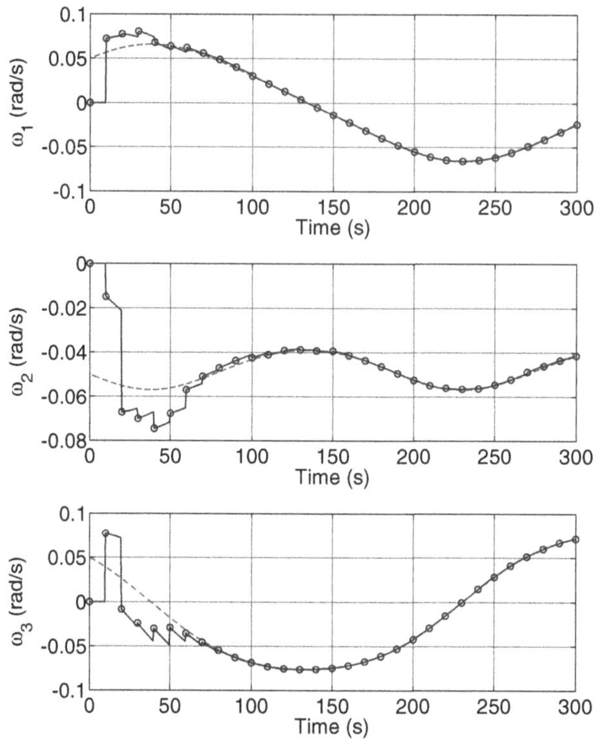

Figure 25.5 Estimate of ω_k (solid line) and the actual value of ω_k (dashed line) versus time. Circles indicate when measurements become available

Shown in Figures 25.4 and 25.5 is the estimate of the vector part of the quaternion, $\hat{\boldsymbol{\epsilon}}_k$, and the estimate of the angular velocity, $\hat{\boldsymbol{\omega}}_k$. Also included is the true values of $\boldsymbol{\epsilon}_k$ and $\boldsymbol{\omega}_k$. Notice that it takes approximately 6 or 7 measurements (which is about one minute) before $\hat{\boldsymbol{\epsilon}}_k$ and $\hat{\boldsymbol{\omega}}_k$ start to match $\boldsymbol{\epsilon}_k$ and $\boldsymbol{\omega}_k$. This is not unreasonable in that our initial estimate of the state was not at all close to the true state.

Notes

This chapter serves as an introduction to spacecraft navigation, and is by no means comprehensive. Spacecraft navigation is a vast and mature subject, and interested readers are encouraged to look at additional books and journal articles.

The material in Section (25.1) is adapted from Jazwinski (1970), Stengel (1994), Simon (2006), and Figliola and Beasley (2011). Wahba's problem described in Section (25.1) is formulated in Wahba (1965). Davenport's q-Method presented in Section (25.1) can be found in the book by Wertz (1978), while the QUEST algorithm presented in Section (25.1) and the TRIAD algorithm discussed in Section (25.1) can be found in Shuster and Oh (1981). A discussion of the TRIAD algorithm in the context of maximum likelihood estimation can be found in Shuster (2006). Davenport's q-Method is also discussed in Crassidis and Junkins (2004). The Kalman filter derivation presented in Section (25.1) is adapted from Jazwinski (1970) and Simon (2006). The norm-constrained Kalman filter derivation in Section (25.1) is adapted from Zanetti et al. (2009). A general derivation of the extended Kalman filter can be found in Jazwinski (1970), Stengel (1994), and Simon (2006); our derivation is based on the derivation in Simon (2006). The discretization and linearization of the spacecraft kinematics and dynamics is motivated by Leung and Damaren (2004), Crassidis and Junkins (2004), and Simon (2006). Additional discussions on observability, discretization, and linearization can be found in Jazwinski (1970), Stengel (1994), Simon (2006), and Crassidis and Junkins (2004). The discussion in this chapter has been restricted to spacecraft attitude determination. The Kalman filter may also be applied to the problem of spacecraft orbit determination. A comprehensive treatment of statistical methods for orbit determination may be found in Tapley et al. (2004).

References

Crassidis JL and Junkins JL 2004 *Optimal Estimation of Dynamic Systems*. Chapman & Hall/CRC, Boca Raton, FL.

Figliola RS and Beasley DE 2011 *Theory and Design for Mechanical Measurements* fifth edn. John Wiley & Sons, Inc., Hoboken, NJ.

Jazwinski AH 1970 *Stochastic Processes and Filtering Theory*. Academic Press, New York, NY.

Leung WS and Damaren CJ 2004 A Comparison of the Pseudo-Linear and Extended Kalman Filters for Spacecraft Attitude Estimation. *AIAA Guidance, Navigation, and Control Conference, Providence, Rhode Island, August 16–19, 2004*.

Shuster MD 2006 The TRIAD Algorithm as Maximum Likelihood Estimation. *Journal of the Astronautical Sciences* **54**(1), 113–123.

Shuster MD and Oh SD 1981 Three-Axis Attitude Determination from Vector Observations. *AIAA Journal of Guidance, Control, and Dynamics (previously the AIAA Journal of Guidance and Control)* **4**(1), 70–77.

Simon D 2006 *Optimal State Estimation*. John Wiley & Sons, Inc., Hoboken, NJ.

Stengel RF 1994 *Optimal Control and Estimation*. Dover, New York, NY.

Tapley BD, Schutz BE and Born GH 2004 *Statistical Orbit Determination*. Elsevier, Burlington, MA.

Wahba G 1965 Problem 65-1, A Least-Squares Estimate of Satellite Attitude. *SIAM Review* **7**(3), 409.

Wertz JR 1978 *Spacecraft Attitude Determination and Control*. Kluwer Academic Publishers, Dordrecht, The Netherlands.

Zanetti R, Majji M, Bishop RH and Mortari D 2009 Norm-Constrained Kalman Filtering. *AIAA Journal of Guidance, Control, and Dynamics* **32**(5), 1458–1465.

26

Practical Spacecraft Attitude Control Design Issues

In Chapters 17 to 25 we have examined some of the fundamental theory required for spacecraft attitude control design. However, on its own this is not enough. There are several associated practical issues that must be addressed if a spacecraft attitude control design is to be successful. In this chapter, we examine some of these issues. The objective is not to provide a comprehensive treatment (that is far beyond the scope of this book). Rather, the objective is to make the reader aware of the issues that need to be addressed when faced with a real spacecraft attitude control design problem.

26.1 Attitude Sensors

As explained in Chapter 11, in order to be able to control the spacecraft attitude, we need to be able to measure (or at least estimate) it. Unfortunately, there is no sensor that can measure the attitude directly, and it must be estimated from available measurements using techniques such as those presented in Chapter 25. In this section, we provide a brief overview of the different sensor types that may be used in spacecraft attitude control, and what measurements they provide.

26.1.1 Sun-Sensors

The two major types of sun-sensors are *analog* and *digital* sun-sensors. Digital sun-sensors are the most accurate, but also the most expensive. Analog sun-sensors are sufficient for many applications.

Spacecraft Dynamics and Control: An Introduction, First Edition.
Anton H.J. de Ruiter, Christopher J. Damaren and James R. Forbes.

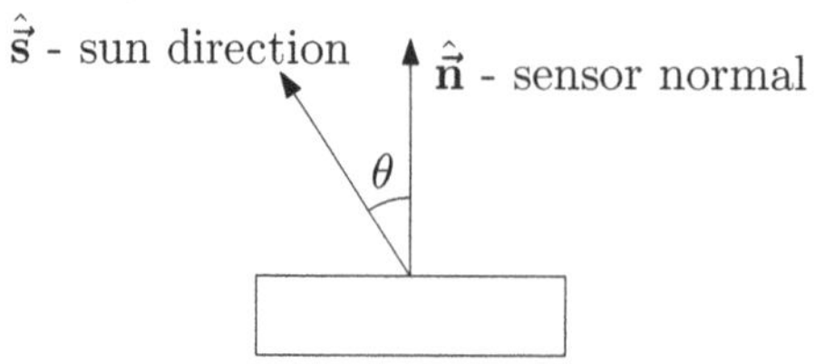

Figure 26.1 Analog sun-sensor

26.1.1.1 Analog Sun-Sensors

Analog sun-sensors are essentially solar cells, whose current output i is proportional to the cosine of the angle θ between the sensor normal $\hat{\mathbf{n}}$ and the incident solar radiation $\hat{\mathbf{s}}$ (refer to Figure 26.1). Mathematically, the sensor output is a current i, given by

$$i(\theta) = i(0)\cos\theta, \tag{26.1}$$

where $i(0)$ is the sensor output when the sun-direction is parallel to the sensor normal. For this reason, analog sun-sensors are sometimes referred to as *cosine sun-sensors*. In reality, the relationship between the current output i and the sun angle θ deviates slightly from the cosine law presented in (26.1), and must be determined by calibration of the sensor.

In any case, an analog sun-sensor provides the angle θ of the sun-vector relative to the sensor normal, and as such, it provides a cone $\mathcal{C}_s$ on which the sun-vector must lie (see Figure 26.2). Combinations of analog sun-sensors (with different sensor normals) can be used to determine more information about the sun vector. For example, a pair of analog sun-sensors with normals $\hat{\mathbf{n}}_1$ and $\hat{\mathbf{n}}_2$, yield a pair of cones on which the sun-vector must lie, which yields a pair of possible sun-vectors, $\hat{\mathbf{s}}_1$ and $\hat{\mathbf{s}}_2$ (see Figure 26.3). Additional measurements are needed to determine which is the true sun-vector. One possibility is a third analog sun-sensor with normal not in the plane of $\hat{\mathbf{n}}_1$ and $\hat{\mathbf{n}}_2$. This leads to three intersecting cones, which in the absence of measurement error will intersect along a single line and give the sun-vector. In the presence of measurement errors, least squares estimation can be used to give an estimate of the sun-vector.

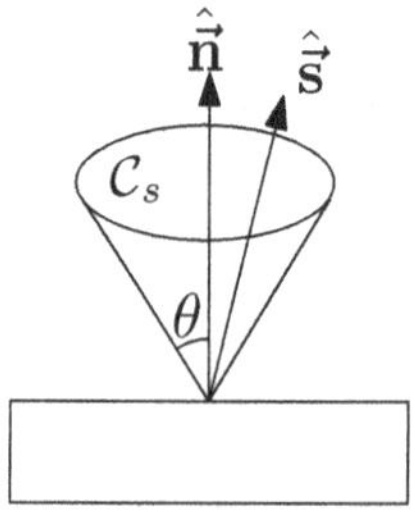

Figure 26.2 Sun-cone provide by an analog sun-sensor

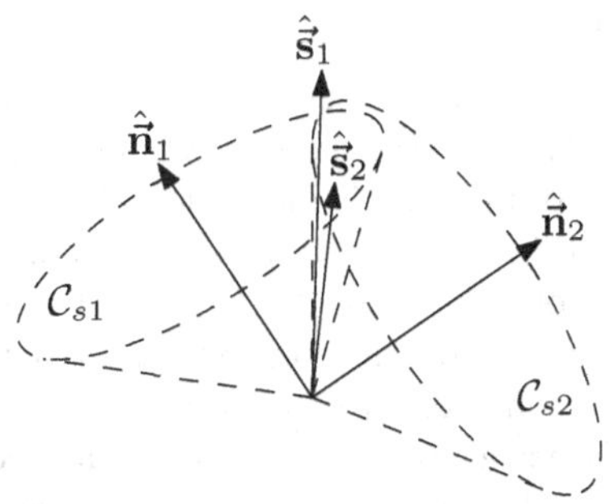

Figure 26.3 Possible sun-vectors from a pair of analog sun-sensors

As an alternative to using a third sun-sensor, if another vector measurement (in spacecraft body coordinates) is available (for example, a magnetometer measurement), then that measurement can be used in combination with knowledge of the angle between that vector and the sun-vector to provide a third cone on which the sun-vector must lie. The sun-vector can then be determined in the same manner as with three analog sun-sensors described in the previous paragraph. Note that this assumes that both the sun-vector and the second vector are known in some coordinate system, such as an inertial coordinate system, from which the angle between them can be determined.

26.1.1.2 Digital Sun-Sensors

Digital sun-sensors are significantly more complex than analog sun-sensors, however they are far more accurate, and a two-axis type digital sun-sensor can provide a full sun-vector in sensor coordinates. Using knowledge of the sensor orientation relative to the spacecraft body frame, the sun-vector can be readily transformed to the body-frame.

26.1.1.3 Sensor Limitations

Both analog and digital sun-sensors have limited useful fields of view, so multiple sensors may be needed to give the required coverage. Additionally, for digital sun-sensors, there may be spacecraft body-rate limitations outside of which the sensor ceases to be useful.

26.1.2 Three-Axis Magnetometers

A three-axis magnetometer measures the local magnetic field vector in sensor coordinates. That is, it gives both the magnitude and direction of the local magnetic field. Using knowledge of the sensor orientation relative to the spacecraft, the local magnetic field vector can be transformed to spacecraft body coordinates.

Magnetometers are relatively inaccurate. However, they do not have the field of view limitations inherent to sun-sensors, Earth-sensors and star-trackers. Therefore, they are very useful for initial attitude determination and initial attitude acquisition before the other more accurate measurements become available.

26.1.2.1 Sensor Limitations

As mentioned, a magnetometer measures the local magnetic field vector. However, this is only useful for the purposes of spacecraft attitude determination and control if the measured magnetic field consists mainly of the Earth's magnetic field. For this reason, magnetometer placement within the spacecraft becomes critical, so as to minimize the corruption of measurements by the spacecraft's own magnetic field due to ferro-magnetic materials (in the spacecraft structure) and spacecraft current loops (in the spacecraft electronics). Some (but not all) of these effects can be removed by calibration. For this reason, magnetometers are sometimes mounted on booms outside of the main spacecraft body. Additional measures are needed if magnetometers are used in conjunction with magnetic torquers so as to ensure that the magnetic torquers do not influence the magnetometer readings (see Section 26.2.2).

26.1.3 Earth Sensors

Earth sensors can be used to measure the nadir vector. That is, they measure the unit vector pointing from the spacecraft toward the Earth's center of mass. They are commonly used on spacecraft that are nominally nadir pointing, such as is the case considered in Chapter 16. In these applications, the nadir vector is a function of the roll and pitch angles of the spacecraft body frame relative to the orbiting frame, and the roll and pitch angles are typically provided directly by the Earth sensor. The yaw angle corresponds to a rotation about the nadir vector, so it cannot be measured.

From the point of view of an Earth orbiting spacecraft, the Earth appears as a large disk, so the direction of the Earth center cannot be measured directly. However, it can be deduced by locating the Earth's horizon as seen from the spacecraft (the boundary of the visible Earth disk). To do this, Earth sensors detect infrared (IR) radiation emitted from the Earth's surface. The reason that the infrared part of the electromagnetic spectrum is used is because the radiation from the Earth's surface is relatively uniform across the entire surface, allowing the horizon to be located accurately. By comparison, the Earth's radiation in the visible part of the spectrum varies greatly depending on whether the surface is in the day or night, which becomes a lot more problematic for location of the horizon.

There are two types of Earth sensors: static Earth sensors and scanning Earth sensors. A static Earth sensor consists of a suite of concentric infrared sensors, as shown in Figure 26.4. The

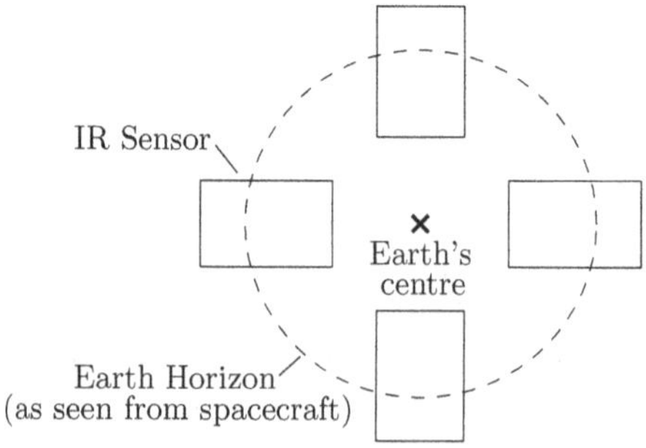

Figure 26.4 Static Earth sensor principle

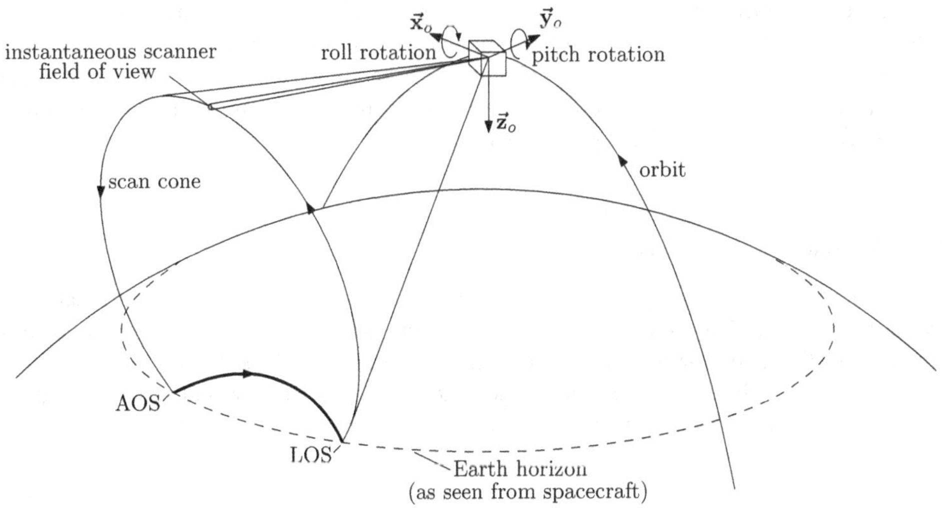

Figure 26.5 Scanning Earth sensor principle

signal from each sensor is proportional to the fraction of the Earth disk contained within its field of view. By combining the signals from each of the sensors, the Earth's horizon can be located, and consequently the nadir vector can be deduced. A scanning Earth sensor has a rotating optical head (with constant angular velocity), sweeping out a cone as shown in Figure 26.5. An infrared sensor then detects when the Earth's surface lies in the scanner's instantaneous field of view. Of particular importance are the horizon crossings by the scanner, that is, when the Earth's surface enters and leaves the scanner's instantaneous field of view. These crossings are called the Acquisition of Signal (AOS) and Loss of Signal (LOS), respectively. As illustrated in Figure 26.6(a), the time between AOS and LOS is dependent on the scanner's roll angle (relative to an orbiting frame), which can consequently be computed. As illustrated in Figure 26.6(b), knowledge of where the Earth's surface passes through the scanner's instantaneous

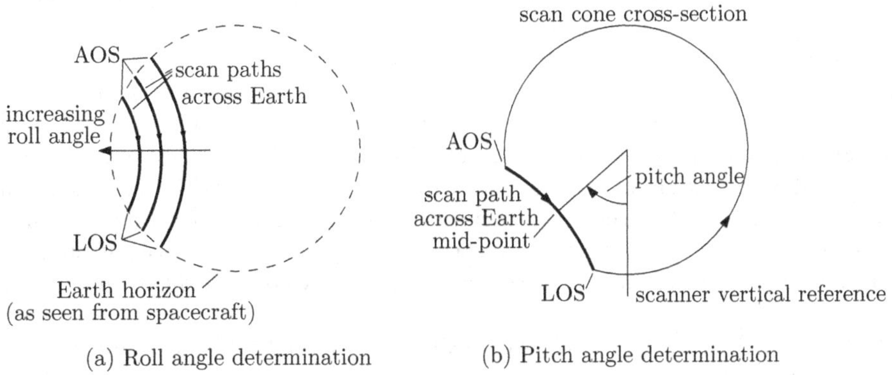

Figure 26.6 Scanning Earth sensor roll and pitch angle determination

field of view is dependent on the scanner's pitch angle (relative to an orbiting frame). Thus, the pitch angle can be determined provided the scanner has a built in vertical reference. Note that for a single scanner, the spacecraft altitude must be known (the time between AOS and LOS depends on the altitude as well as the roll angle). In order to overcome this limitation, scanning Earth sensors often have dual scanners.

26.1.3.1 Sensor Limitations

As is evident from both Figures 26.4 and 26.5, Earth sensors are only useful if the Earth lies within the sensor field of view. As such, Earth sensors are useful only for a limited range of attitudes. Additionally, for scanning Earth sensors, the spacecraft body-rate should be very small compared with the rotation rate of the scanning head. The sun and moon should not come within the sensor field of view.

26.1.4 Star Trackers

Stars provide the most accurate reference for attitude determination. There are two reasons for this. First of all, stars appear very small (compare the apparent size of a star to the apparent size of the sun or Earth). As such, the direction of a star can be measured very accurately. Second, stars are inertially fixed objects. As explained in Chapter 25, given any vector measurements in the spacecraft body frame, it is necessary to know the corresponding vectors in the inertial frame, in order to determine the spacecraft attitude. Considering Earth orbiting spacecraft, the sun vector depends on the orbital position of the Earth around the sun, as well as the orbital position of the spacecraft around the Earth. The nadir vector depends on the spacecraft orbital position around the Earth, and the Earth magnetic field vector depends on spacecraft orbital position as well as the time of day (the magnetic field rotates with the Earth). In all three cases, the reference vectors (in inertial coordinates) must be computed in real-time, and there are errors associated with these computations. Since stars are inertially fixed, there is no real-time computation required, and instead the associated reference vectors may be looked up from a star catalogue, which may be stored onboard.

Star trackers have the ability to lock on, identify and track a star. If the star tracker only tracks a single star, the output will be the measured unit star pointing vector in the star tracker frame. Using knowledge of the sensor orientation relative to the spacecraft, the star pointing vector can be transformed to spacecraft body coordinates. This measurement can then be used in conjunction with other vector measurements to determine the spacecraft attitude (using the techniques presented in Chapter 25). That being said, most modern star trackers have the ability to lock on, identify and track multiple stars at once. The star trackers then process these multiple star vector measurements to output the star tracker attitude directly. Using knowledge of the sensor orientation relative to the spacecraft, the spacecraft attitude can then be determined. Hence, most modern star trackers can be considered to be full three-axis attitude sensors.

For the above reasons, star trackers provide the most accurate attitude information out of all of the sensors. On the other hand, they are also the most complex and as a result the most expensive and least reliable.

26.1.4.1 Sensor Limitations

First of all, the sun and the Earth must not enter the star tracker's field of view. This limits the allowable spacecraft attitude. Additionally, there are typically quite severe spacecraft body-rate limitations if a star tracker is to be used without further compensation for the spacecraft motion. For this reason, star trackers are typically used on three-axis stabilized spacecraft (as opposed to spinning spacecraft). Star trackers are generally designed for specific mission requirements. For example, star trackers typically have a very limited instantaneous field of view. In order to determine the attitude to within requirements, it will need to have at least two stars within the instantaneous field of view at any one time (if not more). If the star-tracker is required for a large range of attitudes, the number of stars stored in the on-board star catalog can become very large, possibly in the thousands. To limit the on-board storage requirements, the star catalog will contain only those stars that fall within the star tracker field of view for the attitudes where the accuracy of a star tracker is required. As a result, the star tracker may be useful only over a very small range of attitudes. Finally, given the significant computational load associated with identifying, locking on and tracking multiple stars, there may be a noticeable time-delay in the star tracker output, which needs to be compensated for in the spacecraft attitude control algorithms.

26.1.5 Rate Sensors

As the name suggests, rate sensors provide measurements of the spacecraft body rate. As we have seen in previous chapters, most spacecraft attitude control laws require some sort of feedback of the spacecraft body rate. If only attitude measurements are available (from one or more of the previously mentioned attitude sensors), then the spacecraft body rate can be determined by differentiation of the attitude measurement, which could lead to very noisy body-rates, or it can be estimated using an Extended Kalman Filter (EKF) as in Chapter 25. Rate sensors can provide very low noise measurements, which can be used directly in the spacecraft attitude control law. This can significantly improve the spacecraft pointing stability. Pointing stability (not to be confused with Lyapunov stability) is a spacecraft attitude control concept, and is defined as the variation in attitude error over a given time interval. Hence, the achievable pointing stability is directly related to the ability to accurately control the spacecraft body rate.

Another advantage of rate sensors is that the measurements do not require any external reference. For example, a sun-sensor is useful only when the sun is in view. As such, rate sensor measurements are available at all times (provided the sensor is operational). Rate sensors can then be used to accurately predict the spacecraft attitude during periods of unavailability of the attitude measurements. This is done by direct numerical integration of the attitude kinematics using the measured body rate $\boldsymbol{\omega}_{bG}$ as input (see Section 11.2 for details of the attitude kinematics).

A limitation of rate sensors is that all rate sensors have an inherent bias. That is, the measured body rate has some small but nearly constant offset (in addition to the measurement noise). This results in a drift in the spacecraft attitude obtained by numerical integration of the spacecraft kinematics. Hence, rate sensors can not be used exclusively to provide attitude estimates, but other attitude measurements must be incorporated to correct the drift. Often, rate sensor measurements are fused together with other attitude sensor measurements in an

EKF to provide optimal estimates of the spacecraft attitude and body rate. By appropriately setting up the EKF, the rate sensor bias can also be estimated and corrected for.

There are a number of different kinds of rate sensor. Traditionally, rate sensors incorporated mechanical gyros. The drawback of this is that it requires moving mechanical parts, which limits the sensor lifetime. Recently, rate sensors such as laser gyros have been developed based on different principles, which require no moving parts.

26.2 Attitude Actuators

As we have already seen, active attitude control requires actuators that can impart a torque to the spacecraft. There are a number of different types of actuators. The actuators can be divided into two separate classes: 1) reaction-type actuators and 2) momentum exchange devices. Reaction-type actuators generate torques that can be considered to be external to the spacecraft. As such, reaction-type actuators have the ability to change the spacecraft angular momentum. Thrusters and magnetic torquers are reaction-type actuators. Momentum exchange devices generate torques that can be considered internal to the spacecraft, and do not change the overall angular momentum of the spacecraft. Reaction wheels, control moment gyros and momentum wheels are momentum exchange devices.

26.2.1 Thrusters

Thrusters eject mass of some form to create a force. A thrust vector that does not pass through the spacecraft center of mass generates a torque. As shown in Figure 26.7, a thruster generating a force F with a moment arm r, creates a torque given by $T = Fr$. The torque can be increased or decreased by either adjusting the amount of force F (which is a characteristic of the thruster), or the moment arm r (which depends on the installation of the thruster). Since a thruster ejects mass, it can only provide force in one direction (it does not inhale mass like a vacuum cleaner). Therefore, a single thruster can only provide a torque about an axis with a single sense (either positive or negative, not both). Referring to Figure 26.8, two thrusters are needed in order to be able to produce both a positive and negative torque about a single axis. Extending this reasoning, a minimum of six thrusters are needed to be able to produce a torque about an arbitrary axis.

Thrusters typically have the characteristic that they operate in an on/off fashion. That is, the force exerted by a thruster (and hence the resulting torque) is always at a constant level when the thruster is switched on (neglecting the short transient period of the thruster force when it

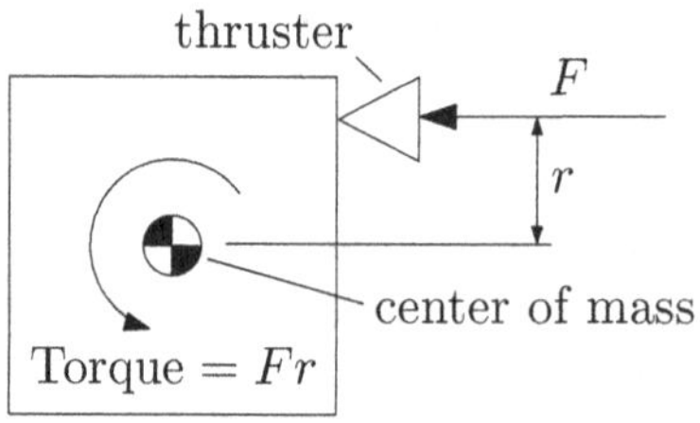

Figure 26.7 Torque on a spacecraft due to a thruster

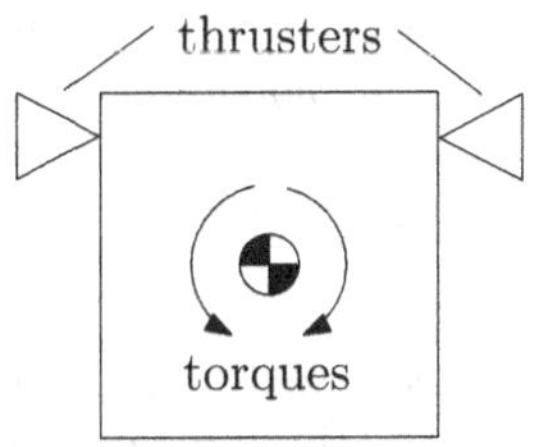

Figure 26.8 Possible torques on a spacecraft due to a pair of thrusters

is first switched on). All of the control laws developed in previous chapters assume that the attitude actuators are capable of creating a continuously variable torque. Therefore, we need some means of being able to approximate a commanded continuously variable control torque with a series of pulses with constant magnitude. As explained in Section 26.3, a real spacecraft attitude control system is implemented digitally. For a digital control implementation using a sample and hold, a measurement is taken at each sample instant, and the computed control torque, T_c is held fixed until the next sample instant, as shown in Figure 26.9. To realize the commanded control torque on average using a thruster that generates a torque with level $\bar{T}_t$ when switched on, the length of time for which the thruster is switched on $t_{p,k}$ (the pulse width) during sample period t_k to t_{k+1}, is computed as

$$t_{p,k} = \frac{T_c(t_k)\Delta t}{\bar{T}_t},$$

where $T_c(t_k)$ is the commanded control torque at sample instant t_k, and $\Delta t = t_{k+1} - t_k$ is the sample period. In this way, by commanding thruster torque pulses with appropriate width (as seen in Figure 26.9), the average torque exerted throughout the sample period is equal to the average commanded control torque. This technique is known as *pulse width modulation* (PWM).

While PWM has the ability to approximate a continuously variable control torque by a series of pulses, it also has its limitations. Problems arise when the commanded control torque becomes very small (in response to a small measured attitude error). Thrusters are not capable

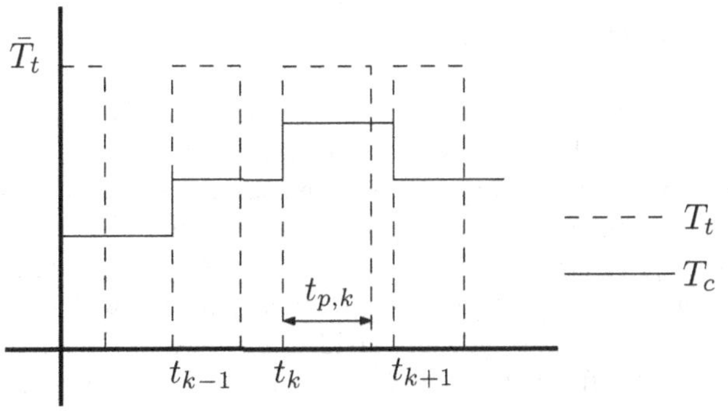

Figure 26.9 Pulse width modulation of a thruster

of creating infinitesimally short pulses. There is a minimum amount of time that a thruster can be turned on for. Related to this is the *minimum torque impulse bit* (MTIB), defined by

$$MTIB = \bar{T}_t t_{p,min},$$

where $\bar{T}_t$ is the torque level of the thruster, and $t_{p,min}$ is the minimum pulse width. So as to avoid unnecessary wasting of propellant, an attitude control system using thrusters should incorporate a dead-zone. That is, the thrusters do not fire when the commanded control torque falls below a certain level (the dead-zone). The dead-zone is directly related to the minimum torque impulse bit. As a result, the achievable accuracy of attitude control systems utilizing thrusters as the means of attitude control is coarse.

The reader can find more detailed treatments of these issues in Sidi (1997) and Wie (2008).

26.2.2 Magnetic Torquers

Magnetic torquers are wire coils attached to the spacecraft. By passing a current through the coils, a magnetic dipole is created. The interaction between the coil dipole $\vec{\mathbf{m}}$ and the Earth's magnetic field $\vec{\mathbf{b}}$ creates a torque according to the law

$$\vec{\mathbf{T}} = \vec{\mathbf{m}} \times \vec{\mathbf{b}}. \tag{26.2}$$

It can be seen from (26.2) that the torque vector $\vec{\mathbf{T}}$ generated by a magnetic torquer is always perpendicular to the instantaneous Earth's magnetic field vector $\vec{\mathbf{b}}$ (a property of the vector cross-product). Therefore, it is impossible to generate a torque about an arbitrary axis using magnetic torquers alone. As such, a spacecraft equipped only with magnetic torquers is *instantaneously underactuated.* However, since the Earth's magnetic field direction varies with orbital position, a full three-axis attitude control law can still be realized on average throughout an orbit using exclusively magnetic control. Note that this is true only for orbits with significant inclination, since there is not much variation in the Earth's magnetic field direction for near equatorial orbits.

Due to the very weak Earth magnetic field, the magnitude of the torque that can be generated by magnetic torquers is also very small, so magnetic torquers have *limited control authority.* For these reasons (instantaneous underactuation and limited control authority), only coarse attitude control is possible when magnetic actuation is employed exclusively. When fine attitude control is required, the actuators of choice are typically a set of reaction wheels providing the attitude control itself. The reaction wheels are often augmented by magnetic torquers for the purpose of unloading excess momentum from the reaction wheels, which builds up due to external disturbances acting on the spacecraft.

When magnetic torquers are used, it is important that they do not operate while magnetometers are being sampled. For a digital control implementation with magnetic torquers and magnetometers, a sample period is typically divided as shown in Figure 26.10. Note that the magnetic torquers are shut off a finite amount of time before the magnetometer is sampled during the next sample period. This is because it takes a very short but finite time for the magnetic torquer dipole moment to dissipate after the current is switched off (it is not instantaneous).

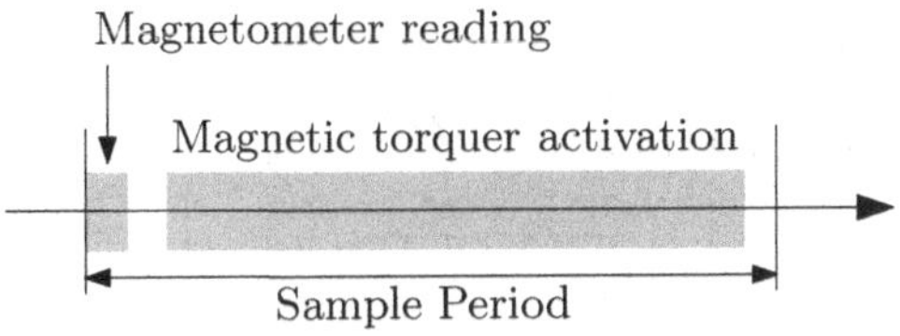

Figure 26.10 Scheduling of magnetometer readings and magnetic torquer activation

26.2.3 Reaction Wheels

A reaction wheel is a nominally non-spinning wheel mounted in the spacecraft. As in Chapter 15, we divide the spacecraft into a platform and the wheels. As shown in Figure 26.11, by accelerating the wheel in one direction about the wheel spin-axis, the wheel applies a reaction torque to the platform in the opposite direction (also about the wheel spin-axis). This is a consequence of conservation of angular momentum. In the absence of an external torque acting on the spacecraft, the total angular momentum of the spacecraft remains constant (see Chapter 2). Therefore, in the absence of external torques, if the angular momentum of the wheels is changed, there must be a corresponding opposite change in angular momentum of the platform. This illustrates why a reaction wheel is considered a momentum exchange device. To control the spacecraft attitude, angular momentum is transferred between the wheels and the platform. By mounting three reaction wheels with spin axes not in the same plane (typically along orthogonal axes), a torque can be created about an arbitrary axis of the spacecraft platform. As actuators, reaction wheels provide the most precise attitude control.

As we have seen in Chapter 12, there are a number of external disturbance torques acting on a spacecraft. These external disturbance torques result in a change in overall spacecraft angular momentum. When the spacecraft attitude is controlled using reaction wheels, this change in spacecraft angular momentum manifests itself as a change in stored angular momentum in the wheels. To illustrate, suppose the control objective is to keep the spacecraft platform attitude inertially fixed. In this case, the controlled platform has zero angular momentum, and the entire change in spacecraft angular momentum occurs in the wheels. Therefore, while reaction wheels provide the most precise attitude control, they cannot be used exclusively, since external disturbances result in a build-up of wheel angular momentum. Therefore, any spacecraft attitude control system utilizing reaction wheels must be augmented with reaction-type actuators (either thrusters or magnetic torquers) capable of creating an

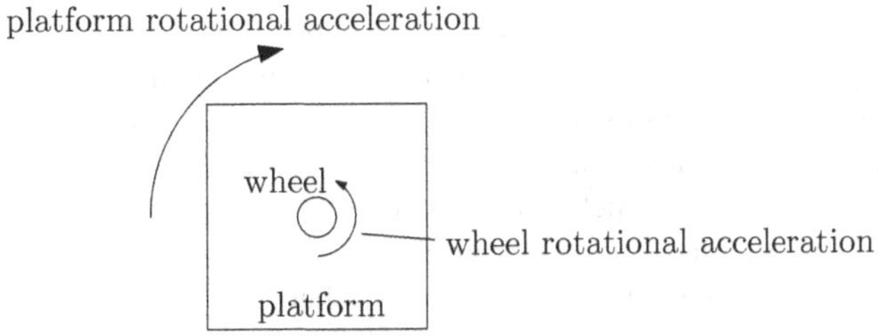

Figure 26.11 Reaction wheel principle

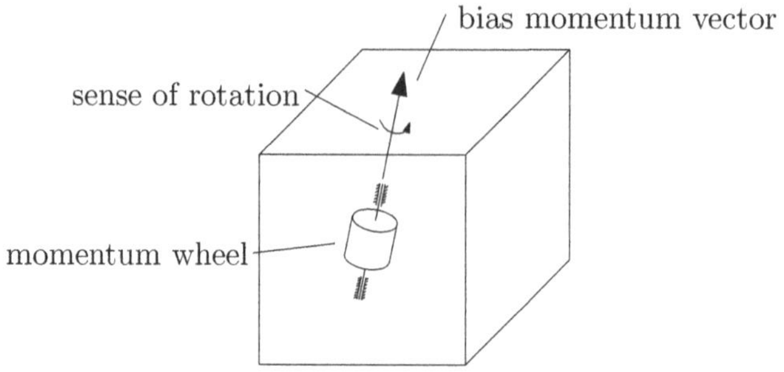

Figure 26.12 Spacecraft with momentum wheel

external torque on the spacecraft to de-load the built up angular momentum in the wheels. This procedure is known as *momentum dumping*.

26.2.4 Momentum Wheels

Momentum wheels have a large non-zero nominal speed. As we have seen in Chapter 15 and Section 18.2, this gives the spacecraft a *bias momentum* which provides gyroscopic stability (see Figure 26.12). What this means is that the spacecraft will resist an external disturbance torque that attempts to turn the bias momentum vector. This is useful in the case when it is desired to keep one of the spacecraft axes inertially fixed (the axis parallel to the wheel spin axis). Momentum wheels can be used in two different ways. In the first, the only purpose for the wheel is to provide a bias momentum to the spacecraft, in which case the momentum wheel speed is kept constant. In the second, the wheel not only provides a bias momentum to the spacecraft, but it also provides attitude control to the spacecraft about the wheel spin axis. For example, if the spacecraft is equipped with a momentum wheel whose spin axis is aligned with the spacecraft pitch axis, then the momentum wheel can also be used to control the spacecraft pitch angle. In the second case, external disturbances on the spacecraft will cause the momentum wheel speed to deviate from the nominal set-point, and momentum management using thrusters or magnetic torquers will be needed just as for reaction wheels.

26.2.5 Control Moment Gyroscopes

Control moment gyroscopes (CMG) are like momentum wheels, except the wheel spin-axis is gimballed. The wheel spins with a constant high speed. By applying a torque to turn the gimbal (which turns the wheel spin-axis), a large gyroscopic reaction torque is generated perpendicular to the gimbal axis (see Figure 26.13). As such, a control moment gyroscope acts as a torque amplifier, and is useful when large control torques are needed. The CMG depicted in Figure 26.13 is a single gimbal control moment gyroscope. There also exist double gimbal control moment gyroscopes, in which case an inner gimbal is mounted in an outer gimbal. In that case, the wheel is mounted in the inner gimbal.

Control moment gyroscopes are also momentum exchange devices. As such, external disturbance torques acting on the spacecraft result in a net change in the gimbal angles. The gimbal angles typically have a limited range. Like reaction wheels, control moment gyroscopes also

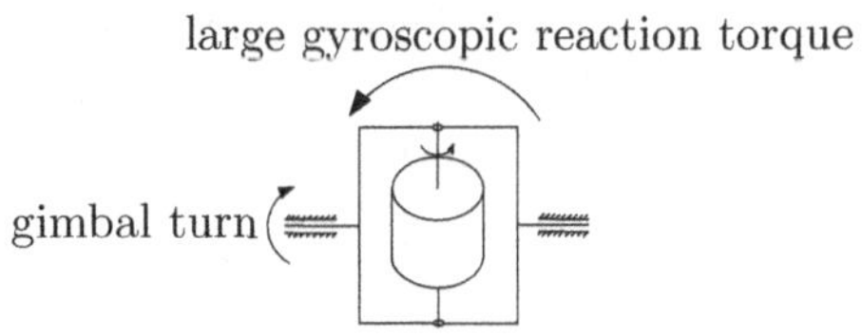

Figure 26.13 Single gimbal control moment gyroscope

must be augmented with reaction-type actuators capable of creating external torques on the spacecraft in order to restore the CMG gimbal angles to their nominal settings.

26.3 Control Law Implementation

In Chapters 17 to 23, all control laws are designed as transfer functions, which are complex functions in the Laplace domain. However, what do these transfer functions mean, given that the control laws must be implemented in the time-domain? How do we implement these control laws? What additional issues must be considered? We shall address these questions in this section.

26.3.1 Time-Domain Representation of a Transfer Function

The control laws designed in Chapters 17 to 23 take the form

$$U(s) = G_c(s)E(s), \tag{26.3}$$

where $E(s)$ is the Laplace transform of the tracking error, $U(s)$ is the Laplace transform of the control command, and $G_c(s)$ is the control law transfer function. We shall now look at what this means in the time-domain. As we have seen in previous chapters, if the control law takes a simple form such as proportional-derivative (PD) or proportional-integral-derivative (PID), then it is a simple matter to obtain the corresponding time-domain representation. However, it is not immediately obvious for more elaborate transfer functions.

Consider the generic control law with transfer function

$$G_c(s) = \frac{a_0 s^n + a_1 s_{n-1} + \cdots a_n}{s^n + b_1 s^{n-1} + \cdots + b_n}. \tag{26.4}$$

We shall now obtain a time-domain representation for (26.4). First, let us rewrite $G_c(s)$ as

$$G_c(s) = a_0 + \bar{G}_s(s), \tag{26.5}$$

where

$$\bar{G}_s(s) = \frac{\bar{a}_1 s^{n-1} + \cdots + \bar{a}_n}{s^n + b_1 s^{n-1} + \cdots + b_n},$$

and $\bar{a}_i = a_i - a_0 b_i$ for $i = 1, \ldots, n$. Substituting (26.5) into (26.3), we find that

$$U(s) = U_1(s) + U_2(s), \tag{26.6}$$

where

$$U_1(s) = a_0(s)E(s), \quad U_2(s) = \bar{G}_c(s)E(s).$$

Taking the inverse Laplace transform of (26.6), we obtain

$$u(t) = u_1(t) + u_2(t). \tag{26.7}$$

We immediately have

$$u_1(t) = a_0 e(t). \tag{26.8}$$

We must now find $u_2(t)$. We first write

$$U_2(s) = \frac{\bar{a}_1 s^{n-1} + \cdots \bar{a}_n}{s^n + b_1 s^{n-1} + \cdots + b_n} E(s). \tag{26.9}$$

A naive approach to finding $u_2(t)$ would be to rearrange (26.9) as

$$\left(s^n + b_1 s^{n-1} + \cdots + b_n\right) U_2(s) = \left(\bar{a}_1 s^{n-1} + \cdots \bar{a}_n\right) E(s),$$

and take inverse Laplace transforms of both sides to obtain

$$\frac{d^n u_2}{dt^n} + b_1 \frac{d^{n-1} u_2}{dt^{n-1}} + \cdots + b_n u_2 = \bar{a}_1 \frac{d^{n-1} e}{dt^{n-1}} + \cdots + \bar{a}_n e(t).$$

This yields a linear ordinary differential equation in the time-domain. However, as seen on the right-hand side, it includes derivatives of the tracking error $e(t)$ up to the order $n - 1$, which are typically not available, so this approach is not very useful for the implementation of the control law in the time-domain.

Instead, let us define

$$Z(s) = \frac{E(s)}{s^n + b_1 s^{n-1} + \cdots + b_n}. \tag{26.10}$$

Then, using (26.9), we find that

$$U_2(s) = \left(\bar{a}_1 s^{n-1} + \cdots \bar{a}_n\right) Z(s). \tag{26.11}$$

Taking inverse Laplace transforms of both sides of (26.11), we obtain

$$u_2(t) = \bar{a}_1 \frac{d^{n-1} z}{dt^{n-1}} + \bar{a}_2 \frac{d^{n-2} z}{dt^{n-2}} + \cdots + \bar{a}_n z(t). \tag{26.12}$$

We now rearrange (26.10) to get

$$\left(s^n + b_1 s^{n-1} + \cdots + b_n\right) Z(s) = E(s).$$

Taking inverse Laplace transforms, we now obtain

$$\frac{d^n z}{dt^n} + b_1 \frac{d^{n-1} z}{dt^{n-1}} + \cdots + b_n z = e(t). \tag{26.13}$$

This is an nth order differential equation in $z(t)$, which does not contain any derivatives of $e(t)$. Let us now put it in first order form. We define

$$x_{c,1} = z, \quad x_{c,2} = \frac{dz}{dt}, \ldots, x_{c,n} = \frac{d^{n-1} z}{dt^{n-1}}.$$

With these definitions we have

$$\frac{dx_{c,1}}{dt} = x_{c,2}, \quad \frac{dx_{c,2}}{dt} = x_{c,3}, \ldots, \quad \frac{dx_{c,n-1}}{dt} = x_{c,n},$$

and from (26.13) we have

$$\frac{dx_{c,n}}{dt} = -b_1 x_{c,n-1} - b_2 x_{c,n-2} - \cdots - b_n x_{c,1} + e(t).$$

Therefore, we can write (26.13) as

$$\dot{\mathbf{x}}_c = \mathbf{A}_c \mathbf{x}_c + \mathbf{B}_c e(t), \tag{26.14}$$

where

$$\mathbf{x}_c = \begin{bmatrix} x_{c,1} \\ x_{c,2} \\ \vdots \\ x_{c,n-1} \\ x_{c,n} \end{bmatrix}, \quad \mathbf{A}_c = \begin{bmatrix} 0 & 1 & 0 & \cdots & 0 \\ 0 & 0 & 1 & \cdots & 0 \\ \vdots & & \ddots & \ddots & \\ 0 & & & 0 & 1 \\ -b_n & -b_{n-1} & -b_{n-2} & \cdots & -b_1 \end{bmatrix}, \quad \mathbf{B}_c = \begin{bmatrix} 0 \\ 0 \\ \vdots \\ 0 \\ 1 \end{bmatrix}.$$

Likewise, equation (26.12) can be written as

$$u_2(t) = \mathbf{C}_c \mathbf{x}_c, \tag{26.15}$$

where

$$\mathbf{C}_c = \begin{bmatrix} \bar{a}_n & \bar{a}_{n-1} & \cdots & \bar{a}_1 \end{bmatrix}.$$

Finally, using (26.7), we can combine equations (26.14), (26.8) and (26.15) to obtain the time-domain representation of (26.3) as

$$\begin{aligned}\dot{\mathbf{x}}_c &= \mathbf{A}_c\mathbf{x}_c + \mathbf{B}_c e(t),\\ u(t) &= \mathbf{C}_c\mathbf{x}_c + D_c e(t),\end{aligned} \tag{26.16}$$

where $D_c = a_0$. The equations (26.16) are called a *state-space realization* of the transfer function relationship (26.3). Furthermore, none of the derivatives of $e(t)$ appear, which is important for the implementation of the control law.

It is interesting to note that $\mathbf{A}_c$ has characteristic equation

$$\det\left[s\mathbf{1} - \mathbf{A}_c\right] = s^n + b_1 s^{n-1} + \cdots + b_n,$$

which is precisely the denominator of the corresponding transfer function $G_c(s)$ in (26.4). Therefore, the poles of the transfer function $G_c(s)$ are equal to the eigenvalues of $\mathbf{A}_c$ in the corresponding state-space realization.

26.3.2 Control Law Digitization

In the previous subsection, we have obtained a time-domain representation of a control law, namely an ordinary differential equation, which must be solved in continuous-time if it is to be implemented directly. However, modern control systems are typically implemented on a digital processor. The processor samples the sensors at discrete sampling instants t_k as shown in Figure 26.14 (a), and computes the corresponding control torque to be applied by the actuators. The control torque is held constant throughout the sample period until the next sensor measurement is processed, as illustrated in Figure 26.14 (b). This is called a *zero-order hold* (ZOH) control implementation.

Therefore, a continuous-time control law as in (26.16) cannot be implemented digitally. However, by sampling fast enough, the digital control implementation can approximate the continuous control design (see Figure 26.14 (b)). The question may be asked as to why the control design is not performed directly in the discrete-time domain. Control design directly in the discrete domain is certainly possible, and tools similar to those presented in Chapters 17 to 23 exist for discrete systems. However, control system design is more common and intuitive in the continuous-time domain, which is why this is the approach taken in this book. Therefore,

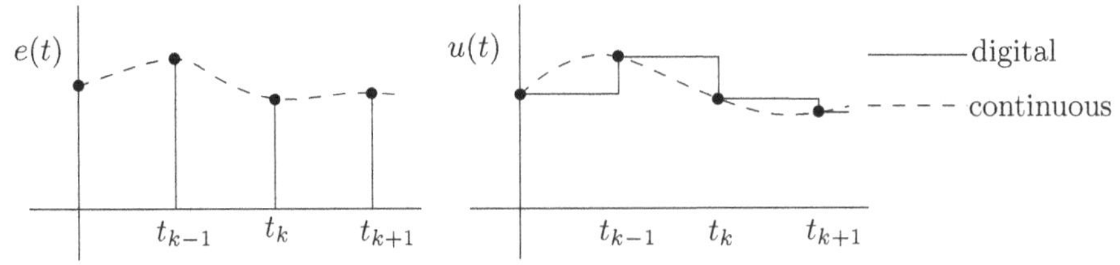

(a) Digital tracking error sampling (b) Digital control input

Figure 26.14 Control law digitization

a typical control design procedure is to initially design a continuous control law, then digitize it and finally check that it performs similarly to the original continuous design.

We now turn our attention to approximating the continuous control law (26.16) digitally. Let us integrate both sides of the differential equation in (26.16) over one sample period from t_k to t_{k+1}. We have

$$\mathbf{x}_c(t_{k+1}) = \mathbf{x}_c(t_k) + \int_{t_k}^{t_{k+1}} [\mathbf{A}_c\mathbf{x}_c(t) + \mathbf{B}_c e(t)]\, dt. \tag{26.17}$$

Since error signal $e(t)$ is sampled and not continuously available, we need to approximate the integral on the right in (26.17). Let us use the trapezoidal rule, which results in

$$\int_{t_k}^{t_{k+1}} \mathbf{x}_c(t)dt \approx \frac{\Delta t}{2}\left(\mathbf{x}_c(t_k) + \mathbf{x}_c(t_{k+1})\right), \quad \int_{t_k}^{t_{k+1}} e(t)dt \approx \frac{\Delta t}{2}\left(e(t_k) + e(t_{k+1})\right),$$

where $\Delta t = t_{k+1} - t_k$ is the sample period. Substituting this into (26.17) and rearranging, we obtain

$$\left[\mathbf{1} - \frac{\Delta t}{2}\mathbf{A}_c\right]\mathbf{x}_c(t_{k+1}) - \frac{\Delta t}{2}\mathbf{B}_c e(t_{k+1}) = \left[\mathbf{1} + \frac{\Delta t}{2}\mathbf{A}_c\right]\mathbf{x}_c(t_k) + \frac{\Delta t}{2}\mathbf{B}_c e(t_k). \tag{26.18}$$

Based on the left-hand side of (26.18), let us define

$$\mathbf{z}_{c,k} = \frac{1}{\sqrt{\Delta t}}\left(\left[\mathbf{1} - \frac{\Delta t}{2}\mathbf{A}_c\right]\mathbf{x}_c(t_k) - \frac{\Delta t}{2}\mathbf{B}_c e(t_k)\right), \tag{26.19}$$

which rearranges to give

$$\mathbf{x}_c(t_k) = \sqrt{\Delta t}\left[\mathbf{1} - \frac{\Delta t}{2}\mathbf{A}_c\right]^{-1}\mathbf{z}_{c,k} + \frac{\Delta t}{2}\left[\mathbf{1} - \frac{\Delta t}{2}\mathbf{A}_c\right]^{-1}\mathbf{B}_c e(t_k). \tag{26.20}$$

Substituting these into (26.18) and rearranging (try this as an exercise), we obtain

$$\begin{aligned}\mathbf{z}_{c,k+1} &= [\mathbf{1} + (\Delta t/2)\mathbf{A}_c]\,[\mathbf{1} - (\Delta t/2)\mathbf{A}_c]^{-1}\,\mathbf{z}_{c,k} \\ &\quad + \sqrt{\Delta t}\,[\mathbf{1} - (\Delta t/2)\mathbf{A}_c]^{-1}\,\mathbf{B}_c e(t_k).\end{aligned} \tag{26.21}$$

From the output equation in (26.17), we have

$$u_k = \mathbf{C}_c\mathbf{x}_c(t_k) + D_c e(t_k), \tag{26.22}$$

where we have defined $u_k = u(t_k)$. Substituting (26.20) into (26.22), and combining with equation (26.21), we obtain a digitized approximation to the controller corresponding to (26.16)

$$\begin{aligned}\mathbf{z}_{c,k+1} &= \mathbf{F}_c\mathbf{z}_{c,k} + \mathbf{G}_c e(t_k), \\ u_k &= \mathbf{H}_c\mathbf{z}_{c,k} + L_c e(t_k),\end{aligned} \tag{26.23}$$

where

$$\mathbf{F}_c = [\mathbf{1} + (\Delta t/2)\mathbf{A}_c][\mathbf{1} - (\Delta t/2)\mathbf{A}_c]^{-1}, \quad \sqrt{\Delta t}\,[\mathbf{1} - (\Delta t/2)\mathbf{A}_c]^{-1}\,\mathbf{B}_c,$$

$$\mathbf{H}_c = \sqrt{\Delta t}\mathbf{C}_c\,[\mathbf{1} - (\Delta t/2)\mathbf{A}_c]^{-1}, \quad L_c = D_c + (\Delta t/2)\mathbf{C}_c\,[\mathbf{1} - (\Delta t/2)\mathbf{A}_c]^{-1}\,\mathbf{B}_c.$$

Note that $\mathbf{z}_{c,k}$ is the state vector of the digitized controller. The digitization procedure we have used to obtain (26.23), is also known as the *bilinear transformation* of (26.16).

We see that when differential equations such as those in (26.16) are digitized, they become difference equations as in (26.23). These difference equations are very suitable for digital implementation. In fact, we have a very simple recursive algorithm corresponding to (26.23), which illustrates how it would be implemented on a digital processor:

1. Set $\mathbf{z}_{c,0} = -(\sqrt{\Delta t}/2)\mathbf{B}_c e(t_0)$ and $k = 0$.
2. Sample $e(t_k)$ and compute the control output $u_k = \mathbf{H}_c\mathbf{z}_{c,k} + L_c e(t_k)$.
3. Update $\mathbf{z}_{c,k+1} = \mathbf{F}_c\mathbf{z}_{c,k} + \mathbf{G}_c e(t_k)$.
4. Store $\mathbf{z}_{c,k+1}$, and set $k = k + 1$.
5. Return to Step 2.

As we have explained, when the controller is implemented digitally, the control input is applied to the plant (which has continuous dynamics) through a zero-order hold as shown in Figure 16.14 (b), namely

$$u(t) = u_k, \quad t_k \leq t < t_{k+1}, \tag{26.24}$$

where u_k is computed at each sample instant using (26.23). Finally, the digital control law implementation with the continuous plant (in this book, the spacecraft attitude dynamics) is shown in Figure 26.15.

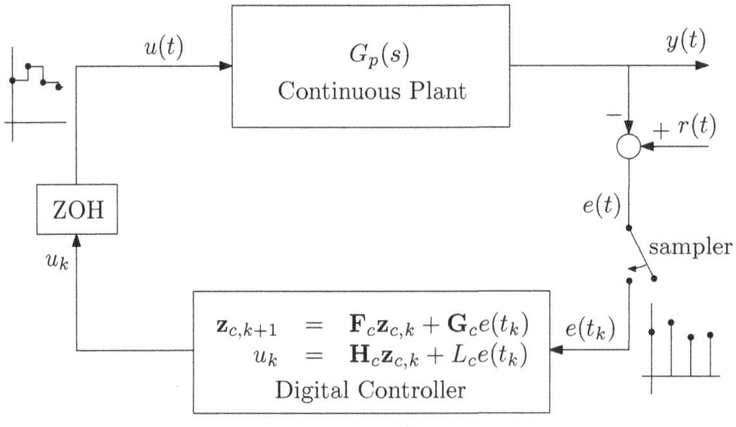

Figure 26.15 Digital control law implementation

26.3.3 Closed-Loop Stability Analysis

It is very important to note that the closed-loop system with the digital controller (as in Figure 26.15), is **not** the same as the originally designed closed-loop system with continuous control. It is therefore important to be able to check the closed-loop system with the digital controller to see that it still conforms to the designed for requirements. In particular, it is important to check that it is stable, which is the subject of this section.

To analyze the closed-loop stability, we first note that similar to the controller with transfer function $G_c(s)$, we can represent the plant with transfer function $G_p(s)$ in state-space form as

$$\begin{aligned} \dot{\mathbf{x}}_p &= \mathbf{A}_p \mathbf{x}_p + \mathbf{B}_p u(t), \\ y(t) &= \mathbf{C}_p \mathbf{x}_p, \end{aligned} \tag{26.25}$$

For simplicity, we assume that the output $y(t)$ does not directly depend on the input $u(t)$. This is certainly the case for spacecraft dynamics. To obtain the state-space representation of the plant in the form (26.25), we can obtain it from the transfer function $G_p(s)$ using the same procedure as in Section 26.3.1, or we could obtain it directly from the differential equations describing the plant dynamics (which are not necessarily first order) using a procedure as in Appendix B. In any case, we assume that a state-space representation of the form (26.25) is available.

Let us now examine the solution of (26.25) when the control (26.24) is applied. We take as initial condition $\mathbf{x}_p(t_k)$ at sampling instant t_k, and examine the resulting solution $\mathbf{x}_p(t_{t+1})$ at the next sampling instant t_{k+1}. To do this, we need the matrix exponential, which is defined as

$$\exp(\mathbf{A}t) = \sum_{n=0}^{\infty} \frac{1}{n!} \mathbf{A}^n t^n, \tag{26.26}$$

where the zeroth power of $\mathbf{A}$ is defined as $\mathbf{A}^0 = \mathbf{1}$. The infinite series in (26.26) converges for all bounded square matrices $\mathbf{A}$ and for all $-\infty < t < \infty$. The matrix exponential has the following properties

$$\begin{aligned} \exp(\mathbf{A}t_1)\exp(\mathbf{A}t_2) &= \exp(\mathbf{A}(t_1 + t_2)), \\ \tfrac{d}{dt}\exp(\mathbf{A}t) &= \mathbf{A}\exp(\mathbf{A}t) = \exp(\mathbf{A}t)\mathbf{A}. \end{aligned} \tag{26.27}$$

Let us now multiply the differential equation in (26.25) by $\exp(-\mathbf{A}_p t)$ to get

$$\exp(-\mathbf{A}_p t)\dot{\mathbf{x}}_p = \exp(-\mathbf{A}_p t)\mathbf{A}_p \mathbf{x}_p + \exp(-\mathbf{A}_p t)\mathbf{B}_p u(t),$$

and making use of the product rule for differentiation as well as the second property in (26.27), this can be rearranged to give

$$\frac{d}{dt}\left[\exp(-\mathbf{A}_p t)\mathbf{x}_p(t)\right] = \exp(-\mathbf{A}_p t)\mathbf{B}_p u(t).$$

We can now integrate both sides to obtain

$$\int_{t_k}^{t_{k+1}} \frac{d}{dt}\left[\exp(-\mathbf{A}_p t)\mathbf{x}_p(t)\right] dt = \int_{t_k}^{t_{k+1}} \exp(-\mathbf{A}_p t)\mathbf{B}_p u(t) dt,$$

Because of the zero order hold in (26.24), this becomes

$$\exp(-\mathbf{A}_p t_{k+1})\mathbf{x}_c(t_{k+1}) - \exp(-\mathbf{A}_c t_k)\mathbf{x}_p(t_k) = \int_{t_k}^{t_{k+1}} \exp(-\mathbf{A}_p t)\mathbf{B}_p dt\, u_k.$$

Pre-multiplying both sides by $\exp(\mathbf{A}_p(t_{k+1}))$, and making use of the first property in (26.27), we obtain the solution

$$\mathbf{x}_p(t_{k+1}) = \exp(\mathbf{A}_p \Delta t)\mathbf{x}_p(t_k) + \int_{t_k}^{t_{k+1}} \exp(\mathbf{A}_p(t_{k+1} - t))\mathbf{B}_c u_k dt. \tag{26.28}$$

We see that when the control (26.24) is applied to the plant (26.25), the plant can be represented by a difference equation of the same form as for the digital controller in (26.23). Note that this discretization of the plant is exact. Therefore, we generically write the discrete representation of the plant as

$$\begin{aligned} \mathbf{x}_p(t_{k+1}) &= \mathbf{F}_p \mathbf{x}_p(t_k) + \mathbf{G}_p u_k, \\ y(t_k) &= \mathbf{C}_p \mathbf{x}_p(t_k), \end{aligned} \tag{26.29}$$

where

$$\mathbf{F}_p = \exp(\mathbf{A}_p \Delta t), \quad \mathbf{G}_p = \int_0^{\Delta t} \exp(\mathbf{A}_p(\Delta t - \tau))\mathbf{B}_c d\tau.$$

Note that a change of variable $t \to t + t_k$ has been made in the integration to obtain the expression for $\mathbf{G}_p$. We can now use (26.29) and (26.23) to form the discrete closed-loop system. Combining (26.29) and (26.23), we obtain (the reader should check this as an exercise)

$$\begin{aligned} \begin{bmatrix} \mathbf{x}_p(t_{k+1}) \\ \mathbf{z}_{c,k+1} \end{bmatrix} &= \begin{bmatrix} \mathbf{F}_p - \mathbf{G}_p L_c \mathbf{C}_p & \mathbf{G}_p \mathbf{H}_c \\ -\mathbf{G}_c \mathbf{C}_p & \mathbf{F}_c \end{bmatrix} \begin{bmatrix} \mathbf{x}_p(t_k) \\ \mathbf{z}_{c,k} \end{bmatrix} + \begin{bmatrix} \mathbf{G}_p L_c \\ \mathbf{G}_c \end{bmatrix} r(t_k), \\ y(t_k) &= \begin{bmatrix} \mathbf{C}_p & \mathbf{0} \end{bmatrix} \begin{bmatrix} \mathbf{x}_p(t_k) \\ \mathbf{z}_{c,k} \end{bmatrix}, \\ u_k &= \begin{bmatrix} -L_c \mathbf{C}_p & \mathbf{H}_c \end{bmatrix} \begin{bmatrix} \mathbf{x}_p(t_k) \\ \mathbf{z}_{c,k} \end{bmatrix} + L_c r(t_k). \end{aligned} \tag{26.30}$$

The closed-loop stability can now be checked. It turns out that the closed-loop system is asymptotically stable if and only if the eigenvalues of the matrix

$$\begin{bmatrix} \mathbf{F}_p - \mathbf{G}_p L_c \mathbf{C}_p & \mathbf{G}_p \mathbf{H}_c \\ -\mathbf{G}_c \mathbf{C}_p & \mathbf{F}_c \end{bmatrix}$$

have magnitude less than 1.

26.3.4 Sampling Considerations

As explained in Section 26.3.2, the sampling rate for the digital controller must be fast enough such that its performance approximates that of the originally designed continuous controller. So the question arises as to how fast is fast enough. The following rule of thumb is often used:

The sampling rate should be at least 20 times the closed-loop bandwidth.

In the remainder of this section, we identify and briefly discuss two consequences of digital control implementation which need to be accounted for.

26.3.4.1 Delay Due to Zero Order Hold

As demonstrated in Figure 26.16, the digital implementation of a control law through a zero order hold as in Figure 26.15, effectively introduces a time-delay of half the sample period $\Delta t/2$. This has the effect of reducing the positive phase margin, and as we have seen in Chapter 23, results in an increase in overshoot.

As already mentioned, if the sampling rate is selected to be at least 20 times the closed-loop bandwidth, the performance with the digital controller can be expected to be similar to the original continuous design. If the sampling rate is at least 30 times the closed-loop bandwidth, the performance can be expected to be almost identical. However, due to the available control hardware, it is not always feasible to have such fast sample rates. One way to mitigate the effect of the effective delay caused by the zero order hold is to account for it in the original continuous control design. This can be accomplished by inserting an approximation of the

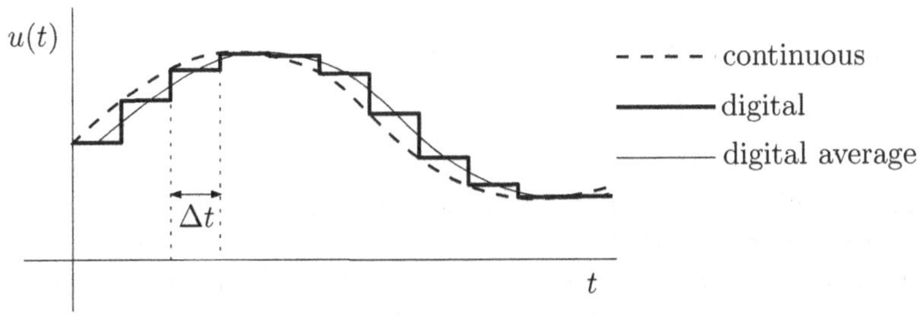

Figure 26.16 Delay due to zero order hold

zero order hold into the feedback loop. One simple continuous approximation of the zero order hold is the first order filter

$$G_{ZOH}(s) = \frac{1}{(\Delta t/2)s + 1}.$$

26.3.4.2 Aliasing

Let us consider a continuous sinusoidal signal of frequency ω, given by

$$y(t) = A\sin(\omega t + \phi), \tag{26.31}$$

where A is the amplitude and ϕ the initial phase. Now, suppose that we will sample this signal at times

$$t_k = k\Delta t, \quad k = 0, 1, \ldots \tag{26.32}$$

where Δt is the sample period. The corresponding sample frequency (in rad/s) is

$$\omega_s = \frac{2\pi}{\Delta t}. \tag{26.33}$$

Correspondingly, the sampled signal has values

$$y(t_k) = A\sin(\omega k\Delta t + \phi). \tag{26.34}$$

Using the simple identity $\sin a = \sin(a + n2\pi)$ for any real number a and any integer n, we can also write (26.34) as

$$\begin{aligned} y(t_k) &= A\sin\left(\omega k\Delta t + \phi + nk2\pi\right), \\ &= A\sin\left(\omega k\Delta t + \phi + nk2\pi(\omega_s/\omega_s)\right). \end{aligned}$$

Recognizing that $\Delta t = 2\pi/\omega_s$ from (26.33), we finally have

$$y(t_k) = A\sin\left((\omega + n\omega_s)k\Delta t + \phi\right), \quad n = \ldots -2, -1, 0, 1, 2, \ldots, \tag{26.35}$$

which is equivalent to sampling any of the signals

$$y(t) = A\sin\left((\omega + n\omega_s)t + \phi\right), \quad n = \ldots -2, -1, 0, 1, 2, \ldots$$

Therefore, the sampling process cannot distinguish between sinusoidal signals of frequency $\omega + n\omega_s$, where ω_s is the sampling frequency. This is called *aliasing*, and is an artefact of the sampling process. Figure 26.17 illustrates the phenomenon with an example. In this example, the primary signal to be sampled has frequency $f = 1.2$ Hz, the sampling frequency is $f = 1.0$ Hz, and the aliased signal has frequency $f_a = f - f_s = 0.2$ Hz. It is clear from Figure 26.17

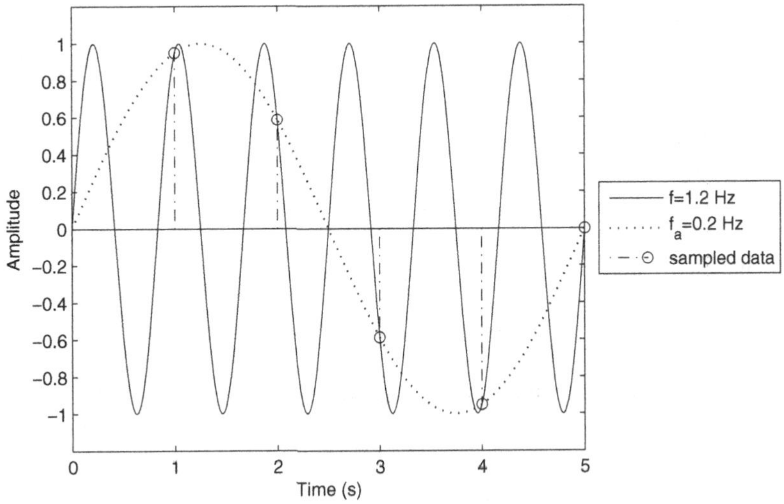

Figure 26.17 Example of aliasing

that both the primary signal and the aliased signal yield the exact same data samples. Hence, the sampler cannot distinguish between the two signals.

Suppose now that we have some additional information. That is, we know that the signal to be sampled has frequency in the range $\omega \in [0, \bar{\omega}]$, where $\bar{\omega} > 0$. Let us now determine what sampling frequency $\omega_s > 0$ is required to be able to uniquely determine the signal.

Before proceeding, suppose that some aliased signal has frequency $\omega + n\omega_s < 0$ for some integer $n \leq -1$, and let $\omega_n = -(\omega + n\omega_s) > 0$. Then, the aliased signal $y_{a,n}(t) = A\sin((\omega + n\omega_s)t + \phi)$ can also be written as

$$y_{a,n}(t) = A\sin(-\omega_n t + \phi) = A\sin\left(\omega_n t + \bar{\phi}\right),$$

where $\bar{\phi} = \pi - \phi$. Therefore, an aliased signal with frequency $\omega + n\omega_s$ in the range $[-\bar{\omega}, 0]$ for some $n \leq -1$, is equivalent to a signal with frequency $\omega_n = -(\omega + n\omega_s)$ in the range $[0, \bar{\omega}]$.

As such, in order to be able to uniquely identify the signal to be sampled, we must ensure that all aliased signals with frequencies $\omega + n\omega_s$ lie outside the range $[-\bar{\omega}, \bar{\omega}]$, for all non-zero integers n. Since $\omega + n\omega_s > \omega$ for all $n \geq 1$, and $\omega + n\omega_s < \omega$ for all $n \leq -1$, and since $0 \leq \omega \leq \bar{\omega}$, this leads to the two requirements that

$$\omega + n\omega_s > \bar{\omega}, \quad n \geq 1, \text{ and } \omega - n\omega_s < -\bar{\omega}, \quad n \geq 1,$$

which reduces to

$$\omega + \omega_s > \bar{\omega}, \text{ and } \omega - \omega_s < -\bar{\omega}. \tag{26.36}$$

The first of these leads to

$$\omega_s > \bar{\omega} - \omega.$$

Since this must hold for all $\omega \in [0, \bar{\omega}]$, this becomes

$$\omega_s > \bar{\omega}. \tag{26.37}$$

The second inequality in (26.36) leads to

$$\omega_s > \bar{\omega} + \omega.$$

Again, since this must hold for all $\omega \in [0, \bar{\omega}]$, this becomes

$$\omega_s > 2\bar{\omega}. \tag{26.38}$$

It is clear that the requirement (26.38) supersedes the requirement (26.37). Therefore, we find that *the sampling frequency ω_s must be at least twice the frequency ω of the sampled signal in order to be able to uniquely reconstruct it.* This is the *sampling theorem of Nyquist and Shannon*, and the frequency

$$\omega_{NS} = 2\omega, \tag{26.39}$$

is called the *Nyquist sampling frequency* corresponding to a signal of frequency ω, and it provides the minimum required sampling frequency necessary to be able to reconstruct the original signal from its samples.

So, why are these issues important? Well first of all, we would like to provide good tracking of reference signals at frequencies within the closed-loop bandwidth. Therefore, it is important that aliases of the signals to be tracked do not appear within the closed-loop bandwidth (aliases outside the closed-loop bandwidth do not matter, since they are attenuated anyway). The Nyquist and Shannon sampling theorem then suggests that the sample rate for the digital controller should be at least twice the closed-loop bandwidth. However, as noted before, we typically try to make the sample rate much faster (at least 20 times the closed-loop bandwidth). Second of all, as explained in Chapter 23, the continuous control law is designed such that high frequency noise has little effect on the closed-loop system. This is done by limiting the closed-loop, open-loop, and controller bandwidths. However, due to aliasing, high frequency noise on the sampled signal can manifest itself at lower frequencies within these bandwidths, which can result in significant degradation in the performance of the digital controller. Therefore, it is important that high frequency sensor noise be attenuated as much as possible **before** sensor sampling. This is accomplished by placing an *anti-alias pre-filter* before the sampler, as shown in Figure 26.18. Obviously, such a filter cannot be digital, and must be implemented using an analog circuit.

For instance, a simple first-order low-pass filter such as

$$H_a(s) = \frac{1}{T_a s + 1}, \tag{26.40}$$

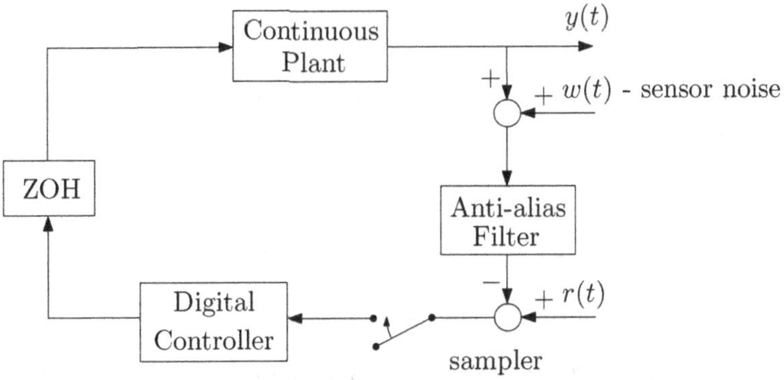

Figure 26.18 Use of anti-aliasing filter

may be used. Now the question arises as to how to select the filter corner frequency $1/T_a$. First of all, the filter bandwidth should be wider than the original controller bandwidth. Next, by the Nyquist and Shannon sampling theorem, we know that any signal with frequency above half the sampling frequency ($\omega_s/2$) will be aliased into the range $[0, \omega_s/2]$ (possibly at very low frequencies within the controller bandwidth). On the other hand, the filter attenuates any signals with frequencies above $1/T_a$. Therefore, it makes sense to select the filter corner frequency as $1/T_a < \omega_s/2$. In this way, any signals with frequencies above $1/T_a$ are significantly attenuated when aliased into the frequency range $[0, 1/T_a]$, and any signals with frequencies in the range $[0, 1/T_a]$ do not have aliases in the frequency range $[0, 1/T_a]$. Therefore, the combined anti-alias filter and sampler essentially pass through an un-altered frequency spectrum in the range $[0, 1/T_a]$, and any aliased signals appear at higher frequencies. The end result is that the digital controller sees the originally designed for spectrum of signals within the controller bandwidth and any alteration of the spectrum of signals due to aliasing occurs at higher frequencies outside the controller bandwidth with little effect on controller performance.

Finally, if the sample rate is chosen to be at least 20 times the closed-loop bandwidth, then the anti-alias pre-filter in (26.40) has a bandwidth that is much wider than the open-loop and controller bandwidths, and therefore does not have much effect on the originally designed continuous closed-loop system (of course, the anti-alias pre-filter could also be incorporated into the original continuous design).

26.4 Unmodeled Dynamics

In all of Chapters 17 to 24, we have considered spacecraft to be rigid bodies for the purposes of attitude control design. This is a useful approximation to make, since it simplifies the attitude control system design. However, as noted in Chapters 14 and 15, no spacecraft structure is perfectly rigid. Additionally, if the spacecraft is equipped with a propulsion system, propellant sloshing in the fuel tanks complicates the spacecraft dynamics as well. Both spacecraft flexibility and propellant sloshing can have significant effects on the performance of a spacecraft attitude control system, and can even lead to instability if not properly accounted for.

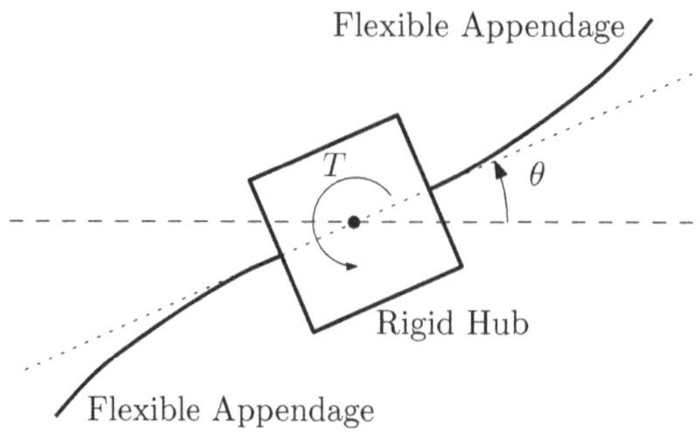

Figure 26.19 Single-axis flexible spacecraft model

26.4.1 Effects of Spacecraft Flexibility

In this subsection, we shall examine the qualitative effects of spacecraft flexibility on the spacecraft attitude dynamics. In keeping with Chapters 17 to 23, we shall limit our discussion to attitude motion (including vibration) about a single axis.

Spacecraft structures may generally be divided into two parts: 1) the main spacecraft hub, and 2) appendages. The main spacecraft hub typically includes all attitude control hardware (sensors and actuators) as well as any payload instrumentation. It is very important that elastic deformations of the spacecraft hub are very small, since the relative orientations of the sensors, actuators and payload need to be known very accurately. As a result, the spacecraft hub structure is generally designed to be very stiff. The appendages on the other hand could include deployable solar panels and antennas. Their orientations relative to the spacecraft hub are generally not as critical, and they could be structurally quite flexible. As such, a reasonable model of a spacecraft is one with a rigid hub and flexible appendages. Since the attitude control hardware is mounted on the hub, the attitude sensors measure attitude motion of the rigid hub, and the attitude actuators apply control torques to the rigid hub.

We shall consider a general single-axis flexible spacecraft such as that illustrated in Figure 26.19. Our development is valid for any number of flexible appendages, we only show two for illustration purposes. Note that θ is the spacecraft attitude, and T is the torque applied to the spacecraft.

Assuming small elastic deformations of the appendages, no internal structural damping, and a finite-dimensional model for the appendages, the equations of motion (which we present without derivation) take the form

$$\mathbf{M}\ddot{\mathbf{q}} + \mathbf{K}\mathbf{q} = \mathbf{F}T, \tag{26.41}$$

where

$$\mathbf{M} = \begin{bmatrix} I & \mathbf{b}^T \\ \mathbf{b} & \mathbf{M}_e \end{bmatrix}, \quad \mathbf{K} = \begin{bmatrix} 0 & \mathbf{0} \\ \mathbf{0} & \mathbf{K}_e \end{bmatrix}, \quad \mathbf{F} = \begin{bmatrix} 1 \\ \mathbf{0} \end{bmatrix}, \quad \mathbf{q} = \begin{bmatrix} \theta \\ \mathbf{q}_e \end{bmatrix}.$$

Note that I is the moment of inertia for the entire undeformed spacecraft about its center of mass (including the hub and appendages in their undeformed (elastically neutral) states). We call $\mathbf{M}$ the *system mass matrix* and $\mathbf{K}$ the *system stiffness matrix*, both of which are constant. Note that $\mathbf{M}$ is symmetric and positive definite. We call the $n \times n$ matrix $\mathbf{M}_e$ the *elastic mass matrix*, and the $n \times n$ matrix $\mathbf{K}_e$ the *elastic stiffness matrix*. Note that both $\mathbf{M}_e$ and $\mathbf{K}_e$ are symmetric and positive definite. The vector $\mathbf{q}_e$ is an n-dimensional vector containing the *elastic coordinates*, and its details depend on the method used to model the appendages. The n-dimensional vector $\mathbf{b}$ couples the spacecraft attitude motion (θ) to the elastic deformations ($\mathbf{q}_e$).

Let us now expand (26.41) into its components as

$$\begin{aligned} I\ddot{\theta} + \mathbf{b}^T\ddot{\mathbf{q}}_e &= T, \\ \mathbf{M}_e\ddot{\mathbf{q}}_e + \mathbf{K}_e\mathbf{q}_e &= -\mathbf{b}\ddot{\theta}. \end{aligned} \tag{26.42}$$

We note that if we fix the spacecraft hub (set $\theta \equiv \dot{\theta} \equiv \ddot{\theta} \equiv 0$), then the equations of motion for the flexible appendages become

$$\mathbf{M}_e\ddot{\mathbf{q}}_e + \mathbf{K}_e\mathbf{q}_e = \mathbf{0}. \tag{26.43}$$

We call these the *constrained flexible spacecraft equations of motion*. By comparison, we call equations (26.42) (equivalently (26.41)) the *unconstrained flexible spacecraft equations of motion*.

We shall now find the transfer function from control torque $\hat{T}(s) = \mathcal{L}(T)$ to spacecraft attitude $\hat{\theta}(s) = \mathcal{L}(\theta)$. First, we shall write equations (26.42) in a simpler form. To do this, we make use of the following result from linear algebra.

Theorem 26.1 *Let* $\mathbf{M}$ *and* $\mathbf{K}$ *be any* $n \times n$ *symmetric real matrices, with the restriction that* $\mathbf{M}$ *be positive definite. Then, there exists a non-singular* $n \times n$ *real matrix* $\mathbf{C}$ *such that*

$$\mathbf{C}^T\mathbf{M}\mathbf{C} = \mathbf{1}, \quad \mathbf{C}^T\mathbf{K}\mathbf{C} = \mathbf{\Lambda},$$

where $\mathbf{\Lambda}$ *is a real diagonal matrix given by*

$$\mathbf{\Lambda} = \text{diag}\,[\lambda_1, \lambda_2, \ldots, \lambda_n],$$

The λ_i *are generalized eigenvalues satisfying the generalized eigenvalue problem*

$$\mathbf{K}\mathbf{x} - \lambda\mathbf{M}\mathbf{x} = \mathbf{0}, \quad \mathbf{x} \neq \mathbf{0}.$$

If K *is also positive definite, then* $\lambda_i > 0$ *for all* $i = 1, \ldots, n$.

Let us now apply Theorem 26.1 to $\mathbf{M}_e$ and $\mathbf{K}_e$. Since they are both positive definite, Theorem 26.1 tells us that there is a non-singular $\mathbf{C}_c$ such that

$$\mathbf{C}_c^T\mathbf{M}_e\mathbf{C}_c = \mathbf{1}, \quad \mathbf{C}_c^T\mathbf{K}_e\mathbf{C}_c = \mathbf{\Lambda}_c, \tag{26.44}$$

where

$$\mathbf{\Lambda}_c = \text{diag}\left[\omega_{c,1}^2, \omega_{c,2}^2, \ldots, \omega_{c,n}^2\right],$$

where $\omega_{c,i} > 0$ for $i = 1, \ldots, n$. From Theorem 26.1, we also see that $\omega_{c,i}$ satisfy

$$\det\left(-\omega_c^2 \mathbf{M}_e + \mathbf{K}_e\right) = 0,$$

which is the characteristic equation associated with the constrained flexible spacecraft equations of motion (26.43). Correspondingly, we call $\omega_{c,i}$ the *constrained natural frequencies* of vibration.

Let us now define

$$\hat{\mathbf{C}}_c = \begin{bmatrix} 1 & \mathbf{0} \\ \mathbf{0} & \mathbf{C}_c \end{bmatrix},$$

and make the change of variables

$$\mathbf{q}_c = \hat{\mathbf{C}}_c^{-1} \mathbf{q}. \tag{26.45}$$

Using the definition of $\mathbf{q}$, we find that

$$\mathbf{q}_c = \begin{bmatrix} \theta \\ \boldsymbol{\eta}_c \end{bmatrix},$$

where $\boldsymbol{\eta}_c = \mathbf{C}_c^{-1} \mathbf{q}_e$. Making the change of variables in (26.41), we obtain

$$\mathbf{M}\hat{\mathbf{C}}_c \ddot{\mathbf{q}}_c + \mathbf{K}\hat{\mathbf{C}}_c \mathbf{q} = \mathbf{F}T.$$

Multiplying both sides on the left by $\hat{\mathbf{C}}_c^T$, this becomes

$$\mathbf{M}_c \ddot{\mathbf{q}}_c + \mathbf{K}_c \mathbf{q}_c = \mathbf{F}_c T, \tag{26.46}$$

where

$$\mathbf{M}_c = \hat{\mathbf{C}}_c^T \mathbf{M} \hat{\mathbf{C}}_c = \begin{bmatrix} I & \bar{\mathbf{b}}^T \\ \bar{\mathbf{b}} & \mathbf{1} \end{bmatrix}, \quad \mathbf{K}_c = \hat{\mathbf{C}}_c^T \mathbf{M} \hat{\mathbf{C}}_c = \begin{bmatrix} 0 & \mathbf{0} \\ \mathbf{0} & \boldsymbol{\Lambda}_c \end{bmatrix}, \quad \mathbf{F}_c = \hat{\mathbf{C}}_c^T \mathbf{F} = \mathbf{F},$$

and $\bar{\mathbf{b}} = \hat{\mathbf{C}}_c^T \mathbf{b}$. Note that we have made use of equations (26.44) to obtain these expressions. Expanding equations (26.46), we obtain

$$\begin{aligned} I\ddot{\theta} + \bar{\mathbf{b}}^T \ddot{\boldsymbol{\eta}}_c &= T, \\ \ddot{\boldsymbol{\eta}}_c + \boldsymbol{\Lambda}_c \boldsymbol{\eta}_c &= -\bar{\mathbf{b}}\ddot{\theta}. \end{aligned} \tag{26.47}$$

Using the notation

$$\boldsymbol{\eta}_c = \left[\eta_{c,1}, \ldots, \eta_{c,n}\right]^T, \quad \bar{\mathbf{b}} = \left[\bar{b}_1, \ldots, \bar{b}_n\right]^T,$$

we can write equations (26.47) as

$$\begin{aligned} I\ddot{\theta} + \textstyle\sum_{i=1}^{n} \bar{b}_i \ddot{\eta}_{c,i} &= T, \\ \ddot{\eta}_{c,i} + \omega_{c,i}^2 \eta_{c,i} &= -\bar{b}_i \ddot{\theta}, \quad i = 1, \ldots, n. \end{aligned} \tag{26.48}$$

It is clear that the equations of motion for $\eta_{c,i}$ for $i = 1, \ldots, n$ become decoupled when the spacecraft hub is fixed ($\theta \equiv 0$). Correspondingly, we call $\eta_{c,i}$ the *constrained modal coordinates*.

To find the transfer function from $\hat{T}(s) = \mathcal{L}(T)$ to $\hat{\theta}(s) = \mathcal{L}(\theta)$, let us take Laplace transforms of equations (26.48). We have

$$Is^2\hat{\theta}(s) + \sum_{i=1}^{n} \bar{b}_i s^2 \hat{\eta}_{c,i} = \hat{T}(s),$$

and

$$\left(s^2 + \omega_{c,i}^2\right) \hat{\eta}_{c,i}(s) = -\bar{b}_i s^2 \hat{\theta}(s).$$

We can rearrange these to obtain

$$Is^2 \left(\hat{\theta}(s) + \sum_{i=1}^{n} (\bar{b}_i / I) \hat{\eta}_{c,i}(s) \right) = \hat{T}(s), \tag{26.49}$$

and

$$\hat{\eta}_{c,i}(s) = -\frac{\bar{b}_i s^2 \hat{\theta}(s)}{s^2 + \omega_{c,i}^2}. \tag{26.50}$$

Substituting (26.50) into (26.49), and rearranging, we obtain

$$\hat{\theta}(s) = G_p(s)\hat{T}(s), \tag{26.51}$$

where

$$G_p(s) = \frac{\prod_{i=1}^{n} \left(s^2 + \omega_{c,i}^2\right)}{Is^2 \prod_{i=1}^{n} \left(s^2 + \omega_{c,i}^2\right) - s^4 \sum_{j=1}^{n} \bar{b}_i^2 \prod_{j=1, j \neq i}^{n} \left(s^2 + \omega_{c,j}^2\right)}. \tag{26.52}$$

Equation (26.52) provides a nice factorization for the numerator of $G_p(s)$, but not the denominator. We now find a factorization for the denominator. To do this, we shall make use of the following result.

Lemma 26.1 *Consider the matrix*

$$\mathbf{A} = \begin{bmatrix} a_{11} & \mathbf{a}_{21}^T \\ \mathbf{a}_{21} & \mathbf{A}_{22} \end{bmatrix},$$

with $\mathbf{A}_{22}$ symmetric. Let $\mathbf{D}_{ij}$ be the matrix formed by deleting the ith row and j^{th} column of $\mathbf{A}_{22}$. Let $\mathbf{G}$ be the co-factor matrix associated with $\mathbf{A}_{22}$, which is defined by $\mathbf{G}_{ij} = (-1)^{i+j} \det\left[\mathbf{D}_{ij}\right]$. Then, the determinant of $\mathbf{A}$ is given by

$$\det[\mathbf{A}] = a_{11} \det[\mathbf{A}_{22}] - \mathbf{a}_{21}^T \mathbf{G} \mathbf{a}_{21}.$$

This can readily be demonstrated by expanding $\det[\mathbf{A}]$ along the first column.

We return to the equations of motion in the form (26.46). Consider

$$s^2\mathbf{M}_c + \mathbf{K}_c = \begin{bmatrix} s^2 I & s^2\bar{\mathbf{b}}^T \\ s^2\bar{\mathbf{b}} & s^2\mathbf{1} + \mathbf{\Lambda}_c \end{bmatrix}.$$

We readily find that the co-factor matrix of $s^2\mathbf{1} + \mathbf{\Lambda}_c$ is given by

$$\operatorname{cof}\left[s^2\mathbf{1} + \mathbf{\Lambda}_c\right] = \operatorname{diag}_{i=1,\ldots,n}\left\{\prod_{j=1,j\neq i}^{n} \left(s^2 + \omega_{c,i}^2\right)\right\}.$$

As a result, by direct application of Lemma 26.1, we find that the denominator of $G_p(s)$ in (26.52) satisfies

$$\det\left[s^2\mathbf{M}_c + \mathbf{K}_c\right] = Is^2 \prod_{i=1}^{n} \left(s^2 + \omega_{c,i}^2\right) - s^4 \sum_{j=1}^{n} \bar{b}_i^2 \prod_{j=1,j\neq i}^{n} \left(s^2 + \omega_{c,j}^2\right). \quad (26.53)$$

Let us now find another expression for $\det\left[s^2\mathbf{M}_c + \mathbf{K}_c\right]$. Since $\mathbf{M}_c = \hat{\mathbf{C}}_c^T \mathbf{M} \hat{\mathbf{C}}_c$ is positive definite (because $\mathbf{M}$ is positive definite and $\hat{\mathbf{C}}_c$ is non-singular), it can be shown that the matrix $\mathbf{1} - (1/I)\bar{\mathbf{b}}\bar{\mathbf{b}}^T$ is positive definite also. Making use of Theorem 26.1, we can then find a non-singular matrix $\mathbf{C}_u$ such that

$$\mathbf{C}_u^T \left[\mathbf{1} - (1/I)\bar{\mathbf{b}}\bar{\mathbf{b}}^T\right] \mathbf{C}_u = \mathbf{1}, \quad \mathbf{C}_u^T \mathbf{\Lambda}_c \mathbf{C}_u = \mathbf{\Lambda}_u, \quad (26.54)$$

where

$$\mathbf{\Lambda}_u = \operatorname{diag}\left[\omega_{u,1}^2, \omega_{u,2}^2, \ldots, \omega_{u,n}^2\right],$$

where $\omega_{u,i} > 0$ for $i = 1, \ldots, n$. Let us now define

$$\hat{\mathbf{C}}_u = \begin{bmatrix} 1 & -(1/I)\bar{\mathbf{b}}^T \mathbf{C}_u \\ \mathbf{0} & \mathbf{C}_u \end{bmatrix}. \quad (26.55)$$

Then, by direct multiplication (and using equations (26.54)), we obtain

$$\hat{\mathbf{C}}_u^T \mathbf{M}_c \hat{\mathbf{C}}_u = \mathbf{M}_u = \begin{bmatrix} I & \mathbf{0} \\ \mathbf{0} & \mathbf{1} \end{bmatrix}, \quad \hat{\mathbf{C}}_u^T \mathbf{K}_c \hat{\mathbf{C}}_u = \mathbf{K}_u = \begin{bmatrix} 0 & \mathbf{0} \\ \mathbf{0} & \mathbf{\Lambda}_u \end{bmatrix}. \quad (26.56)$$

Let us now consider

$$\det\left[s^2\mathbf{M}_u + \mathbf{K}_u\right] = \det\begin{bmatrix} s^2 I & \mathbf{0} \\ \mathbf{0} & s^2\mathbf{1} + \mathbf{\Lambda}_u \end{bmatrix} = Is^2 \prod_{i=1}^{n} \left(s^2 + \omega_{u,i}^2\right).$$

Now,

$$\begin{aligned}\det\left[s^2\mathbf{M}_u + \mathbf{K}_u\right] &= \det\left[\hat{\mathbf{C}}_u^T\left(s^2\mathbf{M}_c + \mathbf{K}_c\right)\hat{\mathbf{C}}_u\right], \\ &= \det\left[\hat{\mathbf{C}}_u\right]^2 \det\left[s^2\mathbf{M}_c + \mathbf{K}_c\right].\end{aligned}$$

Therefore, we find from (26.53) that the denominator of $G_p(s)$ is given by

$$\begin{aligned}&Is^2 \prod_{i=1}^{n}\left(s^2 + \omega_{c,i}^2\right) - s^4 \sum_{j=1}^{n} \bar{b}_i^2 \prod_{j=1,j\neq i}^{n}\left(s^2 + \omega_{c,j}^2\right) \\ &= \frac{1}{\det\left[\hat{\mathbf{C}}_u\right]^2} Is^2 \prod_{i=1}^{n}\left(s^2 + \omega_{u,i}^2\right).\end{aligned} \tag{26.57}$$

Substituting this into the transfer function in (26.52), we obtain

$$G_p(s) = \frac{\det\left[\hat{\mathbf{C}}_u\right]^2 \prod_{i=1}^{n}\left(s^2 + \omega_{c,i}^2\right)}{Is^2 \prod_{i=1}^{n}\left(s^2 + \omega_{u,i}^2\right)}. \tag{26.58}$$

From (26.58), we see that we can write

$$G_p(s) = G_{p,r}(s)G_{p,f}(s), \tag{26.59}$$

where

$$G_{p,r}(s) = \frac{1}{Is^2},$$

and

$$G_{p,f}(s) = K\frac{\prod_{i=1}^{n}\left((s/\omega_{c,i})^2 + 1\right)}{\prod_{i=1}^{n}\left((s/\omega_{u,i})^2 + 1\right)}, \quad K = G_{p,f}(0) = \det\left[\hat{\mathbf{C}}_u\right]^2 \frac{\prod_{i=1}^{n}\omega_{c,i}^2}{\prod_{i=1}^{n}\omega_{u,i}^2}.$$

Rearranging equation (26.59), we have $G_{p,f}(s) = G_p(s)/G_{p,r}(s) = Is^2G_p(s)$. Therefore, from (26.52), we readily find that $K = G_{p,f}(0) = 1$. As seen in (26.59), the transfer function from $\hat{T}(s)$ to $\hat{\theta}(s)$ can be separated into a rigid body component ($G_{p,r}(s)$) and a contribution from the flexible modes ($G_{p,f}(s)$).

Let us now obtain the physical meaning for the frequencies $\omega_{u,i}$ for $i = 1, \ldots, n$. We now show that $(0, \omega_{u,1}^2, \omega_{u,2}^2, \ldots, \omega_{u,n}^2)$ are the generalized eigenvalues of

$$\mathbf{Kx} - \lambda\mathbf{Mx} = \mathbf{0}, \quad \mathbf{x} \neq \mathbf{0}, \tag{26.60}$$

where $\mathbf{M}$ and $\mathbf{K}$ are given in (26.41). The characteristic equation corresponding to (26.60) is

$$\det(\mathbf{K} - \lambda\mathbf{M}) = 0,$$

which can be rewritten as

$$\begin{aligned} 0 &= \det(\mathbf{K} - \lambda\mathbf{M}), \\ &= \det\left(\hat{\mathbf{C}}_c^{-T}\hat{\mathbf{C}}_u^{-T}\left[\mathbf{K}_u - \lambda\mathbf{M}_u\right]\hat{\mathbf{C}}_u^{-1}\hat{\mathbf{C}}_c^{-1}\right), \\ &= \det\left(\hat{\mathbf{C}}_c^{-1}\right)^2 \det\left(\hat{\mathbf{C}}_u^{-1}\right)^2 \det(\mathbf{K}_u - \lambda\mathbf{M}_u). \end{aligned}$$

Therefore, since $\hat{\mathbf{C}}_c$ and $\hat{\mathbf{C}}_u$ are non-singular, the characteristic equation is equivalent to

$$\det(\mathbf{K}_u - \lambda\mathbf{M}_u) = 0.$$

Using the definitions of $\mathbf{M}_u$ and $\mathbf{K}_u$ in (26.56), this becomes

$$I\lambda\prod_{i=1}^{n}(\lambda - \omega_{u,i}^2) = 0,$$

from which the result is obvious. As such, we call $\omega_{u,i}$ the *unconstrained natural frequencies* of vibration.

We now examine the relationship between the constrained and unconstrained natural frequencies. To do this, we make use of the following inclusion theorem from linear algebra.

Theorem 26.2 *Let* $\mathbf{M}$ *and* $\mathbf{K}$ *both be* $n \times n$ *symmetric matrices, with* $\mathbf{M}$ *positive definite, and let* $\hat{\mathbf{M}}$ *and* $\hat{\mathbf{K}}$ *be the matrices obtained by deleting both the first row and column of* $\mathbf{M}$ *and* $\mathbf{K}$*, respectively. Let* $\lambda_1 \le \lambda_2 \le \ldots \le \lambda_n$ *be the generalized eigenvalues corresponding to* $\mathbf{K}\mathbf{x} - \lambda\mathbf{M}\mathbf{x} = \mathbf{0}$ *with* $\mathbf{x} \neq \mathbf{0}$*, and let* $\hat{\lambda}_1 \le \hat{\lambda}_2 \le \ldots \le \hat{\lambda}_{n-1}$ *be the generalized eigenvalues corresponding to* $\hat{\mathbf{K}}\hat{\mathbf{x}} - \lambda\hat{\mathbf{M}}\hat{\mathbf{x}} = \mathbf{0}$ *with* $\hat{\mathbf{x}} \neq \mathbf{0}$*. Then,*

$$\lambda_1 \le \hat{\lambda}_1 \le \lambda_2 \le \ldots \le \lambda_{n-1} \le \hat{\lambda}_{n-1} \le \lambda_n.$$

Let us now order the constrained and unconstrained natural frequencies as $\omega_{c,1} \le \ldots \le \omega_{c,n}$, and $\omega_{u,1} \le \ldots \le \omega_{u,n}$, respectively. Since $\mathbf{M}_e$ and $\mathbf{K}_e$ are obtained by deleting the first row and column of $\mathbf{M}$ and $\mathbf{K}$ in (26.41), we find upon application of Theorem 26.2 that

$$\omega_{c,1} \le \omega_{u,1} \le \omega_{c,2} \le \cdots \le \omega_{c,n} \le \omega_{u,n}. \tag{26.61}$$

The lowest unconstrained natural frequency $\omega_{u,1}$ is given a special name, the *fundamental natural frequency*, corresponding to the *fundamental mode of vibration*.

26.4.1.1 Summary of Findings Thus Far

When there is no structural damping present, the transfer function relationship for the spacecraft attitude is given by

$$\hat{\theta}(s) = G_{p,r}(s)G_{p,f}(s)\hat{T}(s), \tag{26.62}$$

where

$$G_{p,r}(s) = \frac{1}{Is^2},$$

is the transfer function for the rigid spacecraft, and

$$G_{p,f}(s) = \frac{\prod_{i=1}^{n}\left((s/\omega_{c,i})^2 + 1\right)}{\prod_{i=1}^{n}\left((s/\omega_{u,i})^2 + 1\right)}, \tag{26.63}$$

is the contribution of the spacecraft flexibility. Each vibration mode contributes a pair of imaginary poles at the corresponding unconstrained natural frequency $\pm j\omega_{u,i}$ and a pair of imaginary zeros at the corresponding at the corresponding constrained natural frequency $\pm j\omega_{c,i}$, with $\omega_{u,i} \geq \omega_{c,i}$. Figure 26.20 shows the corresponding poles and zeros of the spacecraft attitude transfer function $G_p(s) = G_{p,r}(s)G_{p,f}(s)$. As we have seen in Chapter 22, the presence of open-loop poles on the imaginary axis becomes problematic for Nyquist stability analysis (the Nyquist contour must be modified to pass around them, and encirclements of the complex plane must be accounted for). These poles result in infinite peaks in the gain frequency response, which is problematic for frequency response based control design. Fortunately, every real structure always has some structural damping, so we do not have to worry about poles due to spacecraft flexibility on the imaginary axis (we will still have the poles at the origin due to the rigid body motion).

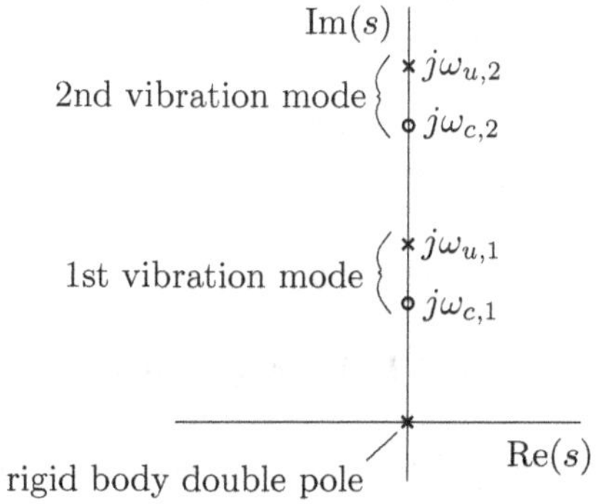

Figure 26.20 Spacecraft attitude transfer function poles and zeros (no structural damping)

26.4.1.2 Effect of Structural Damping

When structural damping is accounted for, the equation of motion in (26.41) becomes

$$\mathbf{M}\ddot{\mathbf{q}} + \mathbf{D}\dot{\mathbf{q}} + \mathbf{K}\mathbf{q} = \mathbf{F}T, \tag{26.64}$$

where

$$\mathbf{D} = \begin{bmatrix} 0 & \mathbf{0} \\ \mathbf{0} & \mathbf{D}_e \end{bmatrix},$$

and all other quantities are the same as before. The matrix $\mathbf{D}_e$ is symmetric and positive definite.

We shall once again seek a transfer function relationship between $\hat{T}(s)$ and $\hat{\theta}(s)$. As before, we expand (26.64) into its components,

$$\begin{aligned} I\ddot{\theta} + \mathbf{b}^T\ddot{\mathbf{q}}_e &= T, \\ \mathbf{M}_e\ddot{\mathbf{q}}_e + \mathbf{D}_e\dot{\mathbf{q}}_e + \mathbf{K}_e\mathbf{q}_e &= -\mathbf{b}\ddot{\theta}. \end{aligned} \tag{26.65}$$

The corresponding constrained equations of motion are

$$\mathbf{M}_e\ddot{\mathbf{q}}_e + \mathbf{D}_e\dot{\mathbf{q}}_e + \mathbf{K}_e\mathbf{q}_e = \mathbf{0}. \tag{26.66}$$

Once again, we make the change of variables in (26.45). This leads to the transformed equations of motion

$$\mathbf{M}_c\ddot{\mathbf{q}}_c + \hat{\mathbf{D}}_c\dot{\mathbf{q}}_c + \mathbf{K}_c\mathbf{q}_c = \mathbf{F}_cT, \tag{26.67}$$

where

$$\hat{\mathbf{D}}_c = \hat{\mathbf{C}}_c^T\mathbf{D}\hat{\mathbf{C}}_c = \begin{bmatrix} 0 & \mathbf{0} \\ \mathbf{0} & \mathbf{D}_c \end{bmatrix}, \quad \mathbf{D}_c = \mathbf{C}_c^T\mathbf{D}_e\mathbf{C}_c,$$

and all other quantities are as in equation (26.46). Note that unlike $\mathbf{\Lambda}_c$ in $\mathbf{K}_c$, the matrix $\mathbf{D}_c$ is not necessarily diagonal. Expanding, we have

$$\begin{aligned} I\ddot{\theta} + \bar{\mathbf{b}}^T\ddot{\eta}_c &= T, \\ \ddot{\eta}_c + \mathbf{D}_c\dot{\eta} + \mathbf{\Lambda}_c\eta_c &= -\bar{\mathbf{b}}\ddot{\theta}, \end{aligned} \tag{26.68}$$

with corresponding constrained equations of motion

$$\ddot{\eta}_c + \mathbf{D}_c\dot{\eta} + \mathbf{\Lambda}_c\eta_c = \mathbf{0}. \tag{26.69}$$

It can be rigorously shown (using LaSalle-type analysis as in Chapter 24) that the constrained equations of motion are asymptotically stable. Since the system is linear and time-invariant, this means that the constrained poles lie strictly in the left-half plane. Because (26.69) is a second-order differential equation, there are an even number of poles ($2n$ poles). Therefore, the constrained characteristic equation can be written as

$$\det\left[s^2\mathbf{1} + s\mathbf{D}_c + \mathbf{\Lambda}_c\right] = \prod_{i=1}^{n}\left(s^2 + 2\zeta_{c,i}\bar{\omega}_{c,i}s + \bar{\omega}_{c,i}^2\right), \tag{26.70}$$

where $\zeta_{c,i} > 0$ is the damping ratio and $\bar{\omega}_{c,i} > 0$ is the undamped natural frequency of the ith constrained mode respectively. Note that $\bar{\omega}_{c,i}^2$ are not necessarily the same as the diagonal entries of $\mathbf{\Lambda}_c$, which are $\omega_{c,i}^2$. This difference is due to the presence of $s\mathbf{D}_c$ in (26.70).

Let us now take Laplace transforms of (26.68). We have

$$s^2 I\hat{\theta}(s) + s^2\bar{\mathbf{b}}^T\hat{\eta}(s) = \hat{T}(s), \tag{26.71}$$

$$\left(s^2\mathbf{1} + s\mathbf{D}_c + \mathbf{\Lambda}_c\right)\hat{\eta}(s) = -s^2\bar{\mathbf{b}}\hat{\theta}(s). \tag{26.72}$$

Noting that

$$\left(s^2\mathbf{1} + s\mathbf{D}_c + \mathbf{\Lambda}_c\right)^{-1} = \frac{1}{\det\left[s^2\mathbf{1} + s\mathbf{D}_c + \mathbf{\Lambda}_c\right]}\mathbf{H}(s),$$

where

$$\mathbf{H}(s) = \operatorname{cof}\left(s^2\mathbf{1} + s\mathbf{D}_c + \mathbf{\Lambda}_c\right),$$

we can rearrange (26.72) to give

$$\hat{\eta} = \frac{-s^2\mathbf{H}(s)\bar{\mathbf{b}}}{\prod_{i=1}^{n}\left(s^2 + 2\zeta_{c,i}\bar{\omega}_{c,i}s + \bar{\omega}_{c,i}^2\right)}\hat{\theta}.$$

Substituting into (26.71) and rearranging, we have

$$\theta(\hat{s}) = G_p(s)\hat{T}(s), \tag{26.73}$$

$$G_p(s) = \frac{\prod_{i=1}^{n}\left(s^2 + 2\zeta_{c,i}\bar{\omega}_{c,i}s + \bar{\omega}_{c,i}^2\right)}{Is^2\prod_{i=1}^{n}\left(s^2 + 2\zeta_{c,i}\bar{\omega}_{c,i}s + \bar{\omega}_{c,i}^2\right) - s^4\bar{\mathbf{b}}^T\mathbf{H}(s)\bar{\mathbf{b}}}. \tag{26.74}$$

As for the undamped case, equation (26.74) provides a nice factorization of the numerator of $G_p(s)$. As before, we shall now find an expression for the denominator.

A direct application of Lemma 26.1 shows that the denominator of $G_p(s)$ in (26.74) satisfies

$$\begin{aligned}\det\left[s^2\mathbf{M}_c + s\hat{\mathbf{D}}_c + \mathbf{K}_c\right] &= \begin{bmatrix} s^2 I & s^2\bar{\mathbf{b}}^T \\ s^2\bar{\mathbf{b}} & s^2\mathbf{1} + s\mathbf{D}_c + \mathbf{\Lambda}_c \end{bmatrix}, \\ &= Is^2\prod_{i=1}^{n}\left(s^2 + 2\zeta_{c,i}\bar{\omega}_{c,i}s + \bar{\omega}_{c,i}^2\right) - s^4\bar{\mathbf{b}}^T\mathbf{H}(s)\bar{\mathbf{b}}. \end{aligned} \tag{26.75}$$

Let us now apply the transformation given in equations (26.54) and (26.55). We consider $\det\left[s^2\mathbf{M}_u + s\hat{\mathbf{D}}_u + \mathbf{K}_u\right]$, where

$$\hat{\mathbf{D}}_u = \hat{\mathbf{C}}_u^T\mathbf{D}_c\hat{\mathbf{C}}_u = \begin{bmatrix} 0 & \mathbf{0} \\ \mathbf{0} & \mathbf{D}_u \end{bmatrix}, \quad \mathbf{D}_u = \mathbf{C}_u^T\mathbf{D}_c\mathbf{C}_u,$$

and all other quantities are as in equation (26.56). Note that unlike $\mathbf{\Lambda}_u$ in $\mathbf{K}_u$, the matrix $\mathbf{D}_u$ is not necessarily diagonal. Then,

$$\begin{aligned}\det\left[s^2\mathbf{M}_u + s\hat{\mathbf{D}}_u + \mathbf{K}_u\right] &= \det\begin{bmatrix} s^2 I & \mathbf{0} \\ \mathbf{0} & s^2\mathbf{1} + s\mathbf{D}_u + \mathbf{\Lambda}_u \end{bmatrix}, \\ &= Is^2\det\left[s^2\mathbf{1} + s\mathbf{D}_u + \mathbf{\Lambda}_u\right].\end{aligned}$$

Using the same reasoning as before, we can write

$$\det\left[s^2\mathbf{1} + s\mathbf{D}_u + \mathbf{\Lambda}_u\right] = \prod_{i=1}^{n}\left(s^2 + 2\zeta_{u,i}\bar{\omega}_{u,i}s + \bar{\omega}_{u,i}^2\right), \tag{26.76}$$

where $\zeta_{u,i} > 0$ is the damping ratio and $\bar{\omega}_{u,i} > 0$ is the undamped natural frequency of the ith unconstrained mode respectively. As before, $\bar{\omega}_{u,i}^2$ are not necessarily the same as the diagonal entries of $\mathbf{\Lambda}_u$. Now,

$$\begin{aligned}\det\left[s^2\mathbf{M}_u + s\mathbf{D}_u + \mathbf{K}_u\right] &= \det\left[\hat{\mathbf{C}}_u^T\left(s^2\mathbf{M}_c + s\mathbf{D}_c + \mathbf{K}_c\right)\hat{\mathbf{C}}_u\right], \\ &= \det\left[\hat{\mathbf{C}}_u\right]^2\det\left[s^2\mathbf{M}_c + s\mathbf{D}_c + \mathbf{K}_c\right].\end{aligned}$$

Therefore, we find from (26.75) that the denominator of $G_p(s)$ is given by

$$\begin{aligned}&Is^2\textstyle\prod_{i=1}^{n}\left(s^2 + 2\zeta_{c,i}\bar{\omega}_{c,i}s + \bar{\omega}_{c,i}^2\right) - s^4\bar{\mathbf{b}}^T\mathbf{H}(s)\bar{\mathbf{b}} \\ &= \left(1/\det\left[\hat{\mathbf{C}}_u\right]^2\right)Is^2\textstyle\prod_{i=1}^{n}\left(s^2 + 2\zeta_{u,i}\bar{\omega}_{u,i}s + \bar{\omega}_{u,i}^2\right).\end{aligned}$$

Substituting this into the transfer function in (26.74), we obtain

$$G_p(s) = G_{p,r}(s)G_{p,f}(s), \tag{26.77}$$

where

$$G_{p,r}(s) = \frac{1}{Is^2},$$

and

$$G_{p,f}(s) = K\frac{\prod_{i=1}^{n}\left((s/\bar{\omega}_{c,i})^2 + 2\zeta_{c,i}(s/\bar{\omega}_{c,i}) + 1\right)}{\prod_{i=1}^{n}\left((s/\bar{\omega}_{u,i})^2 + 2\zeta_{u,i}(s/\bar{\omega}_{u,i}) + 1\right)},$$

where as for the undamped case, $K = \left[Is^2G_p(s)\right]\big|_{s=0}$. From (26.74), we readily find that $K = 1$. Therefore,

$$G_{p,f}(s) = \frac{\prod_{i=1}^{n}\left((s/\bar{\omega}_{c,i})^2 + 2\zeta_{c,i}(s/\bar{\omega}_{c,i}) + 1\right)}{\prod_{i=1}^{n}\left((s/\bar{\omega}_{u,i})^2 + 2\zeta_{u,i}(s/\bar{\omega}_{u,i}) + 1\right)}. \tag{26.78}$$

As seen in (26.77), just as for the undamped case, the transfer function from $\hat{T}(s)$ to $\hat{\theta}(s)$ can be separated into a rigid body component ($G_{p,r}(s)$) and a contribution from the flexible modes ($G_{p,f}(s)$). Note that the damping ratios $\zeta_{c,i}$ and $\zeta_{u,i}$ are typically very small. Each vibration mode contributes a complex conjugate pair of asymptotically stable open-loop zeros and poles, given by $z_i = -\zeta_{c,i}\bar{\omega}_{c,i} \pm j\bar{\omega}_{c,i}\sqrt{1-\zeta_{c,i}^2}$ and $p_i = -\zeta_{u,i}\bar{\omega}_{u,i} \pm j\bar{\omega}_{u,i}\sqrt{1-\zeta_{u,i}^2}$ respectively. Comparing (26.78) to (26.63), we see that the effect of structural damping is to pull both the undamped imaginary zeros and poles slightly into the left-half plane, as shown in Figure 26.21 (compared to Figure 26.20).

26.4.1.3 Effect of Flexibility on the Open-Loop Frequency Response

Let us now look at the effect of spacecraft flexibility on the open-loop frequency response. We shall examine the effect of a single mode of vibration. The contribution of each vibration mode has the form

$$\bar{G}_{p,f}(s) = \frac{\left((s/\omega_c)^2 + 2\zeta_c(s/\omega_c) + 1\right)}{\left((s/\omega_u)^2 + 2\zeta_u(s/\omega_u) + 1\right)}.$$

The corresponding frequency response is given by

$$\bar{G}_{p,f}(j\omega) = \frac{1 - (\omega/\omega_c)^2 + 2j\zeta_c(\omega/\omega_c)}{1 - (\omega/\omega_u)^2 + 2j\zeta_u(\omega/\omega_u)}. \tag{26.79}$$

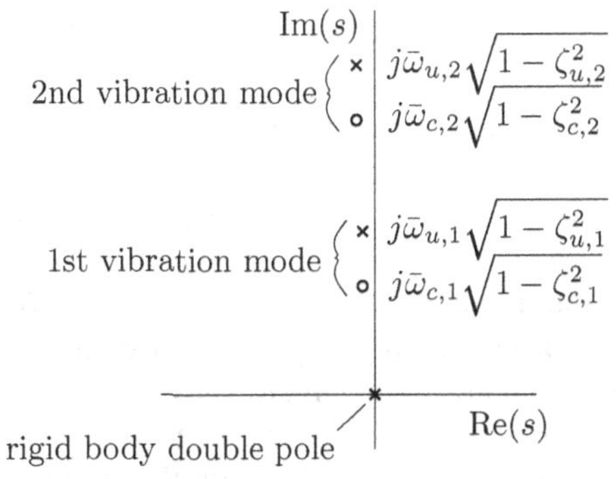

Figure 26.21 Spacecraft attitude transfer function poles and zeros (with structural damping)

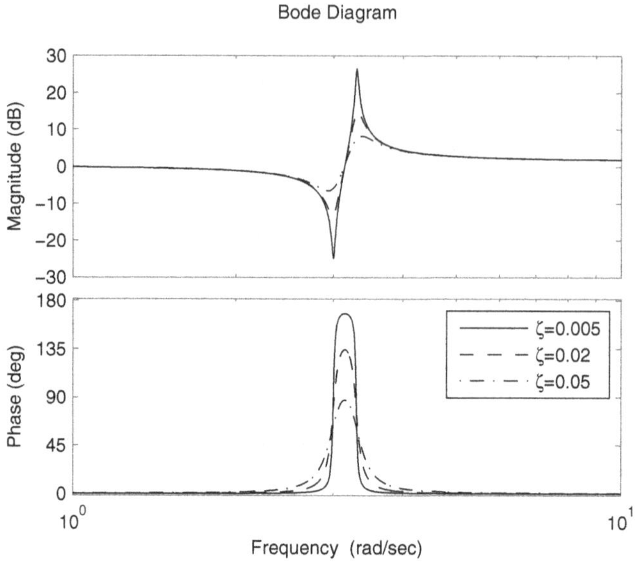

Figure 26.22 Frequency response due to a vibration mode with varying damping ratios

Clearly, for low frequencies $\omega \ll \omega_c, \omega_u$, we have

$$\left|\bar{G}_{p,f}(j\omega)\right| \approx 1, \quad \angle\bar{G}_{p,f}(j\omega) \approx 0^\circ,$$

so each vibration mode has very little effect on the low end of the frequency response. On the other hand, for high frequencies $\omega \gg \omega_c, \omega_u$, we have

$$\left|\bar{G}_{p,f}(j\omega)\right| \approx \left(\frac{\omega_u}{\omega_c}\right)^2, \quad \angle\left|\bar{G}_{p,f}(j\omega)\right| \approx 0^\circ. \tag{26.80}$$

Therefore, the main effect of the vibration mode must be when ω is near ω_c and ω_u. Figure 26.22 shows the frequency response for varying damping ratios (note that in the figure $\zeta = \zeta_c = \zeta_u$). The constrained and unconstrained natural frequencies are taken to be $\omega_c = 3.0$ rad/s and $\omega_u = 3.3$ rad/s, respectively. It can be seen that the vibration mode can contribute a significant phase shift when ω is near ω_c and ω_u. Likewise, a significant peak in the gain occurs when ω is near ω_u. As seen, the size of the peak gain and the phase shift depends significantly on the damping ratio, which unfortunately is one of the most uncertain parameters, and must be typically be determined experimentally. Figure 26.23 shows the frequency response for varying distance between ω_c and ω_u. In this figure, the constrained natural frequency is held fixed at $\omega_c = 3.0$ rad/s, the damping ratios are held fixed at $\zeta_c = \zeta_u = 0.01$. Again, it can be seen that frequency response depends significantly on the spacing between ω_c and ω_u. The closer ω_c and ω_u are, the less influence they have on the open-loop frequency response.

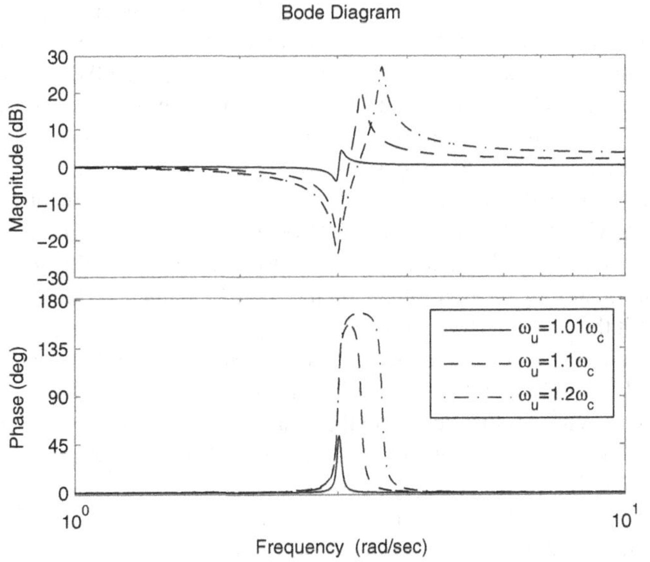

Figure 26.23 Frequency response due to a vibration mode with varying spacing between constrained and unconstrained natural frequencies

26.4.1.4 Accounting for Spacecraft Flexibility in the Attitude Control Design

Now that we understand the effect of spacecraft flexibility on the spacecraft dynamics, we can examine how to account for it in attitude control system design. We have seen that spacecraft flexibility can result in significant peaks and phase shifts in the open-loop gain frequency response near the frequencies of vibration. These can cause significant deviations in the desired open-loop frequency response (refer to Figure 23.6), and if not properly accounted for, they can result in degraded performance, and even lead to instability.

One relatively simple way to ensure that the spacecraft flexibility does not destabilize the closed-loop system is to ensure that the open-loop bandwidth is well below the fundamental vibration frequency, with sufficient cut-off rate. This allows the attitude controller to be

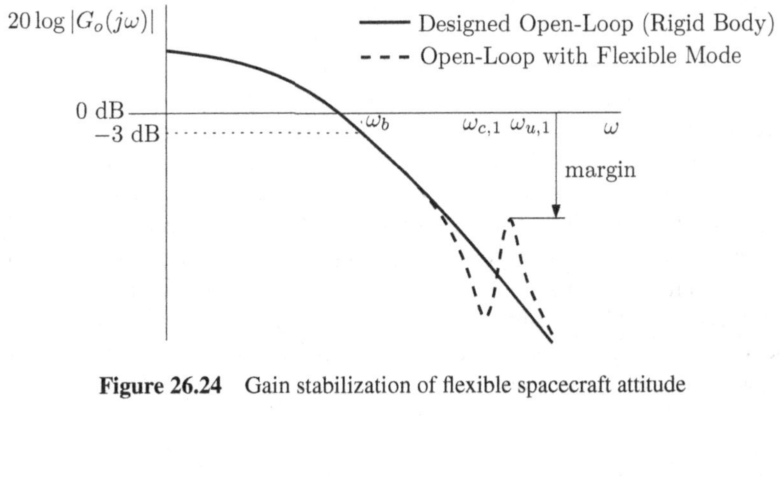

Figure 26.24 Gain stabilization of flexible spacecraft attitude

designed as if the spacecraft is rigid. By ensuring that the designed open-loop gain is sufficiently small at the vibration frequencies, the vibration modes will not cause the open-loop gain to exceed 1 (equivalently that the log-magnitude gain does not exceed 0 dB). As we know from Chapter 22, this ensures that the Nyquist plot stays inside the unit disc at high frequencies, and as such the vibration modes cannot cause additional encirclements of -1. This method of stabilization is called *gain stabilization*, since we are limiting the gain.

Gain stabilization is very useful, especially when the phase component of the frequency response due to the vibration modes is not well known (recall that this depends on the damping ratio of each mode, which is difficult to determine). However, gain stabilization can place severe restrictions on the open-loop bandwidth, especially if the fundamental frequency of vibration is low. This can limit the performance of the closed-loop system (recall from Chapter 23 that the speed of system response is related to the open-loop gain cross-over frequency, which is close to the open-loop bandwidth), and the resulting control design may not meet specifications. In this case, the vibration modes that lie within the desired open-loop bandwidth may be included in the spacecraft model when attitude control design is performed. Alternatively, an attitude controller may be designed to meet specifications assuming a rigid spacecraft. Compensators may then be added to the controller to change the open-loop phase frequency response near the vibration frequencies that lie within the open-loop bandwidth, so as to avoid encirclements of -1. This is called *phase stabilization*. See Wie (2008) for different phase compensator types. The advantage of phase stabilization is that it allows us to keep the open-loop gain above 1 at these frequencies, which means improved closed-loop system performance. The disadvantage is that phase information for the corresponding vibration modes must be known. Vibration modes at frequencies well above the open-loop bandwidth should still be gain stabilized.

Finally, spacecraft flexibility can also be dealt with very effectively by performing control design directly in the time-domain using a state-space representation of the flexible spacecraft dynamics. This approach allows us to consider coupling between the spacecraft axes as well (we limited ourselves to rotation about a single axis).

26.4.2 Effects of Propellant Sloshing

It has been found historically that propellant sloshing has the most significant effect on a spacecraft attitude control during an orbital maneuver when the spacecraft propulsion system is thrusting. Some of the effects of propellant sloshing may be limited by the installation of baffles within the fuel tank, which damp propellant motion.

Propellant sloshing is a very complex phenomenon to model mathematically, since there is the interaction between the fluid dynamics of the propellant and the attitude motion of the spacecraft structure. These models are far too complex to be used in spacecraft attitude control design, and instead simplified dynamically equivalent models have been developed to model the propellant slosh effects. These models generally represent the propellant mass by point masses connected to the fuel tank through springs or pendulums, possibly with some viscous damping. The details of the model used depend on the fuel tank geometry.

The net effect of propellant sloshing on the spacecraft dynamics is the inclusion of vibration modes due to the sloshing. As such, the effects of propellant sloshing may be accounted for in a similar fashion to spacecraft flexibility.

Notes

This chapter has presented a high level overview of some of the important issues that need to be considered for practical spacecraft attitude control. The reader is referred to Wertz (1978) and Sidi (1997) for detailed discussions of different spacecraft attitude sensor and actuator types, including both physical operating principles and mathematical models. The books by Wie (2008) and Sidi (1997) provide more detailed discussions on different modulation schemes for thrusters, and other issues associated with attitude control using thrusters. Section 26.3 discussed some of the issues associated with control law implementation on a digital processor. Much more detailed treatments of these issues may be found in books such as Franklin et al. (1998) and Astrom and Wittenmark (1997). Section 26.4 discussed how spacecraft flexibility and propellant sloshing must be accounted for in the control system design. The flexible spacecraft attitude equations of motion for a single axis (equations (26.41) and (26.64)) were presented without derivation. A detailed treatment of how the flexible spacecraft attitude equations may be derived can be found in Likins (1974) and Junkins and Kim (1993). The underlying assumption in obtaining attitude equations for a single axis is that there be no coupling between the axes. This greatly simplifies the problem, and allows for the excellent analysis presented in Section 26.4.1. In reality however, there may be coupling among the different axes either due to coupled flexible modes, coupled rigid body motion (for example, two axes are gyroscopically coupled for a dual-spin spacecraft, as shown in Chapter 15), or both. If the coupling is insignificant, the approximation of decoupling the axes as in Section 26.4.1 is reasonable to make. Otherwise, the fully coupled equations of motion must be considered, which complicates the analysis. There are several examples in the literature for how to deal with these situations. As a starting point, the reader may consult Wie (2008) and Junkins and Kim (1993). Section 26.4.2 gave a very brief discussion of the effects of propellant sloshing. For detailed models of propellant slosh, the reader is referred to Abrahamson (1966). For examples of how to deal with propellant slosh, the reader may refer to Wie (2008) and Sidi (1997).

For a detailed treatment of spacecraft attitude control from a practical engineering point of view, the reader is referred to Sidi (1997).

References

Abrahamson HN 1966 The dynamic behavior of liquids in moving containers. Technical Report NASA SP−106, National Aeronautics and Space Administration, Washington, DC.

Astrom KJ and Wittenmark B 1997 *Computer Controlled Systems* 3rd edn. Prentice Hall, Upper Saddle River, NJ.

Franklin GF, Powell JD and Workman M 1998 *Digital Control of Dynamic Systems* 3rd edn. Addison-Wesley.

Junkins JL and Kim Y 1993 *Introduction to Dynamics and Control of Flexible Systems*. American Institute of Aeronautics and Astronautics, Washington, DC.

Likins PW 1974 Analytical dynamics and nonrigid spacecraft simulation. Technical Report NASA TR32−1593, Jet Propulsion Laboratory, Pasadena, CA.

Sidi MJ 1997 *Spacecraft Dynamics and Control: A Practical Engineering Approach*. Cambridge University Press, New York, NY.

Wertz JR 1978 *Spacecraft Attitude Determination and Control*. Kluwer Academic Publishers, Dordrecht, The Netherlands.

Wie B 2008 *Space Vehicle Dynamics and Control* 2nd edn. American Institute of Aeronautics and Astronautics, Reston, VA.

Appendix A

Review of Complex Variables

Recall that a complex number, s, may be written as

$$s = a + jb,$$

where a, b are real numbers, called the real and imaginary parts of s, which can be denoted as

$$a = \text{Re}(s), \quad b = \text{Im}(s).$$

Also, j is defined such that

$$j^2 = -1.$$

Since a complex number has two parts, it can be represented by a point in the plane, as shown in Figure A.1.

The magnitude of s is given by

$$r = |s| = \sqrt{a^2 + b^2},$$

and the phase is given by

$$\angle s = \theta = \tan^{-1} \frac{b}{a},$$

where care must be taken in evaluating the inverse tangent so as to obtain the correct quadrant. Therefore, a complex number may be written in polar form as

$$s = r(\cos\theta + j\sin\theta). \tag{A.1}$$

Spacecraft Dynamics and Control: An Introduction, First Edition.
Anton H.J. de Ruiter, Christopher J. Damaren and James R. Forbes.

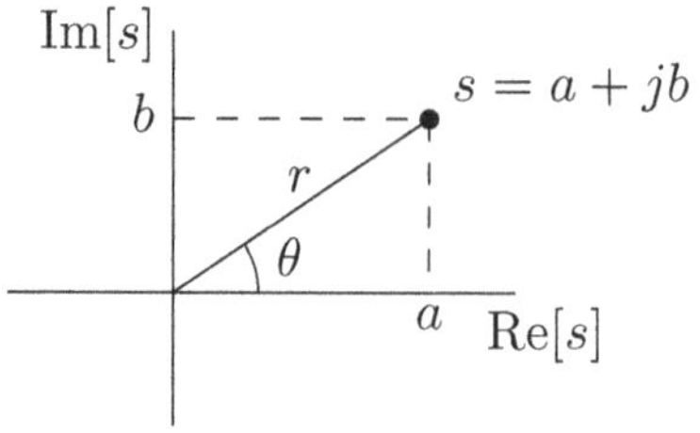

Figure A.1 Complex number representation in the plane

Addition of complex numbers is defined component-wise. If $s_1 = a_1 + jb_1$ and $s_2 = a_2 + jb_2$, then

$$s_1 + s_2 = (a_1 + a_2) + j(b_1 + b_2).$$

Multiplication of complex numbers uses the fact that $j^2 = -1$.

$$\begin{aligned} s_1 s_2 &= (a_1 + jb_1)(a_2 + jb_2), \\ &= a_1 a_2 + j^2 b_1 b_2 + a_1 j b_2 + j b_1 a_2, \\ &= (a_1 a_2 - b_1 b_2) + j(a_1 b_2 + b_1 a_2). \end{aligned}$$

The *complex conjugate* of a complex number $s = a + jb$ is defined as

$$\bar{s} = a - jb.$$

Therefore, using the rule for complex multiplication, we have

$$s\bar{s} = |s|^2 = a^2 + b^2.$$

The *complex exponential* is defined as

$$e^s = e^a(\cos b + j \sin b),$$

(where $s = a + jb$). In particular, this shows that

$$e^{j\theta} = \cos\theta + j \sin\theta.$$

Therefore, from (A.1) we can write

$$s = re^{j\theta}.$$

The inverse of a complex number is now readily found to be

$$\frac{1}{s} = \frac{1}{r}e^{-j\theta} = \frac{1}{re^{j\theta}}.$$

It can be further shown that

$$e^{s_1+s_2} = e^{s_1}e^{s_2}.$$

In particular,

$$e^{j(\theta_1+\theta_2)} = e^{j\theta_1}e^{j\theta_2}.$$

This shows that the product of two complex numbers $s_1 = r_1e^{j\theta_1}$ and $s_2 = r_2e^{j\theta_2}$ is given by

$$s_1s_2 = r_1r_2e^{j(\theta_1+\theta_2)}.$$

Therefore, the magnitude of the product is

$$|s_1s_2| = r_1r_2,$$

and the phase is

$$\angle s_1s_2 = \theta_1 + \theta_2.$$

That is, multiplication by a complex number is just a scaling by its magnitude, and a rotation through its phase, as shown in Figure A.2. Similarly, division of two complex numbers $s_1 = r_1e^{j\theta_1}$ and $s_2 = r_2e^{j\theta_2}$ is given by

$$\frac{s_1}{s_2} = \frac{r_1e^{j\theta_1}}{r_2e^{j\theta_2}} = \frac{r_1}{r_2}e^{j(\theta_1-\theta_2)}.$$

Therefore, the magnitude of the quotient is

$$\left|\frac{s_1}{s_2}\right| = \frac{r_1}{r_2},$$

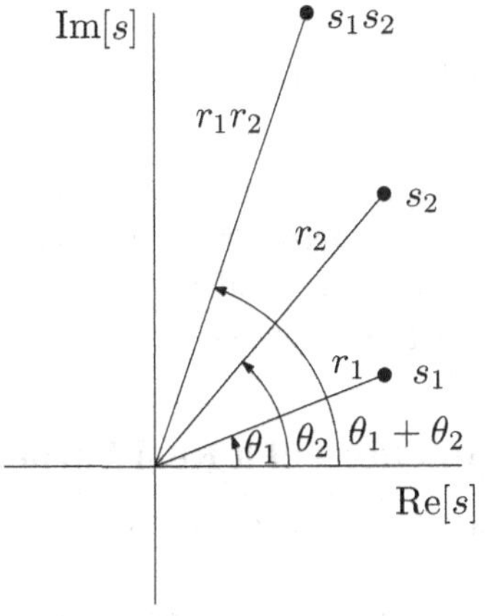

Figure A.2 Complex multiplication

and the phase is

$$\angle \frac{s_1}{s_2} = \theta_1 - \theta_2.$$

A.1 Functions of a Complex Variable

In general, a complex function $G(s)$ of a complex variable $s = a + jb$, has real and imaginary parts, given by

$$G(s) = G_x(a, b) + jG_y(a, b),$$

where $G_x(a, b)$ and $G_y(a, b)$ are real-valued functions. The functions that we deal with in this book are proper rational functions of the form

$$G(s) = \frac{a_0 s^m + a_1 s^{m-1} + \ldots + a_m}{s^n + b_1 s^{n-1} + \ldots b_n} = \frac{N(s)}{D(s)}, \tag{A.2}$$

where a_i, b_i are real coefficients, some of which may be zero, and the degree of the polynomial in the numerator $N(s)$ is less than or equal to the degree of the denominator $D(s)$. That is, $m \leq n$. It turns out that $G(s)$ can be factored into the form

$$G(s) = K \frac{\prod_{i=1}^{m}(s - z_i)}{\prod_{i=1}^{n}(s - p_i)},$$

where z_i are the roots of $N(s) = 0$, called the *zeros* of $G(s)$, and p_i are the roots of $D(s) = 0$, called the *poles* of $G(s)$.

Example A.1 *The function*

$$G(s) = \frac{(s+2)(s+10)}{s(s+5)(s+15)^2},$$

has zeros at $s = -2, -10$ *and poles at* $s = 0, -5, -15, -15$. *Note that the pole at −15 occurs twice.*

Since the coefficients a_i and b_i in (A.2) are real, *any imaginary poles or zeros must occur in complex conjugate pairs*. That is, if the zero z_i has a non-zero imaginary component, then its complex conjugate $\bar{z}_i$ must also be a zero, similarly if the pole p_i has a non-zero imaginary component, then its complex conjugate $\bar{p}_i$ must also be a pole.

A.2 Complex Valued Functions of a Real Variable

A complex valued function of a real variable t, is given by

$$g(t) = a(t) + jb(t),$$

where $a(t)$ and $b(t)$ are real. The derivative of $g(t)$ is defined by

$$\frac{dg}{dt} = \frac{da}{dt} + j\frac{db}{dt}.$$

Likewise, the integral of $g(t)$ is defined by

$$\int_{t_0}^{t_1} g(t)dt = \int_{t_0}^{t_1} a(t)dt + j\int_{t_0}^{t_1} b(t)dt.$$

This means that

$$\begin{aligned}\int_{t_0}^{t_1} \frac{dg}{dt}dt &= \int_{t_0}^{t_1} \frac{da}{dt}dt + j\int_{t_0}^{t_1} \frac{db}{dt}dt,\\ &= a(t)|_{t_0}^{t_1} + jb(t)|_{t_0}^{t_1},\\ &= g(t)|_{t_0}^{t_1}.\end{aligned} \tag{A.3}$$

The product of two complex functions of a real variable is given by

$$g_1(t)g_2(t) = (a_1(t)a_2(t) - b_1(t)b_2(t)) + j\,(a_1(t)b_2(t) + b_1(t)a_2(t)).$$

Differentiating this, making use of the product rule for derivatives of real functions gives

$$\begin{aligned}\frac{d}{dt}(g_1(t)g_2(t)) &= \left(\frac{da_1}{dt}a_2 + a_1\frac{da_2}{dt} - \frac{db_1}{dt}b_2 - b_1\frac{db_2}{dt}\right)\\ &\quad + j\left(\frac{da_1}{dt}b_2 + a_1\frac{db_2}{dt} + \frac{db_1}{dt}a_2 + b_1\frac{da_2}{dt}\right),\\ &= \left(\frac{da_1}{dt}a_2 - \frac{db_1}{dt}b_2\right) + j\left(\frac{da_1}{dt}b_2 + \frac{db_1}{dt}a_2\right)\\ &\quad + \left(a_1\frac{da_2}{dt} - b_1\frac{db_2}{dt}\right) j\left(a_1\frac{db_2}{dt} + b_1\frac{da_2}{dt}\right),\\ &= \frac{dg_1}{dt}g_2 + g_1\frac{dg_2}{dt},\end{aligned} \tag{A.4}$$

which shows that the product rule for derivatives of complex functions of a real variable holds also. Therefore, combining (A.4) and (A.3), we have

$$g_1(t)g_2(t)|_{t_0}^{t_1} = \int_{t_0}^{t_1} \frac{d}{dt}(g_1g_2)\,dt = \int_{t_0}^{t_1} \frac{dg_1}{dt}g_2dt + \int_{t_0}^{t_1} g_1\frac{dg_2}{dt}dt,$$

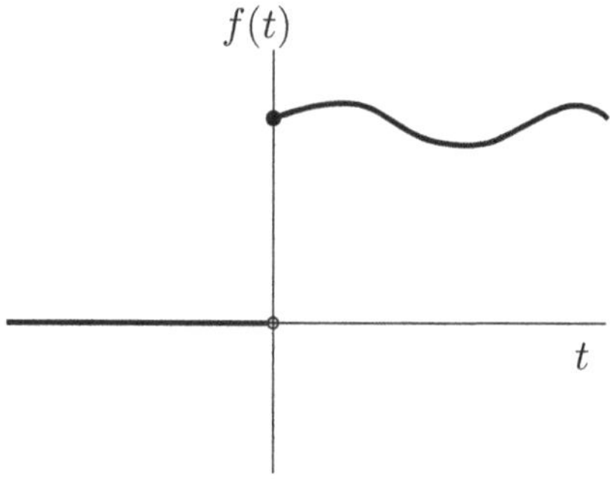

Figure A.3 Functions under consideration in one-sided Laplace transform

which rearranges to give

$$\int_{t_0}^{t_1} \frac{dg_1}{dt} g_2 dt = g_1(t)g_2(t)|_{t_0}^{t_1} - \int_{t_0}^{t_1} g_1 \frac{dg_2}{dt} dt. \tag{A.5}$$

That is, integration by parts holds for complex functions of a real variable.

A.3 The Laplace Transform

We consider only Laplace transforms of (possibly complex) functions $f(t)$ of the real variable t that are zero for all $t < 0$, as shown in Figure A.3. This is called a one-sided Laplace transform. For our purposes, it will be useful to think of t as being time.

In this case, the Laplace transform is given by

$$F(s) = \mathcal{L}(f(t)) = \int_{0^-}^{\infty} e^{-st} f(t) dt. \tag{A.6}$$

The Laplace transform is guaranteed to exist if $f(t)$ is piecewise continuous, and of exponential order, that is $|f(t)| \leq Ae^{tB}$ for all $t \geq 0$, for some $A \geq 0$, and real constant B. The Laplace integral converges if $Re(s) > B$.

Linear Combination

From the definition (A.6), we see that

$$\begin{aligned}\mathcal{L}(\alpha f_1(t) + \beta f_2(t)) &= \int_0^{\infty} e^{-st} (\alpha f_1(t) + \beta f_2(t)) dt, \\ &= \alpha \int_0^{\infty} e^{-st} f_1(t) dt + \beta \int_0^{\infty} e^{-st} f_2(t) dt, \\ &= \alpha \mathcal{L}(f_1(t)) + \beta \mathcal{L}(f_2(t)).\end{aligned}$$

This shows that the Laplace transform is a *linear* operation.

Multiplication by an Exponential

$$\begin{aligned}\mathcal{L}\left(e^{-at}f(t)\right) &= \int_0^\infty f(t)e^{-at}e^{-st}dt,\\ &= \int_0^\infty f(t)e^{-(s+a)t}dt,\\ &= F(s+a),\end{aligned}$$

where $F(s) = \mathcal{L}(f(t))$.

Scaling of Time

$$\mathcal{L}(f(at)) = \int_0^\infty f(at)e^{-st}dt.$$

Making the change of variables $\tau = at$, we have $dt/d\tau = 1/a$, and the integral becomes

$$\mathcal{L}(f(at)) = \frac{1}{a}\int_0^\infty f(\tau)e^{-(s/a)\tau}d\tau = \frac{1}{a}F\left(\frac{s}{a}\right).$$

Laplace Transform of a Derivative

First, we need the derivative of e^{-st} (with $s = a + jb$), which is evaluated as

$$\begin{aligned}\frac{d}{dt}(e^{-st}) &= \frac{d}{dt}\left(e^{-at}(\cos bt - j\sin bt)\right),\\ &= \left(-ae^{-at}\cos bt - be^{-at}\sin bt\right) + j\left(ae^{-at}\sin bt - be^{-at}\cos bt\right),\\ &= -ae^{-at}(\cos bt - j\sin bt) - jbe^{-at}(\cos bt - j\sin bt),\\ &= -(a+jb)e^{-st},\\ &= -se^{-st}.\end{aligned}$$

Now, the Laplace transform of the derivative df/dt can be evaluated using integration by parts as in (A.5).

$$\begin{aligned}\mathcal{L}\left(\frac{df}{dt}\right) &= \int_0^\infty \frac{df}{dt}e^{-st}dt,\\ &= f(t)e^{-st}\Big|_0^\infty - \int_0^\infty f(t)(-s)e^{-st}dt,\\ &= sF(s) - f(0).\end{aligned}$$

Note that we have made use of the fact that $f(t)$ is of exponential order ($|f(t)| \leq Ae^{tB}$), which means that $\lim_{t\to\infty} f(t)e^{-st} = 0$ when $Re(s) > B$.

Continuing the process in general for higher order derivatives, we get

$$\mathcal{L}\left(\frac{d^n}{dt^n}f(t)\right) = s^n F(s) - s^{n-1} f(0) - s^{n-2} \left.\frac{df}{dt}\right|_{t=0} - \cdots - \left.\frac{d^{n-1}f}{dt^{n-1}}\right|_{t=0}.$$

Laplace Transform of an Integral

$$\mathcal{L}\left(\int_0^t f(\tau)d\tau\right) = \int_0^\infty \int_0^t f(\tau)d\tau e^{-st}dt.$$

Integrating by parts gives

$$\mathcal{L}\left(\int_0^t f(\tau)d\tau\right) = \left[-\frac{1}{s}e^{-st}\int_0^t f(\tau)d\tau\right]\Bigg|_0^\infty - \int_0^\infty f(t)\left(\frac{-1}{s}\right)e^{-st}dt = \frac{F(s)}{s}.$$

Note that the first term on the right vanishes because $f(t)$ is of exponential order (which takes care of the upper limit), and $\int_0^t f(\tau)d\tau|_{t=0} = 0$, which takes care of the lower limit.

Convolution

Let two functions $f_1(t)$ and $f_2(t)$ be given. The convolution integral of these is defined by

$$f_1(t) * f_2(t) = \int_0^t f_1(\tau) f_2(t-\tau)d\tau = \int_0^t f_1(t-\tau) f_2(\tau)d\tau.$$

To show that the above two integrals are equal, consider the change of variables $z = t - \tau$, we have

$$\begin{aligned}\int_0^t f_1(\tau)f_2(t-\tau)d\tau &= \int_t^0 f_1(t-z)f_2(z)\frac{d\tau}{dz}dz,\\ &= -\int_t^0 f_1(t-z)f_2(z)dz,\\ &= \int_0^t f_1(t-z)f_2(z)dz.\end{aligned}$$

Replacing z with τ in the last integral shows the equivalence.

The Laplace transform of the convolution integral is then

$$\mathcal{L}(f_1(t) * f_2(t)) = \int_0^\infty \int_0^t f_1(\tau)f_2(t-\tau)d\tau e^{-st}dt.$$

Interchanging the order of integration, this becomes

$$\mathcal{L}(f_1(t) * f_2(t)) = \int_0^\infty \int_\tau^\infty f_1(\tau) f_2(t-\tau) e^{-st} dt d\tau, \text{ multiply by } e^{-s\tau} e^{s\tau} \text{ to get}$$
$$= \int_0^\infty f_1(\tau) e^{-s\tau} \int_\tau^\infty f_2(t-\tau) e^{-s(t-\tau)} dt d\tau.$$

Now, setting $t' = t - \tau$, we have

$$\int_\tau^\infty f_2(t-\tau) e^{-s(t-\tau)} dt = \int_0^\infty f_2(t') e^{-st'} dt' = F_2(s).$$

Substituting this into the double integral above, we have

$$\mathcal{L}(f_1(t) * f_2(t)) = \int_0^\infty f_1(\tau) e^{-s\tau} F_2(s) d\tau = F_1(s) F_2(s).$$

That is, convolution in the time domain becomes multiplication in the Laplace (s) domain.

Final Value Theorem

We can use the Laplace transform to determine limiting behavior in the time domain. This is useful when we do not know the time-domain function, but we do know its Laplace transform. As an example, this situation may arise when we wish to know the limiting behavior of a system governed by an ordinary differential equation.

Consider the limit

$$\lim_{s\to 0} \mathcal{L}\left(\frac{df}{dt}\right) = \lim_{s\to 0} \int_0^\infty \frac{df}{dt} e^{-st} dt.$$

This limit can only exist if the Laplace integral is convergent for $s = 0$. In this case, it turns out that the order of integration and the limit operation may be interchanged, so that

$$\lim_{s\to 0} \mathcal{L}\left(\frac{df}{dt}\right) = \int_0^\infty \lim_{s\to 0} \frac{df}{dt} e^{-st} dt,$$
$$= \int_0^\infty \frac{df}{dt} dt,$$
$$= \lim_{t\to\infty} f(t) - f(0).$$

However, we also know from the Laplace transform of a derivative that

$$\mathcal{L}\left(\frac{df}{dt}\right) = sF(s) - f(0).$$

Combining the above two results leads to

$$\lim_{s\to 0} sF(s) = \lim_{t\to\infty} f(t). \tag{A.7}$$

This is called the *final value theorem.*

It is very important to note that the relationship (A.7) **is only guaranteed if the Laplace integral for** df/dt **is convergent for** $s = 0$**. This means that the final value theorem may be applied if a** $B < 0$ **such that** $|df/dt| \leq Ae^{tB}$ **for all** $t \geq 0$ **for some** $A \geq 0$.

Example A.2 *Consider the function* $f(t) = e^{-2t} + 3$. *Then,* $df/dt = -2e^{-2t}$, *which clearly satisfies* $|df/dt| \leq 2e^{Bt}$ *for any* $B \geq -2$. *Therefore, the final value theorem may be applied. The Laplace transform of* $f(t)$ *is readily found to be*

$$F(s) = \frac{1}{s+2} + \frac{3}{s}.$$

Applying the final value theorem, we have

$$\lim_{t\to\infty} f(t) = \lim_{s\to 0} sF(s) = \lim_{s\to 0} \frac{s}{s+2} + 3 = 3,$$

which is exactly the same as would have been found by taking the limit $\lim_{t\to\infty} e^{-2t} + 3$ *directly.*

Example A.3 *This is an example of an incorrect application of the final value theorem.*

Consider the function $f(t) = e^{-2t} + \cos t$. *Clearly, the limit* $\lim_{t\to\infty} e^{-2t} + \cos t$ *does not exist. The derivative is* $df/dt = -2e^{-2t} - \sin t$, *and since* $\sin t$ *is periodic, and attains magnitude 1 twice per period, it is not possible to find a* $B < 0$ *such that* $|df/dt| \leq Ae^{tB}$ *for all* $t \geq 0$ *for some* $A \geq 0$. *Therefore, the final value theorem cannot be applied. However, let us go ahead and apply it anyway. The Laplace transform is readily found to be*

$$F(s) = \frac{1}{s+2} + \frac{s}{s^2+1}.$$

Applying the final value theorem, we obtain

$$\lim_{s\to 0} sF(s) = 0,$$

Clearly in this case, $\lim_{s\to 0} sF(s) \neq \lim_{t\to\infty} f(t)$.

A.4 Partial Fraction Expansions

There is a complicated Laplace transform inversion formula which yields

$$f(t) = \mathcal{L}^{-1}(F(s)).$$

This is not used often in practise. In practise, look-up tables of known Laplace transforms are used. To do this, the Laplace transform $G(s)$ is decomposed into the sum of simpler known parts. For Laplace transforms of the form,

$$G(s) = K\frac{\prod_{i=1}^{m}(s - z_i)}{\prod_{i=1}^{n}(s - p_i)},$$

with $m \leq n$, this may be expanded in a *partial fraction expansion*. For simplicity, we shall assume that $m < n$, and that the poles p_i are distinct ($p_i \neq p_j$ if $i \neq j$). In this case, $G(s)$ can be written as

$$G(s) = \frac{c_1}{s - p_1} + \frac{c_2}{s - p_2} + \ldots + \frac{c_n}{s - p_n}.$$

The coefficients c_i are called the *residues*. To find the coefficient c_1, multiply $G(s)$ by $(s - p_1)$ to get

$$(s - p_1)G(s) = c_1 + \frac{c_2(s - p_1)}{s - p_2} + \ldots + \frac{c_n(s - p_1)}{s - p_n}.$$

Setting $s = p_1$ gives

$$c_1 = (s - p_1)G(s)|_{s=p_1}.$$

In general,

$$c_i = (s - p_i)G(s)|_{s=p_i}. \tag{A.8}$$

Consider now the case where a pole is repeated k times, say

$$p_1 = p_2 = \ldots = p_k = p',$$

such that

$$G(s) = K\frac{\prod_{i=1}^{m}(s - z_i)}{(s - p')^k \prod_{i=k+1}^{n}(s - p_i)}.$$

In this case, $G(s)$ can be expanded as

$$G(s) = \frac{c_1}{s - p'} + \frac{c_2}{(s - p')^2} + \ldots + \frac{c_k}{(s - p')^k} + \frac{c_{k+1}}{s - p_{k+1}} + \ldots + \frac{c_n}{s - p_n}.$$

Multiplying $G(s)$ by $(s-p')^k$ gives

$$(s-p')^k G(s) = c_1(s-p')^{k-1} + \ldots + c_k + \frac{c_{k+1}(s-p')^k}{s-p_{k+1}} + \ldots + \frac{c_n(s-p')^k}{s-p_n}.$$

Setting $s = p'$ gives c_k, i.e.

$$c_k = (s-p')^k G(s)\big|_{s=p'}.$$

Differentiating with respect to s gives

$$\frac{d}{ds}(s-p')^k G(s) = (k-1)c_1(s-p')^{k-2} + \ldots + c_{k-1} + \frac{d}{ds}\left[\frac{c_{k+1}(s-p')^k}{s-p_{k+1}} + \ldots + \frac{c_n(s-p')^k}{s-p_n}\right].$$

Setting $s = p'$ gives c_{k-1}, i.e.

$$c_{k-1} = \frac{d}{ds}\left[(s-p')^k G(s)\right]\Big|_{s=p'}.$$

We can continue in this manner to obtain

$$c_{k-i} = \frac{1}{i!}\frac{d^i}{ds^i}\left[(s-p')^k G(s)\right]\Big|_{s=p'}.$$

Up to this point, we have assumed that $m < n$. If $m = n$, i.e.

$$G(s) = K\frac{\prod_n^{i=1}(s-z_i)}{\prod_n^{i=1}(s-p_i)} = \frac{a_0 s^n + a_1 s^{n-1} + \ldots + a_n}{s^n + b_1 s^{n-1} + \ldots b_n},$$

then we can write this as

$$G(s) = \frac{a_0\left(s^n + b_1 s^{n-1} + \ldots + b_n\right)}{s^n + b_1 s^{n-1} + \ldots b_n} + \frac{(a_1-b_1)s^{n-1} + \ldots + (a_n - b_n)}{s^n + b_1 s^{n-1} + \ldots b_n},$$

$$= a_0 + \frac{(a_1-b_1)s^{n-1} + \ldots + (a_n - b_n)}{s^n + b_1 s^{n-1} + \ldots b_n}.$$

The second expression has the form required, and we can find the partial fraction expansion of that as before.

A.5 Common Laplace Transforms

The following is a table of common Laplace transforms, which will be useful for finding inverse Laplace transforms and thereby solving ordinary differential equations.

$f(t)$	$F(s)$
1	$\frac{1}{s}$
t^n	$\frac{n!}{s^n}$
e^{-at}	$\frac{1}{s+a}$
$t^n e^{-at}$	$\frac{n!}{(s+a)^n}$
$\sin \omega t$	$\frac{\omega}{s^2+\omega^2}$
$\cos \omega t$	$\frac{s}{s^2+\omega^2}$
$e^{-at} \sin \omega t$	$\frac{\omega}{(s+a)^2+\omega^2}$
$e^{-at} \cos \omega t$	$\frac{s+a}{(s+a)^2+\omega^2}$

A.6 Example of Using Laplace Transforms to Solve a Linear Differential Equation

Consider the linear time-invariant differential equation

$$\ddot{x} + 3\dot{x} + 2x = 0, \quad x(0) = a, \quad \dot{x}(0) = b.$$

Then,

$$\begin{aligned}
\mathcal{L}(x(t)) &= X(s), \\
\mathcal{L}(\dot{x}(t)) &= sX(s) - x(0) = sX(s) - a, \\
\mathcal{L}(\ddot{x}(t)) &= s^2X(s) - sx(0) - \dot{x}(0), = s^2X(s) - sa - b.
\end{aligned}$$

Therefore, taking the Laplace transform of the differential equation gives

$$\mathcal{L}(\ddot{x} + 2\dot{x} + 2x) = \mathcal{L}(0),$$

which leads to

$$s^2X(s) - sa - b + 3(sX(s) - a) + 2X(s) = 0,$$

which can be rewritten as

$$\left(s^2 + 3s + 2\right) X(s) = as + b + 3a.$$

Solving for $X(s)$, we have

$$X(s) = \frac{as + b + 3a}{s^2 + 3s + 2}.$$

Let us now find the poles of $X(s)$, by solving

$$s^2 + 3s + 2 = 0.$$

From the quadratic equation, we have

$$s = \frac{-3 \pm \sqrt{3^2 - 4 \times 2}}{2} = -2, -1.$$

That is, we can write

$$X(s) = \frac{as + b + 3a}{(s + 2)(s + 1)}.$$

Let us now expand this in a partial fraction expansion, i.e.

$$X(s) = \frac{c_1}{s + 2} + \frac{c_2}{s + 1}.$$

Since the poles are distinct, we can use (A.8) to compute c_1 and c_2. We have

$$\begin{aligned} c_1 &= (s + 2)X(s)|_{s=-2}, \\ &= \left.\frac{(s + 2)(as + b + 3a)}{(s + 2)(s + 1)}\right|_{s=-2}, \\ &= \frac{-2a + b + 3a}{-1} = -(a + b), \end{aligned}$$

and likewise for c_2 we obtain

$$\begin{aligned} c_2 &= (s + 1)X(s)|_{s=-2}, \\ &= \left.\frac{(s + 1)(as + b + 3a)}{(s + 2)(s + 1)}\right|_{s=-2}, \\ &= \frac{-a + b + 3a}{1} = 2a + b. \end{aligned}$$

So, we find the partial fraction expansion of $X(s)$ to be

$$X(s) = \frac{-(a+b)}{s+2} + \frac{2a+b}{s+1}.$$

From Section A.5, we find that

$$\mathcal{L}\left(e^{-at}\right) = \frac{1}{s+a}.$$

Making use of this, we can find $x(t)$ by taking the inverse Laplace transform of $X(s)$. We have,

$$\begin{aligned} x(t) = \mathcal{L}^{-1}\left(X(s)\right) &= \mathcal{L}^{-1}\left(\frac{-(a+b)}{s+2}\right) + \mathcal{L}^{-1}\left(\frac{2a+b}{s+1}\right), \\ &= -(a+b)e^{-2t} + (2a+b)e^{-t}, \end{aligned}$$

and the differential equation is solved.

This demonstrates the usefulness of Laplace transforms. They turn differential equations in the time domain into algebraic equations in the Laplace domain which are easier to solve. Finally, taking inverse Laplace transforms yields the solution in the time domain.

Appendix B

Numerical Simulation of Spacecraft Motion

B.1 First Order Ordinary Differential Equations

Numerical simulation of spacecraft motion involves *numerical integration* of the equations of motion. Typically, it is possible to write the equations of motion as first-order differential equations of the form

$$\dot{\mathbf{x}} = \mathbf{f}(\mathbf{x}, t), \quad \mathbf{x}(t_0) = \mathbf{x}_0, \tag{B.1}$$

where the state vector $\mathbf{x}$ may contain the spacecraft orbital position, velocity, attitude variables, angular velocity, and any other state variables (for example controller states) required to fully describe the spacecraft motion. This is called a *state-space* representation of the system. To numerically simulate the spacecraft motion, we need to determine the time history of the state vector $\mathbf{x}(t)$ for $t \geq t_0$ numerically. All numerical integration schemes are based upon a discretization of time with sample times t_k, as shown in Figure B.1. The difference between subsequent sample times $h = t_k - t_{k-1}$ is called the *time step*, which may or may not be constant, and must be sufficiently small.

There are a number of techniques for numerical integration, which can be found in texts on numerical methods. It is not always obvious which technique is the best for a given problem. A possible scheme is the fourth-order Runge-Kutta formula

$$\mathbf{x}(t_{k+1}) = \mathbf{x}(t_k) + \frac{1}{6}\left[\mathbf{k}_1 + 2\mathbf{k}_2 + 2\mathbf{k}_3 + \mathbf{k}_4\right],$$

Spacecraft Dynamics and Control: An Introduction, First Edition.
Anton H.J. de Ruiter, Christopher J. Damaren and James R. Forbes.

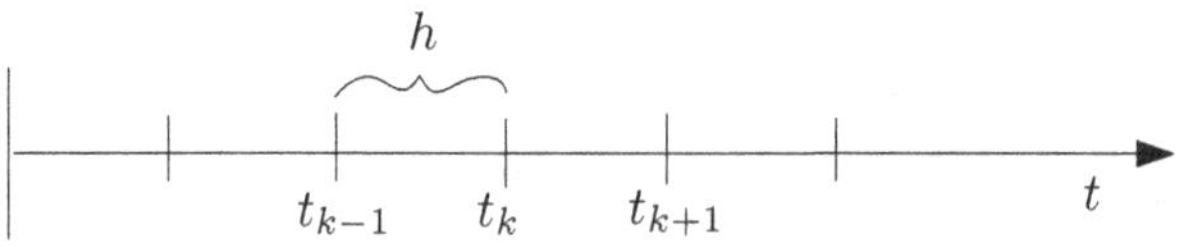

Figure B.1 Time discretization

where

$$\begin{aligned}
\mathbf{k}_1 &= h\mathbf{f}(\mathbf{x}(t_k), t_k),\\
\mathbf{k}_2 &= h\mathbf{f}\left(\mathbf{x}(t_k) + \frac{1}{2}\mathbf{k}_1, t_k + \frac{1}{2}h\right),\\
\mathbf{k}_3 &= h\mathbf{f}\left(\mathbf{x}(t_k) + \frac{1}{2}\mathbf{k}_2, t_k + \frac{1}{2}h\right),\\
\mathbf{k}_4 &= h\mathbf{f}(\mathbf{x}(t_k) + \mathbf{k}_3, t_k + h).
\end{aligned}$$

Alternatively, there are several commercially available numerical packages capable of numerically solving ordinary differential equations of the form (B.1). For example, MATLAB® has a suite of ordinary differential equation solvers.

B.2 Formulation of Coupled Spacecraft Orbital and Attitude Motion Equations

Suppose we wish to numerically simulate the spacecraft attitude motion relative to the Earth-Centered Inertial (ECI) frame $\mathcal{F}_G$, under the influence of a gravity-gradient disturbance torque. The first step will be to identify the required equations.

First, let us parameterize the spacecraft inertial attitude by a 3-2-1 Euler sequence, as summarized in Section 11.2. Then, the attitude kinematics are given by

$$\begin{bmatrix} \dot{\phi} \\ \dot{\theta} \\ \dot{\psi} \end{bmatrix} = \begin{bmatrix} 1 & \sin\phi\tan\theta & \cos\phi\tan\theta \\ 0 & \cos\phi & -\sin\phi \\ 0 & \sin\phi\sec\theta & \cos\phi\sec\theta \end{bmatrix} \boldsymbol{\omega}_{bG}, \tag{B.2}$$

where ϕ is the roll angle, θ is the pitch angle, ψ is the yaw angle, and $\boldsymbol{\omega}_{bG}$ is the spacecraft angular velocity vector relative to $\mathcal{F}_G$, expressed in the spacecraft body-frame $\mathcal{F}_b$. It is clear from equation (B.2), that to solve for the Euler angles we need to know what the angular velocity $\boldsymbol{\omega}_{bG}$ is. This is governed by the attitude dynamics, which as summarized in Section 11.3, are described by Euler's equations

$$\mathbf{I}\dot{\boldsymbol{\omega}}_{bG} + \boldsymbol{\omega}_{bG}^{\times}\mathbf{I}\boldsymbol{\omega}_{bG} = \mathbf{T}_g, \tag{B.3}$$

where $\mathbf{I}$ is the spacecraft inertia matrix, $\mathbf{T}_g$ is the gravity-gradient torque, and all quantities in (B.3) are expressed in the spacecraft body-frame $\mathcal{F}_b$. We see from (B.3) that we need an expression for the gravity-gradient torque. From Section 12.4, we obtain

$$\mathbf{T}_g = \frac{3\mu}{r^5}\mathbf{r}_b^{\times}\mathbf{I}\mathbf{r}_b, \tag{B.4}$$

where μ is the Earth's gravitational constant, $\mathbf{r}_b$ is the spacecraft's orbital position vector expressed in spacecraft body coordinates, and $r = ||\mathbf{r}_b||$ is the orbital radius. From (B.4), it is now clear that we need to simulate the spacecraft orbit as well as the attitude, in order to be able to compute $\mathbf{T}_g$.

Suppose that the spacecraft orbit can be simulated to sufficient accuracy by including only two-body and J_2 gravitational terms. Then, as shown in Chapter 7 the spacecraft orbital motion is described by

$$\ddot{\mathbf{r}} = -\frac{\mu}{r^3}\mathbf{r} + \frac{3\mu J_2 R_e^2}{2r^5}\left[\left(5\frac{(\mathbf{r}^T\mathbf{z}_G)^2}{r^2} - 1\right)\mathbf{r} - 2(\mathbf{r}^T\mathbf{z}_G)\mathbf{z}_G\right], \tag{B.5}$$

where $\mathbf{r}$ is the spacecraft's orbital position expressed in ECI coordinates ($\mathcal{F}_G$), $\mathbf{z}_G = [0 \quad 0 \quad 1]$, R_e is the Earth's equatorial radius and J_2 is the Earth's J_2 constant. Thus, the spacecraft orbit is best simulated in the ECI frame. Referring to equation (B.4), we see that in order to compute $\mathbf{T}_g$, we need to convert the spacecraft orbital position vector to spacecraft body coordinates. From Section 1.3, we find that the required transformation is

$$\mathbf{r}_b = \mathbf{C}_{bG}\mathbf{r}, \tag{B.6}$$

where $\mathbf{C}_{bG}$ is the rotation matrix transforming ECI coordinates to spacecraft body coordinates. From Section 11.2, we find that $\mathbf{C}_{bG}$ may be computed from the Euler angles describing the attitude as

$$\mathbf{C}_{bG}(\phi, \theta, \psi) = \begin{bmatrix} c_\theta c_\psi & c_\theta s_\psi & -s_\theta \\ s_\phi s_\theta c_\psi - c_\phi s_\psi & s_\phi s_\theta s_\psi + c_\phi c_\psi & s_\phi c_\theta \\ c_\phi s_\theta c_\psi + s_\phi s_\psi & c_\phi s_\theta s_\psi - s_\phi c_\psi & c_\phi c_\theta \end{bmatrix}, \tag{B.7}$$

where $s_b = \sin b$ and $c_b = \cos b$.

It can now be readily verified that equations (B.2) to (B.7) are self-contained, and fully describe the coupled spacecraft attitude and orbital motion. As explained in Section B.1, to numerically solve the equations of motion it will be very useful to put the equations into the form of equation (B.1). As a starting point, we examine all of the required equations (equations (B.2) to (B.7)) and identify the variables whose time-derivatives appear. We find all Euler angles ϕ, θ, ψ, the angular velocity vector $\boldsymbol{\omega}_{bG}$, and the inertial orbital position vector $\mathbf{r}$. We see that only the first time-derivatives of ϕ, θ, ψ and $\boldsymbol{\omega}_{bG}$ appear, while only the second time-derivative of $\mathbf{r}$ appears. Since we want to write our equations in the form (B.1), we only want equations with first time-derivatives. To do this, let us introduce the spacecraft inertial velocity vector $\mathbf{v}$, that is

$$\dot{\mathbf{r}} = \mathbf{v} \tag{B.8}$$

Then, the second time-derivative of $\mathbf{r}$ becomes the first time-derivative of $\mathbf{v}$, that is

$$\dot{\mathbf{v}} = \ddot{\mathbf{r}},$$

and (B.5) becomes

$$\dot{\mathbf{v}} = -\frac{\mu}{r^3}\mathbf{r} + \frac{3\mu J_2 R_e^2}{2r^5}\left[\left(5\frac{(\mathbf{r}^T\mathbf{z}_G)^2}{r^2} - 1\right)\mathbf{r} - 2(\mathbf{r}^T\mathbf{z}_G)\mathbf{z}_G\right]. \quad \text{(B.9)}$$

Therefore, by introducing the inertial velocity vector $\mathbf{v}$, we can replace the second-order differential equation (B.6) by the pair of first-order differential equations (B.8) and (B.9). The coupled spacecraft attitude and orbital motion is therefore fully described by the first order differential equations (B.2), (B.3), (B.8) and (B.9) together with equations (B.4) and (B.7). Since the quantities appearing with first time-derivatives are ϕ, θ, ψ, $\boldsymbol{\omega}_{bG}$, $\mathbf{r}$ and $\mathbf{v}$, we can put these equations into the form of equation (B.1) by defining the state vector

$$\mathbf{x} = \begin{bmatrix} \begin{bmatrix} \phi \\ \theta \\ \psi \end{bmatrix} \\ \boldsymbol{\omega}_{bG} \\ \mathbf{r} \\ \mathbf{v} \end{bmatrix}, \quad \text{(B.10)}$$

together with the function

$$\mathbf{f}(\mathbf{x}, t) = \begin{bmatrix} \begin{bmatrix} 1 & \sin\phi\tan\theta & \cos\phi\tan\theta \\ 0 & \cos\phi & -\sin\phi \\ 0 & \sin\phi\sec\theta & \cos\phi\sec\theta \end{bmatrix}\boldsymbol{\omega}_{bG} \\ \mathbf{I}^{-1}\left[-\boldsymbol{\omega}_{bG}^{\times}\mathbf{I}\boldsymbol{\omega}_{bG} + \frac{3\mu}{r^5}\mathbf{r}_b^{\times}\mathbf{I}\mathbf{r}_b\right] \\ \mathbf{v} \\ -\frac{\mu}{r^3}\mathbf{r} + \frac{3\mu J_2 R_e^2}{2r^5}\left[\left(5\frac{(\mathbf{r}^T\mathbf{z}_G)^2}{r^2} - 1\right)\mathbf{r} - 2(\mathbf{r}^T\mathbf{z}_G)\mathbf{z}_G\right] \end{bmatrix}, \quad \text{(B.11)}$$

where $\mathbf{r}_b = \mathbf{C}_{bG}\mathbf{r}$ and

$$\mathbf{C}_{bG}(\phi, \theta, \psi) = \begin{bmatrix} c_\theta c_\psi & c_\theta s_\psi & -s_\theta \\ s_\phi s_\theta c_\psi - c_\phi s_\psi & s_\phi s_\theta s_\psi + c_\phi c_\psi & s_\phi c_\theta \\ c_\phi s_\theta c_\psi + s_\phi s_\psi & c_\phi s_\theta s_\psi - s_\phi c_\psi & c_\phi c_\theta \end{bmatrix}. \quad \text{(B.12)}$$

Equations (B.11) and (B.12) are now in a suitable form to be programmed for use with a numerical ODE solver (such as those available in MATLAB®). It is clear from (B.1), that the initial condition $\mathbf{x}(0)$ must be specified. From (B.10), we see that we need to specify the initial Euler angles $\phi(0)$, $\theta(0)$ and $\psi(0)$, initial angular velocity $\boldsymbol{\omega}_{bG}(0)$, initial orbital position $\mathbf{r}(0)$ and initial orbital velocity $\mathbf{v}(0)$.

Notes

The fourth-order Runge-Kutta method outlined in Section B.1 is but one method of numerically integrating first-order ordinary differential equations. There are several others. For a detailed treatment of different numerical methods, the reader can consult a book such as Burden and Faires (2011).

Reference

Burden RL and Faires JD 2011 *Numerical Analysis* 9th edn. Brooks/Cole, Boston, MA.

Index

Spacecraft Dynamics and Control: An Introduction, First Edition.
Anton H.J. de Ruiter, Christopher J. Damaren and James R. Forbes.